# SteelWall®
## Clutch bars for steel sheet pile walls

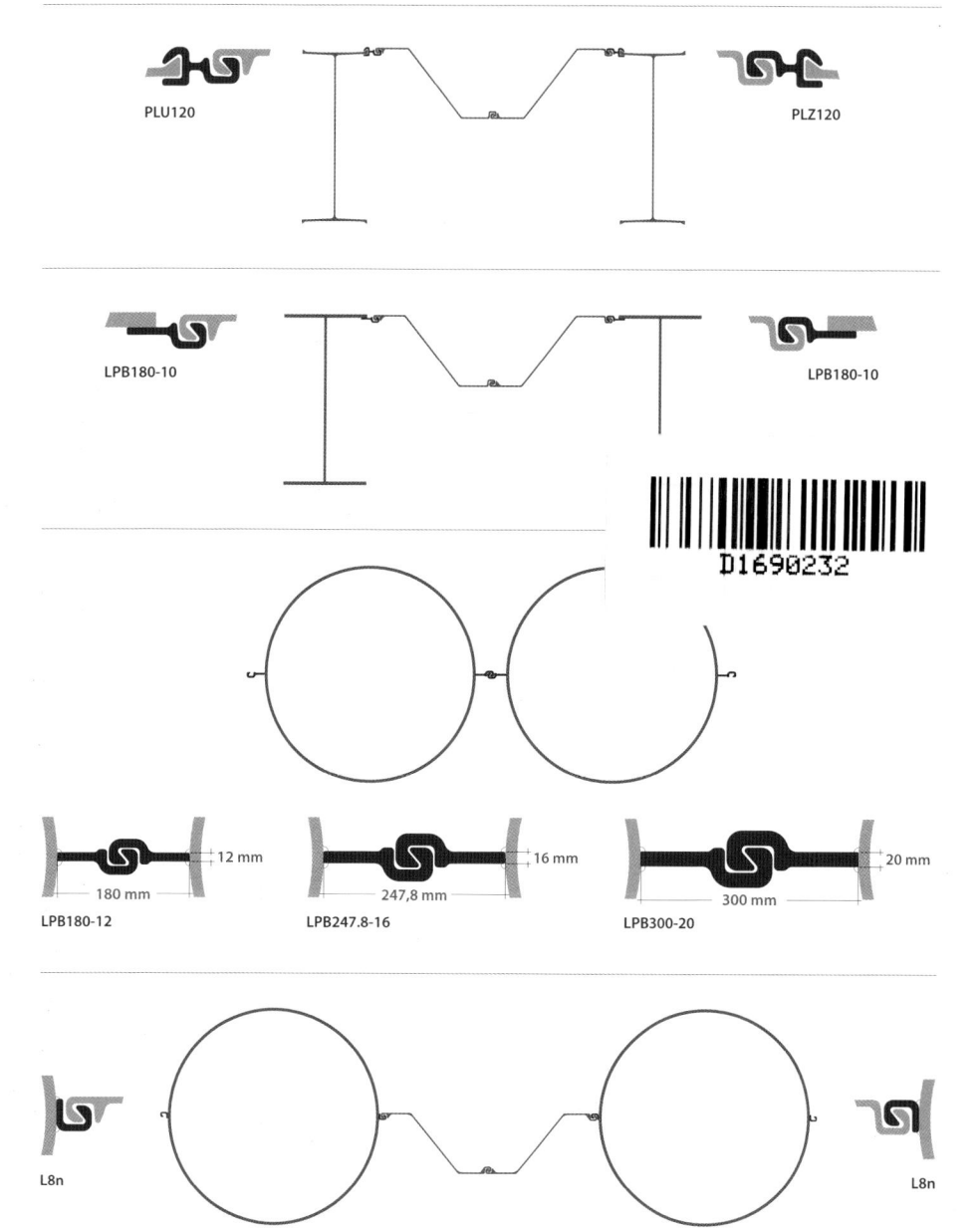

PLU120

PLZ120

LPB180-10

LPB180-10

LPB180-12 — 180 mm, 12 mm

LPB247.8-16 — 247,8 mm, 16 mm

LPB300-20 — 300 mm, 20 mm

L8n

L8n

steelwall.eu

**Ernst & Sohn**
A Wiley Brand

# geotechnik – das Fachgebiet im Überblick

Seit 1978 erscheint die die technischwissenschaftliche Fachzeitschrift geotechnik als Organ der Deutschen Gesellschaft für Geotechnik e.V. (DGGT). Sie behandelt das ganze Fachgebiet der Geotechnik und gibt einen Einblick in die vielfältigen Ziele und Aufgaben der DGGT. Alle Beiträge werden standardmäßig in einem Peer-review Prozess begutachtet.

**DGGT**
Deutsche Gesellschaft
für Geotechnik e. V.
German Geotechnical Society

Hrsg.: Deutsche Gesellschaft
für Geotechnik e.V. (DGGT)
**geotechnik**
38. Jahrgang 2015
4 Hefte / Jahr
ISSN 0172-6145 print
ISSN 2190-6653 online

Auch als **e** journal erhältlich.

Weitere Zeitschriften

- Bautechnik
- Geomechanics and Tunneling
- UnternemerBrief Bauwirtschaft

Probeheft bestellen:
www.ernst-und-sohn.de/geotechnik

---

**Ernst & Sohn**
Verlag für Architektur und technische
Wissenschaften GmbH & Co. KG

Kundenservice: Wiley-VCH
Boschstraße 12
D-69469 Weinheim

Tel. +49 (0)800 1800-536
Fax +49 (0)6201 606-184
cs-germany@wiley.com

1023106_dp

## WORKING WORLDWIDE.

Future-proof quality, convincing profitability, efficient project realisation: With these high service standards Sellhorn has become one of the leading experts for demanding engineering projects worldwide. Hydraulic engineering, port engineering and coastal protection are our core business fields.

Benefit from better solutions. With Sellhorn.

## WELTWEIT IM EINSATZ.

Zukunftssichere Qualität, überzeugende Wirtschaftlichkeit, effiziente Projektrealisierung: Mit diesem hohen Leistungsmaßstab hat sich Sellhorn zu einem der international führenden Experten für anspruchsvollen Ingenieurbau entwickelt. Wasserbau, Hafenbau und Küstenschutz zählen dabei zu unseren Kerngeschäften.

Profitieren Sie von besseren Lösungen. Mit Sellhorn.

www.sellhorn-hamburg.de

- Port planning ■ Infrastructure ■ Railways ■ Bridges ■ Offshore wind energy
- Site supervision ■ FIDIC engineer ■ Project management

**Sellhorn**
INGENIEURGESELLSCHAFT

# Ernst & Sohn
A Wiley Brand

# Schrägkabelbrücken – Cable-Stayed Bridges

### 40 Jahre Erfahrung weltweit – Mit DVD: Vorlesungen live

Holger Svensson
**Schrägkabelbrücken**
40 Jahre Erfahrung weltweit
Mit DVD: Vorlesungen live
2011. 458 S.
€ 79,–*
ISBN 978-3-433-02977-0
Auch als ebook erhältlich

Es werden alle Phasen des Entwurfs, der Montageplanung und der Bauausführung grundsätzlich behandelt und anhand von ca. 250 ausgeführten Beispielen erläutert und illustriert. Die dargestellten Brücken sind nach internationalen Vorschriften bemessen worden, z. B. DIN, Eurocode, AASHTO, British Standard. Besonderes Gewicht wurde auf die Kapitel über Kabel und Montage gelegt, denn hierin liegt der entscheidende Unterschied zu anderen Brückenformen.

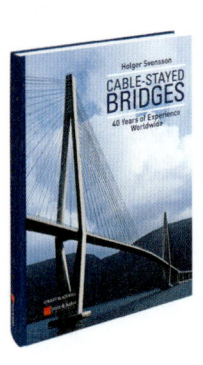

### 40 years Experience Worldwide – 18 lectures on DVD

Holger Svennson
**Cable-Stayed-Bridges**
40 years of Experience Worldwide
18 lectures on DVD
2012. 458 pages
€ 129,–*
ISBN 978-3-433-02992-3
Also available as ebook

The need for large-scale bridges is constantly growing due to the enormous infrastructure development around the world. Since the 1970s many of them have been cable-stayed bridges. In 1975 the largest span length was 404 m, in 1995 it increased to 856 m, and today it is 1104 m. Thus the economically efficient range of cable-stayed bridges is tending to move towards even larger spans, and cable-stayed bridges are increasingly the focus of interest worldwide.

This book describes the fundamentals of design analysis, fabrication and construction, in which the author refers to 250 built examples to illustrate all aspects. International or national codes and technical regulations are referred to only as examples, such as bridges that were designed to German DIN, Eurocode, AASHTO, British Standards. The chapters on cables and erection are a major focus of this work as they represent the most important difference from other types of bridges.

**Order online:**
www.ernst-und-sohn.de

**Ernst & Sohn**
Verlag für Architektur und technische Wissenschaften GmbH & Co. KG

Customer Service: Wiley-VCH
Boschstraße 12
D-69469 Weinheim

Tel. +49 (0)6201 606-400
Fax +49 (0)6201 606-184
service@wiley-vch.de

* Der €-Preis gilt ausschließlich für Deutschland. Inkl. MwSt. zzgl. Versandkosten. Irrtum und Änderungen vorbehalten. 1004146_dp

# Recommendations of the Committee for Waterfront Structures Harbours and Waterways
## EAU 2012

Ernst & Sohn
A Wiley Brand

# Recommendations of the Committee for Waterfront Structures Harbours and Waterways EAU 2012

9th Edition
Translation of the 11th German Edition

Issued by the Committee for Waterfront Structures of the German Port Technology Association and the German Geotechnical Society

Ernst & Sohn
A Wiley Brand

The original German edition was published under the title
*Empfehlungen des Arbeitsausschusses "Ufereinfassungen"*
*Häfen und Wasserstraßen EAU 2012*

Edited by: Arbeitsausschuss "Ufereinfassungen" of HTG and DGGT
Editor: Univ.-Prof. Dr.-Ing. Jürgen Grabe
Hafentechnische Gesellschaft – HTG
Neuer Wandrahm 4
20457 Hamburg
Deutsche Gesellschaft für Geotechnik
Gutenbergstraße 43
45128 Essen

Cover: View across the HHLA Container Terminal Tollerort and the Elbe to the city center of Hamburg (Photo: HHLA/Hampel)

Translated and language polished by Philip Thrift, Quality Engineering Language Services

**Library of Congress Card No.:**
applied for

**British Library Cataloguing-in-Publication Data**
A catalogue record for this book is available from the British Library.

**Bibliographic information published by**
**the Deutsche Nationalbibliothek**
The Deutsche Nationalbibliothek lists this publication in the Deutsche Nationalbibliografie; detailed bibliographic data are available on the Internet at <http://dnb.d-nb.de>.

© 2015 Wilhelm Ernst & Sohn, Verlag für Architektur und technische Wissenschaften GmbH & Co. KG, Rotherstraße 21, 10245 Berlin, Germany

All rights reserved (including those of translation into other languages). No part of this book may be reproduced in any form – by photoprinting, microfilm, or any other means – nor transmitted or translated into a machine language without written permission from the publishers. Registered names, trademarks, etc. used in this book, even when not specifically marked as such, are not to be considered unprotected by law.

Cover design: designpur
Typesetting: Thomson Digital
Printing: Betz-druck GmbH
Binding: Betz-druck GmbH

Printed in the Federal Republic of Germany.
Printed on acid-free paper.

9. revised and up-dated Edition
**Print ISBN:** 978-3-433-03110-0
**ePDF ISBN:** 978-3-433-60520-2
**ePub ISBN:** 978-3-433-60518-9
**eMobi ISBN:** 978-3-433-60519-6
**oBook ISBN:** 978-3-433-60517-2

# Pile instead of anchor?
# Without prestress?
## TITAN Micropile.

- Comparable factor of safety – even without prestressing
- One installation method – for all ground conditions
- Flexible system – also ideal where access is difficult

Further details: www.ischebeck.com

**FRIEDR. ISCHEBECK GMBH**
Loher Str. 31-79 | 58256 Ennepetal | Germany

**ISCHEBECK TITAN**

# DC-SOFTWARE

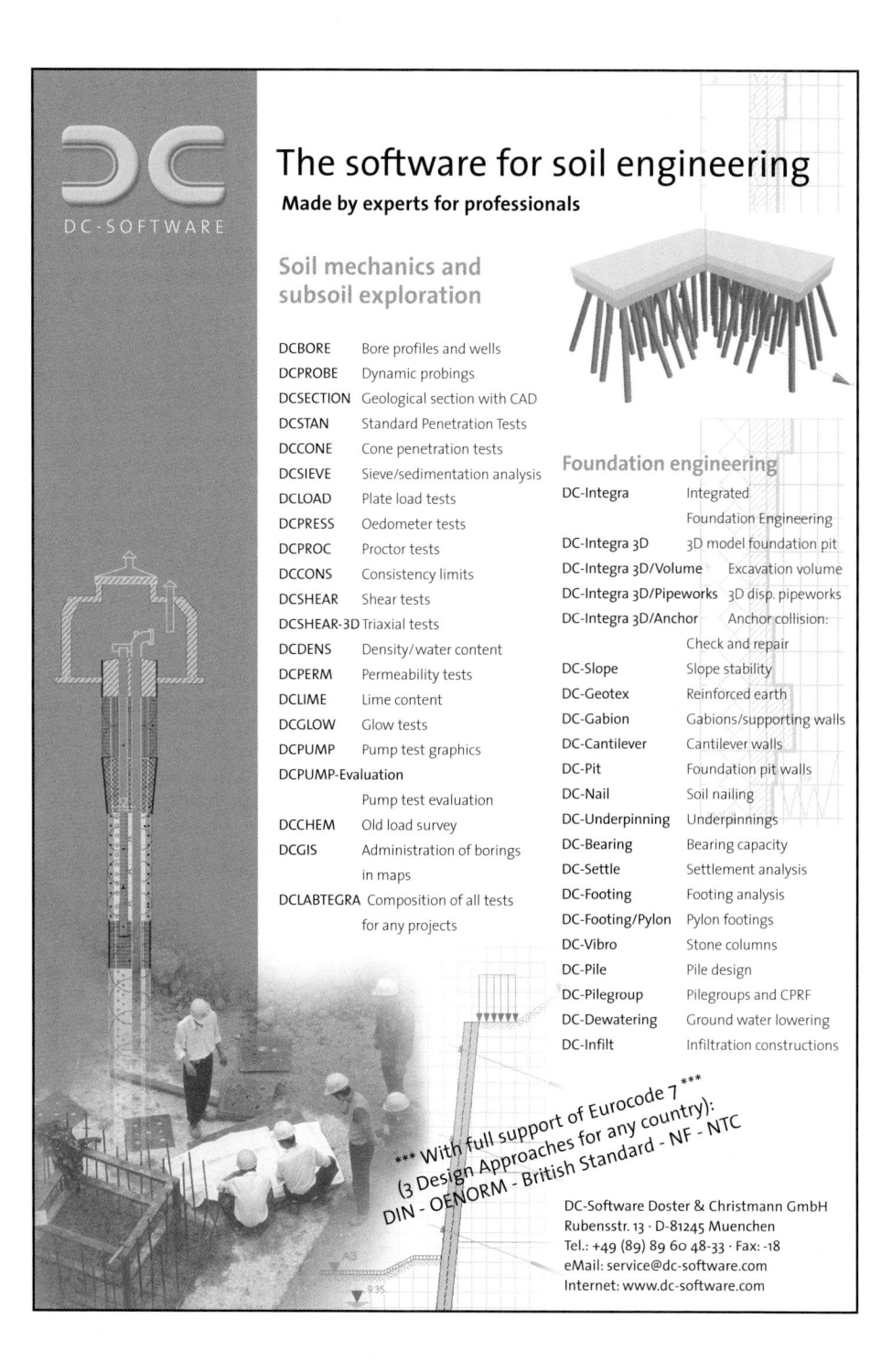

## The software for soil engineering
### Made by experts for professionals

### Soil mechanics and subsoil exploration

| | |
|---|---|
| DCBORE | Bore profiles and wells |
| DCPROBE | Dynamic probings |
| DCSECTION | Geological section with CAD |
| DCSTAN | Standard Penetration Tests |
| DCCONE | Cone penetration tests |
| DCSIEVE | Sieve/sedimentation analysis |
| DCLOAD | Plate load tests |
| DCPRESS | Oedometer tests |
| DCPROC | Proctor tests |
| DCCONS | Consistency limits |
| DCSHEAR | Shear tests |
| DCSHEAR-3D | Triaxial tests |
| DCDENS | Density/water content |
| DCPERM | Permeability tests |
| DCLIME | Lime content |
| DCGLOW | Glow tests |
| DCPUMP | Pump test graphics |
| DCPUMP-Evaluation | Pump test evaluation |
| DCCHEM | Old load survey |
| DCGIS | Administration of borings in maps |
| DCLABTEGRA | Composition of all tests for any projects |

### Foundation engineering

| | |
|---|---|
| DC-Integra | Integrated Foundation Engineering |
| DC-Integra 3D | 3D model foundation pit |
| DC-Integra 3D/Volume | Excavation volume |
| DC-Integra 3D/Pipeworks | 3D disp. pipeworks |
| DC-Integra 3D/Anchor | Anchor collision: Check and repair |
| DC-Slope | Slope stability |
| DC-Geotex | Reinforced earth |
| DC-Gabion | Gabions/supporting walls |
| DC-Cantilever | Cantilever walls |
| DC-Pit | Foundation pit walls |
| DC-Nail | Soil nailing |
| DC-Underpinning | Underpinnings |
| DC-Bearing | Bearing capacity |
| DC-Settle | Settlement analysis |
| DC-Footing | Footing analysis |
| DC-Footing/Pylon | Pylon footings |
| DC-Vibro | Stone columns |
| DC-Pile | Pile design |
| DC-Pilegroup | Pilegroups and CPRF |
| DC-Dewatering | Ground water lowering |
| DC-Infilt | Infiltration constructions |

*** With full support of Eurocode 7 ***
(3 Design Approaches for any country):
DIN - OENORM - British Standard - NF - NTC

DC-Software Doster & Christmann GmbH
Rubensstr. 13 · D-81245 Muenchen
Tel.: +49 (89) 89 60 48-33 · Fax: -18
eMail: service@dc-software.com
Internet: www.dc-software.com

# Contents

List of Recommendations in the 9th Edition ............................. XXI
Preface to 11th, revised edition (9th English edition) of the
*Recommendations of the Committee for Waterfront
Structures – Harbours and Waterways*............................... XXVII

| | | |
|---|---|---|
| **0** | **Structural calculations**............................................... 1 | |
| 0.1 | General ................................................................ 1 | |
| 0.2 | Safety concept........................................................ 2 | |
| 0.2.1 | General ............................................................... 2 | |
| 0.2.2 | Combination factors................................................. 7 | |
| 0.2.3 | Analysis of ultimate limit state..................................... 8 | |
| 0.2.4 | Analysis of serviceability limit state............................... 9 | |
| 0.2.5 | Geotechnical categories ............................................ 9 | |
| 0.2.6 | Probabilistic analysis ............................................... 9 | |
| 0.3 | Calculations for waterfront structures ............................ 10 | |
| **1** | **Subsoil** ............................................................... 11 | |
| 1.1 | Mean characteristic values of soil parameters (R 9) ............ 11 | |
| 1.1.1 | General ............................................................... 11 | |
| 1.2 | Layout and depths of boreholes and penetrometer tests (R 1) .. 11 | |
| 1.2.1 | General ............................................................... 11 | |
| 1.2.2 | Principal boreholes ................................................. 17 | |
| 1.2.3 | Intermediate boreholes ............................................. 17 | |
| 1.2.4 | Penetrometer tests .................................................. 17 | |
| 1.3 | Geotechnical report (R 150)........................................ 18 | |
| 1.4 | Determining the shear strength $c_u$ of saturated, undrained cohesive soils (R 88) ........................................................... 19 | |
| 1.4.1 | Cohesion $c_u$ of undrained soil .................................... 19 | |
| 1.4.2 | Determining the cohesion $c_u$ of an undrained soil............... 20 | |
| 1.4.3 | Determining $c_u$ in laboratory tests ............................... 21 | |
| 1.4.4 | Field tests ........................................................... 22 | |
| 1.4.5 | Correlations......................................................... 22 | |
| 1.5 | Assessing the subsoil for the installation of piles and sheet piles and for selecting the installation method (R 154) ................ 22 | |
| 1.5.1 | General ............................................................... 22 | |
| 1.5.2 | Assessment of soil types with respect to installation methods............................................................... 23 | |

| 2 | **Active and passive earth pressure** | 27 |
|---|---|---|
| 2.1 | General | 27 |
| 2.2 | Considering the cohesion in cohesive soils (R 2) | 27 |
| 2.3 | Considering the apparent cohesion (capillary cohesion) in sand (R 3) | 27 |
| 2.4 | Determining active earth pressure according to the Culmann method (R 171) | 28 |
| 2.4.1 | Solution for uniform soil without cohesion | 28 |
| 2.4.2 | Solution for uniform soil with cohesion | 29 |
| 2.4.3 | Expanded solutions | 29 |
| 2.5 | Active earth pressure in stratified soil (R 219) | 30 |
| 2.6 | Determining active earth pressure for a steep, paved embankment in a partially sloping waterfront structure (R 198) | 32 |
| 2.7 | Determining the active earth pressure shielding on a wall below a relieving platform with average ground surcharges (R 172) | 34 |
| 2.8 | Earth pressure distribution under limited loads (R 220) | 37 |
| 2.9 | Determining active earth pressure in saturated, non- or partially consolidated, soft cohesive soils (R 130) | 38 |
| 2.10 | Effect of artesian water pressure under harbour bottom or river bed on active and passive earth pressures (R 52) | 40 |
| 2.11 | Considering active earth pressure and excess water pressure, and construction guidance for waterfront structures with soil replacement and contaminated or disturbed base of excavation (R 110) | 42 |
| 2.11.1 | General | 42 |
| 2.11.2 | Approach for determining active earth pressure | 42 |
| 2.11.3 | Approaches for determining excess water pressure | 44 |
| 2.11.4 | Guidance for the design of waterfront structures | 44 |
| 2.12 | Effect of groundwater flow on excess water pressure and active and passive earth pressures (R 114) | 46 |
| 2.12.1 | General | 46 |
| 2.12.2 | Determining the excess water pressure | 48 |
| 2.12.3 | Determining the effects on active and passive earth pressures when the flow is mainly vertical | 50 |
| 2.13 | Determining the amount of displacement required for mobilising passive earth pressure in non-cohesive soils (R 174) | 53 |
| 2.14 | Measures for increasing the passive earth pressure in front of waterfront structures (R 164) | 54 |
| 2.14.1 | General | 54 |
| 2.14.2 | Soil replacement | 55 |
| 2.14.3 | Soil compaction | 55 |
| 2.14.4 | Soil surcharge | 56 |
| 2.14.5 | Soil stabilisation | 56 |
| 2.15 | Passive earth pressure in front of abrupt changes in ground level in soft cohesive soils with rapid load application on land side (R 190) | 57 |

| | | |
|---|---|---|
| 2.16 | Waterfront structures in seismic regions (R 124) | 57 |
| 2.16.1 | General | 57 |
| 2.16.2 | Effects of earthquakes on the subsoil | 59 |
| 2.16.3 | Determining the effects of earthquakes on active and passive earth pressures | 59 |
| 2.16.4 | Excess water pressure | 62 |
| 2.16.5 | Transient loads | 62 |
| 2.16.6 | Design situation and partial safety factors | 62 |
| 2.16.7 | Guidance for considering seismic influences on waterfront structures | 62 |
| **3** | **Hydraulic heave failure, ground failure** | **64** |
| 3.1 | Safety against hydraulic heave failure (R 115) | 64 |
| 3.2 | Piping (ground failure due to internal erosion) (R 116) | 70 |
| **4** | **Water levels, water pressure, drainage** | **73** |
| 4.1 | Mean groundwater level (R 58) | 73 |
| 4.2 | Excess water pressure in direction of water side (R 19) | 73 |
| 4.3 | Excess water pressure on sheet piling in front of embankments below elevated platforms in tidal areas (R 65) | 75 |
| 4.3.1 | General | 75 |
| 4.3.2 | Approximation for excess water pressure | 76 |
| 4.4 | Design of weepholes for sheet piling structures (R 51) | 76 |
| 4.5 | Design of drainage systems for waterfront structures in tidal areas (R 32) | 78 |
| 4.5.1 | General | 78 |
| 4.5.2 | Design, installation and maintenance of drainage systems | 79 |
| 4.5.3 | Drainage systems for large waterfront structures | 79 |
| 4.6 | Relieving artesian pressure beneath harbour bottoms (R 53) | 80 |
| 4.6.1 | General | 80 |
| 4.6.2 | Design of relief wells | 80 |
| 4.6.3 | Construction of relief wells | 81 |
| 4.6.4 | Checking the relief installation | 81 |
| 4.7 | Taking account of groundwater flow (R 113) | 81 |
| 4.7.1 | General | 81 |
| 4.7.2 | Principles of groundwater flow | 81 |
| 4.7.3 | Definition of the boundary conditions for a flow net | 83 |
| 4.7.4 | Graphic method for determining a flow net | 83 |
| 4.7.5 | Use of groundwater models to determine flow nets | 84 |
| 4.7.6 | Calculation of individual hydraulic variables | 85 |
| 4.7.7 | Evaluation of examples | 85 |
| 4.8 | Temporary stabilisation of waterfront structures by groundwater lowering (R 166) | 90 |

| | | |
|---|---|---|
| 4.8.1 | General | 90 |
| 4.8.2 | Case with soft, cohesive soil near the ground surface | 90 |
| 4.8.3 | Case as for section 4.8.2 but with high-level aquifer | 92 |
| 4.8.4 | Consideration of intermediate states | 92 |
| **5** | **Ship dimensions and loads on waterfront structures** | **93** |
| 5.1 | Ship dimensions (R 39) | 93 |
| 5.1.1 | Sea-going ships | 93 |
| 5.1.2 | River- and sea-going vessels | 99 |
| 5.1.3 | Inland waterway vessels | 100 |
| 5.1.4 | Displacement | 103 |
| 5.2 | Berthing force of ships at quays (R 38) | 103 |
| 5.3 | Berthing velocities of ships transverse to berth (R 40) | 103 |
| 5.4 | Design situations (R 18) | 105 |
| 5.4.1 | Design situation DS-P | 105 |
| 5.4.2 | Design situation DS-T | 105 |
| 5.4.3 | Design situation DS-A | 105 |
| 5.4.4 | Extreme case | 106 |
| 5.5 | Vertical imposed loads (R 5) | 106 |
| 5.5.1 | General | 106 |
| 5.5.2 | Basic situation 1 | 108 |
| 5.5.3 | Basic situation 2 | 108 |
| 5.5.4 | Basic situation 3 | 109 |
| 5.5.5 | Loading assumptions for quay surfaces | 109 |
| 5.6 | Determining the "design sea state" for maritime and port structures (R 136) | 110 |
| 5.6.1 | General | 110 |
| 5.6.2 | Description of the sea state | 110 |
| 5.6.3 | Determining the sea state parameters | 111 |
| 5.6.4 | Design concepts and specification of design parameters | 116 |
| 5.6.5 | Conversion of the sea state | 117 |
| 5.7 | Wave pressure on vertical quay walls in coastal areas (R 135) | 120 |
| 5.7.1 | General | 120 |
| 5.7.2 | Loads due to non-breaking waves | 120 |
| 5.7.3 | Loads due to waves breaking on structure | 121 |
| 5.7.4 | Loads due to broken waves | 124 |
| 5.7.5 | Additional loads caused by waves | 124 |
| 5.8 | Loads arising from surging and receding waves due to the inflow or outflow of water (R 185) | 125 |
| 5.8.1 | General | 125 |
| 5.8.2 | Determining wave values | 125 |
| 5.8.3 | Load assumptions | 126 |
| 5.9 | Effects of waves due to ship movements (R 186) | 126 |

| 5.9.1 | General | 126 |
|---|---|---|
| 5.9.2 | Wave heights | 128 |
| 5.10 | Wave pressure on piled structures (R 159) | 129 |
| 5.10.1 | General | 129 |
| 5.10.2 | Method of calculation according to Morison et al. | 133 |
| 5.10.3 | Determining the wave loads on a single vertical pile | 134 |
| 5.10.4 | Coefficients $C_D$ and $C_M$ | 135 |
| 5.10.5 | Forces from breaking waves | 136 |
| 5.10.6 | Wave load on a group of piles | 136 |
| 5.10.7 | Raking piles | 137 |
| 5.10.8 | Safety factors | 138 |
| 5.10.9 | Vertical wave load ("wave slamming") | 138 |
| 5.11 | Wind loads on moored ships and their influence on the dimensioning of mooring and fender equipment (R 153) | 145 |
| 5.11.1 | General | 145 |
| 5.11.2 | Critical wind speed | 145 |
| 5.11.3 | Wind loads on moored vessels | 145 |
| 5.11.4 | Loads on mooring and fender equipment | 146 |
| 5.12 | Layout of and loads on bollards for sea-going vessels (R 12) | 147 |
| 5.12.1 | Layout | 147 |
| 5.12.2 | Loads | 148 |
| 5.12.3 | Direction of bollard pull force | 148 |
| 5.13 | Layout, design and loading of bollards for inland facilities (R 102) | 149 |
| 5.13.1 | Layout and design | 149 |
| 5.13.2 | Loads | 150 |
| 5.13.3 | Direction of line pull forces | 151 |
| 5.13.4 | Calculations | 151 |
| 5.14 | Quay loads from cranes and other transhipment equipment (R 84) | 151 |
| 5.14.1 | Typical general cargo port cranes | 151 |
| 5.14.2 | Container cranes | 152 |
| 5.14.3 | Load specifications for port cranes | 153 |
| 5.14.4 | Notes | 155 |
| 5.15 | Impact and pressure of ice on waterfront structures, fenders and dolphins in coastal areas (R 177) | 155 |
| 5.15.1 | General | 155 |
| 5.15.2 | Determining the compressive strength of the ice | 157 |
| 5.15.3 | Ice loads on waterfront structures and other structures of greater extent | 158 |
| 5.15.4 | Ice loads on vertical piles | 161 |
| 5.15.5 | Horizontal ice load on group of piles | 162 |
| 5.15.6 | Ice surcharges | 162 |
| 5.15.7 | Vertical loads with rising or falling water levels | 163 |
| 5.16 | Impact and pressure of ice on waterfront structures, piers and dolphins at inland facilities (R 205) | 164 |
| 5.16.1 | General | 164 |

| | | |
|---|---|---|
| 5.16.2 | Ice thickness | 164 |
| 5.16.3 | Compressive strength of the ice | 165 |
| 5.16.4 | Ice loads on waterfront structures and other structures of greater extent | 165 |
| 5.16.5 | Ice loads on narrow structures (piles, dolphins, bridge and weir piers, ice deflectors) | 166 |
| 5.16.6 | Ice loads on groups of structures | 166 |
| 5.16.7 | Vertical loads with rising or falling water levels | 167 |
| 5.17 | Loads on waterfront structures and dolphins caused by fender reaction forces (R 213) | 167 |
| **6** | **Configuration of cross-sections and equipment for waterfront structures** | **168** |
| 6.1 | Standard cross-section dimensions for waterfront structures in seaports (R 6) | 168 |
| 6.1.1 | Standard cross-sections | 168 |
| 6.1.2 | Walkways (towpaths) | 168 |
| 6.1.3 | Railings, rubbing strips and edge protection | 169 |
| 6.1.4 | Edge bollards | 169 |
| 6.1.5 | Arrangement of tops of quay walls at container terminals | 169 |
| 6.2 | Top edges of waterfront structures in seaports (R 122) | 170 |
| 6.2.1 | General | 170 |
| 6.2.2 | Level of port operations area with regard to water levels | 170 |
| 6.2.3 | Effects of (changing) groundwater levels on the terrain and the level of the port operations area | 171 |
| 6.2.4 | Level of port operations area depending on cargo handling | 171 |
| 6.3 | Standard cross-sections for waterfront structures in inland ports (R 74) | 172 |
| 6.3.1 | Port operations level | 172 |
| 6.3.2 | Waterfront | 172 |
| 6.3.3 | Clearance profile | 173 |
| 6.3.4 | Position of outboard crane rail | 173 |
| 6.3.5 | Mooring equipment | 175 |
| 6.4 | Sheet piling waterfronts on inland waterways (R 106) | 175 |
| 6.4.1 | General | 175 |
| 6.4.2 | Stability analysis | 177 |
| 6.4.3 | Loading assumptions | 177 |
| 6.4.4 | Embedment depth | 178 |
| 6.5 | Upgrading partially sloped waterfronts in inland ports with large water level fluctuations (R 119) | 178 |
| 6.5.1 | Reasons for partially sloped upgrades | 178 |
| 6.5.2 | Design principles | 178 |
| 6.6 | Design of waterfront areas in inland ports according to operational aspects (R 158) | 180 |

| | | |
|---|---|---|
| 6.6.1 | Requirements. | 180 |
| 6.6.2 | Design principles. | 181 |
| 6.6.3 | Waterfront cross-sections. | 181 |
| 6.7 | Nominal depth and design depth of harbour bottom (R 36). | 182 |
| 6.7.1 | Nominal depth in seaports. | 182 |
| 6.7.2 | Nominal depth of harbour bottom for inland ports. | 182 |
| 6.7.3 | Design depth in front of quay wall. | 183 |
| 6.8 | Strengthening waterfront structures for deepening harbour bottoms in seaports (R 200). | 184 |
| 6.8.1 | General. | 184 |
| 6.8.2 | Design of strengthening measures. | 185 |
| 6.9 | Embankments below waterfront wall superstructures behind closed sheet pile walls (R 68). | 189 |
| 6.9.1 | Embankment loads. | 189 |
| 6.9.2 | Risk of silting-up behind sheet pile wall. | 190 |
| 6.10 | Redesign of waterfront structures in inland ports (R 201). | 190 |
| 6.10.1 | General. | 190 |
| 6.10.2 | Redesign options. | 190 |
| 6.10.3 | Construction examples. | 191 |
| 6.11 | Provision of quick-release hooks at berths for large vessels (R 70). | 193 |
| 6.12 | Layout, design and loads of access ladders (R 14). | 194 |
| 6.12.1 | Layout. | 194 |
| 6.12.2 | Design. | 194 |
| 6.13 | Layout and design of stairs in seaports (R 24). | 197 |
| 6.13.1 | Layout of stairs. | 197 |
| 6.13.2 | Practical stair dimensions. | 198 |
| 6.13.3 | Landings. | 198 |
| 6.13.4 | Railings. | 198 |
| 6.13.5 | Mooring equipment. | 198 |
| 6.13.6 | Stairs in sheet pile structures. | 198 |
| 6.14 | Equipment for waterfront structures in seaports with supply and disposal systems (R 173). | 199 |
| 6.14.1 | General. | 199 |
| 6.14.2 | Water supply systems. | 199 |
| 6.14.3 | Electricity supply systems. | 200 |
| 6.14.4 | Other systems. | 201 |
| 6.14.5 | Disposal systems. | 202 |
| 6.15 | Fenders for large vessels (R 60). | 202 |
| 6.15.1 | General. | 202 |
| 6.15.2 | The fendering principle. | 203 |
| 6.15.3 | Design principles for fenders. | 204 |
| 6.15.4 | Required energy absorption capacity. | 205 |
| 6.15.5 | Types of fender system. | 211 |
| 6.15.6 | Construction guidance. | 216 |

| | | |
|---|---|---|
| 6.15.7 | Chains | 217 |
| 6.15.8 | Guiding devices and edge protection | 217 |
| 6.16 | Fenders in inland ports (R 47) | 219 |
| 6.17 | Foundations to craneways on waterfront structures (R 120) | 220 |
| 6.17.1 | General | 220 |
| 6.17.2 | Design of foundations, tolerances | 221 |
| 6.18 | Fixing crane rails to concrete (R 85) | 223 |
| 6.18.1 | Supporting the crane rail on a continuous steel plate on a continuous concrete base | 223 |
| 6.18.2 | Bridge-type arrangement with rail supported centrally on bearing plates | 223 |
| 6.18.3 | Bridge-type arrangement with rail supported on chairs | 227 |
| 6.18.4 | Traversable craneways | 227 |
| 6.18.5 | Note on rail wear | 230 |
| 6.18.6 | Local bearing pressure | 231 |
| 6.19 | Connection of expansion joint seal in reinforced concrete bottom to loadbearing steel sheet pile wall (R 191) | 231 |
| 6.20 | Connecting steel sheet piling to a concrete structure (R 196) | 231 |
| 6.21 | Floating berths in seaports (R 206) | 236 |
| 6.21.1 | General | 236 |
| 6.21.2 | Design principles | 236 |
| 6.21.3 | Loading assumptions and design | 237 |
| **7** | **Earthworks and dredging** | 238 |
| 7.1 | Dredging in front of quay walls in seaports (R 80) | 238 |
| 7.2 | Dredging and hydraulic fill tolerances (R 139) | 240 |
| 7.2.1 | General | 240 |
| 7.2.2 | Dredging tolerances | 240 |
| 7.3 | Hydraulic filling of port areas for planned waterfront structures (R 81) | 242 |
| 7.3.1 | General | 242 |
| 7.3.2 | Hydraulic filling of port above the water table | 245 |
| 7.3.3 | Hydraulic filling of port areas below the water table | 246 |
| 7.4 | Backfilling of waterfront structures (R 73) | 249 |
| 7.4.1 | General | 249 |
| 7.4.2 | Backfilling in the dry | 249 |
| 7.4.3 | Backfilling underwater | 250 |
| 7.4.4 | Additional remarks | 250 |
| 7.5 | In situ density of hydraulically filled non-cohesive soils (R 175) | 251 |
| 7.5.1 | General | 251 |
| 7.5.2 | Empirical values for in situ density | 251 |
| 7.5.3 | In situ density required for port areas | 252 |
| 7.5.4 | Checking the in situ density | 252 |
| 7.6 | In situ density of dumped non-cohesive soils (R 178) | 252 |
| 7.6.1 | General | 252 |

| | | |
|---|---|---|
| 7.6.2 | Influences on the achievable in situ density | 253 |
| 7.7 | Dredging underwater slopes (R 138) | 254 |
| 7.7.1 | General | 254 |
| 7.7.2 | Dredging underwater slopes in loose sand | 254 |
| 7.7.3 | Dredging plant | 255 |
| 7.7.4 | Execution of dredging work | 255 |
| 7.8 | Subsidence of non-cohesive soils (R 168) | 257 |
| 7.9 | Soil replacement along a line of piles for a waterfront structure (R 109) | 258 |
| 7.9.1 | General | 258 |
| 7.9.2 | Dredging | 259 |
| 7.9.3 | Quality and procurement of the fill sand | 261 |
| 7.9.4 | Cleaning the base of the excavation before filling with sand | 262 |
| 7.9.5 | Placing the sand fill | 263 |
| 7.9.6 | Checking the sand fill | 264 |
| 7.10 | Dynamic compaction of the soil (R 188) | 264 |
| 7.11 | Vertical drains to accelerate the consolidation of soft cohesive soils (R 93) | 265 |
| 7.11.1 | General | 265 |
| 7.11.2 | Applications | 265 |
| 7.11.3 | Design | 265 |
| 7.11.4 | Design of plastic drains | 267 |
| 7.11.5 | Installation | 268 |
| 7.12 | Consolidation of soft cohesive soils by preloading (R 179) | 268 |
| 7.12.1 | General | 268 |
| 7.12.2 | Applications | 269 |
| 7.12.3 | Bearing capacity of in situ soil | 270 |
| 7.12.4 | Fill material | 270 |
| 7.12.5 | Determining the depth of preload fill | 270 |
| 7.12.6 | Minimum extent of preload fill | 273 |
| 7.12.7 | Soil improvement through vacuum consolidation with vertical drains | 273 |
| 7.12.8 | Execution of soil improvement through vacuum consolidation with vertical drains | 274 |
| 7.12.9 | Checking the consolidation | 274 |
| 7.12.10 | Secondary settlement | 275 |
| 7.13 | Improving the bearing capacity of soft cohesive soils with vertical elements (R 210) | 275 |
| 7.13.1 | General | 275 |
| 7.13.2 | Methods | 275 |
| 7.13.3 | Construction of pile-type loadbearing elements | 277 |
| 7.13.4 | Design of geotextile-encased columns | 278 |
| 7.13.5 | Construction of geotextile-encased columns | 279 |
| **8** | **Sheet piling structures** | 281 |
| 8.1 | Materials and construction | 281 |

XIII

| | | |
|---|---|---|
| 8.1.1 | Design and installation of timber sheet pile walls (R 22) | 281 |
| 8.1.2 | Design and installation of reinforced concrete sheet pile walls (R 21) | 284 |
| 8.1.3 | Design and installation of steel sheet pile walls (R 34) | 287 |
| 8.1.4 | Combined steel sheet piling (R 7) | 288 |
| 8.1.5 | Shear-resistant interlock connections for steel sheet piling (R 103) | 292 |
| 8.1.6 | Quality requirements for steels and dimensional tolerances for steel sheet piles (R 67) | 296 |
| 8.1.7 | Acceptance conditions for steel sheet piles and steel piles on site (R 98) | 298 |
| 8.1.8 | Corrosion of steel sheet piling, and countermeasures (R 35) | 300 |
| 8.1.9 | Danger of sand abrasion on sheet piling (R 23) | 309 |
| 8.1.10 | Shock blasting to assist the driving of steel sheet piles (R 183) | 309 |
| 8.1.11 | Driving steel sheet piles (R 118) | 312 |
| 8.1.12 | Driving combined steel sheet piling (R 104) | 316 |
| 8.1.13 | Monitoring during the installation of sheet piles, tolerances (R 105) | 321 |
| 8.1.14 | Noise control – low-noise driving (R 149) | 325 |
| 8.1.15 | Driving of steel sheet piles and steel piles at low temperatures (R 90) | 330 |
| 8.1.16 | Repairing interlock declutching on driven steel sheet piling (R 167) | 331 |
| 8.1.17 | Reinforced steel sheet piling (R 176) | 334 |
| 8.1.18 | Design of piling frames (R 140) | 340 |
| 8.1.19 | Design of welded joints in steel piles and steel sheet piles (R 99) | 344 |
| 8.1.20 | Cutting off the tops of driven steel sections for loadbearing welded connections (R 91) | 347 |
| 8.1.21 | Watertightness of steel sheet piling (R 117) | 347 |
| 8.1.22 | Waterfront structures in regions with mining subsidence (R 121) | 350 |
| 8.1.23 | Vibratory driving of U- and Z-section steel sheet piles (R 202) | 353 |
| 8.1.24 | Water-jetting to assist the driving of steel sheet piles (R 203) | 357 |
| 8.1.25 | Pressing of U- and Z-section steel sheet piles (R 212) | 360 |
| 8.2 | Design of sheet piling | 361 |
| 8.2.1 | General | 361 |
| 8.2.2 | Free-standing sheet piling structures (R 161) | 366 |
| 8.2.3 | Design of sheet piling structures with fixity in the ground and a single anchor (R 77) | 367 |
| 8.2.4 | Design of sheet pile walls with double anchors (R 134) | 372 |
| 8.2.5 | Applying the angle of earth pressure and the analysis in the vertical direction (R 4) | 373 |
| 8.2.6 | Taking account of unfavourable groundwater flows in the passive earth pressure zone (R 199) | 386 |
| 8.2.7 | Verifying the loadbearing capacity of the elements of sheet piling structures (R 20) | 386 |
| 8.2.8 | Selection of embedment depth for sheet piling (R 55) | 390 |
| 8.2.9 | Determining the embedment depth for sheet pile walls with full or partial fixity in the soil (R 56) | 391 |
| 8.2.10 | Steel sheet piling with staggered embedment depths (R 41) | 394 |

| | | |
|---|---|---|
| 8.2.11 | Horizontal actions on steel sheet pile walls in the longitudinal direction of the quay (R 132) | 397 |
| 8.2.12 | Design of anchor walls fixed in the ground (R 152) | 400 |
| 8.2.13 | Staggered arrangement of anchor walls (R 42) | 401 |
| 8.2.14 | Steel sheet piling founded on bedrock (R 57) | 401 |
| 8.2.15 | Waterfront sheet piling in unconsolidated, soft cohesive soils, especially in connection with non-sway structures (R 43) | 402 |
| 8.2.16 | Design of single-anchor sheet piling structures in earthquake zones (R 125) | 404 |
| 8.3 | Calculation and design of cofferdams | 405 |
| 8.3.1 | Cellular cofferdams as excavation enclosures and waterfront structures (R 100) | 405 |
| 8.3.2 | Double-wall cofferdams as excavation enclosures and waterfront structures (R 101) | 417 |
| 8.3.3 | Narrow moles in sheet piling (R 162) | 422 |
| 8.4 | Walings, capping beams and anchor connections | 424 |
| 8.4.1 | Design of steel walings for sheet piling (R 29) | 424 |
| 8.4.2 | Verification of steel walings (R 30) | 425 |
| 8.4.3 | Sheet piling walings of reinforced concrete with driven steel anchor piles (R 59) | 427 |
| 8.4.4 | Steel capping beams for sheet piling waterfront structures (R 95) | 431 |
| 8.4.5 | Reinforced concrete capping beams for waterfront structures with steel sheet piling (R 129) | 435 |
| 8.4.6 | Steel nosings to protect reinforced concrete walls and capping beams on waterfront structures (R 94) | 441 |
| 8.4.7 | Auxiliary anchors at the top of steel sheet piling structures (R 133) | 444 |
| 8.4.8 | Screw threads for sheet piling anchors (R 184) | 445 |
| 8.4.9 | Sheet piling anchors in unconsolidated, soft cohesive soils (R 50) | 447 |
| 8.4.10 | Design of protruding quay wall corners with round steel tie rods (R 31) | 450 |
| 8.4.11 | Design and calculation of protruding quay wall corners with raking anchor piles (R 146) | 453 |
| 8.4.12 | High prestressing of high-strength steel anchors for waterfront structures (R 151) | 457 |
| 8.4.13 | Hinged connections between driven steel anchor piles and steel sheet piling structures (R 145) | 458 |
| 8.5 | Verification of stability for anchoring at the lower failure plane (R 10) | 469 |
| 8.5.1 | Stability at the lower failure plane for anchorages with anchor walls | 469 |
| 8.5.2 | Stability at the lower failure plane in unconsolidated, saturated cohesive soils | 471 |
| 8.5.3 | Stability at the lower failure plane with varying soil strata | 471 |
| 8.5.4 | Verification of stability at the lower failure for a quay wall fixed in the soil | 472 |
| 8.5.5 | Stability at the lower failure plane for an anchor wall fixed in the soil | 473 |
| 8.5.6 | Stability at the lower failure plane for anchors with anchor plates | 473 |

| | | |
|---|---|---|
| 8.5.7 | Verification of safety against failure of anchoring soil | 473 |
| 8.5.8 | Stability at the lower failure plane for quay walls anchored with anchor piles or grouted anchors at one level | 474 |
| 8.5.9 | Stability at the lower failure plane for quay walls with anchors at more than one level | 475 |
| 8.5.10 | Safety against slope failure | 477 |
| **9** | **Tension piles and anchors (R 217)** | **478** |
| 9.1 | General | 478 |
| 9.2 | Displacement piles | 478 |
| 9.2.1 | Installation | 478 |
| 9.2.2 | Types | 479 |
| 9.2.3 | Loadbearing capacity of displacement piles | 480 |
| 9.3 | Micropiles | 481 |
| 9.3.1 | Installation | 481 |
| 9.3.2 | Types | 482 |
| 9.3.3 | Loadbearing capacity of micropiles | 482 |
| 9.4 | Special piles | 483 |
| 9.4.1 | General | 483 |
| 9.4.2 | Prefabricated raking piles | 483 |
| 9.5 | Anchors | 484 |
| 9.5.1 | Construction | 484 |
| 9.5.2 | Types | 484 |
| 9.5.3 | Loadbearing capacity of anchors | 485 |
| **10** | **Quay walls and superstructures in concrete** | **486** |
| 10.1 | Design principles for quay walls and superstructures in concrete (R 17) | 486 |
| 10.1.1 | General principles | 486 |
| 10.1.2 | Edge protection | 486 |
| 10.1.3 | Facing | 487 |
| 10.2 | Design and construction of reinforced concrete components in waterfront structures (R 72) | 487 |
| 10.2.1 | Preliminary remarks | 487 |
| 10.2.2 | Concrete | 487 |
| 10.2.3 | Construction joints | 488 |
| 10.2.4 | Structures with large longitudinal dimensions | 489 |
| 10.2.5 | Crack width limitation | 490 |
| 10.3 | Formwork in areas affected by tides and waves (R 169) | 491 |
| 10.4 | Box caissons as waterfront structures in seaports (R 79) | 491 |
| 10.4.1 | General | 491 |
| 10.4.2 | Design | 492 |
| 10.4.3 | Safety against sliding | 492 |

| | | |
|---|---|---|
| 10.4.4 | Construction details | 493 |
| 10.4.5 | Construction work | 493 |
| 10.5 | Compressed-air caissons as waterfront structures (R 87) | 493 |
| 10.5.1 | General | 493 |
| 10.5.2 | Verification | 495 |
| 10.5.3 | Safety against sliding | 495 |
| 10.5.4 | Construction details | 495 |
| 10.5.5 | Work on site | 497 |
| 10.5.6 | Frictional resistance during sinking | 497 |
| 10.6 | Design and construction of block-type quay walls (R 123) | 498 |
| 10.6.1 | Basic principles | 498 |
| 10.6.2 | Forces acting on a block wall | 500 |
| 10.6.3 | Design | 501 |
| 10.7 | Design of quay walls using open caissons (R 147) | 503 |
| 10.7.1 | General | 503 |
| 10.7.2 | Verification | 504 |
| 10.7.3 | Construction details | 504 |
| 10.7.4 | Work on site | 506 |
| 10.7.5 | Frictional resistance during sinking | 506 |
| 10.7.6 | Preparation of the subsoil | 506 |
| 10.8 | Design and construction of solid waterfront structures (e.g. blocks, box caissons, compressed-air caissons) in earthquake zones (R 126) | 507 |
| 10.8.1 | General | 507 |
| 10.8.2 | Active and passive earth pressures, excess water pressure, variable loads | 507 |
| 10.8.3 | Safety | 507 |
| 10.8.4 | Base of the wall | 507 |
| 10.9 | Use and design of bored cast-in-place piles (R 86) | 507 |
| 10.9.1 | General | 507 |
| 10.9.2 | Design | 507 |
| 10.9.3 | Construction of bored cast-in-place pile walls | 509 |
| 10.9.4 | Construction guidance | 509 |
| 10.10 | Use and design of diaphragm walls (R 144) | 510 |
| 10.10.1 | General | 510 |
| 10.10.2 | Verifying the stability of the open trench | 512 |
| 10.10.3 | Composition of the supporting slurry | 512 |
| 10.10.4 | Diaphragm wall construction | 513 |
| 10.10.5 | Concrete and reinforcement | 513 |
| 10.10.6 | Guidance for the design of diaphragm walls | 514 |
| 10.11 | Survey prior to repairing concrete components in hydraulic engineering structures (R 194) | 515 |
| 10.11.1 | General | 515 |
| 10.11.2 | Tests performed on the structure | 516 |
| 10.11.3 | Tests performed in the laboratory | 517 |

| | | |
|---|---|---|
| 10.11.4 | Theoretical investigations. | 518 |
| 10.12 | Repairing concrete components in hydraulic engineering structures (R 195) | 518 |
| 10.12.1 | General | 518 |
| 10.12.2 | Assessing the actual condition | 519 |
| 10.12.3 | Planning the repair works. | 520 |
| 10.12.4 | Execution of the repair works. | 521 |
| | | |
| **11** | **Pile bents and trestles** | **528** |
| 11.1 | General | 528 |
| 11.2 | Calculating subsequently strengthened pile bents/trestles (R 45) | 528 |
| 11.2.1 | General | 528 |
| 11.2.2 | Loads | 529 |
| 11.2.3 | Calculation for cohesive substrata | 530 |
| 11.2.4 | Load from excess water pressure | 530 |
| 11.3 | Design of plane pile bents (R 78). | 531 |
| 11.4 | Design of spatial pile trestles (R 157). | 534 |
| 11.4.1 | Special structures designed as spatial pile trestles | 535 |
| 11.4.2 | Free-standing pile trestles. | 535 |
| 11.4.3 | Structural system and calculations | 537 |
| 11.4.4 | Construction guidance | 537 |
| 11.5 | Design of piled structures in earthquake zones (R 127). | 539 |
| 11.5.1 | General | 539 |
| 11.5.2 | Active and passive earth pressures, excess water pressure, variable loads | 539 |
| 11.5.3 | Resisting the horizontal inertial forces of the superstructure | 539 |
| | | |
| **12** | **Protection and stabilisation structures** | **541** |
| 12.1 | Embankment stabilisation on inland waterways (R 211) | 541 |
| 12.1.1 | General | 541 |
| 12.1.2 | Loads on inland waterways | 541 |
| 12.1.3 | Construction of bank protection | 542 |
| 12.1.4 | Toe protection | 545 |
| 12.1.5 | Junctions | 546 |
| 12.1.6 | Design of revetments | 547 |
| 12.2 | Slopes in seaports and tidal inland ports (R 107) | 547 |
| 12.2.1 | General | 547 |
| 12.2.2 | Examples of impermeable revetments. | 550 |
| 12.3 | Use of geotextile filters in bank and bottom protection (R 189). | 552 |
| 12.3.1 | General | 552 |
| 12.3.2 | Design principles. | 552 |
| 12.3.3 | Requirements. | 553 |
| 12.3.4 | Additional measures | 553 |

| 12.3.5 | General installation guidelines | 554 |
|---|---|---|
| 12.4 | Scour and protection against scour in front of waterfront structures (R 83) | 555 |
| 12.4.1 | General | 555 |
| 12.4.2 | Choosing a greater design depth (allowance for scouring) | 556 |
| 12.4.3 | Covering the bottom (scour protection) | 557 |
| 12.4.4 | Current velocity at revetment due to propeller wash | 560 |
| 12.4.5 | Designing bottom protection | 564 |
| 12.5 | Scour protection at piers and dolphins | 566 |
| 12.6 | Installation of mineral impervious linings underwater and their connection to waterfront structures (R 204) | 567 |
| 12.6.1 | Concept | 567 |
| 12.6.2 | Installation in dry conditions | 567 |
| 12.6.3 | Installation in wet conditions | 567 |
| 12.6.4 | Connections | 568 |
| 12.7 | Flood defence walls in seaports (R 165) | 569 |
| 12.7.1 | General | 569 |
| 12.7.2 | Critical water levels | 569 |
| 12.7.3 | Excess water pressure and unit weight of soil | 570 |
| 12.7.4 | Minimum embedment depths for flood defence walls | 571 |
| 12.7.5 | Special loads on flood defence walls | 571 |
| 12.7.6 | Guidance on designing flood defence walls in slopes | 572 |
| 12.7.7 | Constructional measures | 572 |
| 12.7.8 | Buried services in the region of flood defence walls | 573 |
| 12.8 | Dumped moles and breakwaters (R 137) | 574 |
| 12.8.1 | General | 574 |
| 12.8.2 | Stability analyses, settlement and subsidence, guidance on construction | 574 |
| 12.8.3 | Specifying the geometry of the structure | 575 |
| 12.8.4 | Designing the armour layer | 580 |
| 12.8.5 | Construction of breakwaters | 582 |
| 12.8.6 | Construction and use of plant | 583 |
| 12.8.7 | Settlement and subsidence | 585 |
| 12.8.8 | Invoicing for installed quantities | 586 |
| **13** | **Dolphins (R 218)** | **587** |
| 13.1 | General principles | 587 |
| 13.1.1 | Dolphins – purposes and types | 587 |
| 13.1.2 | Stiffness of the system | 587 |
| 13.1.3 | Loads on dolphins and design principles | 587 |
| 13.1.4 | Actions | 590 |
| 13.1.5 | Safety concept | 592 |
| 13.2 | Design of dolphins | 592 |
| 13.2.1 | Soil–structure interaction and the resulting design variables | 592 |

| 13.2.2 | Required energy absorption capacity of breasting dolphins | 599 |
| 13.2.3 | Other calculations | 600 |
| 13.3 | Construction and arrangement of dolphins (R 128) | 601 |
| 13.3.1 | Type of dolphin structure | 601 |
| 13.3.2 | Layout of dolphins | 601 |
| 13.3.3 | Equipment for dolphins | 602 |
| 13.3.4 | Advice for selecting materials | 603 |

## 14 Inspection and monitoring of waterfront structures (R 193) ........ 604

| 14.1 | General | 604 |
| 14.2 | Documentation | 606 |
| 14.3 | Carrying out structural inspections | 606 |
| 14.3.1 | Structural check/Principle check | 606 |
| 14.3.2 | Structural monitoring/Intermediate inspection | 607 |
| 14.3.3 | Structural survey/Routine inspection | 608 |
| 14.4 | Inspection intervals | 608 |
| 14.5 | Maintenance management systems | 609 |

## Annex I    Bibliography ........................................................ 610

| I.1 | Annual technical reports | 610 |
| I.2 | Books and papers | 611 |
| I.3 | Technical standards | 623 |

## Annex II    Notation ............................................................. 626

| II.1a | Latin lower-case letters | 626 |
| II.1b | Latin upper-case letters | 627 |
| II.1c | Greek letters | 629 |
| II.2 | Subscripts and indexes | 630 |
| II.3 | Abbreviations | 631 |
| II.4 | Designations for water levels and wave heights | 632 |

## Annex III    List of keywords .................................................. 633

# List of Recommendations in the 9th Edition

| | | Section | Page |
|---|---|---|---|
| R 1 | Layout and depths of boreholes and penetrometer tests | 1.2 | 11 |
| R 2 | Considering the cohesion in cohesive soils | 2.2 | 27 |
| R 3 | Considering the apparent cohesion (capillary cohesion) in sand | 2.3 | 27 |
| R 4 | Applying the angle of earth pressure and the analysis in the vertical direction | 8.2.5 | 373 |
| R 5 | Vertical imposed loads | 5.5 | 106 |
| R 6 | Standard cross-section dimensions for waterfront structures in seaports | 6.1 | 168 |
| R 7 | Combined steel sheet piling | 8.1.4 | 288 |
| R 9 | Mean characteristic values of soil parameters | 1.1 | 11 |
| R 10 | Verification of stability for anchoring at the lower failure plane | 8.5 | 469 |
| R 12 | Layout of and loads on bollards for sea-going vessels | 5.12 | 147 |
| R 14 | Layout and design of and loads on access ladders | 6.12 | 194 |
| R 17 | Design principles for quay walls and superstructures in concrete | 10.1 | 486 |
| R 18 | Design situations | 5.4 | 105 |
| R 19 | Excess water pressure in direction of water side | 4.2 | 73 |
| R 20 | Verifying the loadbearing capacity of the elements of sheet piling structures | 8.2.7 | 386 |
| R 21 | Design and installation of reinforced concrete sheet pile walls | 8.1.2 | 284 |
| R 22 | Design and installation of timber sheet pile walls | 8.1.1 | 281 |
| R 23 | Danger of sand abrasion on sheet piling | 8.1.9 | 309 |
| R 24 | Layout and design of stairs in seaports | 6.13 | 197 |
| R 29 | Design of steel walings for sheet piling | 8.4.1 | 424 |
| R 30 | Verification of steel walings | 8.4.2 | 425 |
| R 31 | Design of protruding quay wall corners with round steel tie rods | 8.4.10 | 450 |
| R 32 | Design of drainage systems for waterfront structures in tidal areas | 4.5 | 78 |
| R 34 | Design and installation of steel sheet pile walls | 8.1.3 | 287 |
| R 35 | Corrosion of steel sheet piling, and countermeasures | 8.1.8 | 300 |
| R 36 | Nominal depth and design depth of harbour bottom | 6.7 | 182 |
| R 38 | Berthing force of ships at quays | 5.2 | 103 |
| R 39 | Ship dimensions | 5.1 | 93 |
| R 40 | Berthing velocities of ships transverse to berth | 5.3 | 103 |
| R 41 | Steel sheet piling with staggered embedment depths | 8.2.10 | 394 |
| R 42 | Staggered arrangement of anchor walls | 8.2.13 | 401 |

| R 43 | Waterfront sheet piling in unconsolidated, soft cohesive soils, especially in connection with non-sway structures | 8.2.15 | 402 |
| R 45 | Calculating subsequently strengthened pile bents/ trestles | 11.2 | 528 |
| R 47 | Fenders in inland ports | 6.16 | 219 |
| R 50 | Sheet piling anchors in unconsolidated, soft cohesive soils | 8.4.9 | 447 |
| R 51 | Design of weepholes for sheet piling structures | 4.4 | 76 |
| R 52 | Effect of artesian water pressure under harbour bottom or river bed on active and passive earth pressures | 2.10 | 40 |
| R 53 | Relieving artesian pressure beneath harbour bottoms | 4.6 | 80 |
| R 55 | Selection of embedment depth for sheet piling | 8.2.8 | 390 |
| R 56 | Determining the embedment depth for sheet pile walls with full or partial fixity in the soil | 8.2.9 | 391 |
| R 57 | Steel sheet piling founded on bedrock | 8.2.14 | 401 |
| R 58 | Mean groundwater level | 4.1 | 73 |
| R 59 | Sheet piling walings of reinforced concrete with driven steel anchor piles | 8.4.3 | 427 |
| R 60 | Fenders for large vessels | 6.15 | 202 |
| R 65 | Excess water pressure on sheet piling in front of embankments below elevated platforms in tidal areas | 4.3 | 75 |
| R 67 | Quality requirements for steels and dimensional tolerances for steel sheet piles | 8.1.6 | 296 |
| R 68 | Embankments below waterfront wall superstructures behind closed sheet pile walls | 6.9 | 189 |
| R 70 | Provision of quick-release hooks at berths for large vessels | 6.11 | 193 |
| R 72 | Design and construction of reinforced concrete components in waterfront structures | 10.2 | 487 |
| R 73 | Backfilling of waterfront structures | 7.4 | 249 |
| R 74 | Standard cross-sections for waterfront structures in inland ports | 6.3 | 172 |
| R 77 | Design of sheet piling structures with fixity in the ground and a single anchor | 8.2.3 | 367 |
| R 78 | Design of plane pile bents | 11.3 | 531 |
| R 79 | Box caissons as waterfront structures in seaports | 10.4 | 491 |
| R 80 | Dredging in front of quay walls in seaports | 7.1 | 238 |
| R 81 | Hydraulic filling of port areas for planned waterfront structures | 7.3 | 242 |
| R 83 | Scour and protection against scour in front of waterfront structures | 12.4 | 555 |
| R 84 | Quay loads from cranes and other transhipment equipment | 5.14 | 151 |
| R 85 | Fixing crane rails to concrete | 6.18 | 223 |

| | | | |
|---|---|---|---|
| R 86 | Use and design of bored cast-in-lace piles | 10.9 | 507 |
| R 87 | Compressed-air caissons as waterfront structures | 10.5 | 493 |
| R 88 | Determining the shear strength $c_u$ of saturated, undrained cohesive soils | 1.4 | 19 |
| R 90 | Driving of steel sheet piles and steel piles at low temperatures | 8.1.15 | 330 |
| R 91 | Cutting off the tops of driven steel sections for loadbearing welded connections | 8.1.20 | 347 |
| R 93 | Vertical drains to accelerate the consolidation of soft cohesive soils | 7.11 | 265 |
| R 94 | Steel nosings to protect reinforced concrete walls and capping beams on waterfront structures | 8.4.6 | 441 |
| R 95 | Steel capping beams for sheet piling waterfront structures | 8.4.4 | 431 |
| R 98 | Acceptance conditions for steel sheet piles and steel piles on site | 8.1.7 | 298 |
| R 99 | Design of welded joints in steel piles and steel sheet piles | 8.1.19 | 344 |
| R 100 | Cellular cofferdams as excavation enclosures and waterfront structures | 8.3.1 | 405 |
| R 101 | Double-wall cofferdams as excavation enclosures and waterfront structures | 8.3.2 | 417 |
| R 102 | Layout, design and loading of bollards for inland facilities | 5.13 | 149 |
| R 103 | Shear-resistant interlock connections for steel sheet piling | 8.1.5 | 292 |
| R 104 | Driving combined steel sheet piling | 8.1.12 | 316 |
| R 105 | Monitoring during the installation of sheet piles, tolerances | 8.1.13 | 321 |
| R 106 | Sheet piling waterfronts on inland waterways | 6.4 | 175 |
| R 107 | Slopes in seaports and tidal inland ports | 12.2 | 547 |
| R 109 | Soil replacement along a line of piles for a waterfront structure | 7.9 | 258 |
| R 110 | Considering active earth pressure and excess water pressure, and construction guidance for waterfront structures with soil replacement and contaminated or disturbed base of excavation | 2.11 | 42 |
| R 113 | Taking account of groundwater flow | 4.7 | 81 |
| R 114 | Effect of groundwater flow on excess water pressure and active and passive earth pressures | 2.12 | 46 |
| R 115 | Safety against hydraulic heave failure | 3.1 | 64 |
| R 116 | Piping (ground failure due to internal erosion) | 3.2 | 70 |
| R 117 | Watertightness of steel sheet piling | 8.1.21 | 347 |
| R 118 | Driving steel sheet piles | 8.1.11 | 312 |

| | | | | |
|---|---|---|---|---|
| R 119 | Upgrading partially sloped waterfronts in inland ports with large water level fluctuations | 6.5 | 178 |
| R 120 | Foundations to craneways on waterfront structures | 6.17 | 220 |
| R 121 | Waterfront structures in regions with mining subsidence | 8.1.22 | 350 |
| R 122 | Top edges of waterfront structures in seaports | 6.2 | 170 |
| R 123 | Design and construction of block-type quay walls | 10.6 | 498 |
| R 124 | Waterfront structures in seismic regions | 2.16 | 57 |
| R 125 | Design of single-anchor sheet piling structures in earthquake zones | 8.2.16 | 404 |
| R 126 | Design and construction of solid waterfront structures (e.g. blocks, box caissons, compressed-air caissons) in earthquake zones | 10.8 | 507 |
| R 127 | Design of piled structures in earthquake zones | 11.5 | 539 |
| R 128 | Construction and arrangement of dolphins | 13.3 | 601 |
| R 129 | Reinforced concrete capping beams for waterfront structures with steel sheet piling | 8.4.5 | 435 |
| R 130 | Determining active earth pressure in saturated, non- or partially consolidated, soft cohesive soils | 2.9 | 38 |
| R 132 | Horizontal actions on steel sheet pile walls in the longitudinal direction of the quay | 8.2.11 | 397 |
| R 133 | Auxiliary anchors at the top of steel sheet piling structures | 8.4.7 | 444 |
| R 134 | Design of sheet pile walls with double anchors | 8.2.4 | 372 |
| R 135 | Wave pressure on vertical quay walls in coastal areas | 5.7 | 120 |
| R 136 | Determining the "design sea state" for maritime and port structures | 5.6 | 110 |
| R 137 | Dumped moles and breakwaters | 12.8 | 574 |
| R 138 | Dredging underwater slopes | 7.7 | 254 |
| R 139 | Dredging and hydraulic fill tolerances | 7.2 | 240 |
| R 140 | Design of piling frames | 8.1.18 | 339 |
| R 144 | Use and design of diaphragm walls | 10.10 | 510 |
| R 145 | Hinged connections between driven steel anchor piles and steel sheet piling structures | 8.4.13 | 458 |
| R 146 | Design and calculation of protruding quay wall corners with raking anchor piles | 8.4.11 | 453 |
| R 147 | Design of quay walls using open caissons | 10.7 | 503 |
| R 149 | Noise control – low-noise driving | 8.1.14 | 325 |
| R 150 | Geotechnical report | 1.3 | 18 |
| R 151 | High prestressing of anchors of high-strength steel for waterfront structures | 8.4.12 | 457 |
| R 152 | Design of anchor walls fixed in the ground | 8.2.12 | 400 |
| R 153 | Wind loads on moored ships and their influence on the dimensioning of mooring and fender equipment | 5.11 | 145 |

| | | | |
|---|---|---|---|
| R 154 | Assessing the subsoil for the installation of piles and sheet piles and for selecting the installation method | 1.5 | 22 |
| R 157 | Design of spatial pile trestles | 11.4 | 534 |
| R 158 | Design of waterfront areas in inland ports according to operational aspects | 6.6 | 180 |
| R 159 | Wave pressure on piled structures | 5.10 | 129 |
| R 161 | Free-standing sheet piling structures | 8.2.2 | 366 |
| R 162 | Narrow moles in sheet piling | 8.3.3 | 422 |
| R 164 | Measures for increasing the passive earth pressure in front of waterfront structures | 2.14 | 54 |
| R 165 | Flood defence walls in seaports | 12.7 | 569 |
| R 166 | Temporary stabilisation of waterfront structures by groundwater lowering | 4.8 | 90 |
| R 167 | Repairing interlock declutching on driven steel sheet piling | 8.1.16 | 331 |
| R 168 | Subsidence of non-cohesive soils | 7.8 | 257 |
| R 169 | Formwork in areas affected by tides and waves | 10.3 | 491 |
| R 171 | Determining active earth pressure according to the Culmann method | 2.4 | 28 |
| R 172 | Determining the active earth pressure shielding on a wall below a relieving platform with average ground surcharges | 2.7 | 34 |
| R 173 | Equipment for waterfront structures in seaports with supply and disposal systems | 6.14 | 199 |
| R 174 | Determining the amount of displacement required for mobilising passive earth pressure in non-cohesive soils | 2.13 | 53 |
| R 175 | In situ density of hydraulically filled non-cohesive soils | 7.5 | 251 |
| R 176 | Reinforced steel sheet piling | 8.1.17 | 334 |
| R 177 | Impact and pressure of ice on waterfront structures, fenders and dolphins in coastal areas | 5.15 | 155 |
| R 178 | In situ density of dumped non-cohesive soils | 7.6 | 252 |
| R 179 | Consolidation of soft cohesive soils by preloading | 7.12 | 268 |
| R 183 | Blasting to assist the driving of steel sheet piles | 8.1.10 | 309 |
| R 184 | Screw threads for sheet piling anchors | 8.4.8 | 445 |
| R 185 | Loads arising from surging and receding waves due to the inflow or outflow of water | 5.8 | 125 |
| R 186 | Effects of waves due to ship movements | 5.9 | 126 |
| R 188 | Dynamic compaction of the soil | 7.10 | 264 |
| R 189 | Use of geotextile filters in bank and bottom protection | 12.3 | 552 |
| R 190 | Passive earth pressure in front of abrupt changes in ground level in soft cohesive soils with rapid load application on land side | 2.15 | 57 |

| | | | |
|---|---|---|---|
| R 191 | Connection of expansion joint seal in reinforced concrete bottom to loadbearing steel sheet pile wall | 6.19 | 231 |
| R 193 | Inspection and monitoring of waterfront structures | 14 | 604 |
| R 194 | Survey prior to repairing concrete components in hydraulic engineering structures | 10.11 | 515 |
| R 195 | Repairing concrete components in hydraulic engineering structures | 10.12 | 518 |
| R 196 | Connecting steel sheet piling to a concrete structure | 6.20 | 231 |
| R 198 | Determining active earth pressure for a steep, paved embankment in a partially sloping waterfront structure | 2.6 | 32 |
| R 199 | Taking account of unfavourable groundwater flows in the passive earth pressure zone | 8.2.6 | 386 |
| R 200 | Strengthening waterfront structures for deepening harbour bottoms in seaports | 6.8 | 184 |
| R 201 | Redesign of waterfront structures in inland ports | 6.10 | 190 |
| R 202 | Vibratory driving of U- and Z-section steel sheet piles | 8.1.23 | 353 |
| R 203 | Water-jetting to assist the driving of steel sheet piles | 8.1.24 | 357 |
| R 204 | Installation of mineral impervious linings underwater and their connection to waterfront structures | 12.6 | 567 |
| R 205 | Impact and pressure of ice on waterfront structures, piers and dolphins at inland facilities | 5.16 | 164 |
| R 206 | Floating berths in seaports | 6.21 | 236 |
| R 210 | Improving the bearing capacity of soft cohesive soils with vertical elements | 7.13 | 275 |
| R 211 | Embankment stabilisation on inland waterways | 12.1 | 541 |
| R 212 | Pressing of U- and Z-section steel sheet piles | 8.1.25 | 360 |
| R 213 | Loads on waterfront structures and dolphins caused by fender reaction forces | 5.17 | 167 |
| R 214 | Partial safety factors for loads and resistances | 8.2.1.1 | 363 |
| R 215 | Determining the design values for the bending moments | 8.2.1.2 | 363 |
| R 216 | Partial safety factor for hydrostatic pressure | 8.2.1.3 | 365 |
| R 217 | Tension piles and anchors | 9 | 478 |
| R 218 | Dolphins | 13 | 587 |
| R 219 | Active earth pressure in stratified soil | 2.5 | 30 |
| R 220 | Earth pressure distribution under limited loads | 2.8 | 37 |

# Preface to 11th, revised edition (9th English edition) of the *Recommendations of the Committee for Waterfront Structures – Harbours and Waterways*

Eight years have passed since the 10th German (8th English) edition of the *Recommendations of the Committee for Waterfront Structures* was published. During that period the annual, in some cases six-monthly, technical reports of the years 2005 to 2011 have contained innovations and improvements. This 9th English edition (the translation of the 11th German edition), simply called the "EAU" by those in the know, represents a completely updated version of the recommendations of the Waterfront Structures Committee, a body organised jointly by the German Port Technology Association (HTG) and the German Geotechnical Society (DGGT). I feel sure that this edition, too, will become a standard work of reference for every engineer working on waterfront structures. The main changes to the content are to be found in chapter 1 (production of geotechnical report and calculation of undrained shear strengths), chapter 2 (calculations with total and effective stresses), section 8.1 (installation of sheet pile walls and supervision of such installation work), section 8.2 (verification of vertical load-carrying capacity) and chapter 13 (using the $p$-$y$ method to design dolphins). The previous chapter 14 has been incorporated in other parts of the EAU and the old chapter 15 renumbered accordingly, leaving this edition with just 14 chapters. Furthermore, the notation has been amended to match Eurocode 7 and Germany's National Application Document DIN 1054, which are now valid.

The principle for constituting committees laid down by the German Institute for Standarization (DIN), i.e. appropriate representation of all groups with an interest and availability of the necessary expertise, is followed by the EAU committee. Therefore, the committee is made up of members from all relevant disciplines and drawn from universities, the building departments of large seaports, inland ports and national waterways, the construction industry, the steel industry and consulting engineers.

The following members of the working committee were involved in preparing EAU 2012:

Univ.-Prof. Dr.-Ing. Jürgen Grabe, Hamburg (chair since 2009)
Ir. Tom van Autgaerden, Antwerp
Dipl.-Ing. Dirk Busjaeger, Hamburg
Dr. ir. Jakob Gerrit de Gijt, Rotterdam
Dr.-Ing. Michael Heibaum, Karlsruhe
Dr.-Ing. Stefan Heimann, Berlin
Prof. ir. Aad van der Horst, Delft
Dipl.-Ing. Hans-Uwe Kalle, Hagen

Prof. Dr.-Ing. Roland Krengel, Duisburg
Dipl.-Ing. Karl-Heinz Lambertz, Duisburg
Dr.-Ing. Christoph Miller, Hamburg
Dr.-Ing. Karl Morgen, Hamburg
Dipl.-Ing. Gabriele Peschken, Bonn
Dipl.-Ing. Torsten Retzlaff, Rostock
Dipl.-Ing. Emile Reuter, Luxembourg
Univ.-Prof. Dr.-Ing. Werner Richwien, Essen (chair until 2009)
Dr.-Ing. Peter Ruland, Hamburg
Dr.-Ing. Wolfgang Schwarz, Schrobenhausen
Dr.-Ing. Hartmut Tworuschka, Hamburg
Dr.-Ing. Hans-Werner Vollstedt, Bremerhaven

In a similar way to the work of the DIN when producing a standard, new recommendations are presented for public discussion in the form of provisional recommendations in the annual technical reports. After considering any objections, recommendations are published in their final form in the following annual technical report. Annex I contains a list of the annual technical reports relevant to this edition. The status of the *Recommendations of the Committee for Waterfront Structures – Harbours and Waterways* is therefore equivalent to that of a standard. Seen from the point of view of its relevance to practice and also the dissemination of experience, however, the information provided goes beyond that of a standard; this publication can be seen more as a "code of practice".

As the European standardisation concept is now fully incorporated in the EAU, this edition satisfies the requirements for notification by the European Commission. It is registered with the European Commission under notification No. 2012/426D.

The fundamental revisions in EAU 2012 made in-depth discussions with colleagues outside the committee necessary, even the setting-up of temporary study groups to deal with specific topics. The committee acknowledges the assistance of all colleagues who in this way made a significant contribution to the development of EAU 2012.

In addition, considerable input from experts plus recommendations from other committees and international engineering science bodies have found their way into these recommendations.

So, with such additions and the results of revision work, EAU 2012 corresponds to today's international standards. Experts working in this sector now have at their disposal an updated edition adapted to the European standards which will continue to supply valuable help for issues in design, tendering, award of contract, engineering tasks, economic and environmentally compatible construction, site supervision and contractual procedures. It will therefore be possible to design and build waterfront structures that are in line with the state of the art and have consistent specifications.

The committee would like to thank all those who contributed to and made suggestions for this edition. It is hoped that EAU 2012 will attract the same resonance as earlier editions.

A special vote of thanks goes to my colleague Univ.-Prof. Dr.-Ing. Werner Richwien, who chaired this committee with dedication over many years. He created a working climate that had a positive influence on the motivation of every committee member and will shape the work of the committee in the coming years. I would also like to thank my assistants Dr.-Ing. Hans Mathäus Hügel and Dipl.-Ing. Torben Pichler, who read through the chapters and organised the production process. Only through their efforts it became possible to meet the deadline for printing the 11th German edition in 2012.

I am also grateful to the publishers Ernst & Sohn for the good cooperation, the careful preparation of the many illustrations, tables and equations and the excellent quality of the printing and layout of EAU 2012.

Hamburg, October 2012                    Univ.-Prof. Dr.-Ing. *Jürgen Grabe*

# 0 Structural calculations

## 0.1 General

The recommendations of the "Waterfront Structures" working committee have been repeatedly adjusted in line with the relevant standards. This applied and indeed continues to apply to the safety criteria defined in the standards. Up until the 8th German edition (EAU 1990), the earth pressure calculations were based on reduced soil parameter values, known as "calculation values" with the prefix "cal". The results of calculations using these values then had to fulfil the global safety criteria in accordance with recommendation R 96, section 1.13.2a, of EAU 1990. The publication of EAU 1996 resulted in a changeover to the concept of partial safety factors. It was agreed within the European Union that this safety concept should be pursued in a uniform manner by all Member States.

The "Eurocodes" (EC) – harmonised directives specifying fundamental safety requirements for buildings and structures – were drawn up as part of the realisation of the European Single Market. Those Eurocodes are as follows:

DIN EN 1990, EC 0: Basis of structural design
DIN EN 1991, EC 1: Actions on structures
DIN EN 1992, EC 2: Design of concrete structures
DIN EN 1993, EC 3: Design of steel structures
DIN EN 1994, EC 4: Design of composite steel and concrete structures
DIN EN 1995, EC 5: Design of timber structures
DIN EN 1996, EC 6: Design of masonry structures
DIN EN 1997, EC 7: Geotechnical design
DIN EN 1998, EC 8: Design of structures for earthquake resistance
DIN EN 1999, EC 9: Design of aluminium structures

The Eurocodes "Basis of structural design" (DIN EN 1990) and "Actions on structures" (DIN EN 1991) with their various parts and annexes form the basis of European construction standards, the starting point for building designs throughout Europe. The other eight Eurocodes, along with their respective parts, relate to these two basic standards.

Verification of safety must always be carried out according to European standards. However, in some instances such verification is not possible with these standards alone; a number of parameters, e.g. numerical values for partial safety factors, have to be specified on a national basis. Furthermore, the Eurocodes do not cover the entire range of German standards, meaning that a comprehensive set of national standards has been retained. However, this set of German standards along with its requirements must not contradict the regulations contained in the European standards, which in turn necessitated the revision of national standards.

For proof of stability according to the EAU, the standards DIN EN 1990 to DIN EN 1999, but especially DIN EN 1997 (Geotechnical design), are of particular importance. DIN EN 1997-1 defines a number of terms and describes and stipulates limit state verification procedures. The various earth pressure design models for stability calculations are also included in the annexes for information purposes. A particular feature here is that three methods of verification using the partial safety factor concept are available for use throughout Europe.

The publication of DIN 1054:2010 ensured that any duplication of DIN EN 1997-1 was avoided, but specific German experience has been retained. This standard was combined with DIN EN 1997-1:2010 and the National Annex (DIN EN 1997-1/NA:2010) to create the EC 7-1 manual [85].

DIN EN 1997-2 governs the planning, execution and evaluation of soil investigations. As for part 1, this standard has been published together with DIN 4020:2010 and the National Application Document in the EC 7-2 manual [86].

The existing German execution standards have been replaced by new European standards under the general designation "Execution of special geotechnical works". However, this process is still ongoing.

Likewise, the German calculation standards, containing individual safety stipulations in some instances, have been revised so that all safety criteria are now defined in DIN 1054.

Where standards are cited in the recommendations, then the current version applies, unless stated otherwise. All standards cited are listed in annex I.3.

## 0.2 Safety concept

### 0.2.1 General

A structure can fail as a result of exceeding the ultimate limit state of bearing capacity ("ultimate limit state – ULS", failure of the soil or the structure, loss of static equilibrium) or the limit state of serviceability ("serviceability limit state – SLS", excessive deformations).

In order to verify the ultimate limit state of bearing capacity, three cases were distinguished in the past (five from now on):

| DIN 1054:2005-01 | | EC 7-1 manual | |
|---|---|---|---|
| Loss of static equilibrium | LS 1A | Loss of equilibrium of structure or ground | EQU |
| | | Loss of equilibrium of structure or ground due to uplift by water pressure (buoyancy) | UPL |
| | | Hydraulic heave, internal erosion or piping in the ground due to hydraulic gradients | HYD |
| Failure of structures or components due to failure in the structure or supporting subsoil | LS 1B | Internal failure or very large deformation of the structure or its components | STR |
| | | Failure or very large deformation of the ground | GEO-2 |
| Loss of overall stability | LS 1C | Loss of overall stability | GEO-3 |

DIN EN 1997-1 permits three options for verifying safety, designated "design approaches 1 to 3". For approach 1, two groups of factors are taken into account and used in two separate analyses. For approaches 2 and 3, a single analysis with one group of factors suffices.

In approaches 1 and 2, the factors are applied, in principle, to either actions or action effects and to resistances. However, DIN 1054 stipulates that the characteristic, or representative, effects $E_{Gk,i}$ or $E_{Qrep,i}$ (e.g. shear forces, reactions, bending moments, stresses in the relevant sections of the structure and at interfaces between structure and subsoil) are determined first and then the factors are applied. This is also referred to as design approach $2^*$.

In approach 3, the factors are applied to the soil parameters and to actions or action effects not related to the subsoil. Actions or action effects induced by the subsoil are calculated from the factored soil parameters.

According to DIN 1054, design approach 2 (2*) should be used for the geotechnical analysis of limit states STR and GEO-2, and design approach 3 for analysing limit state GEO-3.

In the Eurocodes, the determination of design situations (DS) has superseded the differentiation between loading cases customary up until now:

- Loading case 1 becomes permanent design situation DS-P
- Loading case 2 becomes transient design situation DS-T
- Loading case 3 becomes accidental design situation DS-A

These design situations are assigned different partial safety factors and combination factors.

Additionally, design situation DS-E has been introduced for earthquakes. According to DIN EN 1990, no partial safety factors are applied in design situation DS-E.

The partial safety factors specified in DIN 1054 are reproduced in Tables R 0-1 to R 0-3.

**Table R 0-1.** Partial safety factors for actions and action effects (to DIN 1054:2010, Table A 2.1, with additions)

| Action or action effect | Symbol | Design situation | | |
|---|---|---|---|---|
| | | DS-P | DS-T | DS-A |
| **HYD and UPL: Limit state of failure due to hydraulic heave and buoyancy** | | | | |
| Destabilising permanent actions[a] | $\gamma_{G,dst}$ | 1.05 | 1.05 | 1.00 |
| Stabilising permanent actions | $\gamma_{G,stb}$ | 0.95 | 0.95 | 0.95 |
| Destabilising variable actions | $\gamma_{Q,dst}$ | 1.50 | 1.30 | 1.00 |
| Stabilising variable actions | $\gamma_{Q,stb}$ | 0 | 0 | 0 |
| Flow force in favourable subsoil | $\gamma_H$ | 1.35 | 1.30 | 1.20 |
| Flow force in unfavourable subsoil | $\gamma_H$ | 1.80 | 1.60 | 1.35 |
| **EQU: Limit state of loss of equilibrium** | | | | |
| Unfavourable permanent actions | $\gamma_{G,dst}$ | 1.10 | 1.05 | 1.00 |
| Favourable permanent actions | $\gamma_{G,stb}$ | 0.90 | 0.90 | 0,95 |
| Unfavourable variable actions | $\gamma_Q$ | 1.50 | 1.25 | 1.00 |
| **STR and GEO-2: Limit state of failure of structures, components and subsoil** | | | | |
| Action effects from permanent actions generally[a] | $\gamma_G$ | 1.35 | 1.20 | 1.00 |

**Table R 0-1.** (*Continued*)

| Action or action effect | Symbol | Design situation | | |
|---|---|---|---|---|
| | | DS-P | DS-T | DS-A |
| Action effects from permanent actions for calculating anchorage[b)] | $\gamma_G$ | 1.35 | 1.20 | 1.10 |
| Action effects from favourable permanent actions[c)] | $\gamma_{G,inf}$ | 1.00 | 1.00 | 1.00 |
| Action effects from permanent actions due to earth pressure at rest | $\gamma_{G,EO}$ | 1.20 | 1.10 | 1.00 |
| Water pressure in certain boundary conditions[d)] | $\gamma_{G,red}$ | 1.20 | 1.10 | 1.00 |
| Water pressure in certain boundary conditions for calculating anchorage[b)] | $\gamma_{G,red}$ | 1.20 | 1.10 | 1.10 |
| Action effects from unfavourable variable actions[e)] | $\gamma_Q$ | 1.50 | 1.30 | 1.00 |
| Action effects from unfavourable variable actions for calculating anchorage[b)] | $\gamma_Q$ | 1.50 | 1.30 | 1.10 |
| Action effects from favourable variable actions | $\gamma_Q$ | 0 | 0 | 0 |

**GEO-3: Limit state of failure due to loss of overall stability**

| | | | | |
|---|---|---|---|---|
| Permanent actions | $\gamma_G$ | 1.00 | 1.00 | 1.00 |
| Unfavourable variable actions | $\gamma_Q$ | 1.30 | 1.20 | 1.00 |

**SLS: Limit state of serviceability**

$\gamma_G = 1.00$ for permanent actions or action effects
$\gamma_Q = 1.00$ for variable actions or action effects

[a)] The permanent actions are understood to include permanent and variable water pressure. Differing from DIN 1054:2010, $\gamma_G = 1.00$ applies in DS-A except when verifying anchorage.
[b)] The design of anchorages (grouted anchors, micropiles, tension piles) also includes verifying stability at the lower failure plane according to R 10 (section 8.5) when dealing with retaining structures.
[c)] If during the determination of the design values of the tensile action effect a characteristic compressive action effect from favourable permanent actions is assumed to act simultaneously, then this should be considered with the partial safety factor $\gamma_{G,inf}$ (DIN 1054, 7.6.3.1, A(2)).
[d)] For waterfront structures in which larger displacements can be accommodated without damage, the partial safety factors $\gamma_{G,red}$ for water pressure may be used if the conditions according to section 8.2.1.3 are complied with (DIN 1054, A 2.4.7.6.1, A(3)).
[e)] Differing from DIN 1054:2010, $\gamma_Q = 1.00$ applies in DS-A except when verifying anchorage.
[f)] The permanent actions are understood to include permanent and variable water pressures.

**Table R 0-2.** Partial safety factors for geotechnical parameters (DIN 1054:2010, Table A 2.2)

| Soil parameter | Symbol | Design situation | | |
|---|---|---|---|---|
| | | DS-P | DS-T | DS-A |
| **HYD and UPL: Limit state of failure due to hydraulic heave and buoyancy** | | | | |
| Friction coefficient $\tan \varphi'$ of drained soil and friction coefficient $\tan \varphi_u$ of undrained soil | $\gamma_{\varphi'}, \gamma_{\varphi u}$ | 1.00 | 1.00 | 1.00 |
| Cohesion $c'$ of drained soil and shear strength $c_u$ of undrained soil | $\gamma_{c'}, \gamma_{cu}$ | 1.00 | 1.00 | 1.00 |
| **GEO-2: Limit state of failure of structures, components and subsoil** | | | | |
| Friction coefficient $\tan \varphi'$ of drained soil and friction coefficient $\tan \varphi_u$ of undrained soil | $\gamma_{\varphi'}, \gamma_{\varphi u}$ | 1.00 | 1.00 | 1.00 |
| Cohesion $c'$ of drained soil and shear strength $c_u$ of undrained soil | $\gamma_{c'}, \gamma_{cu}$ | 1.00 | 1.00 | 1.00 |
| **GEO-3: Limit state of failure due to loss of overall stability** | | | | |
| Friction coefficient $\tan \varphi'$ of drained soil and friction coefficient $\tan \varphi_u$ of undrained soil | $\gamma_{\varphi'}, \gamma_{\varphi u}$ | 1.25 | 1.15 | 1.10 |
| Cohesion $c'$ of drained soil and shear strength $c_u$ of undrained soil | $\gamma_{c'}, \gamma_{cu}$ | 1.25 | 1.15 | 1.10 |

**Table R 0-3.** Partial safety factors for resistances (to DIN 1054:2010, Table A 2.3, with additions)

| Resistance | Symbol | Design situation | | |
|---|---|---|---|---|
| | | DS-P | DS-T | DS-A |
| **STR and GEO-2: Limit state of failure of structures, components and subsoil** | | | | |
| Soil resistances | | | | |
| - Passive earth pressure and ground failure resistance | $\gamma_{R,e}, \gamma_{R,v}$ | 1.40 | 1.30 | 1.20 |
| - Passive earth pressure when determining bending moment[a)] | $\gamma_{R,e,red}$ | 1.20 | 1.15 | 1.10 |
| - Sliding resistance | $\gamma_{R,h}$ | 1.10 | 1.10 | 1.10 |
| Pile resistances from static and dynamic pile loading tests | | | | |
| - Base resistance | $\gamma_b$ | 1.10 | 1.10 | 1.10 |
| - Skin resistance (compression) | $\gamma_s$ | 1.10 | 1.10 | 1.10 |
| - Total resistance (compression) | $\gamma_t$ | 1.10 | 1.10 | 1.10 |
| - Skin resistance (tension) | $\gamma_{s,t}$ | 1.15 | 1.15 | 1.15 |

**Table R 0-3.** (*Continued*)

| Resistance | Symbol | Design situation | | |
|---|---|---|---|---|
| | | DS-P | DS-T | DS-A |
| Pile resistances based on empirical values | | | | |
| - Compression piles | $\gamma_b, \gamma_s, \gamma_t$ | 1.40 | 1.40 | 1.40 |
| - Tension piles (in exceptional cases only) | $\gamma_{s,t}$ | 1.50 | 1.50 | 1.50 |
| Pull-out resistances | | | | |
| - Ground or rock anchors | $\gamma_a$ | 1.40 | 1.30 | 1.20 |
| - Grout body of grouted anchors | $\gamma_a$ | 1.10 | 1.10 | 1.10 |
| - Flexible reinforcing elements | $\gamma_a$ | 1.40 | 1.30 | 1.20 |

[a] Reduction for calculating the bending moment only. For waterfront structures in which larger displacements can be accommodated without damage, the partial safety factors $\gamma_{R,e,red}$ for passive earth pressure may be used if the conditions according to section 8.2.1.2 are complied with (DIN 1054, A 2.4.7.6.1, A(3)).

Remarks:
- For limit state of failure due to loss of overall stability GEO-3, the partial safety factors for shear strength are to be taken from Table R 0-2, and pull-out resistances are multiplied by partial safety factors according to STR and GEO-2.
- The partial safety factor for the material resistance of steel tension members made from reinforced and prestressed steel for limit states GEO-2 and GEO-3 is given in DIN EN 1992-1-1 as $\gamma_M = 1.15$.
- The partial safety factor for the material resistance of flexible reinforcing elements for limit states GEO-2 and GEO-3 is given in *Recommendations for Design and Analysis of Earth Structures using Geosynthetic Reinforcements* [62].

Provided that greater displacements and deformations of the structure do not impair the stability and serviceability of the structure, as can be the case for waterfronts, ports, harbours and waterways, the partial safety factor $\gamma_G$ can be reduced for earth and water pressures in justified cases (DIN 1054, A 2.4.7.6.1, A(3)). This is exploited in EAU by using the factors in the form of $\gamma_{G,red}$ (Table R 0-1) and $\gamma_{R,e,red}$ (Table R 0-3). Furthermore, a partial safety factor $\gamma_G = \gamma_Q = 1.00$ is used for action effects due to permanent and unfavourable variable actions in design situation DS-A.

## 0.2.2 Combination factors

When calculating a design value for actions $F_d$ according to EN 1990, this value must either be stipulated directly or derived from representative values:

$$F_d = \gamma_F \cdot F_{rep}$$

where

$$F_{rep} = \psi \cdot F_k$$

$\gamma_F$ partial safety factor
$\psi$ combination factor

For permanent actions and the leading action of variable actions, then $F_{rep} = F_k$ applies.

In the case of several independent variable characteristic actions $Q_{k,i}$, DIN EN 1990 requires combinations with corresponding coefficients $\psi$ to be investigated for buildings and bridges. In such investigations, one of the independent actions should be taken as the leading action $Q_{k,1}$ on a case-by-case basis.

A combination factor $\psi = 1.00$ is usually used for waterfront structures. Exceptions are discussed in section 5.4.4.

For verifying safety against buoyancy (UPL) and safety against hydraulic heave (HYD), the design values $F_d$ are always calculated without considering combination factors.

## 0.2.3 Analysis of ultimate limit state

Numerical proof of adequate stability is carried out for limit states STR and GEO-2 with the help of design values (index $d$) for actions or action effects and resistances, and for limit state GEO-3 with the help of design values for actions or action effects and soil properties.

Verification of safety is assessed according to the following fundamental equation:

$$E_d \leq R_d$$

where

$E_d$ design value of sum of actions or action effects
$R_d$ design value of resistances derived from sum of resistances of soil or structural elements

When analysing the limit state of loss of equilibrium (EQU) or failure due to hydraulic heave (HYD) or buoyancy (UPL), it is necessary compare the design values for favourable and unfavourable or stabilising or destabilising actions and verify that the respective limit state condition is complied with. Resistances do not play a role in these analyses.

## 0.2.4 Analysis of serviceability limit state

Deformation analyses must be carried out for all structures whose function can be impaired or rendered ineffective through deformations. The deformations are calculated with the characteristic values of actions and soil reactions and must be less than the deformations permissible for correct functioning of the component or whole structure. Where applicable, the calculations should include the upper and lower bounds of the characteristic values.

In particular, deformation analyses must consider the course of actions over time in order to allow for critical deformation states during various operating and construction stages.

## 0.2.5 Geotechnical categories

The minimum requirements in terms of scope and quality of geotechnical investigations, calculations and monitoring measures are described by three geotechnical categories in accordance with EC 7: low (category 1), normal (category 2) and high (category 3) geotechnical difficulty. These are reproduced in DIN 1054, A 2.1.2. Waterfront structures should be allocated to category 2, or category 3 in the case of difficult subsoil conditions. A geotechnical expert should always be consulted.

## 0.2.6 Probabilistic analysis

Even though its origins lie in the idea of a probabilistic concept of verification, the concept of partial safety factors to EC 0 or DIN 1054 is, by its very nature, deterministic. A failure mechanism's compliance with the limit state equation merely shows that there is a sufficient probability that the mechanism investigated will not occur. If, on the other hand, a stability analysis is to include a statement on the likelihood of the occurrence of a limit state, it is necessary to carry out an analysis based on probabilities. If the scatters of the effects and the independent parameters of the resistances are known, which is frequently the case in waterfront engineering, analyses of stability based on probabilities can lead to more economic structures than is possible when using a deterministic analysis concept.

A probability-based analysis assumes that the independent variables describing the actions or action effects and resistances are inserted into the limit state equation as variables of their distribution densities $f(R)$ and $f(E)$ for each limit state being considered. The solution to the limit state equation $f(Z)$ itself then represents a function of a scattered variable:

$$f(Z) = f(R) - f(E).$$

The probability of failure $P_f$, or rather the reliability $1 - P_f$, of a design, is calculated from the integral of the function $f(Z)$ for negative arguments $Z$.

When investigating a larger number of mechanisms, it is expedient to determine the critical mechanism for the failure of a design by means of a tree analysis [8], [173]. Here, the mechanisms not relevant to failure are systematically eliminated, although correlations between the mechanisms are taken into account [191].

## 0.3 Calculations for waterfront structures

Waterfront structures should always be designed as simply as possible in structural terms with clearly defined paths for carrying the loads. The less uniform the subsoil, the greater is the need for statically determinate designs in order to avoid, as far as possible, any additional stresses from unequal deformations which cannot be taken into account properly. Accordingly, wherever possible, stability analyses should be broken down into steps that are as clear and straightforward as possible.

A stability analysis of a waterfront structure must include the following in particular:

- Details of the use of the facility
- Drawings of the structure with all essential, planned structural dimensions
- Brief description of the structure including, in particular, all details that are not readily identifiable from the drawings
- Design value of bottom depth
- Characteristic values of all actions
- Soil strata and associated characteristic values of soil parameters
- Critical levels of bodies of water related to the German NHN height reference system (previously mean sea level) or a local gauge datum, together with corresponding groundwater levels (no high water, no flooding)
- Combinations of actions, i.e. load cases
- Partial safety factors necessary/used
- Intended building materials and their strengths or resistances
- All data regarding construction timetables and construction operations, with critical temporary states
- Description of and justification for the intended verification procedures
- Information about literature used and other calculation aids

The actual analyses of stability and serviceability must take into account the fact that, in foundation and hydraulic engineering, appropriate soil investigations, shear parameters, load assumptions, the ascertaining of hydrodynamic influences and unconsolidated states, a favourable load-bearing system and a realistic computational model are more important than exaggeratedly accurate numerical calculations.

# 1 Subsoil

## 1.1 Mean characteristic values of soil parameters (R 9)

### 1.1.1 General

For preliminary designs, the characteristic values (index $k$) given in Table R 9-1 may be used as empirical values for a larger body of soil. Without verification, the values in the table may only be assumed for low penetration resistance or soft consistency.

Detailed and final designs should always be based on the soil parameter values determined by way of soil investigations and laboratory tests (R 88, section 1.4). Wherever possible, the effective shear parameters $\varphi'$ and $c'$ of cohesive soils should be ascertained in triaxial tests on undisturbed soil samples.

According to Wroth [228], the angle of internal friction $\varphi'$ for non-cohesive, densely bedded, compact soils in the plane strain state amounts to 9/8 of the angle of internal friction measured in a triaxial test. Therefore, this can be increased by up to 10% in calculations for long waterfront structures with the consent of the geotechnical expert. The characteristic values of the shear parameters $\varphi'_k$ and $c'_k$ for cohesive soils apply to calculations for final stability (consolidated state, final strength).

Empirical values for the shear parameters of the undrained, initially loaded soil $c_{u,k}$ are specified in DIN 1055-2:2010-11.

## 1.2 Layout and depths of boreholes and penetrometer tests (R 1)

### 1.2.1 General

The nature and extent of soil investigations, their layout and the depth of any such investigations must be determined by a geotechnical expert according to the provisions of DIN EN 1997-2 and DIN 4020.

The aim of boreholes is to investigate the stratification and obtain soil samples for soil mechanics tests in the laboratory. For investigation and monitoring of groundwater conditions, boreholes can be upgraded to groundwater monitoring wells.

Penetrometer tests allow the strength properties of the in situ soil types to be determined. With the help of empirical correlations, the soil types can be identified and the values of soil properties derived.

---

*Recommendations of the Committee for Waterfront Structures Harbours and Waterways – EAU 2012,*
9th *Edition. Issued by the Committee for Waterfront Structures of the German Port Technology Association and the German Geotechnical Society.*
© 2015 Ernst & Sohn GmbH & Co. KG. Published 2015 by Ernst & Sohn GmbH & Co. KG.

**Table R 9-1.** Characteristic values of soil parameters (empirical values)

| No. | 1 | 2 | 3 | 4 | 5 | | 6 | | 7 | 8 | 9 |
|---|---|---|---|---|---|---|---|---|---|---|---|
| | Soil type | Soil group to DIN 18196[1] | Penetration resistance | Consistency in initial state, i.e. to DIN 14688-1 | Unit weight | | Compressibility[2] Initial loading[3] $E_S = v_e \sigma_{at}(\sigma/\sigma_{at})^{w_e}$ | | Shear parameters of drained soil | | Hydraulic conductivity |
| | | | $q_c$ | | $\gamma_k$ | $\gamma'_k$ | $v_e$ | $w_e$ | $\varphi'_k$ | $c'_k$ | $K_k$ |
| | | | MN/m² | | kN/m³ | kN/m³ | | | degree | kN/m² | m/s |
| 1 | Gravel, uniformly graded | GE $U^{4)} < 6$ | < 7.5 <br> 7.5–15 <br> > 15 | | 16.0 <br> 17.0 <br> 18.0 | 8.5 <br> 9.5 <br> 10.5 | 400 <br> 900 | 0.6 <br> 0.4 | 30.0–32.5 <br> 32.5–37.5 <br> 35.0–40.0 | | $2 \times 10^{-1}$ <br> to <br> $1 \times 10^{-2}$ |
| 2 | Gravel, well- or gap-graded | GW, GI $6 \leq U^{4)} \leq 15$ | < 7.5 <br> 7.5–15 <br> > 15 | | 16.5 <br> 18.0 <br> 19.5 | 9.0 <br> 10.5 <br> 12.0 | 400 <br> 1100 | 0.7 <br> 0.5 | 30.0–32.5 <br> 32.5–37.5 <br> 35.0–40.0 | | $1 \times 10^{-2}$ <br> to <br> $1 \times 10^{-6}$ |
| 3 | Gravel, well- or gap-graded | GW, GI $U^{4)} > 15$ | < 7.5 <br> 7.5–15 <br> > 15 | | 17.0 <br> 19.0 <br> 21.0 | 9.5 <br> 11.5 <br> 13.5 | 400 <br> 1200 | 0.7 <br> 0.5 | 30.0–32.5 <br> 32.5–37.5 <br> 35.0–40.0 | | $1 \times 10^{-2}$ <br> to <br> $1 \times 10^{-6}$ |
| 4 | Sandy gravel with proportion $d < 0.06$ mm $< 15\%$ | GU, GT | < 7.5 <br> 7.5–15 <br> > 15 | | 17.0 <br> 19.0 <br> 21.0 | 9.5 <br> 11.5 <br> 13.5 | 400 <br> 800 <br> 1200 | 0.7 <br> 0.6 <br> 0.5 | 30.0–32.5 <br> 32.5–37.5 <br> 35.0–40.0 | | $1 \times 10^{-5}$ <br> to <br> $1 \times 10^{-6}$ |

| | | | | | | | | | | | | |
|---|---|---|---|---|---|---|---|---|---|---|---|---|
| 5 | Gravel/sand/ fine grain mixture with proportion $d < 0.06$ mm $> 15\%$ | $G\bar{U}, G\bar{T}$ | < 7.5<br>7.5–15<br>> 15 | | 16.5<br>18.0<br>19.5 | 9.0<br>10.5<br>12.0 | 150<br>275<br>400 | 0.9<br>0.8<br>0.7 | | 30.0–32.5<br>32.5–37.5<br>35.0–40.0 | | $1 \times 10^{-7}$<br>to<br>$1 \times 10^{-11}$ |
| 6 | Sand, uniformly graded, coarse sand | SE<br>$U^{+)} < 6$ | < 7.5<br>7.5–15<br>> 15 | | 16.0<br>17.0<br>18.0 | 8.5<br>9.5<br>10.5 | 250<br>475<br>700 | 0.75<br>0.60<br>0.55 | | 30.0–32.5<br>32.5–37.5<br>35.0–40.0 | | $5 \times 10^{-3}$<br>to<br>$1 \times 10^{-4}$ |
| 7 | Sand, uniformly graded, fine sand | SE<br>$U^{+)} < 6$ | < 7.5<br>7.5–15<br>> 15 | | 16.0<br>17.0<br>18.0 | 8.5<br>9.5<br>10.5 | 150<br>225<br>300 | 0.75<br>0.65<br>0.60 | | 30.0–32.5<br>32.5–37.5<br>35.0–40.0 | | $1 \times 10^{-4}$<br>to<br>$2 \times 10^{-5}$ |
| 8 | Sand, well- or gap-graded | SW, SI<br>$6 \leq U^{+)} \leq 15$ | < 7.5<br>7.5–15<br>> 15 | | 16.5<br>18.0<br>19.5 | 9.0<br>10.5<br>12.0 | 200<br>400<br>600 | 0.70<br>0.60<br>0.55 | | 30.0–32.5<br>32.5–37.5<br>35.0–40.0 | | $5 \times 10^{-4}$<br>to<br>$2 \times 10^{-5}$ |
| 9 | Sand, well- or gap-graded | SW, SI<br>$U^{+)} > 15$ | < 7.5<br>7.5–15<br>> 15 | | 17.0<br>19.0<br>21.0 | 9.5<br>11.5<br>13.5 | 200<br>400<br>600 | 0.70<br>0.60<br>0.55 | | 30.0–32.5<br>32.5–37.5<br>35.0–40.0 | | $1 \times 10^{-4}$<br>to<br>$1 \times 10^{-5}$ |
| 10 | Sand $d < 0.06$ mm $< 15\%$ | SU, ST | < 7.5<br>7.5–15<br>>15 | | 16.0<br>17.0<br>18.0 | 8.5<br>9.5<br>10.5 | 150<br>350<br>500 | 0.80<br>0.70<br>0.65 | | 30.0–32.5<br>32.5–37.5<br>35.0–40.0 | | $2 \times 10^{-5}$<br>to<br>$5 \times 10^{-7}$ |
| 11 | Sand $d < 0.06$ mm $> 15\%$ | $S\bar{U}, S\bar{T}$ | < 7.5<br>7.5–15<br>>15 | | 16.5<br>18.0<br>19.5 | 9.0<br>10.5<br>12.0 | 50<br>250 | 0.90<br>0.75 | | 30.0–32.5<br>32.5–37.5<br>35.0–40.0 | | $2 \times 10^{-6}$<br>to<br>$1 \times 10^{-9}$ |

**Table R 9-1.** (*Continued*)

| No. | 1 | 2 | 3 | 4 | 5 | | 6 | | 7 | 8 | 9 |
|---|---|---|---|---|---|---|---|---|---|---|---|
| | Soil type | Soil group to DIN 18196[1)] | Penetration resistance | Consistency in initial state, i.e. to DIN 14688-1 | Unit weight | | Compressibility[2)] Initial loading[3)] $E_S = v_e \sigma_{at}(\sigma/\sigma_{at})^{w_e}$ | | Shear parameters of drained soil | | Hydraulic conductivity |
| | | | $q_c$ | | $\gamma_k$ | $\gamma'_k$ | $v_e$ | $w_e$ | $\varphi'_k$ | $c'_k$ | $K_k$ |
| | | | MN/m² | | kN/m³ | kN/m³ | | | degree | kN/m² | m/s |
| 12 | Inorganic cohesive soils with low plasticity ($w_L < 35\%$) | UL | | soft<br>firm<br>stiff | 17.5<br>18.5<br>19.5 | 9.0<br>10.0<br>11.0 | 40<br>110 | 0.80<br>0.60 | 27.5–32.5 | 0<br>2–5<br>5–10 | $1 \times 10^{-5}$<br>to<br>$1 \times 10^{-7}$ |
| 13 | Inorganic cohesive soils with medium plasticity ($35\% < w_L < 50\%$) | UM | | soft<br>firm<br>stiff | 16.5<br>18.0<br>19.5 | 8.5<br>9.5<br>10.5 | 30<br>70 | 0.90<br>0.70 | 25.0–30.0 | 0<br>5–10<br>10–15 | $2 \times 10^{-6}$<br>to<br>$1 \times 10^{-9}$ |

| | | | | | | | | | | | |
|---|---|---|---|---|---|---|---|---|---|---|---|
| 14 | Inorganic cohesive soils with low plasticity ($w_L < 35\%$) | TL | | soft firm stiff | 19.0 20.0 21.0 | 9.0 10.0 11.0 | 20 50 | 1.0 0.90 | 25.0–30.0 | 0 5–10 10–15 | $1 \times 10^{-7}$ to $2 \times 10^{-9}$ |
| 15 | Inorganic cohesive soils with medium plasticity ($35\% < w_L < 50\%$) | TM | | soft firm stiff | 18.5 19.5 20.5 | 8.5 9.5 10.5 | 10 30 | 1.0 0.95 | 22.5–27.5 | 5–10 10–15 15–20 | $5 \times 10^{-8}$ to $1 \times 10^{-10}$ |
| 16 | Inorganic cohesive soils with high plasticity ($50\% < w_L$) | TA | | soft firm stiff | 17.5 18.5 19.5 | 7.5 8.5 9.5 | 6 20 | 1.0 1.0 | 20.0–25.0 | 5–15 10–20 15–25 | $1 \times 10^{-9}$ to $1 \times 10^{-11}$ |
| 17 | Organic silt, organic clay | OU and OT | | very soft soft firm | 14.0 15.5 17.0 | 4.0 5.5 7.0 | 5 20 | 1.00 0.85 | 17.5–22.5 | 0 2–5 5–10 | $1 \times 10^{-9}$ to $1 \times 10^{-11}$ |
| 18 | Peat[5] | HN, HZ | | very soft soft firm stiff | 10.5 11.0 12.0 13.0 | 0.5 1.0 2.0 3.0 | 5) | 5) | 5) | 5) | $1 \times 10^{-5}$ to $1 \times 10^{-8}$ |

**Table R 9-1.** (*Continued*)

| No. | 1 | 2 | 3 | 4 | 5 | | 6 | | 7 | 8 | 9 |
|---|---|---|---|---|---|---|---|---|---|---|---|
| | Soil type | Soil group to DIN 18196[1] | Penetration resistance | Consistency in initial state, i.e. to DIN 14688-1 | Unit weight | | Compressibility[2] Initial loading[3] $E_S = v_e \sigma_{at}(\sigma/\sigma_{at})^{w_e}$ | | Shear parameters of drained soil | | Hydraulic conductivity |
| | | | $q_c$ | | $\gamma_k$ | $\gamma'_k$ | $v_e$ | $w_e$ | $\varphi'_k$ | $c'_k$ | $K_k$ |
| | | | MN/m² | | kN/m³ | kN/m³ | | | degree | kN/m² | m/s |
| 19 | Mud[6] Digested sludge | F | | very soft soft | 12.5 16.0 | 2.5 6.0 | 4 15 | 1.0 0.9 | [6] | 0 | $1 \times 10^{-7}$ $1 \times 10^{-9}$ |

**Explanatory notes:**

[1] Code letters for primary and secondary components:
Particle size distribution:

F  mud
G  gravel
H  peat (humus)
O  organic inclusions
S  sand
T  clay
U  silt

Code letters for characteristic physical soil properties:
Particle size distribution:

W  wide-graded particle size distribution
E  narrow-graded particle size distribution
I  gap-graded particle size distribution

Plasticity:

L  low plasticity
M  medium plasticity
A  high plasticity

Degree of decomposition of peat:

N  not decomposed or scarcely decomposed peat
Z  decomposed peat

[2] Symbols:

$v_e$  stiffness factor, empirical parameter
$w_e$  stiffness exponent, empirical parameter
$w_e$  empirical parameter
$\sigma$  load in kN/m²
$\sigma_{at}$  atmospheric pressure ($= 100$ kN/m²)

[3] $v_e$ values for repeated load up to 10 times higher, $w_e$ values tend towards 1
[4] $U$ uniformity coefficient
[5] The compressibility and shear parameter values for peat exhibit such a wide scatter that empirical values cannot be given.
[6] The effective angle of internal friction of fully consolidated mud can be very high, but the value corresponding to the true degree of consolidation, which can only be determined reliably in laboratory tests, always governs.

The number and layout of boreholes and penetrometer tests must always be such that all the characteristics of the subsoil relevant to the planning are established and a sufficient number of suitable soil samples is obtained for the laboratory tests. When determining the number and type of boreholes and penetrometer tests, the results of earlier surveys in the form of geological maps and, where applicable, the findings of earlier boreholes and penetrometer tests should also be taken into account.

Geophysical surface measurements in conjunction with the boreholes and penetrometer tests can supply two-dimensional data on the geological profile, groundwater level and indications regarding any large obstacles in the subsoil.

Where major construction projects are involved, it can be useful to begin with principal boreholes and penetrometer tests to gain an overall picture and then to supplement these with intermediate boreholes and further penetrometer tests during the planning phase.

### 1.2.2 Principal boreholes

Principal boreholes should preferably lie on the later axis of the structure (waterfront). For cantilever walls they should be drilled to a depth equal to approximately twice the difference in ground levels or as far as a known geological stratum. As a guide, the recommended borehole spacing is approx. 50 m; recommendations regarding their location and depth are specified in DIN EN 1997-2 (2.4.1.3) and DIN 4020. In specific cases, the positions and spacings of the boreholes must be adapted in line with the geological and constructional boundary conditions. Given that soil samples for soil mechanics tests in the laboratory must be at least grade 2 according to DIN EN ISO 22475-1, the principal boreholes must be designed as boreholes suitable for obtaining samples in liners.

### 1.2.3 Intermediate boreholes

Depending on the findings of the principal boreholes or the earlier penetrometer tests, intermediate boreholes are also sunk to the depth of the principal boreholes, or to a depth at which a known, homogeneous soil stratum is encountered. The typical borehole spacing is again approx. 50 m; in some cases 25 m is necessary.

### 1.2.4 Penetrometer tests

Penetrometer tests are generally executed according to the layout in Fig. R 1-1. As far as possible, they are sunk to the same depth as the principal boreholes. The relevant standards should be consulted regarding details of the equipment for and execution of penetrometer tests, along with their application.

In order to interpret the results of penetrometer tests, individual tests must be carried out directly adjacent to boreholes. In such cases the

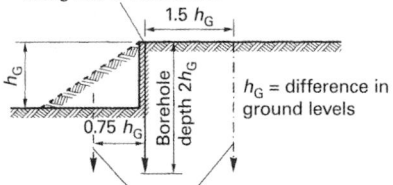

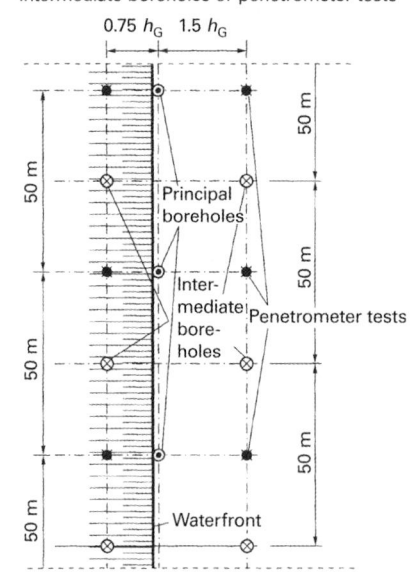

**Fig. R 1-1.** Example of layout of boreholes and penetrometer tests for waterfront structures

penetrometer tests must be performed prior to drilling the boreholes in order to prevent the results of the penetrometer test from being influenced due to any loosening of the soil during drilling.

## 1.3 Geotechnical report (R 150)

The results of the soil investigations are compiled in a geotechnical report according to DIN EN 1997-1 (3.4) or DIN EN 1997-2 (6). The nature and extent of such investigations along with their results must be recorded in that document.

The geotechnical report contains the characteristic design values of the soil parameters, and might also include references to the proposed methods of calculation. An assessment of the subsoil should also include investigations regarding chemical constituents that could damage concrete and/or steel and details of any contamination.

The findings gleaned from the geotechnical report are summarised as foundation recommendations for the specific structure. For waterfront structures, this also includes information on the installation of piles and sheet piles as well as any obstacles to driving.

The soil investigations can be supplemented by loading tests and trial embankments in order to be able to make a proper assessment of the loadbearing behaviour of foundation elements and soil compaction options. If required, a number of model tests can be carried out to assess soil-structure interaction. The execution of and results from loading tests, trial embankments and model tests must be recorded in the geotechnical report.

Together with the verification of stability and serviceability, the aforementioned contents form the basis of the draft geotechnical report according to DIN EN 1997-2 (section 2.8).

## 1.4 Determining the shear strength $c_u$ of saturated, undrained cohesive soils (R 88)

If saturated cohesive soil is loaded without being able to consolidate (undrained conditions), its change in volume is negligible due to the low compressibility of the pore water at loads below its strength. The load generates excess pore water pressure only and no additional effective stresses in the soil skeleton. As a result, the angle of internal friction for saturated cohesive soils in undrained conditions is $\varphi_u = 0$. The strength is only described by the cohesion of the undrained soil $c_u$. In the case of partial saturation, part of the load can generate additional effective stresses in the soil skeleton; in such cases $\varphi' > \varphi_u > 0$.

### 1.4.1 Cohesion $c_u$ of undrained soil

The cohesion $c_u$ of undrained cohesive soils essentially depends on the following conditions:

- For normally consolidated soils, $c_u$ is proportional to the effective vertical stress $\sigma'_v$, i.e. $c_u$ increases linearly with depth:

$$\frac{c_u}{\sigma'_v} = \lambda_{cu}$$

According to Jamiolowski et al. [110], the cohesion constant is $\lambda_{cu} = 0.23 \pm 0.04$, although Gebreselassie [72] states that values as low as $\lambda_{cu} = 0.18$ and even lower are possible.

For north German marine clay, $c_u$ is often very low and any dependency on the vertical stress $\sigma'_v$ is hard to measure with any certainty.

- For overconsolidated soil, $c_u$ is likewise proportional to the effective vertical stress $\sigma'_v$, but is also determined from the stress history:

$$\frac{c_u}{\sigma'_v} = \lambda_{cu} OCR^\alpha$$

The overconsolidation ratio (OCR) is the ratio of the stress $\sigma'_{vc}$, for which the soil is consolidated, and the current stress $\sigma'_v$:

$$OCR = \frac{\sigma'_{vc}}{\sigma'_v}$$

Reference values for exponent $\alpha$ lie between 0.8 and 0.9.
- A number of different authors demonstrate that the cohesion $c_u$ of an undrained soil depends on the stress path. Under triaxial compression (tc), $c_u$ is greater than for triaxial extension (te) Bjerrum [20]; Jamiolowski et al. [110]; Scherzinger [190]; $c_{u,tc}$ can be approx. 50% greater than $c_{u,te}$. Values for direct simple shear $c_{u,dss}$ lie in between for the same pore volume:

$$c_{u,tc} > c_{u,dss} > c_{u,te}$$

- Owing to the viscosity of cohesive soils, the cohesion $c_u$ of an undrained soil depends on the rate of load application. This can be described using the shear rules of Leinenkugel [132] or Randolph [168], for instance. Leinenkugel's relationship is

$$\frac{c_u}{c_{u\alpha}} = \left[1 + I_{v\alpha} \ln\left(\frac{\dot\gamma}{\dot\gamma_\alpha}\right)\right]$$

The viscosity index $I_{v\alpha}$ for the reference strain rate $\dot\gamma_\alpha$ can be determined, for example, by means of CU triaxial tests with an abruptly varying strain rate (step test) or using one-dimensional creep tests. Reference values for $I_{v\alpha}$ can be found, for example, in Leinenkugel [132] and Gudehus [78].

### 1.4.2 Determining the cohesion $c_u$ of an undrained soil

The cohesion $c_u$ of an undrained soil can be determined in laboratory or field tests. There are two essentially different methods for such investigations: the recompression method and the stress history method.

#### 1.4.2.1 Recompression method

In this method, $c_u$ is determined by way of triaxial tests on soil samples that are reconsolidated prior to shearing with the stress that acts on the soil in situ Bjerrum [20]. According to Seah und Lai [197], however, this method overestimates the cohesion $c_u$ of normally consolidated soil. Therefore, the recompression method is preferred for highly structured, brittle soils, e.g. sensitive clays, cemented soils and severely overconsolidated soils. The results of shear tests should always be checked by comparing them with the stress history.

### 1.4.2.2 The stress history and normalised soil engineering properties method (SHANSEP)

This method enables the cohesion $c_u$ to be determined while taking into account the sample disorder, the anisotropy to a limited extent and the rate of load application. It is based on investigations carried out at MIT during the 1960s and was initially published by Ladd and Foott [129]. A revised version can be found in Ladd and DeGroot [128]. The SHANSEP method involves the following steps for specifying the soil model:

- Carrying out a soil investigation, taking special samples (undisturbed soil samples) and compiling a soil profile based on the results of cone penetration tests and field vane shear tests.
- Determining the degree of overconsolidation in the laboratory based on compression tests and deriving the overconsolidation ratio (OCR).
- Determining the effective shear parameters $\varphi'/c'$ and $c_u$ in laboratory tests, normally by way of triaxial tests. Triaxial tests with anisotropic consolidation ($CK_0$) and subsequent undrained triaxial compression (UC with $\sigma_1 > \sigma_3$) and triaxial extension (UE with $\sigma_1 < \sigma_3$) are recommended. Stipulation of the reconsolidation stress corresponding to the calculated OCR.
- Performing shear tests to determine the relationship between the OCR and the normalised shear strength $c_u/\sigma_v'$.
- Stipulating a $c_u$ design profile for the cohesive strata which lies on the safe side.

### 1.4.3 Determining $c_u$ in laboratory tests

The advantage of determining $c_u$ in laboratory tests is that the test conditions can be reproduced in an ideal manner within larger test series. However, the disadvantage is that test specimens can never be obtained from boreholes without disturbing their structure and strength. In addition, test specimens are not continuous, meaning that the distribution of $c_u$ over the stratum thickness is only ascertained at discrete points. The test conditions are best controlled in triaxial tests. CU triaxial tests supply both drained and undrained shear parameters because the pore water pressure is measured. When determining $c_u$ by way of laboratory vane shear tests as well as unconfined compression tests, a distorted capillarity influence cannot be ruled out. For soft soils, $c_u$ can also be determined by way of various pressure and fall cone tests.

### 1.4.4 Field tests

Determining $c_u$ by way of cone penetration tests to DIN 4094-1 and vane shear tests to DIN 4094-4 supplies a profile of the cohesion $c_u$ over the depth. Owing to the high shearing rate, the shear resistance $\tau_{fvt}$ in the vane shear test must be reduced by a factor µ, which depends on the plasticity index $I_P$:

$$c_{u,fvt} = \mu \tau_{fvt}$$

Details of the correction factor µ can be found in DIN EN 1997-2, annex I. The derivation of $c_u$ from the penetration resistance in cone penetration tests requires knowledge of the OCR of the soil. For example, the following applies for the CPTU test:

$$c_{u,cptu} = \frac{q_c - \sigma_v}{N_{kt}}$$

The factor $N_{kt}$ depends on the cone geometry and OCR, and lies between 10 and 20.

The derivation of $c_u$ from borehole-widening tests to DIN 4096 is less common.

Plate load tests to DIN 18134 only supply a $c_u$ value for soils near the surface.

### 1.4.5 Correlations

A number of authors have suggested correlations between $c_u$ and the water content $w$, the consistency index $I_C$, the plasticity index $I_P$ and the liquidity index $I_L$; Gebreselassie [72] provides a detailed overview of this. It should be noted that these correlations apply, at best, to the soils examined and the test conditions, and can therefore only be used as reference values.

## 1.5 Assessing the subsoil for the installation of piles and sheet piles and for selecting the installation method (R 154)

### 1.5.1 General

In the first place, the material, form, size, length and angle of piles and sheet piles play a decisive role with respect to the installation of piles and sheet piles and the selection of the installation method. Important information on this can be found in:

| | |
|---|---|
| R 21, section 8.1.2, | Design and installation of reinforced concrete sheet pile walls |
| R 22, section 8.1.1, | Design and installation of timber sheet pile walls |
| R 34, section 8.1.3, | Design and installation of steel sheet pile walls |
| R 104, section 8.1.12, | Driving combined steel sheet piling |
| R 105, section 8.1.13, | Monitoring during the installation of sheet piles, tolerances |
| R 118, section 8.1.11, | Driving steel sheet piles |
| R 217, section 9.2.2.1, | Tension piles and anchors |

In connection with these recommendations it is especially important to note that when selecting the type of pile section (material, form), it is essential to take into account the stresses due to the installation procedure in the respective subsoil in addition to the structural requirements and economic issues. As a result, the geotechnical report must also include an evaluation of the in situ subsoil with respect to the installation of piles and sheet piles (see also R 150, section 1.3)

## 1.5.2 Assessment of soil types with respect to installation methods

### 1.5.2.1 General

The shear parameters have only a limited significance when it comes to describing the behaviour of the subsoil during the installation of piles and sheet piles. For example, rocky calcareous marl can exhibit relatively low shear parameters due to its fissuring, but may in fact present difficult conditions in terms of piling.

### 1.5.2.2 Impact driving

Easy driving conditions are to be expected in soft or very soft soils such as moorland, peat, silt, marine clay, etc. Easy driving conditions are also to be generally expected in loosely bedded medium and coarse sands and gravels with no rock inclusions, unless there are embedded cemented strata.

Moderate driving conditions are to be expected in moderately densely bedded medium and coarse sands, fine gravel soils and firm clays and loams.

Difficult to very difficult driving conditions are to be expected in most instances of densely bedded medium and coarse gravels, densely bedded fine sandy and silty soils, embedded cemented strata, stiff to very stiff clays, cobbles and moraine strata, glacial till and weathered and soft to medium-hard rock. Earth-moist or dry soils present a greater resistance to penetration during impact driving than those subject to buoyancy. This does not apply to saturated cohesive soils, and silts especially. With a number of blows $N_{10} > 30$ per 10 cm penetration in the heavy dynamic penetration test (DPH, DIN EN ISO 22476-2) or $N_{30} > 50$ per 30 cm penetration in borehole dynamic probing (BDP, DIN 4094-2), an increasingly high penetration resistance during driving must be reckoned with. It can generally be assumed that driving is possible up to a number of blows $N_{10} = 80–100$ per 10 cm penetration (DPH). Driving with a higher number of blows can be possible in individual cases. For more information see Rollberg [176,177].

### 1.5.2.3 Vibratory driving

The skin friction and base resistance of the pile being installed are greatly reduced when using vibratory driving methods. As a result, the piles or sheet piles can quickly reach their required depth compared with impact driving. For more information see R 202, section 8.1.23.

Vibratory driving is particularly successful in sands and gravels with a rounded grain shape and in very soft or soft soil types with low plasticity. Vibrating is much less suitable for highly cohesive soils or sands and gravels with an angular grain shape. Dry fine sands and firm marl and clay soils are particularly critical as they absorb the energy of the vibrator without reducing the skin friction and base resistance.

If the subsoil is compacted during vibration, then its penetration resistance can increase to such an extent that the pile being installed can no longer reach the required depth. This risk arises, in particular, when piles and sheet piles are installed at close spacings and when using vibration in non-cohesive soils. Vibratory driving must be stopped in such cases, see R 202. The use of auxiliary driving measures according to section 1.5.2.5 may represent an option.

Above all, vibratory driving in non-cohesive soils may lead to localised settlement, the magnitude and extent of which depend on the power output of the vibrator, the section being driven, the duration of the vibratory driving and the soil. When working close to existing structures, checks must be made to establish whether any such settlement could cause damage. If required, the installation procedure must be adjusted accordingly.

### 1.5.2.4 Pressing

For pressing to be used, there should be no obstacles in the soil, or if there are any, they must be removed prior to driving.

Slender sections can generally be pressed hydraulically into cohesive soils without obstacles or into loosely bedded non-cohesive soils. Sections can only be pressed into densely bedded non-cohesive soils

**Table R 154-1.** Pressing limits for steel sheet piles

| Soil parameter | | | Without driving aid | With driving aid |
|---|---|---|---|---|
| CPT peak pressure | $q_b$ | MN/m² | < 20 | < 35 |
| CPT skin friction | $q_s$ | MN/m² | < 0.1 | < 0.3 |
| DPH | $N_{10}$ | - | < 25 | < 40 |
| Consistency index | $I_c$ | - | < 1.0 | > 1.0 |
| Plasticity index | $I_P$ | - | > 10 | |
| Ratio[*)] | $I_f$ | - | < 1.0 | > 1.0 |
| Angle of friction | $\varphi'$ | ° | < 35 | < 45 |

[*)] $I_f = (\max e - \min e)/\min e$ (higher compactability in the event of decreasing $I_f$)

if the soil has been loosened beforehand. Empirical values according to Busse [33] are given in Table R 154-1.

### 1.5.2.5 Auxiliary driving measures

Water-jetting can ease driving in densely bedded sands and gravels as well as in firm and stiff clays in particular – indeed, may be essential to enable driving in the first place.

Additional auxiliary driving measures include pre-drilling to loosen the soil or local soil replacement, etc. Rocky soils can be loosened by way of local blasting in such a way that the required depth can be reached using conventional impact driving and appropriate pile sections. For more information see R 183, section 8.1.10.

### 1.5.2.6 Driving plant, pile sections, installation methods

Driving plant, pile sections and installation methods must be suited to the subsoil through which the pile sections are being driven, see: R 104, section 8.1.12; R 118, section 8.1.11; R 202, section 8.1.23; R 210, section 7.13.

Slow-acting drop hammers, diesel hammers and hydraulic hammers are suitable for both cohesive and non-cohesive soils. Rapid-acting hammers and vibration plant place less stress on the pile section, but generally are only effective in non-cohesive soils with a rounded grain shape. Rapid-acting hammers or heavy hammers with short drop heights should be preferred when driving in rocky soils, even when using pre-blasting to loosen the ground.

Interruptions during driving a pile, e.g. between initial and final driving, can make subsequent driving easier or harder depending on the soil type and water saturation as well as the length of the interruption. Any changes to the penetration resistance should generally be identified and quantified by way of tests in advance.

Assessing the subsoil for the installation of piles and sheet piles presumes appropriate experience and specialised knowledge of the installation methods. Experience of construction projects with similar subsoil conditions can indeed be very beneficial.

### 1.5.2.7 Testing installation methods and loadbearing behaviour in difficult conditions

If on construction projects with considerable embedment depths there are concerns that sheet piles cannot be driven to the depth required to satisfy the structural design without being damaged or other piles cannot reach the intended embedment depth to carry the loads, then test piles must be driven and pile loading tests carried out beforehand. At least two test piles should be driven for each installation method in order to obtain accurate information.

Testing the installation method may also be necessary in order to predict any settlement of the soil as well as the spread and impact of vibrations as a result of the installation method.

# 2 Active and passive earth pressure

## 2.1 General

The characteristic earth pressures can be determined according to the graphical and/or analytical methods given in DIN 4085. Other methods, e.g. numerical, are permitted if it can be ensured that they result in the same earth pressures as the methods in that standard. The soil properties are not indexed in the following figures and equations unless the characteristic value (index $k$) or design value (index $d$) is expressly referred to.
The design of sheet pile walls is dealt with in section 8.2. Section 8.2.5 contains information on estimating the active earth pressure inclination $\delta_a$.

## 2.2 Considering the cohesion in cohesive soils (R 2)

The cohesion in cohesive soils may be taken into account when calculating the active and passive earth pressures provided the following conditions are satisfied:

- The soil must be undisturbed. When backfilling with cohesive material, the soil must be placed and compacted without leaving voids.
- The soil should be permanently protected against drying out and freezing.
- The soil should not become very soft when kneaded.

## 2.3 Considering the apparent cohesion (capillary cohesion) in sand (R 3)

The apparent cohesion $cc$ (capillary cohesion according to DIN 18137-1) in sand is caused by the surface tension of the pore water. This cohesion is lost when the soil is fully saturated or dries out completely. It is therefore not normally included in active or passive earth pressure calculations, instead functions as an inherent reserve factor for stability. Apparent cohesion may be taken into account for temporary conditions during construction if it can be ensured that it remains effective throughout the period concerned. Table R 3-1 contains characteristic reference values for the apparent cohesion of strata of medium or greater density.

**Table R 3-1.** Characteristic reference values for the apparent cohesion of strata of medium or greater density [208]

| Soil type | Designation to DIN 4022-1 | Apparent cohesion $c_{c,k}$ [kN/m$^2$] |
|---|---|---|
| Gravelly sand | G, s | $\leq 2$ |
| Coarse sand | gS | $\leq 3$ |
| Medium sand | mS | $\leq 5$ |
| Fine sand | fS | $\leq 9$ |

## 2.4 Determining active earth pressure according to the Culmann method (R 171)

### 2.4.1 Solution for uniform soil without cohesion

In the Culmann method, the Coulomb polygon of forces (Fig. R 171-1) is rotated through an angle 90°-$\varphi'$ to the perpendicular so that the force vectors $G_i$ lie on a straight line at an angle $\varphi'$ to the horizontal ("slope

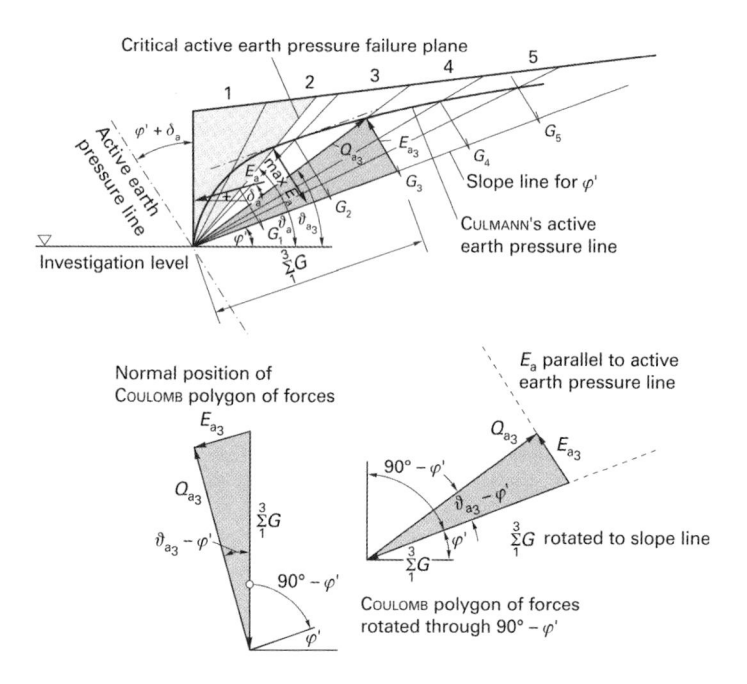

**Fig. R 171-1.** System sketch for determining the active earth pressure according to Culmann for homogeneous soil without cohesion

line"). If we now draw a line parallel with the "active earth pressure line" (straight line passing through base of wall at angle $\varphi' + \delta_a$ to wall) at the start of the dead load $G$, it intersects the associated failure plane at a point on the Culmann active earth pressure line (Fig. R 171-1).

The distance of this intersection from the slope line measured in the direction of the active earth pressure line is the respective active earth pressure for the sliding wedge investigated. This is now repeated for various failure planes. The maximum of Culmann's active earth pressure line represents the unknown design earth pressure.

In homogeneous soil the Culmann method can be used for any ground surface form and any surcharges acting on that surface. The respective groundwater level is also taken into account by determining the sliding wedge loads with $\gamma$ or $\gamma'$. The same also applies to any other changes to the unit weight, provided $\varphi'$ and $\delta_a$ remain unchanged.

The earth pressure loads on a wall are then determined section by section, starting from the top, and are plotted as loads per unit area over the height of the section. For further information on earth pressure distribution see section 8.2.

## 2.4.2 Solution for uniform soil with cohesion

In cases with cohesion, besides the soil reaction $Q$, the cohesion force $C' = c'_k \cdot l$ also acts on the failure plane with length $l$ (Fig. R 171-2). In the Coulomb polygon of forces, $C'$ is applied ahead of the dead load $G$. In the Culmann method, $C'$, rotated through an angle $90°\text{-}\varphi'$, is also placed on the slope line ahead of the dead load $G$. The line parallel to the active earth pressure line is drawn through the starting point of $C'$ and continued to intersect with the associated failure plane, which allows us to find the associated point on the Culmann active earth pressure line. After investigating several failure planes, the critical active earth pressure is the maximum distance of the Culmann active earth pressure line from the line joining the starting points of $C'$ measured in the direction of the active earth pressure line (Fig. R 171-2).

The applicability of straight failure planes must be verified for greater cohesion values, especially for sloping ground. In such cases, curved or polygonal failure planes frequently lead to greater earth pressure loads (R 198, section 2.6).

## 2.4.3 Expanded solutions

The Culmann method can also be used for irregular ground surfaces or when additional loads have to be included in the calculation of the active earth pressure. Failure planes that terminate at the changes of direction of the slope or at the load application point on the surface must be investigated. Concentrated loads lead to an abrupt change in the Culmann active earth pressure line.

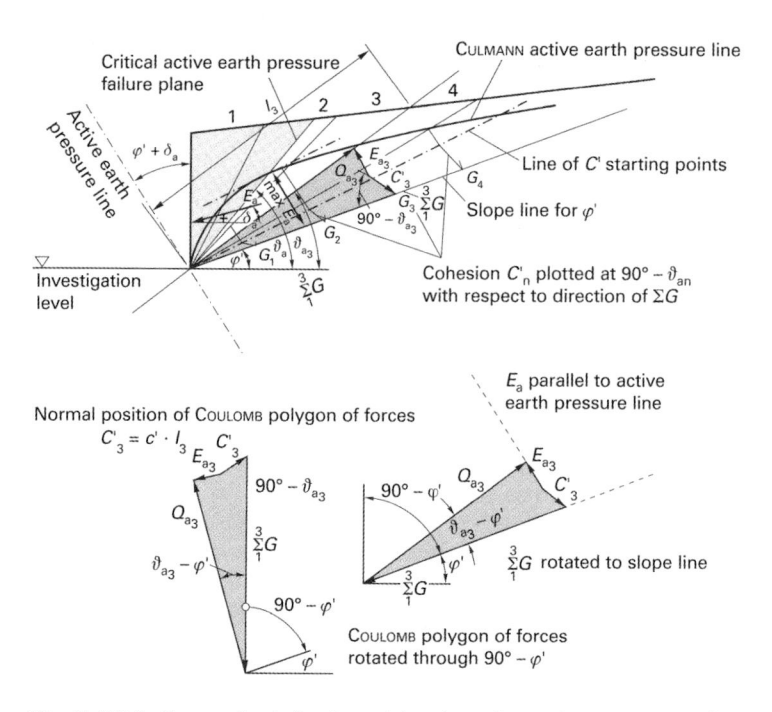

**Fig. R 171-2.** System sketch for determining the active earth pressure according to Culmann for homogeneous soil with cohesion

The Culmann method can only be applied in an approximate manner for stratified soils and straight failure planes. In this case an averaged angle of internal friction is used. A better analysis of the stratification is possible by using the method of Fig. R 219-1.

## 2.5 Active earth pressure in stratified soil (R 219)

Fig. R 219-1 shows an example for determining the resultant of the active earth pressure on a wall for three strata and a straight failure plane. In this example the internal earth pressures are applied horizontally at the boundaries between the lamellas (see Fig. R 219-1b).
The critical failure plane combination is that for which the active earth pressure load $E_a$ is greatest (not examined in Fig. R 219-1). A weighted earth pressure angle or weighted adhesion should be used for the inclination of the earth pressure resultants at the retaining wall. It is easiest to obtain the weighting from a stratum-by-stratum calculation of the earth pressure resultants. For information on earth pressure distribution see section 8.2.

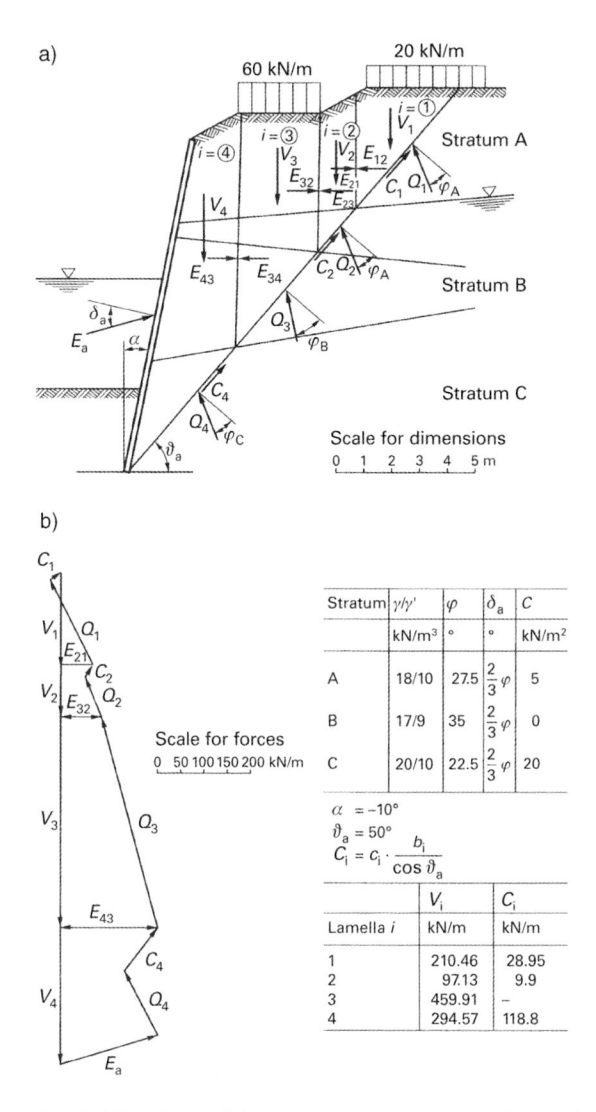

**Fig. R 219-1. Determining the active earth pressure in stratified soil** a) Example for determining the active earth pressure in stratified soil using the lamella method – geometry and forces b) Polygon of forces for the graphical determination of the active earth pressure force $E_a$ with division into lamellas

The analytical solution for straight failure planes according to Fig. R 219-1b is

$$E_a = \left[ \sum_{i=1}^{n} \left( V_i \frac{\sin(\vartheta_a - \varphi_i)}{\cos\varphi_i} - \frac{c_i \cdot b_i}{\cos\vartheta_a} \right) \right] \cdot \frac{\cos\overline{\varphi}}{\cos(\vartheta_a - \overline{\varphi} - \overline{\delta}_a + \alpha)}$$

where

| | |
|---|---|
| $i$ | consecutive number of lamella |
| $n$ | number of lamellas |
| $V_i$ | buoyant unit weight including surcharges of lamellas |
| $\vartheta_a$ | inclination of failure plane with respect to horizontal |
| $\varphi_i$ | angle of friction at failure plane of lamella $i$ |
| $c_i$ | cohesion in lamella $i$ |
| $b_i$ | width of lamella $i$ |
| $\alpha$ | inclination of waterfront structure, defined according to DIN 4085 |
| $\overline{\varphi}$ | mean value of angle of friction along failure plane: |
| $\overline{\delta}_a$ | mean value of earth pressure angle over wall height. For horizontal strata and comparatively low surcharges, $\overline{\delta}_a$ may be approximated with $\overline{\delta}_a \approx \frac{2}{3}\overline{\varphi}$. For more precise examinations, the mean must be formed using the earth pressure calculated stratum by stratum. |

## 2.6 Determining active earth pressure for a steep, paved embankment in a partially sloping waterfront structure (R 198)

A steep embankment case exists if the inclination of the slope $\beta$ is greater than the effective angle of internal friction $\varphi'$ of the in situ soil. The stability of the embankment is only guaranteed if the soil has a permanently effective cohesion $c'$ and is permanently protected against surface erosion, e.g. by means of dense turf or a revetment.

If the cohesion is insufficient to guarantee the stability of the embankment, it requires protection, e.g. in the form of paving, which must be able to resist downslope forces and be structurally connected to the quay wall. The bank protection must be designed in such a way that the resultant of the applied actions always lies within the middle third of its cross-section. Where cohesion does not dominate, i.e. $c'/g \cdot h < 0.1$, the active earth pressure for the embankment area down to the supporting beam for the revetment (active earth pressure reference line in Fig. R 198-1) can be calculated according to R 171, section 2.4, whereby the dead load of the bank protection is not taken into account.

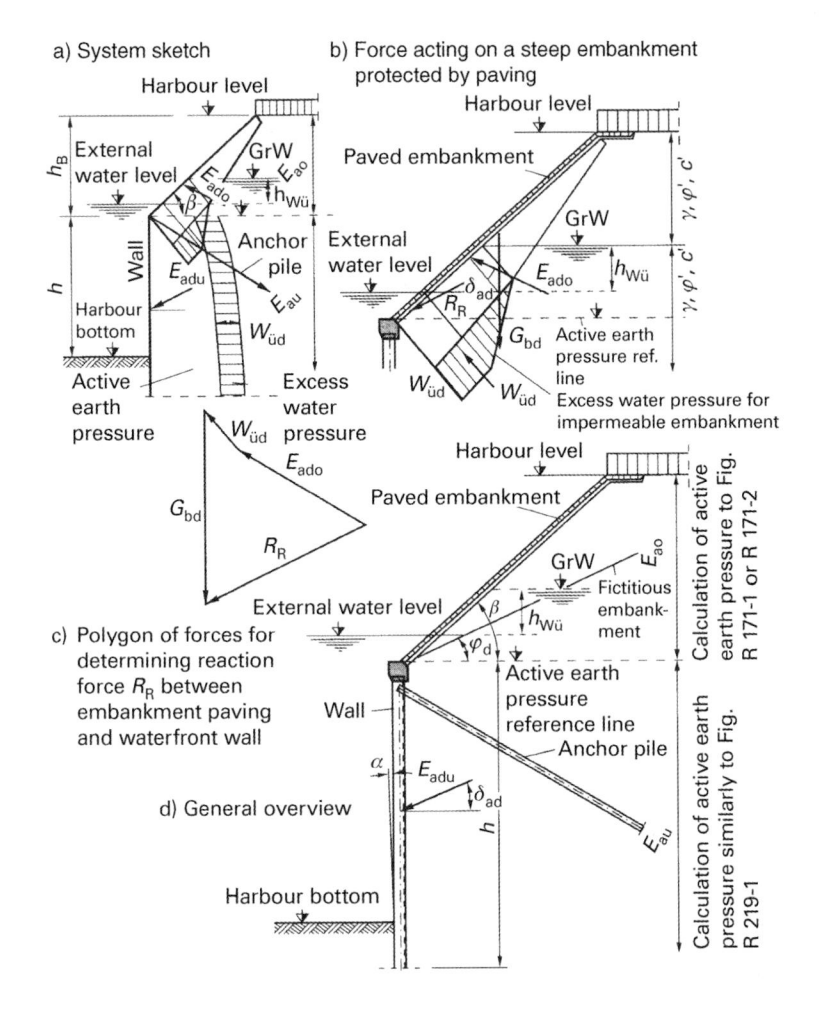

**Fig. R 198-1.** Partially sloping bank with steep, paved embankment

In doing this, any excess water pressure must be allowed for in addition to the active earth pressure $E_a$. This is illustrated in Fig. R 198-1a for impermeable paving. The excess water pressure is lower in the case of permeable paving. The loading assumptions for bank protection are shown in Fig. R 198-1b. The reaction force $R_R$ between the bank protection and the wall is given by the polygon of forces shown in Fig. R 198-1c.

33

The reaction force $R_R$ must be taken into account fully in the calculations for the wall and its anchorages. Below the active earth pressure reference line (theoretical stratum boundary), the active earth pressure $E_{au}$ can be determined similarly to Fig. R 219-1 where cohesion does not dominate (see above). In doing so, it should be noted that the active earth pressure load $E_{ado}$ and the dead load of the bank protection are already included in the reaction force $R_R$ and are carried directly by the wall and its anchorages. Any further calculations are carried out in accordance with R 171, section 2.4. By way of approximation, the active earth pressure load $E_{adu}$ below the active earth pressure reference line of Fig. R 198-1 can also be determined with a wall projecting above the active earth pressure reference line by the fictitious height

$$\Delta h = \frac{1}{2} \cdot h_B \cdot \left(1 - \frac{\tan\varphi'}{\tan\beta}\right)$$

at the same time with a fictitious slope angle $\varphi'$ (Fig. R 198-2). Where cohesion dominates, i.e.

$$\frac{c'}{\gamma \cdot h} \geq 0.1$$

then the calculation with straight failure planes in accordance with Fig. R 171-2 or Fig. R 219-1 underestimates the active earth pressure load $E_a$. In such cases the recommendation is to determine the active earth pressure for the sections both above and below the active earth pressure reference line using curved or polygonal failure planes.

## 2.7 Determining the active earth pressure shielding on a wall below a relieving platform with average ground surcharges (R 172)

A wall can be shielded from the active earth pressure to some extent by building a relieving platform. The shielding effect depends, above all, on the width and position of the platform, but also on the shear strength and compressibility of the soil behind the wall and beneath the structure. A relieving platform can have a favourable influence on the critical active earth pressure distribution used for calculating the internal forces. In homogeneous, non-cohesive soils with average ground surcharges (typically 20–40 kN/m² as uniformly distributed load), the active earth pressure shielding can be determined according to [23], Fig. R 172-1. As can be confirmed by Culmann investigations, the use of the Lohmeyer method works well under the foregoing prerequisites.

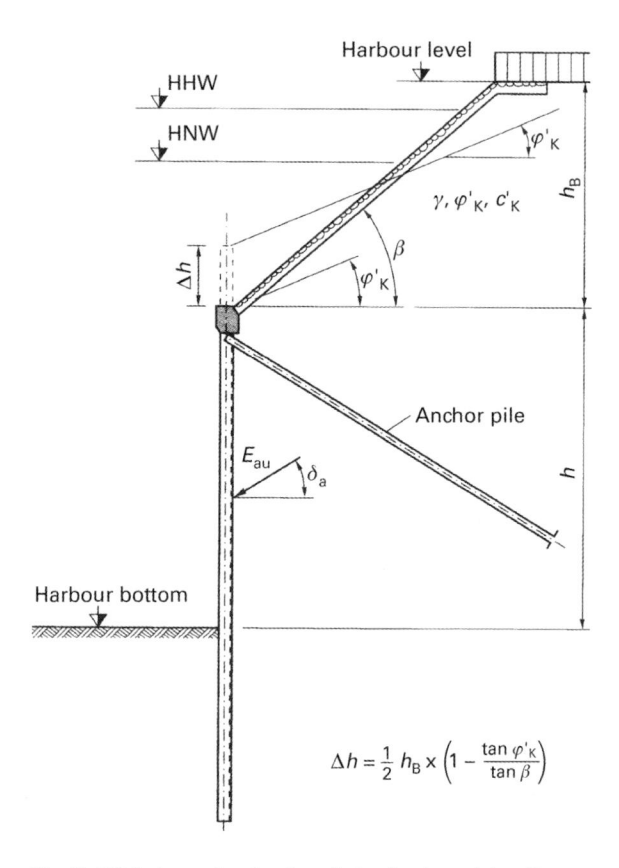

**Fig. R 198-2.** Approximation formulation for determining $E_{au}$

In stratified, non-cohesive soil, the assumptions according to Fig. R 172-2 or Fig. R 172-3 offer approximate solutions. The calculation using Fig. R 172-3 can be carried out easily with commercially available software, even for multiple strata.

If the soil also has a cohesion $c'$, then the shielded active earth pressure can be considered approximately by first calculating the shielded active earth pressure distribution without taking $c'$ into account and then superposing the cohesion on it thus:

$$\Delta e_{ac} = c' \cdot K_{ac}$$

where $K_{ac}$ is the active earth pressure coefficient for cohesion, see DIN 4085. This procedure is, however, only permissible when the degree

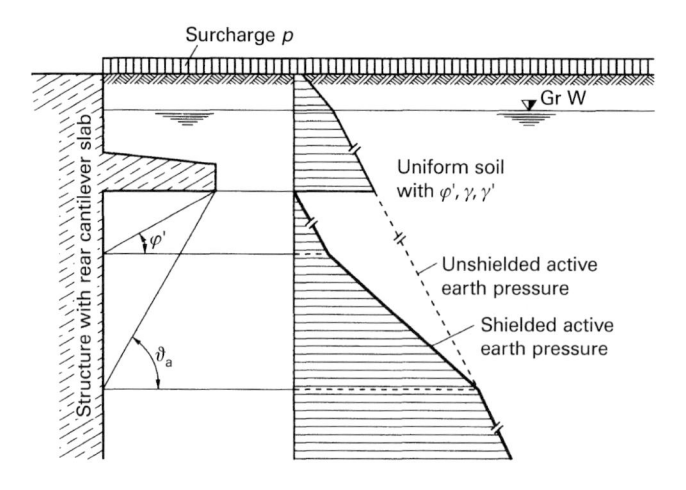

**Fig. R 172-1.** Solution according to Lohmeyer for homogeneous soil

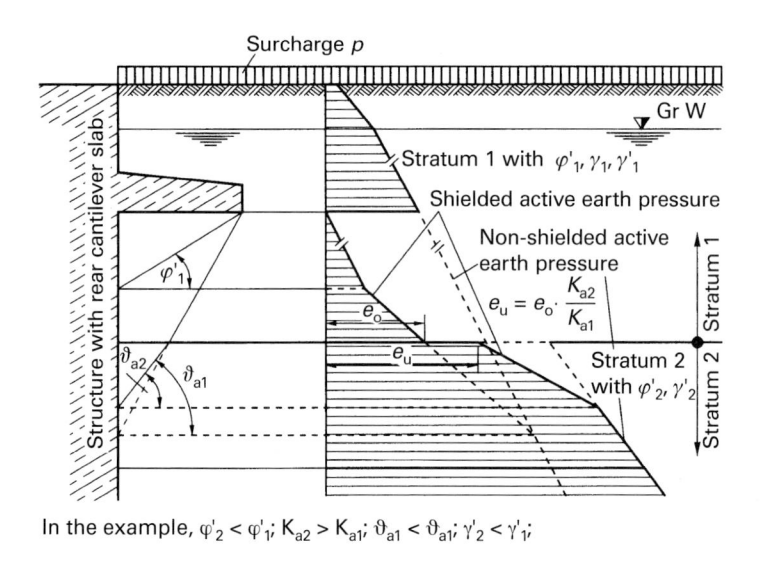

In the example, $\varphi'_2 < \varphi'_1$; $K_{a2} > K_{a1}$; $\vartheta_{a1} < \vartheta_{a1}$; $\gamma'_2 < \gamma'_1$;

**Fig. R 172-2.** Solution according to Lohmeyer expanded for stratified soil (option 1)

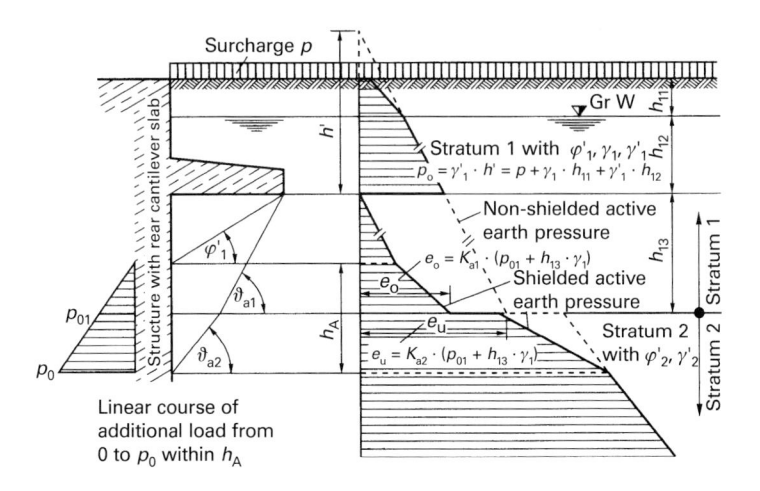

**Fig. R 172-3.** Solution according to Lohmeyer expanded for stratified soil (option 2)

of cohesion is small compared with the total active earth pressure. Here again, a more accurate calculation is possible by using the expanded Culmann method according to R 171, section 2.4.

The same applies to determining the effect of earthquakes, taking into account R 124, section 2.16.

The calculations according to Figs. R 172-1, R 172-2 and R 172-3 may not be used in cases where several relieving platforms are positioned one above the other. Furthermore, irrespective of the shielding, the overall stability of the structure is to be verified for the corresponding limit states according to DIN 1054 – with the full active earth pressure being applied in the critical reference planes.

## 2.8 Earth pressure distribution under limited loads (R 220)

The active earth pressure due to vertical strip or line loads may be applied to the retaining wall via a simplified limited load figure. The load on the wall is spread over an area limited by the angle $\varphi'_k$ from the front edge of the load and the angle $\vartheta_{a,k}$ from the back edge of the load (Fig. R 215-1). The distribution of the load must be selected taking into account any potential deformations. If the conditions are such that active earth pressure redistribution can take place (section 8.2.3.2), then these load components must also be redistributed, in particular to ensure the load concentration via supports is not underestimated. Additional recommendations can be found in *Recommendations on Excavations* (EAB, section 3.5, EB 7; [57]).

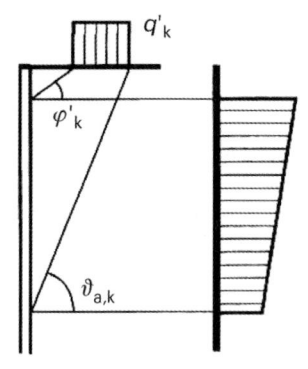

**Fig. R 215-1.** Effective spread of a strip load

When it comes to strip or line loads limited in the longitudinal direction of the wall, the load spreading beyond the end may also be taken into account at an angle of 45°. The associated relief on the wall may be taken into account over an area at an angle ±45° from the end of the load (Fig. R 215-2). If areas overlap in the case of very short loaded areas, then the load is distributed over the section of the wall bounded by the outer spreading lines.

## 2.9 Determining active earth pressure in saturated, non- or partially consolidated, soft cohesive soils (R 130)

In drained conditions, effective stresses and effective shear parameters are to be expected. For undrained conditions, effective stresses and, accordingly, effective shear parameters should be used in the calculations. Performing calculations with total stresses and, accordingly, undrained shear parameters is not recommended because $c_u$ depends on:

- the effective stresses in the initial state,
- the overconsolidation ratio (OCR),
- the chronological order of the loading (stress path), and
- the rate of load application.

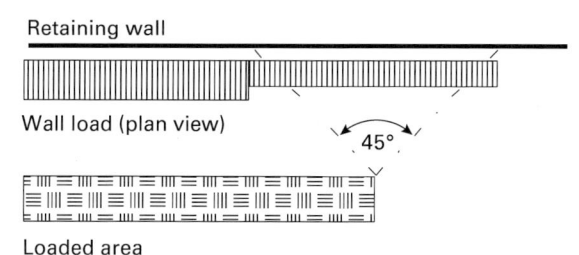

**Fig. R 215-2.** Horizontal load distribution for limited loads

Moreover, these influences can only be fully ascertained by analysing the development of the effective stresses and calculating the active earth pressure with the effective shear parameters.

If a surcharge is applied over such a short period that the soil cannot consolidate, then the surcharge is initially absorbed by the pore water and the effective stress $\sigma'$ remains unchanged. Only after consolidation begins is the surcharge transferred to the soil skeleton. The following applies for the horizontal load on the wall $\sigma_h$ directly after applying the load:

$$\sigma_h = e_{ah} + u$$

where

$e_{ah} = \sigma' \cdot K_{agh} - c' \cdot K_{ach}$ ($K_{agh}$ and $K_{ach}$ to DIN 4085)
$\Delta p = \Delta u$
$u = u_0 + \Delta u$

| | |
|---|---|
| $e_{ah}$ | horizontal active earth pressure due to effective stresses |
| $\sigma'$ | effective stress |
| $c'$ | effective cohesion |
| $\Delta p$ | surcharge applied "quickly" |
| $\Delta u$ | excess pore water pressure due to surcharge |
| $u_0$ | hydrostatic water pressure |
| $u$ | total water pressure |
| $\sigma_h$ | total horizontal load on wall |

If the load is applied slowly, however, consolidation begins during the application of the load. Consequently, the excess pore water pressure $\Delta u$ due to the surcharge $\Delta p$ is smaller than the load, and the effective stress $\sigma'$ increases during the application of the load. Following full consolidation, the surcharge $\Delta p$ is added in full to the effective stress $\sigma'$ (Fig. R 130-1b).

Intermediate states can be taken into account by determining that proportion of $\Delta p$ for which the soil is already consolidated when calculating the degree of consolidation. This is then added to the above equation for the effective stress; the proportion not yet consolidated (residual excess pore water pressure) is superposed in full on the active earth pressure.

Active earth pressure redistribution (section 8.2.3.2) is only permitted for active earth pressures due to consolidated soil strata.

In the example shown in Fig. R 130-1, the active earth pressure distribution is shown for the case where the uniformly distributed surcharge $\Delta p$ extends unrestricted on the land side, i.e. a plane state

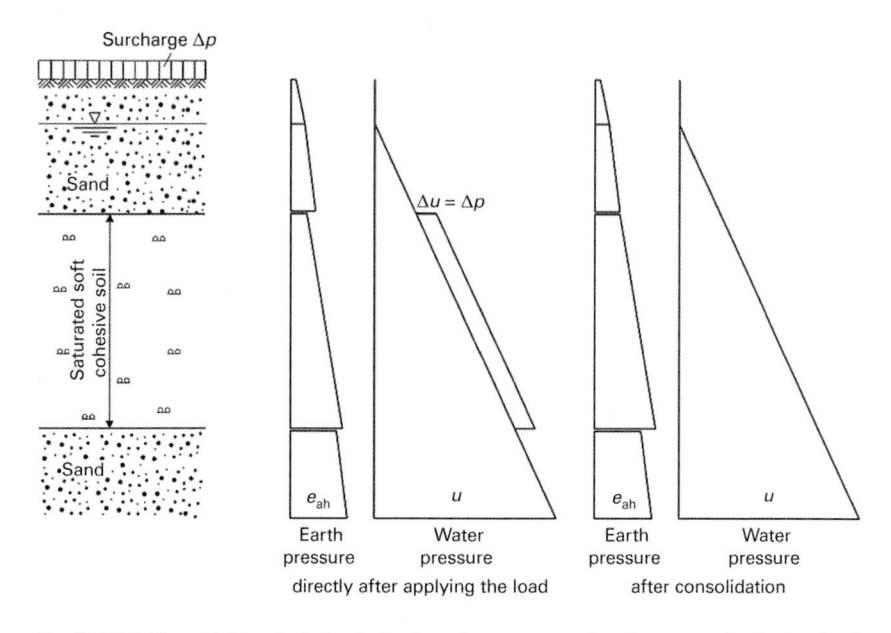

Fig. R 130-1. Example for calculating the horizontal components of earth pressure distribution for the initial state with the shear parameters of the drained soil

of deformation applies. In the soft cohesive stratum, the horizontal load on the wall is due to the active earth pressure as a result of the respective effective stress $\sigma'$ and cohesion $c'$, increased by the excess pore water pressure $\Delta u$ due to the surcharge $\Delta p$ and the hydrostatic water pressure $u_0$.

When it comes to a quick application of the load through hydraulic filling, checks must be performed to see how quickly the sand drains. If this happens immediately, then the surcharge due to the hydraulic fill must be determined using the wet unit weight of the sand. Otherwise, it should be assumed that increased water levels are present in the sand, meaning that part of the fill is still subjected to buoyancy.

## 2.10 Effect of artesian water pressure under harbour bottom or river bed on active and passive earth pressures (R 52)

Artesian water pressure occurs where the harbour bottom or river bed consists of a cohesive stratum of low permeability lying on a groundwater-bearing, non-cohesive stratum and at the same time the low water level of the body of water lies below the concurrent hydraulic head of the groundwater (Fig. R 52-1). The effects of this artesian water pressure on

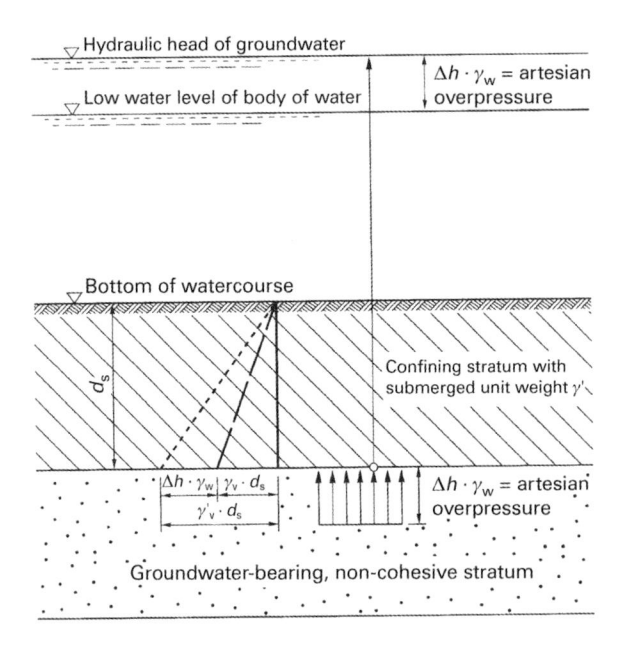

**Fig. R 52-1.** Artesian pressure in groundwater for mainly dead load due to confining stratum

the active and passive earth pressures must be taken into account in the design. The artesian water pressure acts on the confining stratum from below and causes a flow through this, which thus reduces its effective unit weight $\gamma'$. As a result, the active and passive earth pressures also decrease.

In tidal areas, artesian water pressure can be caused by the changing tidal water levels, specifically at low water. The magnitude of the artesian water pressure can then be the same as or greater than the dead load of the confining stratum. At low water, the confining stratum is then lifted off the underlying non-cohesive soil due to the effect of artesian water pressure acting from below and slowly begins to float corresponding to the inflow of pore water.

During the subsequent high water, the confining stratum is then pressed back onto the underlying stratum, thus displacing the pore water again. This process takes place with every tide and is not normally critical under natural conditions. However, if a confining stratum subjected to artesian water pressure depending on the tide is weakened by dredging during the course of construction work, then local heaving of the confining stratum can occur. Such heaving results in local disturbance of the soil in the area around the heave. At the same time, however, the artesian water pressure is relieved, which means that the process is limited to these local disturbances. Similar processes can occur in excavations supported by sheet piling.

The following calculation principles apply to active and passive earth pressures:

1. Passive earth pressure may not be assumed in a cohesive top stratum subjected to artesian water pressure.
2. The passive earth pressure of the soil below the top stratum must be calculated as if the lower stratum is not subjected to any surcharge.

## 2.11 Considering active earth pressure and excess water pressure, and construction guidance for waterfront structures with soil replacement and contaminated or disturbed base of excavation (R 110)

### 2.11.1 General

When soil replacement is used behind waterfront structures as per R 109, section 7.9, it is important to analyse how the contamination in the base of the excavation and slopes to the rear of the excavation can affect active earth pressure and water pressure. This is especially necessary when mud deposits are expected. In addition, in the interests of economic efficiency, soil replacement should be carried out in such a way that intermediate states in which the replacement soil – but primarily any cohesive deposits on the base of the excavation and the rear slope – needs to be considered should not be critical for the design.

### 2.11.2 Approach for determining active earth pressure

Besides the usual design of the structure for the improved soil conditions and ground failure investigations according to DIN 4084, it is necessary to take into account the boundary and disturbing influences arising from a failure plane created by the excavation as per Fig. R 110-1.

The following factors are the most important for the active earth pressure $E_a$:

1. Length and – if applicable – inclination of the restraining section $l_2$ of the failure plane determined by the base of the excavation.
2. Thickness, shear strength $\tau 2$ and effective load of intrusive layer on $l_2$.
3. Any dowelling action for section $l_2$ due to piles or similar.
4. Thickness of the adjacent soft cohesive soil along the inshore edge of the excavation, also its shear strength together with the form and angle of the rear slope.
5. Sand load and imposed load, especially on the rear slope.
6. Properties of the backfilling.

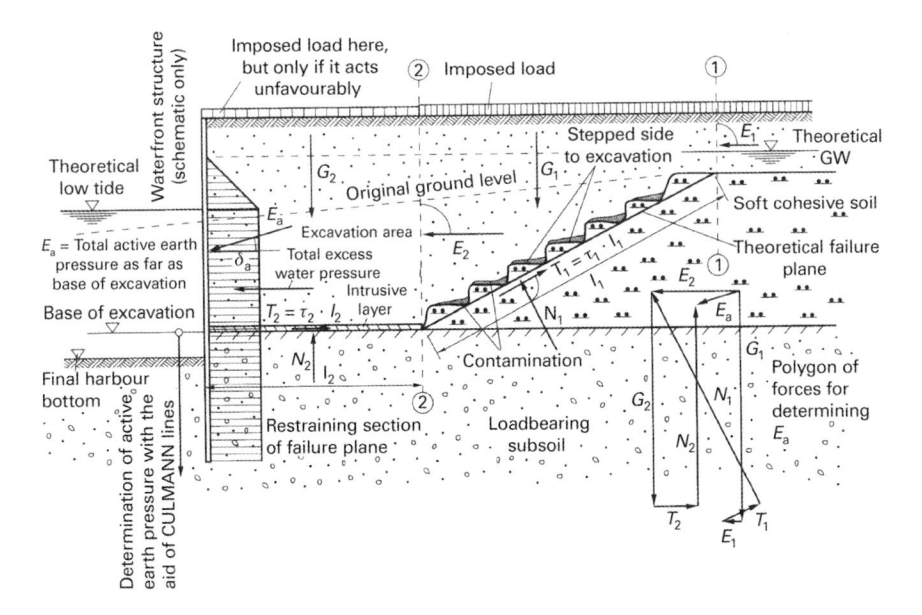

**Fig. R 110-1.** Determination of active earth pressure $E_a$ acting on waterfront structure

The distribution of the active earth pressure $E_a$ as far as the base of the excavation depends on the deformations and the type of waterfront structure.

The active earth pressure below the base of the excavation can be determined, for example, with the aid of the Culmann method. In doing so, the shear forces in section $l_2$, including any dowelling, are to be taken into account.

At all stages of construction, including the original excavation of the harbour bottom or any subsequent deepening of the harbour, the shear stress $\tau_2$ acting at the time in the intrusive layer on section $l_2$ can be calculated as follows:

$$\tau = (\sigma - \Delta u) \cdot \tan\varphi' \approx \sigma' \cdot \tan\varphi'$$

where $\sigma'$ is the effective stress due to vertical loads effective at the place and time of the investigation and $\varphi'$ is the effective angle of friction of the material in the intrusive layer. The final shear strength at the end of the consolidation process is then

$$\tau = \sigma'_a \cdot \tan\varphi'$$

where $\sigma'_a$ is the effective stress due to vertical loads in the area investigated in section $l_2$ at the end of consolidation ($\Delta u = 0$).

Separate calculations are necessary to consider the effects of dowelling section $l_2$ by means of piles [25].

If the rear slope of the excavation has been properly carried out in larger steps, the critical failure plane passes through the rear corners of the steps and thus lies in undisturbed soil (Fig. R 110-1). In this case, owing to the long consolidation periods of soft cohesive soils, the shear strength used for the failure plane is equal to the initial shear strength of the soil prior to excavation. If the soft cohesive soil is made up of strata with different initial shear strengths, these must be taken into account accordingly.

Where a rear slope in soft soil is severely disturbed, has been excavated in small steps or is unusually contaminated, then the shear strength of the disturbed failure plane must be used in the calculations instead of the initial shear strength of the undisturbed soil. This is usually less than the initial shear strength and should therefore be determined in laboratory tests.

The consolidation of soft cohesive soils below the soil replacement area can take a long time owing to the long drainage paths. Considering the increase in the shear strength due to consolidation for such soils is generally only permissible when consolidation is accelerated by closely spaced drains.

## 2.11.3 Approaches for determining excess water pressure

When employing soil replacement, the excess water pressure should be calculated from the difference in levels between the groundwater in the area of reference line 1-1 (Fig. R 110-1) and the lowest simultaneous external water level. A drainage system behind the waterfront structure can lower the groundwater level temporarily, but experience shows that such drainage is not effective over the long term.

The excess water pressure can be applied in the usual approximated form, i.e. increasing linearly over the depth (Fig. R 110-1). A more accurate approach can be derived from an investigation of the flow net around the structure (R 113, section 4.7; R 114, section 2.12).

## 2.11.4 Guidance for the design of waterfront structures

Investigations of excavations have revealed that intrusive layers on the base and sloping sides are fully consolidated within the construction period provided these are no thicker than about 20 cm. Full consolidation during the construction period cannot be assumed for thicker intrusive layers without more detailed investigations. In such cases the expected shear strength of the intrusive layer must be calculated, e.g. from an estimate of the course of consolidation using the shear parameters of the intrusive layer.

However, in order that the reduced shear strength of an only partially consolidated intrusive layer does not become critical for the design, it may be necessary to schedule certain construction measures, e.g. initial dredging or deepening of the harbour, such that consolidation is completed before the loads associated with these measures become effective.

Anchorage forces are transferred via piles or other structural elements into the loadbearing subsoil beneath the base of the excavation because anchorage forces transferred above the base of the excavation would place additional loads on the sliding mass of soil.

Apart from the length required for structural reasons, the length of section $l_2$ in Fig. R 110-1 is to be such that that all structural piles fit within the base of the excavation. This guarantees that the bending stresses in the piles due to settlement of the backfill are kept low.

When mud deposits on the base of the excavation are severe and therefore thicker intrusive layers and/or very loosely compacted sand zones cannot be avoided despite taking every care during soil replacement, the result can be high loads on the piles and, in particular, high bending stresses. To avoid brittle fractures due to these actions, piles made from killed steel must be used (R 67, section 8.1.6.1; R 99, section 8.1.19.2).

If foundation piles have been installed to create a dowelling effect for the failure plane in section $l_2$ according to Fig. R 110-1 [25] and are included when verifying the stability of the overall system according to DIN 4084, the maximum principal stress due to axial loads, shear and bending should in no case exceed 85 % of the yield stress when checking the stresses in these piles. To calculate the dowelling forces, the magnitudes of the pile deformations must be compatible with the other movements of the structure and its components, i.e. generally only a few centimetres. In soft cohesive soils these deformations are not sufficient to ensure effective dowelling (Fig. R 110-1). Piles that are stressed up to yield because of settlement of the subsoil or the backfilling may not be considered as contributing to dowelling.

In order to prevent the properties of an intrusive layer on the base or sloping sides of the excavation becoming critical for the design of the structure, efforts must be made to ensure that the base of the excavation is cleaned immediately prior to backfilling with the replacement soil. In addition, section $l_2$ according to Fig. R 110-1 must be sufficiently long and the sloping side to the excavation must be as shallow as possible (see effects in polygon of forces in Fig. R 110-1).

Where there is only a thin intrusive layer, the shear resistance of the failure plane in section $l_2$ can be substantially improved if it is covered with a layer of rubble. If sufficient time is available, closely spaced drains in the soft cohesive soil behind the slope to the back of the excavation can help to speed up consolidation and hence reduce the active earth pressure. Temporarily reducing the imposed load on the backfill over the slope to the excavation or temporarily lowering the groundwater level as far as a point beyond reference section 1-1 can help to overcome unfavourable initial conditions.

If the restraining section $l_2$ according to Fig. R 110-1 is to be omitted in regions with marine clay, i.e. the slope to the excavation extends up to the back of the wall, then the slope must be as shallow as possible. It should certainly be shallower than about 1:4 in this case because then sliding of the replacement soil on the slope does not result in any additional loads on the waterfront structure. This should in any case be verified by calculation.

## 2.12 Effect of groundwater flow on excess water pressure and active and passive earth pressures (R 114)

### 2.12.1 General

Where groundwater flows around a structure, the flowing water exerts a pressure on the masses of the sliding wedges of soil for the active and passive earth pressures, thus changing the magnitudes of these forces.

The total effects of the groundwater flow on the active earth pressure $E_a$ and the passive earth pressure $E_p$ can be determined with a flow net according to R 113, section 4.7.7 (Fig. R 113-2). To do this, all water pressures acting on the boundary surfaces of the sliding sections of earth are determined and are then taken into account in the Coulomb polygon of forces for the active earth pressure (Fig. R 114-1a) and the passive earth pressure (Fig. R 114-1b). Fig. R 114-1 shows the forces that are to be included for the case of a straight failure plane: $G_a$ and $G_p$ are the dead loads of the sliding wedges for the saturated soil, $W_1$ the resultant water surcharge on the sliding wedges, $W_2$ the resultant water pressure between structure and sliding wedge area and $W_3$ the resultant water pressure acting at the failure plane. Pressures $W_2$ and $W_3$ must be determined using a flow net (R 113, section 4.7.7, Fig. R 113-2). Forces $Q_a$ and $Q_p$ are the soil reactions acting at an angle $\varphi'$ to the failure plane normal and $E_a$ and $E_p$, acting at angles $\delta_a$ and $\delta_p$ respectively to the wall normal, are the total active and passive earth pressures respectively.

The excess water pressure results from the difference between the water pressure $W_2$ acting on the structure from inside and outside.

The accuracy of the water pressure, active earth pressure and passive earth pressure figures determined in this way depend on how well the flow net matches the actual conditions.

The above approach has proved useful for waterfront structures in principle. However, more detailed investigations are required for narrow trenches and trench corners in order to take into account the mutual impact of opposing or abutting sides [230].

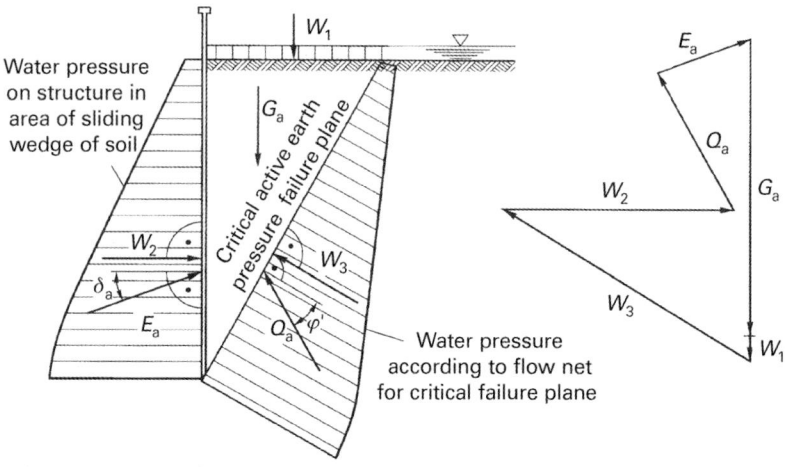

a) Determination of active earth pressure $E_a$

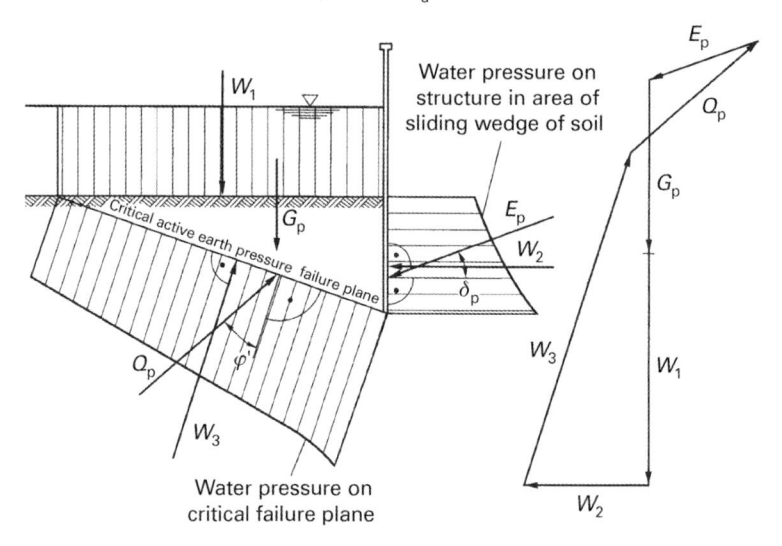

b) Determination of passive earth pressure $E_p$

**Fig. R 114-1.** Determination of active earth pressure $E_a$ and passive earth pressure $E_p$ taking into account the effects of flowing groundwater

The solution according to Fig. R 114-1 only provides the resultants of $E_a$ and $E_p$, not the distribution of the active and passive earth pressures over the height of the wall. Separate consideration of the horizontal and vertical flow forces is therefore worthwhile for practical applications. In this case the horizontal effects are added to the excess water pressure, which is related to the respective failure plane for the active or passive earth pressure (Fig. R 114-2). The vertical effects are added to the vertical soil pressures due to the dead load of the soil, i.e. considered in the unit weight of the soil assumed when calculating the active earth pressure. This approach is dealt with in section 2.12.3.

## 2.12.2 Determining the excess water pressure

The excess water pressure on a waterfront structure can generally be determined according to R 19, section 4.2, or, for predominantly horizontal flow, in accordance with R 65, section 4.3. However, in the event of large differences in water level on the two sides of the structure, the influence of the flow can be so large that it is worthwhile carrying out a more detailed investigation with a flow net.

The flow net according to R 113, section 4.7.7, Fig. R 113-2 will be used to explain the procedure; the calculation sequence can be seen in Fig. 114-2.

The first step is to determine the water pressure at the failure planes for the active and passive earth pressures. This pressure can be calculated from the associated hydraulic head for each intersection between equipotential line and failure plane. The water pressure is the product of the unit weight of the water and the hydraulic head at this intersection (Fig. R 114-2, right side). If the water pressures determined in this way are plotted horizontally from a vertical reference line, the result is the horizontal projection of the water pressures acting at the failure plane. The horizontal excess water pressure is the difference between the calculated water pressure on the failure plane and the external water pressure.

A good approximate solution for the horizontal water pressure is also possible with the approach given in section 2.12.3.2. This takes account of the fact that the potential decreases with the flow around the wall. Here, the potential on the active earth pressure side decreases compared with the hydrostatic distribution and increases on the passive earth pressure side. Both influences can be properly assessed by increasing or decreasing the unit weight $\gamma_w$ of the water by the amount $\Delta\gamma_w$. The excess water pressure is again due to the difference in water pressure distributions to the left and right of the wall.

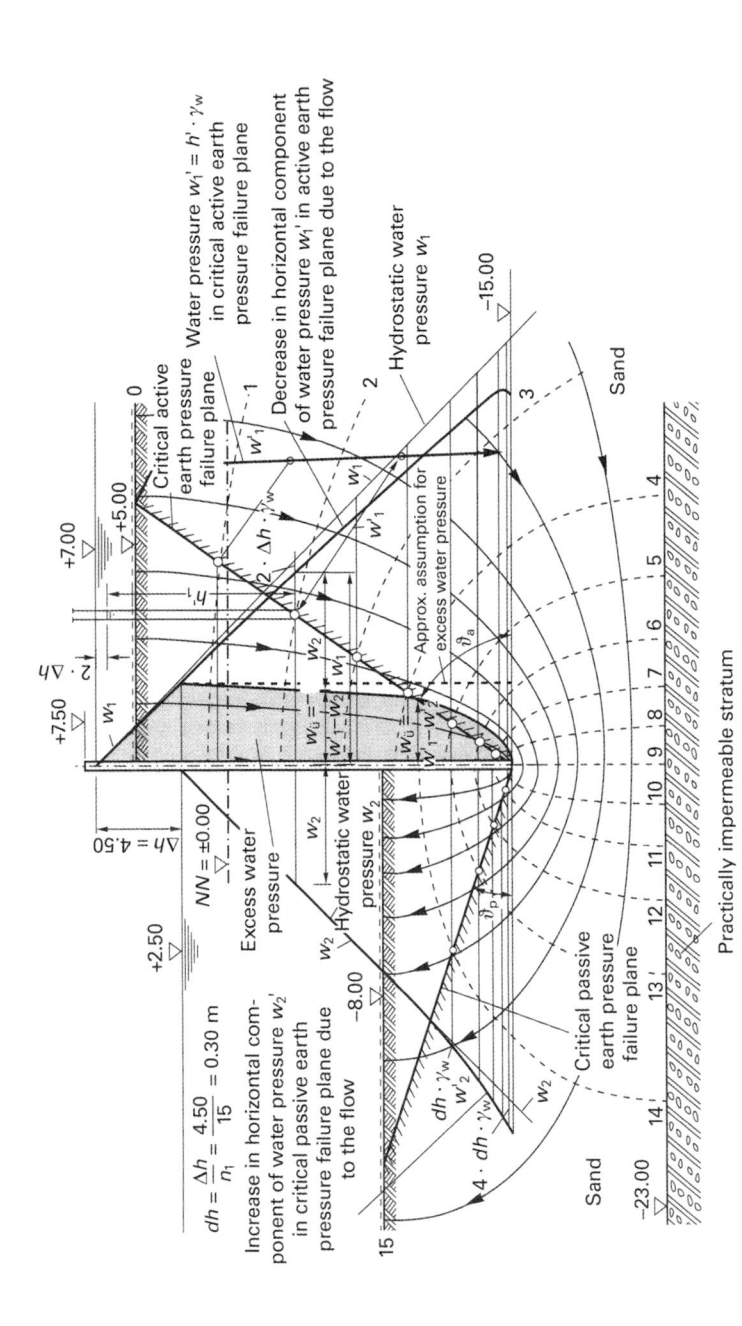

**Fig. R 114-2.** Determination of excess water pressures acting on a sheet pile structure with the flow net according to R 113, section 4.7.7

## 2.12.3 Determining the effects on active and passive earth pressures when the flow is mainly vertical

### 2.12.3.1 Calculations using a flow net

The flow net of R 113, section 4.7.7, Fig. R 113-2 is again used to explain the calculations. The calculation procedure is shown in detail in Fig. R 114-3.

The hydraulic head difference per flow net field is in each case equivalent to a vertical flow force. The flow force increases from top to bottom on the active earth pressure side, and decreases from bottom to top on the passive earth pressure side. If $dh$ is the hydraulic head difference per flow net field and $n$ is the number of fields in the flow direction, the result is an additional vertical stress $n \cdot \gamma_w \cdot dh$ on the active earth pressure side due to the flow force, and that leads to an increase in the horizontal component of the earth pressure stress amounting to

$$\Delta e_{ahn} = +n \cdot \gamma_w \cdot dh \cdot K_{ag} \cdot \cos \delta_a$$

On the passive earth pressure side, the flow is from bottom to top and so the passive earth pressure, and consequently its horizontal component, is reduced accordingly:

$$\Delta e_{phn} = -n \cdot \gamma_w \cdot dh \cdot K_{pg} \cdot \cos \delta_p$$

In contrast to the reduced water pressure on the active earth pressure side, there is usually an increase in the earth pressure amounting to about one-third of the water pressure decrease. On the passive earth pressure side, owing to the much larger $K_p$ value, the decrease in the passive earth pressure is greater than the increase in the water pressure.

The effect of the horizontal component of the flow force on the active or passive earth pressure is taken into account by determining the excess water pressure according to section 2.12.2, Fig. R 114-2, with the water pressure applied to the critical active or passive earth pressure failure plane.

### 2.12.3.2 Approximate solution for the influence of flow on the active and passive earth pressures

With a primarily vertical flow around a waterfront structure, the increase in the active earth pressure and the decrease in the passive earth pressure as a result of the flow can be calculated approximately by increasing the unit weight of the soil on the active earth pressure side and reducing it on the passive earth pressure side.

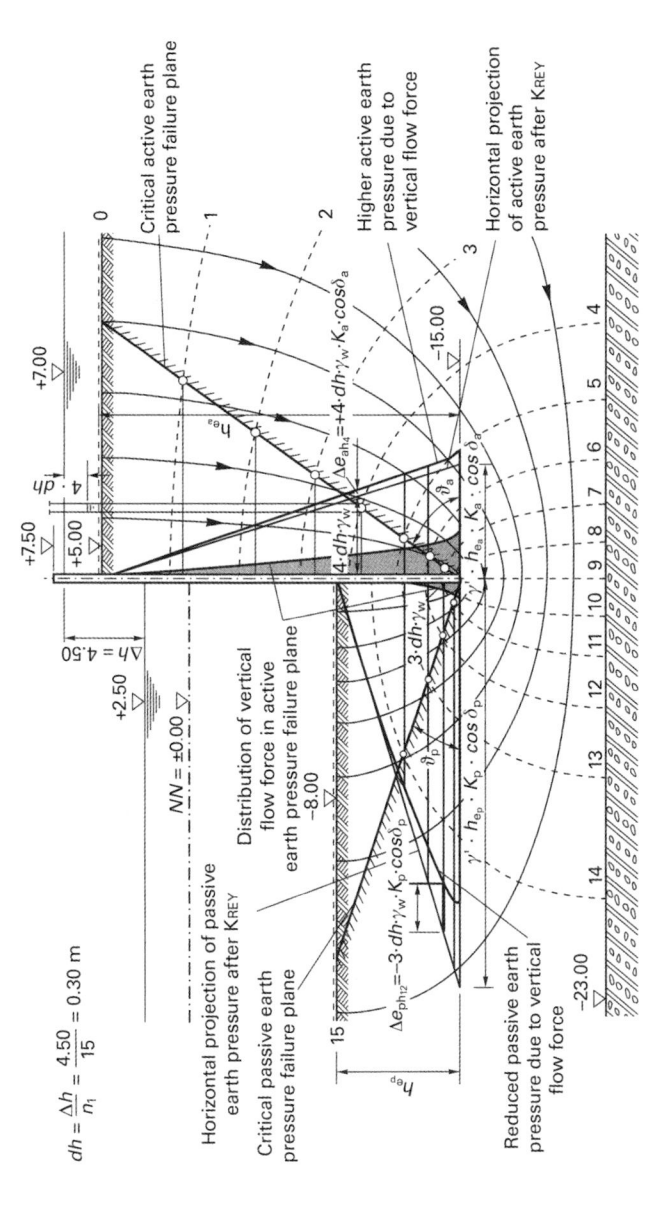

**Fig. R 114-3.** Influence of vertical flow force on the active and passive earth pressures with chiefly vertical flow, determined with the flow net of R 113, section 4.7.7

The increase in the unit weight on the active earth pressure side and the decrease on the passive earth pressure side can be determined approximately for exclusively vertical flow and homogenous subsoil according to [24] using the following equations:

– on active earth pressure side:

$$\Delta \gamma' = \frac{0{,}7 \cdot \Delta h}{h_{so} + \sqrt{h_{so} \cdot h_{su}}} \cdot \gamma_w$$

– on passive earth pressure side:

$$\Delta \gamma' = -\frac{0{,}7 \cdot \Delta h}{h_{su} + \sqrt{h_{so} \cdot h_{su}}} \cdot \gamma_w$$

where

$\Delta h$    difference in water levels on both sides of the wall (difference in hydraulic head)
$h_{so}$    depth of soil affected by flow on land side of sheet piling down to the base of the piling in which a decrease in hydraulic head occurs
$h_{su}$    driving depth or soil stratum thickness on water side of sheet piling in which a decrease in hydraulic head occurs
$\gamma'$    buoyant unit weight of soil
$\gamma_w$    unit weight of water

(The symbols also apply to Fig. R 114-4.)

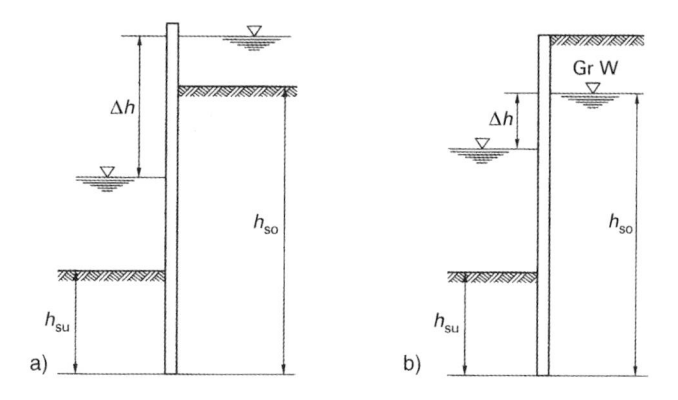

**Fig. R 114-4.** Definitions for the approximate determination of the effective unit weight of the soil in front of and behind a sheet pile structure as changed by flow force

The above equations apply when there is also soil underneath the base of the sheet piling which, in a flow situation, contributes to reducing the potential to the same extent as the soil in front of and behind the sheet piling (see section 4.9.3). Otherwise, section 2.12.3.1 applies similarly. In the case of a horizontal inflow, the residual potential at the base of the sheet piling increases considerably, so the above approximation should not be used in such cases.

Refer to R 219, section 2.5, when determining the active and passive earth pressures in stratified soils.

## 2.13 Determining the amount of displacement required for mobilising passive earth pressure in non-cohesive soils (R 174)

As a rule, a considerable amount of displacement is required to mobilise the full passive earth pressure in front of a waterfront structure. The amount is chiefly dependent on the embedment depth, the in situ density of the soil and the type of movement. According to DIN 4085, the following mobilisation function applies for the horizontal active earth pressure $E'_{pgh}$ reached depending on the displacement $s$:

$$E'_{pgh} = E_{0gh} + (E_{pgh} - E_{0gh}) \cdot \left[1 - \left(1 - \frac{s}{s_p}\right)^b\right]^{0.7}$$

where

$E_{pgh}$    maximum passive earth pressure
$E_{0gh}$    earth pressure at rest due to dead load of soil
$s_p$    displacement required to achieve $E_{pgh}$
$b$    exponent of mobilisation function

The required displacement $s_p$ and the exponent $b$ of the mobilisation function depend on the wall movement, the embedment depth $d$ and the in situ density $D$ of the soil (Table R 174-1):

**Table R 174-1.** Displacement and exponent for mobilisation function

| Wall movement | Displacement | Exponent |
|---|---|---|
| Rotation about base | $s_p = (0.12 - 0.08 \cdot D) \cdot d$ | $b = 1.07$ |
| Parallel displacement | | $b = 1.45$ |
| Rotation about uppermost point of embedment depth | $s_p = (0.09 - 0.05 \cdot D) \cdot d$ | $b = 1.72$ |

According to research by Weißenbach [224], good mobilisation is achieved in front of narrow soil resistance compression zones even with minor displacements. The displacement $s_p^r$ required for the maximum three-dimensional passive earth pressure is calculated as follows:

$$s_p^r = 40 \cdot \frac{1}{1+0.5 I_D} \cdot \frac{d^2}{\sqrt{b_0}}$$

where

$b_0$ width of compression zone ($b_0 < d/3$)
$I_D$ relative density

The test results of [222] are approximated well by the mobilisation function proposed by [97] in the range of values $0.1 \leq s/s_p^r \leq 1$:

$$E_{pgh}^{r\prime} = E_{pgh}^r \cdot \frac{s/s_p^r}{0.12 + 0.88 \cdot s/s_p^r}$$

where

$E_{pgh}^{r\prime}$ mobilised three-dimensional passive earth pressure

$E_{pgh}^r$ maximum three-dimensional passive earth pressure (passive earth pressure in front of narrow soil resistance compression zones)

$s_p^r$ displacement required to achieve $E_{pgh}^r$

## 2.14 Measures for increasing the passive earth pressure in front of waterfront structures (R 164)

### 2.14.1 General

The following are examples of suitable underwater measures for increasing the passive earth pressure in front of waterfront structures:

7. Replacement of in situ soft cohesive soil with non-cohesive material (soil replacement).
8. Compaction of loosely bedded, non-cohesive in situ soil, with additional surcharge if necessary.
9. Consolidation of soft cohesive soils with surcharge.
10. Placing of a fill.
11. Consolidation of the in situ soil.
12. A combination of measures 1 to 5.

These measures differ in terms of their costs; in certain cases they may also prevent the subsequent deepening of a basin or the upgrading of a waterfront structure, e.g. by driving a new wall in front of the old one. In order to retain these options, the various measures taken in specific cases should also be assessed with respect to the further development of the waterfront facilities.

In principle, all the above measures are common in ground engineering. Special aspects of their use in port and harbour engineering are given below.

## 2.14.2 Soil replacement

When replacing soft cohesive subsoil with non-cohesive material, R 109, section 7.9, must be observed in so far as the works themselves are affected. When determining the passive earth pressure, any intrusive layers on the base of the excavation must be taken into account. The remarks of R 110, section 2.11, apply accordingly.

The extent and depth of soil replacement in front of the waterfront structure is normally determined according to earth pressure aspects. In order to exploit fully the higher passive earth pressure achieved by the replacement soil deposited, the passive earth pressure sliding wedge must lie completely within the area of soil replacement.

## 2.14.3 Soil compaction

Loose non-cohesive soils can be compacted by means of vibroflotation. The spacing of the vibration points (grid size) depends on the in situ subsoil and the mean degree of compaction required. The greater the desired improvement in the degree of compaction required and the finer the grains of the soil to be compacted, the closer the spacing must be. An average grid size of 1.80 m is recommended as a guide. In the case of soil replacement, the compaction must encompass all the new soil, i.e. reach as far as the base of the excavation.

Vibroflotation should encompass the entire area of the passive earth pressure sliding wedge in front of the structure and in doing so extend a sufficient distance beyond the critical passive earth pressure failure plane starting from the theoretical base of the sheet piling. If cases of doubt, curved or polygonal failure planes should also be checked to ensure that the area of compaction is sufficiently large.

Compaction by means of vibroflotation causes temporary liquefaction of the soil in the immediate vicinity of the vibration plant, but this soil is then compacted due to the soil surcharge. Therefore, the effect of compaction depends on the soil surcharge, and the soil near the surface (about 2–3 m thick) can only be compacted when a temporary surcharge in the form of fill is introduced during compaction.

Vibroflotation can also be used for subsequent strengthening of waterfront structures. However, when applying this method, appropriate procedures should be chosen to guarantee that the stability of the waterfront wall is not adversely affected by the temporary local liquefaction of the soil. Experience has shown that extensive and prolonged liquid states can occur in loosely bedded, fine-grained, non-cohesive soils and fine sand especially.

Soil compaction can reduce the risk of liquefaction effectively in seismic regions.

### 2.14.4 Soil surcharge

Under certain conditions, e.g. securing an existing waterfront structure, it may be expedient to improve the support for the structure by introducing a fill with a high unit weight and high angle of internal friction in the region of passive earth pressure. Suitable materials here are metal slags or natural stone. The buoyant unit weight of the materials is critical; metal slags can attain values $\gamma' \geq 18\,\text{kN/m}^3$. The characteristic angle of internal friction may be taken as $\varphi'_k = 42.5°$.

In the case of soft in situ subsoil, care must be taken to ensure that the fill material does not subside by choosing a suitable grading for the fill material, including a filter layer between fill and existing soil or limiting the thickness of the fill.

The fill material must be checked constantly for compliance with the conditions and specification. This applies particularly to unit weight.

Sections 2.14.2 and 2.14.3 apply accordingly regarding the necessary extent of the works.

Additional vertical drains can be installed to accelerate the consolidation of soft strata below fill material.

### 2.14.5 Soil stabilisation

If readily permeable, non-cohesive soils (e.g. gravel, gravelly sand or coarse sand) are present in the passive earth pressure area, grout injection represents another method of stabilising the soil. In less permeable, non-cohesive soils, stabilisation by means of high-pressure injection is a potential solution. Stabilisation with chemicals is also possible, provided the chosen stabilising medium can cure properly allowing for the chemical properties of the pore water. Generally, however, chemical stabilisation is too expensive, which restricts it mainly to localised areas with special boundary conditions and tight construction schedules.

It should be noted that stabilisation of the in situ soil constitutes a barrier to any subsequent deepening of the basin and/or driving a new waterfront wall in front of the old one.

The prerequisite for all types of stabilisation through injection is an adequate surcharge, which must be placed in advance and then possibly removed again.

The required dimensions for the stabilisation area can be stipulated in accordance with sections 2.14.2 and 2.14.3. Soil core samples and/or penetrometer tests are always necessary to verify the success of soil stabilisation measures.

## 2.15 Passive earth pressure in front of abrupt changes in ground level in soft cohesive soils with rapid load application on land side (R 190)

When determining the passive earth pressure in front of sheet piling with a rapid load application on the land side, the same principles apply as for determining active earth pressure for this load case (R 130, section 2.9). In the load case "excavating in front of wall", the passive earth pressure should be determined with effective shear parameters ($c'$, $\varphi'$) because excavating does not generally proceed at such a rate that undrained behaviour governs.

In the load case "backfilling behind wall", the load can be applied so rapidly that undrained conditions can occur on the side of the base support, caused by horizontal movement of the wall. A "rapid" surcharge occurs if the rate of load application is significantly greater than the rate of consolidation of the cohesive stratum. In normally consolidated soils, excess pore water pressure can occur in the passive earth pressure area, which results in a weakening of the base support. In overconsolidated soils, negative pore water pressure is the result due to their dilatory material behaviour, but this is generally negligible.

A more accurate analysis of the influence of a rapid loading rate on the passive earth pressure is only possible with numerical modelling. Such an analysis must assess the stress–strain relationship of the soil and the sequence of construction operations as accurately as possible and allow a forecast of the development of the excess pore water pressure over time ([221], *AK Numerik* – Recommendation 1, [4]).

## 2.16 Waterfront structures in seismic regions (R 124)

### 2.16.1 General

In practically all countries where earthquakes can be expected, there are various standards, directives and recommendations – for buildings primarily – which normally contain detailed requirements for earthquake-resistant design and construction. Designers should refer to DIN EN 1998-1:2010-12, DIN EN 1998-1/NA:2011-01, DIN EN

1998-5:2010-12, DIN EN 1998-5/NA:2011-07 for the seismic design of buildings in Germany, but, for example, PIANC [161] can be consulted for port and harbour installations.

The intensity of the earthquakes to be expected in the various regions is generally expressed in such publications by means of the horizontal ground acceleration $a_h$. Any simultaneous vertical acceleration $a_v$ is generally negligible compared with the acceleration due to gravity $g$.

The acceleration $a_h$ causes not only immediate loads on the structure but also has an influence on the active and passive earth pressures, the water pressure and in some cases the shear strength of the soil beneath the foundation as well. In unfavourable circumstances, this shear strength may temporarily disappear completely (liquefaction).

The additional actions during an earthquake are normally taken into account in the actual structure in such a manner that additional horizontal forces

$$\Delta H = \pm k_h \cdot V$$

acting at the centre of gravity of each accelerated mass are applied simultaneously with the other loads.
In the above equation,

$k_h = a_h/g$ seismic coefficient = ratio of horizontal ground acceleration to gravitational acceleration
$V$ dead load of structural member or sliding wedge of soil considered, including pore water

The magnitude of $k_h$ depends on the intensity of the earthquake, the distance from the epicentre and the in situ subsoil. The first two factors are taken into account in most countries by dividing the endangered regions into earthquake zones with appropriate $k_h$ values (DIN EN 1998-1:2010-12, DIN EN 1998-1/NA:2011-01, DIN EN 1998-5:2010-12, DIN EN 1998-5/NA:2011-07). In cases of doubt, agreement on the magnitude of $k_h$ to be used may need to be reached by consulting an experienced seismic expert.

In the case of tall, slender structures at risk of resonance, i.e. when the natural frequency of the structure lies within the earthquake frequency spectrum, inertia forces must also be taken into account in the calculations. This is, however, not generally the case with waterfront structures. Therefore, the principal requirement for the design and construction of an earthquake-resistant waterfront structure is to ensure that it can also accommodate the additional horizontal forces occurring during an earthquake with a reduction in passive earth pressure.

## 2.16.2 Effects of earthquakes on the subsoil

Waterfront structures in earthquake regions must also take account of the soil conditions deeper underground. For example, seismic vibrations are most severe where loose, relatively thin deposits overlie solid rock.

The most sustained effects of an earthquake make themselves felt when the subsoil, especially ground beneath foundations, is liquefied by the earthquake, i.e. loses its shear strength temporarily. This situation (due to settlement flows, liquefaction) arises in loosely bedded, fine-grained, non- or weakly cohesive, saturated and marginally permeable soil (e.g. loose fine sand or coarse silt). The lower the overburden at the depth in question and the greater the intensity and duration of the vibrations, the sooner liquefaction will occur.

If the risk of liquefaction cannot be entirely ruled out, it is advisable to investigate the true liquefaction potential (e.g. [108]). Looser soil strata with a liquefaction tendency can be compacted in advance. Cohesive soils cannot be liquefied by seismic action.

## 2.16.3 Determining the effects of earthquakes on active and passive earth pressures

The influence of earthquakes on active and passive earth pressures is also generally determined according to Coulomb – with the additional forces $\Delta H$ generated by earthquakes according to section 2.16.1 being considered as well. The resultant of the dead loads of the wedges of soil and the additional horizontal forces is no longer vertical. This is taken into account by relating the angle of the active or passive earth pressure reference plane and angle of the ground surface to the new force direction [77]. The result is fictitious angles for the reference plane ($\pm\Delta\alpha$) and the ground surface ($\pm\Delta\beta$).

$$k_h = \tan\Delta\alpha \quad \text{or} \quad k_h = \tan\Delta\beta \quad \text{(Fig. R 124 – 1)}.$$

The active or passive earth pressure is then calculated on the basis of the imaginary system rotated through an angle $\Delta\alpha$ or $\Delta\beta$ respectively (reference plane and ground surface).

An equivalent procedure considers the angle of rotation according to Fig. R 124-1 by calculating the active and passive earth pressures with a wall inclination $\alpha \pm \Delta\alpha$ and a ground inclination $\beta \pm \Delta\beta$.

When determining the active earth pressure below groundwater level, it must be realised that the mass of the soil and the mass of the water trapped in the pores of the soil are accelerated during an earthquake. However, the reduction in the unit weight of the submerged soil remains unchanged due to buoyancy. Therefore, it is expedient to use a larger seismic coefficient – the so-called apparent seismic coefficient $k'_h$ – in calculations for the area below groundwater level.

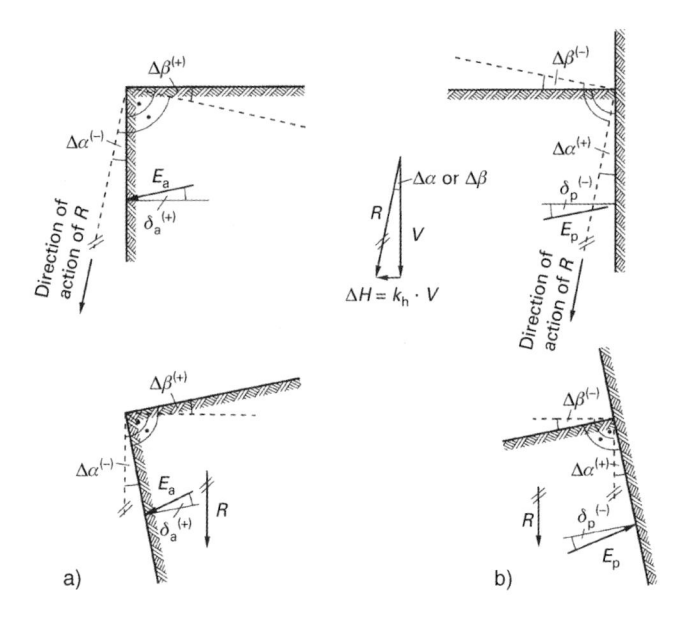

**Fig. R 124-1.** Determining the fictitious angles $\Delta\alpha$ and $\Delta\beta$ and diagrams of systems rotated through angles $\Delta\alpha$ and $\Delta\beta$ (signs after Krey): a) for calculating active earth pressure, b) for calculating passive earth pressure

In the section considered in Fig. R 124-2,

$$\sum p_v = p + \gamma_1 \cdot h_1 + \gamma'_2 \cdot h_2$$

and

$$\sum p_h = k_h \cdot [p + \gamma_1 \cdot h_1 + (\gamma'_2 + \gamma_w) \cdot h_2]$$

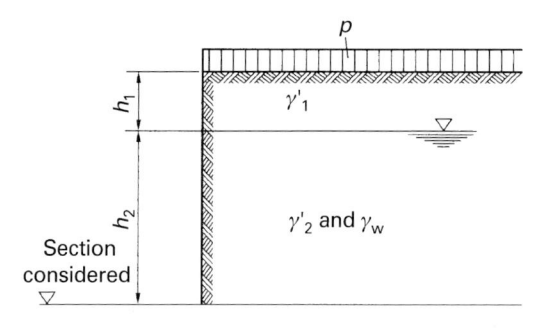

**Fig. R 124-2.** Sketch showing arrangement for calculating $k'_h$

The apparent seismic coefficient $k'_h$ for determining the active earth pressure below groundwater level is thus:

$$k'_h = \frac{\sum p_h}{\sum p_v} = \frac{p + \gamma_1 \cdot h_1 + (\gamma'_2 + \gamma_w) \cdot h_2}{p + \gamma_1 \cdot h_1 + \gamma'_2 \cdot h_2} \cdot k_h$$

A similar procedure can be employed for the passive earth pressure. For the special case where the groundwater level is at the surface and where there is no ground surcharge, the result for the active earth pressure side with $\gamma_w = 10\,\text{kN/m}^3$ is

$$k'_h = \frac{\gamma' + 10}{\gamma'} \cdot k_h = \frac{\gamma_r}{\gamma_r - 10} \cdot k_h \cong 2\,k_h$$

where

$\gamma'$   buoyant unit weight of soil
$\gamma_r$   unit weight of saturated soil

For simplicity, the unfavourable value of $k'_h$ determined in this way for the active earth pressure side is normally also used as the basis for further calculations, even when the groundwater level is lower and for transient loads.

The active earth pressure coefficient $K_{ah}$ determined using $k_h$ and $k'_h$ results in an abrupt change in the active earth pressure at the level of the water table, as shown in Fig. R 124-3. If a more accurate calculation is deemed unnecessary, the active earth pressure can be applied in simplified form according to Fig. R 124-3.

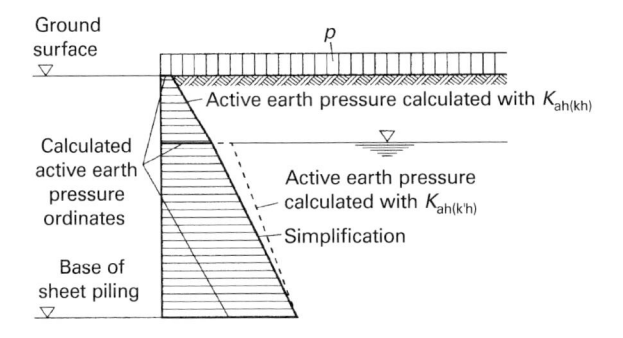

**Fig. R 124-3.** Simplified approach for active earth pressure

In difficult cases for which tabular values are not available for calculating the active and passive earth pressures, it is possible to determine the influences of both the horizontal and – if present – vertical ground accelerations on active and passive earth pressures by applying an extended Culmann method. The forces due to ground accelerations acting on the wedges investigated with the respective seismic coefficient $k_h$ must then be taken into account in the polygons of forces as well. Such an approach is recommended for larger horizontal accelerations as well, especially when the soil lies partly below the groundwater level.

### 2.16.4 Excess water pressure

The excess water pressure for waterfront structures in the seismic loading case may be approximated as for the normal case, i.e. according to R 19, section 4.2, and R 65, section 4.3, because the effects of the earthquake on the pore water have already been taken into account when determining the active earth pressure with the apparent seismic coefficient $k'_h$ according to section 2.12. It should be noted, however, that in the seismic loading case the critical active earth pressure failure plane is inclined at a shallower angle to the horizontal than in the normal case. For this reason, a higher excess water pressure can act on the failure plane.

### 2.16.5 Transient loads

As the simultaneous occurrence of an earthquake, maximum transient load and maximum wind load is unlikely, it suffices to combine the seismic loads with only half the transient load and half the wind load (see also DIN EN 1998-1:2010-12, DIN EN 1998-1/NA:2011-01, DIN EN 1998-5:2010-12, DIN EN 1998-5/NA:2011-07 commentary). Crane wheel loads due to wind and the wind component of line pull should therefore be reduced accordingly as well. Loads due to the travel and slewing movements of cranes do not need to be superposed on earthquake effects.

However, loads that in all probability remain constant for a longer period of time, e.g. loads from filled tanks or silos and from bulk cargo storage, must not be reduced.

### 2.16.6 Design situation and partial safety factors

According to DIN EN 1990, earthquake forces may be taken into account for earthquake design situations (DS-E) without applying partial safety factors to actions and resistances.

### 2.16.7 Guidance for considering seismic influences on waterfront structures

Taking into account the foregoing and other EAU recommendations allows waterfront structures to be designed and constructed systematically and with adequate stability for earthquake regions as well. Supplementary information regarding sheet pile structures can be found in R 125, section 8.2.16, block-type waterfront structures in R 126, section 10.8, and pile bents/trestles in R 127, section 11.5.

Experience gained from the earthquake in Japan in 1995 is outlined in JSCE [112]. For further implications please refer to PIANC [161].

# 3 Hydraulic heave failure, ground failure

## 3.1 Safety against hydraulic heave failure (R 115)

In a hydraulic heave failure, a body of earth in front of a structure is subjected to uplift force due to groundwater flow. This effect reduces the passive earth pressure. Failure occurs when the vertical component $F'_s$ of the flow force is equal to or greater than the dead load $G'$ of the soil body in front of the structure. The soil body is then raised hydraulically (fluidisation) and the passive earth pressure is lost completely.

In homogenous soil, all potential hydraulic heave failure planes start at the base of the structure. The plane with the smallest safety margin, as determined by calculations, is critical for the assessment. Guidance on the relationships and the procedure for stratified subsoil are given in R 113, section 4.7.7.2.

For walls with only a small embedment depth, safety against hydraulic heave can be guaranteed by including a surcharge filter. In this case it is necessary to carry out an iterative analysis of the depth of the soil body for which safety must be verified. The bottom edge of the critical equivalent body lies below the base of the wall because here the vertical gradient of the flow is $i > 1.0$.

For walls with only a very small embedment depth, the degree of utilisation theoretically decreases with the embedment depth in terms of hydraulic heave failure. This range should be avoided for practical construction reasons [148].

DIN 1054 requires that the following condition should apply for limit state HYD for the upward flow within the failure body in front of the base of the wall:

$$F'_{s,k}\gamma_H = G'_k \gamma_{G,stb}$$

where

$F'_{s,k}$    characteristic value of flow force in the soil body subjected to flow

$\gamma_H$    partial safety factor for flow force for limit state HYD to DIN 1054, Table A 2.1

$G'_k$    characteristic value of weight under uplift for the soil body subjected to flow

---

*Recommendations of the Committee for Waterfront Structures Harbours and Waterways – EAU 2012*, 9th Edition. Issued by the Committee for Waterfront Structures of the German Port Technology Association and the German Geotechnical Society.
© 2015 Ernst & Sohn GmbH & Co. KG. Published 2015 by Ernst & Sohn GmbH & Co. KG.

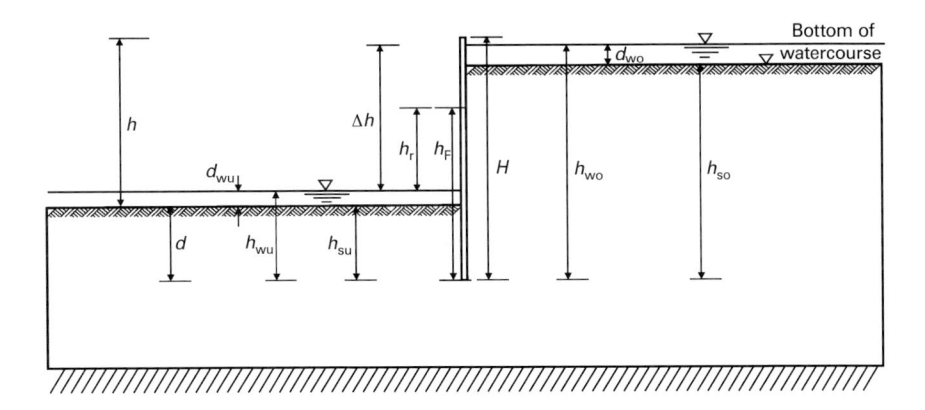

**Fig. R 115-1.** Safety against hydraulic heave failure – relevant dimensions

$\gamma_{G,stb}$  partial safety factor for stabilising permanent actions for limit state HYD according to DIN 1054, Table A 2.1

The flow force $F'_{s,k}$ is the product of the volume of the heave failure body, the unit weight of water $\gamma_w$ and the mean flow gradient measured in this body in vertical direction. It can be calculated using a flow net as per R 113, section 4.7.7, Fig. R 113-2, or R 113, section 4.7.5.

If water flows upwards through the soil in front of the base of the wall, the flow force should be considered in a soil body whose width may generally be assumed to be equal to half the embedment depth of the wall (DIN 1054, 11.5(4)). In critical cases, other boundaries of the soil body should be examined as well.

Fig. R 115-2 shows the method according to Terzaghi/Peck [207], p. 241, and Fig. R 115-3 shows the method according to Baumgart/Davidenkoff [46], p. 61.

In the rectangular failure body with a width equal to half the embedment depth of the structure, the characteristic value of the vertical flow force $F'_{s,k}$ is estimated as follows:

$$F'_{s,k} = \gamma_w \frac{h_l + h_r}{2} \frac{d}{2}$$

where

$\gamma_w$  unit weight of water
$h_r$  effective hydraulic head at base of wall (difference between hydraulic head at base of sheet piling and lower water level)
$h_l$  effective hydraulic head at boundary of failure body, opposite to base of wall
$d$  embedment depth of structure

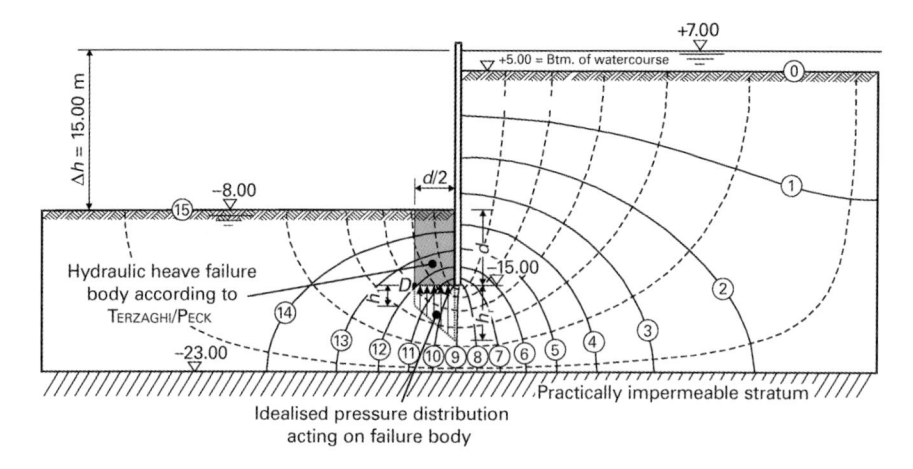

**Fig. R 115-2.** Safety against hydraulic heave failure at the base of an excavation according to the Terzaghi/Peck method, determined with the flow net as per R 113, section 4.7.7

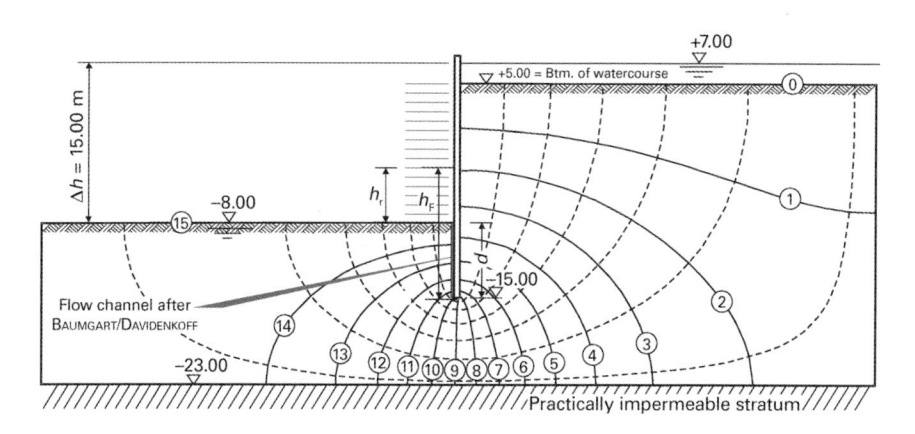

**Fig. R 115-3.** Safety against hydraulic failure heave at the base of an excavation according to the Baumgart/Davidenkoff method

According to Baumgart/Davidenkoff [46], p. 66, verification of safety against hydraulic heave failure can also be calculated in a simplified way as follows (with only one flow channel of the upward flow directly in front of the vertical part of the building element considered):

$$(\gamma_w \, i) \cdot \gamma_H \leq \gamma' \gamma_{G,\text{stb}}$$

where

$\gamma'$      buoyant unit weight of soil
$i$      mean hydraulic gradient in path considered ($i = h_r/d$)
$\gamma_w\, i$      specific flow force

The effective potential difference $h_r$ at the base of the sheet piling can also be determined with this method using a flow net according to R 113, section 4.7.4 or 4.7.5.

For the vertical sheet piling in flowing groundwater, the hydraulic head above the base of the sheet piling $h_F$ according to [24] may be estimated in a simplified form:

$$h_F = \frac{h_{wu}\sqrt{h_{so}} + h_{wo}\sqrt{d}}{\sqrt{h_{so}} + \sqrt{d}}$$

from which we derive

$$h_r = h_F - h_{wu}$$

where (see Fig. R 115-1)

$h_r$      difference between hydraulic head at base of sheet piling and lower water level
$h_F$      hydraulic head at base of sheet piling
$h_{so}$      depth of soil in flow on higher water side of sheet piling
$h_{wo}$      water level above base of sheet piling on higher water side
$h_{wu}$      water level above base of sheet piling on lower water side
$d$      embedment depth of wall

In the case of a horizontal inflow, the potential at the base of the sheet piling increases considerably and therefore in such cases the approximate approach should not be used.

The decrease in hydraulic head is not linear over the height of the sheet piling and therefore it is not correct to calculate $h_r$ from the development of the flow path along the sheet piling! Guidance on determining the hydraulic head at the base of the sheet piling and the margin of safety against hydraulic heave failure in stratified subsoils can be found in R 113, section 4.7.7.2.

In narrow trenches and trench corners the opposing or abutting sides have an impact on each other. This leads to an increased inflow and thus to a decrease in safety. Special investigations are required in such cases [230].

The danger of an impending hydraulic heave failure in an excavation may be indicated by wetting of the base near the sides of the excation; the ground then seems soft and springy. If this occurs, a surcharge in the form of a permeable soil with the appropriate filter properties should be applied immediately in front of the wall supporting the side of excavation and the excavation should be partially flooded at least. Afterwards corrective measures should be taken corresponding to R 116, section 3.2 (fifth and subsequent paragraphs).

Generally, safety against hydraulic heave failure is calculated for the final condition, i.e. once a steady flow around the wall has become established. In soils of low permeability, however, the decrease in the pore water pressure in the soil often takes place at a much slower rate than the external pressure changes acting on the soil due to changes in water level (e.g. tides, waves, lowering of the water level) and excavation work. The lower the permeability $k$ and the saturation $S$ of the soil and the quicker the changes in pressure, the longer is the delay [119]. A process is "quick" when the rate of application $v_{zA}$ of the external pressure reduction (e.g. lowering of the water level in the excavation) is greater than the critical permeability $k$ of the soil ($v_{zA} > k$) [118].

Even below the groundwater table, the soil is still considered to be not fully saturated; in particular, degrees of saturation of the soil amounting to $80\% < S < 99\%$ are frequently encountered directly below the groundwater table for low water depths ($h_w < 4$–$10$ m). Natural pore water contains microscopic gas bubbles, which have a considerable effect on the physical behaviour during pressure changes; such water cannot be considered as an ideal (incompressible) fluid. The pressure compensation in the pore water is therefore always connected with a mass transport, i.e. an (unsteady) flow of pore water in the direction of the lower hydraulic head.

Quick changes in the water level therefore always require verification of the unsteady excess pore water pressure $\Delta u(z)$ for the initial condition in addition to verification of an adequate factor of safety against hydraulic heave failure in the final condition with a steady flow.

The objective of the verification is to avoid limit states for stability and serviceability. Serviceability can be impaired by unacceptable loosening (heave) of the base of the excavation or undesirable base and (lateral) wall deformations [118].

The distribution of the unsteady excess pore water pressure $\Delta u(z)$ over the depth of soil $z$ can be estimated using the following simplified equation according to Köhler/Haarer [119]:

$$\Delta u(z) = \gamma_w \, \Delta h (1 - e^{-bz})$$

where

$\Delta h$   lowering of water level (or depth of excavation)
$\gamma_w$   unit weight of water
$b$   pore water pressure parameter according to Fig. R 115-4
$z$   depth of soil on lower water side

The pore water pressure parameter $b$ (unit: 1/m) can be determined from Fig. R 115-4 depending on the critical time period $t_A$ and the permeability $k$ of the soil.

For a soil depth $z$, the dead load $G_B$ of the soil body with unit width 1, unit length 1 and unit weight $\gamma'$ is

$$G_B = \gamma' z \cdot 1 \cdot 1$$

Due to the excess pore water pressure, a vertical, upward unsteady water pressure develops in the soil and acts on the body of soil:

$$W_{unsteady} = (1 - e^{-bz})\gamma_w \Delta h \cdot 1 \cdot 1$$

The ratio of weight to water pressure is most unfavourable at the section in which the excess pore water pressure reaches a maximum. This

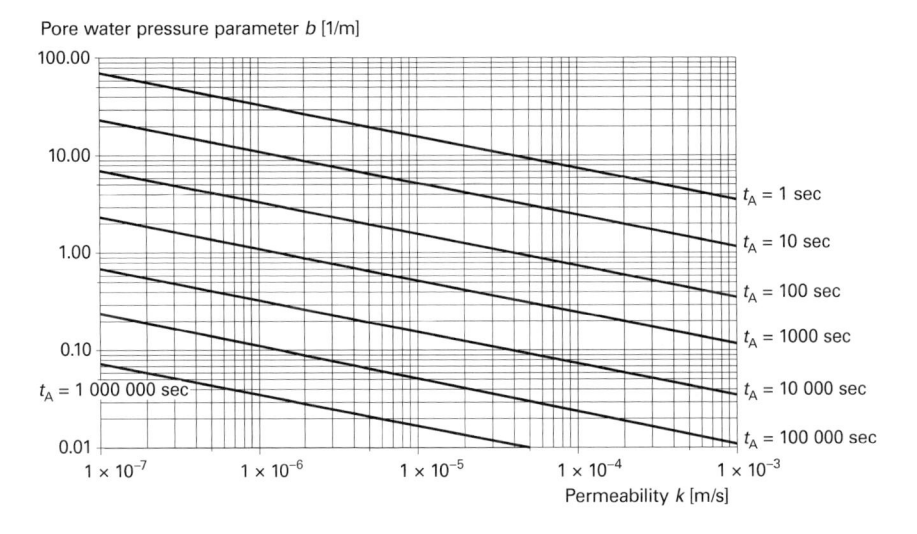

**Fig. R 115-4.** Parameter $b$ for determining the flow force for unsteady flow depending on time $t_A$ and permeability $k$

applies for depth $z = z_{crit}$:

$$z_{crit} = \frac{1}{b} \cdot \ln\left(\frac{\gamma_w \Delta h b}{\gamma'}\right)$$

When $z_{crit} < 0$, no further verification is necessary. If $z_{crit}$ lies below the base of the sheet piling, then the verification should be carried out for the base.
If the condition

$$\gamma_w \cdot \Delta h \cdot b < \gamma'$$

is fulfilled, then adequate safety is guaranteed from the outset.
Verifying adequate safety against hydraulic heave failure for unsteady flow processes requires verification of equilibrium at least:

$$G_B \geq W_{unsteady}$$

## 3.2 Piping (ground failure due to internal erosion) (R 116)

There is a risk of piping if the soil at the bottom of a watercourse or base of an excavation can be washed out through seepage. The process is initiated when the exit hydraulic gradient of the water flowing around the waterfront structure is capable of moving soil particles upwards and out of the soil. This continues in the soil in the opposite direction to the flow of water and is therefore called retrogressive erosion. A channel roughly in the shape of a pipe forms in the ground and propagates along the flow lines with the steepest gradient in the direction of the so-called upstream water level. The maximum gradient always occurs at the interface between the soil and the wall. Once the channel reaches an upstream watercourse, erosion causes widening of the channel within a very short time and leads to a failure similar to hydraulic heave. In doing so, a mixture of water and soil flows into the excavation with a high velocity until equilibrium is achieved between outer water and excavation. A deep crater forms behind the wall.

The presence of loose soils or weak spots (e.g. inadequately sealed boreholes) around the base support and loose zones in the immediate vicinity of the wall-soil interface behind the wall are the conditions that tend to initiate piping, but also a sufficient amount of water (upstream watercourse) and a relatively high hydraulic gradient.

The occurrence of a failure condition in homogenous, non-cohesive soil is shown schematically in Fig. R 116-1.

a) Start of washout of soil at base of excavation

b) Continuation of washout of soil behind sheet piling

c) Breakthrough to upstream watercourse

d) Collapse due to breakthrough to upstream watercourse

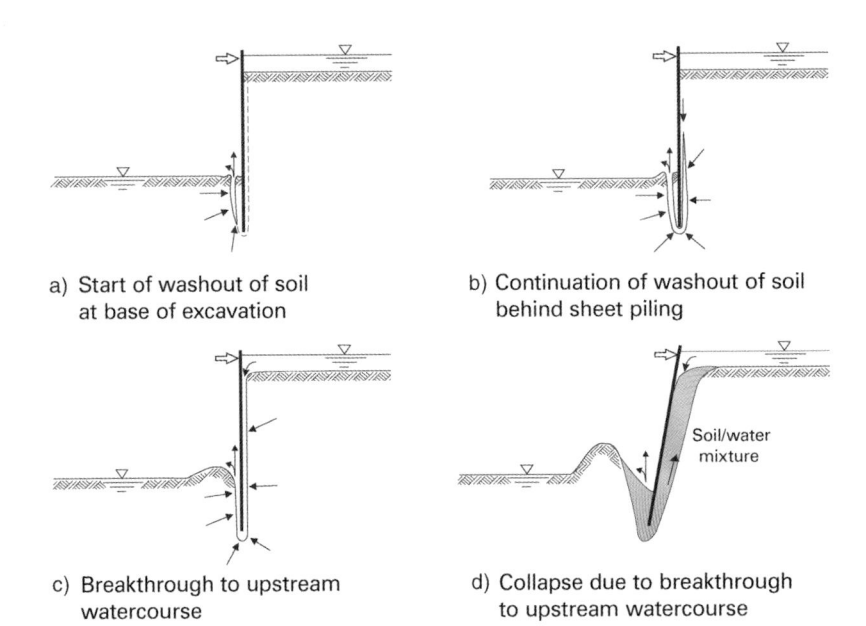

Soil/water mixture

**Fig. R 116-1.** Progression of washout behind sheet piling

Swelling of the ground and the ejection of soil particles on the lower water side or in the base of the excavation is the first indication of a piping failure. At this early stage the impending failure can still be prevented by depositing an adequately thick, graded or mixed gravel filter to prevent further soil from being washed out.

If, however, the condition has already reached an advanced stage (recognisable through the intensity of the water flow in the erosion channels already formed), the breakthrough to the upstream watercourse can no longer be predicted. In such a case immediate equalisation of the water tables must be brought about by raising weir gates, flooding the excavation or similar measures. Only after this has been accomplished is it possible to undertake remedial measures, such as placing a sufficiently thick filter on the lower water side.

The risk of piping cannot generally be ascertained through calculations, and must be assessed for every individual case owing to the diversity of designs and boundary conditions. Other conditions being constant the risk of piping increases in proportion to the increase in the hydraulic head between the upstream and downstream water levels, and is greater when there is looser, more fine-grained material in the subsoil, particularly when there are embedded sand lenses or veins in soils which are otherwise not at risk of erosion. Piping does not normally occur in cohesive soils.

Even if there is no upstream watercourse, erosion channels can still begin from the lower water side; in general, however, these peter out in the subsoil because the flow of water from the groundwater is insufficient to cause a critical erosion impact. If by chance the erosion channel intersects an extraordinarily capacious aquifer, water flowing from this stratum can initiate the erosion process anew.

If the conditions are such that piping seems possible, precautions to prevent this should be planned from the very beginning of work on site so that appropriate countermeasures can be taken immediately if necessary. In particular, in such cases it is important to ensure that works supporting the sides of an excavation are sufficiently deeply embedded in order to minimise the hydraulic gradient. The minimum embedment depth of walls to prevent erosion failures can be determined according to R 113, section 4.7.7.

Defects in the walls (e.g. interlock declutching in sheet piling) shorten the seepage path of the flow around the wall, and hence increase the gradient dramatically. Therefore, such defects are to be assessed with respect to a hydraulic heave failure or internal erosion as far as there is a real risk of these in the actual circumstances.

# 4 Water levels, water pressure, drainage

The loads determined in this section are allocated to the design situations as per R 18, section 5.4. Regarding partial safety factors, chapter 0.2 and section 5.4.4 must be observed.

Allocating the critical hydrostatic loading situation resulting from changing outer and groundwater levels requires an analysis of the geological and hydrological conditions of the area affected. Where available, series of observations over several years should be evaluated.

## 4.1 Mean groundwater level (R 58)

The groundwater level behind a waterfront structure is very much affected by the soil stratification and the design of the waterfront structure. In tidal areas the groundwater level for permeable soils tracks the tides in an attenuated manner to a greater or lesser extent. The relationship between groundwater and tidal water levels can be ascertained by way of measurements. Unless more detailed data is available, a groundwater level 0.3 m above mean half tide (MT½W) in tidal areas or 0.3 m above mean water level (MW) in non-tidal areas may be assumed as an approximation in preliminary designs.

## 4.2 Excess water pressure in direction of water side (R 19)

The excess water pressure $w_e$ in the direction of the water side for a difference in height $\Delta h$ between the critical outer water level and the corresponding groundwater level for the unit weight of water $\gamma_w$ amounts to $w_e = \Delta h \cdot \gamma_w$.

The excess water pressure can be applied according to Fig. R 19-1 in permeable soils in non-tidal areas and Fig. R 19-2 in tidal areas and assigned to design situations DS-P, DS-T and DS-A. If the excess water pressure is applied assuming weepholes, their permanent effect must be guaranteed. The approaches of Figs. R 19-1 and R 19-2 are based on the assumption of a planar flow but without taking into account how waves might influence the excess water pressure.

The water levels associated with Figs. R 19-1 and R 19-2 must be applied as characteristic values.

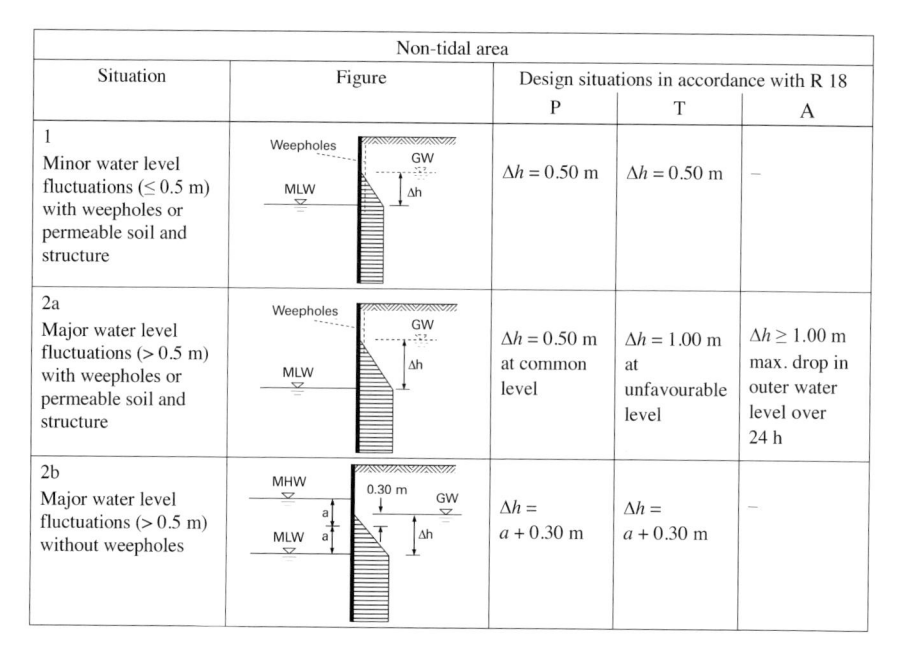

| Situation | Figure | Non-tidal area Design situations in accordance with R 18 | | |
|---|---|---|---|---|
| | | P | T | A |
| 1<br>Minor water level fluctuations ($\leq 0.5$ m) with weepholes or permeable soil and structure | Weepholes, GW, MLW, $\Delta h$ | $\Delta h = 0.50$ m | $\Delta h = 0.50$ m | – |
| 2a<br>Major water level fluctuations ($> 0.5$ m) with weepholes or permeable soil and structure | Weepholes, GW, MLW, $\Delta h$ | $\Delta h = 0.50$ m at common level | $\Delta h = 1.00$ m at unfavourable level | $\Delta h \geq 1.00$ m max. drop in outer water level over 24 h |
| 2b<br>Major water level fluctuations ($> 0.5$ m) without weepholes | MHW, 0.30 m, GW, MLW, a, $\Delta h$ | $\Delta h = a + 0.30$ m | $\Delta h = a + 0.30$ m | – |

**Fig. R 19-1.** Approximate approaches for excess water pressure on waterfront structures for permeable soils in non-tidal areas and standard situations (without significant wave effect)

Hydraulic backfilling to waterfront structures can lead to much higher water levels temporarily which must be taken into account for the respective condition of the building works.

Flooding of the waterfront structure, stratified soils, highly permeable sheet pile interlocks or confined artesian pressure are cases that require special investigations to determine the water levels critical for the excess water pressure (R 52, section 2.10). If a natural groundwater flow is constricted or cut off by a waterfront structure to such an extent that there is a build-up of groundwater in front of the wall, this accumulation can be determined within the scope of a numerical model of the groundwater flow (R 113, section 4.7.5.2).

For details of the excess water pressure on sheet pile walls alongside canals, please refer to R 106, section 6.4.2, Fig. R 106-1.

The relieving effect of drainage systems according to R 32, section 4.5, R 51, section 4.4, and R 53, section 4.6, may only be considered if their effectiveness can be constantly monitored and the system can be restored at any time.

| Tidal area | | | | | |
|---|---|---|---|---|---|
| Situation | Figure | Design situations in accordance with R 18 | | | |
| | | P | T | A | |
| 3a Major water level fluctuations without drainage = normal case | (figure) | $\Delta h = a +$ 0.30 m $+ d$ | – | – | |
| 3b Major water level fluctuations without drainage = limit case for extreme low water level | (figure) | – | – | $\Delta h =$ $a + 2b + d$ | |
| 3c Major water level fluctuations without drainage = limit case for falling high water | (figure) | – | – | $\Delta h =$ 0.30 m $+ 2a$ | |
| 3d Major water level fluctuations with drainage | (figure) | $\Delta h =$ 1.00 m $+ e$ for outer water level at MLWS | $\Delta h = 0.30$ m $+ b + d + e$ | – | |

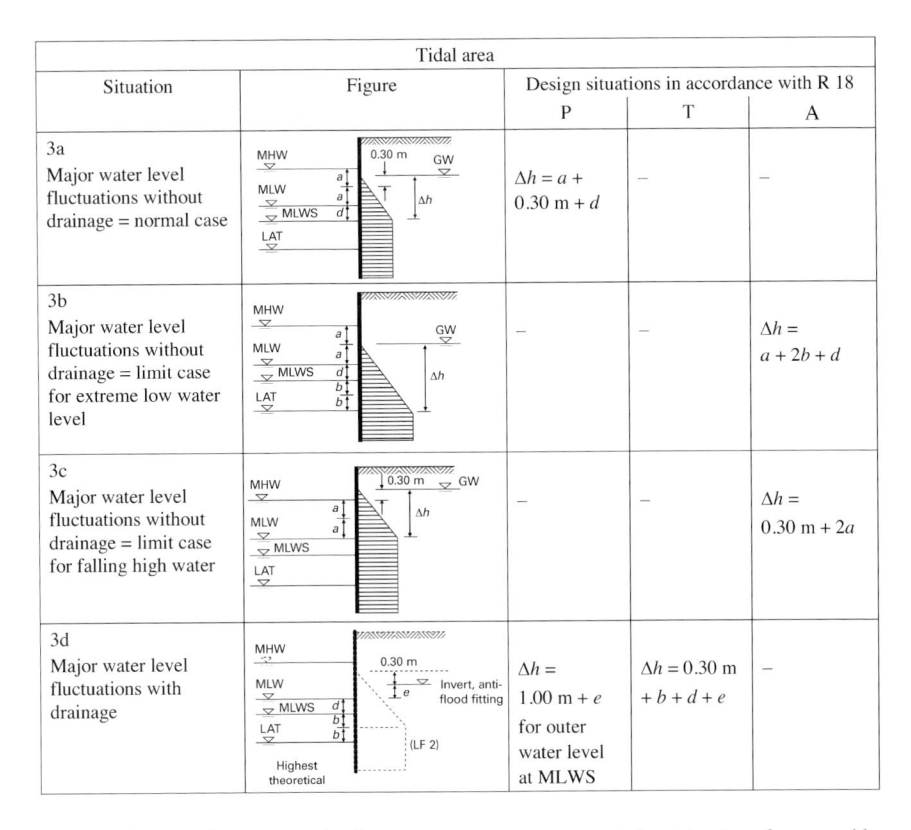

**Fig. R 19-2.** Approximate approaches for excess water pressure on waterfront structures for permeable soils in tidal areas and standard situations (without significant wave effect)

## 4.3 Excess water pressure on sheet piling in front of embankments below elevated platforms in tidal areas (R 65)

### 4.3.1 General

On an embankment with a permeable revetment below an elevated platform in a tidal area, the retained soil drains via the base of the slope at low outer water levels. The position of the seepage line depends on the soil conditions, the size and frequency of the water level fluctuations and the inflow from the land.

The flow through the critical active earth pressure wedge results in an increase in the active earth pressure. The water pressure behind the quay wall can be higher than the hydrostatic water pressure depending on the drainage options and the ensuing flows.

### 4.3.2 Approximation for excess water pressure

In line with experience gained in the tidal areas of Germany's north coast, the approximation for the excess water pressure shown in Fig. R 65-1, which incorporates recommendations R 19, section 4.2, and R 58, section 4.1, can be used in subsoil with a fairly uniform permeability. Fig. 65-1 shows design situation T as an example, but can be used similarly for the other load cases. When approximating the excess water pressure according to Fig. R 65-1, the permeability of the revetment must be such that it does not cause a build-up of groundwater.

## 4.4 Design of weepholes for sheet piling structures (R 51)

Weepholes in sheet piling are only effective over the long term in silt-free water and when the iron content of the groundwater is so low that it is harmless. If these conditions are not met, there is always the possibility that weepholes become silted up or clogged with iron hydroxide particles and thus become ineffective. The iron content in the groundwater of the north German coastal marshlands can be up to 25 mg/l, but in the geest regions it is only 5 mg/l. The risk of iron hydroxide clogging is therefore particularly high where post-ice-age cohesive soils are present in the hinterland.

The long-term efficiency of weepholes may therefore be assumed for silty water and water with a high iron content when their function can be checked and restored as necessary. Weepholes can also lose their efficiency in areas with high shell growth.

In the case of very high temporary outer water levels, water infiltrating from outside can cause a higher water level behind the wall. A margin of safety against uplift for all structures behind the waterfront wall must be guaranteed for this higher water level. It should also be noted that in these cases, rising and falling water levels in backfilling can lead to subsidence.

Weepholes must be located below mean water level so that they do not become blocked. To guarantee their long-term effectiveness, it is expedient to use gravel filters according to R 32, section 4.5.

The weepholes take the form of 1.5 cm wide x approx. 15 cm high slots with rounded ends which are flame-cut in the sheet pile webs (Fig. R 51-1). In contrast to round holes, these slots cannot be blocked by the grains of gravel used for the filter. Please also refer to the last paragraph of R 19, section 4.2.

Weepholes are considerably less expensive than drainage systems with anti-flood fittings (R 32). However, experience shows that in tidal areas they achieve only a minor reduction in the excess water pressure because the water is trapped behind the sheet piling at high tide.

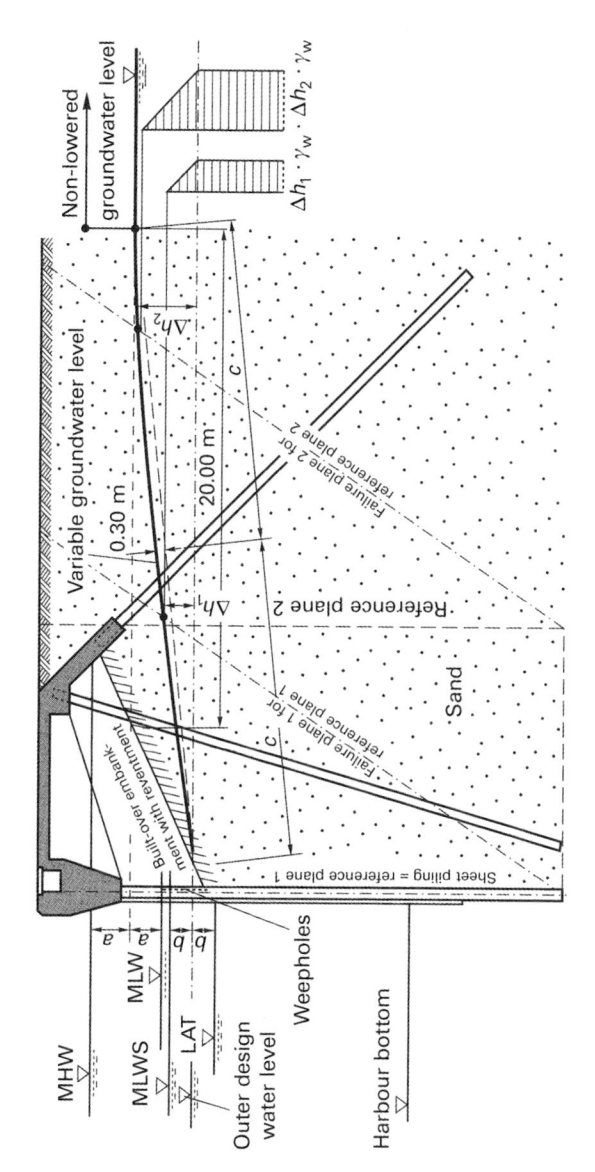

**Fig. R 65-1.** Excess water pressure approach for an embankment below an elevated platform for load case 2

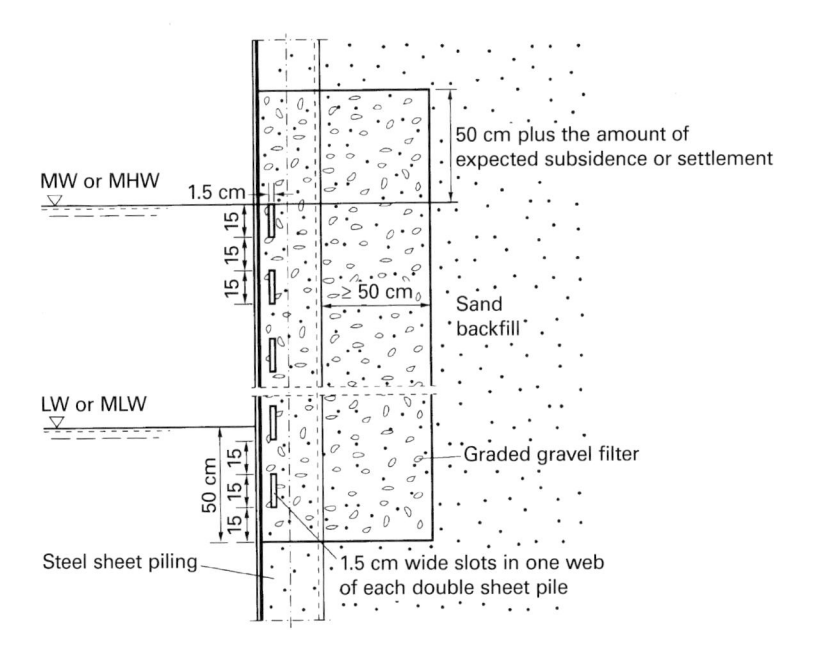

**Fig. R 51-1.** Weepholes in steel sheet piling

Weepholes are not permitted in quay walls that also function as flood defences.

Weepholes are particularly useful in non-tidal areas, at locations where there is a sudden drop in the surface water level, in intense inflows of groundwater or slope seepage water or where the structure can become flooded.

It is also generally necessary to investigate the case of ineffective weepholes within the scope of the structural analysis. According to R 18, section 5.4.3, this case may be allocated to design situation DS-A.

## 4.5 Design of drainage systems for waterfront structures in tidal areas (R 32)

### 4.5.1 General

Effective drainage for waterfront structures is only possible where there is non-cohesive soil behind the structures. The risk of clogging with iron hydroxide particles has already been referred to in section 4.4.

If a drainage system is to remain effective over the long-term in harbour water containing suspended matter and silt and also limit the excess water pressure where there is a larger tidal range, it must include branch drains that discharge into main drains fitted with anti-flood valves or flaps that permit the outflow of water from the system into the harbour water but prevent the ingress of silt-laden water. The drainage system must be designed so that it will continue to function reliably even if the backfilling behind the wall settles.

Experience shows that numerous sheet piling drainage systems become less efficient over time due to silting-up as a result of ineffective anti-flood fittings or clogging with iron hydroxide particles. To ensure this does not occur, the design, installation and maintenance of such drainage systems must be carried out to a high standard.

### 4.5.2 Design, installation and maintenance of drainage systems

Outlets must be fitted with anti-flood valves or flaps to guard against water entering the system from outside and positioned so that they are accessible at mean low tide. Anti-flood fittings must be permanently sealed against wave loads. The spacing between outlets should be about 30 m.

The drains between the outlets should consist of one or more plastic subsoil drain pipes embedded in a gravel filter or ballast sheathed in a geotextile. They must be designed for the surcharge due to the backfilling over the drains.

Connections between pipes and outlets should be designed to resist shearing-through due to settlement of the backfilling behind the sheet piling.

Inspection shafts must be designed and positioned so that the full length of the drainage system can be inspected and, if necessary, cleaned from such shafts.

The drainage system must be inspected and maintained regularly. Inspections and maintenance must be documented.

### 4.5.3 Drainage systems for large waterfront structures

Fig. R 32-1 shows a groundwater relief system for a larger quay in a tidal area. It consists of four DN 350 subsoil drain pipes made from PE-HD (DIN 19666) that run the entire length of the quay. The pipes are encased in gravel wrapped in a non-woven filter material to protect it against the surrounding soil. The depth was chosen so that the pipes are always within the groundwater, which reduces the risk of clogging due to iron hydroxide particles. The pipes are laid without any fall.

Pumps are used to remove the water from the groundwater relief system of Fig. R 32-1. This avoids the need for vulnerable anti-flood fittings and the risk of the connection between drains and anti-flood fittings being sheared through.

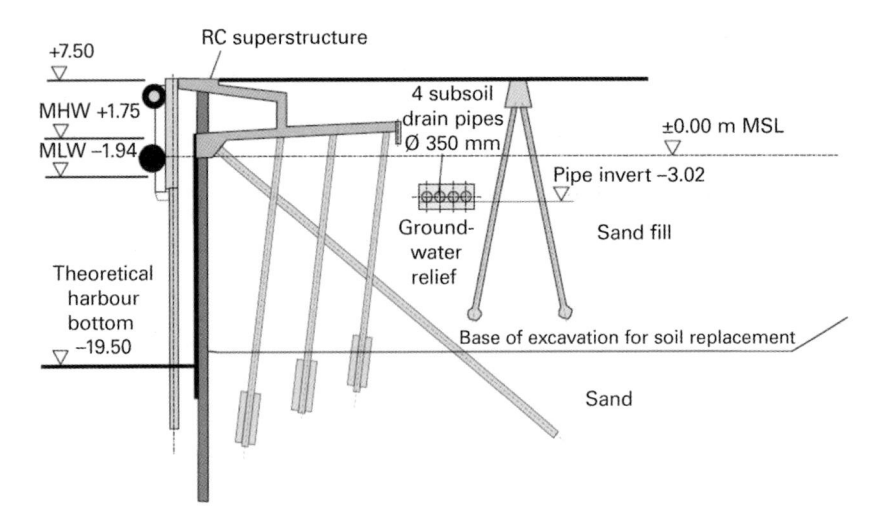

**Fig. R 32-1.** Example of a groundwater relief system for a quay in a tidal area

## 4.6 Relieving artesian pressure beneath harbour bottoms (R 53)

### 4.6.1 General

The most effective way of relieving artesian pressure beneath a harbour bottom is to install efficient relief wells because these work independently of mechanical equipment. Well outlets should always lie below lowest astronomical tide (LAT). A residual artesian pressure – due to the hydraulic head between the depth of the confining stratum and the well outlet – therefore remains below the confining stratum being relieved by the wells. The actual magnitude of this, but a figure of at least $10\,kN/m^2$, should be taken into account when determining the active and passive earth pressures according to R 52, section 2.10.

Gravel-filled trenches in the confining stratum on the harbour bottom do not guarantee permanent artesian pressure relief in silt-laden water because silt accumulations significantly reduce the efficiency of such trenches over time.

### 4.6.2 Design of relief wells

Relief wells are always sized by calculating the groundwater lowering (e.g [98]). The range of such relief and its effect on other parts of the structure must be taken into account (see also R 166, section 4.8.1).

### 4.6.3 Construction of relief wells

Relief wells are best housed in steel box or tubular piles integrated into the outermost sheet piling. This not only simplifies construction and ensures their function, but also places the wells where they are most effective for relief.

In tidal areas the harbour water level at high tide generally lies above the artesian piezometric head of the groundwater. The harbour water can then flow via the wells into the soil behind the wall, thus leading to a rapid decrease in the efficiency of the relief wells in silt-laden water because the flushing effect as the tide ebbs is not sufficient to remove the silt deposits again. In such cases the relief wells must be equipped with permanently effective anti-flood fittings. Ball valves have proved to be ideal for this purpose. They must be easy to remove for inspecting the wells and easy to install again without impairing the watertight seal.

To achieve optimum performance, the filters of the relief wells should be placed in the most permeable stratum.

Filters must be protected against corrosion and clogging with iron hydroxide particles. Wells must be installed by an experienced contractor.

### 4.6.4 Checking the relief installation

The effectiveness of the relief must be checked regularly by way of observation wells behind the waterfront structure which extend below the confining stratum.

If the relief assumed in the structural analyses is no longer being achieved, the relief wells must be cleaned and, if necessary, additional wells must be installed. A sufficient number of steel box piles accessible from the quayside and suitable for subsequent upgrading to relief wells should therefore be installed as a precaution.

Please also refer to the last paragraph of R 19, section 4.2.

## 4.7 Taking account of groundwater flow (R 113)

### 4.7.1 General

Where quay walls and other hydraulic engineering works and their components are to be built in flowing groundwater, the effects of such flowing groundwater on the active and passive earth pressures must be taken into account in the design and detailing.

The groundwater flow can be calculated using Darcy's law if the product of the hydraulic gradient $i$ and the coefficient of permeability $k_f$ is lower than approx. $6 \cdot 10^{-4}$ m/s ($\approx 2$ m/h).

### 4.7.2 Principles of groundwater flow

Laminar groundwater flow is described by potential theory. The solutions to the potential differential equation are orthogonal sets of curves –

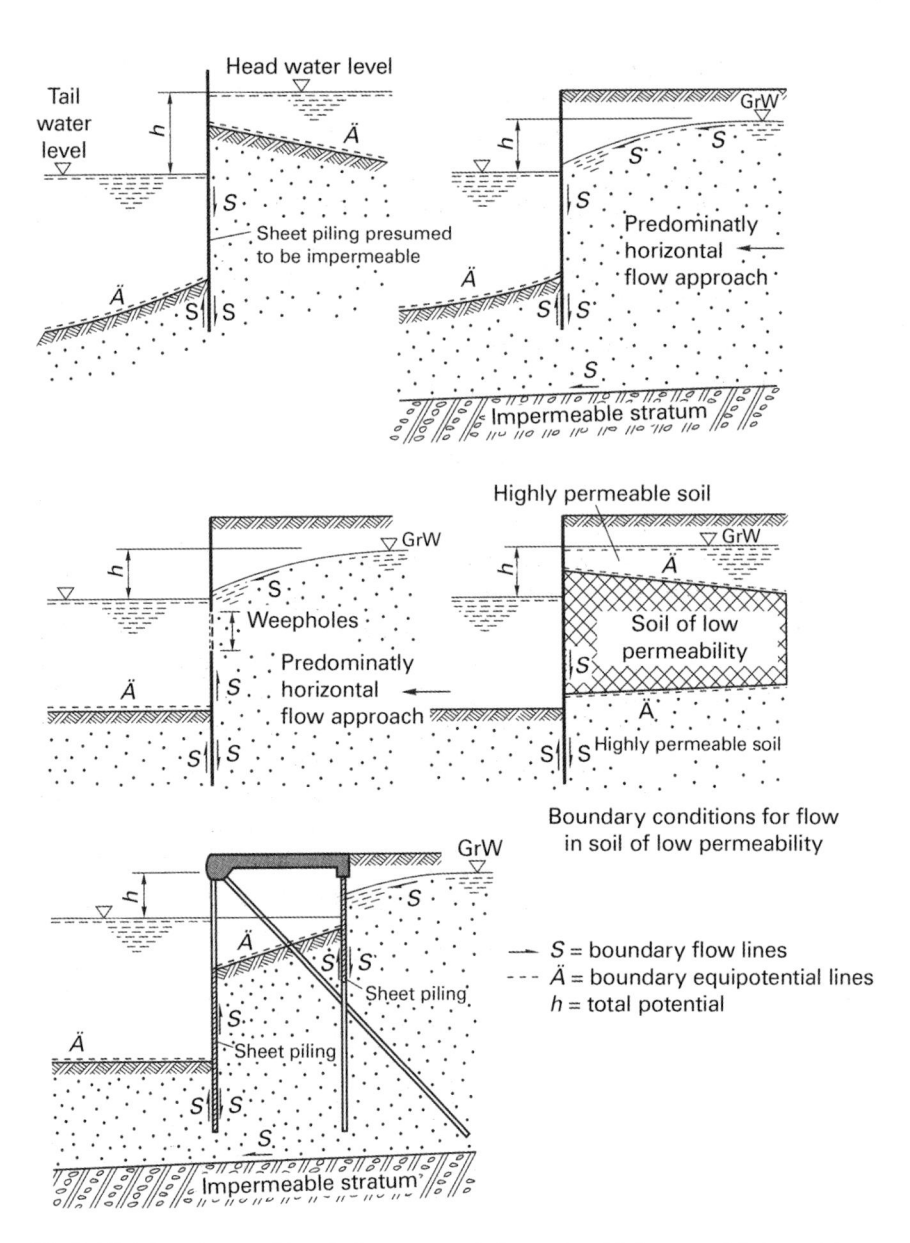

**Fig. R 113-1.** Boundary conditions for flow nets – typical examples with flow around the base of sheet piling

one set representing the flow lines, the other the equipotential lines. The flow net formed by the flow and equipotential lines consists of fields with a constant aspect ratio (Fig. R 113-2).

The flow lines can be interpreted physically as the paths of the water particles, the equipotential lines as lines joining equal hydraulic heads (Fig. R 113-2).

### 4.7.3 Definition of the boundary conditions for a flow net

The boundaries to a flow net can be either boundary flow lines or boundary potential lines.

The following can be boundary flow lines: boundaries between permeable and impermeable soil strata, the surfaces of impermeable structures, an unconfined groundwater level provided it is not horizontal (Fig. R 113-1).

The following can be boundary equipotential lines: horizontal groundwater levels, watercourse bottoms, inlet slopes to permeable barrage structures. Fig. R 113-1 illustrates the boundary conditions for several typical situations.

### 4.7.4 Graphic method for determining a flow net

For simple cases and steady flow conditions, graphical methods represent a quick way of determining flow nets and hence assessing whether and where detailed investigations are necessary. However, this method cannot normally be used to determine flow nets in a stratified subsoil with varying permeabilities.

Once all boundary conditions have been determined, the flow net can be drawn. The following rules apply:

– Flow lines are perpendicular to equipotential lines.
– The total potential difference $\Delta h$ between the highest and lowest hydraulic potentials is divided into equal (equidistant) potential steps $dh$ (15 steps in Fig. R 113-2, i.e. every step is equal to a potential difference of $4.50 \text{ m}/15 = 0.3 \text{ m}$).
– All flow lines pass through the available flow cross-section, i.e. they are closer together in narrower passages and further apart in wider sections.
– The number of equipotential steps and flow lines is chosen such that neighbouring equipotential and flow lines form squares bounded by curved lines in order to guarantee the geometrical similarity across the whole net. The accuracy can be checked by drawing inscribed circles in the squares (Fig. R 113-2).

The procedure continues by trial and error until both the boundary conditions and the requirement for squares bounded by curved lines are

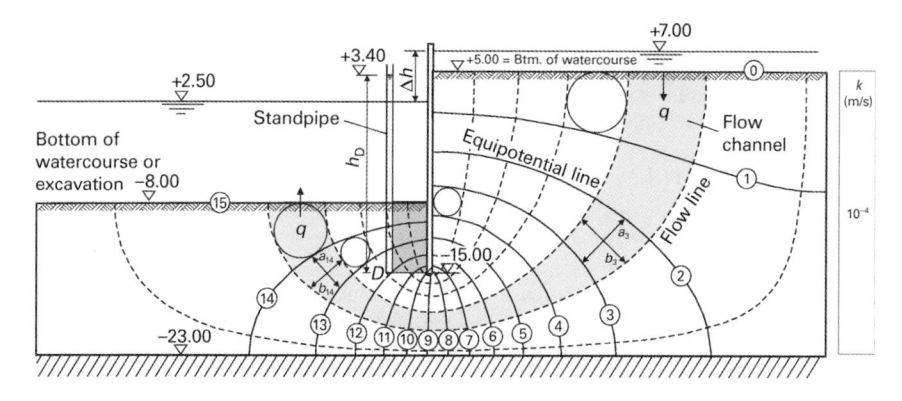

**Fig. R 113-2.** Example of a groundwater flow net in homogenous soil for a vertical flow – case 1

fulfilled with sufficient accuracy throughout the net. Incomplete flow channels or equipotential steps can be accepted along the boundaries. They are included in the calculations according to the proportion of their cross-section (see also section 4.7.7.1).

## 4.7.5 Use of groundwater models to determine flow nets

### 4.7.5.1 Physical and analogue models

Physical models use natural media (water, sands, gravels, clays) on a model scale (selected by the designer). These days they are mainly used for research purposes in 3D analyses and are no longer relevant for practical applications involving waterfront structures.

Analogue models use media whose motion is similar to the groundwater flow. Examples here are the motion of viscous materials between two closely spaced plates (gap models), the flow of electric current through conductive paper or a network of electrical resistances (electrical models); another example is the deformation of thin skins by point loads (seepage line with wellpoint dewatering). To convert the potentials (e.g. electrical voltages) used in these models to the groundwater potentials (hydraulic heads), the kinematic similarity must be considered in addition to the geometrical similarity. However, analogue methods have largely been replaced by numerical groundwater models in modern design practice.

### 4.7.5.2 Numerical models

In numerical groundwater models the entire potential field is divided into individual elements (discretised) and is represented by the hydraulic heads of a sufficient number of support points. These points are the corner points (finite element method, FEM) or the grid points (finite difference method, FDM) of individual small but finite areas. Boundaries and discontinuities (wells, springs, drains, etc.) in the flow field must be represented by nodes or element lines.

The user has to select the correct boundary conditions and geohydraulic parameters in advance when employing numerical groundwater models. This applies in particular to programs that are able to take into account unsteady flows.

### 4.7.6 Calculation of individual hydraulic variables

Whereas groundwater models calculate the entire hydraulic potential field and derive the distribution of gradients, velocities, discharges, etc. from this, simplified methods are available for determining single variables only. These methods are therefore often used successfully within the scope of preliminary planning.

Examples of these methods are:

– Resistance coefficient method after Chugaev for determining gradients and flows in underseepage [46]
– Fragment procedure after Pavlovsky for calculating underseepage [45]
– Diagrams for calculating hydraulic heads for excavations supported by sheet piling [47]

### 4.7.7 Evaluation of examples

#### 4.7.7.1 Sheet piling with underseepage in homogenous subsoil

In the flow net of Fig. R 113-2 the potential difference $\Delta h = 4.50$ m is divided into 15 potential steps ($dh = 0.45/15 = 0.30$ m) between MSL +7.00 m and MSL +2.50 m. At a depth MSL $-23.00$ m the flow net is limited by an impermeable stratum whose upper surface represents the boundary of the model (boundary flow line).

The following flow parameters can be derived from the flow net:

- Hydraulic head at point D (corner point of failure body according to Terzághi – see R 115, section 3.1):

  $h_D = 7.00 - 12/15 \cdot 4.50 \text{ m} = 3.40 \text{ m} (= 2.50 + 3/15 \cdot 4.50 \text{ m})$

- Hydraulic head $h_F$ at base of sheet piling

  $h_F = 7.00 \text{ m} - 9/15 \cdot 4.50 \text{ m} = 4.30 \text{ m} (= 2.50 \text{ m} + 6/15 \cdot 4.5 \text{ m})$

- Hydraulic gradients for two selected cells:

  $i_3 = dh/a_3 = 0.3/6.0 = 0.05$
  $i_{14} = dh/a_{14} = 0.3/4.3 = 0.07$

  Lengths $a_3$ and $a_{14}$ were taken from the drawing.

The discharge $Q$ in every flow channel between two flow lines is the same because all net rectangles have the same aspect ratio $a/b$. The discharge $Q$ between two flow lines is the product of the flow velocity $v$ and the cross-sectional area of the flow $A$. This is equal to the flow channel width $b$ multiplied by the flow channel thickness ($= 1.0$ m in the case of a planar flow net).

The following equation applies for a single flow channel:

$$Q = v \cdot A = k \cdot i \cdot A \quad [\text{m}^3/\text{s}]$$

The following applies for a planar flow net:

$$q_i = k \cdot i \cdot b_i \quad [\text{m}^3/\text{sm}]$$

or

$$q_i = k \cdot dh/a_{14} \cdot b_{14} = k \cdot dh/a_3 \cdot b_3 = k \cdot dh \cdot b/a$$

where $b/a = 1$ for a square net.

The total discharge depends on the number of flow channels. Incomplete flow channels are taken into account according to their cross-section. In Fig. R 113-2 the cross-sectional area of the boundary flow line is only approx. 10 % of that of a complete flow channel. The specific discharge therefore amounts to

$$q = 0.1 \cdot 10^{-4} \cdot 0.3 \cdot 1 = 3.0 \cdot 10^{-6} \text{ m}^3/(\text{s} \cdot \text{m})$$

Verification of safety against hydraulic heave failure at the base of the sheet piling is carried out according to R 115, section 3.1. The following flow parameters are required:

$$h_1 = h_D - h_{wu} = 3.4 - 2.5 = 0.9 \text{ m}$$
$$h_r = h_F - h_{wu} = 4.3 - 2.5 = 1.8 \text{ m}$$

According to Terzághi, the vertical flow force is

$$F'_{s,k} = \gamma_w \cdot (h_1 + h_r)/2 \cdot t/2 = 10 \cdot (0.9 + 1.8)/2 \cdot 7/2 = 47.25 \text{ kN/m}$$

The body of soil has a buoyant unit weight of

$$G'_k = \gamma'_B \cdot t^2/2 = 10 \cdot 24.5 = 245 \text{ kN/m}$$

Adequate safety against hydraulic ground failure, even in unfavourable subsoil, is guaranteed when using the partial safety factors of DIN 1054 (chapter 0) for design situation DS-P:

$$F'_{s,k} \cdot \gamma_H \leq G'_k \cdot \gamma_{G,stb}$$
$$47.25 \cdot 1.8 < 245 \cdot 0.9$$
$$85.05 < 220.5$$

### 4.7.7.2 Sheet piling in flowing groundwater in stratified subsoil

The boundary conditions of R 113-2 are retained, but there is a 2 m thick horizontal stratum at a varying depth whose permeability is much lower than that of the strata above and below. The flow and potential lines are calculated with a groundwater model.

In Fig. R 113-3 we can see the concentration of potential lines in the less permeable stratum, which considerably reduces the safety against hydraulic heave failure for case 2a, but increases it for case 2b. The critical water pressure at the underside of the impermeable stratum must be considered when analysing the hydraulic heave failure.

The prerequisite for this potential distribution is that the stratum with low permeability continues for a sufficient distance in front of and behind the wall. Otherwise, the potential distribution at the base of the wall is determined by the flow around it and not by the flow through this stratum. Impermeable strata on the inflow side in particular must be checked carefully to establish whether they extend for a sufficient distance upstream. Otherwise, groundwater will flow around the stratum, which means that the water pressure below the stratum on the discharge side is considerably higher than the case of a stratum extending far enough into the inflow side. In cases of doubt, the positive influence of a stratum with low permeability on the inflow side should be ignored.

In stratified soil, verifying safety against hydraulic heave failure according to R 115, section 3.1, calls for the designer to find the critical section, or rather failure body, with the lowest margin of safety. Where a less permeable stratum overlies a permeable one, then generally the underside of the less permeable stratum is the underside of the failure body with the least safety against hydraulic heave failure. The critical hydraulic heads or flow gradients can be determined from the flow net: The following applies in Fig. R 113-3 for a high-level stratum with low permeability (case 2a):

- Hydraulic head $h_D$ at point D (underside of stratum with low permeability):

$$h_D = 7.00 - 11/15 \cdot 4.50 \text{ m} = 3.70 \text{ m} (= 2.50 + 4/15 \cdot 4.50 \text{ m})$$

- Mean hydraulic gradient $i$ in stratum with low permeability:

$$i = \Delta h/\Delta l = (3.70 - 2.50)/2.00 = 0.60$$

- Safety:

$$i \cdot \gamma_w \cdot \gamma_H \leq \gamma' \cdot \gamma_{G,stb}$$
$$0.6 \cdot 10 \cdot \gamma_H \leq 10 \cdot 0.95$$

Safety against failure by hydraulic heave is given for favourable subsoil with the partial safety factor $\gamma_H = 1.35$, but not for unfavourable subsoil where $\gamma_H = 1.8$.

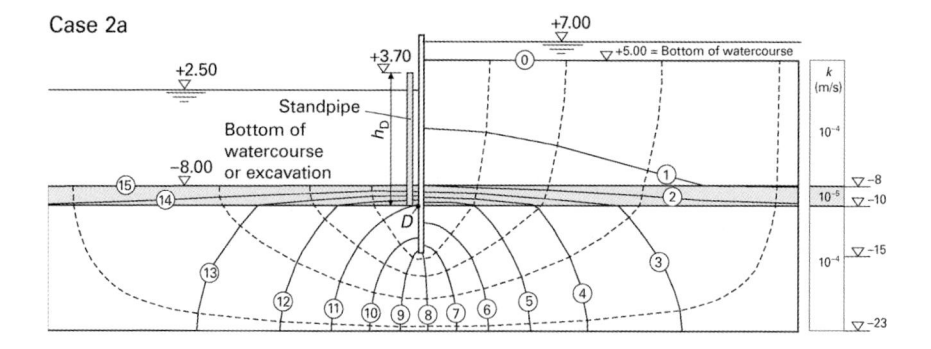

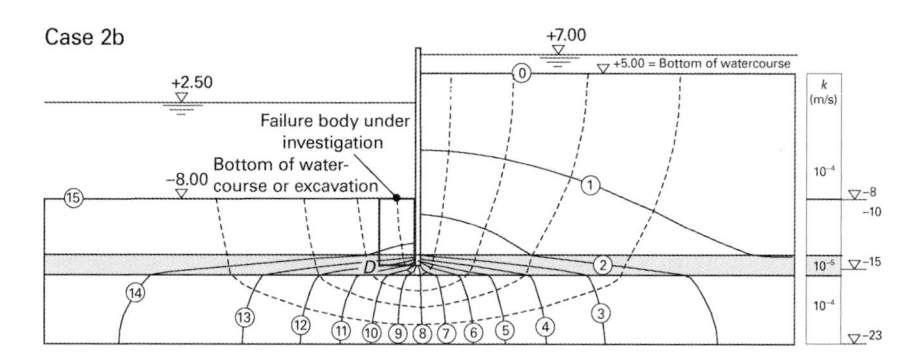

**Fig. R 113-3.** Flow nets in stratified soil with a vertical flow – stratum with low permeability at high level (case 2a) and low level (case 2b)

The following applies in Fig. R 113-3 for a low-level stratum with low permeability (case 2b):

- Hydraulic head $h_D$ at point D (corner point of failure body being investigated):

$$h_D = 7.00 - 12/15 \cdot 4.50 \text{ m} = 3.40 \text{ m} \, (= 2.50 + 3/15 \cdot 4.50 \text{ m})$$

- Hydraulic head $h_F$ at base of wall:

$$h_F = 7.00 - 9/15 \cdot 4.50 \text{ m} = 4.30 \text{ m} \, (= 2.50 + 6/15 \cdot 4.50 \text{ m})$$

- Characteristic value of flow force in body of soil in groundwater flow with a width of 3.0 m and a thickness of 1 m:

$$F'_{s,k} = [(3.4 - 2.5) + (4.3 - 2.5)]/2 \cdot 10 \cdot 3.0 = 40.50 \text{ kN/m}$$

- The body of soil consisting of two strata (with the same unit weight of 10 kN/m³) has a buoyant weight of

$$G'_k = 10 \cdot 3.0 \cdot 1.0 + 10 \cdot 3.0 \cdot 6.0 = 210.0 \, \text{kN/m}$$

- Adequate safety against hydraulic heave failure, even with unfavourable subsoil, is guaranteed when using the partial safety factors of DIN 1054 (chapter 0) for design situation DS-P:

$$F'_{s,k} \cdot \gamma_H \leq G'_k \cdot \gamma_{G,stb}$$
$$40.50 \cdot 1.8 < 210 \cdot 0.95$$
$$72.9 < 199.5$$

For clarity, a relatively wide failure body was considered in Fig. R 113-3. Normally, in accordance with the Terzághi-Peck approach (Fig. R 115-2, section 3.1), a failure body width corresponding to half the embedment depth in the stratum with low permeability (0.5 m here) should be chosen as the most unfavourable failure body.

### 4.7.7.3 Sheet piling in flowing groundwater with horizontal inflow

For a horizontal groundwater inflow (Fig. R 113-4, case 3, ground surface and groundwater on right-hand boundary of model at a level of +7 m MSL instead of surface water), the right-hand vertical boundary of the model is a boundary potential line.

The course of flow and potential lines near the wall is very heavily influenced by the distance between the wall in the flow and the boundary potential line with the maximum water level.

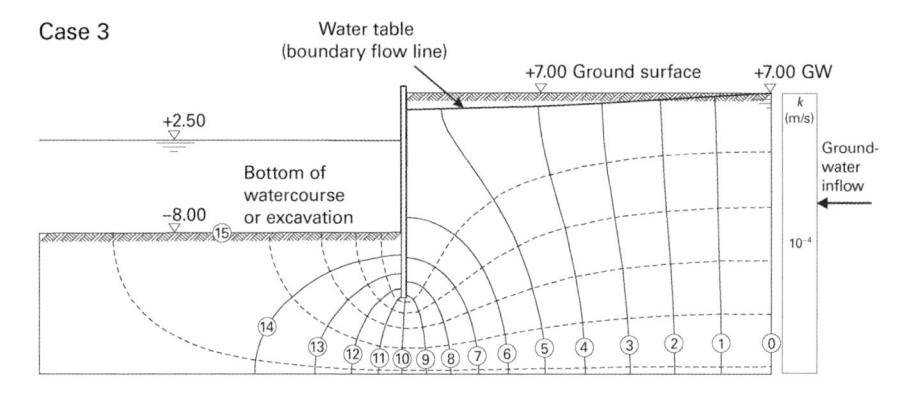

**Fig. R 113-4.** Flow net for a horizontal inflow (case 3)

## 4.8 Temporary stabilisation of waterfront structures by groundwater lowering (R 166)

### 4.8.1 General

The excess water pressure acting on a waterfront structure can be reduced by lowering the groundwater on the land side, which improves the stability of the structure. This method can be used to guarantee the stability of waterfront walls temporarily (states during construction). Where upgrading work on a waterfront structure is necessary, it is therefore possible to schedule the work in such a way that it can be carried out in the most economic way while taking into account the operating conditions of the facility as well.

First of all, however, studies are necessary to ensure that the structure itself or other structures in the area influenced by groundwater lowering will not be endangered by the change in groundwater level. In this context it is especially important to consider the potential increase in negative skin friction for pile foundations.

Lowering the groundwater reduces the excess water pressure, indeed, can even achieve a passive water pressure from the water side. At the same time, the inertial forces in the passive earth pressure zone can be increased by the flow forces.

These positive influences are counteracted on the one hand by an increase in the active earth pressure as a result of the greater inertial forces due to flow force and the loss of buoyancy in the area where the groundwater has been lowered. However, these effects of groundwater lowering have a much smaller influence on the stability of the waterfront wall than the reduction in excess water pressure.

### 4.8.2 Case with soft, cohesive soil near the ground surface

Where soft soil with low permeability extends from the surface to a greater depth and overlies a non-cohesive soil with good permeability (Fig. R 166-1), then the soil is initially not consolidated for the additional inertial forces resulting from the lack of buoyancy due to the depth of lowering $\Delta h$. Since in this state the active earth pressure coefficient $K_{ag} = 1$ for the additional inertial forces, and since with cohesive soil $\gamma - \gamma' \approx \gamma_w$, the additional active earth pressure at the level of the lowered groundwater level at the start of consolidation – without significant water inflow from above – is as follows:

$$\Delta e_{ah} = \Delta h \cdot \gamma_w \cdot 1$$

At the start the reduced excess water pressure is therefore compensated for by the greater active earth pressure in the soft cohesive soil. However,

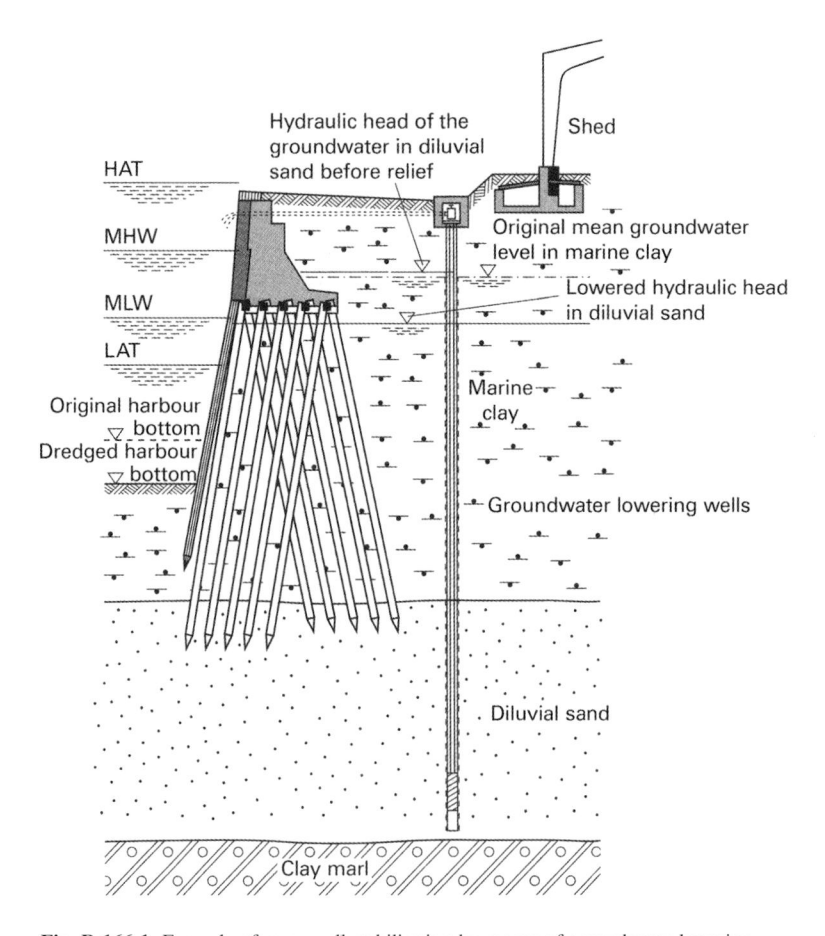

**Fig. R 166-1.** Example of quay wall stabilisation by means of groundwater lowering

as consolidation increases, and when it is concluded, the additional active earth pressure drops to the value

$$\Delta e_a = \Delta h \cdot \gamma_w \cdot K_{ag} \cdot \cos \delta_a$$

here as well. On the passive earth pressure side, an increase in the water surcharge caused by groundwater lowering has a favourable effect, especially in non-cohesive soils (Fig. R 166-1). The state of consolidation in the overlying cohesive soil must also be taken into consideration accordingly.

### 4.8.3 Case as for section 4.8.2 but with high-level aquifer

If in contrast to Fig. R 166-1 there is a water-bearing, non-cohesive stratum above the soft cohesive soil behind the waterfront structure, groundwater lowering sets up a predominantly vertical potential flow to the lower permeable stratum in the underlying uniform, cohesive soil. In this case the hydraulic head of the water in the overlying, highly permeable stratum is critical for the water pressure at the upper surface of the cohesive stratum, and the hydraulic head of the groundwater in the lower, non-cohesive stratum corresponding to the groundwater lowering is critical for the water pressure at the underside of the cohesive stratum. The changes to the active and passive earth pressures depend on the respective flow relationships or water surcharge, and here again the state of consolidation according to section 4.8.2 must be taken into account.

### 4.8.4 Consideration of intermediate states

The effect of groundwater lowering on the stability of waterfront structures is guaranteed for the final state, but is for the initial state heavily dependent on the soil conditions. When groundwater lowering is used for the temporary stabilisation of waterfront structures, the initial state and all intermediate states must also be carefully considered and analysed.

# 5 Ship dimensions and loads on waterfront structures

## 5.1 Ship dimensions (R 39)

### 5.1.1 Sea-going ships

The average ship dimensions given by way of example in Tables R 39-1.1 to R 39-1-11 can be used when designing and detailing waterfront structures, fenders and dolphins. Allowance must be made for the fact that these are average values that can vary by up to 10 % either way. The values have been determined largely statistically from the *Lloyds Register of Ships*, April 2001, and other unpublished analyses from Japan and Germany, and are therefore based on a very extensive database.

Definitions of the conventional data on ship sizes:

- Ship size is based on the gross register tonnage (GRT), a non-dimensional quantity derived from the ship's total volume. The use of the unit of measurement that was the standard previously, the gross registered ton (a registered ton corresponded to 100 cubic feet, i.e. 2.83 m$^3$) has not been permitted since 1994 in accordance with an international agreement.
- The deadweight tonnage (DWT) is given in tonnes and indicates the maximum cargo capacity of a ship fully equipped and ready for operation. There is no mathematical relationship between deadweight tonnage and vessel size.
- Displacement indicates the actual weight of a ship in tonnes including the maximum cargo capacity. There is no mathematical relationship between displacement and cargo capacity and/or vessel size.
- Container vessels are often assessed according to their loading capacity, which is specified in TEUs (**T**wenty feet **E**quivalent **U**nit). A TEU is the smallest available container length of 20 feet, equivalent to 6.10 m.

### 5.1.1.1 Passenger vessels

**Table R 39-1.1.** Passenger vessels

| Tonnage measurement | Cargo capacity | Displacement $G$ | Overall length | Length between perpen-diculars | Beam | Maximum draught |
|---|---|---|---|---|---|---|
| GRT | DWT | t | m | m | m | m |
| 70 000 | – | 37 600 | 260 | 220 | 33.1 | 7.6 |
| 50 000 | – | 27 900 | 231 | 197 | 30.5 | 7.6 |
| 30 000 | – | 17 700 | 194 | 166 | 26.8 | 7.6 |
| 20 000 | – | 12 300 | 169 | 146 | 24.2 | 7.6 |
| 15 000 | – | 9500 | 153 | 132 | 22.5 | 5.6 |
| 10 000 | – | 6600 | 133 | 116 | 20.4 | 4.8 |
| 7000 | – | 4830 | 117 | 103 | 18.6 | 4.1 |
| 5000 | – | 3580 | 104 | 92 | 17.1 | 3.6 |
| 3000 | – | 2270 | 87 | 78 | 15.1 | 3.0 |
| 2000 | – | 1580 | 76 | 68 | 13.6 | 2.5 |
| 1000 | – | 850 | 60 | 54 | 11.4 | 1.9 |

### 5.1.1.2 Bulk carriers

**Table R 39-1.2.** Bulk carriers

| Tonnage measurement | Cargo capacity | Displacement $G$ | Overall length | Length between perpen-diculars | Beam | Maximum draught |
|---|---|---|---|---|---|---|
| | DWT | t | m | m | m | m |
| – | 250 000 | 273 000 | 322 | 314 | 50.4 | 19.4 |
| – | 200 000 | 221 000 | 303 | 294 | 47.1 | 18.2 |
| – | 150 000 | 168 000 | 279 | 270 | 43.0 | 16.7 |
| – | 100 000 | 115 000 | 248 | 239 | 37.9 | 14.8 |
| – | 70 000 | 81 900 | 224 | 215 | 32.3 | 13.3 |
| – | 50 000 | 59 600 | 204 | 194 | 32.3 | 12.0 |
| – | 30 000 | 36 700 | 176 | 167 | 26.1 | 10.3 |
| – | 20 000 | 25 000 | 157 | 148 | 23.0 | 9.2 |
| – | 15 000 | 19 100 | 145 | 135 | 21.0 | 8.4 |
| – | 10 000 | 13 000 | 129 | 120 | 18.5 | 7.5 |

Occasionally, designations for bulk carriers are chosen depending on shipping routes etc. The corresponding sizes are as follows:

| | |
|---|---|
| < 20 000 DWT | Small Bulker |
| 20 000–40 000 DWT | Handysize Bulker |
| 40 000–60 000 DWT | Handymax Bulker |
| 60 000–100 000 DWT | Panamax Bulker |
| > 100 000 DWT | Capesize Bulker |

### 5.1.1.3 General cargo ships

No trend towards larger units is apparent among general cargo ships. If necessary, the dimensions of section 5.1.1.2 can be used.

More and more special ships specifically designed to transport heavy cargo are in use.

**Table R 39-1.3.** General cargo ships

| Tonnage measurement | Cargo capacity | Displacement $G$ | Overall length | Length between perpendiculars | Beam | Maximum draught |
|---|---|---|---|---|---|---|
| | DWT | t | m | m | m | m |
| – | 40 000 | 51 100 | 197 | 186 | 28.6 | 12.0 |
| – | 30 000 | 39 000 | 181 | 170 | 26.4 | 10.9 |
| – | 20 000 | 26 600 | 159 | 149 | 23.6 | 9.6 |
| – | 15 000 | 20 300 | 146 | 136 | 21.8 | 8.7 |
| – | 10 000 | 13 900 | 128 | 120 | 19.5 | 7.6 |
| – | 7000 | 9900 | 115 | 107 | 17.6 | 6.8 |
| – | 5000 | 7210 | 104 | 96 | 16.0 | 6.1 |
| – | 3000 | 4460 | 88 | 82 | 13.9 | 5.1 |
| – | 2000 | 3040 | 78 | 72 | 12.4 | 4.5 |
| – | 1000 | 1580 | 63 | 58 | 10.3 | 3.6 |

### 5.1.1.4 Container vessels

The beam (width) of a container vessel depends on the maximum number of rows of containers that can stand side by side on deck. There has been a very dynamic development in the size of container vessels, and it is hard to predict where this development will end. It is possible that these vessels will reach the sizes of the largest tankers and bulk carriers currently in use, which have reached beams of up to 70 m and draughts of up to 24 m. Design data must therefore be determined meticulously.

**Table R 39-1.4.** Container vessels

| Cargo capacity | Displacement $G$ | Overall length | Length between perpendiculars | Beam | Maximum draught | Number of containers | Generation |
|---|---|---|---|---|---|---|---|
| DWT | t | m | m | m | m | TEU | |
| 200 000 | 260 000 | 399 | 379 | 59.0 | 16.0 | 18 000 | Triple E Maersk |
| 188 000 | 244 400 | 396 | 378 | 53.6 | 16.0 | 16 000 | CMA "Marco Polo" |
| 160 000 | 208 000 | 397 | 379 | 56.4 | 16.0 | 13 700 | Maersk E-class |
| 150 000 | 195 300 | 386 | 369 | 51.0 | 15.5 | 12 900 | |
| 140 000 | 182 700 | 376 | 359 | 48.4 | 15.5 | 12 000 | |
| 130 000 | 170 000 | 365 | 348 | 45.6 | 15.0 | 11 100 | |
| 120 000 | 157 400 | 353 | 337 | 45.6 | 15.0 | 10 200 | |
| 110 000 | 144 700 | 342 | 324 | 42.8 | 14.5 | 9400 | |
| 100 000 | 133 000 | 329 | 312 | 42.8 | 14.5 | 8500 | 6th |
| 90 000 | 120 000 | 315 | 300 | 42.8 | 14.5 | 7600 | 6th |
| 80 000 | 107 000 | 300 | 284 | 40.3 | 14.5 | 6500 | 5th |
| 70 000 | 93 600 | 285 | 270 | 40.3 | 14.0 | 5400 | 5th |
| 60 000 | 80 400 | 268 | 254 | 32.3 | 13.4 | 4400 | 4th |
| 50 000 | 67 200 | 250 | 237 | 32.3 | 12.6 | 3700 | 3rd |
| 40 000 | 53 900 | 230 | 217 | 32.3 | 11.8 | 2900 | 3rd |
| 30 000 | 40 700 | 206 | 194 | 30.2 | 10.8 | 2100 | 2nd |
| 25 000 | 34 100 | 192 | 181 | 28.8 | 10.2 | 1700 | 2nd |
| 20 000 | 27 500 | 177 | 165 | 25.4 | 9.5 | 1300 | 2nd |
| 15 000 | 20 900 | 158 | 148 | 23.3 | 8.7 | 1000 | 1st |
| 10 000 | 14 200 | 135 | 126 | 20.8 | 7.6 | 600 | 1st |
| 7000 | 10 300 | 118 | 109 | 20.1 | 6.8 | 400 | 1st |

### 5.1.1.5 Ferries

The dimensions of ferries are hugely dependent on the areas in which they are used and on their purposes. The dimensions given below should therefore be used for preliminary studies only.

**Table R 39-1.5.** Ferries

| Cargo capacity | Displacement $G$ | Overall length | Length between perpen-diculars | Beam | Maximum draught |
|---|---|---|---|---|---|
| DWT | t | m | m | m | m |
| 30 300 | 40 000 | 223 | 209 | 31.9 | 8.0 |
| 22 800 | 30 000 | 201 | 188 | 29.7 | 7.4 |
| 15 300 | 20 000 | 174 | 162 | 26.8 | 6.5 |
| 11 600 | 15 000 | 157 | 145 | 25.0 | 6.0 |
| 7800 | 10 000 | 135 | 125 | 22.6 | 5.3 |
| 5500 | 7000 | 119 | 110 | 20.6 | 4.8 |
| 3900 | 5000 | 106 | 97 | 19.0 | 4.3 |
| 2390 | 3000 | 88 | 80 | 16.7 | 3.7 |
| 1600 | 2000 | 76 | 69 | 15.1 | 3.3 |
| 810 | 1000 | 59 | 54 | 12.7 | 2.7 |

### 5.1.1.6 Ro-ro vessels

**Table R 39-1.6.** Ro-ro vessels

| Cargo capacity | Displacement $G$ | Overall length | Length between perpen-diculars | Beam | Maximum draught |
|---|---|---|---|---|---|
| DWT | t | m | m | m | m |
| 30 000 | 45 600 | 229 | 211 | 30.3 | 11.3 |
| 20 000 | 31 300 | 198 | 182 | 27.4 | 9.7 |
| 15 000 | 24 000 | 178 | 163 | 25.6 | 8.7 |
| 10 000 | 16 500 | 153 | 141 | 23.1 | 7.5 |
| 7000 | 11 900 | 135 | 123 | 21.2 | 6.6 |
| 5000 | 8710 | 119 | 109 | 19.5 | 5.8 |
| 3000 | 5430 | 99 | 90 | 17.2 | 4.8 |
| 2000 | 3730 | 85 | 78 | 15.6 | 4.1 |
| 1000 | 1970 | 66 | 60 | 13.2 | 3.2 |

### 5.1.1.7 Oil tankers

**Table R 39-1.7.** Oil tankers

| Cargo capacity | Displacement $G$ | Overall length | Length between perpen-diculars | Beam | Maximum draught |
|---|---|---|---|---|---|
| DWT | t | m | m | m | m |
| 300 000 | 337 000 | 354 | 342 | 57.0 | 20.1 |
| 200 000 | 229 000 | 311 | 300 | 50.3 | 17.9 |
| 150 000 | 174 000 | 284 | 273 | 46.0 | 16.4 |
| 100 000 | 118 000 | 250 | 240 | 40.6 | 14.6 |
| 50 000 | 60 800 | 201 | 192 | 32.3 | 11.9 |
| 20 000 | 25 300 | 151 | 143 | 24.6 | 9.1 |
| 10 000 | 13 100 | 121 | 114 | 19.9 | 7.5 |
| 5000 | 6740 | 97 | 91 | 16.0 | 6.1 |
| 2000 | 2810 | 73 | 68 | 12.1 | 4.7 |

### 5.1.1.8 LNG (liquefied natural gas) tankers

**Table R 39-1.8.** LNG (liquefied natural gas) tankers

| Capacity | | Displacement $G$ | Overall length | Length between perpen-diculars | Beam | Maximum draught |
|---|---|---|---|---|---|---|
| DWT | $m^3$ | t | m | m | m | m |
| 100 000 | 155 000 | 125 000 | 305 | 294 | 50.0 | 12.5 |
| 70 000 | 110 000 | 100 000 | 280 | 269 | 45.0 | 11.5 |
| 50 000 | 77 000 | 75 000 | 255 | 245 | 38.0 | 10.5 |
| 20 000 | 30 500 | 34 000 | 195 | 185 | 30.0 | 8.5 |
| 10 000 | 15 000 | 19 000 | 148 | 135 | 26.0 | 7.0 |

### 5.1.1.9 LPG (liquefied petroleum gas) tankers

**Table R 39-1.9.** LPG (liquefied petroleum gas) tankers

| Capacity | | Displacement $G$ | Overall length | Length between perpendiculars | Beam | Maximum draught |
|---|---|---|---|---|---|---|
| DWT | m³ | t | m | m | m | m |
| 70 000 | 105 000 | 90 000 | 260 | 250 | 38.0 | 14.0 |
| 50 000 | 65 000 | 65 000 | 230 | 220 | 35.0 | 13.0 |
| 20 000 | 20 000 | 27 000 | 170 | 160 | 25.0 | 10.5 |
| 10 000 | 10 000 | 15 000 | 130 | 120 | 21.0 | 9.0 |
| 5000 | 5000 | 8000 | 110 | 100 | 18.0 | 6.8 |
| 2000 | 2000 | 3500 | 90 | 75 | 13.0 | 5.5 |

### 5.1.2 River- and sea-going vessels

**Table R 39-1.10.** River- and sea-going vessels

| Tonnage measurement | Cargo capacity | Displacement $G$ | Overall length | Beam | Maximum draught |
|---|---|---|---|---|---|
| GRT | DWT | t | m | m | m |
| 999 | 3200 | 3700 | 94.0 | 12.8 | 4.2 |
| 499 | 1795 | 2600 | 81.0 | 11.3 | 3.6 |
| 299 | 1100 | 1500 | 69.0 | 9.5 | 3.0 |

## 5.1.3 Inland waterway vessels

**Table R 39-1.11.** Inland waterway vessels

| Designation | Cargo capacity | Displacement $G$ | Length | Beam | Draught |
|---|---|---|---|---|---|
| | t | t | m | m | m |
| Motor cargo boats: | | | | | |
|   large Rhine vessel | 4500 | 5200 | 135.0 | 17.2 | 4.5 |
|   2600 t class | 2600 | 2950 | 110.0 | 11.4 | 2.7 |
|   Rhine vessel | 2000 | 2385 | 95.0 | 11.4 | 2.7 |
|   Europa vessel | 1350 | 1650 | 80.0 | 9.5 | 2.5 |
|   Dortmund-Ems canal vessel | 1000 | 1235 | 67.0 | 8.2 | 2.5 |
|   large canal vessel | 950 | 1150 | 82.0 | 9.5 | 2.0 |
|   large Plauer vessel | 700 | 840 | 67.0 | 8.2 | 2.0 |
|   BM 500 vessel | 650 | 780 | 55.0 | 8.0 | 1.8 |
|   Campine barge | 600 | 765 | 50.0 | 6.6 | 2.5 |
|   Lighter | 415 | 505 | 32.5 | 8.2 | 2.0 |
|   Barge | 300 | 405 | 38.5 | 5.0 | 2.2 |
|   large Saale vessel | 300 | 400 | 52.0 | 6.6 | 2.0 |
|   large Finow vessel | 250 | 300 | 41.5 | 5.1 | 1.8 |
| Pushed lighters: | | | | | |
|   Europa IIa | 2940 | 3275 | 76.5 | 11.4 | 4.0 |
| | 1520 | 1885 | | | 2.5 |
|   Europa II | 2520 | 2835 | 76.5 | 11.4 | 3.5 |
| | 1660 | 1990 | | | 2.5 |
|   Europa I | 1880 | 2110 | 70.0 | 9.5 | 3.5 |
| | 1240 | 1480 | | | 2.5 |
| Carrier systems: | | | | | |
|   Sea Bee | 860 | 1020 | 29.7 | 10.7 | 3.2 |
|   LASH | 376 | 488 | 18.8 | 9.5 | 2.7 |
| Push-tow convoys: | | | | | |
|   with 1 Europa IIa lighter | 2940 | 3520[1] | 110.0 | 11.4 | 4.0 |
| | 1520 | 2130[1] | | | 2.5 |
|   with 2 Europa IIa lighters | 5880 | 6795[1] | 185.0 | 11.4 | 4.0 |
| | | | 110.0 | 22.8 | 4.0 |
| | 3040 | 4015[1] | | | 2.5 |
|   with 4 Europa IIa lighters | 11 760 | 13 640[2] | 185.0 | 22.8 | 4.0 |
| | 6080 | 8080[2] | | | 2.5 |

[1] Towboat, 1480 kW, approx. 245 t displacement
[2] Towboat, 2963–3333 kW, approx. 540 t displacement

In accordance with EEC Resolution No. 30 of 12 Nov 1992, TRANS/SC 3/R.153, the classification of Table R 39-3.2 applies to European waterways.

**Table R 39-3.2.** Classification for European inland waterways

| Type of inland waterway | Class of navigable waterway | Motor vessels and barges Type of vessel: general characteristics | | | | | Pushed convoys Type of convoy: general characteristics | | | | | Minimum headroom under bridges [m][2] | Graphic symbol on map |
|---|---|---|---|---|---|---|---|---|---|---|---|---|---|
| | | Designation | Length L [m] | Beam W [m] | Draught d [m][7] | Tonnage T [t] | Formation | Length L [m] | Beam W [m] | Draught d [m][7] | Tonnage T [t] | | |
| 1 | 2 | 3 | 4 | 5 | 6 | 7 | 8 | 9 | 10 | 11 | 12 | 13 | 14 |
| of regional significance west of River Elbe | I | Barge | 38.5 | 5.05 | 1.8–2.2 | 250–400 | | | | | | 4.0 | |
| | II | Campine barge | 50–55 | 6.6 | 2.5 | 400–650 | | | | | | 4.0–5.0 | |
| of regional significance east of River Elbe | I | Large Finow | 41 | 4.7 | 1.4 | 180 | | | | | | 3.0 | |
| | II | BM 500 | 57 | 7.5–9.0 | 1.6 | 500–630 | | | | | | 3.0 | |
| | III | [6] | 67–70 | 8.2–9.0 | 1.6–2.0 | 470–700 | | 118–132[1] | 8.2–9.0[1] | 1.6–2.0 | 1000–1200 | 4.0 | |
| of international significance | IV | Johann Welker | 80–85 | 9.50 | 2.50 | 1000–1500 | | 85 | 9.50[5] | 2.50–2.80 | 1250–1450 | 5.25 or 7.00[4] | |
| | Va | Large Rhine vessel | 95–110 | 11.40 | 2.50–2.80 | 1500–3000 | | 96–110[1] | 11.40 | 2.50–4.50 | 1600–3000 | 5.25 or 7.00 or 9.10[4] | |
| | Vb | | | | | | | 172–185[1] | 11.40 | 2.50–4.50 | 3200–6000 | 9.10[4] | |
| | VIa | | | | | | | 95–110[1] | 22.80 | 2.50–4.50 | 3200–6000 | 7.00 or 9.10[4] | |
| | VIb | [3] | 140 | 15.00 | 3.90 | | | 185–195[1] | 22.80 | 2.50–4.50 | 6400–12000 | 7.00 or 9.10[4] | |
| | VIc | | | | | | | 270–280[1] 195–200[1] | 22.80 33.00–34.20[1] | 2.50–4.50 2.50–4.50 | 9600–18000 9600–18000 | 9.10[4] | |
| | VII | | | | | | | 285 | 33.00–34.20[1] | 2.50–4.50 | 14500–27000 | 9.10[4] | |

Notes to classification table:

1) The first figure takes into account the actual situation, whereas the second corresponds to future developments, although in some cases also existing situations.
2) Takes into account a safety margin of approx. 30 cm between the uppermost point of the ship's structure or cargo on deck.
3) Takes into account the dimensions of self-propelled vessels anticipated in ro-ro and container traffic; the dimensions given are approximate values.
4) Designed for transporting containers:
   5.25 m for vessels carrying two tiers of containers
   7.00 m for vessels carrying three tiers of containers
   9.10 m for vessels carrying four tiers of containers
   50 % of the containers may be empty, otherwise ballasting is required.
5) On account of the greatest permissible lengths of vessels and convoys, a number of existing waterways can be assigned to class IV even though the maximum beam is 11.40 m and the maximum draught 4.00 m.
6) Vessels on the River Oder and on waterways between the River Oder and the River Elbe.
7) The draught for a certain waterway should be defined in accordance with local circumstances.
8) Convoys consisting of a larger number of barges/lighters may also be used on certain class VII waterways. In such cases the horizontal dimensions may exceed the figures given in the table.

### 5.1.4 Displacement

Displacement $G$ [t] is the product of the length between the perpendiculars, the beam (width), draught, the block coefficient $c_B$ and the density of water $\rho_w$ [t/m$^3$]. For sea-going ships the $c_B$ coefficient varies between about 0.50 and 0.80, for inland waterway vessels between about 0.80 and 0.90, and for pushed lighters between 0.90 and 0.93.

### 5.2 Berthing force of ships at quays (R 38)

During preliminary design, no exceptional accident impacts need to be taken into consideration, just the typical berthing forces. The magnitude of these berthing forces depends on each ship's dimensions, the berthing speed, the fenders and the deformation of the ship's side and the structure. In order to provide quays with adequate loading capacity to resist typical berthing forces, at the same time, however, avoiding unnecessarily large dimensions, it is recommended to design the structural components affected by berthing manoeuvres in such a way that a single compression load equal to the critical line pull force can be applied at any point. For quay walls in seaports, the load should be as per R 12, section 5.12.2, with the values of Table R 12-1, and in inland ports as per R 102, section 5.13.2, with a force of 200 kN, but the total load should not exceed the permissible limits.

The single load can be distributed according to the fendering; without fenders, distribution is permissible over an area measuring $0.50 \times 0.50$ m. In the case of sheet pile walls without heavyweight superstructures, only walings and waling bolts need to be designed for this compression load.

Berthing forces on dolphins are dealt with in R 128, section 13.3.

If the failure of the waterfront structure as a result of a collision (e.g. ship impact) poses particular risks, e.g. for another facility situated immediately behind it, further deliberations may be necessary and the measures to be taken then agreed upon between designer, client and authorities.

### 5.3 Berthing velocities of ships transverse to berth (R 40)

When a ship approaches transverse to a berth, in terms of designing the corresponding fenders, the recommendation is to consider the berthing velocities given in Figs. R 40-1 and R 40-2, which correspond to the Spanish recommendations for maritime structures [178].

For inland waterway vessels with a displacement of up to 1500 t, something like the berthing velocities given in Table R 40-3 transverse to the berth can be assumed. These speeds are taken from DIN EN 14504 (2004).

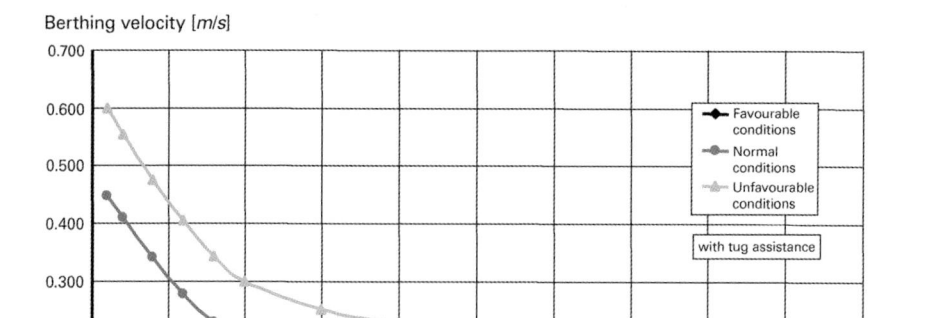

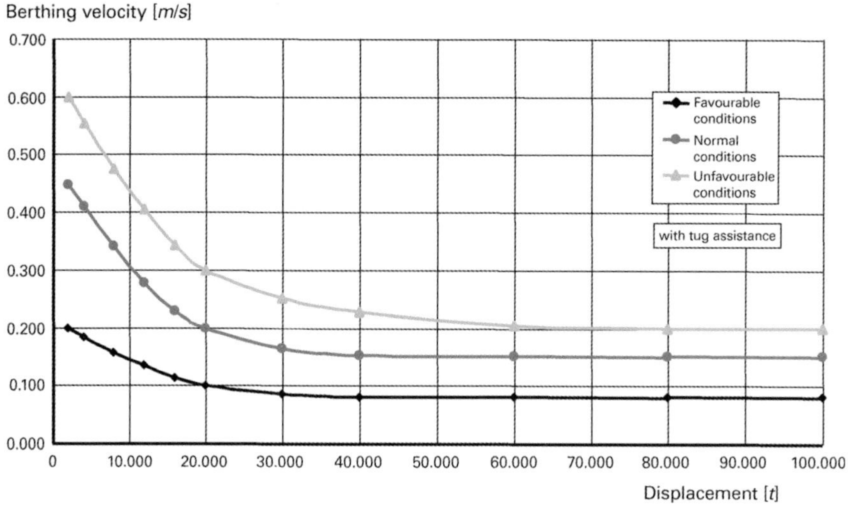

**Fig. R 40-1.** Berthing velocities with tug assistance

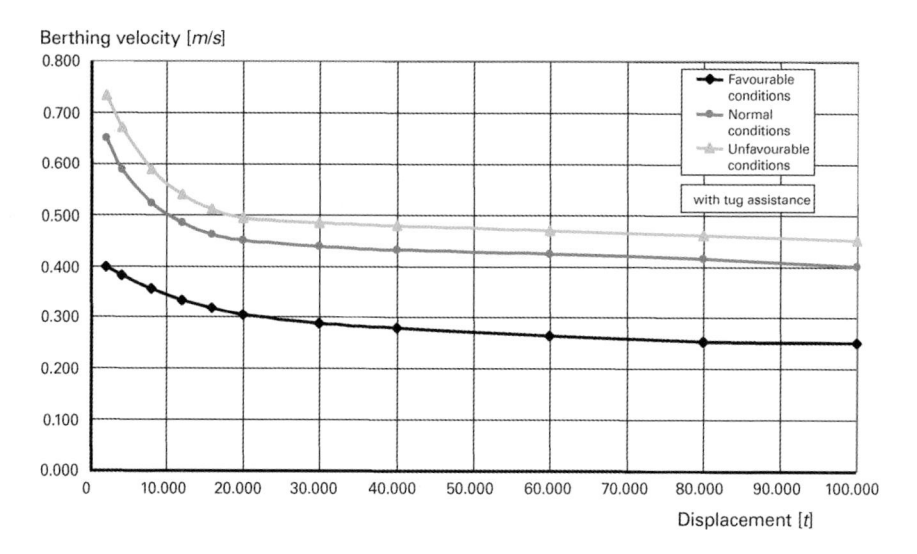

**Fig. R 40-2.** Berthing velocities without tug assistance

**Table R 40-3.** Berthing velocities of inland waterway vessels

| Mass of vessel [t] | Berthing velocity [m/s] |
|---|---|
| 100 | 0.29 |
| 200 | 0.28 |
| 500 | 0.26 |
| 1000 | 0.23 |
| 1500 | 0.21 |
| 2000 | 0.19 |
| 3000 | 0.16 |
| 4000 | 0.14 |
| $\geq 5000$ | 0.13 |

## 5.4 Design situations (R 18)

When it comes to verifying stability and assigning partial safety factors, load cases are defined in DIN 1054, section 6.3.3. These result from the combinations of actions in conjunction with the resistance safety categories. The following classifications apply to waterfront structures:

### 5.4.1 Design situation DS-P
Loads resulting from active earth pressure (separately for the initial and final states in the case of unconsolidated, cohesive soils) and from excess water pressure in the case of the frequent occurrence of unfavourable inner and outer water levels (see R 19, section 4.2), active earth pressure influences from normal imposed loads, normal crane loads, pile loads and instantaneous actions due to self-weight and normal imposed loads.

### 5.4.2 Design situation DS-T
Temporary situations, so-called transient situations, are assigned to design situation DS-T, e.g. situations during construction or maintenance; in hydraulic engineering works the permanent actions and the variable actions of DS-P which occur regularly during the service life of the structure, e.g. limited scour due to currents or ship propellers, or excess water pressure in the case of rare occurrences of unfavourable inner and outer water levels (see R 19, section 4.2), or wave loads according to R 136, section 5.6.4.

### 5.4.3 Design situation DS-A
As for design situation DS-T but with extraordinary design situations such as unscheduled additional loads over a larger area, the unusually large flattening of an underwater slope in front of the base of a sheet pile wall, unusual scour due to currents or ship propellers, excess water

pressure following extreme water levels (see R 19, section 4.2, or R 165, section 12.7), excess water pressure following exceptional flooding of the waterfront structure, combinations of earth and water pressures with wave loads resulting from waves that occur only rarely (see R 136, section 5.6.4); combinations of earth and water pressures with flotsam impact according to section 12.7.5, all load combinations in conjunction with ice states and ice pressures.

### 5.4.4 Extreme case

When extremely improbable combinations of actions occur concurrently, DIN EN 1997-1, section 2.4.6, allows the use of combination factors, and variable actions can be applied as a representative value. Design situation DS-A should be taken as a basis. The largest variable leading action should be applied with the combination factor $\psi = 1$, the others with the combination factor $\psi_2 = 0.5$.

Examples of this include the coincidence of extreme water levels and simultaneous extreme wave loads resulting from plunging breakers according to R 135, section 5.7.3, extreme water levels and the simultaneous, complete failure of a drainage system (see R 165, section 12.7.2), combinations of three short-term events acting simultaneously e.g. high water (highest astronomical tide, see R 165, section 12.7.2), waves that occur rarely (see R 136, section 5.6.4) and flotsam impact (see R 165, section 12.7.5).

## 5.5 Vertical imposed loads (R 5)

In this section, all quantitative loads (actions) are characteristic values.

### 5.5.1 General

Vertical imposed loads (variable loads within the meaning of DIN EN 1991-1) are essentially loads due to stored goods and means of transport. The changing positions of load influences due to mobile cranes (road or rail types) must be considered separately where these affect the waterfront structure. In the case of waterfront structures in inland ports, loads from mobile cranes generally only apply to those waterfront areas expressly intended for heavy-duty loading/unloading with mobile cranes. In seaports, in addition to rail-mounted quayside cranes, mobile cranes are being increasingly used for general cargo handling, i.e. not just for heavy loads. As regards dynamic load influences, a distinction is made between three different basic situations (Table R 5-1):

– In basic situation 1 the loadbearing members of the structures are directly loaded by the means of transport and/or stacked goods, e.g. jetties (Table R 5-1a).

**Table R 5-1.** Vertical imposed loads

| Basic situation | Traffic loads[1] | | |
|---|---|---|---|
| | Rail | Road | |
| | | Vehicle | Crane |
| a) Basic situation 1 | Loading assumptions to German Federal Railways guideline RIL 804 or DIN technical report 101 Dynamic coefficient: Those amounts exceeding 1.0 can be reduced to half their value. | Loading assumptions to DIN 1055 or DIN technical report 101 | Forklift truck loads to DIN 1055; outrigger loads for mobile cranes to sections 5.5.5 and 5.14.3 |
| b) Basic situation 2 | As for basic situation 1 but a further reduction in the dynamic coefficient up to 1.0 for a layer depth $h = 1.00$ m. For a layer depth $h \geq 1.50$ m, use a uniformly distributed surface load of 20 kN/m². | | |
| c) Basic situation 3 | Loads as for basic situation 2 with a layer depth exceeding 1.50 m | | |

[1] Crane loads in accordance with R 84, section 5.14.2.

- In basic situation 2 the means of transport and stacked goods place a load on a layer of a certain thickness which spreads and transmits the loads to the loadbearing members of the waterfront structure. This form of construction is used, for example, for structures built over embankments which include a layer of material above the structure to spread the loads (Table R 5-1b).
- In basic situation 3 the means of transport and stacked goods only place a load on the body of soil behind the waterfront structure, which, consequently, is only loaded indirectly via an active

earth pressure resulting from the imposed loads. Typical examples of this are exclusively sheet pile walls or a partially sloping bank (Table R 5-1c).

There are also intermediate cases between the three basic situations, e.g. a pile trestle supporting a short pile cap.

Provided a complete and reliable basis for calculations is available, the magnitudes of the imposed loads should be assumed to be those expected in normal circumstances. The higher the proportion of dead loads and the better the distribution of loads within the structure, the easier it is to accommodate any increases in the imposed loads which may be necessary at a later date within the scope of the permissible limits. Structural systems complying with basic situation 2 and, in particular, basic situation 3 offer advantages in this respect.

Refer to R 18, section 5.4, regarding the allocation of the respective loads to load cases 1, 2 and 3.

## 5.5.2 Basic situation 1

The railway traffic loads correspond to loading diagram 71 of DIN technical report 101. As regards road traffic, the loading assumptions according to DIN 1055 or DIN technical report 101 should be used, generally load model 1. In the dynamic coefficients specified for railway bridges, used to factor the traffic loads, it is generally possible to reduce amounts exceeding 1.0 to half their value on account of the slow speeds. For road bridges, load model 1 already presumes a slow vehicle travelling speed (congestion situation) and so no reduction is permitted. For jetties in seaports, forklift truck loads to DIN 1055 and mobile crane outrigger loads of 1950 kN should be assumed with an outrigger size of $5.5 \times 1.3$ m, unless higher figures are necessary in particular cases (see Tables R 84-1 and R 84-2, sections 5.14.3 and 5.5.5).

## 5.5.3 Basic situation 2

Essentially the same as basic situation 1. However, depending on the depth of the layer, the dynamic coefficients for railway bridges can be further reduced in a linear manner and eventually completely disregarded if the layer depth is at least 1.00 m (calculated from the top of the rail in the case of tracks embedded in paving). The loading must be considered bay by bay, however.

If the layer is at least 1.50 m deep, the total rail traffic load can be replaced by a uniformly distributed load of $20 \text{ kN/m}^2$.

### 5.5.4 Basic situation 3
Loads as for basic situation 2 with a layer depth exceeding 1.50 m.

### 5.5.5 Loading assumptions for quay surfaces
When heavy road cranes or similar heavy vehicles and heavy construction equipment, such as crawler excavators etc., are operating just behind the front edge of the waterfront structure, the following loads should be assumed as a minimum for designing the waterfront structure and any upper anchorage required:

a) imposed load $= 60$ kN/m$^2$ from rear edge of top of wall landwards over a width of 2.0 m, or
b) imposed load $= 40$ kN/m$^2$ from rear edge of top of wall landwards over a width of 3.50 m.

Both a) and b) include the influences of an outrigger load $P = 500$ kN provided the distance between the axis of the waterfront structure and the axis of the outrigger is at least 2 m. Refer to sections 5.5.2 and 5.14.3 when higher outrigger loads need to be considered.

In accordance with PIANC [154] the following imposed loads are used as a basis outside traffic zones. The container loads take into account gross loads of 300 kN for 40 ft containers and 240 kN for 20 ft containers.

- General cargo  20 kN/m$^2$
- Containers:
  - empty, stored in 4 tiers  15 kN/m$^2$
  - full, stored in 2 tiers  35 kN/m$^2$
  - full, stored in 4 tiers  55 kN/m$^2$
- Ro-ro load  30-50 kN/m$^2$
- Multi-purpose installations  50 kN/m$^2$
- Offshore supply bases  To be agreed with the operator.
- Paper  Characteristic nominal values for unit weights in accordance with EC 1, depending on dumping/stacking height.

- Timber products
- Steel
- Coal
- Ore

Further details of the material parameters of bulk and stacked goods can be found in the tables of [178]. When it comes to calculating active earth pressures on retaining structures, the different loads in traffic and container areas can generally be combined into an average uniformly distributed load of 30–50 kN/m².

Substantial loads can occur at installations intended for cargo handling or supplying offshore facilities, e.g. the construction of offshore wind turbines. These loads must be agreed with the operator in each individual case.

## 5.6 Determining the "design sea state" for maritime and port structures (R 136)

### 5.6.1 General

The wave loads on maritime and port structures are essentially due to the wind-generated sea state, the significance of which for the design has to be checked with reference to the local boundary conditions. In coastal areas it is generally not just the sea state generated locally that is relevant to design, since fetch distances and shallow depths limit the wave energy component. Instead, the local sea state (wind-driven waves) combined with the sea state of the open sea beyond the project area and the sea state approaching the coast (groundswell) should be considered together.

The following descriptions are limited to fundamental processes and simplified approaches to the determination of hydraulic boundary conditions and loads on structures. Detailed information on this can be found, for example, in the coastal protection recommendations [60]. It is recommended that an institute or consulting engineers experienced in coastal engineering be consulted when it comes to investigating the wave conditions in the project area and the specific loads on structures. The need to carry out further physical or numerical studies should be looked at carefully prior to starting detailed design work.

### 5.6.2 Description of the sea state

The natural sea state can basically be described as an irregular chronological sequence of waves of varying height (or amplitude), period (or frequency) and direction. It represents the superposition in space and time of various short- and long-period sea state components. The direct influence of the wind generates an irregular, short-period (short-crested) sea state, also known as wind-driven waves. A long-period, irregular sea state arises through the superposition of wave components with the same direction, with the waves sorted by various interactions and the sea state no longer influenced directly by the wind.

The prevailing natural, irregular sea state in a project area is comprised of the local, short-period sea state (wind-driven waves) and the long-period sea state (groundswell) originally generated as wind-driven waves outside the project area.

In order to take into account the sea state relevant for design which is actually present as a load variable in the existing design methods, it is first necessary to parameterise the irregular sea state because, in general, only individual, characteristic sea state parameters (see section 5.6.3) can be included in the calculations. This parameterisation of the irregular sea state may be carried out both

1. chronologically (direct short-term statistical evaluation of the time series) by determining and presenting characteristic wave parameters (wave heights and wave periods) as arithmetic mean values, and
2. in terms of frequency (Fourier analysis) by determining and presenting the results as a wave spectrum, with the energy content of the sea state being determined as a function of wave frequency [60].

Due to the evaluation, the parameterisation of the sea state relevant for design results in the loss of all the information on the wave time series, its statistics and the wave spectrum.

In the design and detailing of maritime and port structures, the results of the sea state studies from the chronological and frequency analyses must be taken into consideration depending on the location of the project area. In most cases the parameterisation of the sea state relevant for design and the characterisation by way of individual wave height, wave period and wave direction parameters is normally sufficient. Where complex wind and wave conditions prevail and, in particular, in shallow water zones with breakwaters, it may be necessary to determine further local parameters characterising the sea state in order to define reliably the load variables required in the design process [60].

## 5.6.3 Determining the sea state parameters

### 5.6.3.1 General

Sea state parameters are characteristic values that describe and quantify certain properties of the irregular sea state varying with space and time. Depending on the evaluation process (see section 5.6.2), these are:

1. in the time domain, the mean values of individual parameters, e.g. wave heights or wave periods, or combinations thereof, and
2. in the frequency domain, prominent frequencies or integrals from the spectral density of the sea state spectrum.

The wave conditions in the project area must be analysed on the basis of measurements or observations over a sufficiently long period with respect to their theoretical probability of occurrence. To this end, and depending on the particular task in hand, the significant wave parameters resulting from short-term statistical analysis – such as wave heights, wave periods and wave approach directions – must be determined with respect to their seasonal frequencies or maximum long-term values in order to be able to derive data relevant for design. If such measurements are not available, empirical-theoretical or numerical methods must be used to determine the wave parameters from wind data (hindcasting), and these must be verified using any measured wave values available. The parameterisation of the natural sea state is carried out based on the fact that there are statistical correlations between the heights of the individual waves of a natural sea state ascertained by measurements. These correlations can be described with a Rayleigh distribution according to [134] assuming a narrow-band wave spectrum and a large number of different waves (see section 3.7.4 in [60] and [36]).

In deep water ($d \geq L/2$), wave height distributions based on measured data agree very well with the Rayleigh distribution even for wider wave band spectrums.

In shallow water ($d \leq L/20$), various effects (see section 5.6.5) influence the waves and so there are greater deviations between the measured wave height distribution and the theoretical Rayleigh distribution. The wave spectrum in the shallow water zone is no longer a narrow band and the associated wave height distribution may differ substantially from the Rayleigh distribution due to the breaking waves.

Deviations from the Rayleigh distribution increase with larger wave heights and decrease with the narrowing of the spectral bandwidth. The Rayleigh distribution tends to overestimate large wave heights in all water depths. The determination of the sea state parameters in the time and frequency domains and their relationships is explained below.

### 5.6.3.2 Sea state parameters in the time domain

When evaluated in the time domain, the wave heights and wave periods recorded during an observation period are described by stochastic variables of the frequency distribution. Where the wave heights are concerned, the Rayleigh function provides a good approximation of the probability $P(H)$ of the occurrence of a wave of height $H$ (individual probability) and the probability $P(H)$ of the occurrence of a number of waves up to height $H$ (total probability); see also Fig. R 136-1 after Oumeraci [151]:

$$P(H) = 1 - e^{-\frac{\pi}{4}\left(\frac{H}{H_m}\right)^2}$$

with parameters according to Fig. R 136-1.
Fig. R 136-1 is modified by and according to [151]; the notation is as follows:

$N$      frequency of wave heights $H$ in observation period, expressed as a percentage
$H_m$      mean value of all wave heights from sea state records
$H_d$      most frequent wave height

**Probability of falling short of limit** $P(H < H_s) = 1 - \exp\left[-2 \cdot \left(\frac{H}{H_s}\right)^2\right]$

**Frequency distribution**
$$p\left(\frac{H}{H_s}\right) = 4\left(\frac{H}{H_s^2}\right)\exp\left[-2\left(\frac{H}{H_s}\right)^2\right]$$

**Rayleigh distribution of wave heights**
$H_m = 0.63\ H_s$
$H_{1/10} = 1.27\ H_s$
$H_{1/100} = 1.67\ H_s$

$H_{max} = \sqrt{\frac{\ln N}{2}}\ H_s$

$H_{rms} = \sqrt{\frac{1}{N}\sum_{i=1}^{N} H_i^2}$

For $N \approx 1000$ waves:
$H_{max} = 1.86\ H_s$
$H_{rms} = 0.707\ H_s$

33.3 % of highest waves in time series analysed

Relative wave height $[H/H_S]$

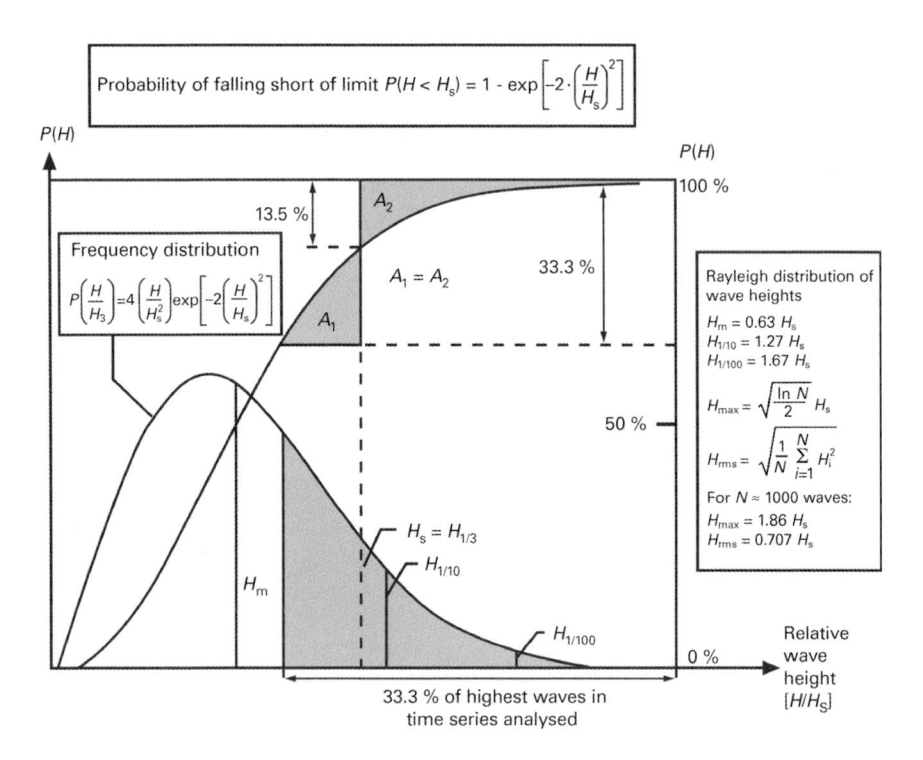

**Fig. R 136-1.** Rayleigh distribution of the wave heights of a natural sea state (schematic)

$H_{1/3}$   mean value of 33 % highest waves
$H_{1/10}$  mean value of 10 % highest waves
$H_{1/100}$ mean value of 1 % highest waves
$H_{max}$   maximum wave height
$H_{rms}$   measurement of mean wave energy; corresponds roughly to $1.13 \cdot H_m$
$H_S$       significant wave height, see section 5.6.3.4

According to Schüttrumpf [194] and Longuet-Higgins [134], the frequency distribution of the wave heights results, approximately, in the following relationships, assuming a theoretical wave height distribution of the sea state corresponding to a Rayleigh distribution. These theoretical ratios agree well with ratios determined from sea state measurements despite the wave spectrum possibly having a greater bandwidth than that presumed by the Rayleigh distribution:

$$H_m = 0.63 \cdot H_{1/3}$$
$$H_{1/10} = 1.27 \cdot H_{1/3}$$
$$H_{1/100} = 1.67 \cdot H_{1/3}$$

The maximum wave height $H_{max}$ depends, in principle, on the number of waves recorded during the measuring period available. According to Longuet-Higgins [134], if we assume

$$H_{max} = 0,707 \cdot \sqrt{\ln(n)} \cdot H_{1/3}$$

for $n = 1000$ waves, the result is a maximum wave height $H_{max} = 1.86 \cdot H_{1/3}$. As far as practical engineering is concerned, it is sufficient to estimate the maximum wave height as

$$H_{max} = 2 \cdot H_{1/3}.$$

A further sea state parameter commonly used in practice is the wave height $H_{rms}$ (rms = root-mean-square). A sea state with Rayleigh distribution results in the relationship $H_{rms} = 0.7 \cdot H_{1/3}$. The value $H_{rms}$, as a measurement of the mean wave energy, gives greater weight to the higher waves in the wave spectrum than the simple mean $H_m$.
Similarly to the ratios of the wave heights, the wave periods in the time domain can be estimated based on measurements from nature [60].
The actual ratios of the wave heights and periods depend on the actual wave height distribution, the specific shape of the sea state spectrum, the duration of measurements, etc. and can deviate from the above theoretical values, especially in shallow water, due to the true distribution of the waves and their asymmetry. Short measurement periods, e.g. 5 or 10 min, can lead to considerable errors when determining ratios, which is why the EAK proposes a period of at least 30 min for measuring and evaluating sea state measurements in order to rule out statistical irregularities.

### 5.6.3.3 Sea state parameters in the frequency domain

When parameterising an irregular sea state in the frequency domain, superposing the individual wave components converts the time series from the sea state records into an energy density spectrum and the associated wave phases into a corresponding phase spectrum (Fourier transformation), see Fig. R 136-2. Presenting the sea state spectrum jointly for all wave directions is known as a "one-dimensional spectrum"; a separate presentation for different wave directions is a "directional spectrum".

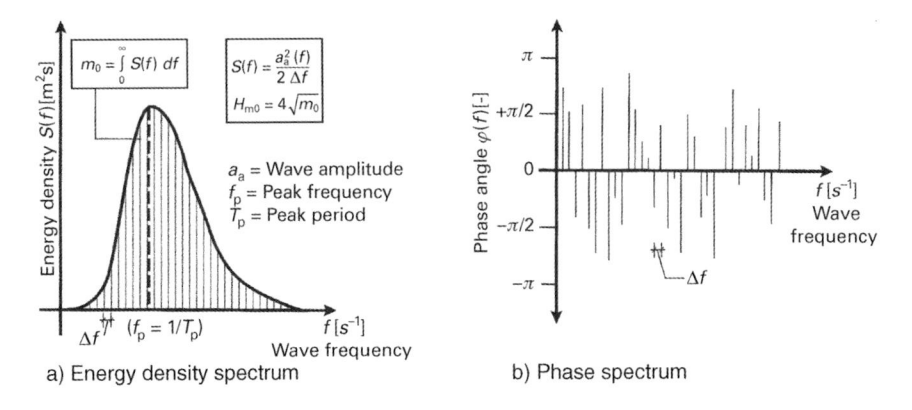

**Fig. R 136-2.** Wave spectrum parameters – definition sketches [150]

The sea state spectrum enables the following and other characteristic sea state parameters to be specified as a function of the frequency $f$ [Hz], taking into account spectral moments of the $n$th order:

$$m_n = \int S(f) \cdot f^n df \quad \text{where } n = 0, 1, 2 \ldots$$

and also as a function of the wave approach direction (see Fig. R 136-2):

$H_{m0}$    characteristic wave height $= 4m_0^{1/2}$, where $m_0$ is the area of the wave spectrum
$T_{01}$    mean period $= m_0/m_1$
$T_{02}$    mean period $= m_0/m_2$
$T_p$    peak period, i.e. wave period at maximum energy density

The wave periods $T_{01}$ and $T_{02}$ have a fixed ratio, which depends on the shape of the wave spectrum, and describe the bandwidth of the spectrum. The wave spectrum allows the identification of long-period wave components in particular, e.g. groundswell waves, wave components transformed by the structure or changes in the spectrum caused by shallow water; these can be significant when defining the hydraulic boundary conditions for the design of maritime and port structures.

### 5.6.3.4 Relationships between sea state parameters in the time and frequency domains

The "significant wave height" $H_s$ was introduced to characterise the irregular sea state for practical engineering applications [156]. Further, presuming a Rayleigh distribution of the sea state, it is assumed for practical applications that $H_s$ can be determined from the wave height $H_{1/3}$ in the time domain or the wave height $H_{m0}$ in the frequency domain:

$$H_s = H_{1/3} = H_{m0}$$

Moreover, theoretically, the wave periods $T_m$ (time domain) and $T_{02}$ (frequency domain) can be equated. As regards further interrelationships between sea state parameters in the time and frequency domains, which can always vary depending on the respective wave spectrum, please refer to EAK [60].

### 5.6.4 Design concepts and specification of design parameters

The design sea state is understood to be the sea state event that leads to the critical load on a structure or part thereof or describes its effect in characteristic terms and is the result of a critical combination of different influencing variables.

With regard to the design of structures, a distinction is made between

- the structural design, i.e. verification of stability for an extreme event, and
- the functional design, which deals with the effect and influence of the structure on its surroundings, see [60], section 3.7.

The irregularity of the sea state and its description as an input variable in the corresponding design procedure is decisive for determining the load variables actually occurring. Depending on the structure to be designed and the design method being used, the design sea state can be

- a characteristic individual wave, which is used to determine a specific load (deterministic method; possible application: load on a flood defence wall), or
- considered as a characteristic wave time series, the result of which is a time series of the loads occurring on the structure which can be evaluated statistically and assessed with respect to the maximum and total loading (stochastic method; possible application: structures resolved in three dimensions, e.g. offshore platforms), or
- integrated into a fully statistical distribution, which enables a failure probability to be determined for the structure taking into account various limit states of the structure (probabilistic method).

Deterministic design methods are predominantly used in practical engineering applications. These will be looked at in more detail below. The EAK contains advice on how the actual irregularity of the sea state can be taken into account in studies, calculations and design work on the basis of regular waves.

**Table R 136-1.** Recommendation for stipulating design wave height

| Structure | $H_d/H_{1/3}$ |
|---|---|
| Breakwaters | 1.0 to 1.5 |
| Sloped moles | 1.5 to 1.8 |
| Vertical moles | 1.8 to 2.0 |
| Flood defence walls | 1.8 to 2.0 |
| Quay walls with wave chambers | 1.8 to 2.0 |
| Excavation enclosures | 1.5 to 2.0 |

Since both wave and wind measurements rarely encompass the planned duration of use or the return periods associated with extreme sea state situations, the available wave data should be extrapolated to a longer period (frequently 50 or 100 years) by using a suitable theoretical distribution (e.g. Weibull). The extrapolation should not exceed three times the measurement period. The theoretical return period, and hence the parameters of the design wave height $H_d$, must be specified taking into account the potential damage or the permissible risk of flooding or destruction of the structure type (type of failure), also the database and other aspects (structural planning).

As far as functional planning is concerned, considerably shorter return periods sometimes have to be used in order to be able to estimate the anticipated restrictions on use and hazard situations as averages.

In the case of high safety requirements, the ratio of the design wave height $H_d$ to the significant wave height $H_{1/3}$ should be taken as 2.0 (see Table R 136-1). However, in order to draw up safe and economic solutions, a more precise analysis of the actual loads and the stability properties of the structure through hydraulic modelling is advisable.

The design wave height is the maximum wave height to be used when designing a structure or structural component. The action effect resulting from the design wave height must be multiplied by the partial safety factors for the critical design situation, which results in the design internal forces.

If the frequency distribution used is based on long observation periods or corresponding extrapolation (approx. 50 years) or corresponding theoretical or numerical studies, the ensuing design wave may be classified as a rare wave as described in section 5.4.3 for design situation DS-A. With shorter observation/study periods, the ensuing design wave should be defined as a frequent wave and classified in design situation DS-T according to section 5.4.2.

## 5.6.5 Conversion of the sea state

Only in exceptional cases are the wave conditions in the immediate vicinity of the proposed structures actually known. As a rule, therefore, the deep water sea state has to be converted for the project area on the coast. Where waves enter shallow water or encounter obstacles, various effects come into play:

1. Shoaling
   When a wave touches bottom, its velocity, and hence wavelength, decreases. Following a local, insignificant reduction, the wave height therefore gradually increases as the wave enters shallow water – for reasons of energy equilibrium (up to breaking point). This effect is called shoaling [226].
2. Bottom friction and percolation
   Frictional losses and exchange processes at the bottom reduce the wave height. This effect is normally negligible for design purposes [220].
3. Refraction
   Refraction occurs where the bottom depth varies and the waves approach the coast obliquely (or, more accurately, oblique to the contour lines of the bottom). The waves tend to turn parallel to the shoreline as a result of local variations in the shoaling effect so that – depending on the shape of the coastline – the effective wave energy is reduced or even increased (e.g. by focusing wave energy on a promontory).
4. Breaking waves
   Generally, waves can break if either the limit steepness is exceeded (parameter $H/L$) or the wave height reaches a certain dimension with respect to the water depth (parameter $H/d$). Once the associated limit water depth is exceeded, the height of deep-water waves entering shallow water is limited by the breaking process. The ratio of breaker height $H_b$ to limit water depth $d_b$ is normally $0.8 < H_b/d_b < 1.0$ (breaker criterion), although higher values have also been observed in special cases [199]. Owing to the different wave heights in a sea state spectrum, the breaking of the waves generally takes place over what is known as a surf zone, the location and extent of which is determined by the underwater topography, the tides and other factors. To be precise, the ratio of breaker height $H_b$ to water depth $d_b$ is a function of the beach slope $\alpha$ and the steepness of the deep-water wave $H_0/L_0$. These parameters are combined in the breaker index $\xi$, which specifies approximately the breaker type of regular waves (i.e. surging, plunging or spilling breakers). Further details can be found in [11], [199], Galvin [70], and [60].
   The breaker index $\xi$ can be related to both the deep-water wave height $H_0$ ($\xi_0$) and the wave height at the breaking point $H_b$ ($\xi_b$) (see Table R 136-2):

$$\xi = \frac{\tan \alpha}{\sqrt{H/L_0}}$$

**Table R 136-2.** Specification of breaker types (the values are based on studies with slope gradients from 1:5 to 1:20)

| Breaker type | $\xi_0$ | $\xi_b$ |
|---|---|---|
| Surging breaker | > 3.3 | > 2.0 |
| Plunging breaker | 0.5 to 3.3 | 0.4 to 2.0 |
| Spilling breaker | < 0.5 | < 0.4 |

where

$\alpha$    angle of inclination of bottom [°]
$H/L_0$    wave steepness
$H$    local wave height
$L_0$    wavelength in deep water

Highly reflective structures and influences due to foreshore geometry can have a considerable effect on the breaking process. Corresponding breaker criteria are then required (see section 5.7.3, for example).

5. Diffraction

   Diffraction occurs when waves encounter obstacles (structures, but also features such as islands lying off the coast). Following diversion around the obstacle, the waves run into the lee of the structure, which results in energy being transported along the crest of the wave and hence normally a reduction in the wave height. At certain points beyond the wave shadow, the superposition of diffraction waves from closely spaced obstacles etc. can result in higher waves [202].

6. Reflections from the structure

   Waves approaching the shore or structures are reflected to a certain extent, the degree of which depends essentially on the properties of the reflecting edges (angle, roughness, porosity, etc.) and the water depth in front of the structure. Non-breaking waves are almost fully reflected where they strike a vertical structure perpendicularly such that, theoretically, a standing wave twice the height of the incoming wave is formed. Moreover, the reflection coefficient of sloping structures depends largely on the steepness of the wave and thus varies for the waves contained in the wave spectrum.

   Since the aforementioned influences depend on many factors, including factors specific to the structure and/or location, a general stipulation is not possible. For more details, see [104], [60], and [150].

## 5.7 Wave pressure on vertical quay walls in coastal areas (R 135)

### 5.7.1 General

The wave pressure on the front of a waterfront structure is to be taken into account for:

- blockwork walls in the uplift pressure beneath the base and pressure in the joints,
- structures above embankments with non-backfilled front wall, taking into account the effective excess water pressure on both sides of the wall,
- non-backfilled sheet pile walls,
- flood defence walls,
- loads during construction, and
- backfilled structures in general, also because of the lowered outer water level in the wave trough.

In addition, quay walls are loaded by line pull forces, vessel impacts and fender pressures resulting from ship movements caused by waves.

When taking into account the wave pressure on vertical quay walls, a distinction must be made between three types of loading:

1. Wall loaded by non-breaking waves
2. Wall loaded by waves breaking on the structure
3. Wall loaded by waves breaking before reaching the structure

Which of these three types of loading applies depends on the water depth, the sea state and the morphological and topographical conditions in the area of the structure.

Load assumptions are explained in the sections below for the various loading types. In addition, reference is made to the fact that the loads resulting from the standing, breaking or already broken waves of a natural sea state can be determined in accordance with Goda [75] and EAK [60]. The dynamic pressure increase coefficient determined empirically according to Takahashi [205] can be used to calculate dynamic pressure loads. The disadvantage of this method is that only landward-facing load components are covered.

### 5.7.2 Loads due to non-breaking waves

A structure with a vertical or approximately vertical front wall in water of such a depth that the highest incoming waves do not break is loaded on the water side by the excess water pressure (which is higher due to reflection) at the crest, or on the land side the higher excess water pressure at the wave trough.

Standing waves are formed when incoming waves are superimposed on the backwash. In reality, true standing waves never occur: the irregularity of the waves creates certain wave impact loads, but these are generally negligible compared with the following load assumptions, and so are considered to be practically static. The wave height doubles as a result of reflection when the waves strike a vertical or approximately vertical wall and if no losses are incurred (coefficient of reflection $\kappa_R = 1.0$). A reduction in the wave height due to partial reflection ($\kappa_R < 0.9$) at vertical walls should only be assumed when verified by large-scale model tests. Otherwise, use the coefficients of reflection listed in EAK [60].

The method according to Sainflou [188] as per Fig. R 135-1 is recommended for calculations when waves impact at 90° to the structure. However, this method supplies loads that are marginally too large if the waves are steep, whereas the loads from very long-period, shallow waves are underestimated. Further details and other design procedures, e.g. according to Miche-Lundgren, can be found in [36] and [60].

Assigning this method to the case of an oblique wave approach is dealt with in Hager [81]. Accordingly, the assumptions for right-angled wave approaches should also be used for acute-angled wave approaches, especially in the case of long structures.

### 5.7.3 Loads due to waves breaking on structure

Waves breaking on a structure can exert extreme impact pressures of 10 000 kN/m² and more. These pressure peaks, however, are very localised and of very brief duration (1/100 to 1/1000 s).

Owing to the huge pressure impulses and dynamic loads that occur, the structure should be suitably arranged and designed to ensure that, as far as possible, high waves do not break directly on the structure. If this is not possible, model studies at the largest possible scale are recommended for the final design. For further information regarding design for pressure impacts, please refer to EAK [60], section 4.3.23, and [36].

The following method of calculation may be used for simple geometries. Tests on a large-scale hydraulic model of a caisson structure on a rip-rap foundation resulted in the following approximation for the impact pressure load on vertical walls [104], [124].

According to Fig. R 135-2, the maximum static horizontal force $F_{max}$ on a quay wall is

$$F_{max} = \varphi \cdot 8.0 \cdot \rho \cdot g \cdot H_b^2 [\text{kN/m}]$$

The point of application of this force lies just below still water level. An approximation for reducing the load as a result of overwash is explained in [104].

**Fig. R 135-1.** Dynamic pressure distribution on a vertical wall for total wave reflection after Sainflou [188] and excess water pressures at the wave crest and trough

Notation for Fig. R 135-1:

$H$   height of incoming wave
$L$   length of incoming wave
$h$   height difference between still water level and mean water level in the reflection area in front of wall:

$$h = \frac{\pi \cdot H^2}{L} \cdot \coth \frac{2 \cdot \pi \cdot d}{L}$$

$\Delta h$ difference between still water level in front of wall and groundwater or inner water level
$d_s$ depth of groundwater or inner water level
$\gamma$ unit weight of water
$p_1$ pressure increase (wave crest) or decrease (wave trough) at base of structure due to wave effect:

$$p_1 = \gamma \cdot H / \cosh \frac{2 \cdot \pi \cdot d}{L}$$

$p_0$ maximum excess water pressure ordinate at land-side water level according to Fig. R 135-1c:

$$P_0 = (p_1 + \gamma \cdot d) \cdot \frac{H + h - \Delta h}{H + h + d}$$

$p_x$ excess water pressure ordinate at level of wave trough according to Fig. R 135-1d:

$$p_x = \gamma \cdot (H - h + \Delta h)$$

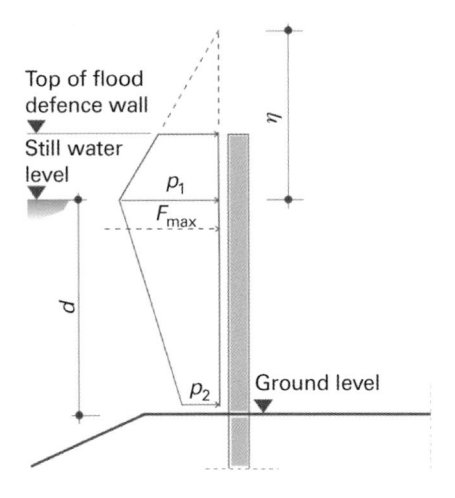

**Fig. R 135-2.** Loads due to plunging breakers [104], [124]

- **Breaking wave height $H_b$:**
  The wave steepness-related breaker criterion developed for relatively steep embankments [104], [151] results in the following equation:

  $$H_b = L_b \cdot [0.1025 + 0.0217(1 - \chi_R)/(1 - \chi_R)]\tan h(2\pi d_b/L_b)]$$

  where

  $\chi_R$    coefficient of reflection of quay wall
  $d_b$    water depth at breaking point
  $L_b$    wavelength of breaking wave, $L_b = L_0 \tan h(2\pi d_b/L_b)$

  With a coefficient of reflection of 0.9 and assuming that the water depth $d_b$ and the wavelength $L_b$ are approximately equal to the corresponding values on the foreshore (water depth $d$ at wall and wavelength $L_d$), the result is

  $$H_b \approx 0.1 \cdot L_0 \cdot [\tan h(2\pi d/L_d)]^2$$

  where

  $L_0$    wavelength in deep water, $L_0 = 1.56 \cdot T_p^2$
  $T_p$    peak period in wave spectrum
  $L_d$    wavelength in water depth, $d \approx L_0 \cdot [\tan h(2\pi d/L_0)^{3/4}]^{2/3}$

- **Impact factor φ:**
  The impact coefficients given below were derived from calculations for the dynamic interaction of impulse-like, wave pressure impact loads varying over time with the stress and deformation conditions of the structure and the subsoil [104].
  The impact coefficient φ depends on the level of the section being checked and is $\varphi = M_{dyn}/M_{stat}$ (wall moment for impact-type load/wall moment for quasi-static load) [104]; [124]; [91].
  Walls with a yielding support in the subsoil, e.g. non-anchored vertical cantilever walls (see Fig. R 135-2, for example) or walls supported deeper than 1.50 m below ground level (Hamburg Guideline 2007 [84]):
  $\varphi = 1.2$ for all analyses above 1.50 m below ground level
  $\varphi = 0.8$ for all analyses below 1.50 m below ground level
  Walls with rigid support (e.g. concrete walls on quay structures) or walls supported at a level higher than 1.50 m below ground level:
  $\varphi = 1.4$ for all analyses above 1.50 m below ground level
  $\varphi = 1.0$ for all analyses below 1.50 m below ground level
- **Pressure ordinate $p_1$** at still water level:

$$p_1 = F_{max}/[0.625 \cdot d_b + 0.65 \cdot H_b]$$

  where η is the level of the pressure figure (difference in height between top of wave pressure load and still water level), $\eta = 1.3 \cdot H_b$
- **Pressure ordinate $p_2$** at ground level:

$$p_2 = 0.25 \cdot p_1$$

## 5.7.4 Loads due to broken waves

SPM [202] provides an approximate calculation of the loads from waves that have already broken. In this calculation it is assumed that the broken wave continues with the same height and velocity after breaking, although this overestimates the actual loads. For a more accurate determination of the actual loads, EAK [60] therefore proposes a correction to the characteristic values based on Camfield's method [35], which is not shown here.

## 5.7.5 Additional loads caused by waves

If a structure supported on a permeable foundation does not have a watertight face, e.g. in the form of an impervious diaphragm, the effect of the waves on the uplift pressure beneath the base must also be taken into account along with the water pressure on the wall surfaces. This also applies to the water pressure in joints between blocks.

## 5.8 Loads arising from surging and receding waves due to the inflow or outflow of water (R 185)

### 5.8.1 General

Surging and receding waves occur in bodies of water as a result of a temporary or temporarily increased inflow or outflow of water. However, surging and receding waves essentially only manifest themselves with wetted watercourse cross-sections that are small in comparison with the inflow/outflow volume per second. Therefore, surging and receding waves and their effects on waterfront structures are generally only significant for navigation channels. In these cases the effects of changes in water levels on embankments, the linings to watercourses, revetments and other facilities should be taken into account.

### 5.8.2 Determining wave values

Surging and receding waves are shallow-water waves in the range

$$\frac{d}{L} < 0.05$$

The wavelength depends on the duration of the water inflow or outflow. The wave propagation velocity can be roughly calculated with

$$c = \sqrt{g \cdot (d \pm 1.5H)}\,[\text{m/s}] \begin{cases} +\text{for surging} \\ -\text{for receding} \end{cases}$$

where

$g$ acceleration due to gravity
$d$ depth of water
$H$ rise for surging or fall for receding compared with still water level

If the $H/d$ ratio is small, then

$$c = \sqrt{g \cdot d}$$

can be used.
The rise or fall in water level is roughly

$$H = \pm \frac{Q}{c \cdot B}$$

where

$Q$ volume of water inflow or outflow per second
$B$ mean width at water level

The wave height can increase or decrease as a result of reflections or subsequent surging or receding waves. With uniform canal cross-sections and smooth canal linings in particular, the wave attenuation is small and so the waves can move back and forth several times, especially with short reaches.

In navigation channels the most frequent cause of surging and receding phenomena is the inflow or outflow of lockage water. In order to prevent extreme surging and receding phenomena, the lockage volume is generally restricted to 70–90 m$^3$/s.

Lockages at intervals of the reflection time or a multiple thereof, particularly in canal stretches, can result in a superposition of the waves and hence to an increase in the degree of surging and receding.

### 5.8.3 Load assumptions

Loading assumptions for waterfront structures must take account of the hydrostatic load due to the height of the surging or receding wave and its possible superposition by reflected or subsequent waves, also any potential simultaneous fluctuations in the water level, e.g. from the raising of the water level by wind, ship waves, etc., in the least favourable configuration each time. Owing to the long-period nature of surging and receding waves, the ensuing effect on the flow gradient of the groundwater must also be checked in the case of permeable revetments.

The dynamic effects of surging and receding waves can be ignored owing to the mostly low flow rates caused by such waves.

The loads calculated in this way are characteristic values that must be multiplied by the partial safety factors for design situation DS-P (see R 18, section 5.4.2) in accordance with DIN 1054.

## 5.9 Effects of waves due to ship movements (R 186)

### 5.9.1 General

Waves of different types are generated by the bow and stern of every moving vessel. Depending on the local circumstances, these cause different loads on waterfronts and/or protective features (Figs. R 186-1 and R 186-2). Initially, an accumulation of water is established ahead of the bow, the size of which may be several ship lengths viewed in the direction of travel. A further local accumulation occurs directly in front of the bow (bow wave). Alongside the ship, the water level dips as a result of the backwash (squat effect), and this tracks the vessel. The ensuing depression extends the length of the ship and extends across the full width of canals and narrow channels. On open water the depression reduces with the distance from the ship; in a first approximation its width corresponds to 1.5–2.0 times the length of the ship [17].

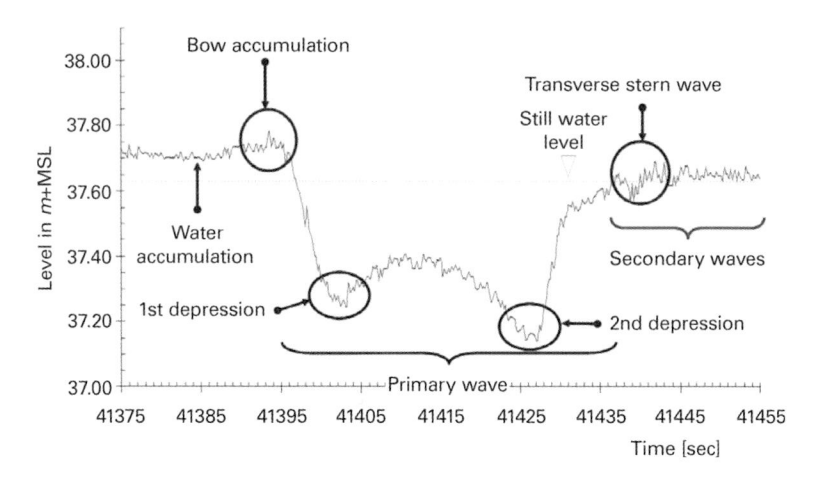

**Fig. R 186-1.** Change in water level for a ship travelling along a confined watercourse

Secondary waves are superposed on this primary wave system and spread out from the ship in a fixed pattern (Fig. R 186-3). Of special significance with regard to the loads imposed on waterfronts are oblique waves, the crests of which are at an angle of 55° to the ship's axis

**Fig. R 186-2.** Large motorised vessel travelling at close to its critical speed along a canal; the water accumulation at the bow, the depression, the breaking stern wave and the secondary wave system of transverse stern waves are all clearly visible.

127

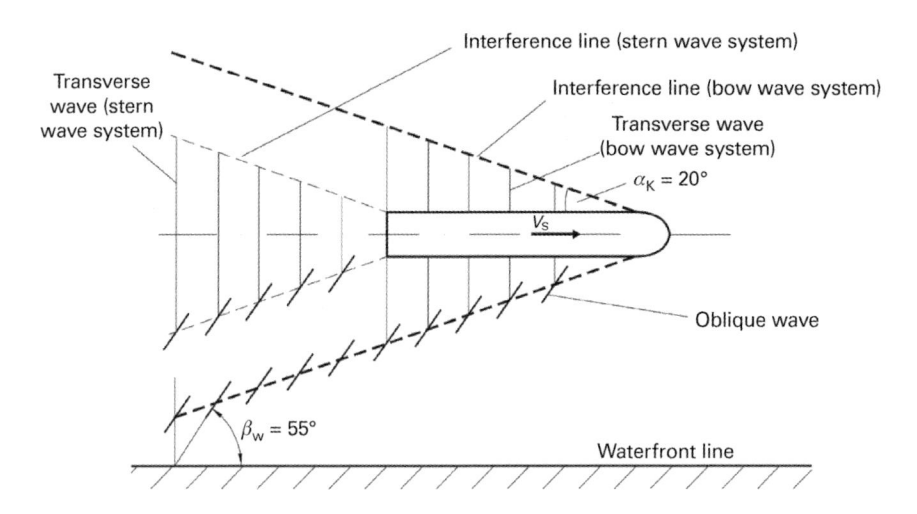

**Fig. R 186-3.** Wave pattern, schematic

### 5.9.2 Wave heights

Particular allowance must be made for the effects of the ship wave system on embankments and waterfront revetments in confined shipping channels. The critical design loads result from the increase and decrease in pressure in the area of the depression in the primary wave system and the breaking of the waves from the primary (stern wave) and secondary (oblique waves and transverse stern waves) wave systems at the transition to the shallow water area on the banks – and depending on the direction of the waves. The stone size required for revetments is usually determined by the load exerted by breaking transverse stern waves or the superposition of these waves on the secondary bow wave. Owing to their hydrostatic pressure changes, the bow wave and depression alongside the ship affect the pore water pressures underground and result – via temporary excess pressure in the subsoil – in a destabilisation of the waterfront protection. They thus generally determine the required thickness of a revetment [17].

Where reflections are possible, e.g. in short branches with a perpendicular termination (lock basins) or along upright waterfront protection, the size of the accumulation or depression can double. More precise values can be determined in model tests.

The chronological sequence of the accumulation or depression might need to be taken into account for permeable waterfront structures, together with its effect on groundwater movement. Attention is drawn to the possible effects on automatically operating closures, e.g. dyke sluices (opening and closing of gates as a result of sudden pressure changes), and on lock gates.

The accumulation in front of a ship can be regarded as a "solitary wave". The height of the accumulation is generally small and rarely exceeds 0.2 m above still water level.

The height of the waves on the interference line of oblique bow and stern waves may be estimated in accordance with BAW [17] as follows:

$$H_{sec} = A_W \frac{v_S^{8/3}}{g^{4/3}(u')^{1/3}} f_{cr}$$

where

$A_W$    wave height factor [–], depending on shape and dimensions of ship, loaded draught and depth of water
$A_W = 0.25$ for conventional inland waterway vessels and tugs
$A_W = 0.35$ for empty pushed lighter convoys, 1 vessel wide
$A_W = 0.80$ for fully laden pushed lighter convoys, > 1 vessel wide

$f_{cr}$    speed factor [–] ($f_{cr} = 1$ applies to $v_S/v_{crit} < 0.8$)
$g$    acceleration due to gravity [m/s²]
$H_{sec}$    secondary wave height [m]
$u'$    distance from ship's side to waterfront line [m]
$v_S$    speed of ship through water [m/s]

The water level depression corresponds to the backwash below and alongside the submerged body of the vessel, and its form and size depend on the vessel's shape, means of propulsion, travelling speed and the conditions of the waterway (ratio $n$ of watercourse cross-section to submerged main frame cross-section of vessel, proximity and shape of waterfront). The maximum depression seldom exceeds approx. 15 % of the depth of the water, even when the vessel attains its critical speed. Figs. R 186-4 and R 186-5 enable the designer to make a safe estimate of the depression $\Delta h$ as a function of $n$ and the ship's critical speed. Depression and wave height change with the distance from the ship, chiefly in the vicinity of the waterfront, see [17], for instance. The maximum values determined can be used as design values.

## 5.10    Wave pressure on piled structures (R 159)

### 5.10.1    General

When designing piled structures, the loads originating from wave motion are to be taken into account with respect to the loads on both a single pile and the entire structure, in so far as this is necessitated by

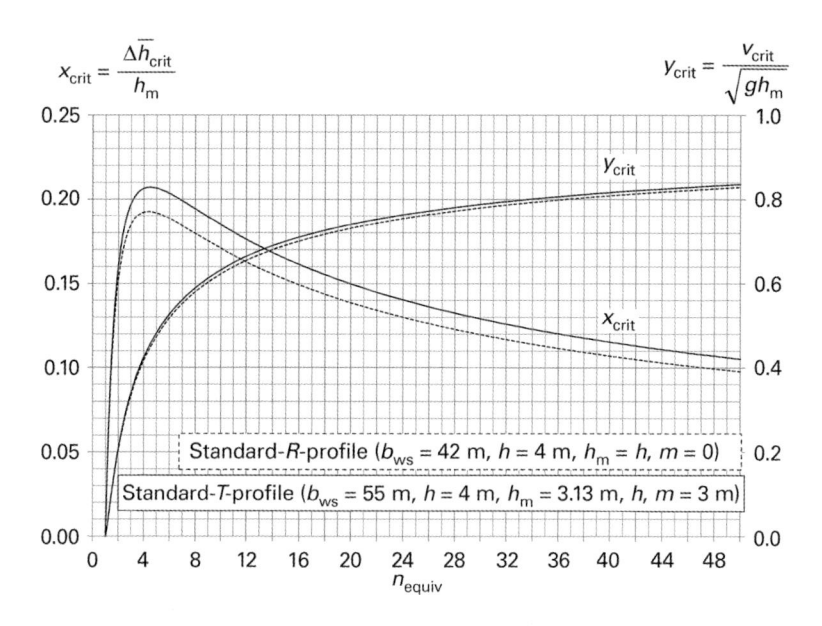

**Fig. R 186-4.** Critical vessel speed $v_{crit}$ in relation to mean depth of water $h_m$, acceleration due to gravity $g$ and cross-section ratio $n$ for watercourses with rectangular (R) and trapezoidal (T) profiles ($n = n_{equiv}$ for typical ship and watercourse dimensions) [13]

Notation for Fig. R 186-4:

| | |
|---|---|
| $b_{ws}$ | width at water level [m] |
| $h$ | maximum depth of water [m] in profile |
| $h_m$ | mean water depth [m] |
| $\Delta\bar{h}_{crit}$ | mean water level depression at $v_{crit}$ |
| $n$ | cross-section ratio [–] |
| $v_{crit}$ | critical speed of vessel [m/s]; maximum feasible hydraulic velocity of vessel travelling in displacement mode; the critical speed of the vessel in shallow water or a navigable channel where the water displaced by the vessel can no longer be fully displaced rearwards in the flowing state opposite to the direction of travel, the transition from subcritical to supercritical flow (Froude number in narrowest cross-section alongside ship = 1). Generally speaking, $v_{crit}$ cannot be exceeded by displacement vessels. |
| $x_{crit}, y_{crit}$ | non-dimensional variables [–] |

local circumstances. Superstructures should be located above the crest of the design wave if possible. Otherwise, large horizontal and vertical loads from direct wave actions can affect superstructures; determining such actions does not fall within the remit of this recommendation since reliable values for such cases can only be obtained from model studies. The level of the crest of the design wave should be determined taking

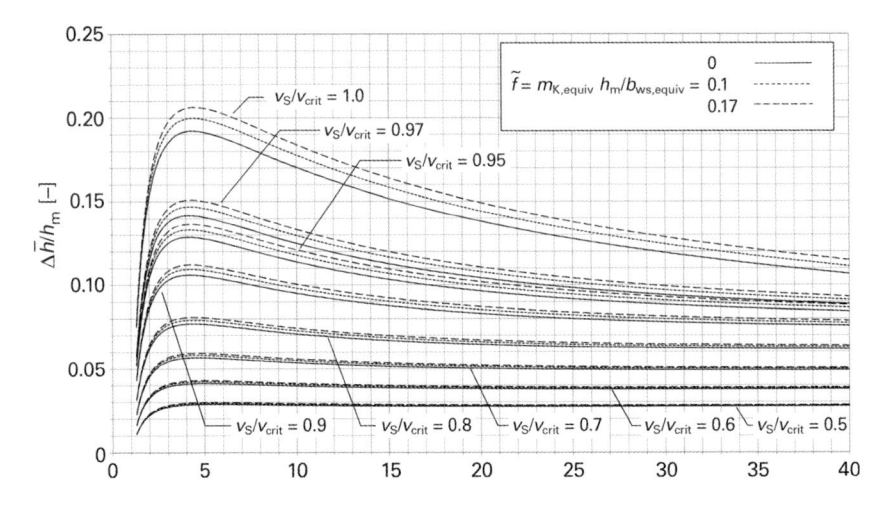

**Fig. R 186-5.** Mean drop in water level $\Delta h$ in relation to mean depth of water $h_m$ and relative – i.e. $v_{crit}$-related – vessel speed $v_s$ for different cross-section ratios $n$ ($n = n_{equiv}$ for standard ship and channel dimensions) [13]

into account the simultaneous occurrence of the highest still water level and, where applicable, also allowing for the raising of the water level by wind, the influence of the tides and the rising and steepening of the waves in shallow water.

The superposition method according to [144] is suitable for slender structural components, whereas the calculation method for wider structures is based on the diffraction theory of [135].

The subject of this recommendation is the superposition method of [144], which applies to non-breaking waves. Owing to a lack of accurate calculation methods for breaking waves, a makeshift method is proposed in section 5.10.5.

The Morison method provides useful values provided the following is satisfied for a single pile:

$$\frac{D}{L} \leq 0.05$$

where

- $D$    pile diameter or, for non-circular piles, characteristic width of structural component (width transverse to direction of wave propagation)
- $L$    length of "design wave" in accordance with R 136, section 5.6, in conjunction with Table R 159-1(3)

**Table R 159-1.** Linear wave theory – physical correlations [226]

| | Shallow water $\frac{d}{L} \leq \frac{1}{20}$ | Transition area $\frac{1}{20} < \frac{d}{L} < \frac{1}{2}$ | Deep water $\frac{d}{L} \geq \frac{1}{2}$ |
|---|---|---|---|
| 1. Profile of free surface | General equation $\eta = \frac{H}{2} \cdot \cos \vartheta$ | | |
| 2. Wave velocity | $c = \frac{L}{T} = \frac{g}{\omega} kd = \sqrt{gd}$ | $c = \frac{L}{T} = \frac{g}{\omega} \tanh(kd) = \sqrt{\frac{g}{k} \tanh(kd)}$ | $c = \frac{L}{T} = \frac{g}{\omega} = \sqrt{\frac{g}{k}}$ |
| 3. Wavelength | $L = c \cdot T = \frac{g}{\omega} kdT = \sqrt{gd} \cdot T$ | $L = c \cdot T = \frac{g}{\omega} \tanh(kd) \cdot T = \sqrt{\frac{g}{k} \tanh(kd)} \cdot T$ | $L = c \cdot T = \frac{g}{\omega} \cdot T = \sqrt{\frac{g}{k}} \cdot T$ |
| 4. Velocity of water particles | | | |
| a) horizontal | $u = \frac{H}{2} \cdot \sqrt{\frac{g}{d}} \cdot \cos \vartheta$ | $u = \frac{H}{2} \cdot \omega \cdot \frac{\cosh[k(z+d)]}{\sinh(kd)} \cdot \cos \vartheta$ | $u = \frac{H}{2} \cdot \omega \cdot e^{kz} \cdot \cos \vartheta$ |
| b) vertical | $w = \frac{H}{2} \cdot \omega \cdot \left(1 + \frac{z}{d}\right) \sin \vartheta$ | $w = \frac{H}{2} \cdot \omega \cdot \frac{\sinh[k(z+d)]}{\sinh(kd)} \cdot \sin \vartheta$ | $w = \frac{H}{2} \cdot \omega \cdot e^{kz} \cdot \sin \vartheta$ |
| 5. Acceleration of water particles | | | |
| a) horizontal | $\frac{\partial u}{\partial t} = \frac{H}{2} \cdot \omega \cdot \sqrt{\frac{g}{d}} \cdot \sin \vartheta$ | $\frac{\partial u}{\partial t} = \frac{H}{2} \cdot \omega^2 \cdot \frac{\cosh[k(z+d)]}{\sinh(kd)} \cdot \sin \vartheta$ | $\frac{\partial u}{\partial t} = \frac{H}{2} \cdot \omega^2 \cdot e^{kz} \cdot \sin \vartheta$ |
| b) vertical | $\frac{\partial w}{\partial t} = -\frac{H}{2} \cdot \omega^2 \cdot \left(1 + \frac{z}{d}\right) \cos \vartheta$ | $\frac{\partial w}{\partial t} = \frac{H}{2} \cdot \omega^2 \cdot \frac{\sinh[k(z+d)]}{\sinh(kd)} \cdot \cos \vartheta$ | $\frac{\partial w}{\partial t} = \frac{H}{2} \cdot \omega^2 \cdot e^{kz} \cdot \cos \vartheta$ |

Definition of symbols used in Table R 159-1:

$$\vartheta = \frac{2\pi \cdot x}{L} - \frac{2\pi \cdot t}{T} = kx - \omega t \text{(phase angle)}$$

$$k = \frac{2\pi}{L}; \omega = \frac{2\pi}{T}, c = \frac{\omega}{k}$$

where
- $t$  duration
- $T$  wave period
- $c$  wave velocity
- $k$  wave number
- $\omega$  wave angular frequency

Otherwise, refer to Fig. R 159-1.

This criterion is generally satisfied.

For the determination of the wave loads, the reader is referred to Hafner [80] and SPM [202], which contain helpful tables and diagrams. The diagrams in [202] are based on stream function theory and may be applied to waves of varying steepness up to the breaking limit, whereas the diagrams in [80] are only applicable under the prerequisites of linear wave theory.

To determine the upward wave load, refer to R 217, section 5.10.9.

Other methods of calculation apply to offshore structures, e.g. according to the API (American Petroleum Institute).

### 5.10.2 Method of calculation according to Morison et al. [144]

The wave load on a single pile is made up of the components of flow force and acceleration force (inertial force), which must be determined separately and superposed to suit the phases.

According to [77], [202], and [201], the total horizontal load per unit length on a vertical pile is

$$p = p_D + p_M = C_D \cdot \frac{1}{2} \cdot \frac{\gamma_W}{g} \cdot D \cdot u \cdot |u| + C_M \cdot \frac{\gamma_W}{g} \cdot F \cdot \frac{\partial u}{\partial t}.$$

For a pile with a circular cross-section:

$$p = C_D \cdot \frac{1}{2} \cdot \frac{\gamma_W}{g} \cdot D \cdot u \cdot |u| + C_M \cdot \frac{\gamma_W}{g} \cdot \frac{D^2 \cdot \pi}{4} \cdot \frac{\partial u}{\partial t}.$$

where

| | |
|---|---|
| $p_D$ | flow force caused by flow resistance per unit length of pile |
| $p_M$ | inertial force due to unsteady wave motion per unit length of pile |
| $p$ | total load per unit length of pile |
| $C_D$ | drag coefficient of flow pressure |
| $C_M$ | drag coefficient of flow acceleration |
| $g$ | acceleration due to gravity |
| $\gamma_W$ | unit weight of water |
| $u$ | horizontal component of velocity of water particles at pile location under consideration |
| $\frac{\partial u}{\partial t} \approx \frac{du}{dt}$ | horizontal component of acceleration of water particles at pile location under consideration |
| $D$ | pile diameter or, for non-circular piles, characteristic width of structural component |
| $F$ | cross-sectional area of the pile within the flow in the direction of flow in area under consideration |

The velocity and acceleration of the water particles are taken from the wave equations, which may be based on various wave theories. For linear wave theory, the correlations required have been compiled in Table R 159-1. [202], [122], and [60] should be consulted regarding the application of theories of a higher order.

### 5.10.3 Determining the wave loads on a single vertical pile

Since the velocities and, accordingly, accelerations of the water particles are a function of, among other factors, the distance of the location considered from the still water level, the wave load diagram of Fig. R 159-1 results from the calculation of the wave pressure load for various values of $z$.

The phase displacement of the components of the wave load max. $p_D$ and max. $p_M$ should not be ignored. The calculation must therefore be performed for different phase angles and the maximum load determined by a phase-adjusted superposition of the components due to flow velocity and flow acceleration. For example, when applying linear

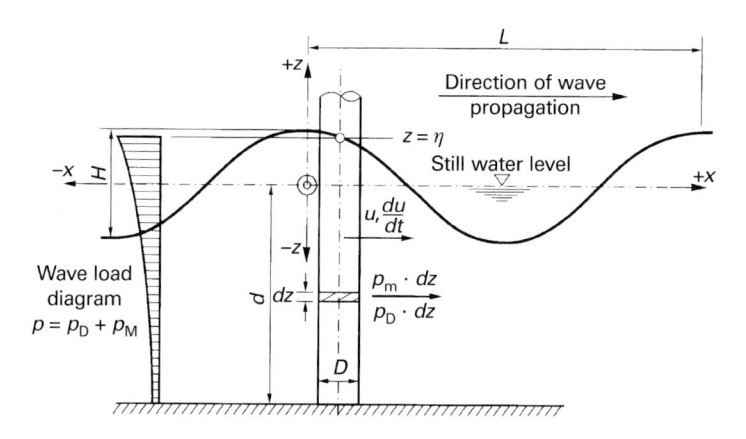

**Fig. R 159-1.** Wave action on a vertical pile
The coordinates zero point lies at still water level, but can be chosen to be at any point on the axis. Definition of symbols used in Fig. R 159-1:

- $z$    ordinate of point investigated ($z = 0 =$ still water level)
- $x$    x coordinate of point investigated
- $\eta$    level of water table varying over time, related to the still water level (water surface displacement)
- $d$    water depth below still water level
- $D$    pile diameter
- $H$    wave height
- $L$    wavelength

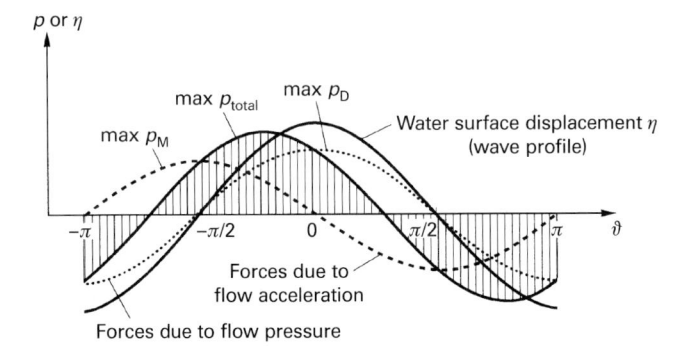

**Fig. R 159-2.** Variation in the forces due to flow pressure and flow acceleration over one wave period

wave theory, the phase of the acceleration force is displaced by 90° ($\pi/2$) with respect to the flow force, which lies in the same phase as the wave profile (Fig. R 159-2).

## 5.10.4 Coefficients $C_D$ and $C_M$

### 5.10.4.1 Drag coefficient for flow pressure $C_D$

The drag coefficient for the flow pressure $C_D$ is determined from measurements. It depends on the shape of the body within the flow, the Reynolds number Re, the surface roughness of the pile and the initial degree of turbulence of the current ([80], [204], [32]). The location of the separation point of the boundary layer is critical for the flow force. In the case of piles where the separation point is defined by corners or flow separation edges, the $C_D$ value is practically constant (Fig. R 159-3). On the other hand, in the case of piles without a stable separation point, e.g. circular cylindrical piles, a distinction is made between a subcritical range of the Reynolds number with a laminar boundary layer and a supercritical range with a turbulent boundary layer.

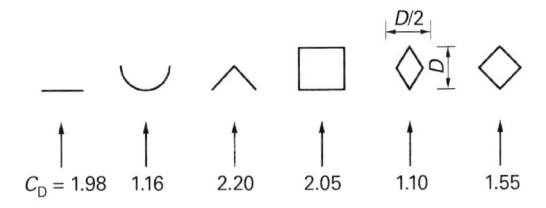

$C_D$ = 1.98   1.16   2.20   2.05   1.10   1.55

**Fig. R 159-3.** $C_D$ values of pile cross-sections with stable separation points [80]

135

Since, however, generally speaking, high Reynolds numbers prevail in nature, in the case of smooth surfaces the recommendation is to assume a uniform value $C_D = 0.7$ ([202], [80]). Further information can be found in Sparboom [201].

Larger $C_D$ values should be expected with rough surfaces; see [48], for example.

#### 5.10.4.2 Drag coefficient $C_M$ for flow acceleration

Using potential flow theory results in a value $C_M = 2.0$ for a circular cylindrical pile, although $C_M$ values of up to 2.5 have been ascertained from tests involving circular cross-sections [49].

Normally, the theoretical value $C_M = 2.0$ can be used. Otherwise, the designer should consult [202], [48], and [201].

### 5.10.5 Forces from breaking waves

At present there is no usable calculation method available for determining the forces from breaking waves correctly. The Morison formula is therefore used again for this range of waves, but under the assumption that the wave acts on the pile as a water mass with high velocity but no acceleration. In this regard, the inertia coefficient $C_M$ is set to 0, while the flow pressure coefficient $C_D$ is increased to 1.75 [202].

### 5.10.6 Wave load on a group of piles

When determining the wave load on a group of piles, the phase angle $\vartheta$ critical for the respective pile location must be taken into account.

Using the designations according to Fig. R 159-4, the total horizontal load on a piled structure consisting of $N$ piles is

$$P_{tot} = \sum_{n=1}^{N} P_n(\vartheta_n).$$

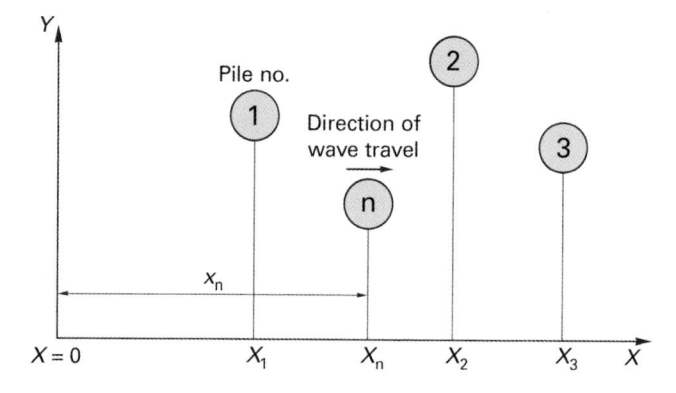

**Fig. R 159-4.** Details for a pile group (plan view) according to [202]

**Table R 159-2.** Correction factors for closely spaced piles [49]

| pile centre-to-centre distance $e$ pile diameter $D$ | 2 | 3 | 4 |
|---|---|---|---|
| for piles in rows parallel with crest of wave | 1.5 | 1.25 | 1.0 |
| for piles in rows perpendicular to crest of wave | 0.7*⁾ | 0.8*⁾ | 1.0 |

*⁾ The reduction does not apply to the front pile directly exposed to the wave action.

where

$N$     number of piles
$P_n(\vartheta_n)$   wave load on a single pile $n$, taking into account the phase angle $\vartheta = k \cdot x_n - \omega \cdot t$
$x_n$     distance of pile $n$ from $y$-$z$ plane

It should be noted that for piles situated closer together than about four pile diameters, there is an increase in load on piles situated side by side in the direction of the wave, but a decrease in load on piles situated one behind the other.

In this situation the correction factors given in Table R 159-2 are proposed for such loads [49].

## 5.10.7 Raking piles

The phase angle $\vartheta$ for the local coordinates $x_0$, $y_0$, $z_0$ differs for the individual pile segments $d_s$ of a raking pile and this must be taken into account.

Consequently, the pressure on the pile at the location under consideration should be determined using the coordinates $x_0$, $y_0$ and $z_0$ in accordance with Fig. R 159-5.

According to SPM [202], the local force $p \cdot d_s$ due to the flow and acceleration of the water particles towards the pile element $d_s$ ($p = f[x_0, y_0, z_0]$) can be equated with the horizontal force on an equivalent vertical pile at position ($x_0$, $y_0$, $z_0$) However, when the pile slopes at a steeper angle, it is necessary to check whether determining the load taking into account the components of the resulting velocity acting vertical to the pile axis

$$v = \sqrt{u^2 + w^2}$$

and the resultant acceleration

$$\frac{\partial v}{\partial t} = \sqrt{\left(\frac{\partial u}{\partial t}\right)^2 + \left(\frac{\partial w}{\partial t}\right)^2}$$

supply less favourable values.

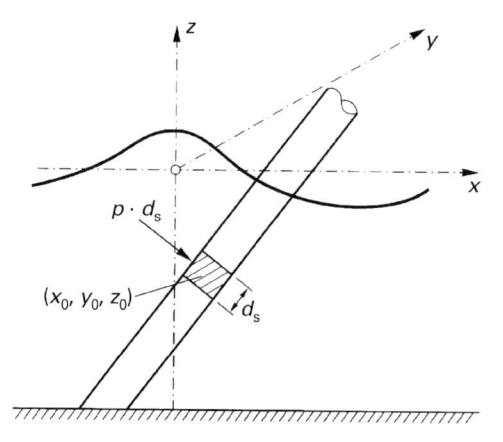

**Fig. R 159-5.** Calculating wave forces on a raking pile [202]

### 5.10.8 Safety factors

Designing piled structures to resist wave action is highly dependent on the choice of "design wave" (R 136, section 5.6, in conjunction with Table R 159-1(3)). The wave theory used, and the values of coefficients $C_D$ and $C_M$, remain influential.

This applies to piled structures in shallow water in particular. In order to allow for such uncertainties, the recommendation is to multiply the calculated loads by increased partial safety factors according to SPM [202]. Consequently, when the "design wave" occurs only rarely, i.e. in the normal case with deep water conditions, the resulting wave load on piles must be increased by a partial safety factor $\gamma_d = 1.5$. When the "design wave" occurs frequently, which is usually the case in shallow water conditions, a partial safety factor $\gamma_d = 2.0$ is recommended.

Readers should refer to Sparboom [201] and EAK [60] regarding the possibility of using coefficients $C_D$ and $C_M$ depending on the Reynolds and Keulegan-Carpenter numbers and a corresponding reduction in the partial safety factor.

Critical vibrations can occur occasionally in piled structures, especially when separation eddies act transverse to the inflow direction, or the eigenfrequency of the structure is close to the wave period, resulting in resonance phenomena. In this case regular waves lower than the "design wave" can have a less favourable effect, which calls for special investigations.

### 5.10.9 Vertical wave load ("wave slamming")

Horizontal structural components located in the vicinity of the water level may experience substantial vertical upward loads if the waves reach the structural component. This process is known as "wave slamming" and induces high impact loads. As these loads are considerable, the general aim is to try to avoid wave slamming completely by raising the structural components or positioning them at an angle.

### 5.10.9.1 Determining the necessary level of the structural components

In order to avoid wave slamming on horizontal slabs, e.g. the decks of jetties or offshore platforms, the level usually chosen for the underside of the deck structure is approx. 1.5 m above the crest of the design wave ("air gap approach"). The crest position is determined for the maximum wave height $H_m$ of the storm on which the design is based (normally $1.6–2.0 \cdot H_s$) while taking account of the design water level. According to Muttray [146], the crest position predicted by Rienecker/Fenton [175] in accordance with Fourier wave theory can be estimated thus:

$$h_{cr} = h_{DWL} + H_{max} \left[ \frac{1}{2} + \frac{1}{3} \frac{\Pi}{3\Pi + 1/2} + \frac{1}{6} \frac{\Pi^2}{\Pi^2 + 1/30} \right] \quad (5.1)$$

where $\Pi = \dfrac{H_{max}}{L} \coth^3\left(\dfrac{2\pi}{L} d\right)$

and

$h_{cr}$ crest [m]
$h_{DWL}$ design water level [m]
$H_{max}$ maximum wave height [m]
$\Pi$ non-linearity parameter [–]
$L$ wavelength [m] (according to linear wave theory)
$d$ water depth [m]

### 5.10.9.2 Estimating the load exerted on structural components due to wave slamming

For functional or economic reasons, e.g. with very high design waves, it can be difficult to arrange the structural components high enough above the crest. In these cases the structural components affected must be designed for vertical wave loads. The calculation methods presented below for the vertical loads due to wave slamming are suitable for preliminary designs. However, model tests at the largest scale possible are recommended for validating these methods or for detailed design work. As with investigations into breaking waves, on account of the reproduction of the proportion of air in the water, the wave height in the model should be at least 0.5 m.

Vertical wave forces acting from below on horizontal surfaces are comparable with horizontal impact loads from breaking waves on vertical walls or piles. As a rule, the action is made up of an impact-type, dynamic load component of very brief duration and a periodic, quasi-static load component. The first contact between wave and structural component leads to high pressure peaks which, however, act only briefly and over a relatively small area. The subsequent periodic wave pressure affects a larger area, the extent of which changes as the wave passes.

### 5.10.9.3 Design approach for horizontal cylindrical structural components

A simple approach for estimating the vertical wave forces was developed for horizontal cylindrical structural components with truss-type offshore platforms. The vertical force $F_s$ is described here as a line load:

$$F_S = \frac{1}{2}\rho_w C_S D w |w| \tag{5.2}$$

where

- $F_s$    vertical force per unit length [kN/m]
- $\rho_w$    unit weight of water [t/m³]
- $C_s$    slamming coefficient [–]
- $D$    diameter of cylindrical structural component [m]
- $w$    vertical velocity of surface of water [m/s]

It is assumed here that the pressure peak is reduced by the three-dimensional movement of the surface of the water in the natural sea state and by air pockets and the entry of air.

As regards circular cylinders, slamming coefficients $C_s$ in the range 0.5–1.7 times the theoretical value $C_s = \pi$ were established in model tests. A slamming coefficient $C_s \geq 3.0$ is recommended for static design. The dynamic (i.e. transient) load peaks can be estimated using a coefficient in the order of magnitude $C_s = 4.5$–6.0. Further information regarding the magnitude of slamming coefficients can be found in BS 6349-1 (2000) and [48]. Surface water movements can be estimated with various (linear or non-linear) wave theories (see also R 159 and Eq. (5.5) below).

### 5.10.9.4 Design approach for horizontal slabs

There is no generally recognised design approach for the vertical wave forces due to wave slamming on horizontal slabs. As regards the preliminary design, the following simple analytical approach according to [206] and [174] can be used for the approximate determination of vertical wave forces on a deck. The dynamic wave pressure (impact

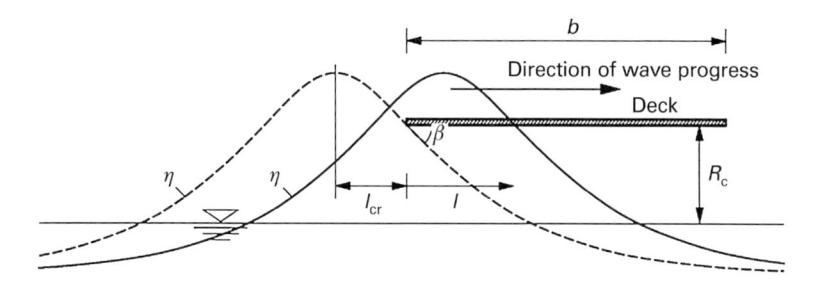

**Fig. R 217-1.** Definitions for slamming

pressure) is then essentially influenced by the contact angle $\beta$ between water level and deck (see Fig. R 217-1 for initial contact between wave and deck). The total wave is the sum of the quasi-static and dynamic pressure components:

$$p = p_0 + p_S \tag{5.3}$$

$$p_0 = C_0 \rho_w g (\eta - R_c) \tag{5.4}$$

$$p_S = \frac{1}{2} \rho_w C_S w^2$$

$$C_S = \min\left\{1 + \left(\frac{\pi}{2} \cot \beta\right)^2 ; 300\right\} \tag{5.5}$$

$$w = c \sin \beta$$

where

- $p$    wave pressure (total) [kN/m²]
- $p_0$    quasi-static pressure component [kN/m²]
- $p_s$    dynamic pressure component [kN/m²]
- $\eta$    water surface displacement [m]
- $R_c$    freeboard of deck (above design water level) [m]
- $\rho_w$    density of water [kg/m³]
- $C_0$    quasi-static pressure coefficient [–] ($C_0 = 1$)
- $w$    velocity component of vertical, upward water particles at wave surface
- $C_s$    slamming coefficient [–] ($C_s \leq 300$)
- $\beta$    contact angle between front of wave and deck [°]
- $c$    velocity of wave [m/s]; $c = L/T$ with wave period $T$ [s] and wavelength $L$ [m] (based on linear wave theory)

The maximum quasi-static pressure load occurs beneath the crest (Eq. (5.4)).

The resulting quasi-static wave force (per unit width) is derived from the integration of the wave pressure $p_0$ (Eq. (5.4)) along the contact area between crest of wave and deck. The length of the contact area $l$ varies during the passage of the wave ($l = ct$) and is limited by the geometry of the wave profile or the deck (with deck length $b$). The water surface displacement $\eta(x, t)$ can be taken from Fig. R 217-2 or Table R 217-1. A coefficient $C_0 = 1$ is recommended for design.

The maximum value of the dynamic load component can be determined using Eq. (5.5). A rapid rise in pressure and an exponential pressure drop are typical. The maximum pressure load migrates along the deck structure; the spatial extent of the peak pressure in this regard is, however, very limited. The dynamic peak pressure $p_s$ therefore only has to be taken into consideration when designing (small-volume) structural components beneath the deck.

The resulting dynamic wave force (per unit width) $F_s$ on the deck can be approximated according to [113] and [27] as follows:

$$F_S = \frac{\pi}{4} \rho_w w \, c \, l'; \quad l' = \min\{l, l_{cr}, b\} \tag{5.6}$$

where

$l$    length of contact area ($l = ct$)
$b$    length of deck (in direction of wave progress)
$l_{cr}$    horizontal distance between crest of wave and original contact point between wave and deck (see Fig. R 217-1)

Once the crest of the wave has passed the front of the deck, a dynamic pressure load no longer occurs. Therefore, $l'$ is always smaller than or equal to $l_{cr}$.

The wave profile and the contact angle (according to Fourier wave theory, [175]) can be determined using the non-dimensional figures in Fig. R 217-2 and Table R 217-1. At the same time, the standardised wave profile varies with the non-linearity parameter $\Pi$; the actual wave profile is obtained by multiplying the values in the table by the actual wavelength $L$. The non-linearity parameter can be calculated according to linear wave theory using the wavelength (Eq. (5.1)).

### 5.10.9.5 Construction details

Vertical wave forces due to wave slamming can be reduced by choosing an open, permeable deck structure (e.g. gratings). Furthermore, damage caused by wave slamming can be limited by a deck structure with loose cover plates, which are simply lifted and displaced by large waves.

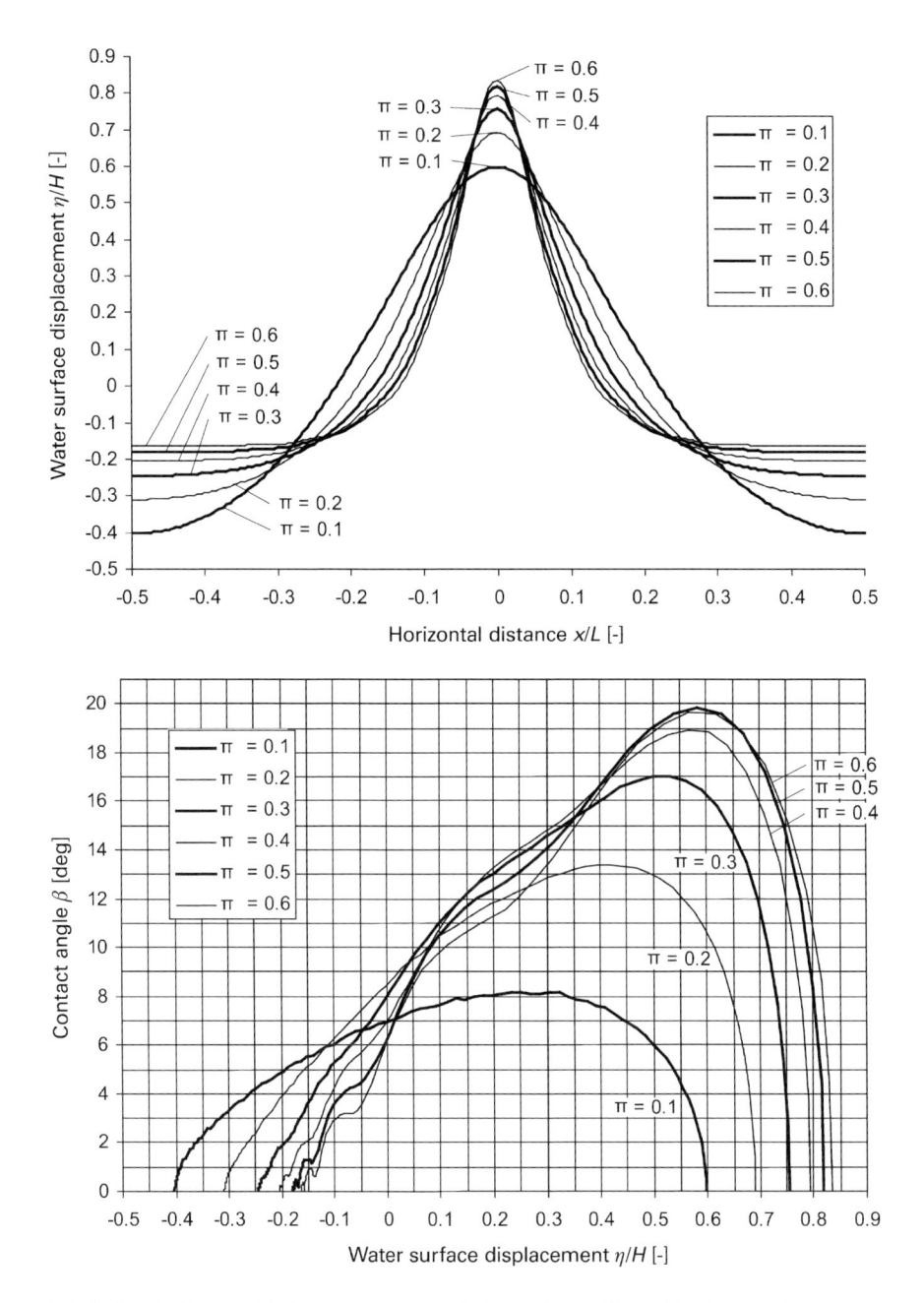

**Fig. R 217-2.** Wave profile (top) and contact angle (bottom) according to Fourier wave theory

**Table R 217-1.** Wave profile (left) and contact angle (right) according to Fourier wave theory

| Distance $x/L$ [-] | Non-linearity parameter $\Pi$ [-] | | | | | | Level $\eta/H$ [-] | Non-linearity parameter $\Pi$ [-] | | | | | |
|---|---|---|---|---|---|---|---|---|---|---|---|---|---|
| | 0.1 | 0.2 | 0.3 | 0.4 | 0.5 | 0.6 | | 0.1 | 0.2 | 0.3 | 0.4 | 0.5 | 0.6 |
| | Relative water surface displacement $\eta/H$ [-] | | | | | | | Contact angle $\beta$ [°] | | | | | |
| -0.5 | -0.40 | -0.31 | -0.25 | -0.21 | -0.18 | -0.16 | 0.8 | | | | | | |
| -0.4 | -0.36 | -0.29 | -0.24 | -0.20 | -0.18 | -0.16 | 0.7 | | | | | | |
| -0.3 | -0.21 | -0.22 | -0.20 | -0.18 | -0.17 | -0.16 | 0.6 | | | | | | |
| -0.2 | 0.06 | -0.03 | -0.08 | -0.10 | -0.11 | -0.12 | 0.5 | | | | | | |
| -0.1 | 0.41 | 0.35 | 0.27 | 0.21 | 0.17 | 0.14 | 0.4 | 0.0 | 0.0 | | 0.0 | 8.4 | 8.7 |
| 0 | 0.60 | 0.69 | 0.75 | 0.79 | 0.82 | 0.84 | 0.3 | 6.0 | 10.7 | 11.9 | 15.6 | 17.2 | 17.5 |
| 0.1 | 0.41 | 0.35 | 0.27 | 0.21 | 0.17 | 0.14 | 0.2 | 7.5 | 13.1 | 16.5 | 18.8 | 19.9 | 19.6 |
| 0.2 | 0.06 | -0.03 | -0.08 | -0.10 | -0.11 | -0.12 | 0.1 | 8.1 | 13.4 | 17.0 | 18.3 | 19.0 | 19.2 |
| 0.3 | -0.21 | -0.22 | -0.20 | -0.18 | -0.17 | -0.16 | 0.0 | 8.1 | 12.9 | 16.2 | 16.4 | 17.1 | 16.0 |
| 0.4 | -0.36 | -0.29 | -0.24 | -0.20 | -0.18 | -0.16 | -0.1 | 7.6 | 11.8 | 14.5 | 14.7 | 14.3 | 13.7 |
| 0.5 | -0.40 | -0.31 | -0.25 | -0.21 | -0.18 | -0.16 | -0.2 | 7.0 | 10.4 | 13.0 | 13.5 | 12.4 | 11.2 |
| | | | | | | | -0.1 | 6.0 | 8.4 | 10.9 | 10.6 | 10.5 | 9.6 |
| | | | | | | | -0.2 | 4.9 | 6.2 | 8.3 | 6.8 | 6.5 | 5.9 |
| | | | | | | | -0.3 | 3.4 | 3.9 | 5.3 | 4.3 | 3.7 | 2.9 |
| | | | | | | | -0.4 | 0.5 | 0.8 | 1.9 | 0.3 | | |

## 5.11 Wind loads on moored ships and their influence on the dimensioning of mooring and fender equipment (R 153)

### 5.11.1 General

This recommendation should be regarded as a supplement to the proposals and advice for the planning, design and detailing of fender and mooring equipment, especially R 12, section 5.12, R 111, section 13.2, and R 128, section 13.3.

The loads for mooring installations – such as bollards or quick-release hooks with their associated anchorages, foundations, support structures, etc. – ensuing according to this recommendation only replace the load variables given in R 12, section 5.12, when the influence of swell, waves and currents on a ship's moorings can be neglected. Otherwise, the latter must be specifically verified and taken into consideration.

R 38, section 5.2, is not affected by this recommendation. When determining the "normal berthing loads" dealt with in that section, the reference to R 12, section 5.12.2, therefore remains applicable without restrictions.

### 5.11.2 Critical wind speed

Owing to the inertia of ships, it is not the short-term (in the order of magnitude of seconds) peak gusts that are critical when determining line pull forces, but rather the average wind over a period $T$. The value of $T$ should be taken as 0.5 min for ships up to 50 000 DWT and 1.0 min for larger ships. The wind strength of the maximum wind speed averaged over a period of one minute is generally about 75 % of that of the value over one second.

It is recommended that wind measurements be used to determine the critical wind speed. If wind data from the immediate vicinity is not available, then wind measurements from more remote measuring stations can be used by means of interpolation or numerical methods of calculation, and taking into account the orography. The time series of the wind measurements should be used to produce statistics of extreme values. A return period of 50 years is recommended for the design value. If no other specific data on the wind conditions is available for the area of the ship's berth, the values given in DIN 1055-4 can be used as critical wind speeds $v$ for all wind directions.

This basic value can be differentiated according to wind direction if more detailed data are available.

### 5.11.3 Wind loads on moored vessels

The loads quoted are characteristic values.

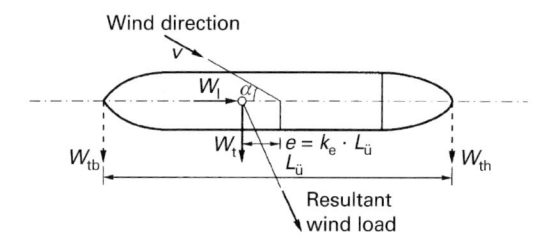

**Fig. R 153-1.** Wind loads acting on moored vessel

Wind load components:

$$W_t = (1 + 3.1 \sin a) \cdot k_t \cdot A_W \cdot v^2 \times \varphi$$
$$W_l = (1 + 3.1 \sin a) \cdot k_l \cdot A_W \cdot v^2 \times \varphi$$

Equivalent loads for $W_t = W_{tb} + W_{th}$:

$$W_{tb} = W_t \cdot (0.50 + k_e)$$
$$W_{th} = W_t \cdot (0.50 - k_e)$$

(see Fig. R 153-1 for force diagram) where

| | |
|---|---|
| $A_W$ | area exposed to wind |
| $v$ | critical wind speed |
| $W_t, W_l$ | wind load components |
| $k_t, k_l$ | wind load coefficients |
| $k_e$ | coefficient of eccentricity |
| $\varphi$ | 1.25; factor comprises dynamic and other non-ascertainable influences |

The area exposed to the wind is derived from the most unfavourable load condition observed for each case, including any cargo on deck that may be present.

As international experience has shown, the load and eccentricity coefficients of Tables R 153-1 and R 153-2 may be used.

Refer to the tables in [162] for more precise data on different types of ship.

### 5.11.4 Loads on mooring and fender equipment

A structural system, consisting of ship, hawsers and mooring and fender structures, must be introduced in order to determine the mooring and fender forces. The elasticity of the hawsers, which depends on their material, cross-section and length, is just as important as the inclination of the hawsers in the horizontal and vertical directions in varying loading and water level conditions. The elasticity of the mooring and fender structures must be ascertained for all support and bearing points of the

**Table R 153-1.** Load and eccentricity coefficients for ships ≤ 50 000 DWT

| \multicolumn{4}{c}{Ships ≤ 50 000 DWT} | | | |
|---|---|---|---|
| $\alpha$ [°] | $k_t$ [kN·s²/m⁴] | $k_e$ [–] | $k_l$ [kN·s²/m⁴] |
| 0   | 0          | 0     | 9.1·10⁻⁵  |
| 30  | 12.1·10⁻⁵  | 0.14  | 3.0·10⁻⁵  |
| 60  | 16.1·10⁻⁵  | 0.08  | 2.0·10⁻⁵  |
| 90  | 18.1·10⁻⁵  | 0     | 0         |
| 120 | 15.1·10⁻⁵  | −0.07 | −2.0·10⁻⁵ |
| 150 | 12.1·10⁻⁵  | −0.15 | −4.1·10⁻⁵ |
| 180 | 0          | 0     | −8.1·10⁻⁵ |

structural system. Anchored sheet piling and structures with raking pile foundations may be considered as rigid elements in this context. Please note that the structural system can alter if individual lines fall slack or fenders remain unloaded under specific load situations.

The wind-shielding effect of structures and facilities may be taken into account to a reasonable extent.

## 5.12 Layout of and loads on bollards for sea-going vessels (R 12)

### 5.12.1 Layout

To ensure maximum simplicity and clarity, a bollard spacing of about 30 m is recommended for plain or reinforced concrete quay walls and quay walls on pile bents/trestles. Where there are joints between blocks, the bollards should be positioned symmetrically within the blocks.

The distance of the bollards from the face of the wall should be as given in R 6, section 6.1.2.

**Table R 153-2.** Load and eccentricity coefficients for ships > 50 000 DWT

| \multicolumn{4}{c}{Ships > 50 000 DWT} | | | |
|---|---|---|---|
| $\alpha$ [°] | $k_t$ [kN·s²/m⁴] | $k_e$ [–] | $k_l$ [kN·s²/m⁴] |
| 0   | 0          | 0     | 9.1·10⁻⁵  |
| 30  | 11.1·10⁻⁵  | 0.13  | 3.0·10⁻⁵  |
| 60  | 14.1·10⁻⁵  | 0.07  | 2.0·10⁻⁵  |
| 90  | 16.1·10⁻⁵  | 0     | 0         |
| 120 | 14.1·10⁻⁵  | −0.08 | −2.0·10⁻⁵ |
| 150 | 11.1·10⁻⁵  | −0.16 | −4.0·10⁻⁵ |
| 180 | 0          | 0     | −8.1·10⁻⁵ |

**Table R 12-1.** Bollard pull forces for sea-going vessels

| Displacement [t] | Bollard pull force [kN] |
|---|---|
| up to 10 000 | 300 |
| up to 20 000 | 600 |
| up to 50 000 | 800 |
| up to 100 000 | 1000 |
| up to 200 000 | 2000 |
| up to 250 000 | 2500 |
| > 250 000 | > 2500 |

Bollards can be designed as single or double bollards and can accommodate several hawsers simultaneously. They should be designed to allow for easy repair or replacement.

All bollards should be clearly marked with their maximum permissible line pull force.

### 5.12.2 Loads

The hawsers placed around a bollard are generally not all fully stressed at the same time, and the effects of line pull forces cancel each other out to some extent. Therefore, the bollard pull forces according to Table R 12-1 may be used for both single and double bollards irrespective of the number of hawsers.

The loads quoted are characteristic values. When designing a bollard and its connections to the structure, use the partial safety factors for load and material strength according to chapter 13. The bollard anchorage in the structure must be designed for 1.5 times the anticipated load to ensure that the quay structure is not damaged should a bollard be torn off. At berths for larger vessels where there is a strong current, the bollard pull forces of Table R 12-1 should be increased by 25 % for ships of 50 000 t or greater displacement.

### 5.12.3 Direction of bollard pull force

The bollard pull force can occur at any angle towards the water side. A bollard pull force towards the landward side is not assumed unless the bollard also serves a waterfront structure situated behind or has a special purpose as a corner bollard. When designing the waterfront structure, the bollard pull force is usually assumed to act horizontally.

When designing the bollard itself and its connections to the waterfront structure, upward inclines of up to 45° with corresponding bollard pull forces must also be taken into consideration.

## 5.13 Layout, design and loading of bollards for inland facilities (R 102)

This recommendation has been brought into line with DIN 19703 "Locks for waterways for inland navigation – Principles for dimensioning and equipment" in so far as the principles of this standard can be transferred to waterfront structures.

The term "bollard" is used to cover all mooring equipment, and includes edge bollards, recessed bollards, dolphin bollards, mooring hooks, mooring rings, etc.

### 5.13.1 Layout and design

In inland ports, ships should be moored to the shore by three hawsers (known as lines): a bow line, a breast line and a stern line. To this end, an adequate number of bollards must be provided on the bank.

Bollards must be arranged on and above the port operating level, with their top edges extending above the highest navigable water level and, if possible, above highest high water. The diameter of such bollards must be > 15 cm. If the bollard does not extend sufficiently above highest high water, slippage of the line must be prevented by a cross-rail. In addition to bollards along the top of the bank, river ports must be equipped with other bollards at various elevations corresponding to the fluctuations in the local water level. Only then can the ship's crew moor the ship without any difficulty at every water level and every freeboard height. In the case of vertical quay walls, the bollards at varying heights should be situated in a vertical line one above the other. The positions of the lines depend on the positions of the vertical access ladders. To avoid mooring lines across ladders, a line of bollards should be located to the left and right of every vertical ladder at a distance of approx. 0.85–1.00 m from the axis of the ladder in the case of concrete or masonry walls and at double the sheet pile spacing in the case of sheet pile walls. The spacing between ladders/lines of bollards should be about 30 m. In the case of steel sheet pile walls, the exact spacing depends on the system dimension of the sheet piles; in the case of concrete or masonry walls the spacing is determined by the block length where there are joints between blocks. The lowest bollard should be approx. 1.50 m above lowest low water, or the mean low water spring level in tidal areas (in the case of inland waterway locks, max. 1.0 m above the lowest tail water level). The vertical distance between the lowest bollard and the upper edge of the quay is divided up by further bollards at a spacing of 1.30–1.50 m (up to 2.00 m in borderline cases).

In the case of waterfront structures of reinforced concrete, the bollards are positioned in recesses, the housings of which are cast into the concrete and provided with anchors. In the case of steel sheet pile walls, the bollards can be bolted or welded in position. The front edge of

149

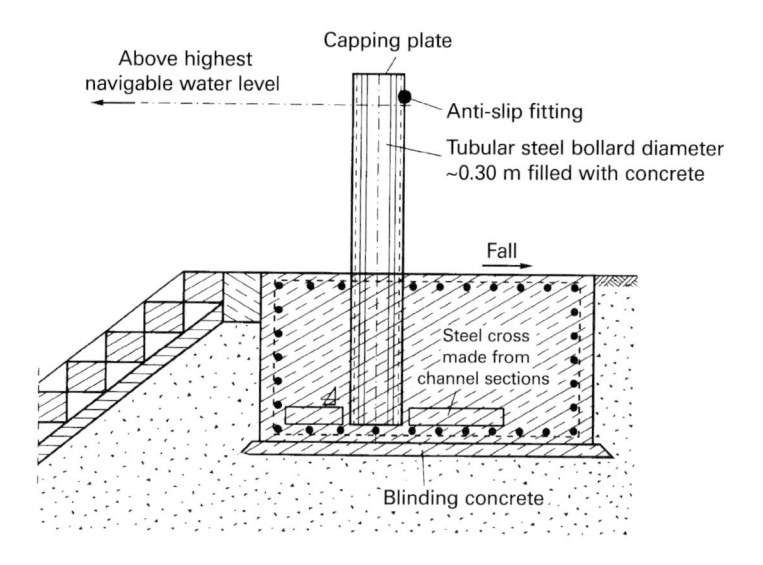

**Fig. R 102-1.** Bollard foundation for a partially sloping bank (drawing shows example of typical layout, actual design must be based on structural requirements)

the bollard post should be 5 cm behind the front edge of the quay. An appropriate gap should be left at the sides, behind and above the bollard post so that ships' hawsers can be easily looped over and removed again. Edges between bollard recess and front of quay must be rounded to prevent any damage to hawsers or waterfront structure.

In the case of (partially) sloping banks, the bollards should be positioned on both sides next to the steps (Fig. R 102-1). Steps should form extensions to ladders.

In this arrangement, it is expedient to continue the bollard foundation beneath the steps to create a joint foundation for both bollards.

## 5.13.2 Loads

The line pull forces mainly depend on the size of the ship, its speed and the distance to passing ships, the flow rate of the water at the berth and the quotient of the water cross-section to the submerged ship's cross-section.

A characteristic load of 200 kN per bollard should be assumed for the load on the wall (i.e. for sheet pile walls, walings, capping beams, anchors, dolphins, etc.). The design for the bollard anchor in the structure must be calculated for a characteristic load of 300 kN in order to ensure that the quay structure is not damaged should a bollard be torn off.

When designing the bollard, the partial safety factors for load and material strength must be selected in accordance with chapter 13.
According to DIN EN 14329, the characteristic bollard load should be increased to 300 kN for a vessel length > 110 m. In this case the anchoring elements should be designed for 400 kN.
Bollards must not be used to brake ships. Therefore, such a load is ignored in the load assumptions (actions).

### 5.13.3 Direction of line pull forces

Line pull forces can only be applied from the water side – mostly at an acute angle and only rarely at 90° to the bank. The calculations must, however, take account of every possible horizontal and vertical angle to the bank.

### 5.13.4 Calculations

Stability analyses must be provided for a single-side line pull force acting at the most unfavourable angle. Proof of stability can also be provided by loading tests.

## 5.14 Quay loads from cranes and other transhipment equipment (R 84)

The following loads are characteristic values that must be multiplied by the partial safety factors for the relevant load cases (see R 18, section 5.4) according to DIN 1054.

### 5.14.1 Typical general cargo port cranes

#### 5.14.1.1 General

In Germany, typical cargo port cranes are mostly constructed as full portal luffing slewing cranes spanning up to three railway tracks, but also occasionally as semi-portal cranes. Safe working loads vary between 7 and 50 t, with jib lengths of 20–45 m.

In the interests of good utilisation of the jib length, counting from the centre of rotation, the axis of rotation of the crane superstructure should lie as close as possible to the outboard crane rail. However, to avoid any collision between the crane and a listing ship, care must be taken to ensure that neither the crane operator's cabin nor the rear counterweight project beyond a plane inclined at approx. 5° landwards from the top edge of the quay.

The distance between the outboard crane rail and the front edge of the quay structure should be in accordance with R 6, section 6.1. The corner distance in the case of small cranes is approx. 6 m, but the minimum corner distance should not be < 5.5 m, otherwise excessive corner loads occur and the cranes must be equipped with excessively large central ballast. The length over buffers is approx. 7–22 m, depending on the size of the crane. If this results in an excessive wheel load, lower wheel loads can be achieved by increasing the number of wheels. Today, however, there are also general cargo handling facilities whose craneways have been built for especially high wheel loads.

As a rule, general cargo port cranes are classified as hoisting class H 2 and load group B 4 or B 5 in accordance with DIN 15018-1. Designers should also refer to F.E.M. 1.001 [65]. The vertical wheel loads due to dead loads, imposed loads, inertial forces and wind loads must be applied when designing the craneway (DIN 15018-1). Allowance should be made for vertical inertial forces arising from the travel or the lifting or setting-down of the loads through the use of a dynamic factor, which is approx. 1.2 for hoisting class H 2. The foundation to the craneway may be designed without taking such a dynamic factor into account. All crane booms can slew through 360°; the respective corner load changes accordingly. With higher wind loads and a crane not in operation, design situation DS-A may be used for designing quay walls and craneways if necessary.

### 5.14.1.2 Full portal cranes

The portals of light-duty port cranes with low safe working loads have either three or four legs, with each leg having between one and four wheels. The number of wheels in each case depends on the permissible wheel load. General cargo heavy-duty cranes have at least six wheels per leg. On straight stretches of quayside, the centre-to-centre spacing of the crane rails is min. 5.5 m, but generally speaking 6.0, 10.0, or 14.5 m depending on whether the portal spans one, two or three railway tracks. The dimensions 10.0 m and 14.5 m result from the theoretical minimum dimension of 5.5 m for one track, to which the track spacing of 4.5 m is added once or twice.

### 5.14.1.3 Semi-portal cranes

The portals of these cranes have only two legs, which run on the outboard crane rail. On the inboard side, the crane is supported by a skid on an elevated craneway, thus allowing unrestricted access to any section of the quay area. The remarks in section 5.14.1.2 apply to the number of wheels under both legs and the skid.

### 5.14.2 Container cranes

Container cranes are constructed as full portal cranes with cantilever beams and trolley (ship-to-shore cranes), the legs of which have 8–13 wheels each as a rule. The crane rails of existing container terminals generally have a centre-to-centre distance of between 15.24 m (50 ft) and 30.48 m (100 ft), in some cases up to 35.00 m. The clear leg spacing

(= uninterrupted space between corners in direction of craneway) is 17–18.5 m, with an overall length over buffers of about 27.5 m (Fig. R 84-1). In this respect it should be assumed that three container cranes work buffer to buffer. If the handling of 20 ft containers requires the use of a shorter length over buffers, a minimum corner spacing of up to 12 m is possible; the length over buffers is then 22.5 m. The corner distance in this case is not the same as the portal leg spacing. As regards the safe working loads of these cranes, a figure of between 60 and 90 t, and up to 110 t for heavy-duty units, including spreader, should be selected. The maximum corner load is particularly affected by crane type and jib length. A jib length of 38–41 m, formerly common and corresponding to the widths of Panamax vessels, is insufficient for post-Panamax vessels, which can no longer pass through the Panama Canal owing to their width. Jib lengths of at least 44.5 m are necessary for this type of vessel. The maximum corner loads for container cranes in operation reach up to 4500 kN for Panamax vessels and up to 15 000 kN for post-Panamax vessels.

The trend in container crane development, however, is towards jib lengths that can handle 22–24 rows of containers on the ships; jib lengths of up to 66 m, measured from the outboard rail, are the result. It is recommended that enquiries be made to the terminal operators in order to obtain more accurate planning data, since the large number of possible approaches does not allow data to be specified in more detail. Particular attention must be paid to developments where several containers are moved simultaneously, e.g. in twin-lift operation, which is already common practice. In addition, advanced systems, including Twin-Forty and others, are being considered or have already been put to the test. Twin-Forty container cranes are capable of loading or unloading two 40 ft containers. Double Hoist employs the same loading system as Twin-Forty, but a different spreader.

### 5.14.3 Load specifications for port cranes

The support structure always consists of a portal-type substructure, either with slewing and height-adjustable jib or with a rigid cantilever beam that can be raised when not in operation. The portal usually stands on four corner points each with several wheels arranged in swing-arm bogies, depending on the magnitude of the corner load. The corner load is distributed as evenly as possible over all the wheels of the corner point. Further to the remarks in sections 5.14.1 and 5.14.2, general load and dimensional data can be found in Tables R 84-1 and R 84-2. These data assist in the general preliminary design of waterfront structures. Final geotechnical and structural calculations must be performed using the specific load specifications of the port cranes envisaged for handling the goods.

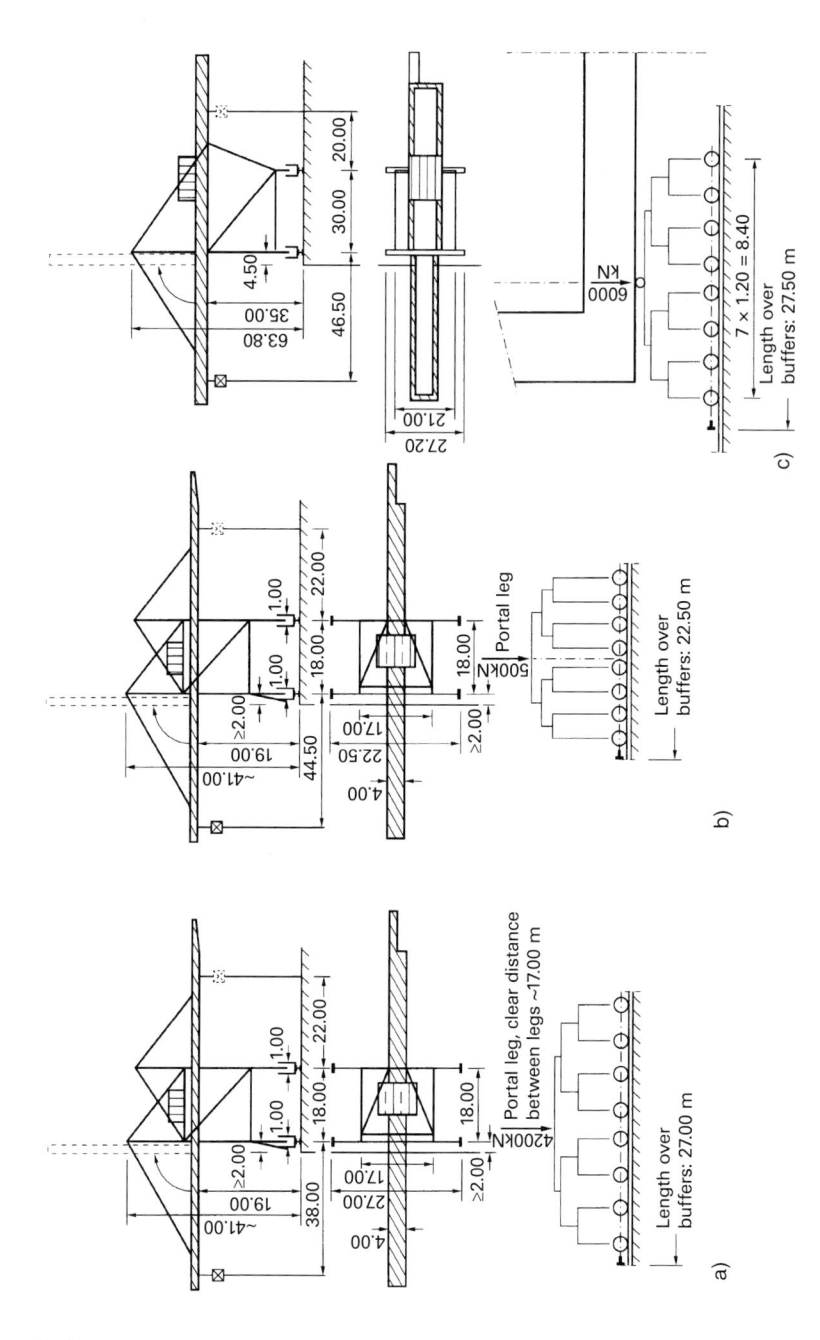

**Fig. R 84-1.** Examples of container cranes: a) with 53 t lifting capacity, 18 m gauge for post-Panamax vessels and 22.5 m length over buffers b) with 53 t lifting capacity, 18 m gauge for Panamax vessels and 27 m length over buffers c) with 53 t lifting capacity, 30 m gauge for post-Panamax vessels and 27.2 m length over buffers

**Table R 84-1.** Dimensions and characteristic loads of slewing and container cranes

|  | Slewing cranes | Container cranes and other goods-handling equipment |
|---|---|---|
| Lifting capacity [t] | 7–50 | 10–110 |
| Self-weight [t] | 180–350 | 200–2000[*)] |
| Portal span [m] | 6–19 | 9–45 |
| Portal clearance [m] | 5–7 | 5–13 |
| Max. vertical corner load [kN] | 800–3000 | 200–15 000 |
| Max. vertical wheel load [kN/m] | 250–600 | 250–1150 |
| Horizontal wheel load transverse to direction of rail in direction of rail | up to approx. 10 % of vertical load up to approx. 15 % of vertical load of braked wheels | |

[*)] Self-weights may be increased further still as a result of more recent developments.

## 5.14.4 Notes

Further details regarding port cranes can be found in AHU (recommendations E 1 and E 9, reports B 6 and B 8), in ETAB (recommendation E 25) and VDI guideline 3576.

Moreover, container cranes with rail gauges of 35 m are in operation.

## 5.15 Impact and pressure of ice on waterfront structures, fenders and dolphins in coastal areas (R 177)

## 5.15.1 General

Loads on hydraulic engineering structures due to the effects of ice can occur in various ways:

a) as ice impact from collisions with ice floes moved by the current or by the wind,

**Table R 84-2.** Dimensions and characteristic loads of mobile cranes

|  | Mobile cranes | | | | |
|---|---|---|---|---|---|
| Max. lifting capacity [t] | 42 | 64 | 84 | 104 | 140 |
| Self-weight [t] | 130 | 170 | 250 | 420 | 460 |
| Associated jib length [m] | 12 | 12 | 15 | 22 | 20 |
| Static outrigger load [kN] | 920 | 1250 | 1660 | 2600 | 3250 |
| Dynamic outrigger load [kN] | 1080 | 1450 | 1950 | 3050 | 3650 |
| Outrigger size [m] | 5.5 × 0.8 | 5.5 × 0.8 | 5.5 × 1.3 | 5.5 × 1.8 | 5.5 × 1.8 |

b) as ice pressure acting through ice thrusting against an ice layer adjacent to the structure or through vessel movements,
c) as ice pressure acting on the structure through an unbroken ice layer as a result of thermal expansion, and
d) as imposed ice loads when ice forms on the structure or as surcharges or uplift loads when water levels fluctuate.

Among other factors, the magnitude of possible actions depends on:

- the shape, size, surface finishes and elasticity of the obstacle with which the ice mass collides,
- the size, shape and rate of advance of the ice masses,
- the nature of the ice and the ice formation,
- the salt content of the ice and the ice strength dependent on this,
- the angle of incidence,
- the critical strength of the ice (compressive, bending, shear),
- the rate of application of the load, and
- the temperature of the ice.

The calculated ice loads are characteristic values. Consequently, according to section 5.4.3, a partial safety factor of 1.0 can generally be used. The calculations presented below apply to sea ice with a salinity (salt content) $S_B \geq 5\%$. Recommendation R 205, section 5.16, applies for lower salinities and freshwater ice. Further, the recommendations mentioned represent rough assumptions for ice loads on structures; they do not apply to extreme ice conditions, as are encountered in Arctic regions, for example. Where possible, the recommendation is to check the load values for waterfront structures, including piled structures, against the assumptions for completed installations that have performed well, or against ice pressures measured in situ or in the laboratory.

As regards protected areas (bays, harbour basins, etc.) where the size of the ice field is small, seaports with clear tidal effects and a substantial amount of maritime traffic and when adopting measures designed to reduce the ice load, e.g.:

- the effect of the current,
- the use of air bubble systems, and
- heating or other thermal transfer systems etc.,

then significantly reduced load assumptions can apply. In any given case where a precise stipulation of the ice loads is required, experts should be called in and, if necessary, model tests carried out.

When positioning port entrances and orienting harbour basins, particular consideration must be given to the wind direction, current and the shear zone formation of the ice when it comes to determining ice loads and ice formation processes.

If the ice loads on dolphins are substantially greater than the loads resulting from vessel impact or line pull, a check should be carried out to determine whether such dolphins should be designed for the higher ice loads or whether, for economic reasons, occasional overloads can be accepted.

The reader is referred to the explanatory notes in Hager [82]. Advice taken from other international regulations (USA, Canada, Russia, etc.) can be found in [83].

### 5.15.2 Determining the compressive strength of the ice

The mean compressive strength of the ice $\sigma_0$ essentially depends on its temperature, salt content and specific rate of expansion, i.e. the ice drift speed. In addition, ice has noticeably anisotropic properties, i.e. the maximum compressive strength depends on the direction of pressure. If no detailed studies of the material properties of the ice are available, the following assumptions apply for the north German coasts:

- Linear temperature distribution over the thickness of the ice, with a temperature on the underside of approx. $-2.0\,°C$ for the German North Sea coast and approx. $-1.0\,°C$ for the German Baltic Sea coast (varies according to the salt content of the water) and the top surface of the ice at air temperature.
- The salinity of the ice in the North Sea and Baltic Sea in accordance with Table R 177-1.

**Table R 177-1.** Guidance values for salt content (salinity $S_B$) of the seawater and sea ice along the German North Sea and Baltic Sea coasts according to [126]

| North Sea | Salinity of water [%] | Salinity of ice [%] | Baltic Sea | Salinity of water [%] | Salinity of ice [%] |
|---|---|---|---|---|---|
| German Bight Estuaries | 32 to 35<br>25 to 30 | 14 to 18<br>12 to 14 | Belt Sea<br>Bay of Kiel<br>Bay of Mecklenburg<br>Arkona Basin and Bornholm Sea<br>Gotland Sea<br>Gulf of Finland and Gulf of Bothnia | 15 to 20<br>15<br>15<br><br>8 to 10<br><br>5 to 7<br>1 to 5 | 10 to 12<br>8 to 10<br>8 to 10<br><br>5 to 7<br><br>*)<br>*) |

*) The compressive strength of ice for salinity < 5% is determined in accordance with R 205, section 5.16

- A specific rate of expansion $\varepsilon = 0.001\ \text{s}^{-1}$ (the compressive strength of the ice depends on the rate of expansion and attains the maximum value in the range between ductile and brittle failure at ($\varepsilon \approx 0.001\ \text{s}^{-1}$).

Based on the salt content and temperature of the ice, the porosity $\phi_B$, i.e. the quantity of salt crystals and air pockets in the ice, according to [126] is as follows:

$$\phi_B = 19.37 + 36.18 S_B^{0.91} \cdot |\vartheta_m|^{-0.69}$$

where

$\phi_B$ porosity [%]
$S_B$ salinity [%]
$\vartheta_m$ mean ice temperature, $(\vartheta_o + \vartheta_u)/2$ [°C]
$\vartheta_u$ temperature on underside of ice ($\vartheta_u = -1\ °C$ in German region of Baltic Sea, $\vartheta_u = -2\ °C$ in German region of North Sea) [°C]
$\vartheta_o$ temperature on top surface of ice (corresponds to air temperature) [°C]

The horizontal uniaxial compressive strength of the ice $\sigma_0$ according to [74] and [126] is derived from the material properties specified above or, in an ideal situation, by means of the material properties determined in situ or through experiments:

$$\sigma_0 = 2700 \dot\varepsilon^{1/3} \cdot \phi_B^{-1}$$

where

$\sigma_0$ horizontal uniaxial compressive strength of the ice [MN/m²]
$\varepsilon$ specific rate of expansion (0.001) [s⁻¹]
$\phi_B$ porosity [%]

If more precise ice strength studies are not available, the flexural strength $\sigma_B$ can be assumed to be roughly $1/3 \sigma_0$ and the shear strength $\tau$ roughly $1/6 \sigma_0$.

### 5.15.3 Ice loads on waterfront structures and other structures of greater extent

#### 5.15.3.1 Mechanical ice pressure

The following design approaches can be used to determine the horizontal ice loads on vertical planar structures on the north German coasts for ice thicknesses $d$ in the order of magnitude of $0.25\ \text{m} \leq d \leq 0.75\ \text{m}$ plus compressive strengths as per section 5.15.2:

a) A horizontal mean line load $p_0$ acting at the most unfavourable height of the water levels under consideration. It is assumed here that the maximum load calculated from the uniaxial compressive strength of the ice $\sigma_0$ is only effective, on average, over one-third of the length of the structure (contact coefficient $k = 0.33$). The mean line load is therefore

$$p_0 = k \cdot h \cdot \sigma_0$$

where

$p_0$    mean line load [MN/m]
$k$    contact coefficient (0.33) [–]
$h$    thickness of ice [m]
$\sigma_0$    compressive strength of the ice [MN/m²]

b) A local load per unit area $p$ acting over the thickness of the ice must be taken into account in local analyses. This results in the following equation:

$$p = \sigma_0$$

where

$p$    local load per unit area [MN/m²]
$\sigma_0$    uniaxial compressive strength of the ice [MN/m²]

c) A reduced horizontal mean line load $p_0'$ acting at the most unfavourable respective height of the water levels under consideration in the case of platforms and revetments in tidal areas when the layer of ice has broken as a result of fluctuating water levels. According to [83], this results in the following equation:

$$p_0' = 0.40 p_0$$

where

$p_0'$    reduced line load [MN/m]
$p_0$    mean line load [MN/m]

It is not necessary to consider the simultaneous effect of ice coupled with wave load and/or vessel impact.

**Table R 177-2.** Measured maximum ice thicknesses as guidance values for design [30]

| North Sea | max. $h$ [cm] | Baltic Sea | max. $h$ [cm] |
|---|---|---|---|
| Helgoland | 30 to 50 | Nord-Ostsee Canal | 60 |
| Wilhelmshaven | 40 | Flensburg (outer fjord) | 32 |
| "Hohe Weg" lighthouse | 60 | Flensburg (inner fjord) | 40 |
| Büsum | 45 | Schleimünde | 35 |
| Meldorf (harbour) | 60 | Kappeln | 50 |
| Tönning | 80 | Eckernförde | 50 |
| Husum | 37 | Kiel (harbour) | 55 |
| Wittdün harbour | 60 | Bay of Lübeck | 50 |
| | | Wismar harbour | 50 |
| | | Bay of Wismar | 60 |
| | | Rostock – Warnemünde | 40 |
| | | Stralsund – Palmer Ort | 65 |
| | | Saßnitz harbour | 40 |
| | | Koserow – Usedom | 50 |

If no maximum ice thickness values specific to the location are available, the maximum values specified in Table R 177-2, which are based on ice observations conducted over many years, can be assumed.

### 5.15.3.2 Thermal ice pressure

Thermal ice pressure, as a further form of static loading on waterfront structures and other planar structures in the water, is caused by rapid changes in temperature and simultaneous restraint to expansion. In confined, iced-up port basins or similar configurations, considerable loads can occur as a result of thermal expansion. A precise determination of the thermal ice pressure on waterfront structures is complex, since the calculation approaches available frequently require numerous input parameters such as air temperature, wind speed, solar radiation, snow covering, etc., which can only be determined with a considerable degree of uncertainty.

As regards the thermal ice pressure of sea ice as a function of the rate of temperature change (°C/h), ISO/FDIS 19906 [109] specifies the values given in Fig. R 177-1 for various ice temperatures and thicknesses.

DIN 19704-1 recommends considering a load of 0.25 MN/m$^2$ uniformly distributed over the ice thickness as a calculation parameter for German

Where

A  $v_m = -30$ °C; $h = 1.0$ m
C  $v_m = -20$ °C; $h = 1.0$ m
B  $v_m = -30$ °C; $h = 0.5$ m
D  $v_m = -20$ °C; $h = 0.5$ m

$p_T$  Thermal ice pressure [kN/m]
$r_T$  Rate of temperature change [°C/h]
$h$   Ice thickness
$v_m$  Mean ice temperature

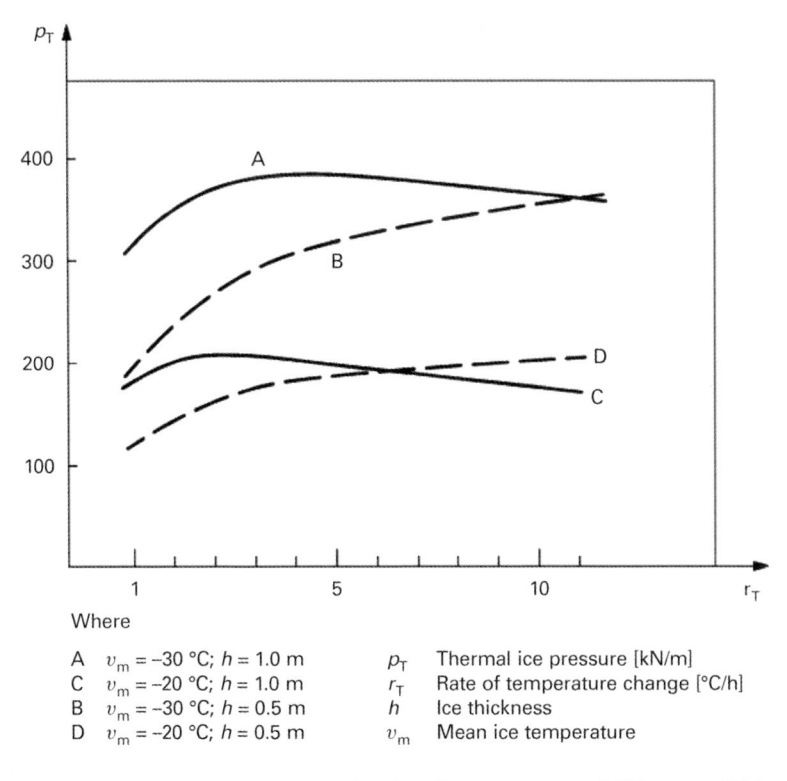

**Fig. R 177-1.** Thermal ice pressure as a function of ice temperature and thickness (to ISO/FDIS 19906 [109])

coasts and ice thicknesses $h < 0.8$ m. The reader is referred to [105] for further information and calculation approaches.

### 5.15.4 Ice loads on vertical piles

The ice loads acting on piles depend on the shape, rake and arrangement of the piles as well as the critical compressive, bending or shear strength of the ice responsible for breaking the ice. In addition, the magnitude of the load depends on the type of load (static or impact load due to ice floes) and the associated rates of expansion and deformation.

In the case of structural components up to 2 m wide with a ratio of structural component width to ice thickness $d/h \leq 12$, and where the ice is flat, the horizontal ice load on vertical piles with a rake steeper than 6:1 ($\beta \geq 80°$) is derived as follows according to [196]:

**Table R 177-3.** Empirical contact coefficients $k$ after [99]

| Load action | $k$ |
|---|---|
| Ice movement with layer of ice *not* tight up against structural component (drifting ice) | 0.564 |
| Ice movement with layer of ice tight up against structural component (icebound structural component) | 0.793 |

$$P_p = k \cdot \sigma_0 \cdot d^{0.5} \cdot h^{1.1}$$

where

- $P_p$ ice load [MN]
- $\sigma_0$ uniaxial compressive strength of the ice at a specific rate of expansion, $\varepsilon = 0.001 \text{ s}^{-1}$ [MN/m²]
- $D$ width of single pile [m]
- $h$ thickness of ice [m]
- $k$ empirical contact coefficient from Table R 177-3 [m$^{0.4}$]

If no studies specific to the location are available, then the ice thicknesses $h$ can be estimated using the guidance values of Table R 177-2 for the German North Sea and Baltic Sea coasts. In the event of ice ridges or ramparts, i.e. banked or humped, compacted sea ice with ice floes on top of each other, the ice loads determined must be doubled.

Along the German North Sea coast, the ice load on free-standing piles is frequently taken to be 0.5–1.5 m above mean high tide, depending on local conditions.

As regards ice loads on tapering and inclined structural components, the guidance and calculation methods of [74] should be taken into account.

### 5.15.5 Horizontal ice load on group of piles

The ice load on a group of piles is derived from the sum of the ice loads on the individual piles. Generally, assuming the sum of the ice loads acting on the piles facing the drifting ice will suffice.

### 5.15.6 Ice surcharges

Ice surcharges must be taken into account depending on local conditions. Without more detailed data, a minimum ice surcharge of 0.9 kN/m² can be regarded as sufficient [73]. In addition to the ice surcharge, an estimate of the typical snow load to DIN 1055-5 should also be considered. Conversely, variable loads that are absent during thicker ice formations do not normally need to be considered to act simultaneously.

### 5.15.7 Vertical loads with rising or falling water levels

Vertical ice forces, which occur when a covering of ice around the pile freezes solid and the water level changes subsequently, are limited by the flexural strength $\sigma_B$ ($\sigma_f$) of the ice. A prerequisite for the transmission of vertical ice forces to piles is a bond (adhesion) between the surface of the pile and the ice.

When determining upward and downward vertical ice loads on piles, the following approach based on the Russian standard (SNiP, [200]) is recommended according to Kohlhase et al. [121] and Weichbrodt [222]:

$$A_V = \left(0,6 + \frac{0,15D}{h}\right) \cdot 0,4 \cdot \sigma_0 h^2$$

where

- $A_V$    vertical ice load [kN]
- $h$    thickness of ice cover [m]
- $D$    pile diameter [m]
- $\sigma_0$    compressive strength of the ice cover [kN/m$^2$]

The vertical ice loads determined using the recommended approach apply to individual vertical piles. If the spacing of the piles or the distance from piles to fixed structures is smaller than the extent of the deformation of the layer of ice under a vertical load (characteristic length of ice layer $\ell_c$), the vertical ice loads per pile are diminished.

The vertical ice load on groups of piles or piles close to fixed structures can be calculated by multiplying the load by a "geometric factor" $f_g$ from the vertical ice load for a single pile.

This factor $f_g$ is calculated from the relationship between the area of deformation, which might be limited at the pile location as a result of neighbouring piles or nearby structures, and the possible area of deformation when assuming an unlimited layer of ice.

The possible deformation area of an unlimited layer of ice is taken to be the area of a circle around the pile with a radius equal to the characteristic length of the ice layer $\ell_c$, which as an approximation can be assumed to be 17 times the thickness of the ice. The limited deformation area is taken to be the mean of the areas of four circles whose radius $r$ in each case is half the distance to the next structure fixed to the ground. The distances are determined in four directions at 90° to each other in accordance with the orientation of the structure. The radii $r$ may not be greater than the characteristic length of the ice layer $\ell_c$. The geometric factor $f_g$ is determined as follows in accordance with Edil et al. [63]:

$$f_g = \frac{r_1^2 + r_2^2 + r_3^2 + r_4^2}{4\ell_c^2}$$

```
                                    Ice thickness = 0.50 m
        Primary pier                Characteristic length
                                    l_c = 0.50 m × 17 = 8.5 m
              Finger pier
                                    Geometric factors
                                    for pile A:
```

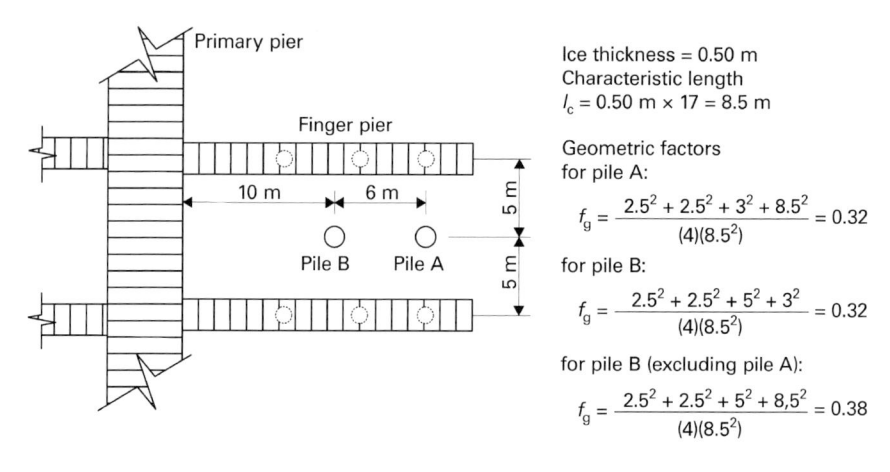

$$f_g = \frac{2.5^2 + 2.5^2 + 3^2 + 8.5^2}{(4)(8.5^2)} = 0.32$$

for pile B:

$$f_g = \frac{2.5^2 + 2.5^2 + 5^2 + 3^2}{(4)(8.5^2)} = 0.32$$

for pile B (excluding pile A):

$$f_g = \frac{2.5^2 + 2.5^2 + 5^2 + 8{,}5^2}{(4)(8.5^2)} = 0.38$$

**Fig. R 177-2.** Examples of calculations for determining the geometric factor $f_g$

where

- $f_g$    geometric factor [–]
- $r_{1-4}$    half the distance from pile to next solid structure [m]
- $\ell_c$    characteristic length of ice layer [m]

Fig. R 177-2 explains the stipulation of the radii $r_1$ to $r_4$.

## 5.16 Impact and pressure of ice on waterfront structures, piers and dolphins at inland facilities (R 205)

### 5.16.1 General

Most of the data provided in recommendation R 177, section 5.15, applies to inland facilities as well. This is true for both the general statements and the loading assumptions, because these depend on the respective dimensions of the structure, the thickness of the ice and the strength properties of the ice.
In accordance with R 177, section 5.15.1, the ice loads determined are characteristic values to which, generally speaking, a partial safety factor of 1.0 should be applied.
The supplementary information provided in R 177, section 5.15.1, must be observed.

### 5.16.2 Ice thickness

According to [125], the ice thickness can be deduced from the sum of degrees of cold occurring on a daily basis during an ice period – the "cold sum". Hence, for example, according to [34]:

$$h = \sqrt{\sum |t_L|}$$

where $h$ is the ice thickness in cm and $\Sigma|t_L|$ the sum of the absolute values of the mean daily subzero air temperatures in °C.
If no more detailed statistics or measurements are available, a theoretical ice thickness $h \leq 30$ cm can generally be assumed provided the conditions given in section 5.16.1 apply.

### 5.16.3 Compressive strength of the ice

The compressive strength of the ice depends on the mean ice temperature $\vartheta_m$, corresponding to half the ice temperature at the surface, as 0 °C is always achieved on the underside (see also section 5.15.2).

The uniaxial compressive strength of the ice vertical to the direction of ice formation can be determined approximately according to [195] as follows:

$$\sigma_0 = 1.10 + 0.35|\vartheta_m| \quad \text{for } 0° < \vartheta_m < -5\,°C$$
$$\sigma_0 = 2.85 + 0.45|\vartheta_m + 5| \quad \text{for } \vartheta_m < -5\,°C$$

where $\vartheta_m$ is the mean temperature of the ice [°C].

### 5.16.4 Ice loads on waterfront structures and other structures of greater extent

#### 5.16.4.1 Mechanical ice pressure

Generally speaking, R 177, section 5.15.3, applies to the force resulting from the pressure of the ice acting as a line load:

$$p_0 = k \cdot h \cdot \sigma_0$$

where

$p_0$    mean line load [MN/m]
$k$    contact coefficient (0.33) [–]
$h$    thickness of the ice [m]
$\sigma_0$    uniaxial compressive strength of the ice [MN/m²]

As regards local structural components, a local load per unit area $p$ acting over the thickness of the ice can be taken as

$$p = \sigma_0$$

where

$p$ local load per unit area [MN/m²]
$\sigma_0$ uniaxial compressive strength of the ice [MN/m²]

On sloping surfaces, the horizontal force due to the pressure of the ice according to [122] can be taken as

$$p'_0 = k \cdot h \cdot \sigma_B \cdot \tan \beta$$

where

$p'_0$ maximum line load [MN/m]
$k$ contact coefficient (0.33) [–]
$\sigma_B$ flexural strength of the ice, 1/3 $\sigma_0$ [MN/m²]
$\tan \beta$ angle of slope [°]

### 5.16.4.2 Thermal ice pressure

Generally speaking, the remarks in R 177, section 5.15.3.2, apply. A load per unit area of 0.15 MN/m² distributed evenly over the ice thickness should be taken as the reference value for the thermal ice pressure resulting from thermal expansion and for ice thicknesses $0.30\,\text{m} \leq h \leq 0.50\,\text{m}$ [105].

### 5.16.5 Ice loads on narrow structures (piles, dolphins, bridge and weir piers, ice deflectors)

The approaches regarding vertical piles according to R 177, section 5.15, are also valid for inland areas taking into consideration the critical ice strengths. They also apply to pier structures and ice deflectors, allowing for their cross-sectional and surface form as well as the angle.

### 5.16.6 Ice loads on groups of structures

The information in R 177, section 5.15.5, applies.
In the case of installations such as bridge piers in watercourses, avoiding ice rafting and the ensuing reduced discharge capacity, the risk of ice rafting can be estimated as shown in Fig. R 205-1.
The structures should be chosen such that the conditions of Fig. R 205-1 are complied with. If this is not possible, additional studies, e.g. physical model tests, are recommended. However, ice rafting does not necessarily increase the ice load in every situation, e.g. when the failure conditions of the banking ice are critical. Additionally, changes to the load distribution and height of load application must be taken into account,

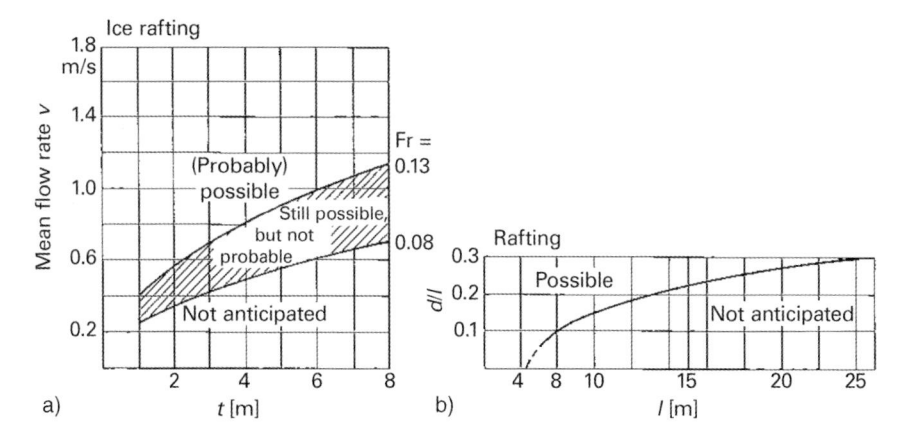

**Fig. R 205-1.** Risk of possible ice rafting according to Hager (2002) [83] in terms of a) water depth $t$ and flow rate $v$, and b) the distance between piers $l$ and the ratio of pier width $d$ to pier spacing $l$

likewise additional loads resulting from accumulated water and changes to currents as a result of cross-sectional restrictions.

### 5.16.7 Vertical loads with rising or falling water levels
The information in R 177, section 5.15.7, applies.

### 5.17 Loads on waterfront structures and dolphins caused by fender reaction forces (R 213)

The energy that can be absorbed by the fenders is determined using the deterministic calculation according to R 60, section 6.15.

The maximum fender reaction force that can act on the waterfront structure or fender dolphin can be determined using the appropriate diagrams/tables of the manufacturer of the type of fender selected and the calculated energy to be absorbed. This reaction force is to be understood as a characteristic value.

Normally, the reaction force does not result in an additional load on the quay and only the local load derivation has to be investigated, unless special structures have been arranged for fenders, e.g. separately suspended fender panels, etc.

# 6 Configuration of cross-sections and equipment for waterfront structures

## 6.1 Standard cross-section dimensions for waterfront structures in seaports (R 6)

### 6.1.1 Standard cross-sections

When building new cargo-handling facilities and extending existing installations, the standard cross-section of Fig. R 6-1 is recommended, taking into account all relevant influences.

The distance of 1.75 m between crane rail and edge of quay should be understood as a minimum. In the case of new construction and watercourse deepening works, it is better to allow 2.50 m as the outboard crane bogies can then be built as wide as required (the crane bogies of modern port cranes are often approx. 0.60–1.20 m wide, see Fig. R 6-1). In addition, it is easier to comply with the health and safety regulations for mooring operations and for embarkation/disembarkation via gangways.

A crane safety clearance must be maintained for railway operations. The centre-line of the first track must be at least 3.00 m from the front crane rail. However, these days, railway tracks are only constructed on quay walls in exceptional circumstances.

### 6.1.2 Walkways (towpaths)

The walkway space (towpath) in front of the outboard crane rail is necessary to provide space for installing bollards and storing gangways, to serve as a path and working space for line handlers, to allow access to berths and for accommodating the outboard portion of the crane gantry. Consequently, this clearance is of special importance for port operations.

Health and safety regulations must also be considered when selecting the walkway width. Furthermore, account must be taken of the fact that ship superstructures often project beyond the hulls of moored vessels and that cargo-handling operations must still be possible on listing ships.

For the reasons outlined above, the walkway must be wide enough so that the outermost edges of the cargo-handling facilities are at least 1.65 m, but preferably 1.80 m, behind the front edge of the quay wall or face of timber fender, fender pile or fender system (Fig. R 6-1).

---

*Recommendations of the Committee for Waterfront Structures Harbours and Waterways – EAU 2012,*
9th Edition. Issued by the Committee for Waterfront Structures of the German Port Technology Association and the German Geotechnical Society.
© 2015 Ernst & Sohn GmbH & Co. KG. Published 2015 by Ernst & Sohn GmbH & Co. KG.

## Interdisciplinary technical know-how in hydraulic and maritime engineering

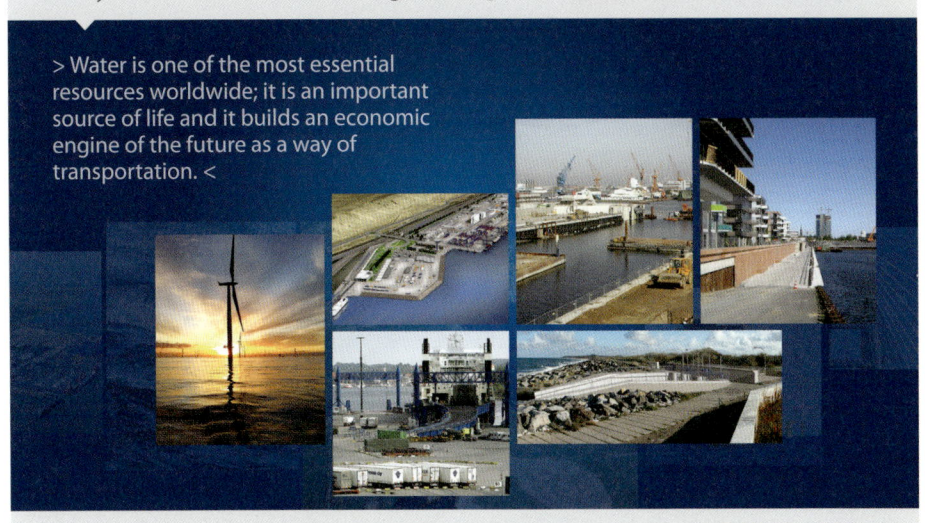

**INROS LACKNER**
Consulting Engineers & Architects

> Water is one of the most essential resources worldwide; it is an important source of life and it builds an economic engine of the future as a way of transportation. <

WWW.INROS-LACKNER.COM

---

**Ernst & Sohn**
A Wiley Brand

**DGGT**
Deutsche Gesellschaft für Geotechnik e. V.
German Geotechnical Society

Ed.: German Geotechnical Society
**Recommendations on Piling (EA Pfähle)**
2013. 496 pages
€ 109,–*
ISBN 978-3-433-03018-9
Also available as **ebook**.

Order online:
www.ernst-und-sohn.de

## Recommendations on Piling (EA Pfähle)

This handbook provides a complete and detailed overview of piling systems and their application. The design and construction of piled foundations is based on Eurocode 7 and DIN 1054 edition 2010 as well as the European construction codes DIN EN 1536 (Bored piles), DIN EN 12699 (Displacement piles) and DIN EN 14199 (Micropiles).

Also available in german:
- Empfehlungen des Arbeitskreises – „EA-Pfähle"

**Ernst & Sohn**
Verlag für Architektur und technische Wissenschaften GmbH & Co. KG

Customer Service: Wiley-VCH
Boschstraße 12
D-69469 Weinheim

Tel. +49 (0)6201 606-400
Fax +49 (0)6201 606-184
service@wiley-vch.de

*€ Prices are valid in Germany, exclusively, and subject to alterations. Prices incl. VAT. excl. shipping. 1013126_dp

# Recommendations in Geotechnical Engineering

**Ernst & Sohn**
A Wiley Brand

Ed.: Deutsche Gesellschaft für Geotechnik e.V.
**Recommendations on Excavations**
3. Edition 2013. 324 pages.
€ 79,–*
ISBN 978-3-433-03036-3
Also available as ebook

For the new 3rd edition, all the recommendations have been completely revised and brought into line with the new generation of codes (EC 7 and DIN 1054), which will become valid soon. The book thus supersedes the 2nd edition from 2008.

Ed.: Deutsche Gesellschaft für Geotechnik e.V.
**Recommendations on Piling (EA Pfähle)**
2013. 496 pages.
€ 109,–*
ISBN 978-3-433-03018-9
Also available as ebook

This handbook provides a complete overview of pile systems and their application and production. It shows their analysis based on the new safety concept providing numerous examples for single piles, pile grids and groups. These recommendations are considered rules of engineering.

Ed.: Deutsche Gesellschaft für Geotechnik e.V.
**Recommendations for Design and Analysis of Earth Structures using Geosynthetic Reinforcements – EBGEO**
2011. 316 pages.
€ 89,90*
ISBN 978-3-433-02983-1
Also available as ebook

The Recommendations deal with analysis principles and the applications of geosynthetics used for reinforcement purposes in a range of foundation systems, ground improvement measures, highways engineering projects, in slopes and retaining structures, and in landfill engineering.

Ed.: HTG
**Recommendations of the Committee for Waterfront Structures Harbours and Waterways EAU 2012**
2015. 676 pages.
€ 129,–*
ISBN 978-3-433-03110-0
Also available as ebook

The "EAU 2012" takes into account the new generation of the Eurocodes. The recommendations apply to the planning, design, specification, tender procedure, construction and monitoring, as well as the handover of and cost accounting for port and waterway systems.

Order online:
www.ernst-und-sohn.de

**Ernst & Sohn**
Verlag für Architektur und technische Wissenschaften GmbH & Co. KG

Customer Service: Wiley-VCH
Boschstraße 12
D-69451 Weinheim

Tel. +49 (0)800 1800-536
Fax +49 (0)6201 606-184
cs-germany@wiley.com

* € Prices are valid in Germany, exclusively, and subject to alterations. Prices incl. VAT. excl. shipping. 1036436_dp

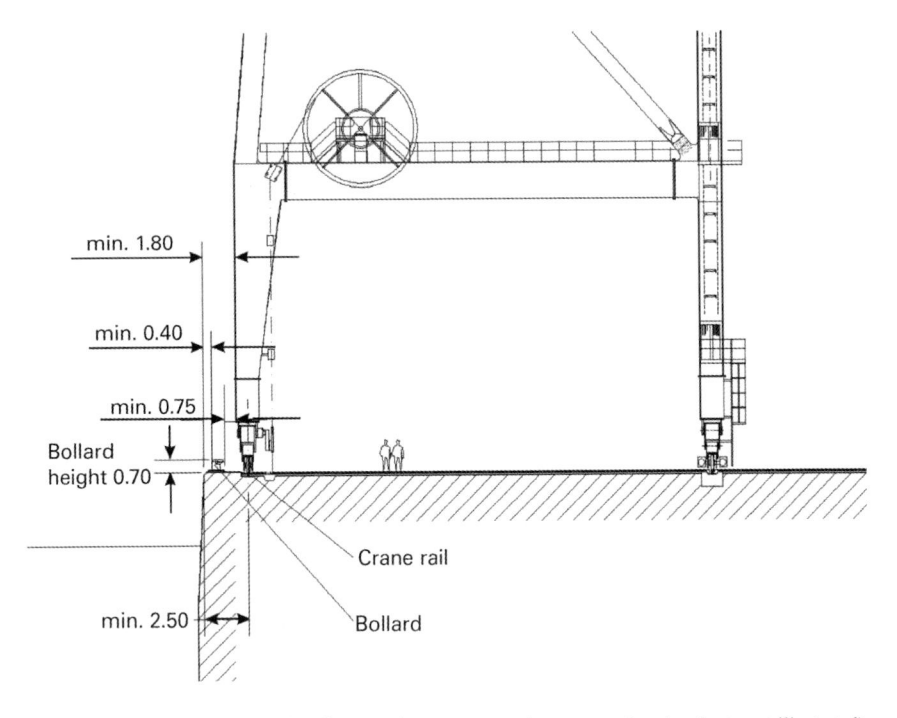

**Fig. R 6-1.** Standard cross-section for waterfront structures in seaports (service ducts not illustrated)

### 6.1.3 Railings, rubbing strips and edge protection

Railings are not required at the edges of quays for mooring and cargo-handling operations. However, the edges of such quays should be provided with adequate edge protection to R 94, section 8.4.6. The edges of quays with public access and those not used for mooring or cargo handling should be provided with railings.

### 6.1.4 Edge bollards

The front edge of an edge bollard must lie 0.15 m behind the front face of the quay wall, because otherwise it is difficult to attach and remove the hawsers of ships moored tight up against the quay wall. The width of the head of a bollard should be taken as 0.50 m.

### 6.1.5 Arrangement of tops of quay walls at container terminals

Owing to the safety requirements and high productivity demands at container terminals, a greater clearance between the front edge of the quay and the axis of the outboard crane rail is recommended, as shown in Fig. R 6-1. It should be possible to store gangways etc. parallel to the ship between the front edge of the quay and the cranes, also park service and delivery vehicles and thus separate the zones for cargo-handling and ship service traffic. It is accepted that container cranes will require longer jibs.

If containers are automatically transported between crane bridge and storage area, it is essential to separate ship service traffic and container-handling areas for safety reasons. In such cases the distance of the front edge of the quay from the outboard crane rail should be such that all service vehicle lanes can be accommodated in this strip. Alternatively, one lane can be placed in this strip and another adjacent to the outboard rail beneath the portal of the crane, separated from the container-handling area by a fence.

## 6.2 Top edges of waterfront structures in seaports (R 122)

### 6.2.1 General

The level of the top edge of a waterfront structure at a seaport is determined by the operations level of the port. When specifying the operations level, the following factors must be observed:

1. Water levels and their fluctuations, especially heights and frequencies of possible storm tides, wind-induced water build-up, tidal waves, the possible effects of flows from high water levels, and maybe other factors mentioned in section 6.2.2.2.
2. Mean groundwater level, with frequency and magnitude of level fluctuations.
3. Shipping operations, port installations and cargo-handling operations, imposed loads.
4. Ground conditions, subsoil, availability of fill material and ground management.
5. Constructional options for waterfront structures.
6. Environmental concerns.

The factors listed above must be assessed in terms of technical and economic aspects according to the requirements of the port or harbour in order to arrive at the optimum level for the operations area.

### 6.2.2 Level of port operations area with regard to water levels

When it comes to the level of the port operations area, a fundamental distinction must be made between wet docks and open harbours with or without tides.

#### 6.2.2.1 Wet docks

In wet docks protected against flooding, the height of the port operations level above the mean operational water level must be such that it is not flooded at the highest possible operational water level in the port. At the same time, it must lie above the highest groundwater level and be suitable for the envisaged cargo-handling operations.

The port operations area should generally be 2.00–2.50 m, but at least 1.50 m, above the mean operational water level in the port.

### 6.2.2.2 Open harbours

The height and frequency of high tides are critical when it comes to selecting a suitable operations level for an open harbour.

As far as possible, planning work should make use of frequency lines to indicate levels above mean high tide. In addition to the main influencing factors stated in section 6.2.1(1), the following influences must be taken into consideration:

- Wind-induced water build-up in the harbour basin
- Oscillating movements of the harbour water due to atmospheric influences (seiching)
- Wave run-up along the shore (so-called Mach effect)
- Resonance of the water level in the harbour basin
- Secular rises in the water level
- Long-term coastal uplift or subsidence

If there are no (meaningful) data records available for the above influences, as many measurements as possible must be taken in situ within the scope of the design work and these linked to high tide levels and wind actions.

### 6.2.3 Effects of (changing) groundwater levels on the terrain and the level of the port operations area

The mean groundwater level and its local seasonal and other changes, as well as their frequency and magnitude, must be taken into consideration when establishing the level of the port operations area, especially with respect to proposed pipes, cables, roads, railways, imposed loads, etc. in conjunction with subsurface conditions. Owing to the need to drain precipitation, the course of the groundwater level with respect to the harbour water must also be given attention.

### 6.2.4 Level of port operations area depending on cargo handling

1. General cargo and container handling
   In general, an operations level not liable to flooding is essential for general cargo and container handling. Exceptions should only be permitted in special cases.
2. Bulk cargo handling
   Owing to the diversity of cargo-handling methods and types of storage, as well as the sensitivity of the goods and vulnerability of

handling gear, it is not possible to give a general recommendation regarding the level of the port operations area for facilities handling bulk cargo. An effort should nevertheless be made to provide an area not liable to flooding, particularly in view of the environmental problems involved.

3. Special cargo handling

   For ships with side doors for truck-to-truck operations, bow or stern doors for roll-on/roll-off operations or other special types of equipment, the top level of the waterfront structure must be compatible with the type of vessel and equipped with either fixed or movable loading/unloading ramps. Such operations do not necessarily require the top of the waterfront structure to be at the same level as the ground. In tidal areas it may be necessary to adjust the levels in the vicinity of ramps. Floating pontoons or similar arrangements may even be necessary. In any case, the requirements of the types of vessel that will use such port facilities must be taken into account.

4. Cargo handling with ship's lifting gear

   In order to achieve adequate working clearances under crane hooks, even for low-lying vessels, the level of a quay where on-board lifting gear is used for handling must generally be lower than one where cargo is handled by quayside cranes.

## 6.3 Standard cross-sections for waterfront structures in inland ports (R 74)

### 6.3.1 Port operations level

The operations level of an inland port should normally be arranged above the highest water level. However, in the case of flowing waters with large water level fluctuations, this is frequently only possible at considerable expense. Occasional flooding can be accepted provided cargoes are not damaged and there is no risk of the watercourse being contaminated by the water lying temporarily on the port operations area.

In ports on inland canals, the operations level should be at least 2.00 m above the normal canal water level.

### 6.3.2 Waterfront

As far as possible, waterfronts in inland ports should be straight and have a front face as smooth as possible (R 158, section 6.6). Sheet piling is ideal for securing a waterfront apart from a few exceptions (R 176, section 8.1.17).

It is important to ensure that the outermost structural parts of cranes do not protrude as far as or beyond the front edge of the waterfront structure. A crane leg width of 0.60–1.00 m should be assumed.

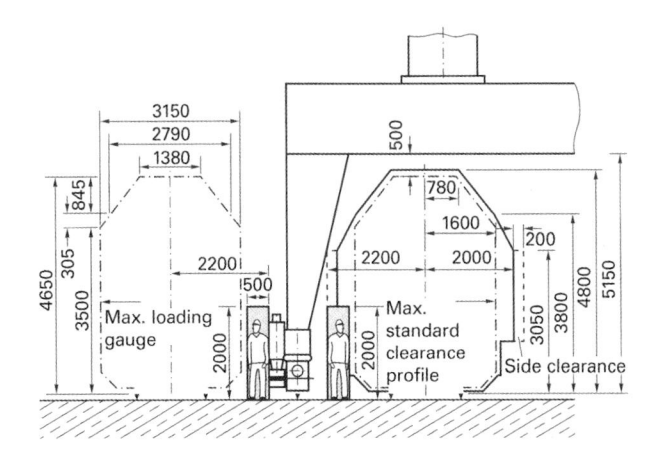

Fig. R 74-1. Side and overhead safety clearances for railways

### 6.3.3 Clearance profile

When positioning crane tracks and designing cargo-handling cranes, care must be taken to comply with the side and overhead safety clearances required by the relevant specifications (in Germany: EBO, BOA, UVV Eisenbahn, ETAB-R 25), see Fig. R 74-1.

As far as roads under crane portals are concerned, in Germany the recommendations given in Fig. R 74-2 by the Association of German Public Inland Ports (EBO, EBA, UVV, E25 in ETAB) apply.

### 6.3.4 Position of outboard crane rail

The aim is to place the outboard crane rail as close as possible to the waterfront edge in order to reduce the length of the crane jib to a minimum and save valuable storage space near the waterfront. The walkway is then positioned inboard of the crane portal leg (Fig. R 74-1). Where a walkway is provided between crane portal and waterfront, it must be at least 0.80 m wide.

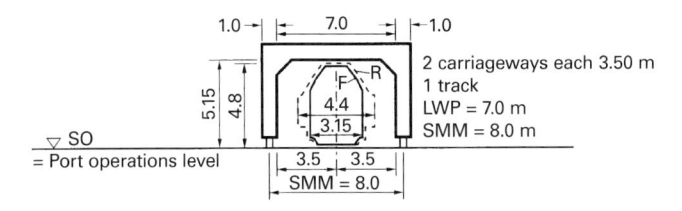

Fig. R 74-2. Recommended track gauge (SMM) and clear width (LWP) for crane portals over roads and covered tracks

173

Depending on the circumstances, the waterfront structure can be in the form of a reinforced concrete wall, a sheet pile wall or a combination of these two forms of construction – a reinforced concrete wall with a facing of sheet piles on bored piles (Fig. R 74-5). The advantage of reinforced concrete retaining walls is that they are easily built with smooth wall surfaces.

In the case of sheet piling, the crane rail should be aligned with the centre-line of the sheet pile wall (Fig. R 74-3). However, the necessary

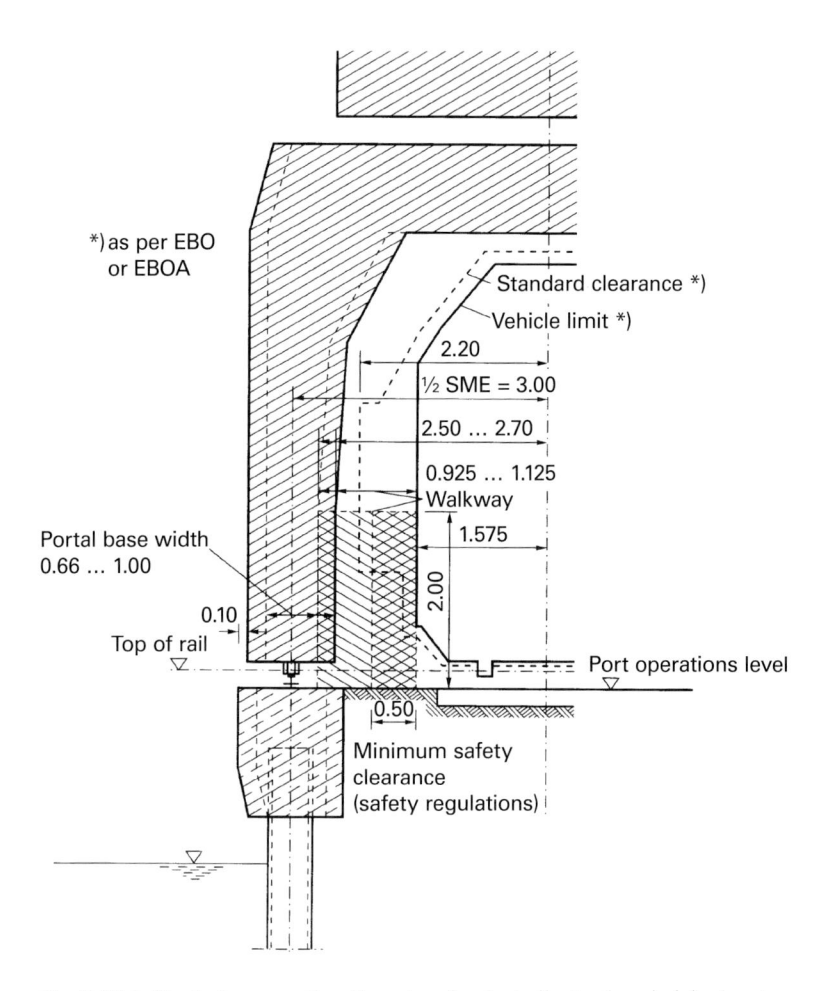

**Fig. R 74-3.** Standard cross-section dimensions for sheet pile structures in inland ports (crane rail on centre-line of sheet piling)

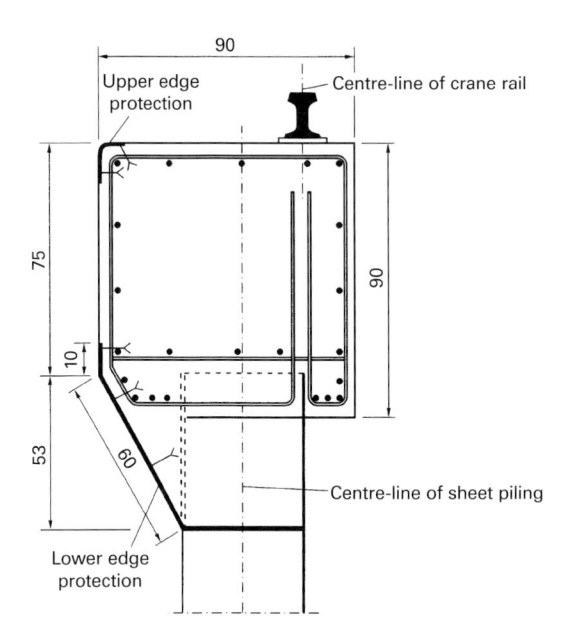

**Fig. R 74-4.** Crane rail eccentric to centre-line of sheet piling (example)

geometrical requirements (section 6.3.2) can make it necessary to support the crane rail off-centre (Fig. R 74-4).

A combined solution (reinforced concrete wall plus facing of sheet piles on bored pile foundations as shown in Fig. R 74-5) has the advantage that the crane loads are transferred via the bored piles to the subsoil. In addition, the crane rail can be routed close to the edge without vessel impacts having any influence on the crane. Ladders for access to vessels can be placed in ideal positions behind mooring piles (Fig. R 74-5 and recommendation R 42 of the Association of German Public Inland Ports).

### 6.3.5 Mooring equipment

Sufficient mooring equipment for vessels must be installed on the water side of the waterfront structure (R 102, section 5.13).

## 6.4 Sheet piling waterfronts on inland waterways (R 106)

### 6.4.1 General

Where canals are to be constructed or extended in areas where only limited space is available, waterfront structures of anchored steel sheet

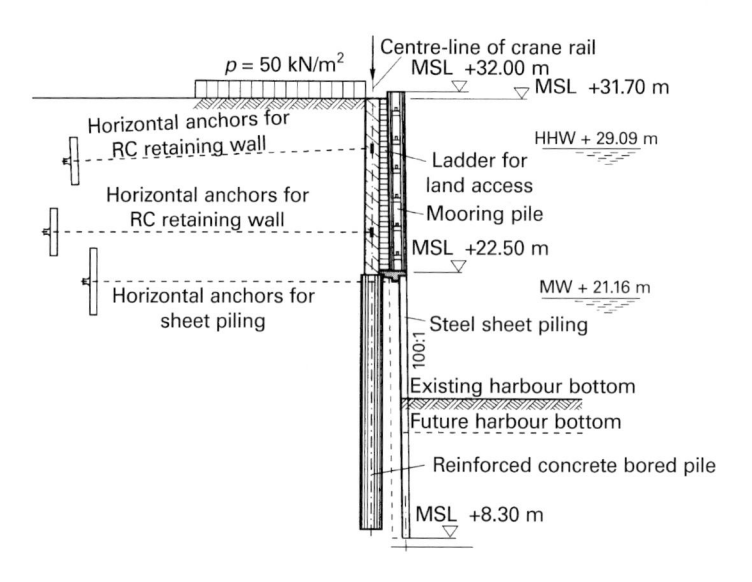

**Fig. R 74-5.** Anchored sheet pile wall with mooring piles/reinforced concrete wall (example)

piles are frequently the best engineering solution and – after considering the land acquisition and maintenance costs – also the most economical. This is especially true for stretches requiring an impervious bottom. If necessary, sheet pile interlocks can be sealed, see R 117, section 8.1.21. Fig. R 106-1 shows a typical example of such a structure.

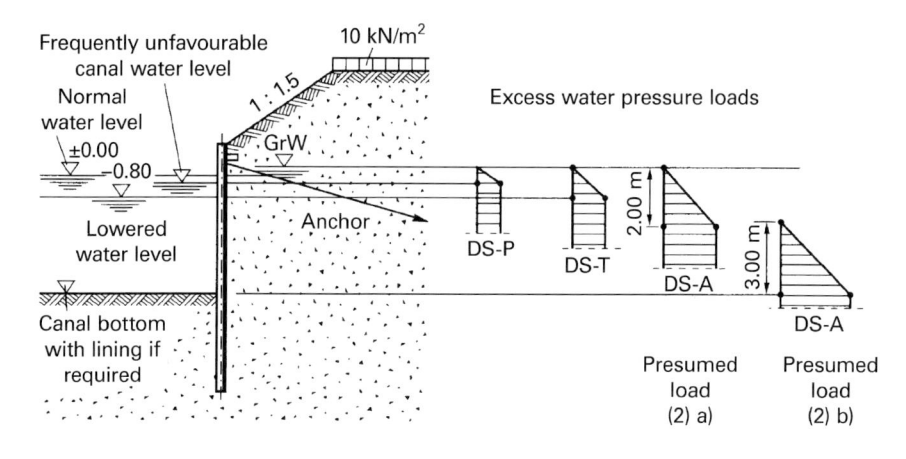

**Fig. R 106-1.** Section through a sheet piling waterfront on a normal stretch of inland waterway showing the most important loading assumptions

Where shipping conditions allow, the top edge of the sheet piling should remain below the water level for reasons of corrosion protection and landscaping. Please refer to the Federal Ministry of Transport, Planning, and Housing for more information on standard cross-sections.

### 6.4.2 Stability analysis

To verify stability, the waterfront structure and its components are analysed and designed according to the pertinent recommendations. Please refer to R 19, section 4.2, and R 18, section 5.4, for water pressures. In the case of vertical imposed loads, in contrast to R 5, section 5.5, a uniformly distributed surcharge of $10\,kN/m^2$ (characteristic value) should to be assumed (Fig. R 106-1).

In addition, please also refer to R 41, section 8.2.10, for guidance on staggered embedment depths and R 55, section 8.2.8, for guidance on selecting embedment depths.

### 6.4.3 Loading assumptions

The loads assigned to the design situations are characteristic values. The excess water pressure to be used in design situation DS-P is the pressure that can be expected to occur frequently due to unfavourable canal and groundwater levels. This also includes a lowering of the canal water level by 0.80 m in front of the sheet piling due to passing vessels. The groundwater level is often assumed to be at the top of the sheet piling. Design situation DS-A considers the following loads:

If there has been a rapid and severe fall in the canal water level, the following load cases must be investigated (Fig. R 106-1):

a) Canal water level 2.00 m below groundwater level
b) Canal water level at canal invert with groundwater level 3.00 m higher

Where the failure of a waterfront structure would cause bridges and loading installations etc. to collapse, then sheet piling must be designed for the load case "canal empty", or must be adequately secured by means of special structural measures.

In the structural investigations, the design canal invert level or base of excavation (e.g. underside of bottom protection) may be assumed to be the theoretical level for calculation purposes. Excavation down to 0.30 m below the design bottom level is generally permissible without special calculations when the wall is fully fixed in the ground (see R 36, section 6.7). However, this does not apply to walls without anchors and anchored walls with a simply supported base. If in exceptional cases greater deviations are to be expected and severe scour damage due to ship propeller action is likely, the calculations should assume a bottom depth at least 0.50 m below the design bottom level.

### 6.4.4 Embedment depth

If soil with a low permeability is encountered at an attainable depth in dam or dyke stretches that are to be made watertight, the sheet piling should be driven to such a depth that it is embedded in this stratum. A lining to the canal bottom is then unnecessary. However, this course of action must not have a negative impact on the wider flow of groundwater.

## 6.5 Upgrading partially sloped waterfronts in inland ports with large water level fluctuations (R 119)

### 6.5.1 Reasons for partially sloped upgrades

The berthing, mooring, lying and casting-off of vessels in inland ports must be possible at every water level without the use of anchors, and port and operations personnel should have safe access to vessels at all water levels. This is only possible with vertical waterfronts.

If in inland ports with fluctuating water levels a port operations area free from flood water is also required, this can result in very high quay walls. In these cases it is appropriate to design cargo-handling facilities with a vertical quay wall and an adjoining upper slope (Figs. R 119-1 and R 119-2).

### 6.5.2 Design principles

For cargo-handling operations on a partially sloping waterfront, the level of transition from vertical wall to sloping bank should be such that it is not below water for more than 60 days (long-term mean).

On the Lower Rhine, for example, this corresponds to a transition level about 1.00 m above mean water level (Fig. R 119-1). The level of this vertical/sloping transition should remain the same throughout the port basin.

For waterfronts with a port operations area at a very high level, the transition from the quay wall to the bank should be positioned so that – for operating and structural reasons – the bank is max. 6.00 m high (Fig. R 119-2).

Guide piles about 40 m apart are advisable along the vertical bank section at berths and push-tow coupling quays for unmanned vessels without cargo-handling operations. These piles serve for marking, safe mooring and protecting the bank. They should extend 1.00 m above highest high water but not project over the water (Fig. R 119-1).

The vertical waterfront is generally constructed of sheet piles with a fixed-end support at the base and a single row of anchors at or near the top.

The top of the piles should be finished with a 0.70 m wide steel or reinforced concrete capping beam, which also can also function as a safe

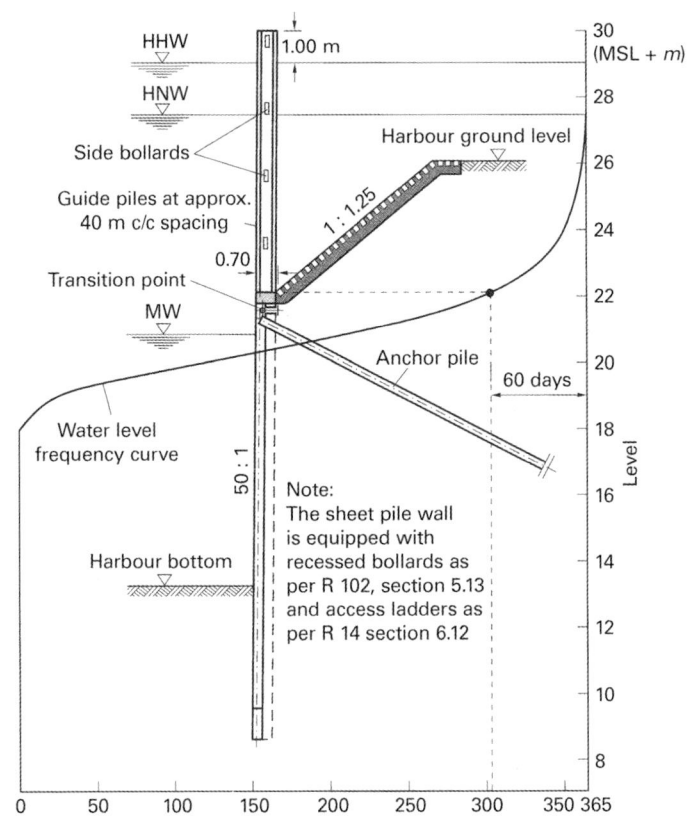

**Fig. R 119-1.** Partially sloped bank for berths, particularly for pushed lighters where the port operations level is subject to flooding

walkway between the guide piles (Figs. R 119-1 and R 119-2). It must be possible to pass behind the guide piles.

Ensuring that ships are moored correctly will prevent them grounding in the event of falling water levels.

The outboard edge of a reinforced concrete capping beam is to be protected against damage by a steel plate as described in R 94, section 8.4.6.

To ensure safe access via steps, the slope above the wall must not be steeper than 1:1.25. Slopes of 1.25 to 1:1.5 are chiefly used.

Bollards should be in accordance with R 102, section 5.13.

179

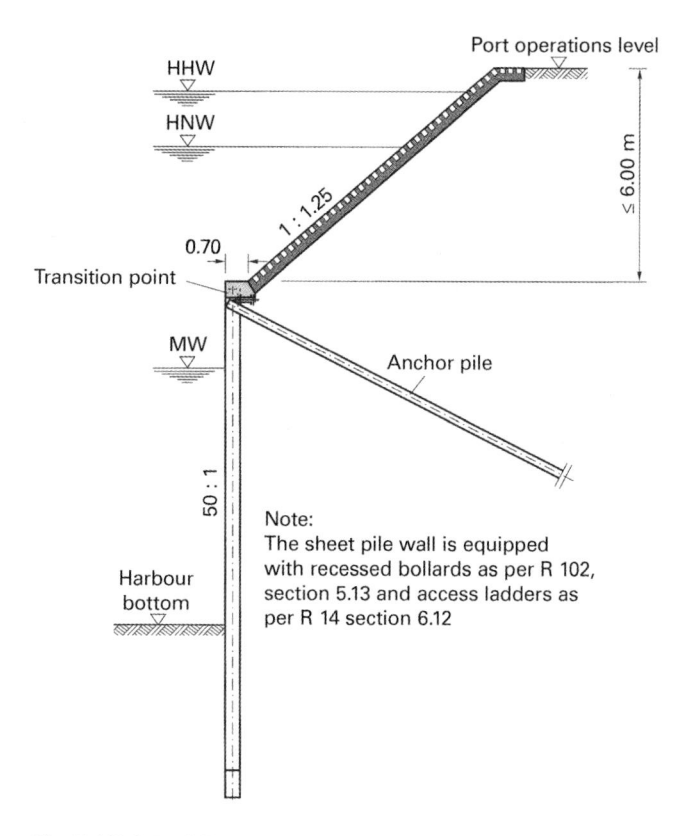

**Fig. R 119-2.** Partially sloped bank with flood-free port operations level

## 6.6 Design of waterfront areas in inland ports according to operational aspects (R 158)

### 6.6.1 Requirements

Designs for embankment cross-sections at inland ports are primarily influenced by economic and operational considerations. When handling cargoes with cranes, a clearly laid out facility is important.

Trouble-free shipping operations are ensured when vessels can tie-up and cast-off easily at the waterfront, and passing ships or push-tow convoys do not create disadvantageous ship movements. Mooring cables and ropes must be able to slacken as the water level changes.

The waterfront structure should also act as a guide when berthing ships.

Pushed lighters have relatively large masses and are box-shaped with sharp corners and edges. Therefore, waterfront structures for cargo-handling operations involving pushed lighters should be as flat as possible.

Vessels should move as little as possible during loading/unloading to ensure fast and safe operations. On the other hand, if necessary it should be possible to warp the vessel without difficulty.

Regarding the carriage of passengers between land and ship, it must be possible to transfer them directly or use a gangway safely (ETAB [64], E 42).

### 6.6.2 Design principles

In principle, long straight waterfronts are preferable. If changes in direction are unavoidable, they should be designed in the form of angled, not rounded, turns. Distances between changes of direction should be such that the intermediate straight stretches match the lengths of the ships or push-tow convoys using the facility. Trouble-free shipping operations are easiest to achieve with smooth quay walls without recesses or protruding structures. The front faces of the waterfront walls can be sloped, partially sloped or vertical.

### 6.6.3 Waterfront cross-sections

1. Embankments
   Embankment surfaces should be designed as flat as possible. Intermediate landings should be avoided if feasible. Steps and stairs should be installed in the direction of the slope, i.e. at a right-angle to the shoreline. Bollards and mooring rings must not protrude above the embankment surface. If intermediate berms are unavoidable on high embankments, they must not be located in the area of frequent water level fluctuations. Instead, they should be situated in the high water zone.
   See R 119, section 6.5.2, for details of the transition from the sloping to the vertical waterfront.
   On sloping and partially sloping waterfronts, safe guiding of vessels can only be guaranteed in conjunction with closely spaced mooring dolphins.
2. Vertical waterfronts
   Vertical or only slightly inclined waterfront structures in concrete or masonry must be built with a smooth front face. This condition is easily fulfilled when constructing a new structure in a dry excavation. If built in the form of a diaphragm wall or contiguous piling, the side of the wall on the water side usually requires further work to meet operational requirements regarding a flat wall.
   Using sheet piles for vertical waterfront structures represents a proven, economical solution. Should the shipping operations require

it, sheet pile walls can be reinforced as in R 176, section 8.1.17, in order to achieve a flat outer surface.

## 6.7 Nominal depth and design depth of harbour bottom (R 36)

### 6.7.1 Nominal depth in seaports

The nominal depth is the water depth below a defined reference level. When stipulating the nominal depth of the harbour bottom in front of a quay wall, the following factors should be considered:

1. The draught of the largest, fully laden ship berthing in the port. The draught must be calculated taking into account the salinity of the water and the listing of the ship.
2. A safety clearance between ship's keel and nominal depth. The safety clearance depends on the regulations of the local harbour authorities, but should be at least 0.50 m.

The reference level for the nominal depth is generally a statistically based low water level.

In regions without tides, e.g. the Baltic Sea, the low water level is derived from data collected over many years.

When determining the reference level and hence the nominal depth in front of waterfront structures in tidal regions, appropriate consideration of the tide-related changes to the water level is necessary so that an adequate water depth is available with adequate statistical frequency. In this case the chart datum (CD) is frequently chosen as the reference level. Up until the end of 2004 the chart datum in Germany was derived from the mean low water spring level. Since 2005 chart datum has been defined as the lowest astronomical tide (LAT). This is a standardised reference level, which is also common internationally, for all countries bordering the North Sea.

LAT defines the lowest possible water level caused by astronomical influences. For the German North Sea, LAT is about 0.50 m below mean low water spring.

The reference level has to be fixed with respect to local requirements and can also be different from chart datum LAT when a lower water level due to exceptional meteorological or astronomical conditions with a higher statistical frequency is acceptable. Therefore, the reference level must be agreed unanimously by all those involved prior to fixing the nominal depth for the harbour bottom.

### 6.7.2 Nominal depth of harbour bottom for inland ports

The nominal depth of the harbour bottom in inland ports and harbour entrances should be selected so that ships can reach their destinations with the greatest possible loaded draught. In inland ports on rivers, the

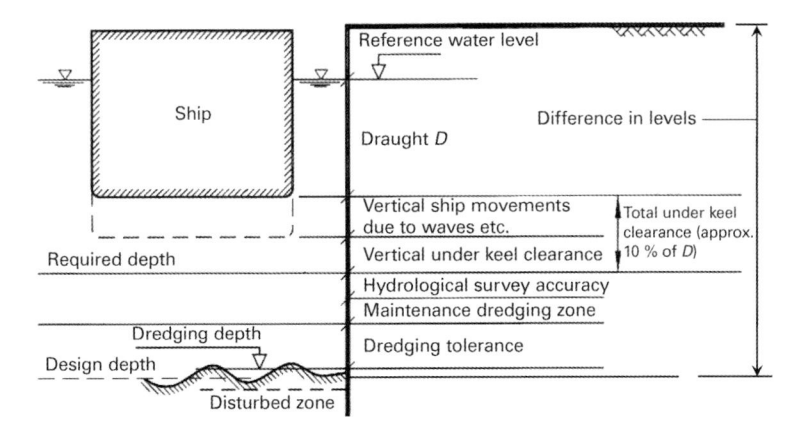

**Fig. R 36-1.** Calculating the design depth in accordance with CUR [42]

water depth should generally be 0.30 m deeper than that of the adjoining waterway in order to rule out any dangers for ships in ports at low water levels.

### 6.7.3 Design depth in front of quay wall

If dredging is to be carried out in front of a quay wall because of silt, sand, gravel or rubble deposits, the dredging must be deeper than the intended nominal depth of the harbour bottom stipulated in sections 6.7.1 and 6.7.2 (Fig. R 36-1).

The design depth is made up of the nominal depth of the harbour bottom, the maintenance margin down to the planned dredging depth plus dredging tolerances and other allowances, e.g. possible loosening of the floor while dredging (disturbance zone). The design depth can therefore be any value selected by the design team below the dredging depth.

The dredging depth is determined using the following factors:

1. Extent of the silt mass, sand drift, gravel or rubble deposits per dredging period.
2. Depth below the nominal depth of the harbour bottom to which the soil may be removed or disturbed.
3. Costs of every interruption to cargo-handling operations caused by dredging works.
4. Availability of the required dredging plant.
5. Costs of dredging work with regard to the depth of the maintenance margin.
6. Extra costs of a quay wall with a deeper harbour bottom.

**Table R 36-1.** Maintenance margins and minimum dredging tolerances, reference values [m]

| Depth of water below lowest water level [m] | Depth of maintenance margin [m] | Minimum dredging tolerance*⁾ [m] |
|---|---|---|
| 5  | 0.5 | 0.2 |
| 10 | 0.5 | 0.3 |
| 15 | 0.5 | 0.4 |
| 20 | 0.5 | 0.5 |
| 25 | 0.5 | 0.7 |

*⁾ Depends on dredging plant

Additional information can be found in R 139, section 7.2.

Owing to the importance of the above factors, the maintenance margin, representing an addition to the nominal depth, must be stipulated with care. On the one hand, an inadequate margin can lead to high costs for frequent maintenance dredging and the ensuing interruptions to operations. On the other hand, an excessive margin results in higher construction costs and may encourage sedimentation.

It is practical to attain the harbour bottom depth first in at least two dredging operations executed at intervals. A maximum dredging depth of 3.00 m must be observed.

Table R 36-1 provides a general guide to the depth of maintenance margins and minimum dredging tolerances to be used for different water depths. More information on dredging tolerances can be found in R 139, section 7.2.2.

The reference values in Table R 36-1 already take account of the allowances required by DIN EN 1997-1 (see also section 2.12).

If erosion of the harbour bottom is expected in front of the wall, the design depth must be increased, or suitable measures taken to prevent erosion.

## 6.8 Strengthening waterfront structures for deepening harbour bottoms in seaports (R 200)

### 6.8.1 General

Developments in ship dimensions mean that occasionally it is necessary to deepen the harbour bottom in front of an existing quay wall. In these cases, larger crane and imposed loads are often necessary, meaning that the waterfront structures must also be strengthened.

Whether in any individual case it is both possible and economical to deepen the harbour bottom and reinforce the waterfront structure depends on various factors:

First, checks must be carried out to establish whether the required deepening of the harbour bottom is even possible given its current depth. Next, checks must be carried out to establish whether the design and condition of the waterfront structure is suitable for use with a deeper harbour bottom. To this end, design drawings and structural calculations of the wall and the results of soil investigations from the construction phase can be studied, provided they are available. If required, additional soil investigations may reveal more favourable soil properties, resulting from the improvement of the subsoil characteristics due to consolidation since construction.

Finally, it is necessary to check whether the required strengthening is economical with respect to the remaining useful life of the waterfront structure compared with the costs of building a new one.

Stability and serviceability must then be verified for the actions due to the new loads and the increased theoretical depth of the harbour bottom. If the stability analyses of the present wall and/or its design drawings are no longer available, compensating for the loads on the wall due to the increase in the depth of the harbour bottom by reducing the imposed loads is one possible approach.

### 6.8.2 Design of strengthening measures

There are numerous possibilities for reinforcing quay walls to withstand greater loads resulting from deepening the harbour bottom. It is crucial to ensure that the embedment depth of the wall is still sufficient after the deepening. This also applies to the intermediate piles of combined walls. A few typical solutions are given below.

### 6.8.2.1 Measures to increase the passive earth pressure

The load-carrying capacity of waterfront walls can be improved by increasing the passive earth pressure at the base of the wall.

If the in situ soil is soft and cohesive with little strength, it can be replaced by a non-cohesive material with a high unit weight and high shear strength down to the required depth as shown in Fig. R 200-1.

The transition to the in situ soil must ensure a stable filter action. The soil replacement may only be carried out in stages and it is important to observe deformations of the wall. No cargo-handling operations should be carried out in the area affected during soil replacement activities. If required, the load on the wall can be relieved temporarily by removing the backfill behind the wall. Details on soil replacement can be found in R 164, section 2.11, and R 109, section 7.9.

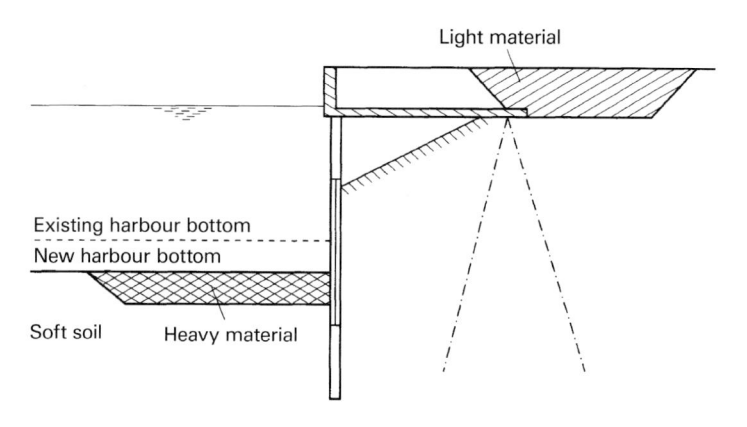

**Fig. R 200-1.** Soil replacement in front of and/or behind the structure

If the soil in front of the waterfront wall is non-cohesive and can be compacted, the passive earth pressure can be increased by compaction and, if necessary, by adding gravel or ballast (Fig. R 200-2). Permeable non-cohesive soils can be stabilised by grout injection.

### 6.8.2.2 Measures to reduce active earth pressure

The active earth pressure on a waterfront wall can be reduced, for example, by means of a relieving slab supported on piles (Fig. R 200-3). Further options include the partial replacement of the backfill with a lighter material (Fig. R 200-1) or employing grout injection to stabilise the backfill.

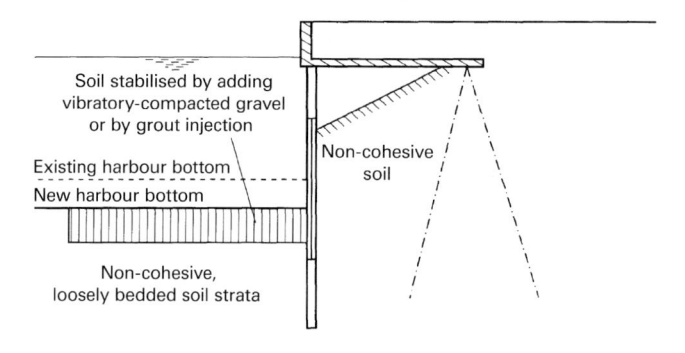

**Fig. R 200-2.** Soil stabilisation or soil compaction in front of the structure

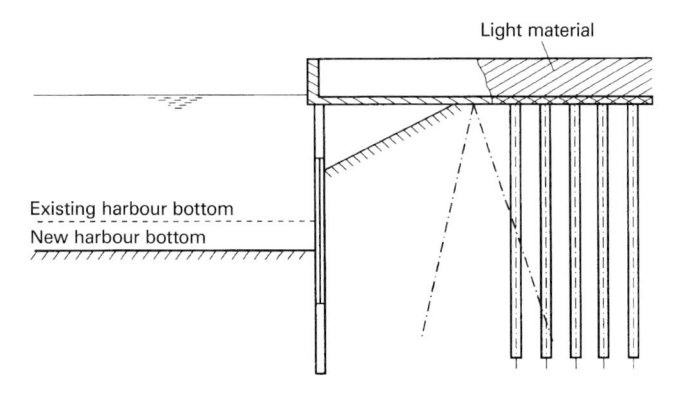

**Fig. R 200-3.** Stabilisation with a relieving slab supported on piles

### 6.8.2.3 Measures involving the quay wall

The waterfront structure itself can be upgraded using additional anchors for carrying increased loads (Fig. R 200-4). If required, it is also possible to drive the existing waterfront structure deeper and extend it (Fig. R 200-5). This option requires the existing anchors to be temporarily detached and then reattached or replaced by new anchors.

However, in most cases it will be necessary to build a new wall in front of the old one and to anchor it either with a new superstructure (Fig. R 200-6) or with raking or horizontal anchors (Fig. R 200-7).

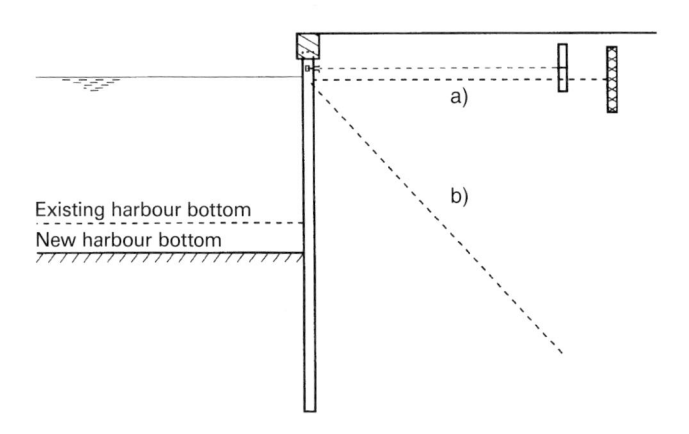

**Fig. R 200-4.** Use of additional anchors: a) horizontal, b) raking

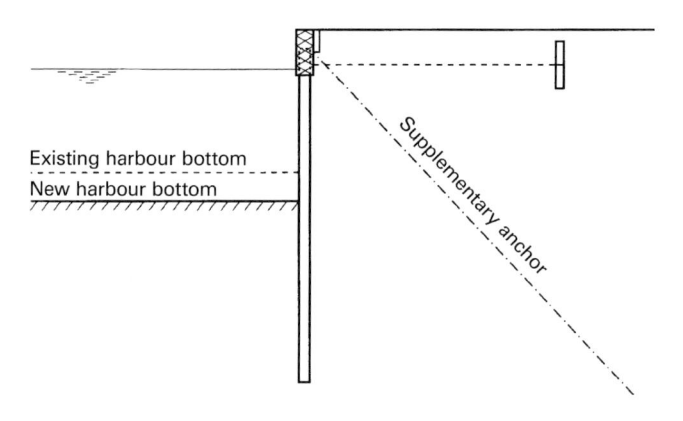

**Fig. R 200-5.** Driving the existing waterfront structure deeper and extending it upwards, plus additional anchors

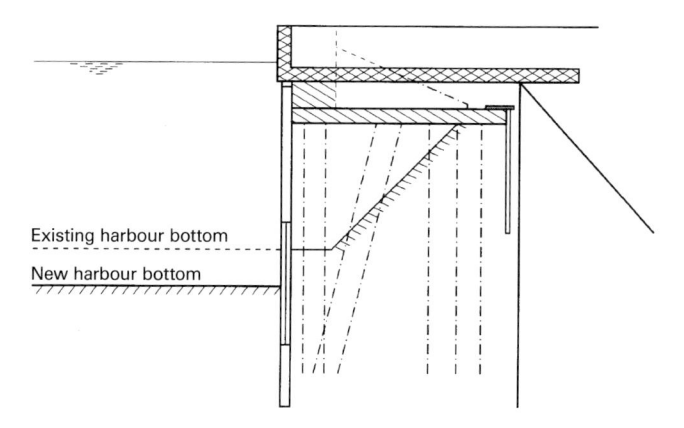

**Fig. R 200-6.** New wall in front of existing one plus a new superstructure

Provided there is sufficient space, an entirely new pile cap can be built to provide extra support for the loads from cargo-handling operations (Fig. R 200-8). This solution also creates a larger cargo-handling area. For more information on the design and calculation of pile caps see R 157, section 11.4.

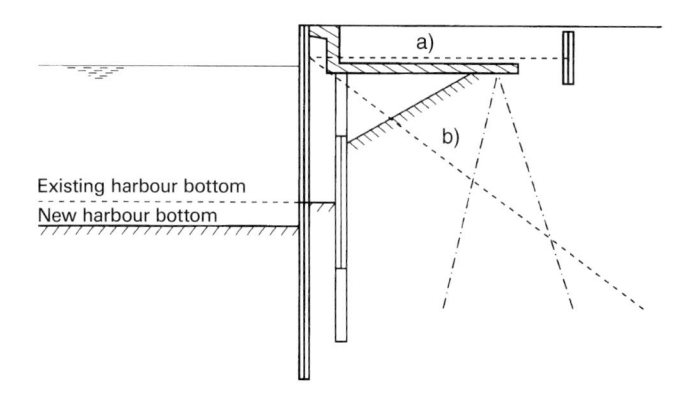

**Fig. R 200-7.** New wall in front of existing one plus additional anchors: a) horizontal, b) raking

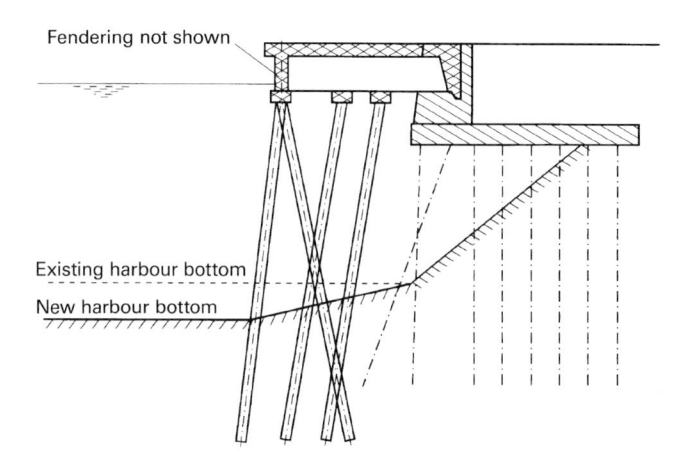

**Fig. R 200-8.** Forward extension on piles with underwater embankment

## 6.9 Embankments below waterfront wall superstructures behind closed sheet pile walls (R 68)

### 6.9.1 Embankment loads

Slopes below waterfront wall superstructures can be loaded by currents along the waterfront wall and by flow forces from groundwater flows, in addition to active earth pressure. Regarding the stability of slopes, flow forces are disadvantageous when the groundwater level behind the slope is higher than the outer water level. Consequently, the flow forces are directed away from the slope (see R 65, section 4.3).

Therefore, the angle of the slope must always be such that the slope is stable for all water levels. In addition, the surface of the slope must be protected against erosion due to currents in the outer water.

### 6.9.2 Risk of silting-up behind sheet pile wall

In tidal areas there is a risk that silt will accumulate on slopes located beneath superstructures. The ensuing additional loads on the superstructure piles can be substantial.

Silt deposits can only be prevented permanently when no silt-laden water can penetrate the area beneath the superstructure. However, this usually entails considerable extra costs for the waterfront structure.

It is therefore generally accepted that outer water will penetrate the area beneath the superstructure. Silt deposits are therefore avoided by including outlets at regular intervals along the sheet pile wall just above the base of the slope.

The effect of such measures should be monitored on an individual basis. Should they not achieve the desired results, the silt deposits must be removed regularly.

## 6.10 Redesign of waterfront structures in inland ports (R 201)

### 6.10.1 General

In canals and rivers with controlled levels, expansion work for a deeper draught may require port facilities to be deepened. In some cases increasing the crane and imposed load capacities may require a redesign of the waterfront structure.

In the case of river ports, an increase in water depth is required when the port is located in a side basin and the river bed is deepened by erosion. In order to ensure access to the port facilities, the harbour bottom must then be deepened.

Where port facilities have sloping banks, deepening the bottom results in a reduction in the width of the basin and water cross-section. If this is unacceptable, a partial slope or vertical quayside are the options. This is then also advantageous for cargo handling as the outreach of cargo cranes on partially sloping or vertical expansion projects is shorter than on sloping waterfronts.

Deepening the harbour bottom and/or higher imposed loads results in higher stresses on individual structural components which then, under certain circumstances, are no longer adequately designed.

### 6.10.2 Redesign options

Given the aforementioned circumstances, it is generally always possible to construct a new waterfront structure in front of or instead of the old one. However, it is often sufficient to renew or reinforce certain parts of

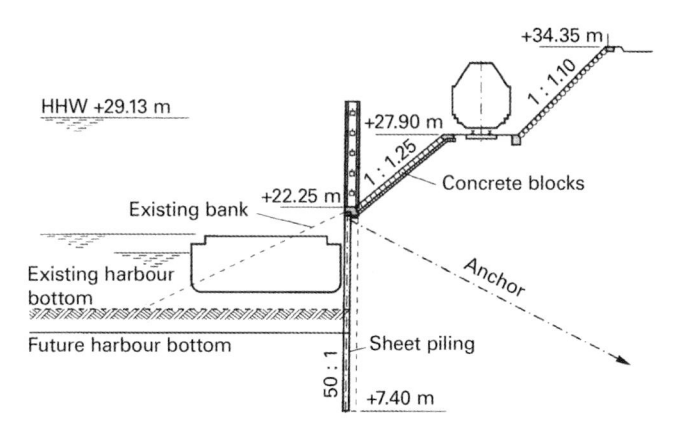

**Fig. R 201-1.** Waterfront upgrade by replacing a sloping bank by a partially sloping bank

the waterfront structure, or to implement other upgrading measures as per R 200, section 6.8.

Thus, for example, sheet piles can be driven deeper and a new superstructure built off these. Higher anchor forces can be accommodated with additional anchors. Non-cohesive soils on the harbour bottom can be compacted, which leads to an increase in passive earth pressure. Soil nailing can improve the stability of an embankment.

### 6.10.3 Construction examples

Figs. R 201-1 to Fig. R 201-6 show typical examples of redesigns for waterfront structures for inland ports. All levels are related to MSL.

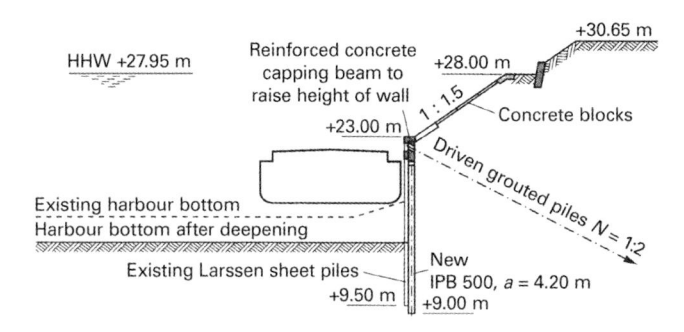

**Fig. R 201-2.** Waterfront upgrade achieved by driving anchored IPB 500 beam sections behind the existing sheet piles and raising the height of the sheet piling

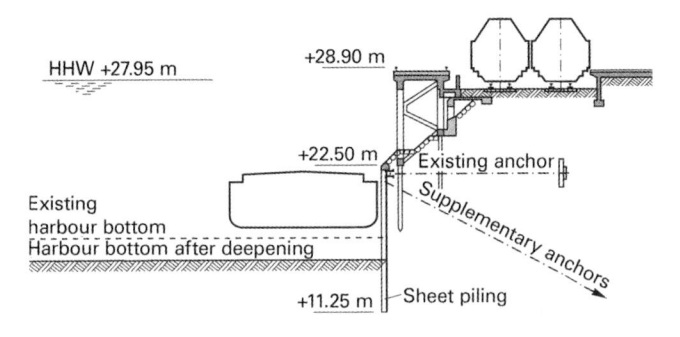

**Fig. R 201-3.** Waterfront upgrade by means of additional anchorages to existing sheet piling

Fig. R 201-1 shows an example of increasing the water depth by redesigning a sloping bank as a partially sloped bank with sheet piling.
In the example shown in Fig. R 201-2, IPB 500 steel beam sections were driven 4.2 m apart behind existing waterfront sheet piling. The sheet pile wall was raised with a reinforced concrete capping beam and re-anchored with driven grouted piles.
In the example shown in Fig. R 201-3, the existing sheet pile wall was given additional anchorage in the form of driven grouted piles, whereas in Fig. R 201-4 a new anchored sheet pile wall has been driven in front of the existing wall.

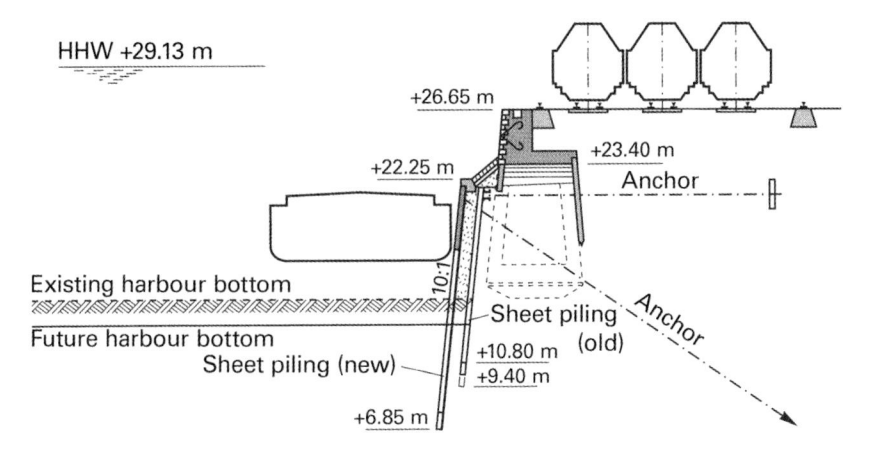

**Fig. R 201-4.** Waterfront upgrade by driving new anchored sheet piles in front of the old wall

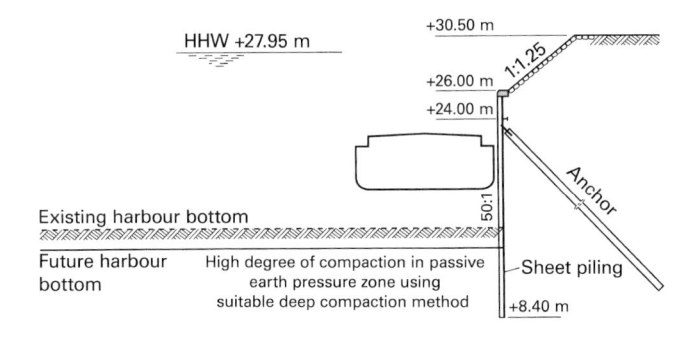

**Fig. R 201-5.** Waterfront upgrade by compacting non-cohesive soil in the passive earth pressure area in front of the sheet piling

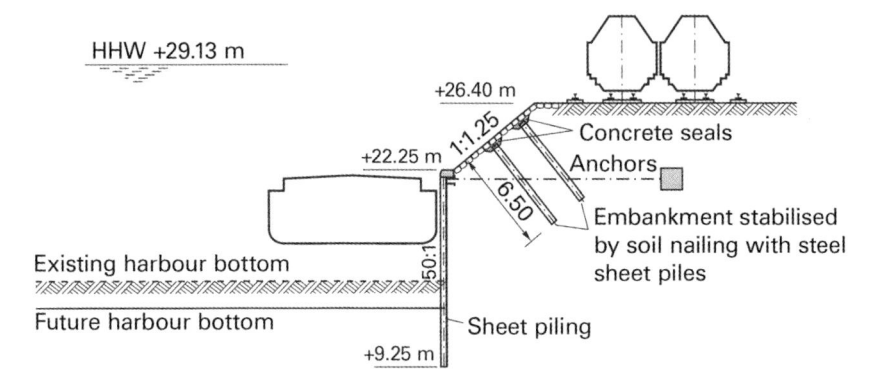

**Fig. R 201-6.** Waterfront upgrade with slope secured with soil nailing

Fig. R 201-5 shows a waterfront wall upgraded by increasing the passive earth pressure in front of it – achieved by compacting the in situ soil. In Fig. R 201-6, soil nailing has been used to stabilise the slope above the waterfront.

## 6.11 Provision of quick-release hooks at berths for large vessels (R 70)

Quick-release hooks are provided instead of bollards only in exceptional cases at special berths for large vessels where mooring takes place according to a defined mooring system. The range of movement of the quick-release hook is defined according to the mooring system. Manual and hydraulic release mechanisms with remote control enable simple tying-up and swift casting-off of hawsers, even in the case of heavy hawsers with loads of up to 3000 kN.

Fig. R 70-1 shows an example of a quick-release hook for 1250 kN maximum load with manual release mechanism. It can be used with several hawsers and it takes little effort to release the hook whether hawsers are at full or lower loads.

A quick-release hook is attached to its base via a universal joint. The number of quick-release hooks depends on the line pull to be considered according to R 12, section 5.12, and the directions from which the principal line pulls can occur simultaneously. Several quick-release hooks can be installed on one base. The range of movement must be chosen so that the hook can cover all anticipated operational requirements without jamming. The swivel range is max. 180° horizontally and 45° vertically.

It is easier to attach heavy-duty towing hawsers when the quick-release hook is combined with a capstan.

## 6.12 Layout, design and loads of access ladders (R 14)

### 6.12.1 Layout

Vertical ladders are used for vessel embarkation/disembarkation in exceptional circumstances only. They are primarily intended to provide access to mooring equipment and, in emergencies, to enable persons who have fallen into the water to climb onto the quayside. Trained and experienced shipping and operations personnel may also be expected to use the ladders when there are very large water level fluctuations, even in the case of great differences in water levels.

Vertical ladders in waterfront structures of reinforced concrete should be placed at approx. 30 m intervals. The position of the ladder depends on the position of the bollard because ladders must not be obstructed by mooring lines. If joints between sections have been included in quay walls of reinforced concrete, it is advisable to place the ladders near those joints. In the case of sheet piling waterfront structures, positioning the vertical ladders in the pile troughs is recommended.

Mooring equipment should be installed on both sides of each ladder (R 102, section 5.13.1).

### 6.12.2 Design

In order to be accessible from the water at all times, even at low water levels, each ladder must extend down 1.00 m below the lowest low water or lowest astronomical tide. For easy installation and replacement, the lowest ladder mountings are designed as plug-in items into which the stiles can be inserted from above.

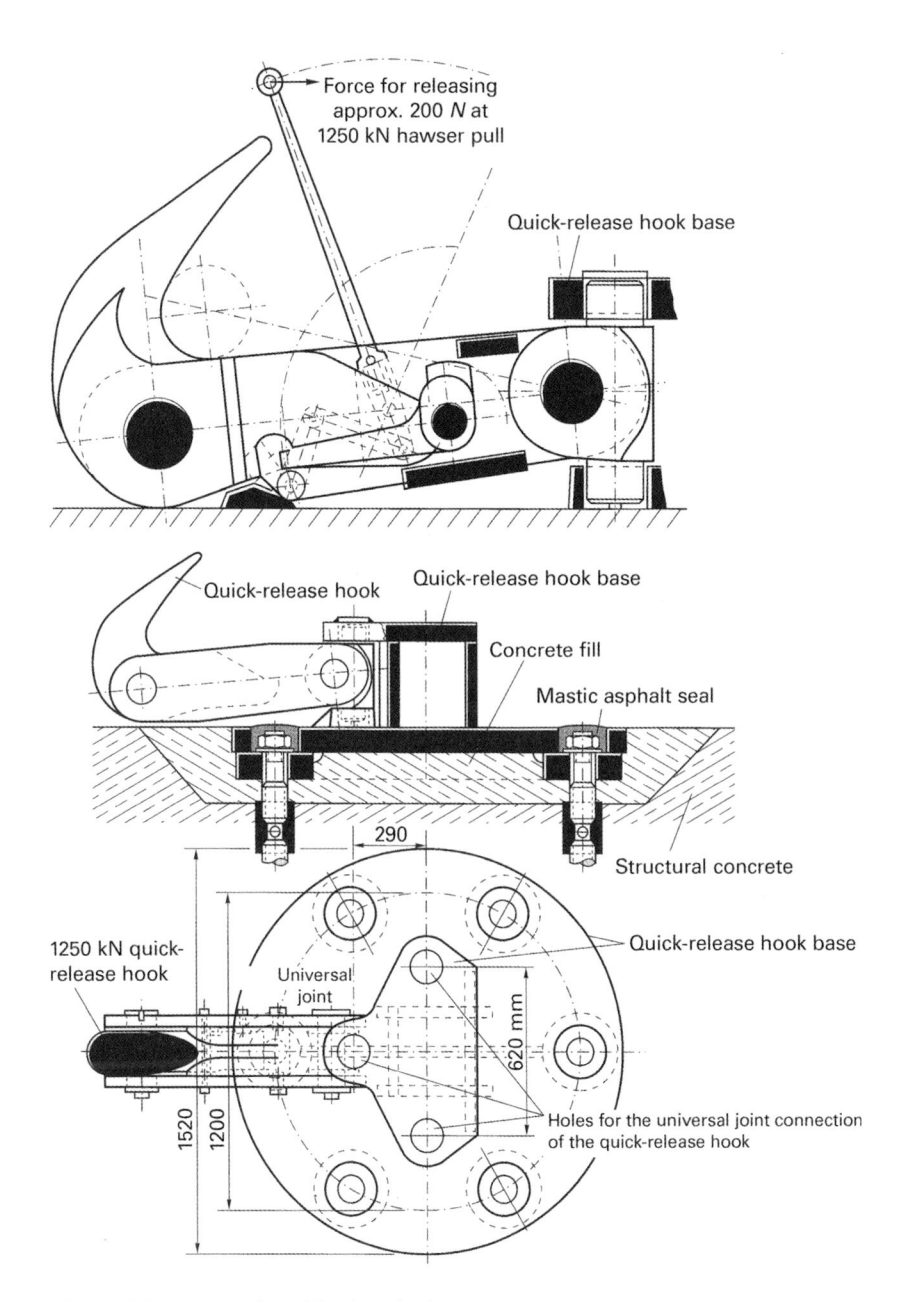

**Fig. R 70-1.** Example of a quick-release hook

Transitions between top of ladder and quayside must be designed to ensure that ascending and descending the ladder can be accomplished safely. At the same time, the ladder must not be a hazard to traffic on the quayside.

Fig. R 14-1 shows a tried-and-tested design that satisfies these two requirements. Here, the edge protection is dished at each ladder. The

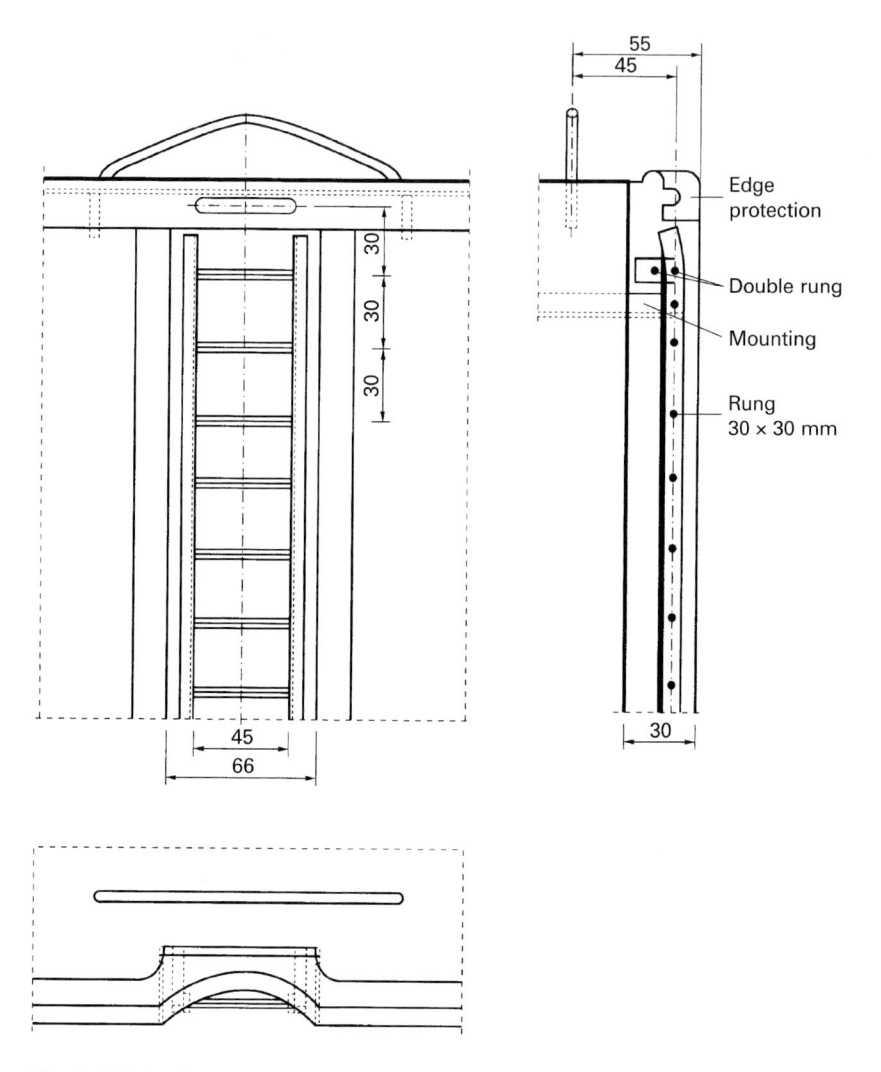

**Fig. R 14-1.** Vertical ladder in steel capping beam (dims. in cm)

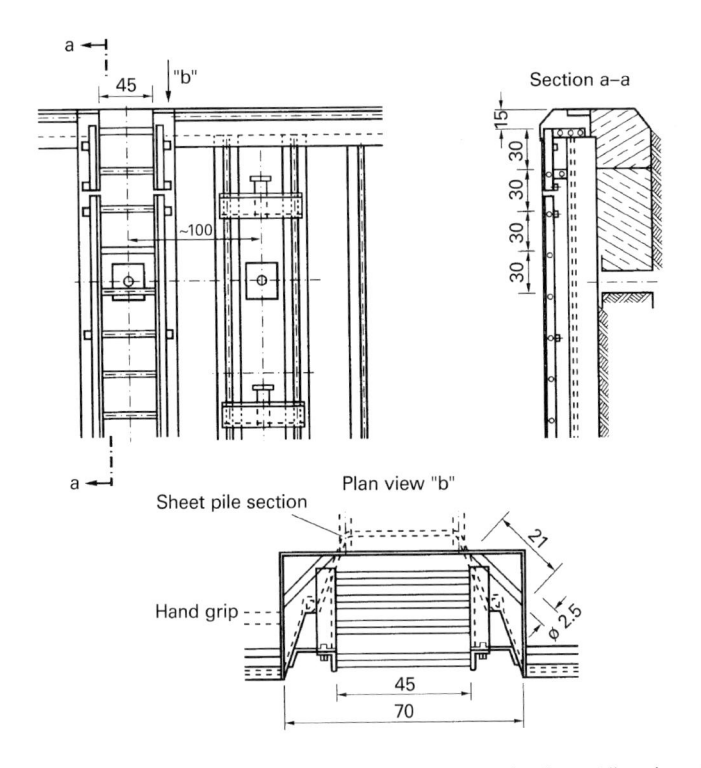

**Fig. R 14-2.** Vertical ladder in reinforced concrete capping beam (dims. in cm)

detail includes a handrail, made of 40 mm dia. material about 30 cm above the quayside and its longitudinal axis about 55 cm from the face of the wall. If the handrail proves to be an obstacle during cargo handling, other suitable aids for climbing the ladder must be provided.

Fig. R 14-2 shows a proven design of this type. The topmost rung of the ladder in this solution is 15 cm below the top of the quay wall.

Ladder rungs should be installed with their centre min. 10 cm behind the face of the quay wall and should be made from $30 \times 30$ mm square steel bars installed so that one edge points upwards. This reduces the risk of slipping due to ice or dirt. Rungs should be fastened to the stiles at a centre-to-centre distance of 28–30 cm, with a clear width between stiles of 45 cm.

## 6.13 Layout and design of stairs in seaports (R 24)

### 6.13.1 Layout of stairs

Stairs are used in seaports where persons not acquainted with the conditions in ports require access and where such persons cannot be expected to use vertical ladders to climb from ships to waterfront structures. The upper end of the stairs should be placed so that there is little or no interference with foot traffic and cargo handling. The approach to the stairs must be clearly visible and thus permit the smooth flow of foot traffic. The lower end of stairs should be positioned so that ships can berth easily and safely with safe passage between ship and stairs.

### 6.13.2 Practical stair dimensions

Stairs should be max. 1.50 m wide so that they can be positioned in front of the outboard crane rail on quay walls for seagoing vessels, without projecting into the area of the fixings for the crane rail, which are 1.75–2.50 m from the edge of the quay. The pitch of the stairs should be determined using the well-known formula $2s + a = 59$ to 65 cm (rise $s$, going $a$). Concrete steps should have rough, granolithic concrete finish and the edges of the steps should be fitted with steel nosing for protection.

### 6.13.3 Landings

For larger tidal ranges, landings should be positioned at 0.75 m above mean low tide, mean tide level and mean high tide respectively. Depending on the height of the structure, further landings may be necessary. Intermediate landings are to be positioned after max. 18 steps; the length of the landing should be 1.50 m or equal to the stair width.

### 6.13.4 Railings

Stairs should be fitted with a handrail whose upper edge is 1.10 m above the front edge of the pitch line. Where port operations permit, stairs should be enclosed by a 1.10 m high railing, which can be removable if necessary.

### 6.13.5 Mooring equipment

The quay wall next to the lowest landing should be equipped with mooring hooks (R 102, section 5.13). In addition, a recessed bollard or mooring hook should be positioned below each landing. Recessed bollards are used in concrete or masonry quay walls or wall components, mooring hooks generally for steel sheet pile structures.

### 6.13.6 Stairs in sheet pile structures

Stairs in sheet pile wall structures are made from steel. The sheet piles at stairs are set back to create a recess large enough to contain the stairs. The stairs must be protected against underrunning by suitable means (e.g. fender piles).

## 6.14 Equipment for waterfront structures in seaports with supply and disposal systems (R 173)

### 6.14.1 General

Supply systems provide public installations and facilities, the businesses in the port and moored ships etc. with the operational resources, power, etc. they require. Disposal systems serve to drain any wastewater and operational resources.

The supply and disposal systems must be located in the immediate vicinity of a waterfront structure, sometimes directly in the structure itself.

Adequate openings for these services must be provided in the structural members of waterfront structures, e.g. in craneway beams. Therefore, consultation among all participants must take place during the planning of supply and disposal systems. Spare openings must be included to allow for any later expansion.

Supply systems include:

- water supplies
- electric power
- communication and remote control systems
- other systems

Disposal systems include:

- rainwater drainage
- wastewater drainage
- fuel and oil interceptors

The respective disposal regulations must be observed.

### 6.14.2 Water supply systems

Water supply systems provide drinking and process water and normally can also be used for extinguishing fires.

#### 6.14.2.1 Drinking and process water supplies

In order to safeguard the drinking and process water supply systems in the port, at least two independent supply points are required for each port section, with the lines laid out as ring systems to guarantee a permanent flow.

Hydrants should be installed at approx. 100–200 m intervals; every 60 m along the quayside is a typical spacing for water hydrants for supplying ships. Underground hydrants are placed on quay walls and in paved crane and rail areas so as not to hinder operations. The hydrants must be arranged so that there is no danger of them being crushed by railborne cranes and vehicles, even when standpipes are fitted.

When using underground hydrants, special attention must be paid to protecting the connection coupling against contamination, even in case of any possible flooding of the quay wall. An additional shut-off valve is required to isolate the hydrant from the supply line. Hydrants must be accessible at all times. They must be situated in areas where operations prevent goods being stored.

The pipes are normally laid with an earth cover of 1.50–1.80 m. To protect them against frost, they are also placed at least 1.50 m from the front face of the quay wall. In loaded zones with railway tracks, the lines should be placed in protective ducts.

In quay walls with concrete superstructures, the lines may be placed in the concrete structure. Here, the different deformation behaviour of adjacent structural sections must be taken into account, together with the differing settlement behaviour of structures on deep or shallow foundations. Drinking water is typically supplied via inboard ring mains and branches to the hydrants located at the front of the quay. It must be possible to drain the branches so that, for hygiene reasons, water can be drained from those branches not constantly in use.

To reduce expenditure in the case of a burst pipe, pipes used for water supplies should not be laid under areas covered by concrete. Typically, they are located under strips reserved for such services.

### 6.14.2.2 Separate fire-fighting water

When there is a high fire risk in a certain section of a port, the recommendation is to supplement the drinking and process water supplies system with an independent fire-fighting system. Water for fighting fires is pumped directly from the harbour basin. The associated pumping stations can be located within the quay wall below ground so as to not disturb cargo handling.

It is also possible to feed fire-fighting water into the system from the pumps of the fireboats via special connection points.

In sheet pile quay walls, the suction pipes may be placed in the sheet piling troughs, where they are adequately protected against vessel impacts. In concrete superstructures such pipes should be positioned in slots for protection.

The routing of the pipes of the fire-fighting system must satisfy the same requirements as the drinking and process water supplies.

### 6.14.3  Electricity supply systems

Office buildings, port installations, cranes, lighting to railway tracks, roads, operations areas, open areas, quays, berths, dolphins, etc. must all be provided with electric power. As a contribution to avoiding air and noise pollution, it may be necessary in future to provide ships moored in the port with electricity from the public grid. Although up until now this has been done in individual cases only, it is nevertheless recommended that consideration be given to the possibility of supplying ships with electric power from onshore when building new or converting existing facilities.

Only buried cables are used for the high- and low-voltage supply systems in ports, except during construction. The cables should be laid in the ground in plastic ducts with an earth cover of approx. 0.80–1.00 m in quay walls and operations areas; concrete cable drawpits designed to accept vehicular traffic are also required. The advantage of such a duct system is that the cable installations can be augmented/modified without interrupting port operations.

When there is a risk of frequent flooding of the quayside, power sockets must be mounted on posts raised above the flood level.

Power sockets are generally installed in the top of the quay wall at intervals of approx. 100–200 m. They must be capable of accepting vehicular traffic and be fitted with a drain pipe. These sockets are used to provide the power for welding equipment when carrying out minor repairs on ships and cranes as well as the power for emergency lighting and other purposes.

Ducts for conductor rails, cable troughs and crane power feeding points must be provided in the quayside for power supplies to cranes. The drainage and ventilation of these facilities is particularly important. In quay walls with concrete superstructures, these facilities can be incorporated in the concrete structure.

Special attention is drawn to the fact that electricity supply networks require equipotential bonding facilities. This is to prevent unduly high voltages occurring in crane rails, sheet piling or other conductive components of the quay wall due to a fault in any electrical systems (e.g. a crane). Such equipotential bonding systems should be installed about every 60 m.

In the case of craneways integrated into the quay wall superstructure, the equipotential bonding lines are normally concreted in during construction of the superstructure, for reasons of cost. However, they must be laid in protective ducts with sufficient freedom of movement in areas in which differential settlement might be expected.

### 6.14.4 Other systems

These include all supply systems not mentioned in sections 6.14.2 and 6.14.3 but are required, for example, in the quay walls to shipyards. These include: gas, oxygen, compressed air, acetylene, steam and condensate lines. The layout and installation of such facilities must comply with pertinent regulations, particularly safety regulations.

Connections for telephones are usually placed at a spacing of 70–80 m along the front edge of the quay. Although the growing availability and use of mobile phones means that such telephone points are being relied on less and less, port authorities nevertheless usually require that a hazardous goods landline telephone be installed.

### 6.14.5 Disposal systems

#### 6.14.5.1 Rainwater drainage

The rainwater falling in the quay wall area and also on its land side is drained into the harbour directly via the quay wall. To do this, the quay and operations areas are provided with a drainage system consisting of inlets, transverse and longitudinal channels and a main drain with outlet into the harbour. The sizes of drainage catchment areas depend on local circumstances. As few outlets as possible should be installed in the waterfront structure. They should be designed and installed in such a way that they are not damaged by moored ships or vessel impacts.

To prevent pollution of the harbour water, the outlets are to be provided with gate valves in quay and operational areas with a risk of leakage of dangerous or toxic substances or contaminated fire-fighting water into the drainage system.

#### 6.14.5.2 Wastewater disposal

It is currently not customary for the port operator to accept the wastewater from seagoing vessels. The wastewater occurring in the port itself is fed through a special waste disposal system into the municipal sewer network. It may not be drained into the harbour water. Wastewater drains are therefore only found in waterfront structures in exceptional cases.

#### 6.14.5.3 Fuel and oil interceptors

Fuel and oil interceptors must be included wherever they are required, in the same way as non-port facilities.

#### 6.14.5.4 Disposal regulations for ship waste

Corresponding to the MARPOL convention, ports should provide facilities for the disposal of ship waste, such as liquids containing oils and chemicals, solid ship waste (galley waste and packaging refuse) and sanitary wastewater.

### 6.15 Fenders for large vessels (R 60)

#### 6.15.1 General

In order to enable vessels to berth safely alongside waterfront structures, it is customary these days to provide fenders. They absorb the impact of vessels during berthing and prevent damage to ship and structure while the vessel is moored. For large vessels in particular, fenders are indispensable. Although timber baulks, rubber tyres, etc. are still common, other, modern forms of fender are becoming more and more established. The main reasons for this are:

- The use of fenders increases the service life of the waterfront structure (see R 35, section 8.1.8.4).
- The cost of vessels is on the increase and so ships demand good fenders.
- Ships are growing in size – hence, the surface area exposed to the wind as well.
- The demands placed on moored ships by cargo-handling equipment are on the increase.
- The strength of the outer hull is being reduced further and further.

Fenders are used not only on waterfront structures, but also frequently on dolphins, and work together with the elastic dolphins to absorb the energy (see also R 218, section 13).

## 6.15.2 The fendering principle

A fender is, in principle, an intermediate layer between vessel and waterfront structure which absorbs part of the kinetic energy of a berthing ship; indeed, energy-absorbing fenders absorb most of this energy. In the case of fenders attached to waterfront structures, the energy absorbed by the fender is transferred to the structure. A portion of the berthing energy is absorbed by the ship's hull by means of elastic deformations.

The energy absorption $E_f$ of a fender is shown by its characteristic load–deflection curve (Fig. R 60-1), which illustrates the relationship between fender deflection $s$ and fender reaction force $F_R$. The area beneath the curve represents the energy absorption $E_f$. The energy absorbed at maximum deflection $s_{max}$ is denoted the energy absorption capacity.

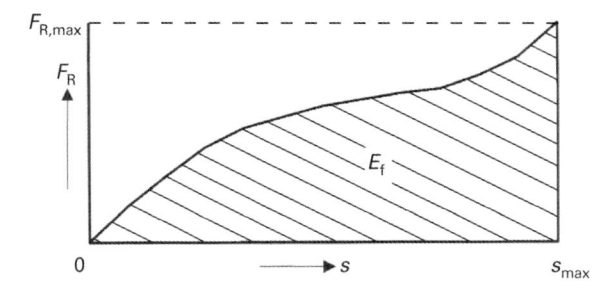

**Fig. R 60-1.** Load–deflection curve for a fender

All fender designs braced against a rigid waterfront structure are generally characterised by an abrupt increase in the reaction force of the structure once the energy absorption capacity of the fender has been reached. These fender reaction forces must be taken into account when designing the structure.

The dimensions and properties (e.g. characteristic force/load–deflection curve) of the various fenders available can be found in the publications of the fender manufacturers. However, it should be noted that these curves apply only when lateral buckling of the fender is ruled out and when creep deformations under permanent load are not excessive.

Where fenders are used on flexible components and other supporting structures (e.g. dolphins), then it is not only the energy absorption of the fender, but also that of those components which must be taken into account.

### 6.15.3 Design principles for fenders

A fender system must be designed with the same level of care and attention as the entire waterfront structure. The fender system must be considered at the planning stage. Comprehensive guidance on the design and detailing of fenders can be found in [162].

A fender system has to satisfy the following requirements:

- Ships must be able to berth without being damaged
- Ships must be able to moor without being damaged
- Fenders should retain their effectiveness for as long as possible
- Damage to the waterfront structure must be avoided
- Damage to dolphins must be avoided

Therefore, the following steps must be incorporated into the design and detailing of fender systems:

- Compilation of functional requirements
- Compilation of operational requirements
- Assessment of the local conditions
- Assessment of the boundary conditions for the design
- Calculation of the energy to be absorbed by the fender system
- Selection of a suitable fender system
- Calculation of the reaction force and possible friction forces
- Checking whether the forces transferred to the waterfront structure and the ship's hull can be accommodated
- Ensuring that all constructional details in the waterfront structure or dolphin can be accommodated, especially fixings, built-in parts, chains, etc., without any damage being caused to the ship or the waterfront structure due to projecting fixings or other parts of the construction

Many different fenders and fender systems are available from numerous manufacturers. The manufacturers frequently offer not only standard products, but also systems tailor-made to suit particular situations.

In order to compare the various fenders available, the quality and system data of the manufacturers should comply with the test methods in the [162].

Some fenders require considerable maintenance. Therefore, before selecting and installing fenders, the designer is recommended to check carefully whether and to what extent vessels and/or structures are really at risk and which special requirements a fender system will have to fulfil.

When designing a quay wall, pier, dolphin, etc. and the fender support structures, it is not only berthing loads that must be taken into account. The horizontal and vertical movements of the ship during berthing and departure, loading and unloading procedures, swell or fluctuations in the water level, etc. can lead to friction forces in the horizontal and/or vertical direction (provided these movements are not accommodated by the rotation of suitable cylindrical fenders) which are additional to the berthing forces. If lower values cannot be verified, to be on the safe side a friction coefficient $\mu = 0.9$ should be assumed for dry elastomeric fenders. Polyethylene surfaces result in less friction on the ship's hull; a friction coefficient $\mu = 0.3$ should be assumed in such cases.

The permissible pressure between a fender and the hull of a modern large ship is max. 200 kN/m$^2$. All customary fender types take account of this pressure. This should be borne in mind when selecting a type of fender. Certain ships, e.g. naval vessels, require softer fenders.

### 6.15.4 Required energy absorption capacity

#### 6.15.4.1 General

The required energy absorption capacity of a fender describes the energy that can be absorbed under specified boundary conditions. It must be at least as great as the kinetic energy produced by a berthing ship.

Typically, during berthing, a ship moves transversely and/or longitudinally and at the same time rotates about its centre of mass. At the moment of mooring, contact is generally initially with a single dolphin or fender only (Fig. R 60-2). The berthing velocity $v$ of the ship at the fender is the critical factor for calculating the energy transferred to the dolphin, the size and direction of which can be calculated from the vectorial addition of the velocity components $\upsilon$ and $\omega \cdot r$. In the case of a full frictional connection between ship and fender, during the berthing procedure the berthing velocity of the ship, which is then identical to the deformation rate of the fender, is reduced to $\upsilon = 0$. The ship's centre of mass will then,

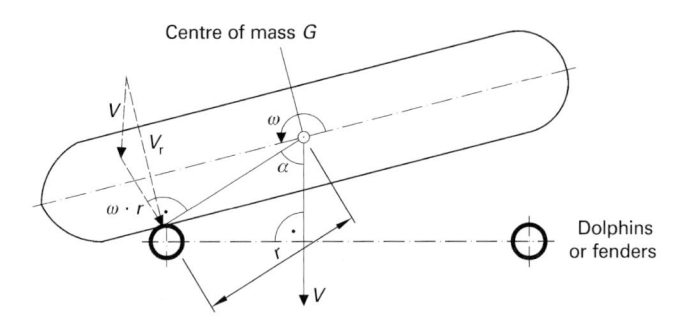

**Fig. R 60-2.** Explanatory diagram for calculating the energy to be absorbed due to a berthing manoeuvre

however, usually remain in motion albeit with a different velocity and, potentially, also in a different rotational direction.

The ship therefore retains a portion of its original kinetic energy at the moment of maximum fender deformation. Under certain circumstances this can result in the ship turning to make contact with a second fender after striking the first one which then, at the moment of contact, causes a greater berthing force.

The deterministic method of analysis is normally used for designing fenders. This is based on the energy equation

$$E = \tfrac{1}{2} \cdot G \cdot v^2$$

where

$E$ kinetic energy of ship [kNm]
$G$ mass of ship, i.e. displacement [t] according to section 5.1
$v$ berthing velocity of ship [m/s]

The amount of a ship's kinetic energy that a structure (fender and/or dolphin) has to absorb during a berthing manoeuvre represents the energy absorption capacity required to prevent damage to ship and/or structure. This energy absorption capacity for the example shown in Fig. R 60-2 is

$$E_d = \frac{G \cdot C_m \cdot C_s \cdot C_c}{2 \cdot (k^2 + r^2)} \cdot \left[ v^2 \cdot (k^2 + r^2 \cdot \cos^2 \alpha) + 2 \cdot v \cdot \omega \cdot r \cdot k^2 \cdot \sin \alpha + \omega^2 \cdot k^2 \cdot r^2 \right]$$

When $\omega = 0$ (no rotation of ship), the equation simplifies to

$$E_d = \frac{1}{2} \cdot G \cdot v^2 \cdot \frac{k^2 + r^2 \cos^2\alpha}{k^2 + r^2} \cdot C_m \cdot C_s \cdot C_c$$
$$= \frac{1}{2} \cdot G \cdot v^2 \cdot C_e \cdot C_m \cdot C_s \cdot C_c$$

In both equations, the following definitions apply:

| | |
|---|---|
| $E_d$ | berthing energy to be absorbed [kNm] |
| $G$ | mass of ship, i.e. displacement [t] according to R 39, section 5.1 [t]. The mass of the fully laden ship should always be used – even when operational conditions dictate that only unloaded ships usually berth at the dolphin concerned – in order to cover the case of an unscheduled re-berthing of a ship. |
| $k$ | radius of gyration of ship [m], generally taken as 0.25 $l$ for large ships with a high block coefficient |
| $l$ | length of ship between perpendiculars [m] |
| $r$ | distance of ship's centre of mass from point of impact on fender/dolphin [m] |
| $v$ | berthing velocity, i.e. translational movement speed of centre of mass at time of first contact with fender/dolphin [m/s] |
| $\omega$ | ship's rotational speed at time of first contact with fender/dolphin [rad/s] |
| $\alpha$ | angle between velocity vector $v$ and distance $r$ [°] |
| $C_m$ | virtual mass factor [1] |
| $C_s$ | ship flexibility factor [1] |
| $C_e$ | eccentricity factor [1] |
| $C_c$ | waterfront structure attenuation factor [1] |

### 6.15.4.2 Information on dolphin berths

If a ship is manoeuvred to a dolphin berth with the help of tugs, it can be assumed that it is hardly moving in the direction of its longitudinal axis and that its side is virtually parallel to the line of the dolphins while it is berthing. Therefore, when designing the inner dolphins within a row of dolphins, the velocity vector $v$ can be assumed to be perpendicular to the distance $r$, i.e. angle $\alpha = 90°$.

When designing the outer dolphins of a row of dolphins, however, this will not be necessary because in this case the ship's centre of gravity in the direction of the line of dolphins can also approach close to the centre of the dolphins.

The individual factors of the aforementioned equations are defined as follows:

1. Mass of ship/displacement $G$
   The mass of the ship, i.e. its displacement, is required for calculating the energy to be absorbed.
   Recommendation R 39, section 5.1, contains reference values for the displacement of different types of ship. Provided no particular values have been specified for the port design, the values in the tables can be used for displacement calculations.
2. Berthing velocity $v$
   The square of the berthing velocity $v$ is included in the equation for calculating the berthing energy to be absorbed and is therefore one of the main parameters to consider when designing fenders and dolphins. The berthing velocity is specified at a right-angle to the waterfront structure or row of dolphins. Measured values for the berthing velocity are not usually available. As a rule, the figures given in R 40, section 5.3, can be assumed.
3. Angle $\alpha$
   Measurements carried out in Japan resulted in a berthing angle of, generally, <5° for ships with DWT > 50 000 t (accordingly $\alpha > 85°$). To remain on the safe side in calculations, the designer is recommended to assume a berthing angle of 6° for such ships (accordingly $\alpha = 84°$).
   For smaller vessels, and primarily when berthing without tug assistance, an angle of 10–15° should be assumed (accordingly $75° \leq \alpha \leq 80°$).
4. Eccentricity factor $C_e$
   The eccentricity factor $C_e$ takes into account the fact that the first contact between ship and fender is not normally in the middle of the ship's side and therefore not in line with the vessel's centre of mass either. According to [162], the eccentricity factor is calculated as follows (using the factors explained above for the energy equation):

$$C_e = (k^2 + r^2 \cos^2 \alpha)/(k^2 + r^2)$$

Assuming a berthing angle of 0°, i.e. $\alpha = 90°$, is sufficiently accurate for fenders and the inner dolphins in a row of dolphins. The eccentricity factor is therefore

$$C_e = k^2/(k^2 + r^2)$$

The radius of gyration $k$ for large ships with a high block coefficient can usually be taken as $0.25\,l$, where $l$ is the length between perpendiculars.

When designing fenders alongside quay walls, $C_e = 0.5$ can be assumed if more accurate data is not available and for rough calculations, or $C_e = 0.7$ for dolphin fenders. At ro-ro berths, $C_e = 1.0$ should be assumed for the end fenders for ro-ro ships that dock with bow or stern.

5. Virtual mass factor $C_m$

   The virtual mass factor takes into account the fact that a considerable quantity of water is moved together with the ship, and this must be included in the mass of the ship in the energy calculation. Various approaches have been used to determine the $C_m$ factor, see [162]. Assessing this and other approaches in the literature results in $C_m$ values between 1.45 and 2.18.

   PIANC (2002, [162]) recommends using the following values:
   - for a large clearance under the keel $(0.5 \cdot d)$: $C_m = 1.5$
   - for a small clearance under the keel $(0.1 \cdot d)$: $C_m = 1.8$
   
   where $d$ is the draught of the ship [m].

   Mass factor values for a clearance under the keel between $0.1 \cdot d$ and $0.5 \cdot d$ can be found by linear interpolation.

6. Ship flexibility $C_s$

   The factor for the flexibility of a ship $C_s$ takes into account the ratio of elasticity of the fender system to that of the ship's hull because part of the berthing energy is absorbed by the latter. The following $C_s$ values are normally used:
   - for soft fenders and small vessels: $C_s = 1.0$
   - for hard fenders and larger vessels: $0.9 < C_s < 1.0$
   
   Generally, $C_s = 1.0$ can be assumed, which lies on the safe side.

7. Waterfront structure attenuation factor $C_c$

   The attenuation factor $C_c$ takes into account the type of waterfront structure. With a closed structure (e.g. vertical sheet pile wall), the water between ship and wall already absorbs a considerable portion of the berthing energy as it is accelerated and displaced laterally as the ship approaches. This waterfront structure attenuation factor $C_c$ depends on various influences, e.g.:
   - arrangement of the waterfront structure
   - clearance under keel
   - berthing velocity
   - berthing angle
   - depth of fender system
   - ship cross-section

   Experience has shown that the following values can be assumed for $C_c$:
   - open waterfront structure: $C_c = 1.0$
   - closed waterfront structure and parallel berthing $(\alpha \approx 90°)$: $C_c = 0.9$

**Table R 60-1.** Additional factors for exceptional berthing manoeuvres depending on size and type of vessel

| Type of vessel | Size of vessel | Additional factor |
|---|---|---|
| Tanker, bulk cargo | large | 1.25 |
| | small | 1.75 |
| Container | large | 1.5 |
| | small | 2.0 |
| General cargo | | 1.75 |
| Ro-ro vessel, ferry | | $\geq 2.0$ |
| Tug, workboat | | 2.0 |

Values $< C_c = 0.9$ should not be used.
The attenuation can be considerably less at a berthing angle of just $\alpha = 85°$, i.e. in such cases assume $C_c = 1.0$.

8. Fender design programs
   Manufacturers can provide computer programs for designing their fenders. Depending on the manufacturer, these programs require influencing parameters to be entered in either metric or non-metric units. Consequently, the results of the calculations may have to be converted for comparisons.
9. Additional factors for exceptional berthing manoeuvres
   Using the detailed information given here allows the designer to calculate the required energy absorption of fenders and dolphins with sufficient accuracy. It is left to the discretion of the design engineer to take account of any potential difficulties caused by exceptional berthing manoeuvres, e.g. by assuming higher berthing velocities or general additional factors when calculating the energy. Exceptional conditions could be, for example, the frequent handling of hazardous goods. The designer should consult the PIANC Report [162]. Generally, additional factors between about 1.1 and max. 2.0 are recommended. Table R 60-1 provides guidance on additional factors depending on the type and size of vessel in the case of an extraordinary berthing manoeuvre.
10. Selection of fenders
    Once the energy has been calculated, the fenders required can be selected from the relevant manufacturers' publications. However, for detailed planning the designer is advised to consult the manufacturer because many of the construction details cannot be gleaned from the manufacturers' publications. This concerns, in particular, the construction details regarding the mounting of the fenders.

### 6.15.5 Types of fender system

Diverse fender systems are available on the international market. The various types and models can be seen in the catalogues of the manufacturers.

Cylindrical fenders are the most common type, and these are available in many different sizes. Floating fenders have proved to be worthwhile for quay walls exposed to considerable water level fluctuations as a result of tides. Berths for ferries are frequently custom solutions with polyethylene-coated low-friction panels on conical fenders or cylindrical fenders loaded along their longitudinal axis.

For a comparison of various types of fender, please refer to [162]. Further information on the advantages and disadvantages of different fender systems can also be found in the report.

Please note that the designations of the manufacturers can vary for the same type of fender. Test methods for materials and fenders should comply with the data given in [162] in order to be able to assess the equivalence or otherwise of the products of different manufacturers.

Materials for fenders are these days almost exclusively elastomer or other synthetic products. With the exception of dolphins and rare custom designs, these products guarantee that the energy that occurs during berthing can be transferred to the loadbearing structure in accordance with the calculations and without damage.

For this reason, types of fenders common in the past, e.g. brushwood, vehicle tyres, or timber (rubbing strips, timber fenders, fender piles), cannot be designated as fenders because their insufficiently defined material properties prevents them from being included in the energy absorption calculation. Fenders made from wood or vehicle tyres can therefore only be used in construction details, e.g. as nosings or guides.

#### 6.15.5.1 Elastomer fenders

##### 6.15.5.1.1 General

Elastomer fenders are used in many ports for absorbing the impacts of vessels and berthing pressures. These fenders are generally made from a material resistant to seawater, oil and ageing and are not destroyed by occasional overloads, so a long service life can be expected.

Elastomer fender elements are manufactured in various shapes, dimensions and with specific performance characteristics. They meet every requirement, from simple fendering for small vessels to fender structures for large tankers and bulk cargo freighters. Special attention must be given to the particular loads on fenders at ferry terminals, locks, dry docks, etc.

Elastomers are used either alone as a fender material, against which the ships berth directly, or as suitably designed buffers between fender piles or fender panels and the structure. Occasionally, both types of usage are combined. In such cases it is possible to attain the energy absorption capacity and spring constants best suited to any specific requirement using elements made from commercially available elastomers.

#### 6.15.5.1.2 Cylindrical fenders

Thick-walled elastomer cylinders are frequently used. They can have various diameters from 0.125 to >2.00 m. They have variable spring characteristics depending on the application. Cylinders with smaller diameters are attached with ropes, chains or rods in horizontal, vertical or, where applicable, diagonal positions in front of quay walls.

In the latter case they are frequently suspended as a protective "garland" in front of a quay wall, mole head, etc.

Cylindrical fenders are usually installed in a horizontal position. To avoid the risk of deflection and tearing at the ends, they must not be hung directly in front of quay walls with ropes or chains, instead should be threaded over rigid steel tubes or steel trusses made from tubular sections. These are then suspended from the quay wall with chains or wire ropes, or mounted on steel brackets located next to the fenders (Fig. R 60-3).

#### 6.15.5.1.3 Axially loaded cylindrical fenders and conical fenders

Cylindrical fenders can also be installed to carry loads in their longitudinal direction. However, owing to the risk of buckling, only shorter cylinders are possible in this case. If the deformation to absorb the

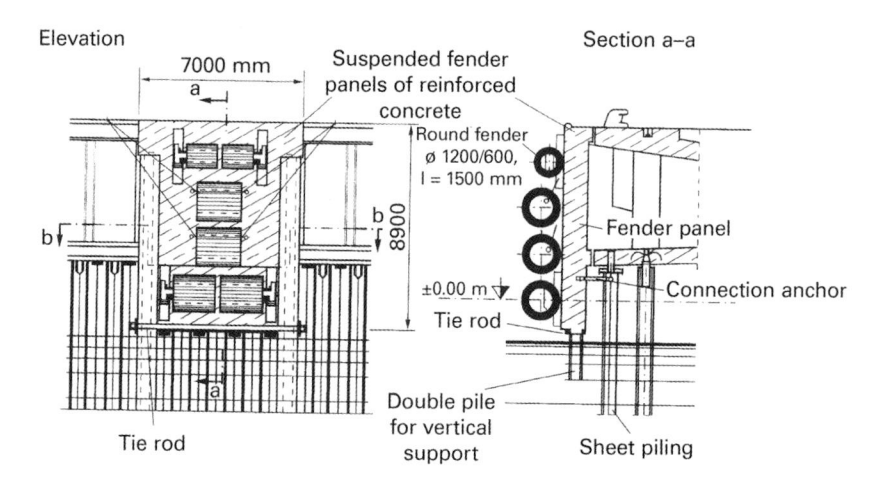

**Fig. R 60-3.** Example of a fender arrangement with cylindrical fenders

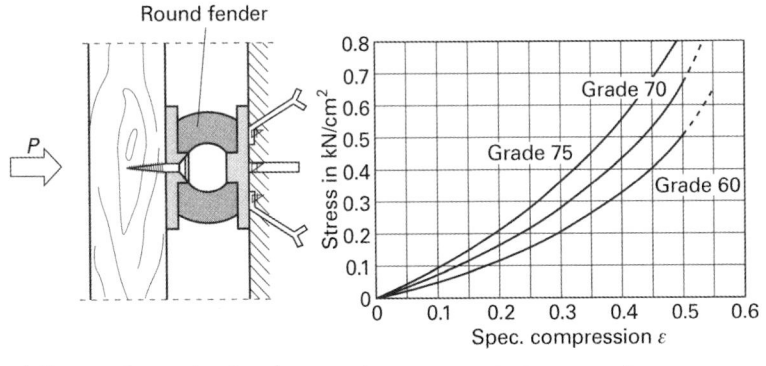

a) Example under load   b) Characteristic stress/
                        compression diagram

**Fig. R 60-4.** General data for round elastomer fenders loaded in the longitudinal direction in Shore A grades 60, 70 and 75 to DIN 53505

berthing energy is not sufficient, several cylindrical fender elements can be arranged adjacent to each other. To prevent buckling of such a row of elements, steel plates in suitable guides can be fitted between the individual elements (Fig. R 60-4). The reaction force of this type of fender initially climbs rapidly to the buckling load and then drops again as the fender deforms.

The conical fender is a special form that is far less vulnerable to buckling. The energy and deformation characteristics are similar to those of an axially loaded cylindrical fender.

#### 6.15.5.1.4 Trapezoidal fenders

In order to obtain a more favourable load–deflection curve, special shapes have been developed using special inlays, e.g. textiles, spring steels or steel plates vulcanised into the elements. Metal inlays must be blasted to a bright surface finish and must be completely dry before vulcanising. These elements are frequently made in a trapezoidal form with a height of approx. 0.2–1.3 m. They are attached to the quay wall with dowels and bolts (Fig. R 60-5).

#### 6.15.5.1.5 Floating fenders

The great advantage of floating fenders, primarily in tidal waters, is that ships are fendered practically exactly on the waterline and hence also roughly in line with their centre of gravity. Floating fenders are available in foam- or air-filled versions.

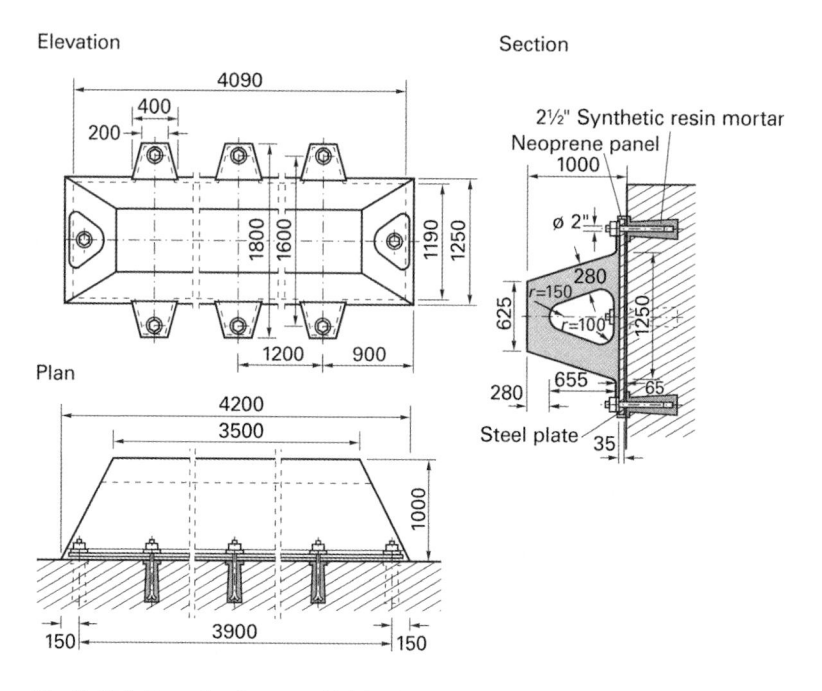

**Fig. R 60-5.** Example of a trapezoidal fender

Air-filled fenders are fitted with a blow-off valve which prevents the fender from bursting if it is overloaded. The valve must be serviced regularly.

Owing to their method of manufacture, foam-filled fenders can be produced in virtually any size and with virtually any properties. They have a core of closed-cell polyethylene foam and a jacket of polyurethane reinforced with a textile. The jacket is easily repaired. Special attention should be paid to the material properties of the jacket because the stresses during deformation are very high.

#### 6.15.5.1.6 Fenders made from car tyres and rubber waste

Various seaports use old car tyres, mostly filled with rubber waste, as fenders, suspended flat against the face of quays. They have a cushioning effect, but do not possess any appreciable energy absorption capacity. More frequent is the use of several stuffed truck tyres – usually between 5 and 12 – threaded onto a steel shaft with a pipe collar welded on at each end for attaching the suspension and retaining ropes. The ropes hold the fender so that it can rotate on the face of the quay wall. The tyres are filled with elastomer sheets placed crosswise which brace the tyres

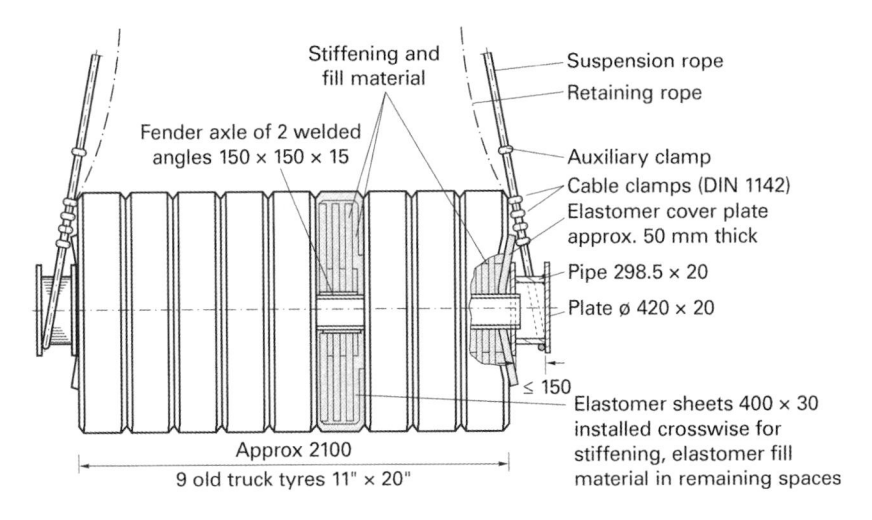

**Fig. R 60-6.** Example of a fender made from truck tyres

against the steel shaft. Any remaining voids are filled with elastomer material (Fig. R 60-6). Fenders of this type – occasionally in an even simpler design with a wooden shaft – are inexpensive and have generally performed well where the impact energy of berthing ships has not been severe. However, their energy absorption capacity cannot be determined reliably.

Not to be confused with these improvised measures are the accurately designed fenders free to rotate on an axle. These fenders are fabricated mostly from very large special tyres which are either stuffed with rubber waste or inflated with compressed air. Fenders of this type are used successfully at exposed positions, such as the entrances to locks or dry docks as well as at narrow harbour or port entrances in tidal areas, where they are suspended horizontally and/or vertically and successfully serve as guides for ships.

Tyres from road vehicles are occasionally used as fenders at ore loading/unloading facilities in the vicinity of open-cast mines. In such cases the energy absorption capacity should be determined in tests.

### 6.15.5.2 Fenders made from natural materials

Suspended brushwood fenders are still used in countries in which suitable materials are available and/or funding is limited. However, if elastomer fenders are available, then brushwood fenders should be rejected because they involve a higher capital outlay and also higher maintenance costs than elastomer fenders. Brushwood fenders are

**Table R 60-2.** Fender dimensions for brushwood fenders

| Size of vessel [DWT] | Fender length [m] | Fender diameter [m] |
|---|---|---|
| up to 10 000 | 3.0 | 1.5 |
| up to 20 000 | 3.0 | 2.0 |
| up to 50 000 | 4.0 | 2.5 |

subject to natural wear and tear due to weather and wave conditions. Their dimensions are adapted to the largest vessels berthed. Unless special circumstances require larger dimensions, fender sizes may be taken from Table R 60-2.

### 6.15.6 Construction guidance

Fenders should be located at regular intervals along the quay. The spacing of the fenders depends on the design of the fender system and the vessels anticipated. One important criterion here is the radius of the ship between the bow and the flat side in the middle of the ship. This radius defines – for a given fender spacing – the projecting dimension of completely compressed fenders with the bow making contact with the quay wall between two fenders. To avoid that, the fender spacing must be adjusted to suit.

As a rule, fenders should not be spaced more than 30 m apart.

The fender projection from the structure should not be too large. The maximum load moment of the cranes frequently influences the projection of the fenders.

It is difficult to design a fender system that is equally suitable for both large and small vessels. Whereas a fender designed for a large vessel is sufficiently "soft", this might be too inflexible for a small vessel, which can result in damage to the ship. Furthermore, the level of the fenders with respect to the water level is more significant for small vessels than for large ones. In tidal waters, floating fenders can offer considerable advantages.

If container feeder ships or inland vessels are handled at berths for large ships, there is a danger of such vessels becoming caught beneath fixed fenders. In addition, the listing of small ships during loading/unloading procedures at low water can lead to high-level fenders damaging the superstructure and the cargo.

In a new development for the container terminal at Bremerhaven, Germany, floating fenders have been installed in front of fender panels, and these can move up and down with the tide between lateral guide

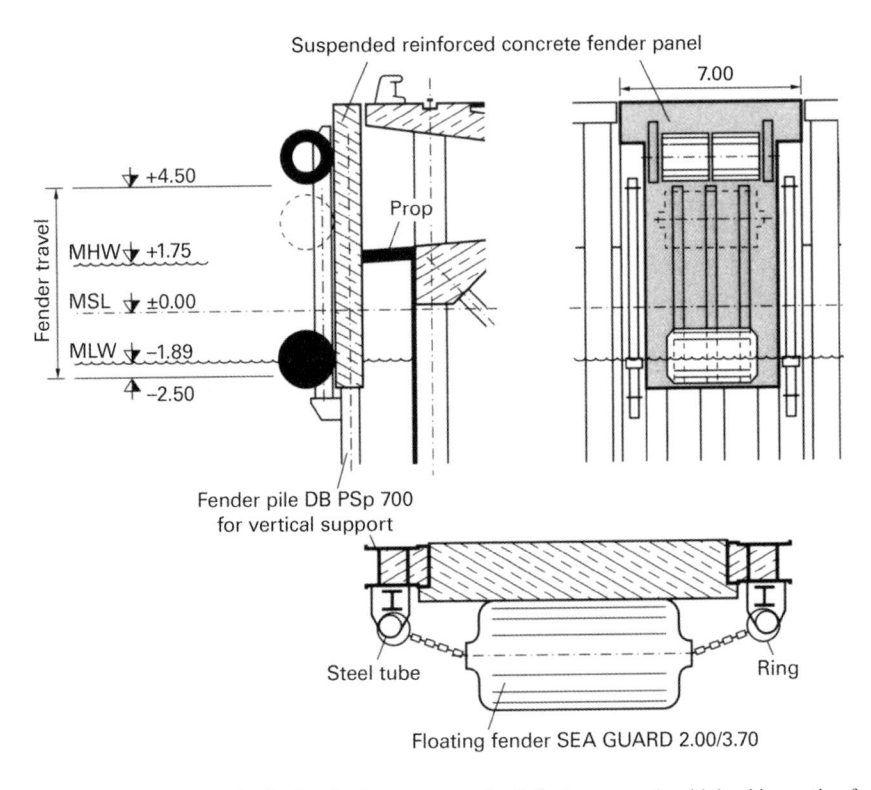

**Fig. R 60-7.** Example of a floating fender system at a berth for large vessels with berthing option for feeder and inland vessels

tubes. This solution is shown in Fig. R 60-7. The fender construction here consists of a fixed upper cylindrical fender (⌀1.75 m) and a moving floating fender (⌀2.00 m) in front of a fixed fender panel. These diameters were chosen to ensure that sufficient listing of smaller vessels is possible at low water levels.

### 6.15.7 Chains

Chains in fender systems should be designed for at least three to five times the theoretical load.

### 6.15.8 Guiding devices and edge protection

#### 6.15.8.1 General

Besides the actual fenders, which are specifically designed to absorb energy, there are many elements that are provided merely for constructional reasons, e.g. guiding devices in channels and locks, edge protection, or berthing equipment for smaller vessels not designed for specific situations. Such devices include fender piles, timber fenders, rubbing strips and nosings.

### 6.15.8.2 Timber fenders and fender piles

Timber used in seawater and brackish water can be attacked by the so-called naval shipworm (*Teredo navalis*). Such an attack can lead to total destruction of the timber in a port facility within just a few years. The destruction of the timber is practically invisible from the outside because the worm bores into the wood radially and then extends its path horizontally. The shipworm attacks mainly softwoods, but European hardwoods such as oak, and even tropical woods, are also at risk.

The use of timber in loadbearing structures cannot be recommended in seawater and brackish water with a salt content >5%. When, for example, timber fender piles are used, replacement after infestation by the naval shipworm must be considered.

### 6.15.8.3 Edge protection

Edge protection is made from special sections and fender sections. Owing to their small size and their shape, they possess no significant energy absorption capacity.

### 6.15.8.4 Rubbing strips of polyethylene

In addition to other components such as timber fenders, fender piles, etc., rubbing strips made of plastic, frequently polyethylene (PE), are used in order to reduce the friction forces between waterfront structures and berthing/moored vessels. These components must absorb the loads arising from pressure and friction without fracture and be capable of transmitting them to the waterfront structure via their mountings. To do this they will need to be supported by supplementary loadbearing members in certain cases.

Polyethylene compounds of medium density to DIN EN ISO 1872 (HDPE) and high density to DIN 16972 (UHMW-PE) have proved suitable for use as rubbing strips in hydraulic engineering and seaport construction. Standard forms are rectangular solid profiles with cross-sections between $50 \times 100$ mm and $200 \times 300$ mm and lengths up to 6000 mm. Custom sections and lengths can also be supplied. HDPE is cast in moulds and is vulnerable to brittle fracture at low temperatures ($<-6\,°C$). UHMW-PE sections are cut to suit the profile required and therefore have smooth edges.

In order to minimise friction forces, rubbing strips should be made of a material that has a very low coefficient of friction together with low

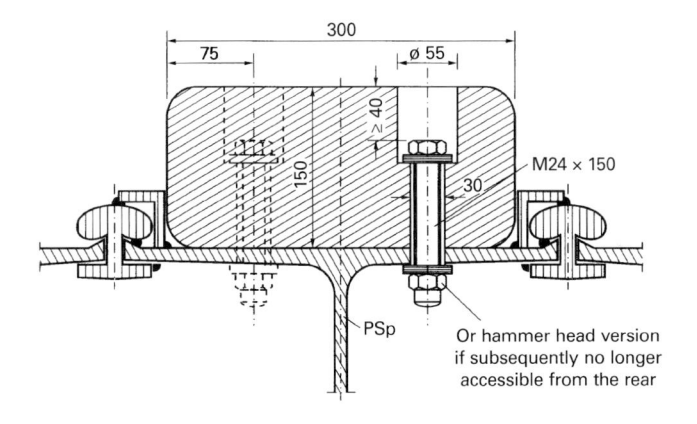

**Fig. R 60-8.** Rubbing strip fixed directly to a Peiner sheet pile wall

abrasion and wear rates, e.g. ultra-high molecular weight polyethylene (UHMW-PE).
The shaped parts must always be free from voids and must be produced and processed in such a way that they are free from distortion and inherent stresses. The quality of the processing can be checked by acceptance tests to verify the properties and by additional hot storage tests on samples cut from sections.
Regenerated PE compounds of medium density may not be used because of their reduced material properties.
Figs. R 60-8 and R 60-9 show fixing and construction details for typical rubbing strips. The heads of the fixing bolts should be recessed at least 40 mm below the contact surface of the strips. Replaceable bolts should be at least $\varnothing 22$ mm, cast-in bolts hot-dip galvanised and at least $\varnothing 24$ mm.

## 6.16    Fenders in inland ports (R 47)

The berthing areas of waterfront structures at inland ports generally consist of concrete, steel sheet piles or faced natural stone. They are constructed either vertical or with a minimal landward batter (1:20 to 1:50).
To protect the waterfront structure and the hull of the ship, the ship's crew will normally suspend rubbing strips about 1 m long between quay wall and side of ship.
The designer is recommended to refrain from equipping waterfront structures at inland ports with fender piles or timber fenders.

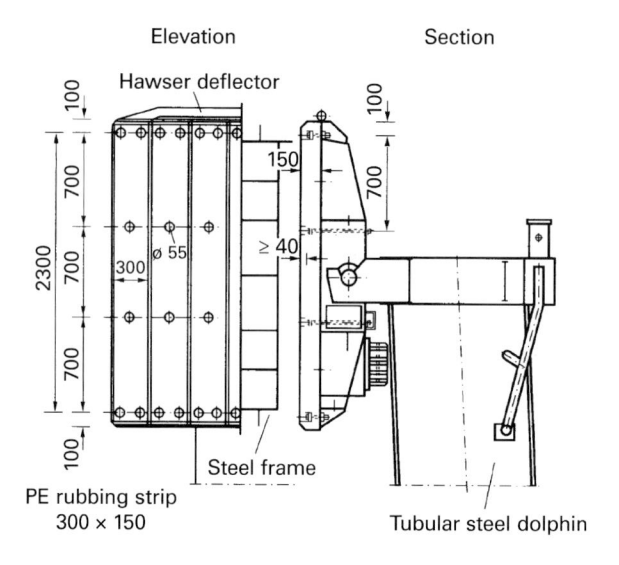

**Fig. R 60-9.** Rubbing strips on a fender panel to a tubular steel dolphin

## 6.17 Foundations to craneways on waterfront structures (R 120)

### 6.17.1 General

In many cases constructional requirements make it practical to construct a deep foundation for the outboard crane rail as an integral part of the quay wall, whereas the foundation for the land-side rail is generally independent of the waterfront structure. In inland ports the outboard crane rail is also frequently on a foundation separate from the waterfront structure. This facilitates any later modifications that might be required, e.g. different operating conditions due to new cranes or alterations to the waterfront structure. Separate foundations for waterfront structure and craneways may also be necessary in the case of different responsibilities and ownership.

Whether the outboard and/or inboard craneway requires a deep foundation depends on the in situ subsoil and whether the unavoidable settlements in the case of shallow foundations can be permitted.

Owing to the fact that the settlement of the ground also affects the decision regarding the type of foundation for the craneway, an appropriate assessment must be carried out in (see R 1, section 1.2).

Please refer to R 72, section 10.2.4, for details of designing long craneway beams without joints. Craneway beams over 2000 m long have been constructed successfully without joints.

## 6.17.2 Design of foundations, tolerances

Craneway foundations may be shallow or deep depending on the local subsoil conditions, the sensitivity of the cranes to settlement and displacements, the crane loads, etc.

The permissible dimensional deviations of the craneway must be taken into account here, distinguishing between tolerances during installation and tolerances during operation.

The installation tolerances mainly relate to the permissible dimensional deviations during the laying and fixing of crane rails and are therefore not usually relevant when selecting the type of foundation. Operational tolerances on the other hand relate to permissible settlement and differential settlement during operation and are therefore critical when deciding on the type of foundation.

Depending on the design of the crane portal, the following values can be taken as a guide for the operational tolerances:

- Level of one rail (longitudinal gradient): 2–4‰,
- Level of rails in relation to each other (cross-level): max. 6‰ of gauge
- Inclination of rails in relation to each other (offset): 3–6‰

Considerably tighter operational tolerances apply, e.g. ≤1‰ for the longitudinal gradient, when using special cargo-handling equipment, e.g. container cranes. Operational tolerances should be specified together with the crane manufacturer in every single case.

Please refer to HTG [102] for more information on the relationship between craneways and crane systems.

### 6.17.2.1 Shallow craneway foundations

1. Strip foundations of reinforced concrete
   In soils not sensitive to settlement, the craneway beams may be constructed as shallow strip foundations in reinforced concrete. The craneway beam is then calculated as an elastic beam on an elastic foundation. An analysis should be carried out to verify the soil pressures beneath the beam. Settlement and differential settlement must also be analysed and compared with the agreed operational tolerances.
   DIN 1045 applies for the design of the beam cross-section. The action effects due to vertical and horizontal wheel loads – also due to braking along the craneway axis – must be verified.
   Craneways with a narrow gauge, e.g. gantry cranes spanning only one track, require the gauge to be maintained with tie beams or tie bars installed at a spacing roughly equal to the gauge. With wide gauges, both crane rails are designed separately on individual foundations. In this case the cranes must be designed with a pinned leg on

one side. Please refer to R 85, section 6.18, for the design of the rail fastening.

Settlement of the craneway beam of up to 3 cm can still be accommodated, generally by installing rail bearing plates or by exploiting the adjustment options of rail chairs – work that can even take place during operations without causing serious disruptions. In the case of greater settlement and settlement that abates only slowly during operations, a deep foundation will generally be more economical because it will not be possible to compensate for the settlement merely by inserting bearing plates or adjusting rail chairs. That means costs and longer downtimes.

2. Sleeper foundations

Crane rails on sleepers on a ballast bed are comparatively easy to realign and so are used primarily in mining subsidence regions and where excessive settlement is expected. Even substantial movements can be corrected quickly by realigning level, lateral position and gauge. Sleepers, sleeper spacing and crane rails are calculated according to the theory of an elastic beam on an elastic foundation and in line with permanent way standards. Timber, steel, reinforced concrete and prestressed concrete sleepers can be used. Timber sleepers are preferred at facilities for loading/unloading lump ores, scrap and similar cargoes because of the reduced risk of damage from falling pieces.

### 6.17.2.2 Deep craneway foundations

On soil sensitive to settlement or deep fill, craneway beams should be founded on piles. If the piles are installed deep enough, the deep foundation to the craneway beam also relieves the loads on the waterfront structure because the loads from the craneway beam are no longer carried by the structure itself.

Basically, all customary types of pile can be used for deep foundations under craneways. The piles beneath the outboard craneway in particular are loaded in bending as well due to the deflection of the quay wall. Likewise, larger asymmetric imposed loads can lead to considerable additional, horizontal loads on the piles.

All horizontal forces due to crane operations must either be resisted by the mobilised passive earth pressure in front of the craneway beam, by raking piles or by anchors.

The craneway beam is designed as an elastic beam on an elastic foundation.

Instead of deep foundations on piles, craneway beams can also be mounted on shallow foundations on soil that has been improved or introduced through soil replacement measures.

## 6.18 Fixing crane rails to concrete (R 85)

Crane rails are to be attached free from stresses but allowing longitudinal movement. Crane rail fixings tested for the respective type of use are available. A number of options that can be used for mounting crane rails on concrete are given below.

### 6.18.1 Supporting the crane rail on a continuous steel plate on a continuous concrete base

When supporting the crane rail on a continuous steel plate, the steel plate is first aligned as flat as possible on the craneway beam and then suitably grouted or bedded on earth-damp, compacted single-sized aggregate concrete. The crane rail is only guided longitudinally on the continuous steel plate, but is anchored vertically in such a way so that even upload forces due to the interaction between bedding and rail can be accommodated. When calculating the maximum moment, anchorage force and maximum concrete compressive stress, the modulus of subgrade reaction method may be used.

Fig. R 85-1 shows an example for a heavy crane rail. Here, the concrete base was tamped in between the steel angles, levelled and given a levelling coat $\geq 1$ mm of synthetic resin or a thin bituminous coating.

If an elastic intermediate layer is placed between the concrete and the continuous steel plate, both rail and anchorage must be calculated for this softer support, which can lead to larger dimensions. The rails should be welded to minimise the number of joints. Short pieces of rail are used to bridge over expansion joints between quay wall sections.

### 6.18.2 Bridge-type arrangement with rail supported centrally on bearing plates

In this arrangement special bearing plates are used which assure a centred transfer of the vertical forces into the craneway beams and also guide the rails, which are able to move longitudinally. Furthermore, they must prevent overturning of the rail, which with this type of support is necessarily quite tall. They must accommodate both uplift and horizontal forces.

This type of crane rail fixing is used for normal general cargo cranes and in inland ports is also preferred for bulk cargo cranes. The heavy-duty design is to be recommended, above all, for the craneways of heavy-duty cranes, very heavy unloaders, unloading gantries, etc. Rail sections S 49 and S 64 are used as running rails in lightweight systems. Heavy-duty installations require PRI 85 or MRS 125 in accordance with parts 1 and 2 of DIN 536, or very heavy special rails made from steel grade St 70 or St 90.

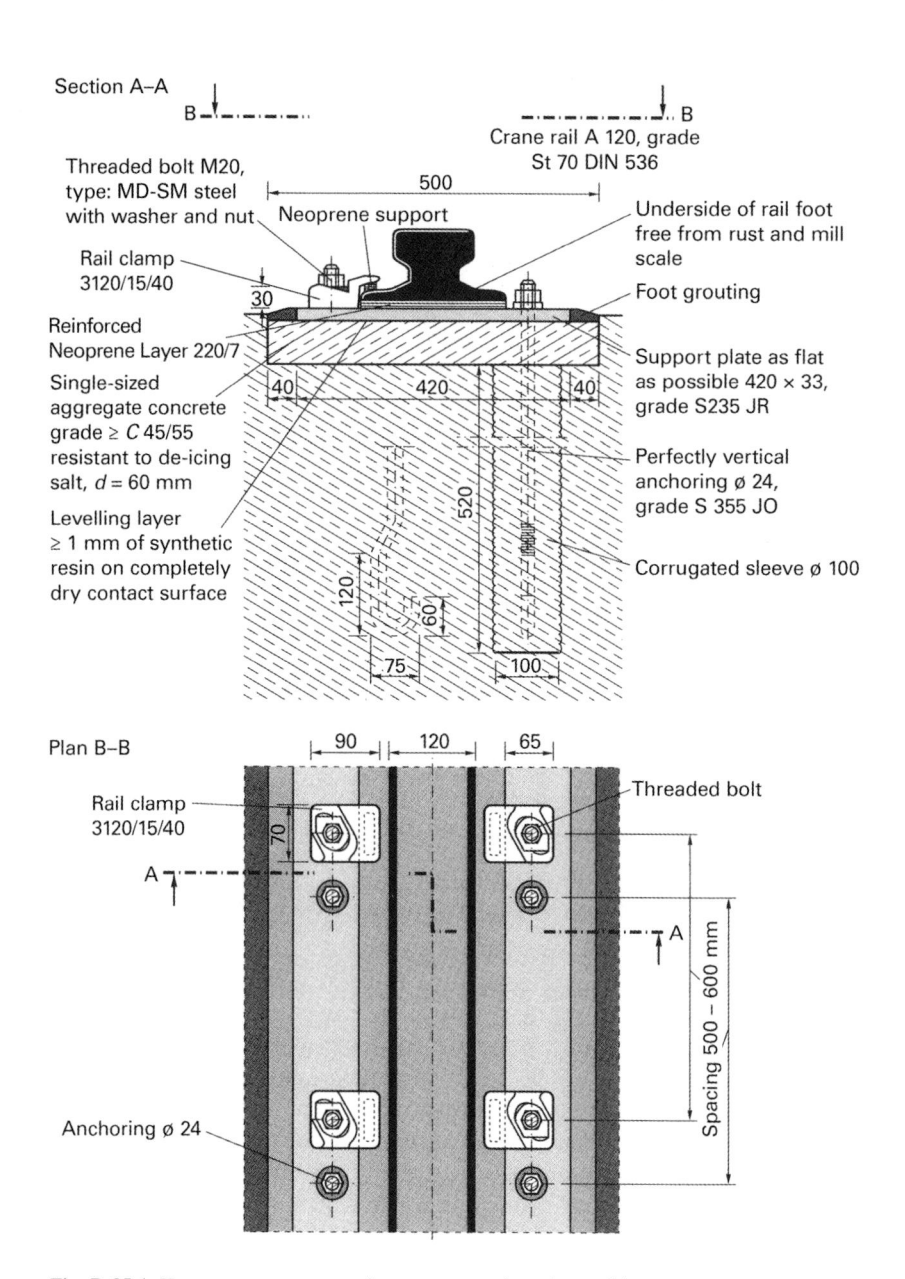

**Fig. R 85-1.** Heavy craneway on continuous concrete base (example)

S 49 or S 64, grade St 70

Plastic intermediate layer

Fastened according to *K*-type permanent way of Deutsche Bahn AG with horizontal washers at 600 mm *c/c* spacing

≥18 cm

Special anchor or anchor bolt

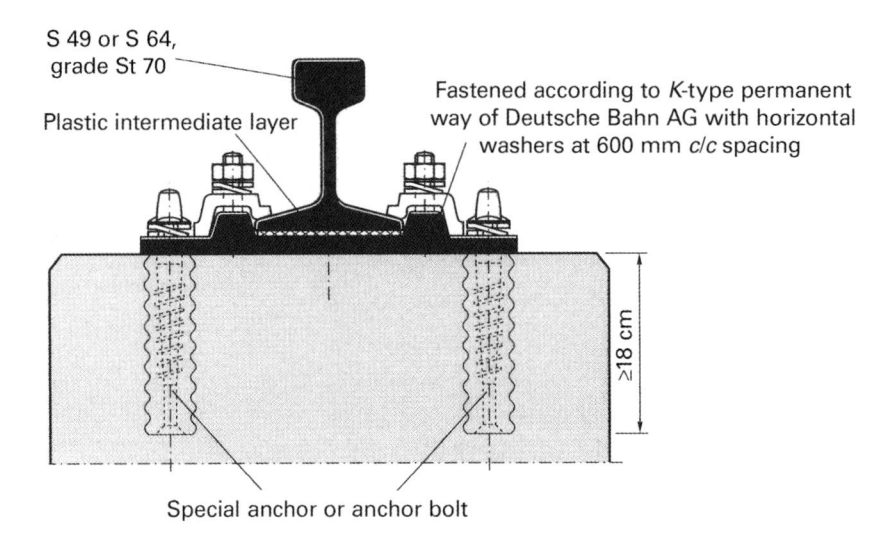

**Fig. R 85-2.** Lightweight craneway on individual supports

A typical example of a lightweight installation is shown in Fig. R 85-2. Here, S 49 or S 64 rails are supported on bearing plates according to the K-type permanent way of Deutsche Bahn AG. Rail, bearing plates, anchors and special anchors are fully assembled on the formwork or a special adjustable steel support which can be rigidly mounted. The concrete is then placed with the aid of vibration so that the bearing plates are supported across their entire area. Occasionally, an intermediate layer of approx. 4 mm thick plastic is placed between bearing plate and underside of rail (Fig. R 85-2). With cambered bearing plates, the detailing must ensure that the plastic interlayer cannot slide off.

Fig. R 85-3 shows a heavy craneway in which the support for the rail is cambered upwards in the longitudinal direction so that the rail is supported on the camber. A non-shrink material is placed or packed beneath the bearing plates. The bearing plates are also provided with elongated holes in the transverse direction so that changes to the gauge can be corrected if need be. This type of support must be provided primarily for rails with a deep web.

Non-continuous support is also possible for very high wheel loads. For crane rails with small section moduli, e.g. A 75 to A 120 or S 49, however, continuous support is recommended for loads exceeding approx. 350 kN because otherwise the spacing between the plates or chairs becomes too small.

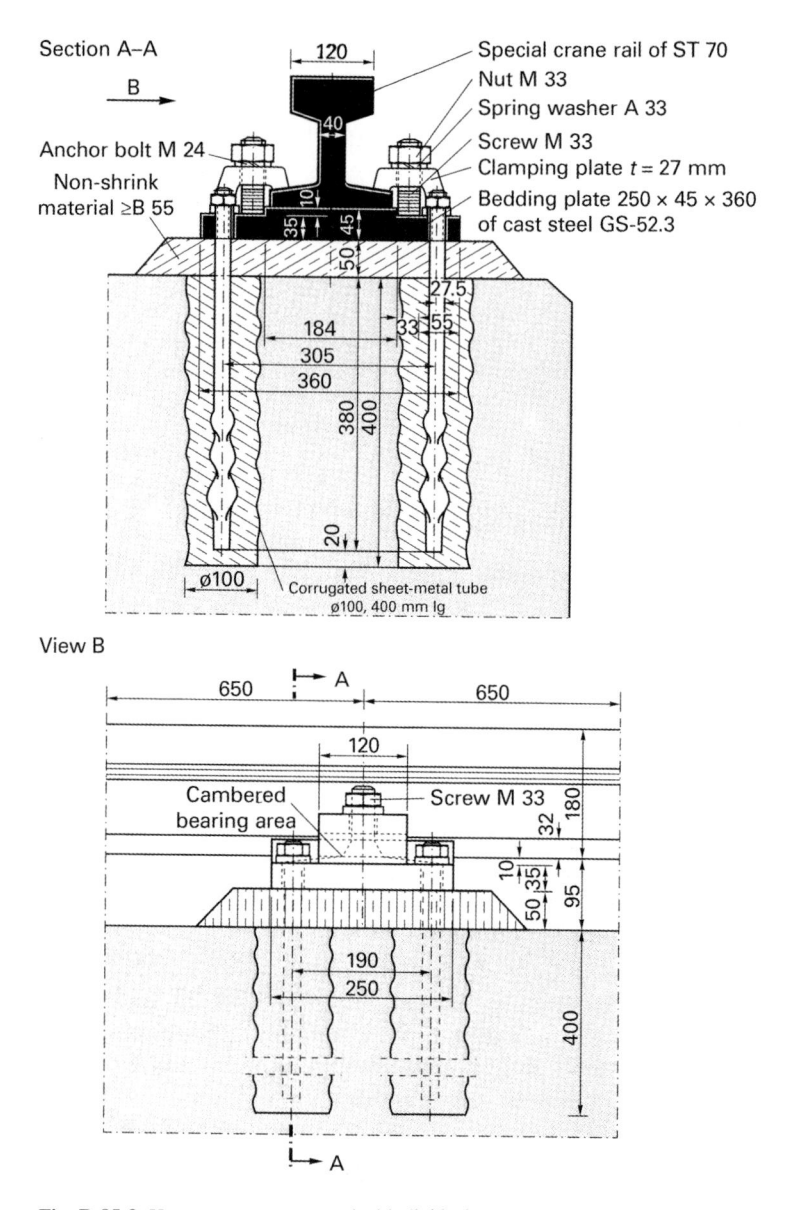

**Fig. R 85-3.** Heavy craneway on packed individual supports

### 6.18.3 Bridge-type arrangement with rail supported on chairs

When rails are supported on chairs, the rail – from a structural point of view – becomes a continuous beam on an infinite number of supports. In order to use the elasticity of the rail, an elastic interlayer is inserted between rail and chair. This layer is up to 8 mm thick and can be made from, for example, neoprene (for bearing pressures up to 12 N/mm$^2$) or a textile-reinforced rubber. It also cushions the crane wheels and chassis against impacts and shocks.

The top of the rail chair is cambered, which results in a centralised transfer of the support reaction into the concrete. This cambered bearing lies some distance above the concrete, and the flexibility of the spring washers in the rail fixings allow the rail to expand in the longitudinal direction. Changes in the length of the rails due to temperature changes and rocking movements can thus be accommodated (Fig. R 85-4). Through flexible shaping, the rail chairs can be adapted to any desired requirements. For example, the chairs permit subsequent rail realignment of, for example, $\Delta s = \pm 20$ mm in the transverse direction and $\Delta h = +50$ mm in the vertical direction.

In addition, lateral pockets can be included to accommodate the edge protection angles at crossing points.

The chairs are mounted together with the rails. Following alignment and fixing in position, additional longitudinal reinforcement is inserted through special openings in the chairs and connected to the projecting bars of the substructure (Fig. R 85-4). The concrete grade depends on structural requirements, but should be grade C20/25 at least.

If settlement and/or horizontal displacement of the crane rail necessitate subsequent realignment of the rail, this must be taken into account right at the planning stage by choosing chairs with appropriate adjustment options.

### 6.18.4 Traversable craneways

The demands of port operations frequently require the crane rails to be installed sunk into the quay surface so that they can be crossed without difficulty by vehicular traffic and port cargo-handling gear. Crane rails must therefore be installed so that they are flush with the quay surface.

1. Traversable heavy crane rails

    Fig. R 85-5 shows an example of a proven form of construction for a traversable heavy crane rail. The rail is supported on a craneway beam on a bedding of single-sized aggregate concrete, grade >C 45/55, levelled off horizontally by means of a flat steel bar. To distribute the loads, the rail, which has a thin coat of bitumen on its underside, is supported on a continuous bearing plate that is bedded on a >1 mm synthetic resin levelling layer. To prevent loads from the

**Section A–A**

Elastic intermediate layer — PRI 85$^R$ of St 70 — Rail chair of GG 25
Spring washer — T-bolt
Spring washer
Seal

**Section B–B**

Spacing according to rail and loads

PRI 85$^R$

Longitudinal reinforcement inserted after installation

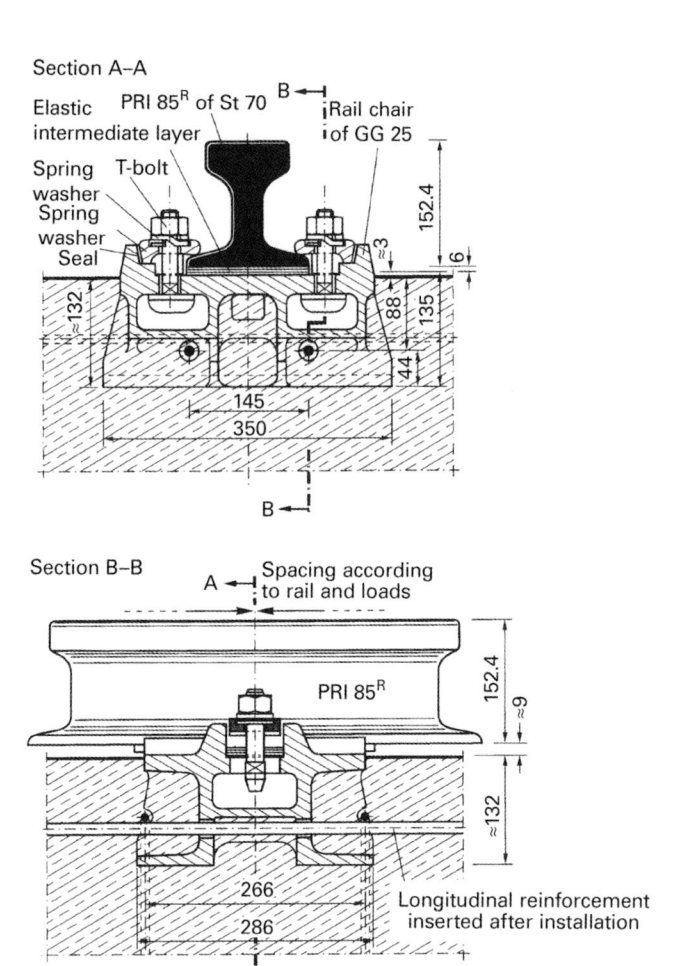

**Fig. R 85-4.** Example of a heavy craneway on chairs

longitudinal movement of the rail and bearing plate being transferred to the bolts holding the rail in position, the bearing plate is not connected to the rail fixings. Subsequent installation of the bolts is preferable because this helps to position the bolts exactly. However, this approach must be allowed for when placing the reinforcement in the craneway beam so that enough space is left between the reinforcing bars for the cast-in sheet metal or plastic sleeves for the bolts. If necessary, the holes for the bolts can also be drilled subsequently.

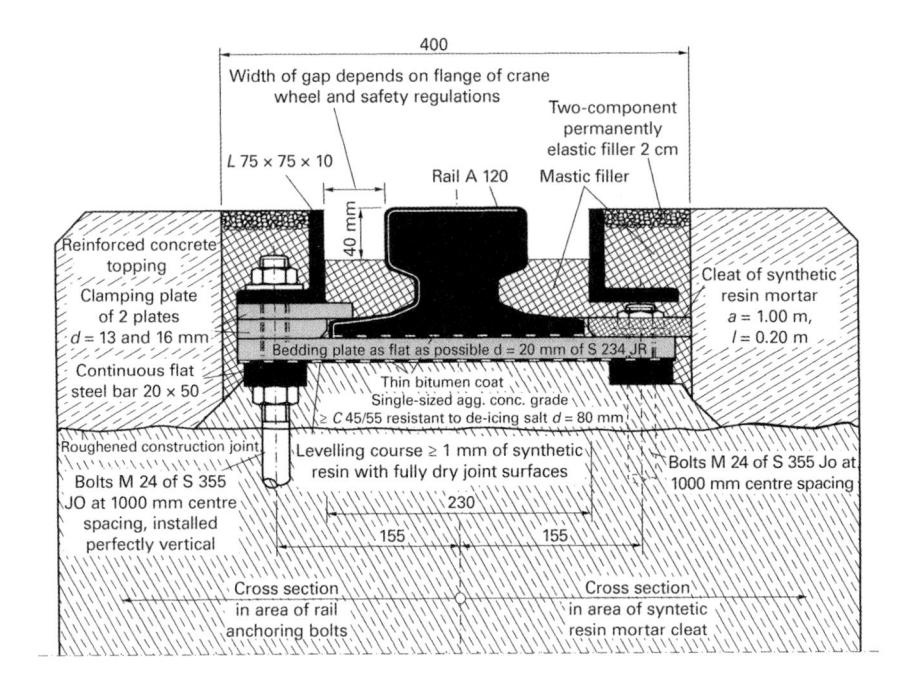

**Fig. R 85-5.** Example of traversable heavy craneway (reinforcement omitted for clarity)

In order to transfer horizontal forces transverse to the rail axis and hold the rail exactly in position, approx. 20 cm wide cleats of synthetic resin mortar are inserted between the foot of the rail and the side of the concrete topping every approx. 1 m.

It is expedient to use a permanently elastic, two-part filler for the top 2 cm of the mastic compound in the reinforced concrete topping joined with stirrups to the rest of the craneway beam. See Fig. R 85-5 for further details.

2. Traversable light crane rails
   A tried-and-tested example of this is shown in Fig. R 85-6.
   Horizontal ribbed plates are fastened to the flat, levelled craneway beam with anchors and screw spikes at a pitch of approx. 60 cm. The crane rail, e.g. S 49, is connected to the ribbed plates with clamping plates and T-bolts in accordance with Deutsche Bahn AG specifications. Levelling plates of steel, plastic, etc. can be installed beneath the foot of the rail to correct minor differences in the level of the concrete surface.
   A continuous steel angle ($80 \times 65 \times 8$ mm, grade S 235 JR) is installed to form an abutment for any adjoining reinforced concrete

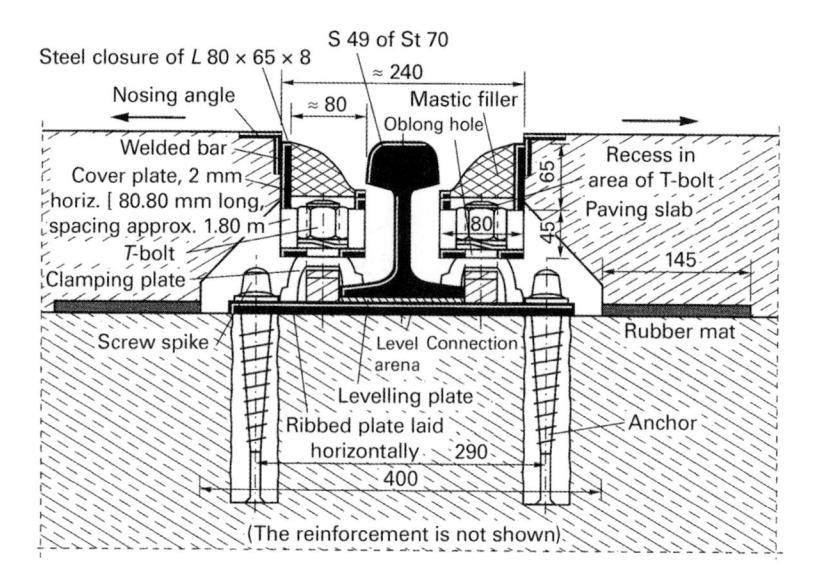

**Fig. R 85-6.** Example of a traversable light craneway

ground slabs. The angle is fitted parallel with the head of the rail, beneath which 80 mm channel sections 80 mm long are welded at every third ribbed plate. These sections have elongated holes in the (horizontal) web for fastening with T-bolts. At the intermediate ribbed plates, the angle is stiffened by 8 mm thick steel plates. Holes are cut in the horizontal leg of the angle above the fastening nuts which are covered by 2 mm thick plates after the nuts have been tightened. A bar is welded alongside the rail on the toe of each angle to hold the subsequent mastic filler in place.

To pave the port area, large reinforced concrete slabs, e.g. Stelcon, are laid loose up against the steel angle. It is recommended to lay rubber sheets underneath to prevent tilting of the slabs and at the same time to create falls that drain away from the crane rails.

### 6.18.5  Note on rail wear

The wear to be expected for the foreseeable service life of all crane rails must be taken into account in the design. As a rule, a height reduction of 5 mm with good rail support is adequate. Furthermore, more or less frequent maintenance and, depending on the type, checks of the fixings, are recommended during operation to prolong the service life.

### 6.18.6 Local bearing pressure

If local bearing pressure between the rail fastening and the mortar underneath due to travelling loads cannot be prevented by detailing, then this local bearing pressure must be considered when specifying the mortar. If there are sharp parts of screws beneath the bearing plate, then it is essential to verify that the mortar is not damaged by notch effects or elastic deformations.

### 6.19 Connection of expansion joint seal in reinforced concrete bottom to loadbearing steel sheet pile wall (R 191)

Expansion joints in reinforced concrete bottoms, e.g. in a dry dock or similar, are protected against large mutual vertical displacements by means of a joggle joint. Only minor mutual vertical displacements are then possible. The transition between the bottom and the vertical loadbearing steel sheet pile wall is formed via a relatively narrow reinforced concrete beam fixed to the sheet piling. The bottom plates separated by the expansion joint are flexibly connected to the beam, also by way of a joggle joint.

The joggle joint also continues in the connection beam.

The expansion joint in the bottom slab is sealed from below with a waterstop with loop. The waterstop ends at the U-section sheet piling, at the crest of a single sheet pile specially installed for this purpose, see Fig. R 191-1.

When using Z-section sheet piles, the waterstop ends at a connecting plate welded over the entire trough of the sheet pile as shown in Fig. R 191-2. The waterstop is turned up and clamped to this.

The interlocks of the connection piles (single piles for U-section, double piles for Z-section) are to be generously greased with a lubricant before installation. See Figs. R 191-1 and R 191-2 for further details.

### 6.20 Connecting steel sheet piling to a concrete structure (R 196)

The connection of a steel sheet pile wall to a concrete structure is always a one-off detail and must be designed to suit the actual geometric situation. Workmanship on site is especially important and all work must be carried out properly. The aim should always be to provide the simplest and most robust solution.

The connection between the sheet piling and the concrete structure must allow mutual vertical movements of the overall structure, but must also remain permanently watertight in hydraulic engineering applications, for instance.

Fig. R 196-1 shows examples of the connection between U-section sheet piles and a concrete structure.

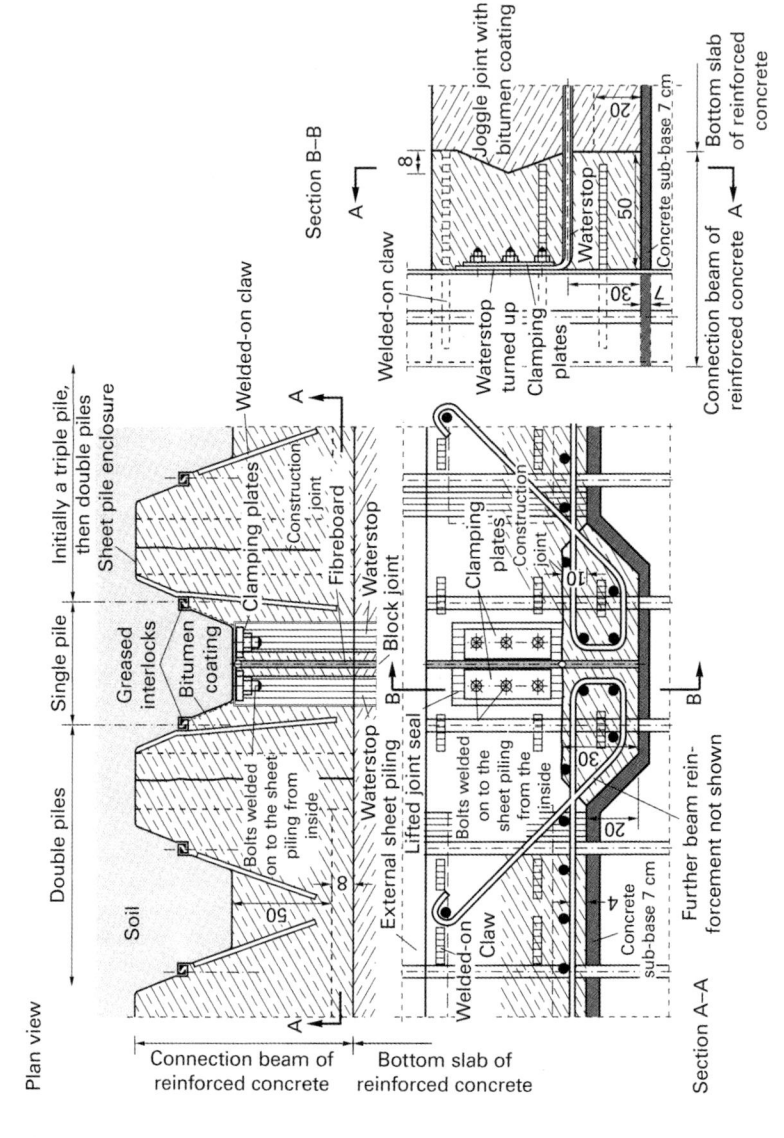

**Fig. R 191-1.** Connection of bottom seal of an expansion joint to U-section sheet piles (example)

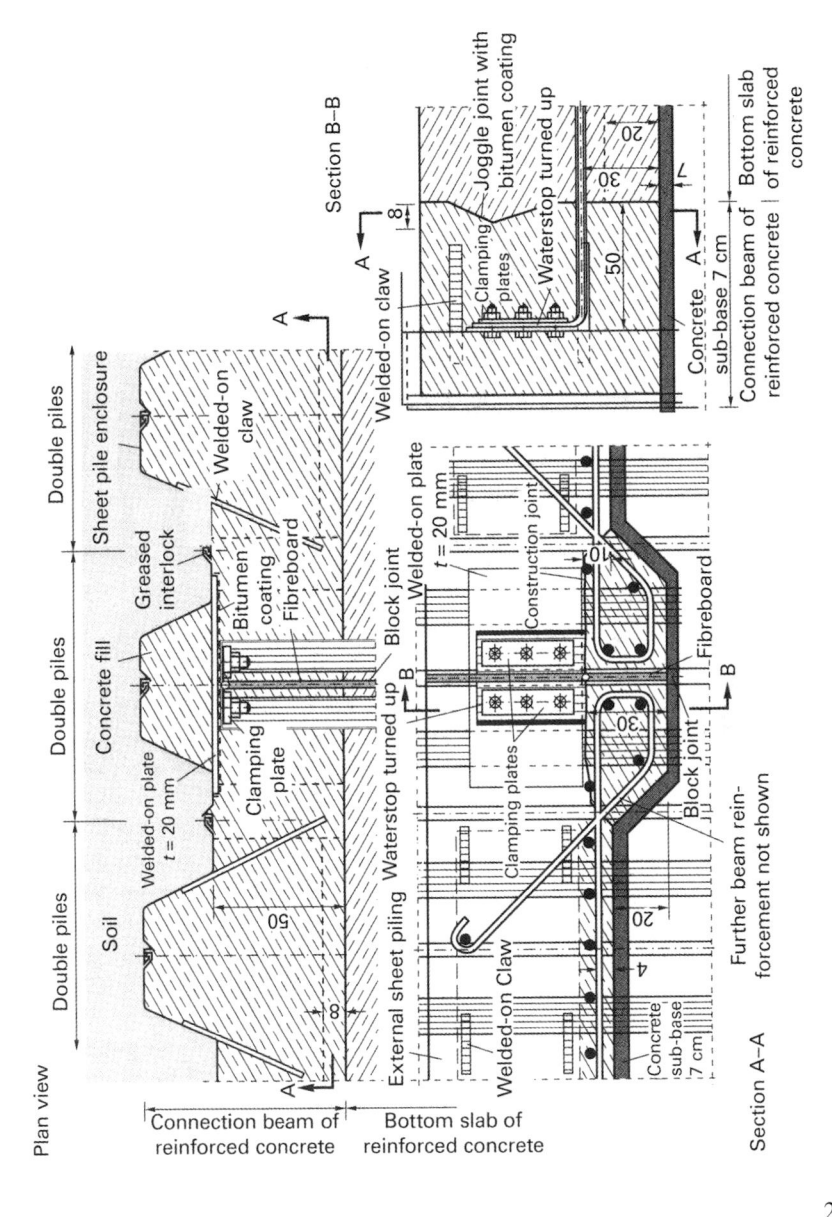

**Fig. R 191-2.** Connection of bottom seal of an expansion joint to Z-section sheet piles (example)

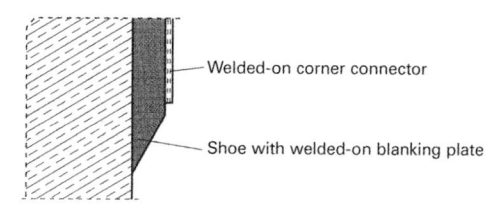

**Fig. R 196-1.** Connection of U-section sheet piling to a concrete structure

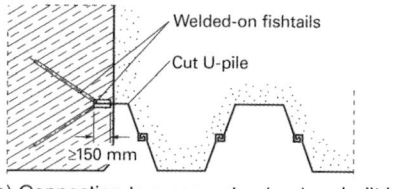

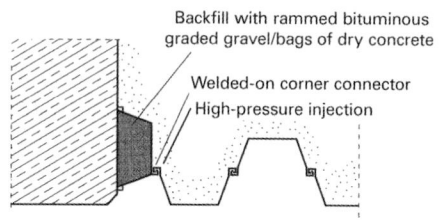

If the concrete structure is being newly constructed, a cut individual pile with welded fishtails can be inserted through the formwork and cast into the concrete. The adjoining sheet piling is then threaded into the interlock of the cast-in pile (Fig. R 196-1a). The connecting interlock must be filled with a plastic compound so that threading remains possible (see R 117, section 8.1.21).

When steel sheet piling is to be connected to an existing concrete structure, a solution such as the one shown in Fig. R 196-1b is recommended. Here, a U-section sheet pile with an interlock welded to the back is bolted to the concrete structure and the intermediate space filled. Instead of bituminous graded gravel, a backfill with small sacks of dry concrete (like for bagwork) has proved successful. The next sheet pile is then threaded into the interlock welded to the back of the filled pile. High-pressure grout injection behind the wall may be advisable in order to ensure the permanent watertightness of the connection.

Similar examples for Z-section sheet piles are shown in Fig. R 196-2.

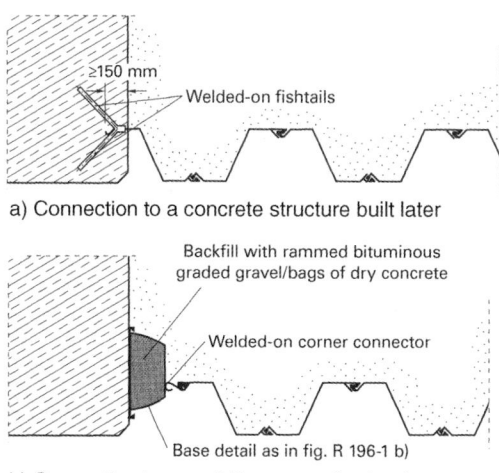

a) Connection to a concrete structure built later

b) Connection to an existing concrete structure

**Fig. R 196-2.** Connection of Z-section sheet piling to a concrete structure

If the watertightness and/or flexibility of the connection must satisfy a demanding specification, then waterstops must be included in the connection, fixed to the sheet pile with clamping plates and to a steel flat cast into the concrete (Fig. R 196-3). The embedment depths of the sheet piles must be such that when embedded in a low-permeability soil stratum, groundwater flow is either stopped or the length of the seepage path is long enough to limit the flow to a permissible level and hydraulic heave is ruled out. Movements of the components with respect to each other must be checked carefully when assessing the sealing effect. Please refer to DIN 18195 parts 1– 4, 6, 8–10.

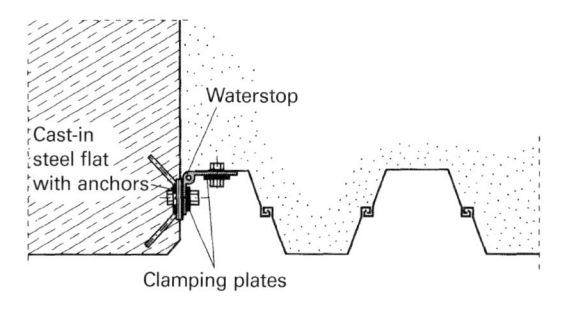

**Fig. R 196-3.** Connection of U-section sheet piling to a concrete structure with high demands on the watertightness of the junction

## 6.21 Floating berths in seaports (R 206)

The "Floating Berths" specification of the German Federal Ministry of Transport, Building and Urban Development applies to floating berths on federal waterways. It can be applied correspondingly to seaports as well by taking into account the advice given below.

### 6.21.1 General

In seaports, floating berths are reserved for passenger ferries, port vessels and pleasure craft. They consist of one or more pontoons and are connected to the shore by a bridge or permanent stairs. The pontoons are generally held in place by driven guide piles and the access bridge/stair has a fixed support at the land end, a movable one at the pontoon. If the floating berth consists of several pontoons, interconnecting walkways ensure that it is possible to move from one pontoon to the next.

### 6.21.2 Design principles

Stipulating the location of a floating berth must take account of current directions and velocities together with wave influences.

In tidal areas, the highest and lowest astronomical tides should be used as the design water levels. The incline of the access jetty should not be steeper than 1:6 at mean tide and not steeper than 1:4 at extreme water levels.

Especially when used by the public, the facility must comply with strict requirements, e.g. even under icy conditions. To do this, suitable constructional and organisational measures will be necessary.

The bulkhead divisions of the pontoons must be chosen such that failure of one single cell through an accident or other circumstances will not cause the pontoon to sink. The cells should be vented individually, e.g. with swan-neck pipes. Cells with sounding pipes accessible from the deck are recommended for simplifying the checking of the watertightness. In certain cases it may be advisable to include an alarm system that warns of an undetected ingress of water. For industrial safety reasons, every cell should be accessible from the deck or through no more than one bulkhead.

Filling the cells with a non-porous foam can also be considered.

A cambered pontoon deck must be provided to ensure surface water run-off.

It is advisable to provide a disconnecting option for the access bridge, e.g. by two piles driven next to the bridge with suspended cross-member, to guarantee that pontoons can float away rapidly in the event of an accident.

The minimum freeboard required for a pontoon depends on the permissible listing, anticipated wave heights and intended use. For smaller facilities, e.g. for pleasure craft, a minimum freeboard of 0.20 m is adequate for one-sided use, whereas large pontoons require far greater freeboard heights. As a guide, freeboard heights for pontoons up to 30 m long and 3–6 m wide should be about 0.8–1.0 m, whereas pontoons 30–60 m long and up to 12 m wide should have a freeboard height of approx. 1.2–1.5 m.

The freeboard heights must be adjusted to suit the embarkation and disembarkation heights of the vessels, particularly when the facility is used by the public.

### 6.21.3 Loading assumptions and design

As a basic rule, the position of the pontoon should be verified with an even keel, with ballast balance being provided where necessary.

An imposed load of 5 kN/m$^2$ should be assumed when checking floating stability and listing (one-sided load).

Floating stability verification should also include hydrodynamic loads such as banking-up pressure, flow force and waves, with calculations being confirmed by tests if necessary. Listing of pontoons and the angles of walkways between pontoons must be checked.

Depending on the pontoon dimensions, listing acceleration and the mutual offset of several pontoons, listing may not exceed 5°; the upper limit is 0.25–0.30 m. Greater listing angles are to be checked on a case-by-case basis.

The ship's berthing force as a load from moored ships is to be taken as 300 kN and 0.30 m/s, or 300 kN and 0.5 m/s for larger facilities (pontoon length >30 m).

A cushioning effect to reduce the ship's berthing force on pontoons by way of fenders on the outer surface, spring-mounted brackets, rubbing strips and guide dolphins can be considered if these are verified in appropriate investigations. The cushioning effect of guide dolphins can be increased when they are constructed in the form of coupled tubular piles.

# 7 Earthworks and dredging

## 7.1 Dredging in front of quay walls in seaports (R 80)

This recommendation deals with the technical options and conditions to be taken into account when planning and executing dredging work in front of quay walls in ports and harbours.

A distinction must always be made between new dredging and maintenance dredging when it comes to the necessary authorisations, approvals, etc.

Dredging down to the design depth according to R 36, section 6.7, should be carried out by grab dredgers, bucket-ladder dredgers, cutter-suction dredgers, cutter-wheel suction dredgers, plain suction dredgers or hopper suction dredgers. Harrows and water-jetting machines can be employed as well.

When using cutter-suction dredgers, plain suction dredgers or hopper suction dredgers in areas of passive earth resistance in front of quay walls, they must be equipped with devices that ensure exact adherence to the planned dredging depth. Cutter-suction dredgers with high capacity and high suction force are, however, less suitable because of the danger of overdigging and disturbing the soil below the cutter. Dredging by suction dredgers without cutter equipment should not be allowed in areas of soil that constitute passive earth resistance.

It should also be noted that, even under favourable conditions and with careful workmanship, bucket-ladder, cutter-suction and hopper suction dredgers cannot achieve the exact theoretical nominal depth when making the final dredging cut, because a wedge of soil 3–5 m wide remains unless the soil is able to slide down. The need to remove this residual wedge depends on the type of fender on the quay wall and on the block coefficients of the ships that will berth there. Only grab dredgers can remove any residual wedge of soil. Additionally, the troughs of the sheet piles must be flushed free of cohesive soil in certain circumstances.

When dredging a port or harbour with floating plant, the work is usually divided into dredging cuts of between 2 and 5 m depending on the type and capacity of the plant. The intended use of the excavated soil can also be relevant to the choice of dredging plant in the case of varying soil types. It is advisable to survey the front face of the quay prior to and at intervals throughout the dredging operations and not only afterwards in order to ensure early detection of any structural deformation.

---

*Recommendations of the Committee for Waterfront Structures Harbours and Waterways – EAU 2012*, 9th Edition. Issued by the Committee for Waterfront Structures of the German Port Technology Association and the German Geotechnical Society.
© 2015 Ernst & Sohn GmbH & Co. KG. Published 2015 by Ernst & Sohn GmbH & Co. KG.

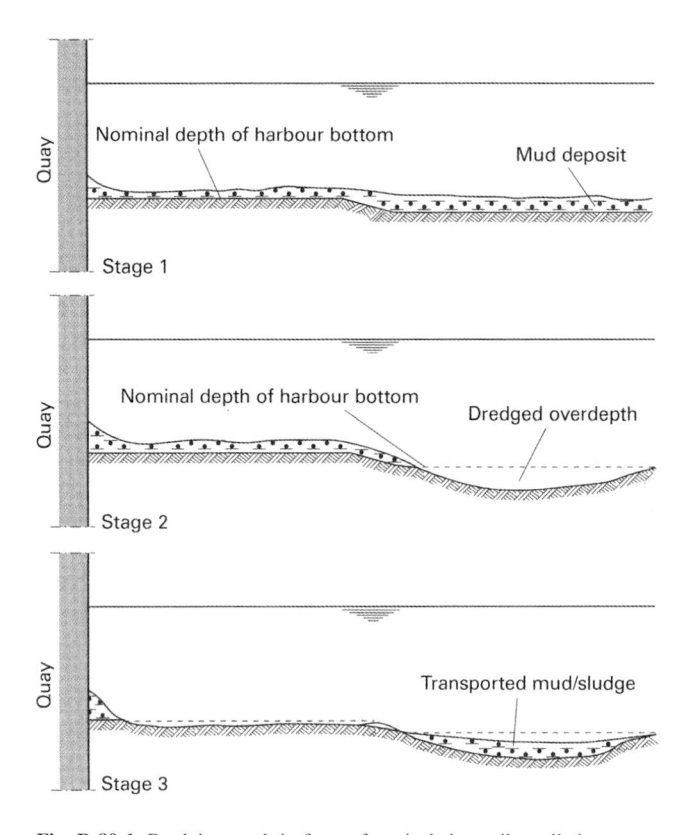

**Fig. R 80-1.** Dredging work in front of vertical sheet pile walls in seaports.
Stage 1: Existing situation
Stage 2: Situation after dredging
Stage 3: Situation after using harrow (hedgehog) or grab dredger

Inspections by divers are required to detect any damage to sheet pile interlocks that might have occurred in wall areas exposed by dredging (R 73, section 7.4.4).

An approach such as that depicted in Fig. R 80-1 can be economical and less disruptive to port operations.

Following overdepth dredging (stage 2), the mud/sludge lying in front of the quay wall is moved into the overdepth area of the harbour bottom with grab dredgers or a harrow (stage 3). Wherever possible, the overdepth area should be created during the new dredging works.

It is essential to check the stability of the quay wall before every dredging operation that might fully exploit the theoretical total depth beneath the nominal depth of the harbour bottom. In addition, the behaviour of the quay wall must be monitored before, during and after dredging.

## 7.2 Dredging and hydraulic fill tolerances (R 139)

### 7.2.1 General

The specified dredging depths and hydraulic fill heights are to be achieved within clearly defined permissible tolerances. If the difference in level between individual points of the base of an excavation or a filled area exceed those tolerances, then supplementary measures will be necessary. Specifying excessively tight tolerances can lead to disproportionately high extra costs for dredging works. Stipulating tolerances for dredging and hydraulic filling work should therefore be based on technical but, first and foremost, economic aspects. The client must consider how important it is to achieve a certain accuracy for the base of an excavation or a height of fill.

In addition to deviations from the intended level (vertical tolerances), horizontal tolerances must also be specified when trenches have to be dug for soil replacement measures, inverted siphons and tunnels, for instance. Here, too, it is almost always necessary to reach an optimum balance between the extra costs involved in more extensive dredging and filling quantities with more generous tolerances, and the extra costs due to reduced performance of the plant because of more precise and hence slower working plus the costs of any potential additional measures.

The achievable accuracy of dredging work for inland waterways is generally better than that in waterways for ocean-going vessels, where tides, waves, shoaling by sand, and/or sludge play a major role. For nautical reasons, a minimum depth is normally required in waterways. Where dredging work is followed by construction measures such as profiling in front of embankments, installing revetments or concreting bottom slabs, dredging tolerances must be tailored to meet the requirements of subsequent operations.

### 7.2.2 Dredging tolerances

Dredging tolerances are to be specified taking into account the following factors:

a) Quality demands regarding the accuracy of the depths to be achieved resulting from the objective of the dredging work, e.g.:
- regular, recurring maintenance dredging to eliminate all sediments and maintain navigability,
- the creation or deepening of a berth in front of a quay wall or a navigable channel to improve navigability,

- the creation of a watercourse bottom to accept a structure (inverted siphon or tunnel structures, bottom protection measures, etc.),
- dredging to remove non-loadbearing subsoils within the scope of soil replacement, and
- dredging to remove contaminated soil sediments.

Each of these objectives calls for specific and distinctly different dredging accuracies. This affects the choice of plant and so has an influence on the cost of dredging work.

b) In addition, structural boundary conditions, the extent of dredging and the properties of the soil to be dredged must all be taken into account, e.g.:
- the stability of nearby underwater slopes, moles, quay walls, etc.,
- the depth below the planned harbour bottom to which disturbance of the subsoil may be accepted,
- the horizontal or vertical dimensions of the dredging work, the length and width of the excavation, the thickness of the stratum to be dredged and the overall volume of dredging works,
- the soil types and properties, particle sizes and distribution and shear strengths of the soils to be dredged,
- the use of dredged soil, and
- any possible contamination, with specific requirements for dealing with the dredged soil.

This last point in particular is becoming increasingly significant due to the, generally, extremely high cost of treating and disposing of contaminated soils, and can therefore make it necessary to work to very tight tolerances.

c) Local circumstances continue to play a decisive role, influencing the use and control of dredging plant and the general strategy of the dredging works. Examples of this are:
- the depth of the water,
- accessibility for dredging plant,
- tide-related changes in water levels with changing currents,
- any seawater/freshwater changes,
- weather conditions (wind and current conditions),
- waves, sea conditions, swell,
- interruptions to the dredging works by waterborne traffic,
- constraints due to the proximity of berths or ships at anchor, and
- the extent of regularly recurring sediments (sand or silt), possibly even during the dredging works.

d) Finally, the dredging plant itself and its equipment also play their part in terms of:
- the size and technology of the plant as well as its dredging accuracy depending on soil and depth,

- the instrumentation on board the operation (positioning instruments, depth measurement, performance measurement, quality of monitoring and logging technology for the entire dredging process),
- the experience and skills of dredging crews, and
- the magnitude of the specific drop in performance of dredging operation owing to the tolerances that have to be maintained, and the ensuing costs.

It must also be taken into account that the outcome of dredging work and the accuracy achieved is generally checked and recorded by sounding. Therefore, the actual results determined through sounding are always affected by the accuracy of the sounding equipment itself. So, when specifying dredging tolerances, the resolution of the sounding system employed to check the work must also be considered.

Specifying technically optimum and hence also economic dredging tolerances is therefore a complex issue. Before commencing major dredging operations it is therefore imperative to weigh up carefully the many factors that influence dredging accuracy. At the time of inviting bids it is often unclear as to which type of dredger will be used. It is therefore wise to ask bidders to submit their price based on the specified accuracy along with their price for the accuracy they themselves propose and will guarantee (special proposal and special bid for dredging work). For general guidance, reasonable vertical dredging tolerances for various dredger types are given in Table R 139-1. These are based mostly on experience gained in the Netherlands (International Association of Dredging Companies [41]).

Horizontal tolerances have been deliberately omitted because in the case of slopes these ensue from the vertical tolerance due to the required angle of the slope (R 138, section 7.7). Furthermore, the horizontal tolerances should be specified for each individual case in conjunction with the specific requirements and the equipment available for determining the position of the dredger. Soundings should be carried out with instruments that measure the actual depth to the bottom and do not give a false reading to the top of an overlying suspended layer.

## 7.3 Hydraulic filling of port areas for planned waterfront structures (R 81)

### 7.3.1 General

Recommendation R 73, section 7.4, covers the direct backfilling of waterfront structures.

Port areas with a good bearing capacity behind waterfront structures can be created by hydraulic filling, provided well-graded sand is available. In hydraulic filling over water, a greater degree of compaction is achieved

**Table R 139-1.** Guide values for vertical dredging tolerances in cm (International Association of Dredging Companies [41])

| Dredger | Non-cohesive soils | | | Cohesive soils | | | Surcharges for ... | | |
|---|---|---|---|---|---|---|---|---|---|
| | Sand | Gravel | Rock | Silt | Clay | water depth 10–20 m | current 0.5–0.1 m/s | unprotected watercourse |
| Grab dredger | 40–50 | 40–50 | – | 30–45 | 50–60 | 10 | 10 | 20 |
| Bucket-ladder dredger | 20–30 | 20–30 | – | 20–30 | 20–30 | 5 | 10 | 10 |
| Cutter-suction dredger | 30–40 | 30–40 | 40–50 | 25–40 | 30–40 | 5 | 10 | 10 |
| Cutter-wheel suction dredger | 30–40 | 30–40 | 40–50 | 25–40 | 30–40 | 5 | 10 | 10 |
| Suction dredger with reduced turbidity | 10–20 | – | – | 10–20 | – | 5 | 5 | – |
| Backhoe dredger | 25–50 | 25–50 | 40–60 | 20–40 | 35–50 | 10 | 10 | 10 |
| Hopper suction dredger | 40–50 | 40–50 | – | 30–40 | 50–60 | 10 | 10 | 10 |

Notes to Table R 139-1:
Guide values for positive and negative deviations (in cm) for normal conditions (e.g. 50 = ±0.5 m)
The lower value of each pair applies to work in which maximum accuracy is required.
The upper value of each pair applies to work for which large plant would seem to be suitable.
Normally, there is a probability of the values being 5% higher or lower than the values given.
Greater accuracy is possible with correspondingly greater effort.

without additional measures than when filling underwater for the same conditions (R 175, section 7.5).

In all hydraulic filling work, but especially in tidal areas, measures must be taken to ensure good run-off of the filling water as well as the water flowing in with the tide.

The fill sand should contain as few fine particles as possible (silt and clay <0.06 mm). The allowable volume of fine particles in any specific case depends not only on the loadbearing capacity of the intended waterfront structure and the required quality of the planned port area, but also on the magnitude of the anticipated residual settlement or when port operations are due to start.

The upper 2 m of hydraulic fill must be easy to compact and exhibit a loadbearing capacity suitable for setting up manoeuvring and storage areas.

If the port area will include settlement-sensitive facilities, silt and clay deposits must be avoided and the proportion of fine particles in the hydraulic fill in loadbearing areas should be <10%.

For economic reasons it is frequently necessary to obtain the fill material locally, or to use material obtained, for example, from dredging work in the port itself. The dredged soil is often loosened with cutter-suction or plain suction dredgers and pumped via pipelines directly to the planned port area.

In such cases, meaningful soil investigations beforehand at the source are indispensable to establish the types of soil and their properties, especially the proportion of fine particles. Continuous sampling of the sand obtained using core samples is expedient because these also reveal thin inclusions of cohesive material. Employed in conjunction with supplementary cone penetration tests, these investigations provide worthwhile knowledge of the proportion of fine particles in the sand intended for the fill and how it is distributed vertically and horizontally over the source site.

If the fill sand is pumped via pipelines directly into the port area from the area from which is being sourced, the content of fine particles is completely restored at the filling site, possibly even concentrated in layers around the outlets. Where hopper suction dredging is used or the fill sand is transported to the filling site in barges, some of the fine particles are washed out with the overflow. In this respect, this type of backfill for sand deposits with a larger silt and clay content is more advantageous than direct pumping. However, turbidity in the watercourse resulting from the overflow of fines must be acceptable.

If mud or clay has been deposited locally on the surface of the hydraulic fill, e.g. around the outlet, then this material must be removed to a depth of 1.5–2.0 m below the future ground level and replaced by sand (see also R 175, section 7.5).

Silt or clay deposited in layers in the hydraulic fill area prevents rapid run-off of the hydraulic fill water, which means that the sand fill is compacted to a lesser degree than when the water can drain away quickly. Appropriate maintenance of the hydraulic fill area and controlling the flow of hydraulic fill material can ensure that silt or clay is removed from the filling area.

If cohesive inclusions still cannot be prevented, the consolidation of these layers can be promoted and accelerated by using vertical drains, for example (R 93, section 7.11).

Embankments of hydraulic fill material with a medium sand have slopes of 1:3 to 1:4 above the water table, up to 1:2 in some cases in depths $\geq 2$ m below the water table. Such embankments are unstable and currents can flatten the slopes.

### 7.3.2 Hydraulic filling of port above the water table

Hydraulic filling of port areas in the dry is shown schematically in Fig. R 81-1. Clearly defining the width and length of the area of fill and the positions of the outlets can facilitate the removal of fine particles with the hydraulic fill flow.

The width and length of the filling area and the outlets must be specified in such a way that the water carrying suspended solids and fine particles can drain away as quickly as possible and, in particular, that no eddies ensue. In addition, hydraulic filling operations must continue without interruption if at all possible.

Where interruptions cannot be avoided (e.g. weekends), checks must be carried out after every interruption to establish whether layers of fine particles have settled anywhere. Any such layers must be removed before filling resumes.

If the water containing suspended particles and fine material is to be returned to the watercourse, settlement basins may need to be included to ensure compliance with the corresponding regulations of the authorities with regard to water turbidity and the introduction of suspended particles. Sediments separated off in this way must be disposed of separately.

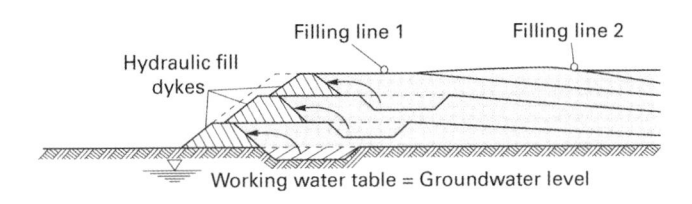

**Fig. R 81-1.** Hydraulic filling of port areas above the water table (schematic)

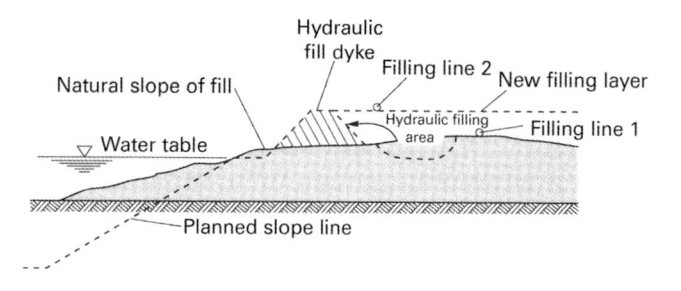

Fig. R 81-2. Hydraulic filling of port areas on a surface below the water table

When the hydraulic fill dyke is to form the subsequent boundary to a port or harbour, e.g. the waterfront to a watercourse, it is recommended to construct this as a sand dyke with a covering of plastic sheeting. In order that sand with the coarsest possible grain size settles in the area of the dyke, the filling lines are laid on the hydraulic fill dyke or at the base of the dyke on the backfill side. The coarse sand deposited in front of the hydraulic fill dyke can then be used for raising the dyke further (Fig. R 81-1). This increase in height must be limited so that the bearing capacity of the subsoil is not exceeded.

## 7.3.3  Hydraulic filling of port areas below the water table

### 7.3.3.1  Filling with coarse-grained sand

Coarse-grained sand can be used as hydraulic fill without any further measures (Fig. R 81-2). The angle of the natural hydraulic fill slope depends on the coarseness of the fill sand and on the currents. The fill material deposited outside the intended underwater embankment will be dredged away later (R 138, section 7.7).

The sand deposited as fill in the first stage should reach a level of about 0.50 m above the relevant working water level for coarse sand and at least 1.0 m for coarse to medium sand. Above this, work continues between hydraulic fill dykes according to section 7.3.2. Filling in tidal areas may need to be carried out to suit the tides.

### 7.3.3.2  Filling with fine-grained sand

Fine-grained fill sand is placed underwater by pumping or dumping between hydraulic fill dykes of rock fill material (Fig. R 81-3). This method can also be recommended when, for instance, waterborne traffic requirements do not leave sufficient space for a natural hydraulic fill slope.

The fill material used for the hydraulic fill dykes underwater should create a stable filter with respect to the fill sand.

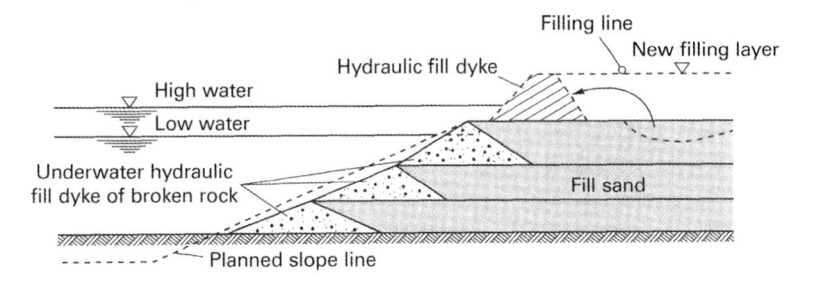

**Fig. R 81-3.** Underwater hydraulic fill dyke of broken rock with fine-grained sand deposited by pumping or dumping

If the hydraulic fill dykes form the boundary embankments to a port or harbour area, they must be able to withstand the effects of currents and waves. If necessary, protection must be provided in the form of revetments. Hydraulic fill dykes of rock fill material are problematic where driving works are to be carried out later.

It is also possible to build up the shore in advance with dumped sand (Fig. R 81-4) and then backfill behind this. The coarsest sand should be used for this method in order to avoid drifting caused by strong currents. Excess dumped sand outside the theoretical underwater slope line is dredged away later (R 138, section 7.7).

An alternative to dumped sand is the so-called rainbow method. In this method the filling material is deposited through a jet. This method is only possible if the fill material is not carried away by currents.

### 7.3.3.3 Hydraulic filling of port areas above soft sediment deposits

If existing idle harbour or port basins are to be filled as part of restructuring work for new uses, it can be economical for existing

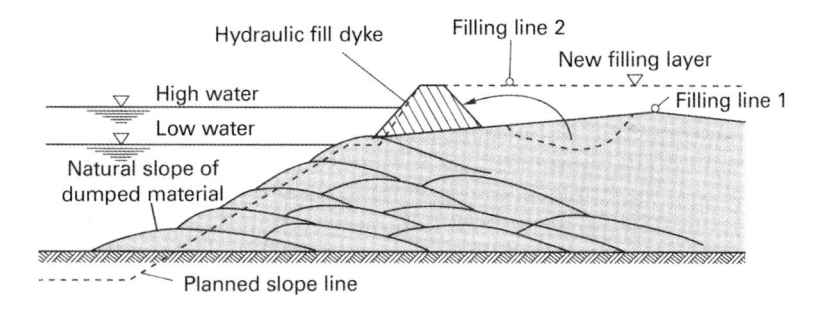

**Fig. R 81-4.** Underwater construction of dykes of coarse sand by dumping

sediments to be left on the bottom and covered by fill material. For structural reasons, however, it is generally necessary to replace the sediment layers in front of and behind waterfront structures and embankments.

In the basin the sediment remains on the bottom and is covered using the hydraulic filling method. The fill sand must be placed in layers, the thickness of which is limited by the bearing capacity of the in situ sediments. Where sediments have very little strength, then in order to rule out local deformations and ground failure, the first layer of hydraulic fill material can often be only a few decimetres thick. Only the rainbow method is suitable for placing such thin layers. The plant customarily available these days enables the placing of layers 10 cm thick [142]. Maintaining the permissible thickness of the layers of hydraulic fill material must be constantly monitored in order to avoid the fill sand collapsing into the soft in situ sediments. If efforts to do this fail, large differential settlement can occur later in the finished surface.

The in situ sediments consolidate under the load of the newly deposited fill sand. Thus, the second layer of sand can be correspondingly thicker – to suit the increased strength of the sediments. The required duration of the consolidation process can be estimated beforehand with settlement analyses. Obtaining a reliable figure for the required consolidation time, however, is only possible by using metrological techniques accompanying the work (see R 179, section 7.12.5.1).

Hydraulic filling of port areas above soft sediment deposits therefore requires careful prior soil mechanics investigations of the strength and consolidation characteristics of the soft in situ sediments, plus exhaustive monitoring of the building measures by way of measurements during the work, e.g. settlement and the development of excess pore water pressure as well as shear strength.

Once the hydraulic fill is raised to such a height that the surface is suitable for vehicles, consolidation can be accelerated using vertical drains, possibly also in conjunction with preloading (R 93, section 7.11). When developing port areas by covering soft sediments below bodies of water, the area is first enclosed with a dyke or pile wall. Within this polder it is then possible to backfill over in situ soft sediments up to the intended height.

### 7.3.3.4 Camber during hydraulic filling to account for settlement

The amount of camber during hydraulic filling is largely dependent on the accuracy with which the settlement of the subsoil and the settlement and subsidence of the fill material can be predicted. An appropriate estimate of the settlement requires a sufficient number of good-quality field and laboratory soil mechanics investigations. It will certainly be necessary, however, to carry out profiling work to establish the target height. Moreover, the fill tolerances are determined in relation to the height of the fill.

When greater settlement is expected, the amount of settlement – and the camber derived from that – should be indicated in the tender and taken into account in the hydraulic fill specification. If this is not possible, separate remuneration for the additional filling volume due to settlement can be agreed upon.

## 7.4 Backfilling of waterfront structures (R 73)

### 7.4.1 General

Where waterfront structures are to be erected and subsequently backfilled in open water, it can be useful to remove soft in situ sediments from areas where pile driving will take place and from regions of active and passive earth pressure prior to driving sheet pile walls and other piles and to replace the material with good, compactable soil. Subsequent backfill can thus be supported directly on subsoil with sufficient bearing capacity. This avoids construction stages during which the quay wall is affected by the surcharge due to backfill on top of soft cohesive layers that have not been consolidated for these loads. These measures prevent actions due to active earth pressure in the unconsolidated state, and the full passive earth pressure in front of the wall can be assumed from the beginning. In addition, differential settlement between waterfront structure and backfilling can be minimised.

### 7.4.2 Backfilling in the dry

Waterfront structures erected in the dry should also be backfilled in the dry wherever possible. The backfill must be placed in horizontal layers in depths to suit the compaction plant used and then well compacted. Sand or gravel should be used as the backfill material wherever possible. Non-cohesive backfill must have a minimum in situ density $D \geq 0.5$, particularly near the top of the backfill. Otherwise, maintenance of roads, rail tracks, etc. can be expected.

The in situ density of the backfill can be checked with cone penetration tests. In a backfilling of inhomogeneous sand in which the content of fines <0.06 mm dia. is <10% by wt., the toe resistance $q_c$ should be >6 MN/m$^2$. Following compaction, $q_c$ can reach >10 MN/m$^2$ at depths below approx. 0.6 m.

For backfilling in the dry, cohesive soil types such as boulder clay, sandy loam, loamy sand and, in exceptional cases, even firm or silty clays may be suitable. Cohesive backfill must be placed in thin layers and well distributed and compacted to achieve a maximum density and rule out voids in the backfilling. A sufficiently deep layer of sand must be added over a cohesive backfill to protect it from the direct effects of vehicular traffic.

### 7.4.3 Backfilling underwater

Only sand and gravel or other suitable, non-cohesive soil may be used as fill underwater. A moderate in situ density ($0.3 < D < 0.5$) can usually be achieved if inhomogeneous sand is hydraulically pumped in such a manner that it is deposited without segregation. When using homogenous sand as fill, hydraulic pumping alone can generally achieve a loose in situ density only ($D < 0.3$). Higher in situ densities can only be achieved through additional compaction by means of vibroflotation. When using vibroflotation methods in an area directly affecting the quay wall, always ensure that the structure can accommodate the vibrations generated and the pressure from the local liquefied soil.

When the sand for hydraulic fill contains fine particles or mud is to be expected, then backfilling must proceed in such a manner that no continuous sediment horizon is created in the body of soil behind the quay wall. Such horizons can form failure planes with reduced strength in which a higher active earth pressure and/or lower passive earth pressure might have to be assumed.

If during backfilling the upper layers are deposited by hydraulic methods above the water table, the hydraulic filling is to be controlled such that the water can drain away quickly. Otherwise, a higher excess water pressure must be assumed for the quay wall. Any existing drainage system for a waterfront structure may not be used for draining the hydraulic fill water because it could become irreparably clogged and therefore ineffective.

As a rule, subsoil settlement beneath a backfill surcharge and settlement of the backfill itself do not abate before dredging works are carried out in front of the waterfront structure and the area becomes operational. Therefore, residual settlement can occur in the first years of port operation.

### 7.4.4 Additional remarks

Sheet pile interlocks are occasionally damaged during driving (declutching), which results in a considerable flow of water at these points when there is a difference in water pressure. Backfill can then be washed out locally and the harbour bottom possibly eroded in front of the sheet piling.

Voids ensue in the backfill, which owing to the arching effect can remain stable for some time. However, sooner or later the voids become visible at the surface in the form of subsidence. Washout at the harbour bottom can be identified using divers and/or sounding methods. Where declutching results in only a very small opening, subsidence often only occurs after many years of port operation, but can then cause substantial damage to persons and property.

To prevent this, immediately after dredging, divers should inspect the structure for driving damage between the water level and the base of the excavation or harbour bottom.

## 7.5 In situ density of hydraulically filled non-cohesive soils (R 175)

### 7.5.1 General

The bearing capacity of port areas with hydraulic fill is essentially determined by the in situ density and strength of the uppermost 1.5–2.0 m of the hydraulic fill. The in situ density of hydraulically filled ground depends primarily on the following factors:

- The granulometric composition, especially the silt content of hydraulic fill material. To achieve the highest in situ density it is important to limit the proportion of fine particles < 0.06 mm to max. 10%. This can be guaranteed through, for example, correct barge loading (R 81, section 7.3).
- The type of extraction and further processing of the fill material.
- The shape and setup of the hydraulic fill site.
- The positioning and type of drainage for hydraulic fill water.

During hydraulic filling above water, a higher in situ density is generally achieved without additional measures than is the case below water. The influence of tides and waves often compacts the hydraulically filled sand within a short time, which therefore achieves a very high in situ density.

### 7.5.2 Empirical values for in situ density

Experience has shown that hydraulic filling below water results in the following in situ densities $D$:

- Fine sand with different uniformity coefficients and a mean grain size $d_{50} < 0.15$ mm:

  $D = 0.35 - 0.55$.

- Medium sand with different uniformity coefficients and a mean grain size $d_{50} = 0.25 - 0.50$ mm:

  $D = 0.15 - 0.35$.

As the granulometric composition and silt content of the material do not remain constant during hydraulic filling work, the aforementioned empirical values represent only a rough guide to the actual in situ density that can be achieved.

**Table R 175-1.** Use-related in situ density $D$ of non-cohesive soils for port areas

| Type of use | In situ density $D$ | |
|---|---|---|
| | Fine sand $d_{50} < 0.15$ mm | Medium sand $d_{50} = 0.25$–$0.50$ mm |
| Storage areas | 0.35–0.45 | 0.20–0.35 |
| Traffic areas | 0.45–0.55 | 0.25–0.45 |
| Structure areas | 0.55–0.75 | 0.45–0.65 |

### 7.5.3 In situ density required for port areas

The upper 1.5–2.0 m of a port area should exhibit the following in situ densities $D$ depending on the respective use and particle sizes in the fill (Table R 175-1).

Therefore, for the same loads, fine sand always requires a higher in situ density than medium sand.

### 7.5.4 Checking the in situ density

The in situ density in the upper part of hydraulic fill can be determined with the customary tests for density determination to DIN EN 22475-1, normally using equivalent methods, and by plate bearing tests to DIN 18134 or radiometric penetration sounding equipment. However, these can only measure the in situ density or load-carrying capacity of the upper areas (max. 1 m). At greater depths, the in situ density can be determined through cone or dynamic penetration tests to DIN EN 22476-2, or with a radiometric depth sounder.

The cone penetration test (CPT) is ideal for checking the in situ density of hydraulically filled sands. However, the heavy dynamic penetration test (DPH) is also suitable when, for instance, surfaces are inaccessible for the cone penetration test. The light dynamic penetration test (DPL) represents a further option for depths <3 m. The values given in Table R 175-2 are empirical values for the correlation between the respective test findings in fine and medium sands and the in situ density. They only apply from about 1.0 m below the application point of the test.

### 7.6 In situ density of dumped non-cohesive soils (R 178)

### 7.6.1 General

This recommendation essentially supplements recommendations R 81, R 73 and R 175.

The dumping of non-cohesive soils generally leads to a more or less pronounced segregation of the material. The result is that the composition

**Table R 175-2.** Correlation between in situ density $D$, toe resistance $q_c$ of cone penetration test and dynamic penetration test for number of blows $N_{10}$ in hydraulically filled sands (empirical values for non-uniform fine sand and uniform medium sand)

| Type of utilisation | | Storage areas | Traffic areas | Structure areas |
|---|---|---|---|---|
| In situ density $D$ | Fine sand | 0.35–0.45 | 0.45–0.55 | 0.55–0.75 |
| | Medium sand | 0.20–0.35 | 0.25–0.45 | 0.45–0.65 |
| Cone penetration test CPT 15 $q_c$ in MN/M² | Fine sand | 2–5 | 5–10 | 10–15 |
| | Medium sand | 3–6 | 6–10 | >15 |
| Dynamic penetration test DPH, $N_{10}$ | Fine sand | 2–5 | 5–10 | 10–15 |
| | Medium sand | 3–6 | 6–15 | >15 |
| Light dynamic penetration test DPL, $N_{10}$ | Fine sand | 6–15 | 15–30 | 30–45 |
| | Medium sand | 9–18 | 18–45 | >45 |
| Light dynamic penetration test DPL-5, $N_{10}$ | Fine sand | 4–10 | 10–20 | 20–30 |
| | Medium sand | 6–12 | 12–30 | >30 |

of the backfill changes considerably, particularly as currents can also wash the fine constituents out of the soil. Embankments formed by dumping non-cohesive soils are initially relatively steep. But they are not stable and repeated slope failures cause them to become shallower. Such movements also loosen the soil. Embankments with a slope of 1:5 or less are stable over a longer period of time.

Dumped non-cohesive soils can be further compacted by tide and wave effects, too.

### 7.6.2 Influences on the achievable in situ density

The in situ density of dumped non-cohesive soils depends primarily on the following factors:

a) Generally, a non-uniform granulometric composition produces a higher in situ density than a uniform one. The silt content should not exceed 10%.

b) Segregation increases with the depth of the water, especially for non-cohesive soils with a uniformity coefficient $U > 5$. This changes the particle size distribution, with the coarse grain fractions reaching a higher in situ density than the fine grain ones. On the whole this results in a body of soil with an inhomogeneous in situ density.

c) The greater the flow, the greater is the segregation and the more irregular the settlement of the soil.
d) In general, a higher in situ density is achieved with split-hopper barges than with bottom-dump barges.

Owing to the fact that these influences sometimes have opposing effects on the in situ density, the density of dumped non-cohesive soils can vary considerably.

The earth surcharge has only a minor influence on the in situ density of dumped non-cohesive soils. Even with increasing overburden pressure, there is generally scarcely any change to the in situ density of the dumped sand. Therefore, only a loose in situ density should be assumed in dumped sand without additional compaction.

## 7.7 Dredging underwater slopes (R 138)

### 7.7.1 General

In many cases underwater slopes are constructed as steep as stability requirements will allow. As regards the long-term stability of dredged underwater slopes, the influences of wave impact and currents are taken into account plus the effects of the dredging work itself. Waterborne traffic must also be evaluated. Experience has shown that slope failures frequently take place during and shortly after the dredging work.

The high costs of restoring failed slopes justify careful soil surveys and soil mechanics investigations in advance as a basis for specifying the intended angle of the dredged embankment and the tender for the dredging work.

In general, the recommendation is to withdraw groundwater by means of wells installed immediately behind the slope and thus create a flow gradient towards the fill. This results in the flow gradient during dredging work – which would otherwise be away from the fill – no longer being critical for the stability of the slope. Moreover, these measures compensate for the effects of the dredging work, which can compromise stability.

### 7.7.2 Dredging underwater slopes in loose sand

Special problems can arise when dredging underwater slopes in loose sand. Effects such as vibrations and local stress changes in the soil due to dredging operations can mobilise large quantities of sand. The sand then behaves temporarily like a heavy liquid, and so this type of slope failure is also referred to as a "flow failure". The latent flow sensitivity of the in situ soil must be investigated prior to dredging in order to initiate countermeasures such as compacting the soil. Failing this, the slope must be kept shallow from the beginning so that a flow failure is ruled out.

However, this is often not possible in practice because even a thin layer of loosely bedded sand in the mass of soil to be dredged can trigger a flow failure.

### 7.7.3 Dredging plant

In principle, underwater slopes can be excavated by all common dredgers. The dredger must be selected to suit the operating conditions.

Slopes down to a depth of approx. 30 m can be successfully excavated with large cutter-suction dredgers and cutter-wheel suction dredgers, approx. 35 m with large bucket-ladder dredgers. Backhoe dredgers can currently excavate max. 20 m deep.

Backhoe dredgers are preferably employed for dredging in heavy soils. Grab dredgers can be used when dredging only small quantities, or when dredging is to be carried out in accordance with R 80, section 7.1.

Plain suction dredgers are only advisable for forming slopes with low requirements regarding accuracy. Their operating mode means that they are apt to cause slope failures. They are therefore not generally considered for specifically dredging underwater slopes.

Undercutting must be avoided at all costs when forming underwater slopes.

### 7.7.4 Execution of dredging work

#### 7.7.4.1 Rough dredging

Preliminary dredging is carried out from above to just below the water level, where, for example, the profile of this part of the slope can be properly formed with a grab dredger. Before dredging the remainder of the underwater slope, dredging is carried out at such a distance from the slope that the dredger can operate as close as possible to full capacity without causing any risk of a slope failure in the future slope.

Indications as to the safe distance to be maintained between the dredger and the planned embankment are gained by observing the in situ soil during the preliminary dredging work.

After concluding the rough dredging work, a strip of soil remains at the top of the underwater slope. This must be removed by a suitable method chosen such that the stability of the slope is not at risk (Figs. R 138-1 and R 138-2).

#### 7.7.4.2 Slope dredging

Dredging to form underwater slopes must be carried out carefully so that slope failures are kept within bounds and under control. The following types of plant are acceptable. Their suitability in each case must be researched and assessed.

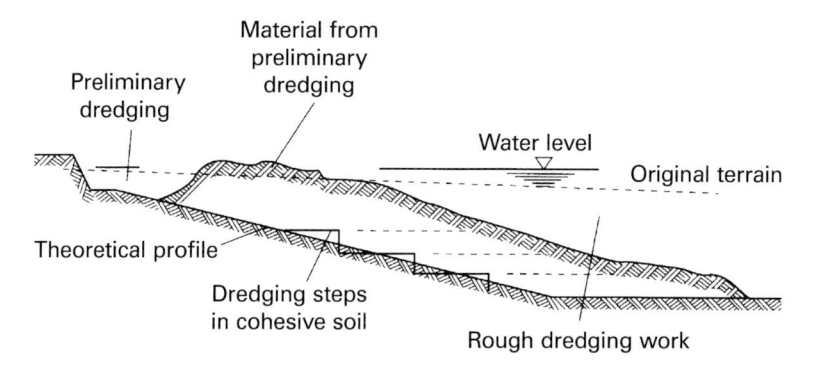

**Fig. R 138-1.** Dredging an underwater slope with a bucket-ladder dredger

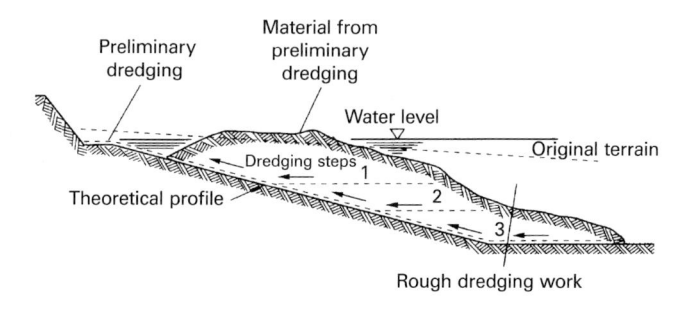

**Fig. R 138-2.** Dredging an underwater slope with a cutter-suction or cutter-wheel suction dredger

### 7.7.4.2.1 Bucket-ladder dredger

In the past, bucket-ladder and grab dredgers were employed without exception for both the rough dredging and slope dredging work. Small bucket-ladder dredgers can be employed for dredging from approx. 3 m below the water level.

For practical reasons, the bucket-ladder dredger operates parallel to the slope, generally dredging layer by layer. Full or semi-automatic control of movements of the dredger ladder is possible and is to be recommended. The slope is dredged in steps. The type of soil determines the extent to which the steps may intrude into the theoretical slope line (Fig. R 138-1). In cohesive soils the steps are generally dredged in the intended slope line. In non-cohesive soils, however, intruding into the intended slope line is not permitted. The potential removal of the excess soil depends on the tolerances stipulated contingent on the soil conditions and the boundary conditions listed in R 139, section 7.2.

The height of the steps depends on the soil conditions and other factors and is generally between 1.0 and 2.5 m.

The planned slope inclination, the type of soil and the capabilities and experience of the dredger crew are just some of the factors affecting the precision with which slopes can be formed in this manner.

With slopes of 1:3 to 1:4 in cohesive soils, it is possible to achieve an accuracy of $\pm 50$ cm measured perpendicular to the theoretical slope line. The tolerance in non-cohesive soils should be $+25$ to $+75$ cm, depending on the dredging depth.

### 7.7.4.2.2 Cutter-suction or cutter-wheel suction dredger

For economic reasons, cutter suction and cutter-wheel suction dredgers are also used these days for forming underwater slopes.

When dredging, the cutter-suction dredger preferably moves along the slope, dredging layer by layer like the bucket-ladder dredger. Computerised control of the dredger and dredger ladder is recommended.

Fig. R 138-2 shows how the cutter-wheel works upwards, parallel to the theoretical slope line, after having made a horizontal cut. Highly accurate underwater slopes can be produced in this way. In the case of computerised dredger control, tolerances $T_h$ of $+25$ cm measured transverse to the slope can be achieved with small cutter-suction dredgers, and $+50$ cm for larger models. If dredging is carried out without special control, the same tolerances apply as for bucket-ladder dredging, provided flow failures can be ruled out.

## 7.8  Subsidence of non-cohesive soils (R 168)

Subsidence, in a narrower geotechnical sense, is a non-loading-related deformation of a non-cohesive soil which leads to a higher in situ density. Subsidence is caused by the cessation of capillary cohesion (apparent cohesion) due to saturation or dynamic actions. Subsidence also occurs in cemented soils if the bonds between soil aggregates disintegrate or cohesion breaks down as a result of chemical bonds caused by external influences.

Fig. R 168-1 shows subsidence as a sudden occurrence of settlement when an earth-moist sand is saturated during a compression test.

Subsidence is especially likely in the first few years following earthworks. It is most likely in soils with a low in situ density; in loosely bedded sands, subsidence can amount to 10% of the stratum thickness. In densely bedded soils, subsidence can reach an order of magnitude of up to 0.5% of the stratum thickness. However, subsidence in densely bedded soils is only triggered by particular actions, e.g. vibrations due to pile driving or other operations.

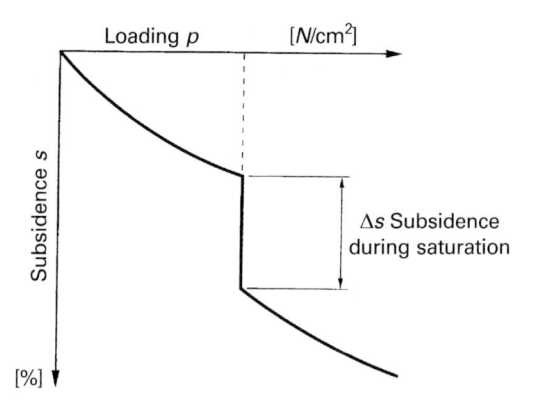

**Fig. R 168-1.** Load–settlement diagram of an earth-moist sand during saturation

In coarse-grained, and therefore sufficiently permeable, soil, the possibility of subsidence is reduced by adding water during compaction (the "wet branch" of the Proctor curve). In fine-grained soils, i.e. with low permeability, adding water can prevent compaction.

Subsidence can occur in rock fill when it is dammed for the first time. In this case subsidence is mainly triggered by weaknesses in the rock structure due to water absorption and the resulting material fracture at very highly stressed contact points.

In general, subsidence is greater in round-grained than in angular-grained soils. Uniform sands show a greater degree of subsidence than non-uniform sands. The difference, however, is only discernible in loose and very loose deposits.

In a broader sense, subsidence can also occur as a result of removing material in deeper strata. This can be caused by hydraulic material transport (suffosion, contact erosion, etc.), by damaged areas in the supporting walls, by leaching (sinkholes) or by organic decomposition processes.

## 7.9 Soil replacement along a line of piles for a waterfront structure (R 109)

### 7.9.1 General

Soil replacement can be useful where soils presenting difficult driving conditions and/or obstacles are present along a line of piles for a waterfront structure. In this case there is a risk that driving damage cannot be avoided, especially with combined sheet piling. The cost of the soil replacement is compensated for by the greater certainty for design and construction and by the fact that damage is much less likely.

Soil replacement is also helpful where low-strength cohesive soils are encountered in thicker layers along the line of the waterfront structure. In such a case, soil replacement is the prerequisite for guaranteeing the stability of the quay wall during all stages of construction and in its final state.

An economic prerequisite for soil replacement is that the new soil, normally a sand that can be readily installed and compacted, can be obtained inexpensively.

The design basis for soil replacement is a meaningful soil investigation during the preliminary design phase. The results should allow exact determination of the replacement area and enable an optimised plan to be drawn up for the necessary plant.

To optimise plant usage, it can be advisable to excavate a trial pit prior to extensive dredging. Moreover, observing the pit allows the slope stability to be observed under the effect of waves, currents and mud deposits.

## 7.9.2 Dredging

### 7.9.2.1 Dredger selection

Only bucket-ladder dredgers, cutter-suction dredgers, cutter-wheel suction dredgers or dipper dredgers can be used for excavating cohesive soil. If soil containing obstacles has to be dredged (e.g. soil interlaced with rubble), the risk when using a suction dredger is that rubble not picked up by the dredger remains on the base of the excavation and forms a layer that is almost impossible to penetrate during later driving work. In such instances, soil must be replaced to such a depth that obstacles brought up by dredging lie beneath the embedment depth of all pile sections.

When dredging cohesive soils underwater, it is impossible to avoid soil from the bucket becoming deposited on the base of the excavation as a layer of sludge. Owing to overfilled buckets, incomplete emptying of buckets and barges overflowing, large deposits of sediment must be reckoned with when excavating with bucket-ladder dredgers in particular (Fig. R 109-1). The ensuing layer of sediment, possibly in conjunction with sludge from the watercourse, has only a very low strength and should therefore be removed prior to filling. To ensure that the layer is removed as thoroughly as possible, a shallower cut must be used when reaching the base of the excavation (Figs. R 109-1 and R 109-2).

To do this, a slack lower bucket chain must be used together with low bucket and cutting speeds. When loading barges, overflowing soil must be absolutely avoided to prevent deposits of mud.

Using cutter-suction and cutter-wheel suction dredgers results in an undulating base to the excavation (see Fig. R 109-2), and the layer of

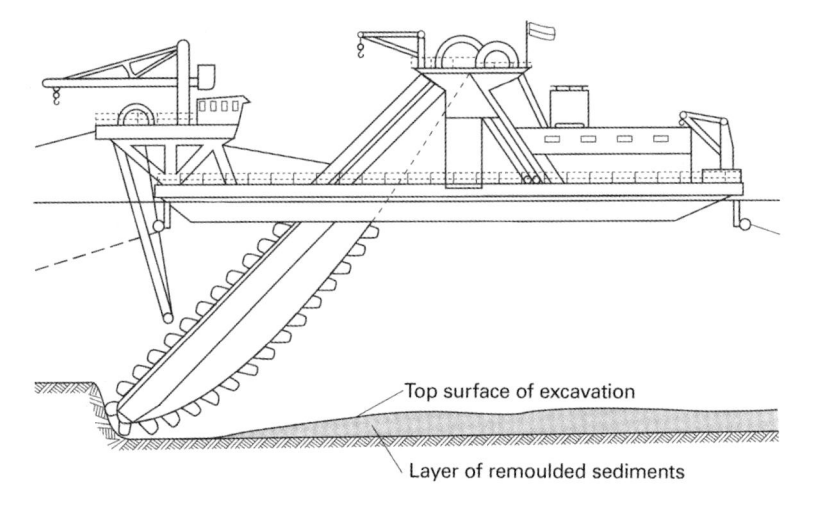

**Fig. R 109-1.** Formation of layer of remoulded sediments when dredging with a bucket-ladder dredger

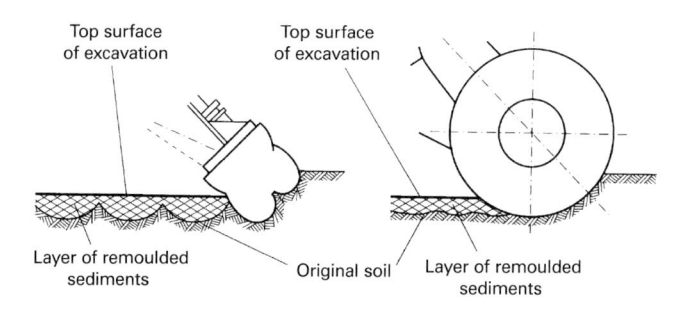

**Fig. R 109-2.** Formation of layer of remoulded sediments when excavating with a cutter-suction or cutter-wheel suction dredger

remoulded sediments is thicker than that produced by a bucket-ladder dredger.
The thickness of the layer of remoulded sediments can be reduced by using a special cutting head shape, slower rotation, short thrusts and a slow cutting speed.

### 7.9.2.2 Executing and checking dredging work

Excavation is performed in steps corresponding to the mean profile inclination at the edge of the base of the excavation. The height of the steps depends on the type and size of the plant and on the type of soil. Strict control of cut widths must be maintained because cuts that are too wide can cause an excessively steep slope in places, possibly resulting in slope failures.

Proper dredging progress can be monitored by modern surveying methods (e.g. depth sounder in combination with the global positioning system, GPS). It is also possible to detect any profile changes, possibly caused by underwater slope failures, in good time. Merely marking the dredger cutting width solely on the side lines of the dredger is not adequate for checking dredging works.

Measurements with inclinometers around the edge of the excavation have also proven successful for monitoring underwater slope failures.

The last sounding should be taken immediately before the sand fill is placed. In order to obtain information on the characteristics of the base of the excavation, soil samples should be taken for testing. A hinged sounding tube (sediment core drill) with a minimum diameter of 100 mm and a gripping attachment (core catcher) has proved effective. Depending on requirements, this tube is driven 0.5–1.0 m or even deeper into the base of the excavation. After extracting and opening the tube, the sample inside permits an assessment of the soil strata at the base of the excavation.

### 7.9.3 Quality and procurement of the fill sand

When planning soil replacement works, it is advisable to investigate the sand procurement areas by means of boreholes and penetrometer tests before starting work. The fill sand should contain only very little silt and clay, and no major accumulations of stones.

If the in situ fill sand contains thin silt and clay lenses and/or includes stone layers but the proportion of fines, related to the total layer thickness, remains below the 10% limit, hydraulic filling methods may not be used; instead the material must be dumped. This avoids local concentrations of fine particles and stones in the fill.

In the interests of continuous, economic filling, adequate deposits of suitable sand must be reserved within a reasonable distance of the site. When determining the amount of fill required, losses due to current washout must be taken into account. The finer the sand and the stronger the current, the greater are the losses. The percentage loss of soil is greater for slower replacement rates than for faster rates. Greater losses (as a percentage) are also to be expected in deeper waters than in shallower waters.

Powerful hopper suction dredgers or barge loaders and large barges are recommended for sand procurement. Overflow from the barges washes out some of the silt and clay inclusions present in the soil; this cleaning effect can be intensified by loading the barges in a certain way and using longer overflow times. However, this method is restricted by the (usually) strict limits imposed on the discharge of materials into watercourses. Samples of the fill sand are to be taken frequently from the barges and tested for compliance with the requirements in the design, especially with regard to the amount of fine particles.

### 7.9.4 Cleaning the base of the excavation before filling with sand

All cohesive soils must be removed from the base of the excavation immediately prior to starting filling work. Suitable suction plant can be used if the deposits are not too firm. However, if a period of several days or longer has elapsed between the conclusion of dredging work and the start of silt removal, the deposits may already be so solid that suction is impossible and another cleaning cut may be necessary.

Water-jetting has proved effective here. Large quantities of water are pumped at low pressure (approx. 1 bar) through nozzles directed at the base of the excavation, using a movable cross-member suspended beneath floating plant. The clearance to the bottom is kept to a minimum – between 0.3 and 0.5 m. Deposited sediments are thus turned back into a full suspension. Water-jetting must be carried out immediately prior to placing the sand fill and continued until the bottom is proved to be free from mud.

The cleanliness of the base of the excavation must be checked regularly. The sounding tube described in section 7.9.2.2 can be used for this. A suitably designed grab (hand grab) can be used for taking the samples in soft deposits. A combination of silt soundings at discrete positions and depth soundings (with differing frequencies to ensure reliable detection of the base of the excavation) is a good evaluation option.

If it cannot be guaranteed that no mud deposits remain on the base of the excavation, then it must be ensured that the replacement soil interlocks with the in situ, loadbearing soil beneath the base of the excavation once it has been installed so that the layer of mud cannot form a continuous sliding surface. For example, the first layer of the fill can be of crushed rock, which displaces the sediment and therefore interlocks with the soil beneath the base of the excavation. The thickness of the layer of crushed rock must be chosen such that the sediment can be accommodated in the pores of the crushed rock without the mineral contact between the individual stones being lost. If it can be ensured that there will be no pile driving in nor deepening of the harbour bottom at a later date, keying between in situ soil and fill sand can be achieved with coarse rubble rather than crushed rock.

A sediment-free base to the excavation is especially important in the filling area in front of quay walls because a layer of sediment remaining on the base of the excavation could drastically reduce the passive earth pressure there. With non-cohesive soils and a shallow excavation, keying between the fill material and the subsoil can also be achieved by the "dowelling" effect achieved with vibroflotation with a unit of 2–4 vibrators.

## 7.9.5 Placing the sand fill

The excavation can be filled hydraulically or by dumping. Backfilling operations must be carried out without interruption and around the clock. This is particularly important when the risk of sediments is high. Winter operations with working days lost to cold weather, drifting ice, storms and fog should therefore be avoided.

Placing the sand fill should follow removal of the poor soil as soon as possible to minimise the inevitable deposits of sediment accumulating in the meantime. On the other hand, mixing of the soil to be replaced with that being filled is not permitted, and so sufficient time must be allowed between excavating and filling operations. This danger is particularly prevalent in waters with strong, changing currents (tidal areas especially).

Contamination of the sand fill by ongoing sedimentation can be minimised by installing the fill rapidly. How the soil mechanics characteristics of the fill sand are affected by the anticipated inclusion of fine material in the soil matrix must be taken into account in the analyses. The sand filling operation must be carried out in such a way that no continuous layers of sludge can build up. In the case of severe deposits, this can be achieved through continuous, efficient operations, not even interrupted at weekends.

The use of water-jetting on the surfaces of layers of sand already deposited during the filling operations has proved to be effective. Any deposits are then turned into suspensions again and layers of mud are prevented from the very outset. Sand losses due to this water-jetting must be taken into account when determining the quantities.

Should interruptions during filling be unavoidable, then the silt that has built up on sand already deposited in the meantime must be removed before further sand filling takes place. If embedded silt layers are detected after filling has been completed, these must be assessed with regard to the serviceability of the soil replacement measures. If necessary, these areas must be clearly defined using specific soil investigations and treated using, for example, preloading or other soil improvement measures. During any interruptions to filling, a check must be made as to whether and where the surface level of the fill has changed.

In order to avoid causing active earth pressure on the waterfront structure in excess of the design load, the excavation must be filled in such a way that silted-up slopes occurring during the filling have an inclination opposite to that of the failure plane of the active earth pressure wedge

that will later act on the waterfront structure. The same applies similarly to soil replacement on the passive earth pressure side.
For additional information, please refer to [155].

### 7.9.6 Checking the sand fill

Soundings should be taken constantly during the sand filling operation and the results logged. This enables the filling processes itself and the effects of the currents to be checked to a certain extent. At the same time, these records show clearly how long surfaces have remained stable and whether sediments have accumulated.

The taking of samples from the fill area can be dispensed with only when there is fast, uninterrupted hydraulic filling and/or dumping. If sand filling was interrupted, however, the surface must be checked for silt deposits, as described above, before further filling can take place.

After completing the filling work, samples of the new soil must be taken and tested randomly but systematically by means of core samples and cone penetration tests. Sampling and testing are to be carried out as far as the soil beneath the base of the excavation.

An acceptance certificate forms the basis for the final design of the waterfront structure and any measures that may be required to adapt the design to conditions at the site.

## 7.10 Dynamic compaction of the soil (R 188)

Effective compaction with heavy weights dropped from a height is primarily suitable for soils with good water permeability. This method can also be used on weakly cohesive and non-saturated cohesive soils because the impacts increase the pressure in the compressible water and air in the pores to such an extent that it tears open the soil structure. The decrease in the pressure in the pores leads to a relatively rapid consolidation rate.

The use of very heavy weights and high drop heights enables even saturated cohesive soils to be compacted with this method.

In order to assess the success of dynamic compaction measures reliably, the soil to be compacted must be examined for its suitability beforehand using soil mechanics methods. It is also necessary to define any areas where dynamic compaction cannot be used; soil replacement is then necessary in such areas.

The drop height required, the size of the weight and the number of compaction passes should be determined in advance in tests. This also permits an assessment of the side-effects of compaction such as noise and the vibration of neighbouring structures. The compaction achieved can be checked using the methods outlined in R 175.

## 7.11 Vertical drains to accelerate the consolidation of soft cohesive soils (R 93)

### 7.11.1 General

Vertical drains have an accelerating effect on consolidation because, during the consolidation process, they enable radial drainage of the strata instead of drainage just upwards and downwards. The radial drainage path is half the spacing of the drains and, in contrast to upward drainage (= stratum thickness) or upwards and downwards (= half the stratum thickness), it can be influenced. Consolidation settlement (primary settlement) of soft cohesive, relatively impermeable strata can be considerably speeded up with vertical drains. At the same time, the settlement due to consolidation, especially in soft cohesive strata, is greater with vertical drains than without. The reason for this is that with shorter drainage paths, the stagnation pore water pressure (at which the consolidation stops) is lower than when using longer drainage paths.

Secondary settlement (creep settlement), however, is not influenced by vertical drains. Such settlement can be relatively large and continue for some time in soft cohesive and organic soils in particular.

Vertical drains accelerate consolidation primarily in horizontal strata of soft cohesive soils such as marine clay because they exploit the horizontal permeability of these soils. This is also true for stratified soils with alternating permeability (e.g. layers of marine clay and mudflats sand). The strata of low permeability are drained via the adjacent layers of higher permeability, which feed the water into the vertical drains.

### 7.11.2 Applications

Vertical drains are used where bulk materials, dykes, dams or fill are dumped on soft cohesive soils. The period of consolidation is shortened and the in situ soil attains the bearing capacity required for the intended use at an earlier date. Vertical drains are also used to stabilise embankments or terraces and when it is necessary to limit lateral flow movements from fill.

There are limits to the use of vertical drains, e.g. when contamination in the soil could be mobilised unacceptably.

### 7.11.3 Design

When designing a vertical drain system, take the following factors into account:

- The consolidation of a soil requires a surcharge in every case. Vertical drains can speed up consolidation. In an ideal situation, consolidation is finished before the structure gets into operation. However,

secondary settlement still takes place, the course of which cannot be influenced by vertical drains, and in any event occurs during the operational phase.
- The application of a surcharge greater than the total of all intended loads can compensate for a part of the secondary settlement. If the surcharge in excess of the intended load is removed again, the soil is overconsolidated.
- The settlement to be expected from consolidation can be estimated using Terzaghi's consolidation theory. Approaches for estimating the consultation and secondary settlement simultaneously can be found in [123]. However, as those methods were developed for very much simplified boundary conditions, the results they provide might have limited applicability within the scope of detailed design work. In any case, settlement should always be calculated for the probable bandwidths of the soil mechanics parameters so that the inhomogeneity of the in situ soils is also taken into account.
- Owing to the different consolidation theories, the chronological sequence of consolidation cannot be derived for certain; at best, the theories provide rough estimates. However, in conjunction with settlement measurements, the input parameters for the calculations can be calibrated.
- Normally, the settlement due to consolidation with vertical drains is greater than that without vertical drains for the same conditions because the stagnation gradient of the consolidation is lower with vertical drains than it is without.
- If there is a confined water table below the stratum to be consolidated, the drains should terminate about 1 m above the lower stratum. Otherwise, groundwater is forced upwards through the vertical drains.
- The structure of the subsoil must be very carefully investigated beforehand in order to specify the optimum drain spacing. It is primarily the permeability of the in situ soil types that can only be determined reliably by means of trial pumping.

When setting up vertical drains it is important to avoid contaminating the surface in contact with the soil, which might increase the resistance to water entering the drain to such an extent that drainage of the soil is prevented. Moreover, it must be ensured that mechanical overstresses (e.g. due to a local ground failure) do not limit or even inhibit the effect of the vertical drains.

Vertical drains are covered with a layer of sand or gravel in which the water escaping from the soil is held and which in turn discharges the water into an outfall.

Vertical drains suffer from the same settlements as the soil strata in which they are installed. Large settlements can bend plastic drains, which can seriously impair their function. Nowadays, vertical drains are mainly made from plastic.

## 7.11.4 Design of plastic drains

The aim of the design is to determine the spacing of the drains so that the consolidation is completed within a period specified in the construction schedule (usually less than two years), where this is called for.

The time $T$ needed to achieve a degree of consolidation $U_h$ is calculated on the basis of the one-dimensional consolidation theory according to Kjellmann [116] and Barron [10]. Hansbo [87] has simplified Kjellmann's approach by assuming uniform soil deformation and undisturbed soil conditions:

$$t = \frac{D_e^2 \alpha}{8 c_h} \cdot \ln\left(\frac{1}{1 - U_h}\right)$$

where

$$\alpha = \frac{n^2}{n^2 - 1} \cdot \left(\ln(n) - \frac{3}{4} + \frac{1}{n^2} \cdot \left(1 - \frac{1}{4 \cdot n^2}\right)\right)$$

$$n = \frac{D_e}{d_w}$$

- $t$    time available for consolidation [s]
- $D_e$    diameter of drained soil cylinder [m], area of influence of the drain
- $c_h$    horizontal consolidation coefficient [m²/s]
- $d_w$    equivalent diameter of vertical drain (drain circumference/π) [m]
- $U_h$    average degree of consolidation [-]

With a symmetrical arrangement of the drain starting points and partial overlap of the drained soil cylinders, the distance $s$ between each drain is as follows:

- for equilateral triangle grids: $s = \dfrac{D_e}{1.05}$
- for square grids: $s = \dfrac{D_e}{1.13}$

For the design, the drain spacing $s$ is chosen first and then $D_e$ calculated from this. Thereafter, $n$ is calculated as the ratio of the diameter of the drained soil cylinder $D_e$ to the equivalent diameter $d_w$ of the selected drain and the coefficient α. Finally, the duration of consolidation $t$ can be calculated with these values and the target value of the degree of consolidation $U_h$. If $t$ is too long, the drain spacing $s$ must be reduced and the calculation repeated.

The design can also be carried out with nomograms and computer programs, which are available from the drain manufacturers. Further calculation approaches can be found in [88] and [115].

Since the required service life of plastic vertical drains is usually less than two years, there are no special requirements in terms of durability. Only in the case of large settlements is it necessary to prove that the drains remain functional even with kinks. Guidelines for the assessment of water flow with kinked drains can be found in the Dutch recommendations BRL 1120 [26].

### 7.11.5 Installation

Sand drains have diameters of approx. 25–35 cm and are installed using driving, drilling or water-jetting methods. Thin layers of fine cohesive sediments at the contact surface between drain and soil can only be reliably avoided by using driven vertical drains.

Plastic drains are supplied in widths of approx. 10 cm. Their walls are 5–10 mm thick and their water drainage capacity lies between 0.01 and 0.05 l/s with a ground pressure of 350 kN/m². This capacity is therefore relatively large in comparison to the volume to be drained from the soil. Plastic vertical drains are pressed or vibrated into the ground using special equipment; however, in many cases a flat working area must be set up for this equipment in advance. The material for this must be at least 0.5 m deep, but an excessively thick working level can make installing the drains more difficult.

In the event of contaminated subsoil, avoid installation procedures that produce drilling or jetting debris. Moreover, any existing sealing stratum should not be penetrated. In order to gain time for consolidation, it is advisable to install the drains at an early stage and initiate consolidation by means of preloading.

## 7.12 Consolidation of soft cohesive soils by preloading (R 179)

### 7.12.1 General

Frequently, only areas with soft cohesive soil types of inadequate bearing capacity are available for extensions to ports and harbours. However, in many cases the bearing capacity of these areas can be improved by preloading. At the same time, settlement is pre-empted such that residual settlement due to port operations does not exceed the permissible tolerances.

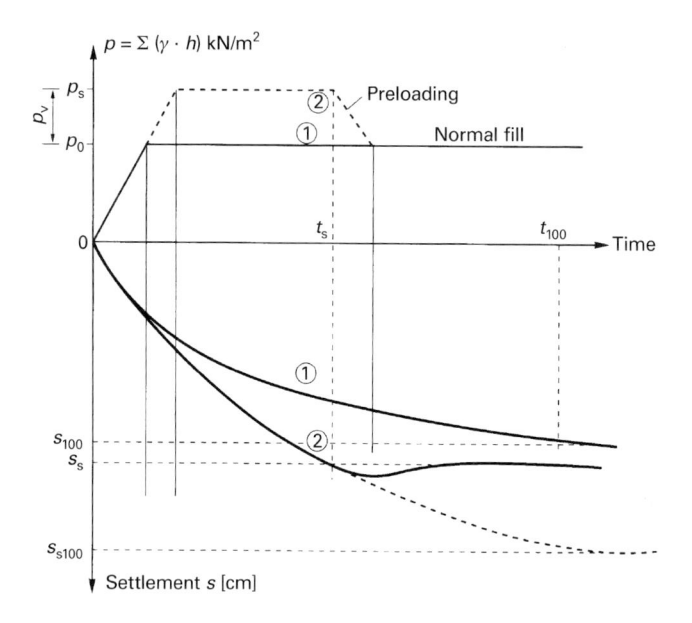

**Fig. R 179-1.** Relationship between settlement, time and surcharge (principle)

Apart from preloading by applying fill material, the bearing capacity can also be improved by vacuum consolidation.

### 7.12.2 Applications

The aim of preloading is to pre-empt the settlement of a soft soil stratum which would otherwise take place during the later utilisation of the area (Fig. R 179-1). The time needed for this depends on the thickness of the soft strata, their permeability and the magnitude of the preload. However, preloading is only effective when there is sufficient time for the consolidation to take place. It is advisable to use the maximum possible preload and to apply this at such an early stage that all primary settlement is completed before commencing any construction work for the actual waterfront structure.

Preloading in accordance with Fig. R 179-1 is divided into:

a)  That part of the preload fill constructed in the form of earthworks (permanent filling). It generates the surcharge stress $p_0$.
b)  The excess preload fill, which temporarily acts as an additional surcharge with the preload stress $p_v$.
c)  The sum of both fills (total fill), which results in the overall stress $p_s = p_0 + p_v$.

### 7.12.3 Bearing capacity of in situ soil

The magnitude of the initial preload is limited by the bearing capacity of the in situ subsoil. The maximum depth of a preload fill $h$ can be estimated with

$$h = \frac{4c_u}{\gamma}$$

The shear strength $c_u$ – and thus the permissible depth $h$ of further preload fill – increases with the consolidation.

When filling underwater, it must be expected that there is a soft layer of sediment on the bottom of the watercourse. If this remains in the soil, it affects the settlement behaviour of the future port area. Where this is unacceptable, it must be removed (R 109, section 7.9.4).

Occasionally, attempts are made to displace the layer of soft soils by pushing the fill material ahead in one single thick layer. Experience has shown that this method is only successful for very soft sediments. It is much more often the case that the displacement is not completed or not at all successful. Sediment layer residue then remains in the subsoil, with the risk of differential settlement.

### 7.12.4 Fill material

The permanent fill material must constitute a stable filter with respect to the soft subsoil. If applicable, filter layers or geotextiles should be laid before the permanent fill is applied. Otherwise, the required quality of the permanent fill material is governed by the intended use.

### 7.12.5 Determining the depth of preload fill

#### 7.12.5.1 Soil mechanics principles

The requirements with respect to the preload fill depth are essentially due to the consolidation time available. The dimensioning is based on the consolidation coefficient $c_v$. If $c_v$ values are derived from the time–settlement curves of compression tests, experience shows that only a rough estimate of the consolidation time is possible. Therefore, such values should be used for preliminary appraisals only. This also applies to consolidation coefficients, which can be calculated using

$$c_v = \frac{k \cdot E_s}{\gamma_w}$$

from which the stiffness modulus $E_s$ and the permeability $k$ can be derived. It should be remembered that considerable scatter can affect the permeability coefficient $k$.

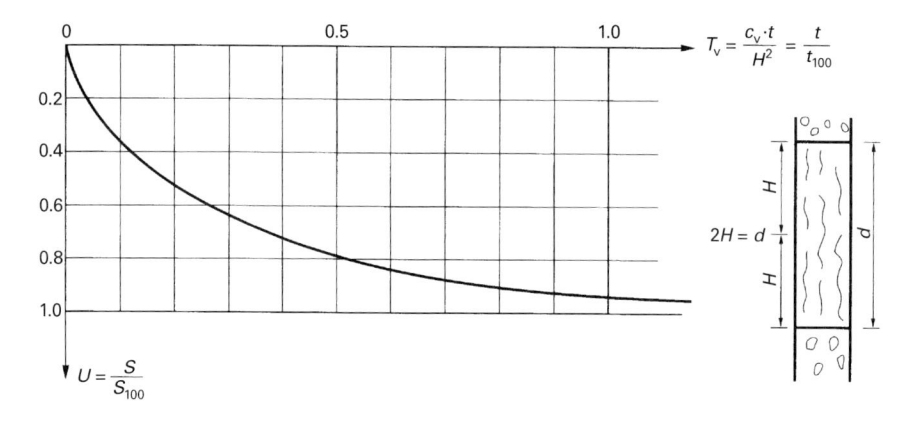

**Fig. R 179-2.** Relationship between time factor $T_V$ and degree of consolidation $U$

A reliable estimate of the course of settlement is possible, however, when the consolidation coefficient $c_v$ is determined in advance from a trial fill. The settlement and, if possible, the pore water pressure should also be measured. Using Fig. R 179-2, $c_v$ is then calculated with the following equation:

$$c_v = \frac{H^2 \cdot T_v}{t}$$

where

$T_v$ specific consolidation time [-]
$t$ consolidation time for trial fill [s]
$H$ drainage length of soft soil stratum [m]

For a degree of consolidation $U = 0.95$, i.e. virtually complete consolidation, the specific consolidation time $T_V = $ approx. 1.0. With a consolidation time $t_{100}$ corresponding to the degree of consolidation $U = 0.95$, the consolidation coefficient can thus be calculated:

$$c_v = \frac{H^2}{t_{100}}$$

### 7.12.5.2 Determining the preload fill

The drainage length of the soft soil stratum and the $c_v$ value must be known for determining the preload fill. Additionally, the consolidation

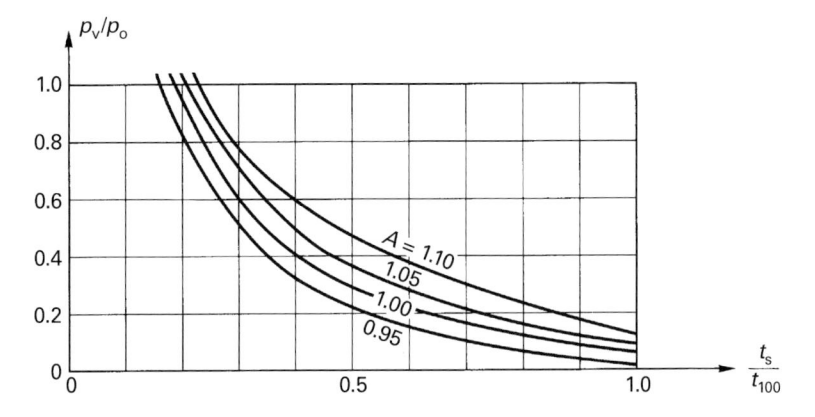

**Fig. R 179-3.** Determination of preload $p_v$ as a function of time $t_s$

time $t_S$ (Fig. R 179-1) must be specified (construction schedule). The figure $t_S/t_{100} = T_V$ is determined with $t_{100} = H^2/c_v$, and the required degree of consolidation $U = s/s_{100}$ under preloading is determined with the aid of Fig. R 179-2. The 100% settlement $s_{100}$ of the permanent fill $p_0$ is determined with the aid of a settlement calculation in accordance with DIN 4019.

The magnitude of the preload $p_v$ (Fig. R 179-1) is then derived in accordance with Horn [111] as

$$p_v = p_0 \cdot \left(\frac{A}{U} - 1\right)$$

where $A$ is the ratio of the settlement $s_s$ after removing the preload to the settlement $s_{100}$ of the permanent fill: $A = s_s/s_{100}$. The ratio $A$ must be $\geq 1.0$ if complete elimination of the settlement is to be achieved (Fig. R 179-3).

Thus, for example, with a stratum drained on both sides, a depth $d = 2H = 5$ m, $c_v = 3$ m²/year and a given consolidation time $t_s = 1$ year, the time necessary for complete consolidation of this stratum at $t_{100} = 2.5$ m²/3m²/year $= 2.08$ years. Thus $t_s/t_{100} = 1$ year/2.08 years $= 0.48$. The degree of consolidation as shown in Fig. R 179-2 after one year $U =$ approx. $0.78$, i.e. 80% of the consolidation settlement has taken place. Should the settlement $s_s$ due to the preload be approx. 5% greater than the settlement from the later permanent load $s_{100}$, then $A = 1.05$, and therefore the magnitude of the preload can be estimated as

$$p_v = p_0(A/U - 1) = p_0(1.05/0.78 - 1) = 0.35 p_0$$

As stated above, $p_v$ is limited by the bearing capacity of the in situ soil. Thus, it may be necessary to apply preload fill in several stages.

**7.12.6 Minimum extent of preload fill**

In order to save preload material, the soil stratum to be stabilised is generally preloaded in sections. The sequence of preloading is governed by the construction schedule. To maximise the even distribution of stress in the subsoil, the area of the preload fill should not be too small. As a guide, the smallest side length of the preload fill area should two to three times the sum of the depths of the soft stratum and the permanent fill.

**7.12.7 Soil improvement through vacuum consolidation with vertical drains**

With conventional preloading, the increase in the shear strength is achieved through an additional surcharge (total stresses). In contrast to this, the vacuum method achieves the consolidation by reducing the pore water pressure – the total stresses remain unchanged.

First of all, a layer of sand at least approx. 0.8 m deep must be placed on the soil to be consolidated to serve as a working platform for installing the vertical drains and also to function as a drainage layer. The strength of very soft cohesive soils is usually insufficient for applying the drainage layer in the necessary thickness in one operation.

The water within the drainage layer is collected and drained away with the help of a horizontal drainage system. The drainage layer is covered with a layer of plastic waterproofing material.

Special pumps that can pump, or rather extract, water and air simultaneously are connected to the drainage layer; these generally achieve a maximum vacuum of 75% of atmospheric pressure (approx. 0.75 bar). The pumping capacity $n < 1.0$ denotes the ratio of the applied vacuum to atmospheric pressure.

The stresses in the soil result from the atmospheric pressure $P_a$, the density $\gamma$ of the moist soil in the drainage layer, the density $\gamma_r$ of the in situ soil under the drainage layer und the density $\gamma_w$ of the water as well as the depth $z$ and thickness $h$ of the drainage layer in the following relationships:

total stress: $\quad \sigma = z \cdot \gamma_r + h \cdot \gamma + P_a$
pore water pressure: $\quad u = z \cdot \gamma_w + P_a$
effective stress: $\quad \sigma' = \sigma - u = z \cdot \gamma' + h \cdot \gamma$

Once the vacuum pumps are in operation, the atmospheric pressure decreases by $P_a$, which increases the effective stress by this amount to

$$\sigma'_{vacuum} = \sigma' + n \cdot P_a$$

From the consolidation with the additional stress $\Delta\sigma = n \cdot P_a$, it follows that the degree of consolidation $U_t$ leads to an increase in the shear strength $\Delta\tau$ amounting to

$$\Delta\tau = U_t \cdot (\tan\varphi' \cdot \Delta\sigma)$$

The final settlement and the progression of the settlement can be calculated using the consolidation theory of Terzaghi and Barron. The shear strength after consolidation corresponds to that due to an equivalent sand surcharge. The temporary increase in strength of non-cohesive soils due to the negative pressure is lost again once the vacuum pump is switched off.

### 7.12.8 Execution of soil improvement through vacuum consolidation with vertical drains

Firstly, a working platform (thickness > approx. 0.8 m) is installed. Vertical drains (a size equivalent to 5 cm dia. is usual but flat and round drains can be used in combination) are then installed from here extending down to about 0.5–1.0 m above the underlying non-cohesive strata to prevent hydraulic contact with the strata below reliably. Therefore, in inhomogeneous subsoil conditions, indirect investigations (e.g. penetration tests) during the work are necessary in order to satisfy this condition reliably and establish the depth quickly beforehand.

In subsoil conditions with major variations, test installations without drains can be carried out on a 10×10 m grid, for example. The test installations enable the plant operator to detect increased penetration resistance of intermediate sand strata or the topside of in situ sands.

The success of soil improvement using vacuum consolidation depends on the covering of plastic waterproofing material maintaining a seal during the consolidation. Any defects in this sheeting are difficult to locate and repair. It should therefore not be covered with stony, angular material. Flooding with water can be used to protect the sheeting.

In special cases the in situ soft stratum can itself function as the sealing layer maintaining the vacuum. To do this, the vertical drains are interconnected hydraulically by a horizontal drain approx. 1.0 m below the top of the soft stratum.

Owing to the limited depth of vertical drains, the vacuum process is currently limited to layer thicknesses of approx. 40 m of soil to be improved.

### 7.12.9 Checking the consolidation

As, generally, settlement and degree of consolidation are difficult to predict accurately in soil improvement measures involving soft cohesive soils, soil improvement using vacuum consolidation must be monitored through observations of settlement and pore water pressure. The results can be used to calibrate the calculations.

At the edges of the fill, inclinometers can measure whether the bearing capacity of the subsoil has been exceeded.

A limiting value should be specified for the rate of settlement at which the preload fill can be removed, e.g. in mm/day or cm/month.

### 7.12.10 Secondary settlement

It should be noted that preloading can pre-empt secondary settlement, which is independent of consolidation, to only a very limited extent (in highly plastic clays, for example). If secondary settlement is expected on a fairly large scale, special supplementary investigations are necessary.

## 7.13 Improving the bearing capacity of soft cohesive soils with vertical elements (R 210)

### 7.13.1 General

Grouted or non-grouted ballast or sand columns, supported on low-lying loadbearing strata, are frequently constructed for the foundations to earthworks on soft cohesive soils.

The columns carry the vertical loads and transfer them to the loadbearing subsoil, and in doing so they are supported by the surrounding soil. Therefore, the soil must exhibit a strength $c_u > 15 \, kN/m^2$ at least. Grouted vibro-replacement stone columns only need such support during installation.

Very soft, organic cohesive soils cannot guarantee the necessary degree of support and therefore such foundation systems can be built in this soil type only when the lateral support is achieved by other means, e.g. geotextile jackets.

Ballast and sand columns exhibit drainage properties similar to those of vertical drains (see R 93, section 7.11) and hence also increase the shear strength of the in situ soil through consolidation.

### 7.13.2 Methods

Vibroflotation is a method of improving the subsoil with vertical elements which has been used and accepted for a long time. Using vibrations, the soil is compacted in the sphere of influence of the vertical element. The resulting loss of volume is compensated for by adding soil. Positioning the compaction points on a triangular or square grid improves the bearing capacity over a wide area. However, the use of this method in non-cohesive, compactable soils is limited. Even just small proportions of silt can prevent compaction by vibration because the fine soil particles cannot be separated from each other by vibration.

Vibro-displacement compaction was developed for these soils. Soil improvement by this method is achieved, on the one hand, by displacing the in situ soil and, on the other, by introducing columns of compacted, coarse-grained material into the in situ soil. But even vibro-displacement compaction requires the support of the surrounding soil. As a guide to the applicability of this method, the undrained shear strength should generally be $c_u > 15 \text{ kN/m}^2$.

The loadbearing elements arranged on a grid are embedded in the low-lying loadbearing soil strata. The load transfer into the loadbearing elements takes place via a layer of sand above the elements. This concentrates the stress over the loadbearing elements and relieves the surrounding cohesive soil.

The loadbearing effect is enhanced by including geotextile reinforcement above the loadbearing elements, spanning across the soft strata like a membrane.

When designing geotextile-reinforced earthworks on pile-type foundation elements, the following boundary conditions and dimensions must be taken into account (see Fig. R 210-1):

- The diameter of the pile-type loadbearing element is generally 0.6 m and the spacing between these, arranged on a regular grid, is approx. 1.0–2.5 m.
- The geotextile reinforcement is generally placed 0.2–0.5 m above the tops of the columns. Additional layers of reinforcement are placed at a distance of 0.2–0.3 m above the first layer. The sand between the reinforcement layers prevents a failure in the form of one geotextile sliding on another.

$$E = \frac{F_P}{F_{AE}} = 1 - \frac{\sigma_{zo} \cdot (A_E - A_P)}{(\gamma \cdot h) \cdot A_E}$$

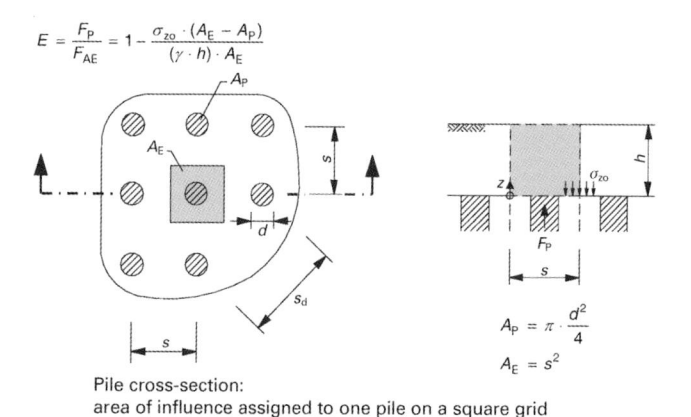

$$A_P = \pi \cdot \frac{d^2}{4}$$

$$A_E = s^2$$

Pile cross-section:
area of influence assigned to one pile on a square grid

**Fig. R 210-1.** Loading on the piles [229]

- Analysing the stability for the construction phases and the final condition can be carried out according to DIN 4084 using curved failure planes or rigid body failure mechanisms. In doing so, the three-dimensional system should be converted into an equivalent planar system with wall-like plates but maintaining the area ratios. Resistances due to the "truncated" piles and the geotextile reinforcement may be taken into account.
- The allocation of the loads to the pile elements and the surrounding, settlement-sensitive soft soil is expressed by the load redistribution $E$ in the loadbearing layer. The redistribution $E$ is the force $F_P$ that must be carried by one pile related to the area of influence $F_{AE}$:

$$E = \frac{F_P}{F_{AE}} = 1 - \frac{\sigma_{zo} \cdot (A_E - A_P)}{(\gamma \cdot h) \cdot A_E}$$

The load redistribution $E$, and hence the stress $\sigma_{zo}$ acting on the soil between the columns, can be determined numerically with the help of a vaulting model in which it is assumed that a redistribution of stress takes place only within a limited zone above the soft stratum [229].

The load redistribution in the loadbearing layer is directly dependent on the shear strength of the material of the loadbearing layer. This is valid for:

rectangular grid with axis dimensions $s_x$ und $s_y$:

$$F_p = E \cdot (\gamma \cdot h + p) \cdot s_x \cdot s_y$$

triangular grid with axis dimensions $s_x$ und $s_y$:

$$F_p = E \cdot (\gamma \cdot h + p) \cdot \frac{1}{2} s_x \cdot s_y$$

The membrane effect of the geotextile reinforcement is improved by laying the reinforcement as close as possible over the almost rigid piles. To prevent shearing of the geotextile, provide a levelling layer over the heads of the piles so that the material does not rest directly on the piles. The membrane effect of the geotextile layer can also relieve the soft strata yet further. Please refer to [229] and [115] for more details of the calculations.

### 7.13.3 Construction of pile-type loadbearing elements

To construct the pile elements, a working platform with adequate bearing capacity for the necessary plant is essential. This can be achieved through improving the soil near the surface or by replacing the topmost layer of soil.

Vibro-displacement stone columns are produced with a bottom-feed vibrator. After reaching the design depth or the loadbearing subsoil, the bottom-feed vibrator is raised by a few decimetres and coarse-grained material is driven out of a chamber in the vibrator by compressed air or water. Afterwards, the vibrator is lowered again, which compacts the material just added. The in situ soil is thus displaced and also compacted. This process is repeated in several stages until a compacted pile element is created from the bottom upwards.

To prevent the pile material escaping into the surrounding soil, vibro-displacement stone columns are only recommended for cohesive soils with an undrained shear strength $c_u \geq 15\,kN/m^2$.

Grouted or partly grouted vibrated stone columns make use of the same method as the vibro-displacement stone column. However, the coarse-grained material is mixed with a cement suspension as it is installed; alternatively, the column can be entirely of concrete. The same rule applies here: an undrained shear strength of cohesive strata $c_u > 15\,kN/m^2$ is recommended. The shear strength may be lower in intermediate strata less than approx. 1.0 m thick, but must be $c_u > 8\,kN/m^2$.

Soil displacement using vibration gives rise to an excess pore water pressure which in the case of vibro-displacement stone columns is dispersed relatively quickly owing to the drainage effect of the loadbearing elements. This excess pore water pressure is therefore not relevant for the design. In grouted or partly grouted vibrated stone columns or concrete columns, the drainage effect is at best severely limited. In these cases it must be verified that the excess pore water pressure does not affect load-carrying capacity or serviceability.

### 7.13.4 Design of geotextile-encased columns

In cases of very soft cohesive soils with $c_u < 15\,kN/m^2$, a geotextile sleeve must be used to compensate for the lack of lateral support from the soil. When designing geotextile-encased columns, the following factors must be taken into account:

- The geotextile-encased column is a flexible loadbearing element that can adapt to horizontal deformations.
- Practical experience on site and in the laboratory has shown that there is no punching shear risk in low-strength strata beneath the columns because the settlement of the geotextile-encased column is equal to that of the surrounding soil.
- Residual settlement can be pre-empted by applying an overload greater than the sum of all the later loads (excess fill).
- The geotextile-encased columns are arranged on a uniform grid (usually triangular). The spacing between the columns is approx. 1.5 m to approx. 2.0 m depending on the grid.

- The diameter of the columns is generally about 0.8 m.
- Installing a layer of horizontal geotextile reinforcement above the geotextile-encased columns is recommended. This increases the stability during critical construction phases and reduces expansion deformations.
- The stress concentration above the columns leads to an increase in the overall shear strength in the columns and to a decrease in the shear strength of the surrounding soft strata. The shear strength effect should be taken into account by introducing so-called equivalent shear parameters and using the pore water pressure due to the surcharge [166].
- The design of the geotextile jacket can be carried out in such a way that the short-term strength $F_K$ of the geotextile is reduced by various factors $A_i$ and the factor of safety $\gamma$, and compared with the calculated tensile forces in the geotextile:

$$F_d = \frac{F_K}{\prod A_i \cdot \gamma}$$

The calculation of the respective tensile force in the geotextile is based on the compression, or rather settlement, of an individual column as a result of the effective stress concentration above the column and the horizontal support of the segment of surrounding soft soil. This results in a volume-based bulging of the column over the depth, which in turn causes stretching of the geotextile. Further details and methods of calculation can be found in [166] and [114].

- Measurements of the pore excess water pressure and stresses in and above the soft strata during and after construction must be carried out to check the true state of consolidation. In terms of consolidation, geotextile-encased columns behave similarly to a large-diameter vertical drain.

### 7.13.5 Construction of geotextile-encased columns

Soil displacement or soil excavation within a casing is used to install geotextile-encased columns.

During installation via soil excavation, an open casing down to the level of the loadbearing subsoil is introduced by vibration. The soil is then excavated from within the casing and a prefabricated geotextile jacket is suspended in the casing and filled with sand or gravel. Afterwards, the casing is extracted while being vibrated and the geotextile jacket filled with sand or gravel remains in the ground.

With the displacement method, flaps at the bottom of the drive pipe are closed and the geotextile sleeve is driven to its final depth with the pipe, filling simultaneously. Afterwards, the pipe is extracted.

The prefabricated geotextile jacket is either woven directly as a tube or factory-sewn to form a tube.

The displacement method is more economical than excavating. In addition, soil displacement can already increase the initial shear strength of the soft soil during the installation of the geotextile-encased column. This may need to be checked by measurements before and after installing the columns.

A brief increase in the excess pore water pressure after constructing the column is to be expected when using soil displacement. However, this is quickly dispersed by the drainage effect of the column. Soil displacement causes a temporary lifting of the top level of the soft stratum.

# 8 Sheet piling structures

## 8.1 Materials and construction

### 8.1.1 Design and installation of timber sheet pile walls (R 22)

#### 8.1.1.1 Range of applications

Timber sheet pile walls are advisable only where the existing subsoil is favourable for their installation. In addition, the thickness of the piles is limited, meaning the bending stresses must not be too great. For structures designed for a long working life, the tops of the piles must be at such a level that they remain continuously wet, and steps must be taken to prevent infestation by wood-boring insects. In these conditions, and when other building materials are not available due to local circumstances, the use of timber sheeting for waterfront walls can be considered. DIN EN 1995 is to be applied accordingly when planning and designing timber structures. Connecting elements of steel must be hot-dip galvanised at least or be provided with equivalent corrosion protection.

#### 8.1.1.2 Wood species and dimensions

Timber sheet piles are usually made from highly resinous pine wood, but also spruce and fir. They can also be made from tropical hardwoods (Table R 22-1).

Standard dimensions and forms for piles and their joints are shown in Fig. R 22-1. For straight- or chamfer-tongued joints the tongue is normally made several millimetres longer than the depth of the groove so that a tight fit develops as the pile is driven.

Corner piles are required at the corners of timber sheet pile walls. These are thick square timbers in which the grooves for the adjoining piles are cut corresponding to the angle at the corner.

#### 8.1.1.3 Driving

Timber sheet piles are mostly driven as double piles joined together with pointed cramps. Staggered driving or driving in panels is always necessary (see R 118, section 8.1.11) to protect the piles and achieve a watertight wall. The tongue side faces in the driving direction and is bevelled at the toe of the pile so that, as it is driven, the pile presses against the adjacent, already driven, pile. The toe of the pile is shaped like a cutting edge – and reinforced by an approx. 3 mm thick steel plate

# Steel Sheet Piling | 2015

**ArcelorMittal**

## Z Section

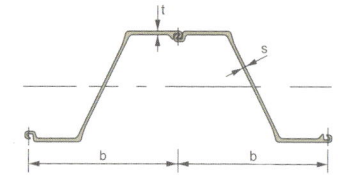

| | | | | |
|---|---|---|---|---|
| Mass | from | 67.7 | to | 157.7 kg/m |
| Thickness t | from | 8.5 | to | 20.0 mm |
| Thickness s | from | 8.5 | to | 16.0 mm |
| Width b | from | 580 | to | 800 mm |
| $W_x$ | from | 1 205 | to | 5 015 cm³/m |

## U Section

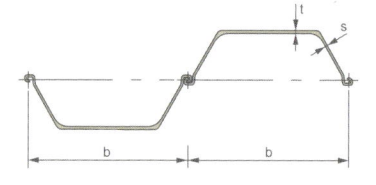

| | | | | |
|---|---|---|---|---|
| Mass | from | 41.9 | to | 118.4 kg/m |
| Thickness t | from | 6.0 | to | 20.5 mm |
| Thickness s | from | 6.0 | to | 11.4 mm |
| Width b | from | 400 | to | 750 mm |
| $W_x$ | from | 625 | to | 3 340 cm³/m |

## AS Section

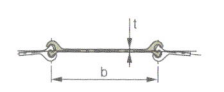

| | | | | |
|---|---|---|---|---|
| Mass | from | 63.8 | to | 77.1 kg/m |
| Thickness t | from | 9.5 | to | 12.7 mm |
| Width b | | | | 500 mm |
| $F_{max}$ | from | 3 000 | to | 5 500 kN/m |

## HZ®-M Section

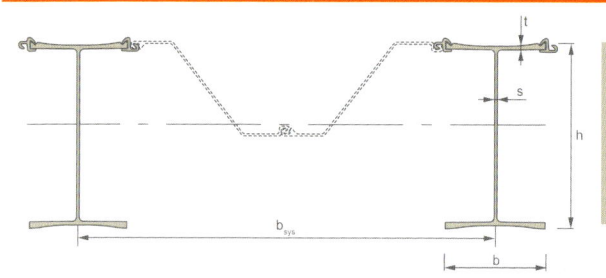

| | | | | |
|---|---|---|---|---|
| Mass | from | 234.7 | to | 999.6 kg/m |
| Thickness t | from | 16.9 | to | 37.0 mm |
| Thickness s | from | 13.0 | to | 22.0 mm |
| Height h | from | 631.8 | to | 1 087.4 mm |
| Width b | from | 454 | to | 460 mm |
| $W_x$ | from | 3 500 | to | 48 570 cm³/m |

sheetpiling.arcelormittal.com

# SteelWall®

## Clutch bars for steel sheet pile walls

**steelwall.eu**

SteelWall ISH GmbH · Tassilostr. 21 · 82166 Graefelfing · Germany

# NEWSLETTER

**Ernst & Sohn**
A Wiley Brand

Anmelden / Subscribe to:
www.ernst-und-sohn.de/newsletter

# Ernst & Sohn journals online
easy to search – easy to access – easy to archive

**More details:**
www.ernst-und-sohn.de/wol

Journals online – a product of
**Wiley Online Library**

**Ernst & Sohn**
Verlag für Architektur und technische
Wissenschaften GmbH & Co. KG

Customer Service: Wiley-VCH
Boschstraße 12
D-69469 Weinheim

Tel. +49 (0)800 1800 536
Fax +49 (0)6201 606-184
cs-germany@wiley.com

## a) Sections

Corner details
Corner pile
Sheet pile with 2 tongues
($\leftarrow$ = driving direction)
Corner pile

Overlapping plank sheeting, $t < 6$ cm
Overlapping plank sheeting, $t < 6$ cm
V-jointing, $t \geq 6$ cm, $t/3$
Straight-tongued jointing, $t = 10...30$ cm
Chamfer-tongued jointing, $t = 10...30$ cm
Joint details, $t/3$, $a = t/3$, but $\leq 5$ cm

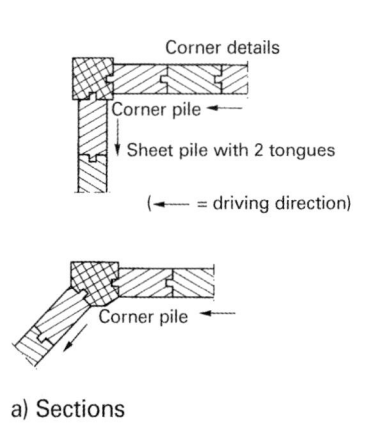

Rule of the thumb for pile thickness $t$
$t$ (cm) = $2 \times l$ (m)
Pile width $b$ = approx. 25 cm
Pile length $l \leq 15$ m

## b) Toe detail

$2 - 3\ t$ in easy driving conditions
$1 - 1.5\ t$ in difficult driving conditions
Driving direction

## c) Double pile

Forged ring approx. 2 cm thick
Pointed cramps
Driving direction

**Fig. R 22-1.** Timber sheet piles

on all sides in soil that presents difficult driving conditions. The top of the pile is always protected from splitting by a conical steel flat ring about 20 mm thick.

#### 8.1.1.4 Watertightness

To a certain extent, timber sheet pile walls seal themselves as a result of the swelling of the wood. The watertightness of excavation enclosures in open water can be improved by sprinkling fine, uncontaminated suspensions (ash, sawdust, finely ground slag) into the water on the outside of the piling while pumping out the excavation (for more information see R 117, section 8.1.21). These suspensions become trapped in the tongue and groove joints. Large leaks in the wall, e.g. due to joint damage, can be temporarily repaired by attaching canvas over them, but must be permanently sealed by divers with wooden laths and caulking.

#### 8.1.1.5 Protecting the wood

Wood species indigenous to Germany are usually only protected against rotting when underwater. Timber sheet pile walls made from indigenous species with permanent loadbearing roles within the structure must therefore be placed above the water level (above low water in open waters and above mid-tide level in tidal waters) and protected with an environmentally friendly impregnation treatment. In waters with a salt content >9‰ in which attack by shipworm (*Teredo navalis*) is a risk, the sheet piling must be protected over its entire length with an impregnation treatment. Tropical hardwoods are more resistant under such circumstances. The most important characteristic values of tropical hardwoods often used in hydraulic engineering are listed in Table R 22-1.

### 8.1.2 Design and installation of reinforced concrete sheet pile walls (R 21)

#### 8.1.2.1 Range of applications

Reinforced concrete sheet piles may be used only with the assurance that the sheet piles can be installed the soil without damage and with tight joints. Their use should, however, be restricted to structures in which requirements in respect of watertightness are not high, e.g. groynes etc.

#### 8.1.2.2 Concrete

When selecting a type of concrete for reinforced concrete sheet piles, take into account the respective exposure classes in relation to the ambient conditions (DIN EN 1992-1-1).

#### 8.1.2.3 Reinforcement

Reinforced concrete sheet piles are designed in accordance with DIN EN 1992-1-1, although DIN EN 12699 must also be observed for the loads

# momentum
MAGAZIN

Das Online-Magazin für Bauingenieure
www.momentum-magazin.de

**Ernst & Sohn**
A Wiley Brand

## TAGU® Tiefbau GmbH Unterweser

- Marine Construction
- Seacable/Conduit Laying
- Landfall Installation
- Piling Works
- Quay Construction
- Offshore Services

Foundations
·
Marine Logistics
·
Rental Equipment
·
Planning
·
Hydrographic Surveys
·
Hydraulic Engineering

**Tiefbau GmbH Unterweser**
Ammerländer Heerstr. 368 • D- 26129 Oldenburg
Tel. +49 (0) 441 9704-500 • Fax +49 (0) 441 9704-510
E-mail: info@tagu.de • www.tagu.de

A Member of the Ludwig Freytag group

**Ernst & Sohn**
A Wiley Brand

# Soil Dynamics with Applications in Vibration and Earthquake Protection

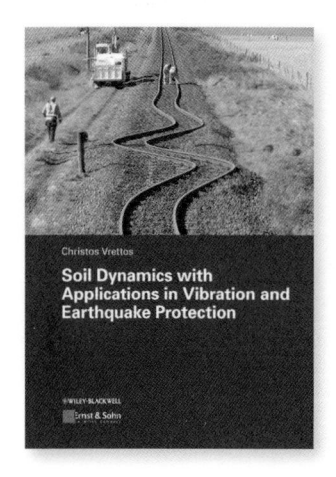

For numerous geotechnical applications soil dynamics are of special importance. In seismic engineering this affects the stability of dams, slopes, foundations, retaining walls and tunnels, while vibrations due to traffic and construction equipment represent a significant aspect in environmental protection. Foundations for mechanical equipment and cyclically loaded offshore structures are also part of the spectrum of application. This book covers the basics of soil dynamics and building thereon the practical applications in vibration protection and seismic engineering.

About the author:
Dr. Christos Vrettos is Professor of Soil Mechanics and Foundation Engineering at the Technical University of Kaiserslautern in Germany. He holds a Dipl.-Ing. and a Dr.-Ing. degree from the University of Karlsruhe and a Habilitation from the Technical University of Berlin. He spent several years in construction industry and geotechnical consulting being involved in notable projects. His expertise covers soil dynamics, railroad track dynamics, geotechnical earthquake engineering, numerical methods in geomechanics, deep foundations, unsaturated soils, and deep excavations. He is member of various international and national committees, and author of numerous publications.

Christos Vrettos
**Soil Dynamics with Applications in Vibration and Earthquake Protection**
2015.
approx. 200 pages.
approx. € 59,–*
ISBN 978-3-433-02999-2
Auch als **ebook** erhältlich

Recommendations:

- Geomechanics and Tunneling
- Rock Mechanics Based on an Anisotropic Jointed Rock Model (AJRM)

Order online:
www.ernst-und-sohn.de

**Ernst & Sohn**
Verlag für Architektur und technische Wissenschaften GmbH & Co. KG

Customer Service: Wiley-VCH
Boschstraße 12
D-69469 Weinheim

Tel. +49 (0)6201 606-400
Fax +49 (0)6201 606-184
service@wiley-vch.de

* € Prices are valid in Germany, exclusively, and subject to alterations. Prices incl. VAT. excl. shipping. 1066116_dp

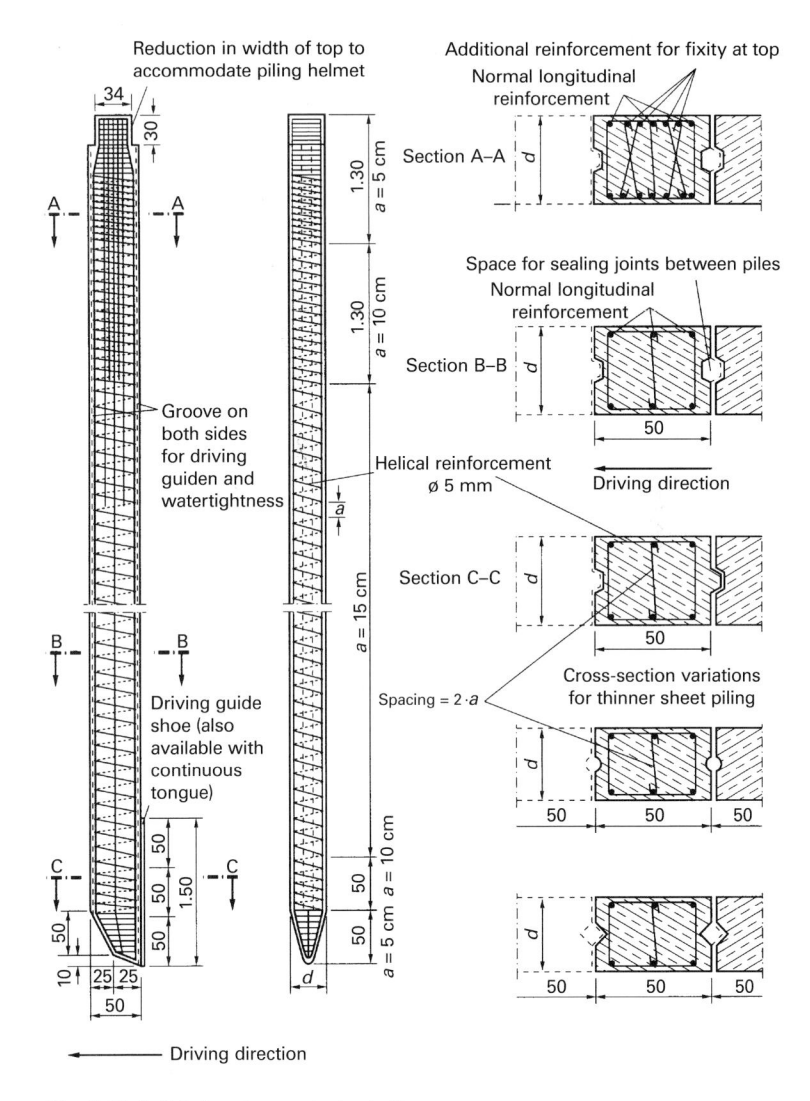

**Fig. R 21-1.** Reinforced concrete sheet piles

caused when removing precast concrete piles from their moulds and pitching piles prior to driving. Sheet piles generally have longitudinal structural reinforcement of grade BSt 500 S. In addition, the sheet piles have helical reinforcement made from grade BSt 500 S or M or Ø5 mm wire. Details are shown in Fig. R 21-1.

285

The concrete cover to the main structural reinforcing bars in freshwater and seawater should be at least $c_{min} = 50$ mm, with a nominal dimension $c_{nom} = 60$ mm, and hence larger than the values given in DIN 1992-1-1.

### 8.1.2.4 Dimensions

Reinforced concrete sheet piles must be min. 14 cm thick but generally should not be thicker than 40 cm for weight reasons. The thickness depends on structural and constructional needs as well as driving requirements. The normal pile width is 50 cm, but the width at the top should be reduced to 34 cm if possible so that it fits standard piling helmets. Reinforced concrete sheet piles can be supplied in lengths of up to about 15 m, although up to about 20 m is possible in exceptional cases. Normal groove forms for reinforced concrete sheet piles are shown in Fig. R 21-1. The width of the groove can be up to 1/3 of the sheet pile thickness, but no larger than 10 cm. The groove runs continuously to the bottom of the pile on the leading side. On the opposite side, the toe has a tongue about 1.50 m long which fits into the groove. A groove runs upward from this tongue (Fig. R 21-1). The tongue guides the toe as the pile is driven. It may extend from the toe to the top of the pile and thus help to seal the wall. In the case of quay walls backfilled with non-cohesive soil and subjected to excess water pressure from behind, this type of groove may only be used when a filter can form naturally behind each joint so that soil can only escape through the joints immediately after the quay wall is completed.

### 8.1.2.5 Driving reinforced concrete sheet piles

The toe of the pile is given a bevel a little less than 2 : 1 on the leading side so that, during driving, the pile presses against the piles already driven. This form is also retained for jetted piles. Driving is made easier if the piles have a wedge-shaped point in the transverse direction, too. The piles are always driven individually, and tongue and groove types with the groove side forward. A piling helmet should be used when driving with a drop hammer. To protect them, piles should be driven with the heaviest (slow-stroke) hammers possible and with a small drop. When driving with hydraulic hammers, the impact energy must be regulated so that the top of the pile is not damaged.

When driving reinforced concrete sheet piles into soils of fine sand and silt, better progress will be made if jetting is used.

### 8.1.2.6 Sealing to prevent loss of soil

If the sheet pile has only a short tongue, sufficient space for the insertion of joint seals is available. Before these are fitted, each groove should be cleaned with a jet of water. The space is then filled with a concrete mix using a tremie pipe. Sacks filled with a plastic concrete mix can be used

Based on the innovative engineering performance combined with its highly qualified team of employees and an extensive modern equipment pool, Colcrete-von Essen covers the whole spectrum of hydraulic engineering.

**Colcrete-von Essen GmbH & Co. KG**
Am Waldrand 9c · 26180 Rastede · Germany
Tel. +49 4402 / 9787-0
Fax +49 4402 / 97948

info@colcrete-von-essen.de
www.colcrete-von-essen.de

**Colcrete-von Essen Wasserbau GmbH & Co. KG**
Ziegeleistraße 4 · 17373 Ueckermünde · Germany
Tel. +49 39721 / 5417-0
Fax +49 39721 / 5417-20

ueckermuende@colcrete-von-essen.de
www.colcrete-von-essen.de

**Colcrete - von Essen**
Spezial-Wasserbau
Küstenschutz

# ebooks @ Ernst & Sohn

**Ernst & Sohn**
A Wiley Brand

- Over 35.000 ebooks available as PDF, ePUB and mobi
- Fast and efficient search results

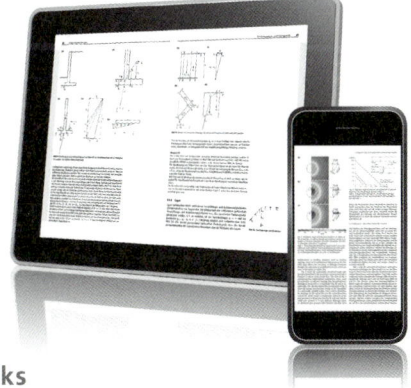

www.ernst-und-sohn.de/ebooks

1008436_dp

# Otl Aicher (1922–1991)
## an outstanding personality in modern design

**Ernst & Sohn**
A Wiley Brand

Otl Aicher was a co-founder of the legendary Hochschule für Gestaltung (HfG), the Ulm School of Design, Germany. His works since the fifties of the last century in the field of corporate design, e.g. Lufthansa, and his pictograms for the 1972 Summer Olympics in Munich are major achievements in the visual communication of our times.

**analogue and digital**
2nd Edition 2015. 188 pages.
€ 24,90
ISBN 978-3-433-03119-3

**the world as design**
2nd Edition 2015. 198 pages
€ 24,90
ISBN: 978-3-433-03117-9

"An integral component of Aicher's work is that it is anchored in a "philosophy of making" inspired by such thinkers as Ockham, Kant or Wittgenstein, a philosophy concerned with the prerequisites and aims, the objects and claims, of design. Aicher's complete theoretical and practical writings on design (which include all other aspects of visual creativity, such as architecture) are available with this new edition of the classic work."

*Wilhelm Vossenkuhl*

"Otl Aicher likes a dispute. For this reason, the volume contains polemical statements on cultural and political subjects as well as practical reports and historical exposition. He fights with productive obstinacy, above all for the renewal of Modernism, which he claims has largely exhausted itself in aesthetic visions; he insists the ordinary working day is still more important than the "cultural Sunday"."

*Wolfgang Jean Stock*

Order online:
www.ernst-und-sohn.de

**Ernst & Sohn**
Verlag für Architektur und technische
Wissenschaften GmbH & Co. KG

Customer Service: Wiley-VCH
Boschstraße 12
D-69469 Weinheim

Tel. +49 (0)6201 606-400
Fax +49 (0)6201 606-184
service@wiley-vch.de

\* € Prices are valid in Germany, exclusively, and subject to alterations. Prices incl. VAT. excl. shipping. 1107146_dp

in large grooves. A seal of bituminised sand and fine gravel can also be used. In any case, the sealing measures must be fitted so that they fill the entire groove without leaving any gaps. This type of seal between the sheet piling is especially suited to C-shaped grooves as shown in Fig. R 21-1. Such joints are also suitable with prestressing in the plane of the wall.

However, the degree of sealing achievable with the methods described above is limited. Subsequent resealing is only possible at great expense.

## 8.1.3 Design and installation of steel sheet pile walls (R 34)

### 8.1.3.1 General

With regard to structural and driving requirements, steel sheet piles can be used everywhere. Damage is easy to repair in most cases. The interlocks of sheet piles can be sealed with constructional measures (R 117, section 8.1.21). Walls made from steel sheet piles can be protected against corrosion according to R 35, section 8.1.8.

### 8.1.3.2 Selecting type of section and grade of steel

When selecting the type of section, its dimensions and the steel grade for steel sheet piles, then the structural requirements, the expectations placed on serviceability, economic considerations and the stresses that will arise while driving the section into the subsoil are all vital considerations. It must be ensured that the section chosen can be installed without damaging the interlocks in particular. In addition, the required wall thickness of the section can be determined by the mechanical stresses imposed on the wall during the berthing manoeuvres of ships and sand abrasion effects (R 23, section 8.1.9).

In the case of large excavation depths and large actions due to earth pressure and excess water pressure, the required standard cross-section values (section modulus) can no longer be achieved with U- and Z-sections. In these cases, beam-type sections (bearing piles) with large section moduli are combined with U- and Z-sections (intermediate piles), which form a closed wall when connected. This combined steel sheet piling is covered in more detail in R 7, section 8.1.4, and often represents an economical solution. The sections can be strengthened further by welding on plates or additional interlocks so that they can accommodate additional bending stresses.

Draft standard DIN EN 10248 lists seven grades for steel sheet pile walls, up to steel grade S 460 GP. The choice of higher steel strengths is to be agreed with the manufacturers on placing an order. In addition, R 67, section 8.1.6, is to be observed.

### 8.1.4 Combined steel sheet piling (R 7)

#### 8.1.4.1 General

Combined steel sheet piling is a type of wall in which long bearing piles alternate with shorter, lighter intermediate piles. After installation, the bearing piles must stand vertically so that the intermediate piles can also be driven without exposing the interlocks between the bearing and intermediate piles to excessive stresses that damage the joints. For more information on this and on the most common wall forms and elements used in combined steel sheet piling see R 104, section 8.1.12.

#### 8.1.4.2 Structural system for combined sheet piling

In the case of combined sheet piling, the vertical and horizontal loads are transferred into the subsoil exclusively by the bearing piles. The intermediate piles complete the wall and transfer the direct excess water pressure and part of the earth pressure to the bearing piles. DIN EN 1993-5, annex D, section D.1.2, can be used to calculate the transfer of loads from the intermediate to the bearing piles through local bending of the flanges.

The bearing piles must be designed in accordance with DIN EN 1993-5, sections 5.5.1(2) and 5.5.4(1)P, for the horizontal and vertical loads acting on the system dimension (= width of intermediate pile + width of bearing pile). At the bottom, the bearing piles are embedded in the soil, at the top they are held by anchors.

When backfill of at least medium density is used behind the wall, the intermediate piles are primarily stressed by excess water pressure as the majority of the earth pressure is resisted directly by the bearing piles through a horizontal arching effect. If this condition is satisfied, experience has shown that intermediate piles in the form of unwelded Z-sections with a wall thickness of 10 mm and a clear spacing between the bearing piles of up to 1.50 m, or U-sections with a clear spacing of up to 1.80 m, can transfer an excess water pressure load of up to $40 \, kN/m^2$ to the bearing piles without the need for further checks.

Where bearing piles are spaced further apart and/or higher loads from excess water pressure are to be expected, or if a horizontal arching effect cannot be assumed behind the bearing piles, the transfer of the loads from the intermediate to the bearing piles must be verified.

For Z-sections with a clear bearing pile spacing between 1.50 and 1.80 m, this check can be deemed to be satisfied according to DIN EN 1990, provided the sheet pile interlocks have been verified for the excess water pressure acting or local experience with such walls is available. In all other situations, horizontal intermediate walings can be used as supplementary support components for accommodating excess water pressure.

With bearing pile spacings of up to 1.80 m and/or embedment depths of at least 5.00 m, the full passive earth pressure in front of the bearing piles can, for simplicity, be assumed even if the intermediate piles are not driven as deep as the bearing piles.

If the spacing of the bearing piles exceeds 1.80 m and/or the embedment depth is less than 5.00 m, checks must be carried out to ascertain whether, instead of the full passive earth pressure in front of the continuous wall, the volumetric passive earth resistance in front of the narrow compression areas of the bearing piles according to DIN 4085, section 6.5.2, is critical.

In combined sheet piling with an off-centre arrangement of the intermediate piles, considering them as part of a composite cross-section according to R 103, section 8.1.5, is only appropriate if the displacement of the centroid axis is compensated for by strengthening the bearing piles on the opposite side.

### 8.1.4.3 Design guidance for combined sheet piling

The properties of the bearing piles in line with section 8.1.4.1 are to be certified with inspection certificate 3.1 to DIN EN 10204. For the other components in a combined steel sheet pile wall, steel grades with a yield stress $<355 \text{ N/mm}^2$ require test certificate 2.2, steel grades with a yield stress $\geq 355 \text{ N/mm}^2$ certificate 3.1 to DIN EN 10204. For special requirements, the specification of the alloying and accompanying elements (C, Si, Mn, P, S, Nb, V, Ti, Cr, Ni, Mo, Cu, N, Al) can also be agreed. If required, forms and sizes deviating from the standard can be agreed separately. Notwithstanding, national technical approvals must still be taken into account with regard to the intended purpose when it comes to tubes made from fine-grained structural steels to DIN EN 10219 and thermo-mechanically treated steels.

Tubular bearing piles must have corresponding interlock sections welded on to guarantee adequate connection of the intermediate piles. For this purpose, the interlock connection must comply with the tolerances in R 67, section 8.1.6, and be able to transmit reliably the loads from the intermediate piles to the tubular sections. At the intersection points between girth and helical welds, interlock sections must fit tightly against the tube. Suitable tube weld seams and interlock section details are required at the intersection point. A welding method test to DIN EN ISO 15614-1 must be performed for interlock welds on tubular sections made from fine-grained steels.

Tubular sections with internal interlocks are particularly useful as bearing piles in combined sheet piling walls if they can be installed using a rotary drilling system, which reduces noise and avoids vibrations. The steel grade of the tubes should comply with DIN EN 10219 and fulfil all other demands placed on sheet piling steels.

Tubes as bearing piles in combined steel sheet piling are fabricated in the works in full lengths spirally welded, or as individual lengths longitudinally welded and connected by girth welds, using full or semi-automatic welding equipment. Differing pile thicknesses, stepped on the inside, are possible with longitudinally welded tubes.

Longitudinal and helical line welds must be subjected to ultrasonic testing. Any flaws identified must be colour-coded and repaired manually. The re-testing of repaired weld seams should be recorded within the scope of a new ultrasonic test with the help of a printed log.

Girth welds between the individual tubes lengths and transverse welds between the coil ends of spirally welded tubes must be checked by X-ray methods.

The intermediate piles generally consist of Z-section double piles or U-section triple piles. In the case of tubular bearing piles they are usually arranged on the wall axis, which means that such walls do not achieve a plain berthing surface. In combined walls with bearing piles made from box or H-sections, the intermediate piles are usually fitted in the waterside interlock connection.

### 8.1.4.4 Installing bearing and intermediate piles for combined steel sheet piling

Impact driving is normally used for installing the bearing piles of combined sheet piling (box piles, H-piles or tubular piles) (R 104, section 8.1.12). Intermediate piles can be installed by means of impact or vibratory driving (R 202, section 8.1.23). To ensure that the intermediate piles can be installed without overloading the interlock connections, the bearing piles must stand parallel to each other within the permissible tolerances at the planned spacing and without any distortion.

When driving tubular piles, any obstacles can be removed by excavating within the tube. However, to do this, the inside diameter of the pile must be at least 1200 mm to allow the operation of suitable excavation plant and there can be no construction elements, e.g. internal interlock elements, inside the tube itself (section 8.1.4.3).

With vibratory driving, obstacles in the soil can usually be detected immediately.

The risk with impact driving of steel tubes is that the pile heads will buckle, especially in the case of tubes with relatively thin walls. This can mean that the piles cannot achieve their intended depth. In order to avoid buckling in these cases, the pile head must be stiffened. There are various proven methods for this:

1. Welding several steel angles approx. 0.80 m long vertically to the outside of the tube (Fig. R 7-1). This method is comparatively simple

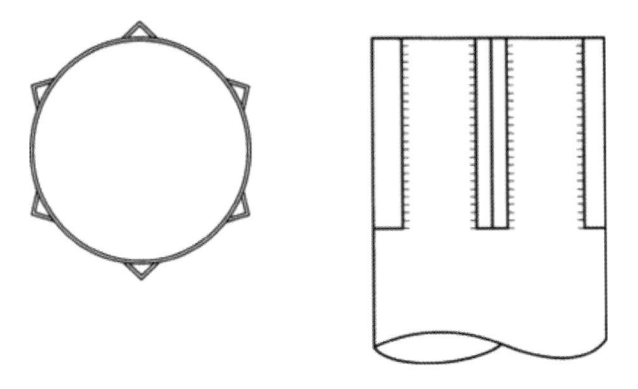

**Fig. R 7-1.** Stiffening by welding on external angles

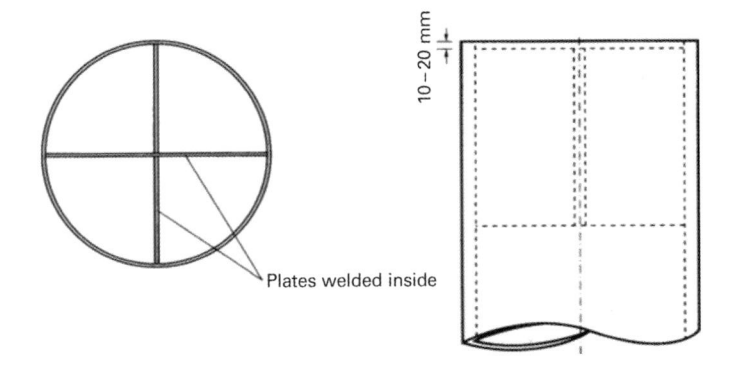

**Fig. R 7-2.** Stiffening by welding plates inside

and cost-effective as welding work is only performed on the outside of the tube.
2. Welding steel plates, approx. 0.80 m long, into the head of the pile cross-wise (Fig. R 7-2). This involves more work than method 1.

### 8.1.4.5 Structural analyses

Please refer to section 8.2 for the transmission of the axial loads from a tubular bearing pile into the subsoil. Where large axial loads and large tube diameters are concerned, it may be necessary to weld metal plates in the form of a cross into the base of the tube to stiffen it so that the base resistance is mobilised. The prerequisite for mobilising this base resistance is the prestressing of the soil within the tube (plug formation) and hence the compactability of the soil around the base of the pile. Any obstacles to driving should be removed in advance, e.g. by soil replacement. Furthermore, it is also possible to use internal wedge, adapted to fit the tube walls, which are braced against a concrete tube fitted into the toe of the tube to generate the required end bearing pressure. The end bearing pressure that can be mobilised may need to be verified by loading tests.

Where using tubular bearing piles in combined steel sheet piling, an analysis of the risk of buckling is unnecessary when the bearing piles are filled with compacted, non-cohesive soil over their entire length.

### 8.1.5 Shear-resistant interlock connections for steel sheet piling (R 103)

#### 8.1.5.1 General

From the structural viewpoint, we distinguish between walls without shear force transmission across the interlocks and those with shear force transmission, the so-called jagged walls. In the case of the latter, every sheet pile makes a full contribution to the section modulus of the wall. When designing steel sheet piling as a jagged wall, it must be ensured that the interlocks can accommodate the shear forces.

In a steel sheet pile wall made from individual Z-section piles, all interlocks are located on the outside and can guarantee the required interlocking effect by means of friction alone.

For steel sheet pile walls made from U-sections, the interlocks are all situated on the wall axis, which means that proper transmission of the shear forces can only be assumed when the factory-assembled interlocks are crimped or welded and the threading interlocks are welded after the wall has been driven.

Crimping the interlocks can only achieve a partial connection, because shear forces can displace the interlocks by several millimetres at the crimping points. The number of crimping points per interlock has a positive effect on the resistance against pile movement in the interlocks and therefore increases shear force transmission and interaction of piles.

#### 8.1.5.2 Verifying the interlocking effect using welding

The shear flow in welded interlock connections can be calculated using the following formula:

$$T_d = V_d \cdot \frac{S}{I}$$

where

$V_d$  design value of the shear force [kN]

$S$     static moment of the cross-section portion to be connected, related to the centroid axis of the connected jagged wall [m³]
$I$     moment of inertia of jagged wall [m⁴]

For intermittent welds, the shear stress should be set higher correspondingly to DIN EN 1993-1-8, section 4.9.

Verification of the welds is to be carried out according to EN 1993-1-8, section 4.5, where a plastic analysis – assuming a uniform shear flow – is permitted. For steel grades with yield stresses not covered by DIN EN 1993-1-8, Table 4.1, the correlation coefficient $\beta_w$ may be obtained through linear interpolation.

### 8.1.5.3 Layout and execution of weld seams

Interlock welds should be designed and executed so that, as far as possible, the shear forces are transferred continuously. A continuous seam is best for this. If an intermittent seam is chosen, the length of each seam should be >200 mm, provided section 8.1.5.2 does not call for longer welds. In order to keep the secondary stresses within limits, the interruptions in the seam should be ≤800 mm.

Continuous weld seams should always be used in areas where the sheet piling is subjected to heavy loads, especially, for example, near anchor connections and at the point where the equivalent force $C$ is introduced at the base of the wall (Fig. R 103-1).

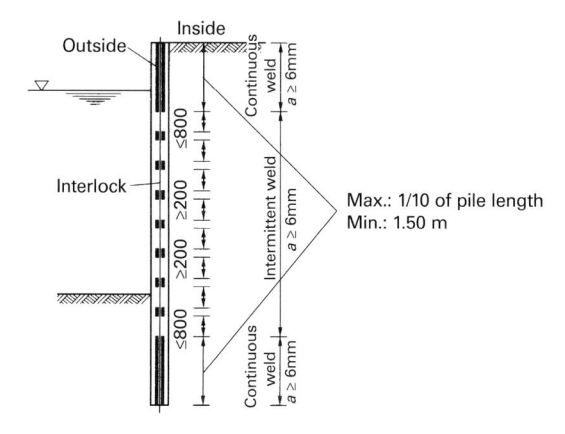

**Fig. R 103-1.** Interlock welding principle for walls in easy driving conditions and only minor corrosion due to harbour water and groundwater

In addition to structural requirements, the loads of driving stresses and corrosion must be considered, too. In order to be able to cope with the driving stresses, the following measures are necessary:

1. Interlocks are to be welded on both sides at head and toe.
2. The lengths of the welds on both sides at head and toe depend on the length of the sheet pile and on the difficulties that might be encountered when driving. For easy to moderate driving conditions, they should be at least 1/10 of the length of the pile but not less than 1.50 m.
3. For waterfront structures, these weld seam lengths should be ≥3000 mm for sheet pile walls in difficult driving conditions.
4. Moreover, additional seams as per Fig. R 103-1 are necessary for easy driving conditions, seams as per Fig. R 103-2 for difficult driving.

In areas where the harbour water is severely corrosive, a continuous weld seam with a thickness $a \geq 6$ mm (Fig. R 103-2) is required on the outside down to the sheet pile toe.

If both the harbour water and the groundwater can cause severe corrosion, a continuous seam with a thickness $a \geq 6$ mm is required on the inside of the wall as well.

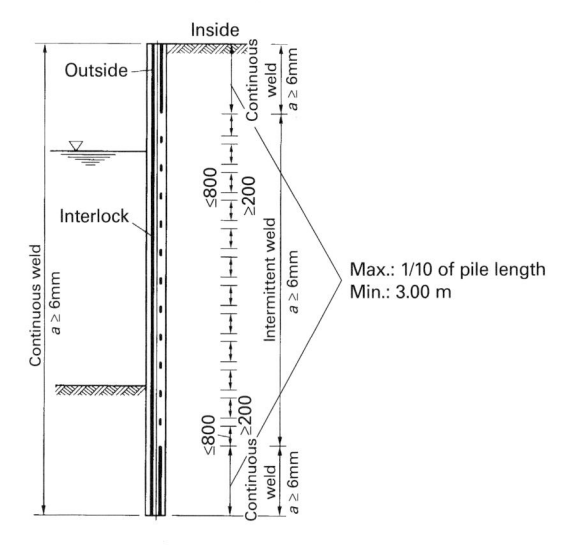

**Fig. R 103-2.** Interlock welding principle for walls in difficult driving conditions or severe corrosion from outside in harbour water area

If, as part of the contract, additional technical contractual term for sheet piling, piles and anchoring (ZTV-W (LB 214)) is agreed upon, the values given there for minimum weld seam thicknesses and additions for corrosion are to be used.

### 8.1.5.4 Choice of steel grade
As the amount of welding work on jagged walls is comparatively extensive, the sheet piles must be manufactured from steel grades that are fully suitable for fusion welding. In view of the starting points of welds, not only for intermittent welds, killed steels without brittle fracture tendencies should be used as per R 99, section 8.1.19.2.

### 8.1.5.5 Verification of interlocking action with crimped interlocks
The design value of the shear flow at the crimping points given by the primary structural system and the loads and supporting influences are determined according to section 8.1.5.2. The analysis of the transfer of shear forces is carried out in accordance with DIN EN 1993-5, sections 5.2.2 and 6.4. The resistances of the crimped points are determined in accordance with draft standard DIN EN 10248:2006.

### 8.1.5.6 Crimping points
Crimping points can be fabricated as single, double or triple crimps. The spacing of the crimping points should comply with DIN EN 1993-5, section 6.4(5), at least. Check in each case whether the number of crimping points per unit width of pile is sufficient for the entire shear force at a certain elevation (shear force region with the same sign). To ensure this, the spacing of the crimping points can be reduced to suit the shear flow. This aspect should be discussed and agreed with the supplier prior to fabricating the piles.

### 8.1.5.7 Welding on strengthening plates
Strengthening plates are used to increase the flexural strength of the bearing piles and must always be welded to the piles around their full perimeter to prevent corrosion between plate and pile. The ends of the plates should be tapered to reduce the abrupt change in the moment of inertia (see R 99, section 8.1.19.5). The weld seam thickness $a$ should be at least 5 mm where there is no risk of corrosion, at least 6 mm where corrosion is likely, or should comply with the requirements set out in ZTV-W (LB 214).

Where a strengthening plate spans an interlock located on the flange of a sheet pile, this interlock must be provided with a continuous weld beneath the plate plus an extension of at least 500 mm on either side. This weld must be on the side opposite to that on which the plate is located and the weld thickness should be $a \geq 6$ mm. Under the plate, the weld thickness should be such that the plate makes contact with the pile without the need for any further machining. If this is not done, the welds attaching the plate may be seriously damaged during driving.

If welding of the interlock is not desired, the strengthening plates must be cut in half and welded on either side of the interlock according to the procedure above.

### 8.1.6 Quality requirements for steels and dimensional tolerances for steel sheet piles (R 67)

This recommendation applies to steel sheet piles, trench sheeting and driven steel piles, which are all called steel sheet piles in the following. DIN EN 10248 parts 1 and 2 as well as DIN EN 10249 parts 1 and 2 apply.

If steel sheet piles are stressed in the direction of their thickness (normal to the rolling direction), e.g. junction piles for cofferdams with circular and diaphragm cells, then steel grades with appropriate properties must be ordered from the sheet pile supplier in order to avoid lamellar tearing, see DIN EN 1993-1-10.

#### 8.1.6.1 Designation of steel grades

Steel grades with designations S 240 GP to S 430 GP are used for hot-rolled steel sheet piles in normal cases, as indicated in sections 8.1.6.2 and 8.1.6.3.

The grades of steels with yield stresses of 240, 270 and 320 N/mm² in accordance with DIN EN 10248 should be verified with test certificate 2.2 as per DIN EN 10204. Higher-quality steel in accordance with DIN EN 10248 should be verified with inspection certificate 3.1 to DIN EN 10204. For specific requirements, specifying the alloying and accompanying elements (C, Si, Mn, P, S, Nb, V, Ti, Cr, Ni, Mo, Cu, N, Al) can also be agreed.

In special cases, e.g. for accommodating greater bending moments, steel grades with higher minimum yield stresses up to 500 N/mm² can be used, in accordance with R 34, section 8.1.3. When using steel grades with a minimum yield stress >430 N/mm², a national technical approval for sheet piles of this grade of steel is required in Germany.

Steel grades S 235 JRC, S 275 JRC and 355 J0C to DIN EN 10249 can be considered for cold-rolled steel sheet piles.

In special cases, e.g. as stated in section 8.1.6.4, steels to DIN EN 10025 are used.

#### 8.1.6.2 Requirements regarding the mechanical and technological properties of sheet pile steels

Table R 67-1 contains information on the requirements regarding the mechanical properties of hot-rolled steel sheet piles. The mechanical

**Table R 67-1.** Requirements for the mechanical properties of steel grades for hot-rolled steel sheet piles

| Steel grade | Minimum tensile strength $R_m$ | Minimum yield stress $R_{eH}$ | Minimum fracture elongation for measuring length $Lo = 5.65\sqrt{So}$ |
|---|---|---|---|
| | [N/mm$^2$] | [N/mm$^2$] | [%] |
| S 240 GP | 340 | 240 | 26 |
| S 270 GP | 410 | 270 | 24 |
| S 320 GP | 440 | 320 | 23 |
| S 355 GP | 480 | 355 | 22 |
| S 390 GP | 490 | 390 | 20 |
| S 430 GP | 510 | 430 | 19 |
| S 460 GP [*] | *530* | *460* | *17* |

[*] As per Table 2 of draft standard DIN EN 10248-1:2006

properties of cold-rolled sheet piling made from grades S 235 JRC, S 275 JRC and S 355 J0C are specified in DIN EN 10025 and DIN EN 10249.

### 8.1.6.3 Chemical composition

The ladle analysis is binding for the verification of the chemical composition of steel sheet piles (see Table R 67-2). The analysis of single bars can be used as an additional test in cases of doubt. If verification of the chemical composition of individual bars is wanted, a separate agreement must be made for the testing.

**Table R 67-2.** Chemical composition of ladle/bar analysis for hot-rolled steel sheet piles

| Steel grade | Chemical composition, max. %, for ladle/bar | | | | |
|---|---|---|---|---|---|
| | C | Mn | Si | P and S | N[*][**] |
| S 240 GP | 0.20/0.25 | –/– | –/– | 0.045/0.055 | 0.009/0.011 |
| S 270 GP | 0.24/0.27 | –/– | –/– | 0.045/0.055 | 0.009/0.011 |
| S 320 GP | 0.24/0.27 | 1.60/1.70 | 0.55/0.60 | 0.045/0.055 | 0.009/0.011 |
| S 355 GP | 0.24/0.27 | 1.60/1.70 | 0.55/0.60 | 0.045/0.055 | 0.009/0.011 |
| S 390 GP | 0.24/0.27 | 1.60/1.70 | 0.55/0.60 | 0.040/0.050 | 0.009/0.011 |
| S 430 GP | 0.24/0.27 | 1.60/1.70 | 0.55/0.60 | 0.040/0.050 | 0.009/0.011 |
| *S 460 GP* [***] | *0.24/0.27* | *1.70/1.80* | *0.55/0.60* | *0.035/0.045* | *0.012/0.014* |

[*] The values stipulated may be exceeded on condition that for every increase by 0.001% N, the max. P content is reduced by 0.005%. However, the N content of the ladle analysis may not exceed 0.012%.
[**] The maximum N value does not apply when the chemical composition has a minimum total aluminium content of 0.020%, or when there are sufficient N-binding elements. The N-binding elements are to be specified in the test certificate.
[***] As per Table 1 of draft standard DIN EN 10248-1:2006

#### 8.1.6.4 Weldability, special cases

Unlimited suitability of steels for welding cannot be presumed, since the properties of a steel after welding depend not only on the material, but also on the dimensions and the shape plus the fabrication and service conditions of the structural member. Generally, killed steels are preferred for welding (R 99, section 8.1.19.2).

All sheet pile steel grades can be assumed to be suitable for arc welding, provided general welding standards are observed. When selecting higher-strength steel grades S 390 GP and S 430 GP, the welding stipulations of the national technical approval must be adhered to. To ensure weldability, the carbon equivalent (CEV) should not exceed the values for steel grade S 355 given in DIN EN 10025, Table 6. Rimming steels should not be used.

In cases with unfavourable welding conditions and stresses due to installation (e.g. welding in low temperatures and difficult driving conditions), or three-dimensional stresses and/or predominantly changing loads as per R 20, section 8.2.6.1, fully killed steels are to be used according to DIN EN 10025, grade groups J2 or K2, in order to satisfy the brittle fracture resistance and ageing resistance required.

Filler metals are to be selected according to DIN EN ISO 2560, DIN EN 756 and DIN EN ISO 14341 or according to data provided by the supplier (R 99, section 8.1.19.2).

#### 8.1.6.5 Interlock forms

Examples of proven forms of interlock for steel sheet piling are shown in Fig. R 67-1. The nominal dimensions $a$ and $b$, which can be obtained from the suppliers, are measured at right-angles to the least favourable direction of displacement. The minimum interlock hook connection, calculated from $a - b$, must correspond to the values in the figure. Over short sections, the values may not fall below these minimum values by more than 1 mm. In forms 1, 3, 5 and 6, the required coupling must be present on both sides of the interlock.

#### 8.1.6.6 Permissible dimensional deviations for interlocks

Deviations from the nominal dimensions are unavoidable during the rolling of sheet piles and interlocks. The permissible deviations are summarised in Table R 67-3.

### 8.1.7 Acceptance conditions for steel sheet piles and steel piles on site (R 98)

Although careful and workmanlike construction methods are hugely important for the usability of structures made from steel sheet piling or steel piles, it is also essential that the sections delivered to the building site correspond to delivery requirements and comply with certain tolerances regarding dimensions and form. To achieve this, materials

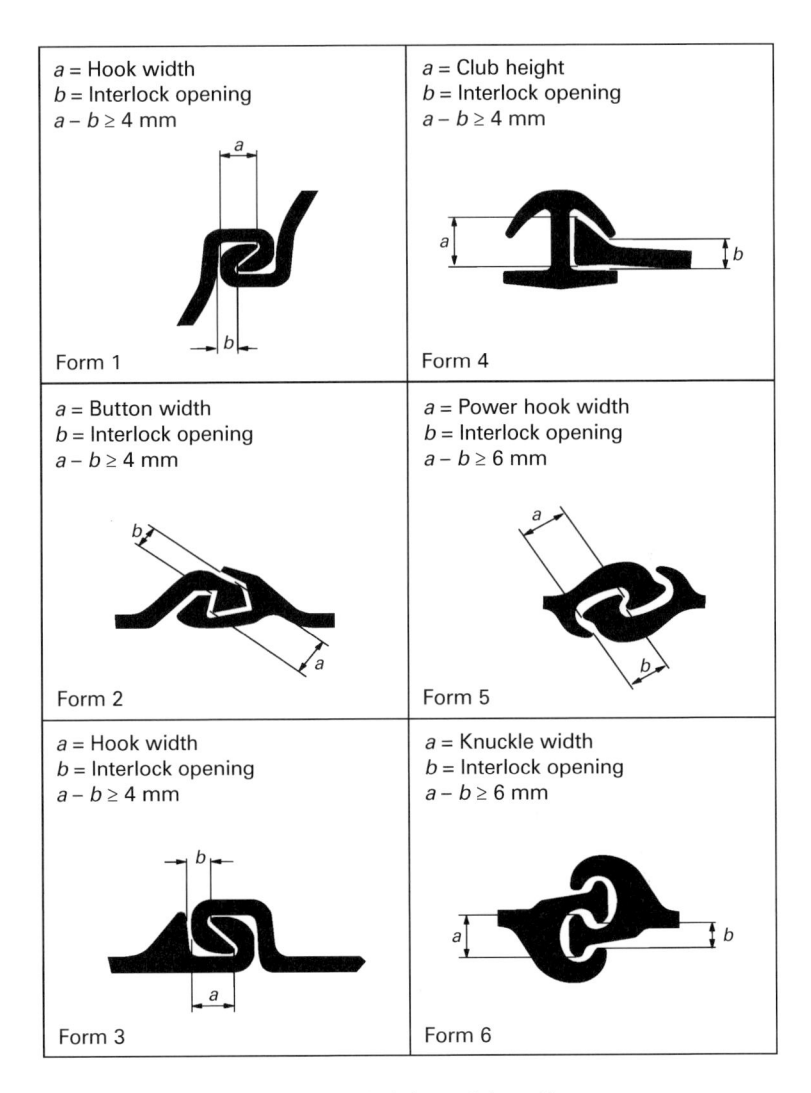

| | |
|---|---|
| $a$ = Hook width<br>$b$ = Interlock opening<br>$a - b \geq 4$ mm<br><br>Form 1 | $a$ = Club height<br>$b$ = Interlock opening<br>$a - b \geq 4$ mm<br><br>Form 4 |
| $a$ = Button width<br>$b$ = Interlock opening<br>$a - b \geq 4$ mm<br><br>Form 2 | $a$ = Power hook width<br>$b$ = Interlock opening<br>$a - b \geq 6$ mm<br><br>Form 5 |
| $a$ = Hook width<br>$b$ = Interlock opening<br>$a - b \geq 4$ mm<br><br>Form 3 | $a$ = Knuckle width<br>$b$ = Interlock opening<br>$a - b \geq 6$ mm<br><br>Form 6 |

**Fig. R 67-1.** Established types of interlock for steel sheet piles

are inspected on site to ensure that dimensional and form tolerances have been adhered to. The results are documented. As a supplement to the manufacturer's own internal inspections, a works acceptance can be agreed in each case. For shipments overseas, inspection is frequently carried out prior to shipping.

**Table R 67-3.** Permissible interlock deviations as per Fig. R 67-1

| Form | Nominal dimensions (according to section drawings) | Deviations from nominal dimensions | | |
|---|---|---|---|---|
| | | Designation | plus [mm] | minus [mm] |
| 1 | Hook width $a$<br>Interlock opening $b$ | $\Delta a$<br>$\Delta b$ | 2.5<br>2 | 2.5<br>2 |
| 2 | Button width $a$<br>Interlock opening $b$ | $\Delta a$<br>$\Delta b$ | 1<br>3 | 3<br>1 |
| 3 | Button width $a$<br>Interlock opening $b$ | $\Delta a$<br>$\Delta b$ | $(1.5–2.5^{*)})$<br>4 | 0.5<br>0.5 |
| 4 | Club height $a$<br>Interlock opening $b$ | $\Delta a$<br>$\Delta b$ | 1<br>2 | 3<br>1 |
| 5 | Power hook width $a$<br>Interlock opening $b$ | $\Delta a$<br>$\Delta b$ | 1.5<br>3 | 4.5<br>1.5 |
| 6 | Knuckle width $a$<br>Interlock opening $b$ | $\Delta a$<br>$\Delta b$ | 2<br>3 | 3<br>2 |

$^{*)}$ depends on section

The acceptance procedure on site should specify that every unsuitable pile will be rejected until it has been reworked to a suitable standard, unless it is rejected outright. Acceptance of sections on the building site is based on:

DIN EN 10248 parts 1 and 2 for hot-rolled sheet piles,
DIN EN 10249 parts 1 and 2 for cold-formed sheet piles, and
DIN EN 10219 parts 1 and 2 for cold-formed welded hollow sections.

In addition, the values of Table R 67-3, section 8.1.6.6, apply to permissible dimensional deviations for interlocks.
R 104, section 8.1.12.4, applies for the limit deviation for straightness of combined sheet pile walls.
Please refer to DIN EN 12063, section 8.3, for more detailed information on the handling and storage of the sections on site.

## 8.1.8 Corrosion of steel sheet piling, and countermeasures (R 35)

### 8.1.8.1 General

Steel in contact with water undergoes a natural process of corrosion, which is influenced by numerous chemical, physical and, occasionally, biological parameters. There are different corrosion zones (Fig. R 35-1) over the exposed height of a sheet pile wall, which are characterised by the type of corrosion (surface, pitting or tuberculation) and its intensity.

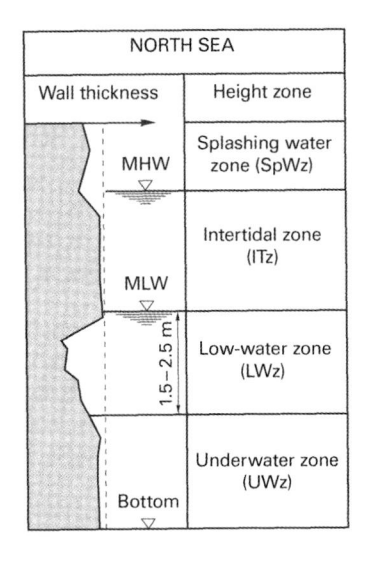

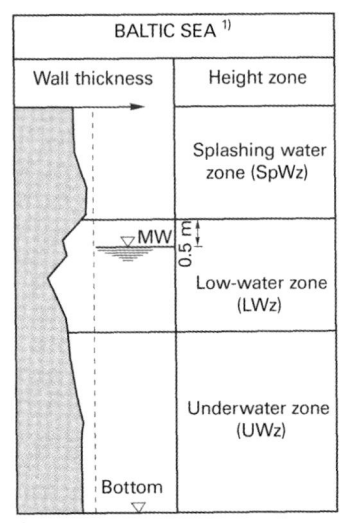

| NORTH SEA | |
|---|---|
| Wall thickness | Height zone |
| MHW | Splashing water zone (SpWz) |
| | Intertidal zone (ITz) |
| MLW | |
| 1.5–2.5 m | Low-water zone (LWz) |
| | Underwater zone (UWz) |
| Bottom | |

| BALTIC SEA [1] | |
|---|---|
| Wall thickness | Height zone |
| | Splashing water zone (SpWz) |
| MW 0.5 m | Low-water zone (LWz) |
| | Underwater zone (UWz) |
| Bottom | |

[1] Can be transferred to inland waters

**Fig. R 35-1.** Qualitative depiction of corrosion zones for steel sheet piling using the North and Baltic Seas as examples

To measure the degree of corrosion, we can assess the decrease in wall thickness [mm] or – related to the time in use – the rate of corrosion [mm/a].

Typical mean and maximum values for wall thickness reduction are shown in Figs. R 35-3 and R 35-4. These diagrams are based on numerous wall thickness measurements on sheet pile walls, piles and dolphins in the North Sea and Baltic Sea as well as inland waters and can be assigned to the corrosion zones shown in Fig. R 35-1. Owing to the multitude of factors that influence corrosion, the measured values are subjected to a very wide scatter.

**8.1.8.2 How corrosion influences the loadbearing capacity, serviceability and durability of steel sheet piling**

Corrosion influences the stability, serviceability and durability of unprotected steel sheet piling and so the following aspects must be considered:

1. Corrosion decreases the design value of the component resistance corresponding to the different reductions in wall thickness in the various corrosion zones (Fig. R 35-1). Depending on the local bending stresses, this can reduce the loadbearing capacity and serviceability of the structure (DIN EN 1993-5, sections 4–6).

2. In analyses of loadbearing capacity and serviceability according to EN 1993-5, the section modulus and the cross-sectional area of the sheet piles are to be reduced in proportion to the mean values of the wall thickness losses according to Fig. R 35-3a (freshwater) or Fig. R 35-4a (seawater).

   Unprotected sheet piling should be designed in such a way that the maximum bending moment lies outside the zone of maximum corrosion.

3. In the light of experience gained over the past few decades, it would seem that corrosion can limit the durability of sheet pile walls (DIN EN 1993-5, section 4) to a useful life of 20–30 years, especially in the seawater of the North Sea and Baltic Sea [5]. Once the steel is rusted through, the soil behind the wall can be washed out, thus leaving voids that can suddenly collapse, causing subsidence at the surface. This is linked with considerable safety risks and restrictions on port operations. U-sections frequently rust through in the middle of the flange of the front pile, Z-sections frequently at the junction between flange and front web (see Fig. R 35-2).

The basis for assessing the durability of sheet pile structures (estimating the useful life until the first occurrence of rusting) are the maximum values for wall thickness losses in accordance with Fig. R 35-3b for freshwater and Fig. R 35-4b for seawater.

Unprotected sheet pile walls should be planned and designed taking into account wall thickness losses as per the regression curves in Figs. R 35-3 and R 35-4 unless other values are available for the location. A decision must be made as to whether the design is to be based on the mean or maximum regression curve values. To avoid uneconomic designs, the recommendation is to use the corrosion values above the regression curve only if local experience suggests this is necessary.

For older, unprotected sheet pile walls, stability assessments should always be based on ultrasound measurements of the wall thickness in

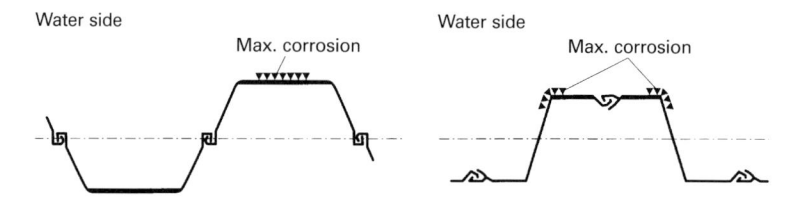

**Fig. R 35-2.** Zones where U- and Z-section sheet piles can rust through in the low-water zone (seawater)

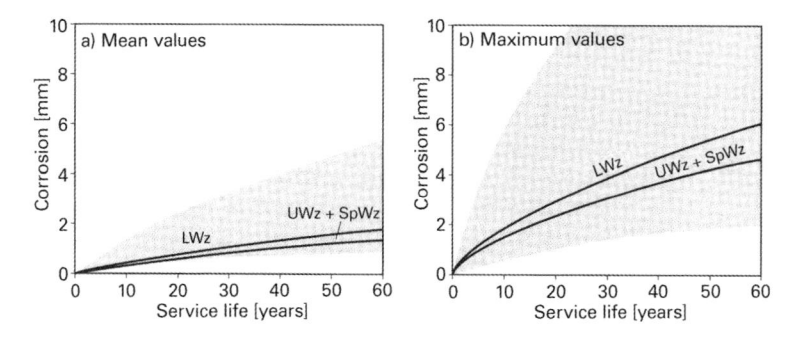

**Fig. R 35-3.** Decrease in thickness as a result of corrosion in freshwater

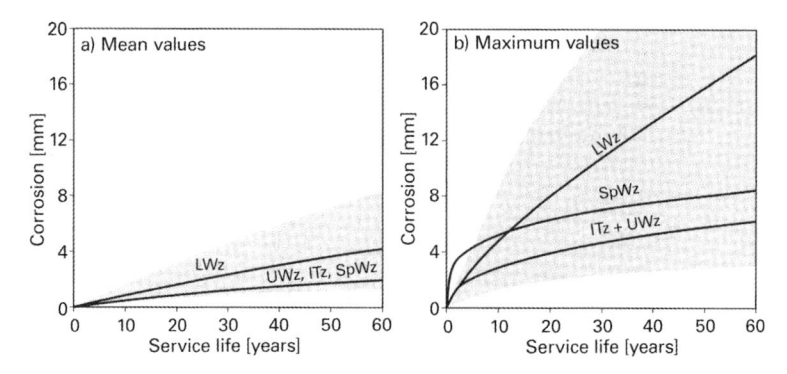

**Fig. R 35-4.** Decrease in thickness as a result of corrosion in seawater

order to assess the local corrosion influences. Information on performing, analysing and troubleshooting such ultrasound measurements can be found in [7] and [6].

### 8.1.8.3 Design values for loss of wall thickness in various media

The following empirical figures for wall thickness loss through corrosion in different media can be used as design values for new sheet piling structures and for checking existing sheet pile walls, unless other values are available for the location.

1. Freshwater:
   The design values for the reduction in wall thickness for sheet pile walls in freshwater can be taken from the regression curves

of Fig. R 35-3 depending on the age of the structure. The area shaded grey represents the scatter of those structures investigated.

2. Seawater of the North and Baltic Seas
   The design values for the reduction in wall thickness for sheet pile walls in the seawater of the North and Baltic Seas can be taken from Fig. R 35-4 depending on the age of the structure. The area shaded grey represents the scatter. The measured values in Figs. R 35-3 and R 35-4 are comparable with those from international literature on the subject [92]. They are, however, on the whole somewhat higher.

3. Brackish Water
   Brackish water zones are a mixture of freshwater from inland areas and salty seawater. The design values for the reduction in wall thickness can be estimated on the basis of the figures for seawater (Fig. R 35-4) and freshwater (Fig. R 35-3) depending on the location of the structure within the briny water zone.

4. Corrosion above the splashing water zone (atmospheric corrosion)
   The rate of corrosion above the splashing water zone (Fig. R 35-1) on waterfront structures is low, amounting to a corrosion rate of about 0.01 mm/a (C1 to C5 as per DIN EN ISO 12944). Higher values can be expected where de-icing salts are in use or in areas involving the storage and handling of substances known to attack steel.

5. Corrosion in the soil
   The aggressiveness of soils and groundwater can be roughly assessed using DIN 50929. In soils the corrosion can be further aggravated by the activity of bacteria that attack steel (microbiologically induced corrosion, MIC) [19, 76]. Microbiologically induced corrosion can be expected where organic substances reach the rear face of the sheet pile wall, either through circulating water (e.g. in the vicinity of landfill sites for domestic waste or in wastewater percolation areas) or through soils with a high organic content. In such cases high corrosion rates and, typically, non-uniform degradation can be expected. Aggressive soils such as humus, carbonaceous soils, waste washings and slag must, as a matter of principle, be avoided whenever possible when backfilling sheet pile structures. This also applies to aggressive water.

6. If steel sheet piling is embedded in non-aggressive, natural soil, then the expected corrosion rate on both sides of the piling is very low (0.01 mm/a). Corrosion rates in the same order of magnitude are to be expected when the sand filling behind a sheet pile wall is such that the troughs of the piling are also completely embedded.

### 8.1.8.4 Corrosion protection

The corrosion protection required for a specific sheet piling structure is ascertained by assessing the following boundary conditions and usage requirements:

- Intended purpose and design life of the structure
- General and specific corrosion loads at the location of the structure
- Experience with corrosion phenomena in adjacent structures
- Options for configuring and designing the structure to inhibit corrosion
- Costs of premature repairs to unprotected sheet pile walls, e.g. adding plates [18]

Subsequent protection measures or complete renewal are extremely difficult so particular care must be taken when planning and applying protective systems. Additional, specific protection measures may be required depending on the nature and intensity of the corrosion and also on the demands placed on the corrosion protection system.

In essence, we distinguish between the following methods of corrosion protection:

1. Corrosion protection with coatings

   According to experience so far, coatings can increase the working life of sheet piling structures by more than 20 years.

   The prerequisite is that the surfaces to be protected are blast-cleaned to standard Sa 2½ before applying the coating and that the coating system selected is suitable for the specific application.

   Guidance on selecting coating systems depending on the local conditions plus surface preparation, laboratory tests, workmanship, supervision of coating work and maintenance of coating systems can be found in the "List of Recommended Coating Systems for the Corrosion Protection of Hydraulic Steel Structures" published by the Federal Waterways Engineering and Research Institute [www.baw.de]. ZTW-W 218 contains additional guidance on contractual matters.

   When it comes to the *serviceability* of sheet piling structures, it is the rate of corrosion in the low-water zone (LWz) that governs. Fig. R 35-4b (maximum values) shows that a 12 mm thick sheet pile in seawater without a coating is completely rusted through after 35 years (Fig. R 35-5, curve 1). If a coating that is assumed to last 25 years is applied, the first cases of rusting will appear after a working life of 60 years (Fig. R 35-5, curve 2).

   When it comes to the loadbearing capacity, the mean corrosion rate should be selected. Maximum bending stresses in sheet piling structures in ports and harbours are mostly found in the underwater

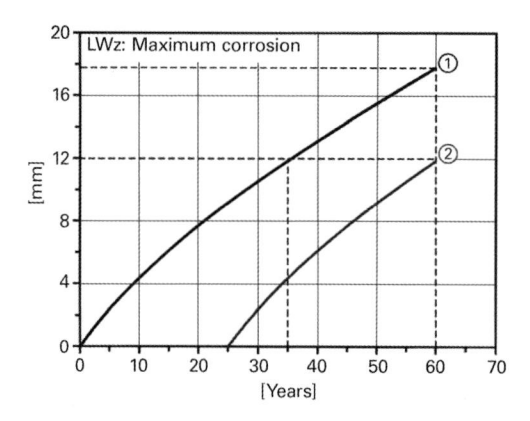

**Fig. R 35-5.** Corrosion in seawater in the low-water zone, maximum corrosion rate; curve 1: uncoated steel, curve 2: coated steel

zone (UWz). Therefore, when assessing the effect of corrosion on the stability of sheet piling structures, the (uniform) mean corrosion rate according to Fig. R 35-4a should be used. Accordingly, a structure suffers 2.0 mm of corrosion during a 60-year working life. Coating reduces the rusting to 1.4 mm over the same period.

In the low-water zone (LWz), according to Fig. R 35-6b, an uncoated structure suffers 4.0 mm of corrosion, a coated structure 2.6 mm.

Coatings should be applied to steel sheet piling completely at the works so that only transport and installation damage needs to be repaired on site.

The choice of coating system should make allowance for the possibility that cathodic corrosion protection (CCP) might be installed subsequently. Attention must be paid to the compatibility of the coating materials with the CCP. More information on BAW-approved systems can be found at [www.baw.de]. This list is based on laboratory tests, so local experience should be taken into account. Where coatings have to protect steel sheet piling against sand abrasion, the abrasion value $A_w$ of the coating material governs [184], Guidelines for the Testing of Coating Systems for the Corrosion Protection of Hydraulic Steel Structures [184]). Fenders must be installed to protect the coating from vessel impacts.

2. Cathodic corrosion protection (CCP)

   The corrosion of steel sheet piling below the waterline can be substantially eliminated by cathodic corrosion protection (CCP) with stray current or sacrificial anodes. Additional coating or partial coating is an economical measure and usually indispensable in

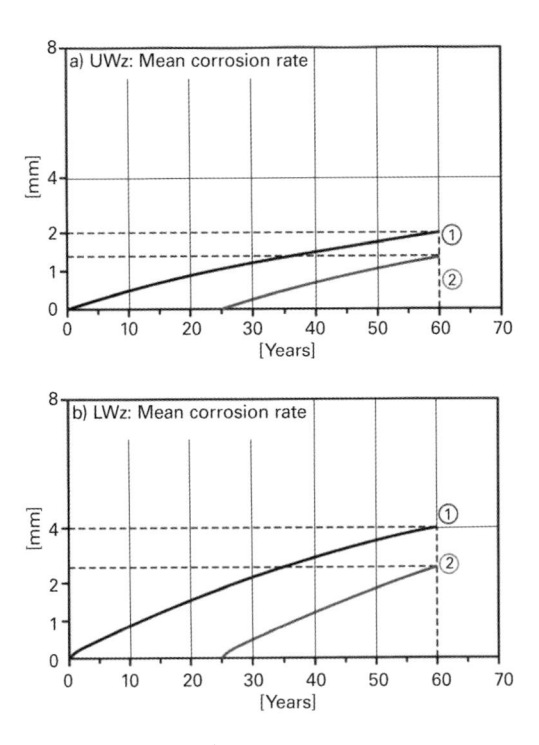

**Fig. R 35-6.** Reduction in wall thickness in seawater: a) underwater zone (UWz), b) low-water zone (LWz); curve 1: uncoated steel, curve 2: coated steel

the interests of good current distribution and lower power requirements.

CCP systems are ideal for protecting those sections of sheet piling, e.g. a tidal low water zone, where renewing any protective coatings or repairing corrosion damage on unprotected sheet piling is impossible or at best complicated and costly.

Sheet piling structures with CCP require special constructional measures [227]. Therefore, any potential subsequent expansion of the CCP system must be taken into account during the planning phase of the sheet piling structure (ZTV-W, [232, 233, 234]). Combined CCP and coating guarantees permanent protection against the soil behind sheet piling structures, too.

3. Alloying additives for steel sheet pile steels

The sheet pile steels according to DIN EN 10248 and the steels given in DIN EN 10025 (structural steels), DIN EN 10028 and DIN EN 10113 (higher-strength fine-grained structural steels) have not been observed to exhibit any differences in corrosion behaviour. Furthermore, the much publicised positive effect of adding small amounts of copper together with nickel and chromium have not been confirmed to date.

4. Corrosion protection by oversizing
   The useful life of a steel sheet pile structure can be extended by oversizing the thickness of the pile sections. In these cases, when analysing the ultimate limit state (ULS) and serviceability limit state (SLS), the mean wall thickness loss values as per Figs. R 35-3 and R 35-4 are to be assumed unless experience of local corrosion rates is available. Unless the client specifies otherwise, the ULS analyses for wall thicknesses expected at the end of the wall's service life which take into account rusting can be assigned to design situation DS-A.
   When checking corrosion, the maximum wall thickness loss values as per Figs. R 35-3 and R 35-4 can be assumed unless values based on local experience are available.

Further information on corrosion protection:

- As far as corrosion attack is concerned, structures in which the back of the sheet piling is only partially backfilled, or not at all, are at a disadvantage.
- Surface water should be collected and drained clear of the rear face of a sheet pile wall. This particularly applies to quays handling aggressive substances (fertilisers, cereals, salts, etc.).
- Free-standing, open piles are exposed to corrosion over their whole periphery, whereas essentially only the outer surfaces of closed piles, e.g. box piles, are exposed. Experience has shown that the inner surfaces of closed piles are protected against corrosion when the pile is filled with sand.
- In the case of free-standing sheet piling, e.g. flood defence walls, the coating of the sheet piling must be extended deep enough into the ground. Subsequent settlement of the soil must be taken into account.
- A sand bed is recommended around round steel tie rods; inhomogeneous fill material must be avoided. A coating on or other protection for the tie rod is essentially superfluous as the design according to R 20, section 8.2.7.3, includes a reserve for corrosion. Once a coating has been applied, it must not be damaged, because such damage encourages pitting. The connections of round steel tie rods must be carefully sealed.
- When backfilling coated sheet piling structures, damage to the coating cannot be completely avoided. However, this risk can be reduced to a negligible level in most cases by using backfill material free from stones.

- Ship berths should always be equipped with timber rubbing strips, fender piles and permanent fender systems to protect sheet piling against constant chafing and scuffing caused by pontoons, ships or their fenders. This protection should ensure that there is never any direct contact between the surface of the sheet pile wall and a ship or pontoon. Otherwise, greater decreases in thickness and further corrosion must be expected – clearly exceeding the values given in Figs. R 35-3 and R 35-4.
- Steel in concrete is a particularly active cathode and can therefore increase corrosion. Increased corrosion should therefore always be expected at the transition from steel to reinforced concrete (e.g. at concrete capping beams); see [93] for details. R 95, section 8.4.4, is to be taken into account for steel capping beams.

## 8.1.9 Danger of sand abrasion on sheet piling (R 23)

When steel sheet piling is used, it must be coated with a system that can permanently withstand the sand abrasion at the location. Assessment of the necessary abrasion resistance of the coating is performed in accordance with RPB 2011 (Guidelines for the Testing of Coating Systems for the Corrosion Protection of Hydraulic Steel Structures, [184]).

## 8.1.10 Shock blasting to assist the driving of steel sheet piles (R 183)

### 8.1.10.1 General

If difficult driving is anticipated, then it is always necessary to check which driving aids can be used in order to prepare the subsoil in such a way that driving progress is economic and pile sections or plant are not overstressed. This will ensure that the required driving depths are reached. Driving aids reduce energy consumption when installing sections and result in the desired reductions in noise and vibrations.

Blasting is frequently used to loosen up rocky soils in advance. The blasting fragments the rock along the planned line of the sheet pile wall in such a way that a vertical, ballast-filled trench is created into which the sheet piles can be – preferably – vibrated. The loosening should extend as deep as the intended base of the sheet pile wall and should be wide enough to accommodate the sheet pile wall section. On either side of the trench the rock remains stable.

In principle, every type of rock can be blasted. But critical for achieving the aims of the blasting are the choice of suitable blasting method, the detonation sequence, the layout of the charges, the type of explosive (waterproof, highly explosive) and, in particular, accurate positioning, spacing and inclination of the blast-holes.

### 8.1.10.2 Blasting method

Trench blasting with short-period detonators and inclined blast-holes, as shown in Fig. R 183-1, has proved to be a successful way of achieving the aims of blasting while causing minimum vibration. Trenches up to 1 m wide can be produced by this method [54].

The blasting sequence begins by drilling stab-holes in a V-arrangement (shown on the right of the longitudinal section in Fig. R 183-1) to relieve the stress in the rock by creating a second free face in addition to the surface of the trench, against which the rock can be thrown. This improves the effect of the blasting and leads to lower vibration than is the case with vertical blast-holes.

Following the V-cut, the other charges detonate in succession at intervals of, usually, 25 ms. The first round of detonations creates space for the rock to be loosened up by the subsequent detonations. In addition, the detonations bounce off each other so that the rock along the line of the blast-holes in the trench is thrown backwards and forwards several times, thus reducing the size of the fragments.

The explosive acts in a V-form (included angle 90°) in the direction of the free face (see cross-section in Fig. R 183-1). In order that the sheet pile wall can be driven to the intended depth despite the narrow tip of the cone of debris, the blast-holes for the explosives must extend below the planned toe of the sheet pile wall. Above the cone of debris, a ballast-

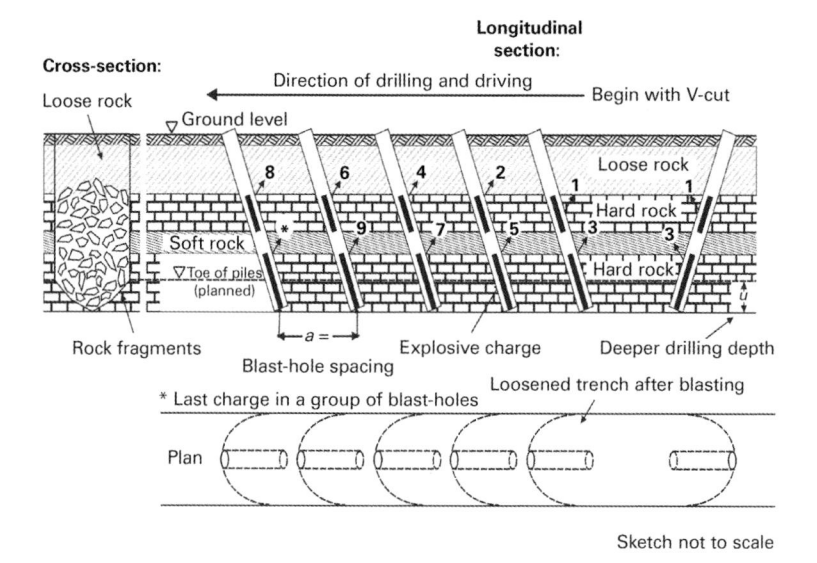

**Fig. R 183-1.** Principle of trench blasting with inclined blast-holes

filled trench ensues into which the sheet piles can be subsequently vibrated without damage.

The spacing of the blast-holes $a$ in Fig. R 183-1 roughly corresponds to the average width of the trench. Common blast-hole spacings for inclined drilling are 0.5–1.0 m

The trench should be only a few decimetres wider than the sheet pile sections being used. Trenches should not be too wide because the base of the sheet piling is assumed to be fixed in the structural analyses.

In rocks with changing strengths, the individual charges should be placed in the hard rock segments in order to achieve an optimum blasting effect.

### 8.1.10.3 Blasting guidance

The following basic advice should be taken into account when planning and carrying out blasting work:

1. Before beginning work it is necessary to investigate the subsoil by means of trial blast-holes and trial blasting. The aim of these trials is to determine the optimum blast-hole spacing and the explosives required. Ultrasound measurements taken prior to and after blasting can be used, for example, to estimate the volume loosened.
2. As vibratory driving of sheet piles causes less damage than impact driving, the sections should be vibrated into the trenches of loosened material. Compaction of the loosened rock during the vibration process is not generally a risk. Impact driving is only permissible in exceptional cases.
3. Sheet pile sections should be vibratory driven immediately after blasting because the geostatic surcharge and, possibly, groundwater flow forces (hydraulic compaction) can partially reverse the loosening effect in the trench.
4. The driveability of the blasted rock and the possibility of installing sections with vibration can be assessed with the results of dynamic penetration tests (DPH). A high number of blows ($N_{10} > 100$) indicates that difficulties will be experienced when driving the sheet pile wall.
5. If difficulties occur while driving the sections, additional blasting will be necessary. The sheet pile sections must be extracted prior to this.
6. Simply splitting the rock permits impact driving only.
7. Evidence of the vibrations due to blasting occurring at the nearest structures should be comprehensively collected with vibration measuring instruments to DIN 45669. These measurements can be carried out within the scope of the contractor's own monitoring procedures.

8. Data on the drilled blast-holes must be documented in blast-hole logbooks. These should record the boundaries of the strata, contact pressure during drilling, flushing losses and water-bearing strata. This information provides details of clefts, voids and narrowing of the blast-hole. The angle and depth of every blast-hole must be recorded. The logbooks should be made available to the chief blasting engineer prior to placing the charges in the holes to ensure optimum placement of the explosive charges in each blast-hole.
9. Spot-checks should be carried out to assess the accuracy of the blast-holes, especially at the start of the drilling. Precise blast-hole measurement systems are available (e.g. BoreTrack). A drilling accuracy of 2% is feasible. The axes of the blast-holes and the sheet pile wall must lie in one plane throughout the intended depth so that the sheet piles are always driven into the rock loosened by the blasting.
10. Only water should be used for flushing because the changing colour of the water and the drilling debris allows changes of strata to be identified, whereas flushing with air produces only a uniform dirty cloud of dust. Another problem is that the air introduced under high pressure is forced into softer strata and existing clefts, which may create undesirable paths. This can, in turn, lead to blowouts during blasting that are remote from the desired blasting point and thus diminish the success of the blasting.
11. Casings should be used when drilling in soft cohesive soils and non-cohesive soils because otherwise there is a risk of material collapsing into the inclined blast-holes.
12. The blasting parameters should be systematically optimised on the basis of the records of the first blasts. Separate blasting trials are then unnecessary. Strict coordination of the drilling and driving works and a constant exchange of information on the work carried out so far are necessary and improve the success of the operations.
13. The maximum amount of explosive per detonation stage should be established on the basis of the vibration calculations prior to any blasting. On no account should this amount be exceeded.

## 8.1.11 Driving steel sheet piles (R 118)

### 8.1.11.1 General

Impact driving of steel sheet piles represents a widely used and proven installation method for sheet piling. However, the work must always be based on sufficient expert knowledge and carried out carefully so that the sections are driven to create a closed wall with the necessary driving accuracy. The more difficult the soil conditions, the longer the piles, the greater the embedment depth and the deeper the later dredging in front of the sheet piling, the greater will be the need to insist on a high degree of workmanship during construction.

Poor results can be expected if long pile sections are driven one after the other to their final embedment depth because the exposed length of interlock is then too short and so adequate initial guidance for the next pile is lacking.

A working basis for judging the behaviour of the ground with respect to pile driving is obtained from boreholes and soil mechanics investigations as well as cone and dynamic penetration tests (see R 154, section 1.5). In critical cases, driving trials at specific locations are recommended to assess the driveability of the sheet piles and the possible deviations of the sections from their intended positions.

The success and quality of the sheet piling installation depend largely on the driving. This presumes that, in addition to suitable, reliable construction plant, the contractor has the necessary experience and is therefore capable of making the best use of the plant and the skills of qualified technical and supervisory personnel. All plant and methods for installing sheet piles must comply with DIN EN 996.

### 8.1.11.2 Pile sections

Steel sheet piles in the form of U- or Z-sections are generally driven in pairs (double piles). Triple or quadruple piles may also have technical and economic advantages in specific cases. The driving of single piles should be avoided if possible.

Sheet piles joined in pairs should, as far as possible, be securely connected to form a unit by crimping or welding the middle interlock. This facilitates handling and driving of the double piles so that piles already in place are hardly dragged down.

In difficult conditions, rocky soils and/or with deep embedment depths, technical reasons might demand the selection of sheet piles with thicker walls or a higher grade of steel than required by the structural calculations. The pile toe and, if necessary, also the pile head may have to be strengthened occasionally, which is particularly recommended for driving in rocky soils or soils with stony inclusions.

Please refer to R 98, section 8.1.7, for the acceptance conditions for steel sheet piles on site.

### 8.1.11.3 Driving plant

The driving plant must be designed so that the pile sections can be driven safely and carefully, at the same time being guided adequately. This guiding is particularly important when driving long piles and for deep embedment depths so as to avoid unacceptably large deviations during driving. The size and efficiency of the driving plant depend on the dimensions and weights of the pile sections, their steel grade, the embedment depth, the subsoil conditions and the driving method selected.

Drop hammers, diesel hammers, hydraulic hammers and rapid-action hammers can all be used for pile driving, also vibratory hammers (R 202, section 8.1.23) or pressing systems (R 212, section 8.1.25).

In the case of drop hammers, the ratio of hammer weight to weight of pile section with driving helmet should be about 1 : 1 in order to achieve the best degree of efficiency. Slow-action, heavy hammers can be used for all applications, especially in cohesive soils. Rapid-action hammers tend not to damage the pile section and are especially well-suited to driving in non-cohesive soils. Hydraulic hammers are also suitable for all applications; their energy per blow can be carefully controlled to suit the respective driving resistance and the subsoil (including rock).

The following factors determine the degree of efficiency and driving performance of pile hammers:

- Total weight of hammer
- Weight of piston, energy of single blow, type of acceleration
- Energy transmission, force transmission (driving helmet and guide)
- Pile sections: weight, length, angle, cross-section, form
- Subsoil (see R 154, section 1.5)

A good degree of efficiency for impact driving is achieved through optimum coordination of these factors.

A driving helmet between pile and hammer is essential for impact driving. Its size and form must be chosen to match the requirements of the plant and the pile.

Suitable preliminary investigations (e.g. numerical prognosis models) are definitely recommended for estimating the effects of driving vibrations and noise expected in each specific case. If necessary, driving trials might prove useful for calibration. DIN 4150 parts 2 and 3 contain guidance on the values for permissible oscillation velocities when driving.

Please refer to R 149, section 8.1.14, with regard to noise control.

### 8.1.11.4 Driving sheet piles

As far as possible, the driving blow is to be transferred to the driving element in such a way that, as regards the resistances, the force is introduced symmetrically and axially. The effect of the interlock friction, acting on one side only, can be countered by adjusting the point of impact.

All pile sections must be guided according to their stiffness and the driving stresses so that their final position is the intended design position. To guarantee this, the pile driver itself must be adequately stable and set up on firm ground, and the leader must always be parallel to the inclination of the pile section. Ensuring the required driving accuracy calls for guiding the pile sections at two points at least, spaced as far apart as possible. A strong lower guide plus spacer blocks for the pile section in this guide are especially important for the accuracy of the driving. The leading interlock of each pile section must also be well guided.

Suitable crane-supported guidance is required where a leader is not being used. Where pile sections are being driven from floating plant, the motion of the plant must be minimised to ensure there is no impact on driving accuracy.

The first sheet pile section for a quay wall must be positioned with special care so that good interlock engagement is ensured when driving the following elements. This is especially important for accuracy when driving in deep water.

In the case of difficult subsoil conditions and greater embedment depths, a driving method with two-sided interlock guidance of the pile sections is required to guarantee accuracy. With continuous driving, if the driving resistance along the line of the piles increases due to soil compaction and the sections therefore deviate from their intended positions, staggered driving should be used (e.g. initial driving with a light hammer and redriving with a heavy one). This should be carried out in panels, with several pile sections being pitched and then driven in a leap-frog sequence (1-3-5-2-4).

Staggered driving is also recommended for constructing sheet piling enclosures.

The heads of U-section sheet piles tend to lean forwards in the driving direction, those of Z-section piles tend to lean backwards. For continuous driving of U-sections, the leading web (in the driving direction) can be bent outwards by a few millimetres to increase the system dimension somewhat. In the case of Z-sections, the leading web can be slightly pressed in a little towards the trough.

In many cases this leaning can be prevented by using staggered or panel driving. If this does not work, taper piles must be used. These must be designed so that the pile section has the same form at both ends and the connecting flange with a welded-in taper section is on the driving direction side (Fig. R 118-1a). This prevents ploughing up the webs in the soil. Chamfering the toes of either U- or Z-section sheet piles can lead to damage to the interlocks and is therefore not permitted.

Maintaining the width tolerance of the pile sections is essential where the grid-line dimensions of certain stretches of sheet pile walls must be ensured with great accuracy. If necessary, make-up pile sections must be driven (Fig. R 118-1b).

Driving can be assisted by blasting to R 183, section 8.1.10, drilling to loosen the soil, soil replacement or water-jetting to R 203, section 8.1.24. Rocky subsoils can be perforated by drilling boreholes and thus relieved to such an extent that the sheet piles can be driven.

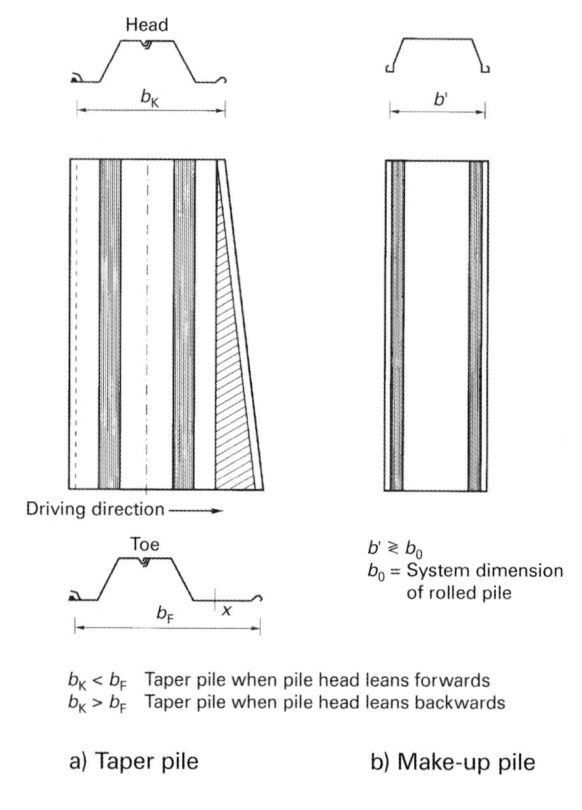

**Fig. R 118-1.** Taper and make-up piles

$b_K < b_F$ Taper pile when pile head leans forwards
$b_K > b_F$ Taper pile when pile head leans backwards

a) Taper pile            b) Make-up pile

$b' \geq b_0$
$b_0$ = System dimension of rolled pile

Choosing hammers and driving methods to match local circumstances and taking greater care when pitching and guiding pile sections reduces the energy required for driving and improves driving progress. The minimum penetration per blow for impact driving should be in accordance with the manufacturer's specification.

## 8.1.12 Driving combined steel sheet piling (R 104)

### 8.1.12.1 General

Combined sheet piling is generally used when dealing with large differences in ground levels. Heavy combined sheet pile walls employing bearing piles with webs > 800 mm deep are often used in quay walls for ships with a very large draught. These quay walls are frequently located in exposed positions as it is in these locations that the natural water conditions enable ships with large draughts to berth. Exposed locations are typically associated with difficult environmental conditions (swell, groundswell, waves, wind, etc.), which must be taken into account during design, preparatory work and construction.

In view of the (usually) long lengths needed for such structures, primarily the bearing piles of steel sheet piling, the greatest possible care must be exercised when driving. This is the only way to ensure that the bearing piles achieve their intended positions and the intermediate piles can be installed without damaging the interlocks.

### 8.1.12.2 Wall forms

Combined steel sheet piling consists of bearing piles between which intermediate piles are inserted as double or triple piles (R 7, section 8.1.4).

Rolled or welded I-sections are frequently used as bearing piles, either as a single pile or welded together to form a double pile with a box section. Additional plates or connectors can be welded on to increase the section modulus. In addition, the bearing piles, e.g. in the form of box piles, can be made from U- or Z-sections welded together with web plates.

LN or SN welded tubes, with welded corner sections or single piles, are also suitable for bearing piles (see R 7, section 8.1.4). In special cases, interlocks are welded flush with the outer edge of the tubes in the tube wall to create an interior interlock.

The use of box piles or double piles made from broad flange or box sheet pile sections, which have a sufficiently high flexural stiffness in both directions and a high torsional stiffness, or tubes should be used when bearing piles >20 m long are required. The increased driving work for such cross-sections must be accepted.

Intermediate piles are generally U- or Z-sections in the form of double or triple piles. Triple piles may require adequate stiffeners for structural and constructional reasons. Other suitable sections may also be considered for intermediate piles if they can properly transmit the loads to the bearing piles and can be installed without damaging interlock connections. This presumes flexibility of the intermediate piles during driving.

### 8.1.12.3 Sheet pile sections for walls

If intermediate piles with interlock types 1, 2, 3, 5 or 6 as per R 67, section 8.1.6, or DIN EN 10248-2 are used, matching interlocks or sheet pile sections are to be attached to the bearing piles by shear-resistant welded joints. The thickness of the inner and outer weld seams of these joints should $a > 6$ mm. The sections of the intermediate piles are to be secured against displacement by welding or crimping their interlocks together.

The use of intermediate piles with interlock type 4 as per R 67, section 8.1.6, calls for bearing piles with this type of interlock. Type 4 interlocks are mounted on either the intermediate or bearing piles prior to driving. When mounting interlocks on bearing piles, there is a risk that they will fill with soil during driving. Declutching is then more likely when driving the intermediate piles. Therefore, in this case the interlocks are welded at the bottom end before driving and filled, for example, with soft bitumen.

If the interlocks are mounted on the intermediate piles, only the upper end is welded in the case of greater embedment depths so that the rotational flexibility of the connection between bearing and intermediate pile is maintained during driving and interlock friction during driving is reduced. The length of the weld must suit the pile length, embedment depth, soil conditions and anticipated driving difficulties. Generally, the length of the weld will be 200–500 mm/m. With especially long piles and/or difficult driving, an additional safety weld at the base of the pile is recommended. With short embedment depths, a shorter transport weld at the top of the piles is generally sufficient.

Care must be taken to ensure that the driving helmet covers the outer connectors when driving the intermediate piles deeper than the top edge of the bearing piles, but leaving enough clearance to bearing piles.

Shear-resistant weld seams ($a \geq 6$ mm) are required if the interlocks are mounted on the bearing piles, which achieves a higher moment of inertia and section modulus at the expense of reduced rotational flexibility of the connection.

If the bearing piles are made from U- or Z-section sections connected to each other via web plates, the plates are to be welded continuously to the U- or Z-section sections on the outside, and inside at the ends of the bearing piles for at least 1000 mm. Weld seam thickness must be $a > 8$ mm. Furthermore, the bearing piles must be stiffened at toe and head with wide plates between the web plates so the driving energy can be transmitted without damaging the bearing pile.

### 8.1.12.4 General requirements placed on sheet pile wall sections

In addition to the otherwise customary requirements in accordance with R 98, section 8.1.7, the bearing piles must be straight with, as a rule, a perpendicular offset no greater than 2‰ of the pile length as per DIN EN 10248-2. They must be free of warp and, in the case of long piles plus deep embedment, must possess adequate flexural and torsional rigidity. Permissible twisting tolerances for bearing piles in combined sheet piling are not given in DIN EN 10248 or comparable standards. Therefore, these must be agreed separately with suppliers or stipulated by the customer upon placing the order.

The top of a bearing pile must be flat and at a right-angle and be so shaped that the hammer blow is transferred over the entire cross-section by means of a well-fitting driving helmet. When driving bearing piles it is important to ensure that the resultant forces of the driving energy and resistance to penetration will act along the centroid axis of the bearing pile in order to prevent it creeping out of alignment due to an eccentric load.

Elements to strengthen the toe of the sheet pile, e.g. "wings", to increase the axial loadbearing capacity must be positioned such that the resultant force of the resistance to penetration acts along the centroid axis of the pile. Otherwise, the pile could creep out of alignment during driving. Furthermore, wings should continue far enough up the pile to provide guidance during driving.

Intermediate piles are to be designed in such a way that they can follow acceptable deviations of the bearing piles from their intended position. Owing to their external interlocks, intermediate piles made from Z-sections can adapt to changes in the position of the bearing piles to a limited extent. Intermediate piles with interlocks on the axis of the wall (U-sections) can only cope with bearing pile driving deviations by lengthening or shortening the section.

The interlocks between intermediate and bearing piles must allow easy threading and must have an adequate load-carrying capacity (see R 67, section 8.1.6). Interlocks must fit together properly and should not be twisted with respect to each other.

### 8.1.12.5 Driving procedure

Bearing piles must be installed such that, after driving, they fulfil the following requirements:

- *Parallelism*: Bearing piles must be installed essentially parallel, i.e. every pile must stand vertically or adhere to the stipulated inclination.
- *Alignment*: The required driving alignment must be achieved.
- *Distortion/twisting*: Distorting and twisting increase the risk of interlock declutching and therefore must be prevented as far as possible.
- *Spacing*: The distance between the piles must be equal over their entire length, matching the system dimension.

These requirements can only be fulfilled by accurately guiding the bearing piles, with double guidance proving expedient. This guidance must be ensured during both the pitching and the driving of the bearing piles. Suitable, heavy, adequately rigid driving plant, suitable for the length and weight of the piles, set up on a firm base and sufficiently stable in itself, should be used when pitching and driving bearing piles.

The following procedures are recommended for driving combined sheet piling in exposed locations such as those with extreme weather conditions and high waves, e.g. on sites in river estuaries:

- Jack-up platforms can be used to support driving plant and driving guides or sufficiently stiff piling frames on supporting piles, which can be extracted afterwards. Floating plant, e.g. working pontoons or so-called half divers are less suitable; it must be assumed that this type of plant will move in the groundswell and waves – if only marginally. Floating plant is only suited to driving combined sheet piling walls in protected locations.
- Leader-guided driving.
- Guide the bearing pile via the piling helmet on the leader so that it is held in the intended position above the horizontal guide. The play between bearing pile and helmet as well as between helmet and leader must be as small as possible during driving.
- The position of jack-up platforms must be checked constantly during driving because driving vibrations can shift the position (e.g. angle).
- Avoidance of distortion in the bearing piles by means of stiff guidance which is mounted independently of tide, swell, groundswell and waves. Guidance also improves the parallelism and direction of the bearing piles (vertical or raking).
- In deep waters, bearing piles can be secured beneath the driving guidance by an accompanying underwater parallel guide or cages. Cages are fixed guide frames placed as deep as possible. It should be noted that extracting cages sometimes requires considerable lifting capacity greatly exceeding the weight of the bearing piles.

All equipment and methods for installing sheet piles must comply with the safety requirements of DIN EN 996.

Bearing piles are initially positioned using vibration plant and then driven to their final depth using heavy driving equipment.

When working in shallow water, guidance can be improved by digging a trench in the base of the watercourse before driving so that the guidance can be used at the greatest depth possible and the driving depth is reduced.

Bearing piles are not driven in a continual order but instead in a "leap-frog" sequence. This ensures that the base of each pile is never driven in soil compacted on one side only. Typically, a driving unit comprising seven bearing piles is driven in the order 1-7-5-3-2-4-6 (large leap-frog sequence). However, the following sequence should be observed at least: 1-3-2-5-4-7-6 (small leap-frog sequence).

Generally, all the bearing piles should be driven to their full depth without interruption.

Subsequently, the intermediate piles can be pitched and driven in succession. If they are installed (partially) with vibration, it is essential to make sure that the piles are always driven deeper. If progress during vibration suggests that an intermediate pile is making only marginal progress, driving must be halted immediately in order to avoid any damage to the threading interlocks and the interlock welds. Therefore, as stated in section 8.1.23.5, driving rates should not be less than 0.5 m/min.

A sufficient wall thickness should be selected depending on the length of the intermediate piles. In the case of intermediate piles $\geq 20$ m long, walls should not be less than 12 mm thick.

In soil free from rocks and suitable for jetting, the bearing piles – and possibly the intermediate piles, too – can be driven with the aid of water-jetting. Jetting equipment is to be installed symmetrically and properly guided in order to counter the lateral deviation of bearing piles from their intended position.

In boulders and hard soil strata, soil replacement is recommended. For building sites on land, a trench can be excavated along the planned line of the wall, and the sheet piling subsequently placed and driven in this trench.

Special measures, especially in the case of difficult soil conditions, may be required in order to prevent damage to the combined sheet piling during driving. For example, soil replacement, pilot holes, etc. may be appropriate.

The design and construction of combined sheet piling calls for planners, design engineers, fabricators and contractors with considerable experience, with personnel skilled in preparatory works and execution of the works in particular.

## 8.1.13 Monitoring during the installation of sheet piles, tolerances (R 105)

### 8.1.13.1 General

During the installation of steel sheet piling, the position and condition of the pile sections must be constantly checked and suitable measurements carried out to ascertain when the intended embedment depth has been reached. Together with the correct starting position, adherence to tolerances must also be checked in intermediate phases, especially after the first few metres of penetration. This should make it possible to detect even small deviations from the design position (angle, out-of-plumb, distortion) or deformations of the pile head so that early corrections can be made and, if necessary, suitable countermeasures initiated.

In difficult driving conditions with obstacles, the penetration, line and position of the pile sections are to be observed frequently and with particular care. If a pile section no longer moves, i.e. unusually slow penetration, stop the driving/vibration immediately. Subsequent piles can then be inserted first. Later, a second attempt can be made to drive the protruding pile deeper.

Observations when driving bearing and intermediate piles are to be carried out for each pile section (see section 8.1.13.5). All records should be made available immediately so that decisions regarding further work or other measures can be made promptly.

Individual pile sections that become very difficult to drive just before reaching their design depth should not be forcibly driven further as there is a risk of damaging the section, interlocks and weld seams. In individual cases, a shorter embedment depth can be accepted if this avoids damage to the section. However, it is important to ensure that, due to the shallower depth of an individual pile section, neither the stability (passive earth pressure, hydraulic heave failure) nor serviceability (e.g. water penetration) of the structure as a whole are compromised.

In the case of unusually large embedment depths for the bearing piles of combined sheet piling, and when the bearing piles are not driven to a final set, it may be necessary to remove the piles, compact the soil and redrive them so that they can transmit the vertical loads assigned to them. The piles may need to be lengthened in some cases.

If observations during driving, such as distortion or out-of-plumb sections, indicate that sections have been damaged, an attempt should be made to inspect the piles by digging them partially free or extracting them to investigate the cause of the distortion or misalignment, e.g. by examining the subsoil for obstacles.

### 8.1.13.2 Declutching, signal transmitters

Declutching on sheet piling arises when the interlock of a section being driven becomes detached from that of a section already driven. Possible causes of declutching include obstacles in the soil and, particularly in the case of combined sheet piling, deviation of driven bearing piles from the intended position both on plan and in the plane of the wall. Therefore, adhering to driving tolerances is the most important precautionary measure for preventing declutching. However, declutching, particularly in the case of combined sheet piling walls, cannot be completely ruled out even when driving is performed carefully and driving tolerances are observed.

Only in exceptional cases is it possible to identify a case of declutching from the driving data (driving energy, driving progress). It is therefore necessary, after the driven pile has been revealed through excavation, to inspect it for declutching, possibly with the help of divers, and to carry out any necessary repairs.

Where the stability or serviceability of a structure could be compromised by declutching (e.g. excavations in open water), declutching detectors

a) Proximity switch    b) Electric contact pin    c) Mechanical spring pin

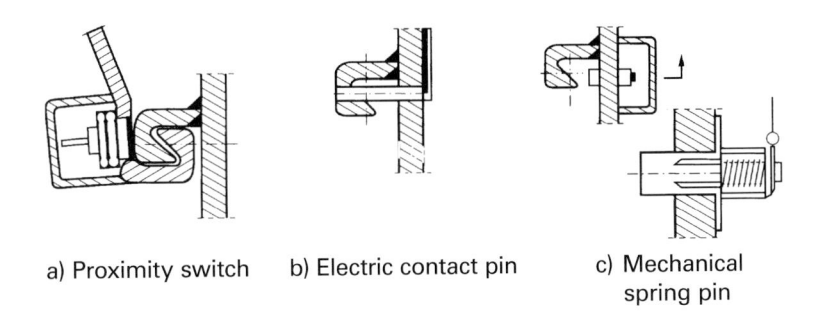

**Fig. R 105-1.** Signal transmitters

(signal transmitters) can be used which make it possible to identify declutching during driving (Fig. R 105-1).

A proximity switch such as the one shown in Fig. R 105-1a can continually measure whether the interlock connection is still intact during driving. If the interlock of the threading pile makes contact with the electric contacting pin shown in Fig. R 105-1b, it is sheared off, which then indicates that the threading interlock is still in the interlock of the driven pile. Similarly, the mechanical spring pin shown in Fig. R 105-1c indicates that the interlock connection is still intact.

In addition, there is also the robust option – tried and tested in practice – of driving a short piece of interlock with a welded-on cable ahead of the interlock to the required depth which thus indicates that the threading interlock is still running in the driven interlock.

## 8.1.13.3 Driving deviations and tolerances

The following tolerances for sheet piling should be included in the calculations at the planning stage for deviations of the piles from the design position in accordance with DIN EN 12063:

±1.0% of the embedment depth for normal soil conditions and driving on land
±1.5% of the embedment depth for driving in water
±2.0% of the embedment depth for difficult subsoil

The deviation is to be measured in the top metre of the pile section.
The deviation of the top of the sheet piling perpendicular to the axis of the wall must not exceed 75 mm for driving on land and 100 mm for driving in water.
The tolerances given in DIN 12063 for sheet piling are not suitable as a measure of the required accuracy of bearing piles in combined sheet piling walls. The driving of bearing piles in a combined sheet piling wall must be

significantly more accurate than the tolerances for sheet piling indicate in DIN EN 12063. Consequently, the tolerances given above for driving deviations do not apply to bearing piles in combined sheet piling walls. The tolerances for such piles must be agreed on an individual basis. Refer to Fig. 6 in DIN EN 12063 for details of establishing tolerances for bearing piles in combined sheet piling walls.

Owing to the risk of declutching, bearing piles in combined sheet pile walls must be straight, vertical (or at the stipulated inclination), parallel to each other, undistorted and at the planned spacings (R 104, section 8.1.12.5).

### 8.1.13.4 Measuring driving deviations

The correct starting position, and also intermediate positions of the pile sections, can be checked by using two measuring devices: one checking the position in the $y$ direction, the other in the $z$ direction. These measurements should generally be prescribed for driving bearing piles in combined sheet piling walls. The tolerances to be maintained when driving combined sheet piling must always be determined and agreed between designer, client and contractor as per DIN EN 12063, taking into account the rolling tolerances stated in DIN EN 10248 and, if required, the additional section tolerances stated by the product supplier, e.g. increase/decrease in width, limits for interlock twist, etc. When constructing quay walls in exposed locations, the agreed tolerances for each bearing pile must be checked after driving and removing guides, not only at the pile head and directly above the lowest possible waterline, but also by divers at the depth of the watercourse bottom.

If the permissible tolerances have been exceeded, the bearing piles must be extracted and redriven. Alternatively, make-up piles must be installed as intermediate piles. The make-up piles are fabricated either in accordance with the measurements or as particularly flexible elements (spring piles). This flexibility can be achieved by removing the middle interlock of an intermediate pile and welding on a half shell.

If the vertical positioning of pile sections is checked with spirit levels, these must be long enough (at least 2.0 m) and used with a straightedge if necessary. The checks are to be repeated at various points to compensate for local irregularities.

### 8.1.13.5 Records

Records of the driving observations are to be kept according to DIN EN 12699, section 10. The list in this standard corresponds to the preprinted forms of DIN 4026:1975, which has been withdrawn. Under difficult driving conditions, the driving energy for the whole driving procedure should be recorded for the first three pile sections and thereafter for every 20th section.

Modern pile drivers record the driving data fully on data media so that the information can be quickly evaluated using special software. When working in difficult driving conditions in changing soils, complete records of all the driving data are recommended.

When installing pile sections using vibratory driving, the penetration achieved should be continually recorded and documented in order to identify any irregularities.

## 8.1.14 Noise control – low-noise driving (R 149)

### 8.1.14.1 General remarks on sound level and sound propagation

The sound emissions of a sound source (e.g. a machine) are characterised by the sound power or sound pressure level measured at a defined distance. The sound power is the sound energy emitted per unit of time. It does not depend on the ambient conditions and hence is a parameter of the sound source, which is assessed according to subjective criteria (e.g. human hearing).

The principal method of assessment is the A-weighted sound power level, which is 10 times the logarithm of the ratio of the sound power to the reference power ($P_0 = 1 \text{ pW} = 1 \cdot 10^{-12} \text{ W}$). The A-weighting reflects a filter for the frequency response of the human ear. The type of assessment is identified either by using the index $L_{WA}$ or by adding a suffix to the unit of measurement dB(A).

The sound power level, being 10 times the logarithm of the ratio of the sound pressure of the sound source to a reference pressure (in air: $p_0 = 20 \text{ μPa}$), is a variable that depends on the measuring distance and the ambient acoustic conditions.

Other important parameters that can be used for estimating the sound emissions of a sound source are the sound distribution over various frequency bands (third-octave, octave, narrow) and possible chronological fluctuations and directional characteristics.

Owing to the logarithmic calculation in acoustics, it is easy to appreciate that doubling the number of sound sources raises the sound pressure level by 3 db(A). However, studies have revealed that the sound pressure level must rise by 10 dB(A) in order for the noise to be perceived by the human ear as "twice as loud". Fig. R 149-1a shows the increase in the total sound pressure level upon superimposing a number of equally loud sources.

Sources with different sound levels have a totally different influence on the overall sound level. If the difference between the sound pressure levels of two sources is greater than 10 dB(A), the quieter source actually has no influence on the overall sound level (see Fig. R 149-1b).

Consequently, measures to control noise can only be effective when, as a first measure, the loudest individual noise levels are reduced. Eliminating

$$L_{PGes} = 10 \cdot \lg \sum_{i=1}^{n} 10^{0.1 L_{p1}}$$

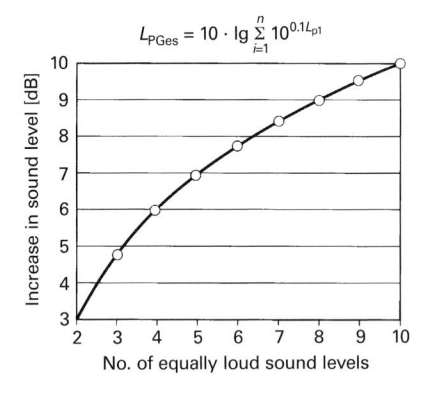

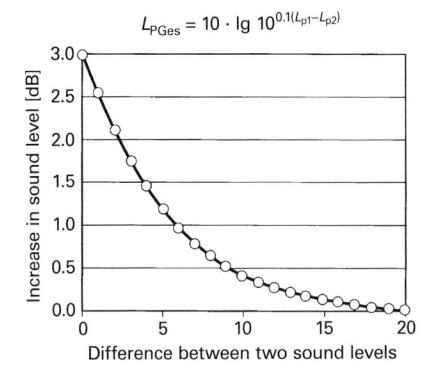

$$L_{PGes} = 10 \cdot \lg 10^{0.1(L_{p1}-L_{p2})}$$

**Fig. R 149-1.** (a) Increase in sound level when several equally loud sound levels are heard simultaneously. (b) Increase in sound level for two levels of different loudness

the weaker individual noise levels only makes a minor contribution to reducing noise.

Given ideal free-field propagation in an infinite semi-spherical space, and on account of the geometric propagation of sound energy, the sound pressure of a point-like sound source diminishes with the distance from the source by

$$\Delta L_P = -20 \cdot \lg \frac{S}{S_0}$$

where

$\Delta L_P$   change in sound pressure [dB]
$S$   distance 1 to sound source [m]
$S_0$   distance 2 to sound source [m]

Therefore, doubling the distance reduces the sound pressure by 6 dB(A). In addition, the sound is attenuated by up to 5 dB(A) over larger distances over natural, uneven terrain on account of air and ground absorption, vegetation and buildings.

On the other hand, it must be taken into account that a single sound reflection on a structure in the vicinity of the sound source or concrete or asphalt surfaces can increase the sound level by up to 3 dB(A) depending on the extent to which surfaces absorb or scatter sound. When there are several reflecting surfaces, each can be substituted by a theoretical mirror sound source with the same loudness level as the original sound source and the resulting increase in level calculated taking account of the calculation rules for the interaction of several sound sources (see Fig. R 149-1a).

When sound propagates over larger distances, it is also necessary to take into account the fact that the decrease in sound level due to meteorological influences, e.g. wind currents and temperature stratification, can have both a positive effect, i.e. in the sense of a greater decrease in sound level, and a negative effect. For example, a positive temperature gradient (increase in air temperature with altitude = inversion) amplifies the sound level because the sound waves are deflected back to the ground at locations more than approx. 200 m away from the sound source. This effect is particularly noticeable over areas of water, which are generally colder than the ambient air (which heats up more quickly) and thus give rise to a positive temperature gradient. A rapid cooling of the land after sunset also results in a positive temperature gradient.

In interaction with ground reflections, the curvature of the sound waves can also have the effect that the propagation of the sound remains limited to a corridor between the ground and the inversion layer, so that the geometric propagation attenuation is reduced by 50%.

The influence of the wind is comparable with that of temperature. Here again, the smaller reduction in sound level in the wind direction is caused by a change in the horizontal wind speed with increasing altitude and the resulting downward deflection of the sound waves. This effect is particularly noticeable on cloudy or foggy days, when the wind with a speed of up to 5 m/s exhibits an essentially laminar current. On the other hand, turbulence and vertical air circulation, caused mainly during the day by solar radiation, can result in a greater reduction in sound level due to the scatter and refraction of the sound waves.

### 8.1.14.2 Regulations and directives for noise control

The following regulations and directives apply to noise control:

- *Allgemeine Verwaltungsvorschrift zum Schutz gegen Baulärm – Geräuschimmissionen* (regulation for protection against construction

noise – noise immissions). Carl Heymanns Verlag KG, Cologne, 1971.
- *Allgemeine Verwaltungsvorschrift zum Schutz gegen Baulärm – Emissionsmessverfahren* (regulation for protection against construction noise – emission measuring methods). Carl Heymanns Verlag KG, Cologne, 1971.
- Council Directive 79/113/EEC of 19 December 1978 on the approximation of the laws of the Member States relating to the determination of the noise emission of construction plant and equipment (OJ L 33, 8 Feb 1979, pp. 15–30, [171]).
- 15th Regulation for Enforcement of the *Bundesimmissionsschutzgesetz* (federal imission control act) of 10 November 1986 (*Baumaschinen-LärmVO*, construction plant noise act, [12]).
- Directive 2000/14/EC of the European Parliament and of the Council of 8 May 2000 on the approximation of the laws of the Member States relating to the noise emission in the environment by equipment for use outdoors (OJ L 162, 3 Jul 2000 pp. 1–78; corr.: OJ L 311, 12 Dec 2000, p. 50).
- ISO 9613-1 (1993): Acoustics – attenuation of sound during propagation outdoors – Part 1: Calculation of the absorption of sound by the atmosphere.
- DIN ISO 9613-2 (1999): Acoustics – attenuation of sound during propagation outdoors – Part 2: General method of calculation (ISO 9613-2:1996).
- 3rd regulation for *Gerätesicherheitsgesetz* (equipment safety act) – *Maschinenlärminformationsverordnung* (machine noise information act) of 18 Jan 1991 (*Federal Gazette* 15. 146; 1992, p. 1564; 1993, p. 704).
- VDI Directive 2714 (Jan 1988): Outdoor sound propagation.

Furthermore, the regulations of federal state legislation regarding noise levels to be observed at night and on public holidays must be complied with, together with the laws and regulations regarding occupational safety (safety regulations UVV, GDG in conjunction with equipment safety act).

The permissible noise exposure in the area of influence of a sound source is stipulated graduated according to the need to protect the surrounding areas from construction site noise. The need for protection results from the actual use of the areas as stipulated in the local development plan. If no local development plan exists or the actual use deviates considerably from the use intended in the plan, the need for protection results from the actual use of the areas.

According to the regulation covering construction noise (*AVV Baulärm*), the effective level generated by the construction machine at the place of exposure may be reduced by 5 or 10 dB(A) if the average daily operating period is less than 8 or 2.5 hours respectively. On the other hand, a nuisance surcharge of 5 dB(A) is to be added when the noise includes clearly audible sounds such as whistling, singing, whining or screeching. If the rating level of the noise caused by construction plant determined in this way exceeds the permissible recommended exposure value by more than 5 dB(A), measures must be initiated to reduce the noise. However, this is not necessary when the operation of the construction plant does not cause any additional dangers, disadvantages or nuisances as a result of not merely occasional effective extraneous noises.

The emissions measuring procedure serves to ascertain and compare the noise levels of construction plant. For this purpose, each item of construction plant undergoes a minutely prescribed measuring procedure during various operating procedures under defined boundary conditions. As part of the standardisation of EU regulations, the noise emissions of construction plant are now stated as sound power levels $L_{WA}$ related to a semi-spherical surface of $1\,m^2$. The sound pressure level $L_{PA}$ is still frequently used, and relates to a radius of 10 m around the centre of the sound source or – in combination with occupational safety regulations – at the position of the plant operator.

Recommended emissions levels have been defined for the certification or use of various items of construction plant. Using the latest equipment prevents these levels from being exceeded. Up until now, no mandatory values have been stipulated for pile drivers.

### 8.1.14.3 Passive noise control measures

Screens represent a method of passive noise control which prevents the propagation of sound waves in certain directions. The screen is lined with sound-absorbent material on the side facing the sound source in order to avoid reflections and so-called standing waves. The effectiveness of a screen depends on its effective height and width and the distance from the source that needs to be screened off. Basically, the screen should be erected as close as possible to the sound source. To remain effective, there should not be any gaps (e.g. open joints) in the screen.

So-called encapsulated solutions employ enclosures or baffles to surround the sound source completely with soundproofing material.

A soundproof enclosure around pile driver and pile can reduce the noise level during driving. However, such enclosures complicate working procedures considerably. As encapsulation makes it impossible to monitor the driving procedure and thus increases the risk of accidents, the use of this type of passive noise control is very limited. In addition, encapsulation is expensive, adds considerable extra weight to the plant and is susceptible to damage.

If some form of a screen cannot be avoided, a U-shaped mat of textured sheeting suspended over the hammer and piling element is preferable and still enables the pile driver operator to watch the driving procedure. A reduction of up to 8 dB(A) can be achieved in the screened direction. Up to now, so-called sound chimneys have been used for smaller pile drivers only.

#### 8.1.14.4 Active noise control measures

The most effective and, as a rule, also least expensive way of reducing noise levels both on and around pile-driving sites is to use low-noise construction plant. For example, compared with hammers, vibratory driving can reduce the sound level considerably. Hydraulic pressing of sheet piles and the installation of bearing piles in pre-drilled holes can certainly be classed as low-noise methods. But the scope for using these low-noise driving methods essentially depends on the properties of the in situ subsoil.

Active noise control measures also include construction methods that ease the driving of piles or sheet piles into the subsoil, thus reducing the energy required for the driving procedure. Alongside water-jetting or drilling or blasting to loosen the soil, these methods also include limited soil replacement in the area of the pile sections plus threading sheet piles into pre-cut, suspension-supported trenches. However, such noise control measures can be used only if they are suitable for the in situ subsoil and construction conditions.

#### 8.1.14.5 Planning a driving site

The aim – right from the planning phase of the driving works – must be to keep the anticipated environmental nuisance of the planned construction site to a minimum. This includes shifting the times of construction work involving high levels of noise to times of the day when it will be less of a nuisance and strictly observing driving breaks during the early morning, midday rest periods and the evenings. If this means accepting a reduction in daily output, then this must be taken into consideration in the tender.

Screening measures are particularly cost-intensive and frequently ignored in the pricing.

### 8.1.15 Driving of steel sheet piles and steel piles at low temperatures (R 90)

At temperatures above 0 °C steel sheet piles of all steel grades can be driven without hesitation. If driving has to be carried out at lower temperatures, then special care is required when handling and driving the pile sections.

In easy driving conditions, driving is still possible down to temperatures of about $-10\,°C$, particularly when using S 355 GP and higher steel grades. However, fully killed steels to DIN EN 10025 should be used when difficult driving with a high energy input is expected, and when working with thick-walled sections or welded pile sections.

At temperatures below −10 °C, steel grades with enhanced cold workability must be used.

## 8.1.16 Repairing interlock declutching on driven steel sheet piling (R 167)

### 8.1.16.1 General

Interlock declutching can occur during driving of steel sheet piles or might be due to other external actions. However, the better the recommendations for the design and construction of sheet piling structures are observed with care and diligence, so the more the risk is reduced. Please refer to the following recommendations: R 34, section 8.1.3, R 73, section 7.4, R 67, section 8.1.6, R 98, section 8.1.7, R 104, section 8.1.12, and R 118, section 8.1.11.

However, interlock declutching, especially on combined sheet piling, is still possible even when strictly observing these recommendations. Interlock declutching in quay walls in tidal areas is particularly critical as these walls are regularly exposed to excess water pressures during low water, which can wash out the backfilling via the gaps. Cavities then form in the backfill which, due to the arching effect, can remain stable for long periods of time, but then collapse unexpectedly, causing subsidence in the port operations area.

Therefore, sheet piling in water must be exposed by excavating and checked by divers for interlock declutching. Any damage found must be repaired. Repairs with steel offer many options in line with the respective boundary conditions.

### 8.1.16.2 Repairing interlock declutching

If observations during driving suggest interlock declutching over an extended area of the wall and the sheet piling cannot be extracted due to lack of time, repairs in the form of grouting a large area of the soil behind the wall is a popular option. This prevents the soil from escaping through the wall and thus causing subsidence behind the wall. High-pressure injection has proved particularly effective for grouting soil (see Fig. R 167-5). Afterwards, the damaged areas can be permanently sealed with steel sections or plates attached to the front of the wall.

Individual cases of interlock declutching are usually only discovered during assessments by divers after the soil in front of the wall has been excavated. Declutching must be repaired in such a way that not only is the damaged area sealed, but that the wall can continue serving its purpose without any restrictions. This is especially important for sheet pile walls consisting of only U- or Z-piles.

The sealing of interlock declutching depends, above all, on the size of the opening and on the sheet pile section.

Repair work is generally carried out on the water side. Smaller interlock openings can be closed with wooden wedges. Large openings can be temporarily sealed, e.g. with a rapid-setting material, such as rapid-hardening cement or a two-part mortar, both placed in sacks. However, in order to seal the damaged areas permanently, it is necessary to close off the damaged area down to at least 0.5 m, but preferably 1.0 m, below the bottom of the harbour floor (R 36, section 6.7) by attaching steel sections or plates to the water side of the sheet piling. In addition, the damaged area must be concreted in order to prevent sand from being washed out at a later date, resulting in subsidence behind the wall. The concrete used may need to be reinforced to protect it against vessel impacts. Installing an additional protective layer, e.g. of ballast, on the bottom of the watercourse is recommended in front of the damaged section when the soil in this area is at risk of scouring. Otherwise, the repairs should extend below the calculated scouring depth.

Most of this work has to be carried out underwater and therefore always requires diver assistance. Very high demands are therefore placed on the technical abilities and reliability of the divers. In difficult conditions the plates can be fitted in the dry using underwater enclosures, which are held against the wall by water pressure.

The smoothest possible sheet piling surface on the water side is to be strived for. For this reason, protruding bolts, for example, must be burned off after the damaged area has been concreted and they have fulfilled their function as a formwork element. The steel plates and the sheet piling are to be joined to the concrete with anchors in the form of splayed fishtails.

The solutions shown in Figs. R 167-1 to R 167-5 for repairing underwater declutching have proved successful in practice, but do not claim to be exhaustive.

Interlock declutching on combined sheet piling is repaired, if possible, by driving further intermediate piles behind the wall as per Fig. R 167-4 or on the water side as per Fig. R 167-3. In the former, e.g. in accordance with Fig. R 167-4, the excess pressure behind the wall presses the pile section against the intact parts of the sheet piling during dredging on the water side.

The repairs shown in Figs. R 167-1 to R 167-3 presume that no soil is washed out through the damaged area, e.g. because there is no excess water pressure behind the wall or the damaged area has already been

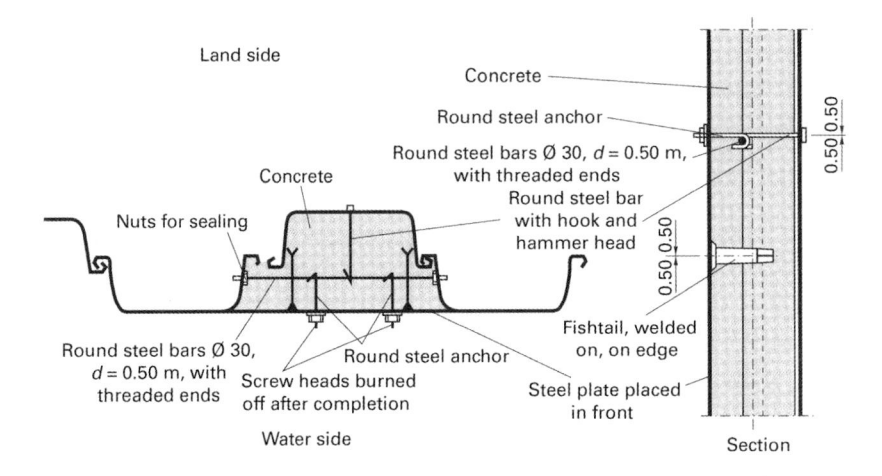

**Fig. R 167-1.** Example of repairing damage at a small opening in a wall

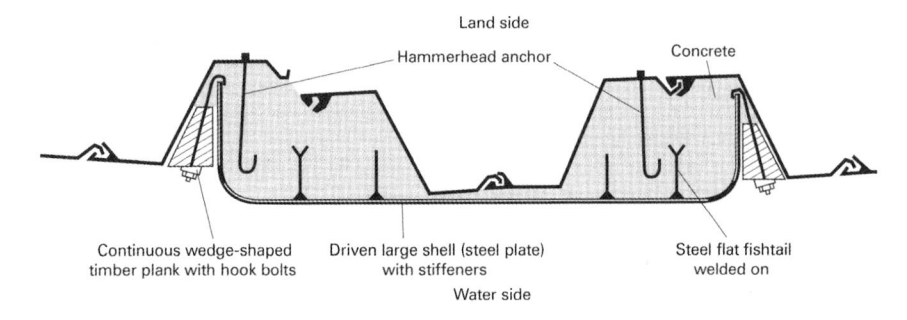

**Fig. R 167-2.** Example of repairing damage at a large opening in a wall

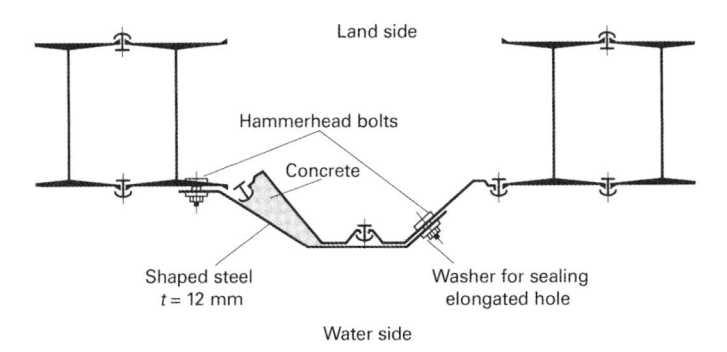

**Fig. R 167-3.** Example of repairing damage at an opening in combined sheet piling

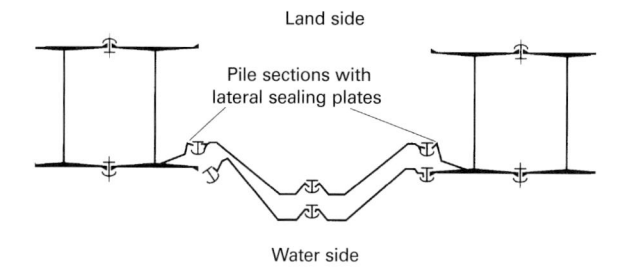

**Fig. R 167-4.** Example of repairing damage by driving a sheet pile section behind the wall to close off an opening in combined piling

successfully sealed temporarily. However, this is normally not the case, meaning that the repair shown in Fig. R 167-5 must take place in two steps. The soil behind the wall must first be grouted using high-pressure injection, with the damaged area plugged using a generously dimensioned frontal embankment of dumped material so that the high-pressure injection suspension does not escape through the damaged area. Afterwards, the dumped material can be dredged away and the declutching permanently sealed with concrete and tailored steel parts reaching below the design harbour bottom.

When driving in front of the wall, suitable measures must be taken to ensure that the base of the pile section is always pressed against the bearing pile; in other words, it has to be located as close to the bearing pile as possible. Before dredging takes place, the head of the pile section must be fixed to the bearing pile, e.g. with hammerhead bolts. The elevation of the first dredging cut must be coordinated with the bearing capacity of the pile section between this upper fixed support and the lower, more flexible earth support. After driving, the pile section must be fixed to the bearing pile in the exposed area etc.

Constant investigations of the repaired areas by divers are necessary during dredging operations. Any local leaks can be sealed by additional injections.

### 8.1.17  Armoured steel sheet piling (R 176)

#### 8.1.17.1  Necessity

Ever larger vessels, convoy traffic and more powerful engines have resulted in increased operational requirements for waterfronts in inland ports and on waterways. To prevent damage, sheet pile waterfront structures must therefore have front faces that are as flat as possible

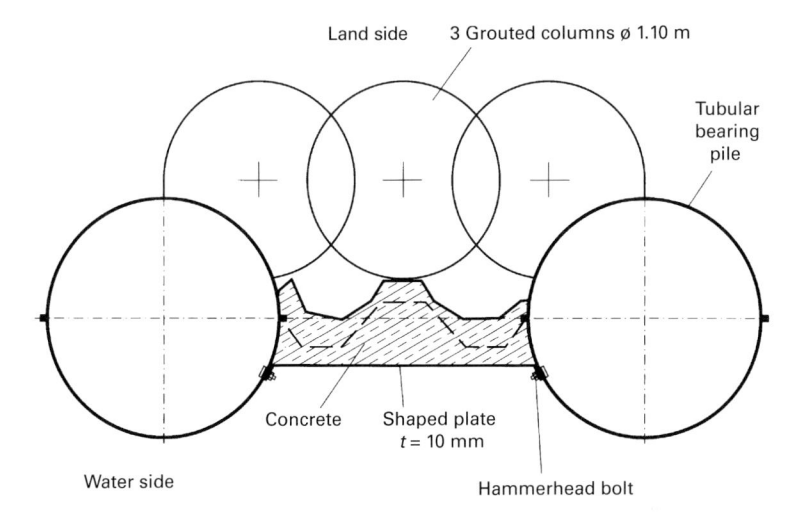

**Fig. R 167-5.** Example of repairing declutching in combined sheet piling with high-pressure injection and a plate attached in front of the wall

(R 158, section 6.6). This can be achieved by welding plates in or across the troughs of the sheet piles to strengthen the structure (Fig. R 176-1). The reinforcement creates a uniformly flat waterfront structure with less elasticity than a wall without such strengthening.

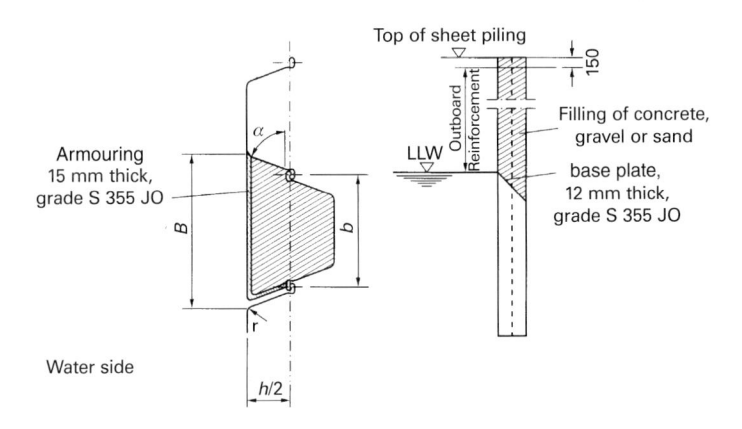

**Fig. R 176-1.** Armouring to a U-section sheet pile

335

### 8.1.17.2 Applications

Owing to the technical input and the costs, armouring is only recommended for stretches of waterfront that are particularly exposed to waterborne traffic loads. In inland ports, these are waterfronts with very heavy traffic with push lighters and large motor vessels, waterfronts at changes of direction and the guidance structures at lock entrances.

### 8.1.17.3 Elevation

Sheet piling reinforcement is required in the area between the lowest and highest possible water levels where vessel impacts are likely (Fig. R 176-1).

### 8.1.17.4 Design

Armouring dimensions depend primarily on the width $B$ of the sheet pile trough. This is determined by the system dimension $b$ of the wall, the angle $\alpha$ of the web of the pile, the section depth $h$ and the radius $r$ between the web and flange of the sheet pile (Fig. R 176-1).

### 8.1.17.5 Pile shape and method of armouring

Armouring differs according to section form (U or Z). In addition, it is necessary to distinguish between plating factory-set before driving and armouring attached on site by welding on after driving.

In the case of walls made from Z-sections and armouring attached on site, the plates extend over the full width of two sections right up to the interlocks. This reinforcement projects beyond the section and so also protects the interlocks (Fig. R 176-2). Welding on the armouring in the workshop cannot be recommended for Z-sections because the armouring stiffens the piles to such an extent that, during driving, they can no longer compensate for unavoidable driving deviations between neighbouring double piles. With walls made from U-sections, the back of the piles and the free leg of the forward pile can deform elastically.

When installing the armouring on site, pressing and adjustment work will be unavoidable with both types of section. Walls made from U-sections with armouring fitted in the shop are better when it comes to driving and shipping operations. This type of wall armouring creates a completely flat surface (Fig. R 176-2).

### 8.1.17.6 Factory-attached armouring for U-sections

When attaching armouring to U-sections in the workshop, the straight or bent strengthening plates (Fig. R 176-3) are welded to the rearward pile interlock or on the flange of the forward pile (Fig. R 176-2). In addition, the interlock of the double pile must be welded to create a rigid connection. Only then can the connecting welds of the armouring survive the driving procedure without being damaged.

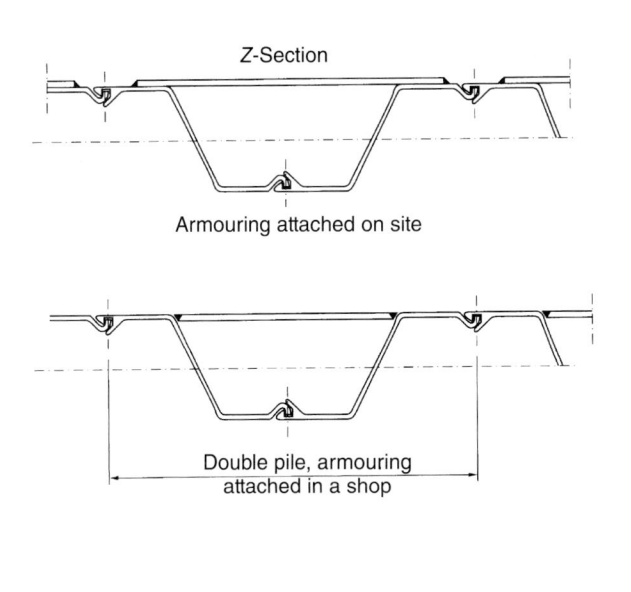

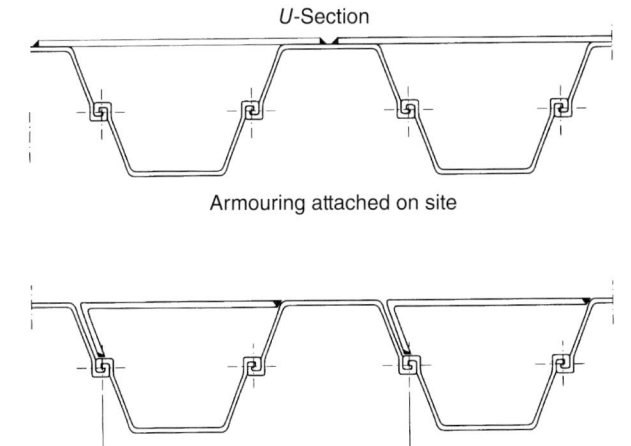

**Fig. R 176-2.** Armouring for U- and Z-section sheet piles

Strengthening plates can be straight or bent (Fig. R 176-3). The gap between the piles is approx. 20 mm wide with straight welded plates so that, related to a system dimension of 1.0 m, up to about 98% of the wall is closed. In the case of walls with bent plates, the gap is

337

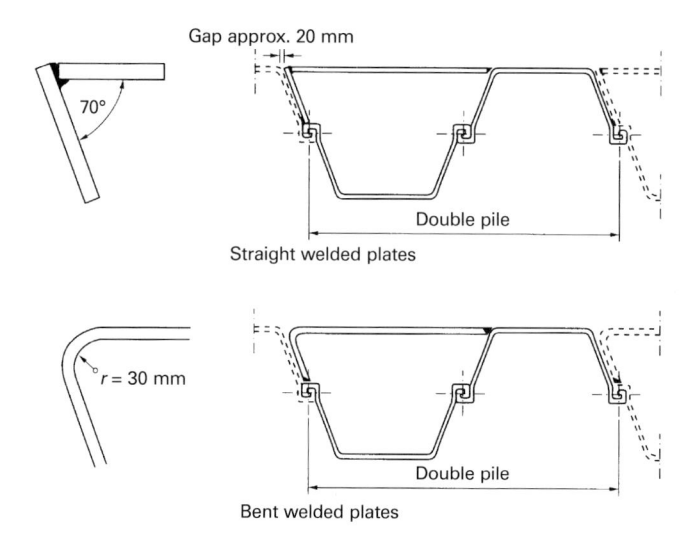

**Fig. R 176-3.** Factory-attached armouring using straight or bent welded plates

wider due to the minimum radius that must be observed during cold forming.

### 8.1.17.7 Dimensions of the armouring

The plate thickness required for the armouring is derived from the width of the sheet pile trough. As this is always larger than the flange width of the forward pile, the strengthening plates must be thicker than the flange of the pile. To avoid plate thicknesses $\geq 15$ mm, the space between armouring and pile can be filled.

Armouring is not taken into account in the structural design of the wall.

### 8.1.17.8 Filling behind the armouring

When filling the space between armouring and pile, a base plate is generally welded in to retain the filling (Fig. R 176-1). Sand, gravel or concrete are used as the filling material.

### 8.1.17.9 Ladders and mooring equipment

Sheet pile armouring is generally interrupted around ladders and recessed bollards. If a largely flat surface is also required here, the ladder pile can be armed with recessed footholds as per Fig. R 176-4, or a continuous recessed foothold box can be included as per Fig. R 176-5.

**Fig. R 176-4.** Armouring with recessed footholes

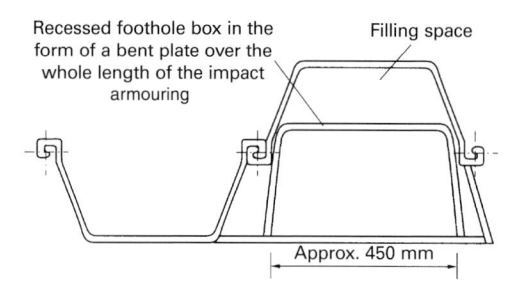

**Fig. R 176-5.** Impact armouring with recessed foothole box

The positioning of recessed bollards in an armed wall is shown in Fig. R 176-6. This solution complies with the requirements of R 14, section 6.11, and R 102, section 5.13.

#### 8.1.17.10 Cost of reinforcement

The additional costs of armed compared with not armed sheet piling depend primarily on the ratio of the length of the reinforcement to the entire length of the sheet piles and on the sheep pile section. Reinforcement increases the cost of supplying wall materials by 25–40%. In the case of U-sections, it is less expensive and technically better to plan armouring from the very beginning and have it attached in the works.

### 8.1.18  Design of piling frames (R 140)

#### 8.1.18.1  General

Driving operations in water can be carried out from either a jack-up platform or a pontoon. Both solutions are mobile and can be moved to match construction progress. However, their usage requires a navigable water depth. If the water is too deep, a driving platform can be created with dumped materials. One alternative involves piling frames that provide the pile driver with a stable base and can be moved to match the progress of the works.

#### 8.1.18.2  Construction and planning boundary conditions

Piling frames are working platforms erected on steel, timber or concrete piles (Fig. R 140-1). They must be designed and built in such a way that they can accommodate the plant and equipment required for driving operations but are still economical. The following boundary conditions should be borne in mind for their design and construction:

Elevation – double pile with impact armouring and recessed bollards

Top of double pile

Section A–A

Mount and weld retaining plate after driving

Section B–B

Recessed bollard area

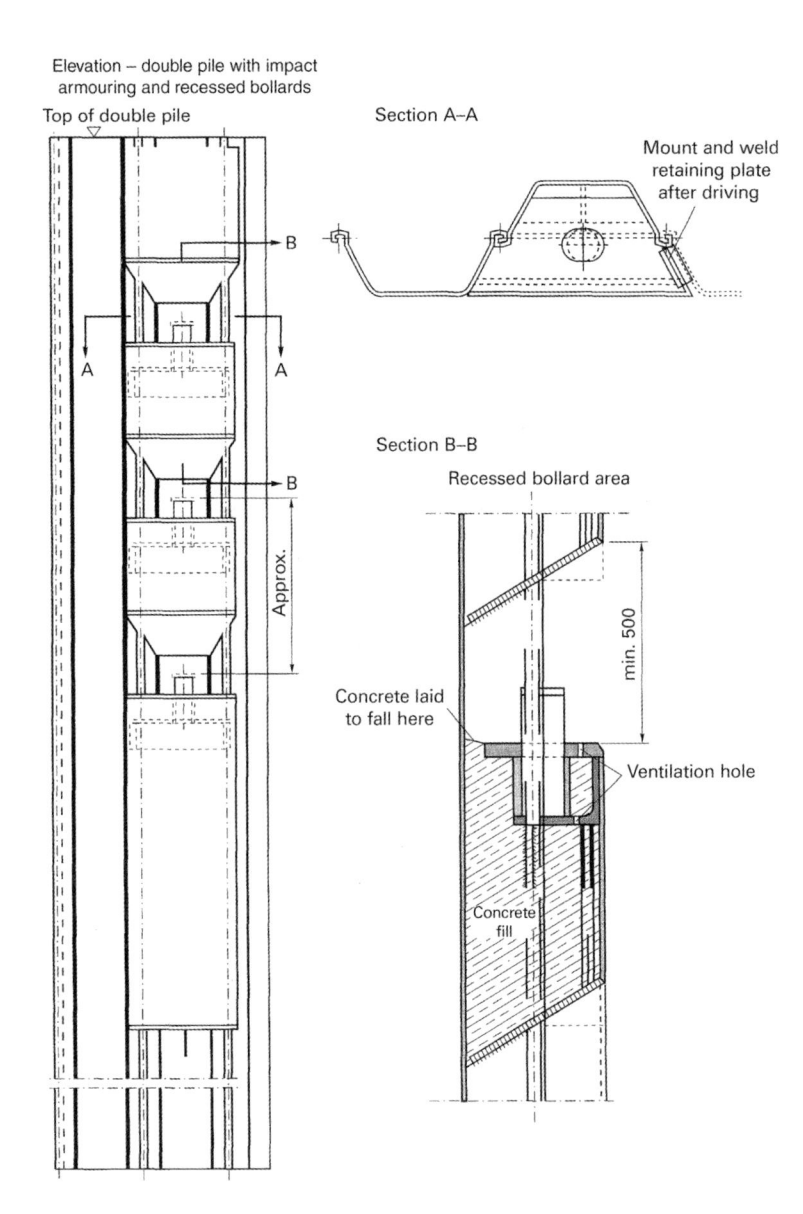

min. 500

Concrete laid to fall here

Ventilation hole

Concrete fill

Approx.

**Fig. R 176-6.** Reinforcement with recessed bollards

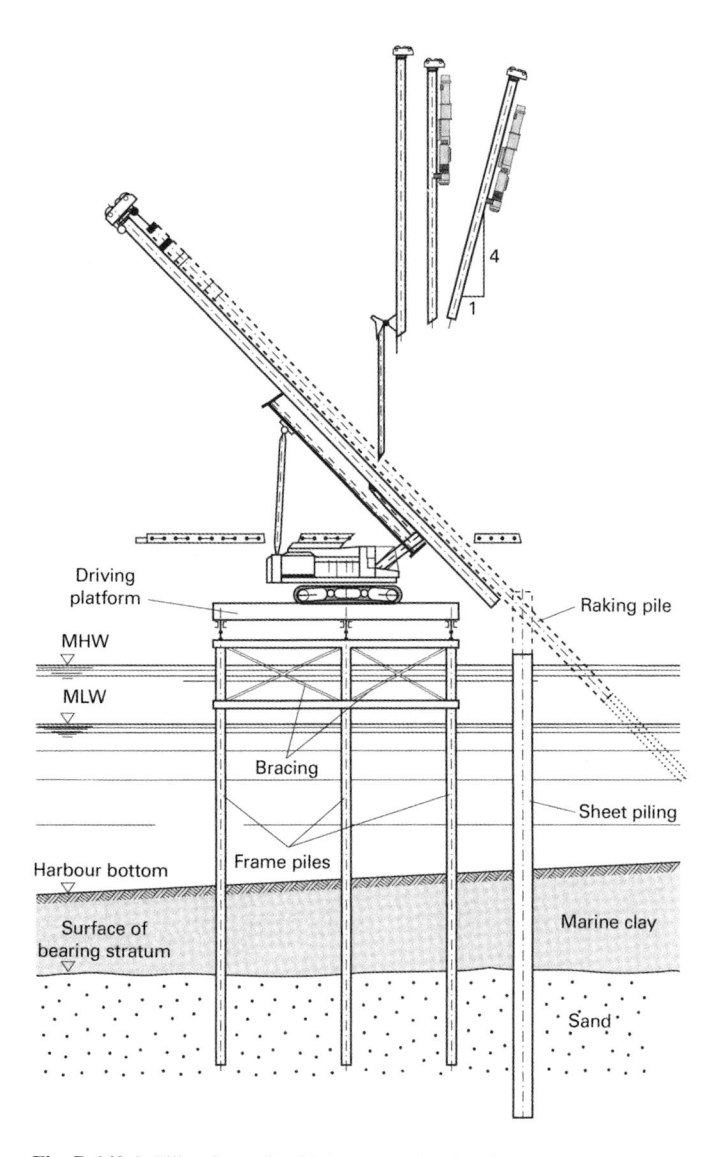

**Fig. R 140-1.** Piling frame for driving vertical and raking piles

1. Piling frames can be driven from a floating pontoon or barge although the driving deviations to be expected under such conditions must be taken into account during design.
2. The length of the frame is to be selected such that it can be dismantled from an area of the waterfront structure already driven and re-erected ahead of the driving work without interrupting driving operations.
3. The frame foundation piles are extracted in the dismantling area and redriven ahead of the current driving operations. If foundation piles cannot be extracted, even with the help of jetting, they must cut off beneath the planned harbour bottom depth (taking into account dredging tolerances and later deepening work). The frame must be designed to handle the loads arising from the extraction of the frame piles.
4. Simple structural systems and configurations should be used for the piling frame so that the components can be reused many times.
5. If the harbour bottom consists of soil at risk of scouring (sand, silt), potential scouring must be taken into account when determining the embedment depth of the frame piles. As the scouring depth around pile groups in tidal currents is particularly difficult to estimate in advance, the scouring must be regularly monitored during construction by means of sounding. If scouring is found to be deeper than that assumed when determining the embedment depth of the frame piles, the scouring must be filled with soil not susceptible to scour (e.g. gravel, stones, cohesive soil).
6. The frame piles should be driven well clear of the permanent structure so that they are not dragged down by the driving of the structural piles. The clearance required depends on the strata of the subsoil and the structural pile types to be used. In cohesive soils, cavities may remain after the frame piles have been extracted. These cavities are to be filled if it is thought that they could compromise the stability or serviceability of the final structure.
7. Piling frames built close to the waterfront can consist of a row of piles in the water and supports on the bank for the platform beams. Piling frames in water stand on two or more rows of piles. Piles that form part of the final structure can also be used.

### 8.1.18.3 Loading assumptions

The safety measures and requirements and operating conditions, etc. as per DIN EN 996, section 4, apply to piling frames.

Together with the loads from the pile driver and the platform, and crane if present, the loads due to currents, waves and ice must be taken into account when designing the supporting piles and frame bracing. If the piling frame is not safeguarded by additional measures (safety dolphin) against, for example, contact with pontoons or other floating construction plant, additional loads due to vessel impact and, where applicable, line pull forces of 100 kN each applied in the most unfavourable position must also be considered in the design.

## 8.1.19 Design of welded joints in steel piles and steel sheet piles (R 99)

This recommendation applies to welded joints in steel sheet piles and all types of driven steel pile.

### 8.1.19.1 General

The design of the joints is carried out according to DIN EN 1993-1-8. Design and fabrication must comply with the requirements of DIN EN 12063 and DIN EN 12699, or DIN EN ISO 3834 in the case of more demanding requirements. Working areas for welding on site must be protected from wind and weather. Welded joints are to be thoroughly cleaned, kept dry and, if necessary, preheated beforehand.

### 8.1.19.2 Materials

Sheet piling grades according to R 67 and steels according to DIN EN 10025 are suitable for welding. Welding suitability is covered in R 67, section 8.1.6.4, and must always be verified by inspection certificate 3.1B to DIN EN 10204, which indicates both mechanical and technological properties as well as the chemical composition (R 67, section 8.1.6.1).

Filler metals are to be selected by the welding engineer of the welding contractor licensed to perform the work, taking into account the recommendations of the sheet piling and steel pile supplier. General basic electrodes or filler metals with a high degree of basicity should be used (filler wire, powder).

### 8.1.19.3 Classification of the welded joints

A butt joint is intended to replace the steel cross-section of a pile or sheet pile as fully as possible. The percentage of effective butt weld coverage is, however, contingent on the type of section, the offset of edges at the joint ends and on the prevailing conditions on site (Table R 99-1).

If the butt weld cross-section steel does not match the cross-section of the pile or sheet pile, and if the full cross-section is required for structural reasons, splice plates or additional sections must be used to achieve the full cross-section.

The effective butt coverage is expressed as a percentage and is the ratio between the butt weld cross-section and the steel cross-section of the pile or sheet pile. Possible butt coverage values can be found in Table R 99-1.

**Table R 99-1.** Effective butt joint coverage expressed as %

| Type of pile of sheet pile | | Effective butt coverage as % allowance | |
|---|---|---|---|
| | | in the workshop | under the pile driver |
| a) Tubular sections, calibrated joint ends, root welded through | | 100 | 100 |
| b) Piles of I-sections, box sheet piles, cross-section reduction with material removed from throats | | 80–90 | 80–90 |
| c) Sheet piles | Single sheet piles | 100 | 100 |
| | Double piles, interlock with one-sided welding only U-sections Z-sections | 90 80 | ~80 ~70 |
| d) Box piles made from individual sections Individual sections jointed then assembled Box pile to be jointed | | 100 70–80 | 50–70 |

### 8.1.19.4 Making weld joints

1. Preparation of joint ends
   Each section to be welded should be cut in one plane at right-angles to the axis; an offset in the joint is to be avoided.
   Special attention is to be paid to ensuring a good fit between the cross-sections and, in the case of steel sheet piling, to preserving free movement in the interlocks as well. Differences in width and height between pieces to be welded should not exceed ±2 mm so that offsets in the welded edges will not exceed 4 mm.
   It is recommended that hollow piles assembled from several sections should first be fabricated in the full length required and then cut into working lengths after having been suitably coded (e.g. for transportation, driving, etc.).
   The ends intended for each butt joint are to be checked for laminations over a length of about 500 mm.
2. Edge preparation for welding
   In the shop, butt welds are generally in the form of V- or Y-groove welds. Both edges of the butt joint are to be suitably prepared.

If a butt joint must be welded in the field on driven steel sheet piles or steel piles, the top of the driven section must first be trimmed as required by R 91, section 8.1.20. The extension piece is to be prepared for a butt weld with or without backing weld.

3. Welding procedure

All accessible sides of the butted sections are to be fully connected. Wherever possible, the roots should be gouged and sealed with backing welds.

Root positions that are inaccessible require a high degree of accuracy when fitting together the sections and careful edge preparation.

The proper welding sequence is to be determined so that loads from the welding process do not overlap with those occurring when the structure is in service.

### 8.1.19.5 Special details

1. Wherever possible, joints are to be positioned in a lowly stressed part of the cross-section. The joints of adjacent sections must be offset by at least 1 m.
2. When preparing I-sections for welding, the throat areas of the web are to be drilled out in the shape of a semi-circle facing the flange with a diameter of 35–40 mm so that the flange can be fully welded through with a backing weld. After welding, the edges of these openings must be machined to remove any notches. Run-on and run-off plates must be provided for the flange welds in the vicinity of the butt joint in order to achieve a clean termination at the flange. After removing the plates, grind the edge of the flange to remove any notches.
3. If flange splice plates are required to achieve the butt coverage for structural reasons, the following rules must be observed:
   - Splice plates should be no more than 20% thicker than the parts of the section being spliced, and should never exceed 25 mm thick.
   - The width of the plates should be such that they can be welded to the flanges on all sides without end craters.
   - The ends of the plates should be tapered to 1/3 the plate width at a slope of 1 : 3.
   - Before the plate is positioned, the butt weld is to be ground flush.
   - Non-destructive tests must be completed before the splice plates are mounted.
4. If butt joints in service are not subjected to predominantly static loads within the meaning of R 20, section 8.2.7, splice plates over the joints are to be avoided.
5. Only killed steels should be used where butt joints are planned, e.g. for reasons of transportation or driving methods.

6. Butt joints under the pile driver are to be avoided as far as possible for economic reasons and because unfavourable weather conditions could have a negative effect on the weld.
7. If sheet piling structures include openings for welding through which the soil could be washed out, such openings must be sealed in a suitable manner (see R 117, section 8.1.21).

## 8.1.20 Cutting off the tops of driven steel sections for loadbearing welded connections (R 91)

If structural welded connections (e.g. butt joints, welds for fittings, etc.) are required at the tops of driven steel sheet piles or steel piles, these may not be located in areas with driving deformation so as to avoid any embrittlement having an effect on the load-carrying capacity of the welded connections. Therefore, the top ends of the piles must be cut off at a point below the extent of the deformation. Alternatively, the welded joints should be positioned outside the deformed area.

## 8.1.21 Watertightness of steel sheet piling (R 117)

### 8.1.21.1 General

Walls made from steel sheet piles are not completely watertight due to the necessary play in the interlocks. Interlocks threaded in the workshop (W interlocks) are typically more watertight than those threaded on site (B interlocks), which are at least partly filled with soil during driving. Generally, a self-sealing effect due to corrosion with incrustation plus an accumulation of fine particles in the interlocks in sediment-laden water can usually be expected over the course of time. If this is not sufficient, the interlocks can be sealed (section 8.1.21.3).

Annex E of DIN EN 12063 can be used to assess the permeability of sheet pile interlocks. The suppliers of steel sheet piles can provide information on the permeability for the various synthetic seals according to section 8.1.21.3. If very strict requirements have been specified for the watertightness, the interlock seals must be verified in practical tests. Should there be a wide scatter of results, this must be taken into account when determining permeability.

In soil with low permeability, unsealed interlocks function like vertical drains.

If there is fine-grained, non-cohesive soil such as fine sand or coarse silt behind the sheet piling, this can be easily washed out through unsealed interlocks. This is especially the case when, behind enclosures to excavations, there is a high excess water pressure and/or the wall is exposed to varying loads due to waves. In these cases, specific measures are usually required for sealing the interlocks.

### 8.1.21.2 Assisting the self-sealing process

The self-sealing effect of sheet pile interlocks can be encouraged for walls standing in open water when, e.g. for walls around construction sites, excess water pressure acts intermittently on one side of the wall. To this end, an environmentally friendly suspended substance, e.g. boiler ash, is poured into the water outside the excavation.

When the maximum possible pumping capacity is used to pump water out of an excavation, there is a difference in water levels between inside and outside. The resulting excess water pressure presses the interlocks together. As the water flows through the interlocks, so the suspended matter is deposited in the interlocks, thus reducing permeability.

If sheet piling can move due to wave action or swell, however, the sealing effect is not permanent because the sealant is crushed between the interlocks and washed out. Similarly, a permanent seal cannot be created using suspended matter when the direction of the excess water pressure alternates.

### 8.1.21.3 Artificial seals

Sheet pile interlocks can be sealed with artificial seals both before and after installation.

1. Methods for sealing sheet piles before driving:
    a) Filling the interlocks with a durable, environmentally compatible, sufficiently plastic compound – for W interlocks in the shop and B interlocks on site. A noticeable improvement in watertightness is achieved by factory-fitting an extruded polyurethane seal to B interlocks.
    b) The use of this method allows B interlocks that are no longer accessible after driving, e.g. areas below the bottom of an excavation or watercourse, to be sealed as well. Please refer to c) concerning the position of the sealed joint.
    c) With both sealing methods, the interlock watertightness achievable depends on the excess water pressure and the method of driving. Impact driving places little stress on the seal, since the movement of the pile in the interlock takes place in one direction only. Vibratory driving places greater stresses on the seal. Therefore, a complete loss of watertightness due to friction and heat cannot be ruled out.
    d) Interlock joints of W interlocks are welded watertight, either at the works or on site. In order to avoid cracks in the watertight seam during driving, additional seams are required, e.g. on both sides at the head and base of the pile section, as well as counter-seams in the area of the sealing seam. The sealing seam must be placed on

the correct side of the sheet piling, e.g. on the air/water side for dry docks and navigation locks.

e) One method known from the rehabilitation of contaminated sites can be used for sealing dykes along waterways with the highest demands on watertightness. In this method, lined holes with a diameter of about 0.1–0.3 m are drilled at a spacing to match the system dimension of the section being used and then filled with a slurry. After the casing has been removed, and before the slurry hardens, the sections welded to the W interlocks are driven.

2. Methods for sealing sheet piles after driving:
    a) Caulking the interlock joints with wooden wedges (swelling effect), rubber or plastic cords (round or profiled) or a caulking compound capable of swelling and setting, e.g. fibres mixed with cement.
    The cords are tamped in with lightweight pneumatic hammers and a blunt chisel. The caulking work can also be carried out on water-bearing interlocks. Caulking generally works better on B interlocks rather than crimped W interlocks.
    Soil particles must be removed from interlock joints prior to caulking.
    b) The interlock joints are welded watertight. As a rule, W interlocks are already welded tight at the works (see section 8.1.21.3, 1c), B interlocks on site.
    Only dry and properly cleaned interlocks can be welded directly. Water-bearing interlocks must therefore be covered with steel plates or sections, which are welded to the sheet piling with two fillet welds. Using this method, a fully watertight sheet pile wall can be achieved.
    c) On the completed structure, plastic sealing compounds may be placed in accessible joints above the water level at any time, or PU foam injected into the interlocks. Plastic sealing compounds can only be applied to dry surfaces. To achieve this, interlocks must be sealed provisionally beforehand.
    In the case of box sheet piling with double interlocks, sealing can also be achieved by filling the emptied cells with a suitable sealing material, e.g. underwater concrete.

### 8.1.21.4 Sealing of penetrations

Aside from the watertightness of the interlocks, special attention must be paid to adequate sealing at the points where anchors, waling bolts, etc. penetrate the sheet piling.

Lead or rubber washers can be placed between sheet pile and plate washer, also between plate washer and nut. To prevent damage to the sealing washers, anchors must be tensioned by means of a turnbuckle, and waling bolts with the nuts on the waling side.

Holes in sheet piles for waling bolts and anchors must be properly deburred so that plate washers fit flush.

### 8.1.22 Waterfront structures in regions with mining subsidence (R 121)

#### 8.1.22.1 General

Waterfront structures in regions liable to mining subsidence must be built to withstand the ground movements expected during their operational life. With regard to mining subsidence, a differentiation is made between vertical ground movements (subsidence) and horizontal ground movements (tensile and compressive ground strains).

As movements in mining subsidence regions generally occur at different times, waterfront structures can be exposed to vertical subsidence and horizontal ground strains in changing sequences.

Local subsidence does not usually influence the groundwater level. Water levels in canals also stay the same and are therefore not affected by local subsidence either.

Before beginning to construct waterfront and other structures in mining subsidence areas, the plans must be submitted to the mining company operating in the area. It is then left to the discretion of the mining company whether to propose safety measures against the effects of mining and to cover the ensuing costs or, alternatively, to cover the costs of rectifying any potential damage resulting from mine subsidence. However, it is actually normally impossible to limit or foresee the extent of any damage in advance.

If the mining company responsible is not willing to undertake measures to prevent damage arising from mine subsidence, or does not consider such measures necessary, and the client is not prepared to provide for such measures, the method of construction selected should be planned and executed so that mine subsidence can be accommodated without severe damage and any damage can be easily repaired.

Experience has shown that concrete and masonry waterfront structures are frequently seriously damaged by tensile and compressive horizontal ground strains and torsional movements caused by mine subsidence. By contrast, no appreciable damage due to mine subsidence has been seen so far in structures made from steel sheet piles.

For waterfront structures in mining subsidence regions, steel sheet piling can therefore be generally classed as suitable, provided several fundamental rules are observed during planning, design, calculations and construction.

### 8.1.22.2 Guidance for planning waterfront structures in mining subsidence regions

The magnitude of the ground movements to be expected must be ascertained from the mining company responsible. Levels and load assumptions are determined on the basis of this data.

Anticipated subsidence, on canals, for example, can be compensated for by providing a taller waterfront structure. That is generally more economical than raising the height of a wall once subsidence has occurred. If different amounts of subsidence are expected along the length of a waterfront structure, the top of the wall can be raised to match the expected subsidence so that, once subsidence has occurred, the height of the wall is more or less uniform.

In specific cases it may be appropriate, for reasons of appearance, to increase the heights of quay walls only after subsidence has taken place. However, in these cases the correspondingly greater loads from active earth pressure and hydrostatic pressure plus the ensuing increase in anchor load must be considered during planning in order to avoid subsequent, often very involved, strengthening measures.

Tensile and compressive horizontal ground strains at the level of the waterfront structure do not usually damage U- or Z-section sheet piles because the deformation potential of such structures (concertina effect) enables them to withstand ground movements without being overloaded. Compressive horizontal ground strains perpendicular to the waterfront structure displace the wall towards the water. Tensile horizontal ground strains perpendicular to the waterfront structure can increase the load on anchors when they are very long. However, additional loads on the anchors due to such horizontal strains can usually be accommodated without the anchors failing.

### 8.1.22.3 Guidance for design, calculations and construction

With regard to the loads due to mining subsidence, waterfront structures do not usually have to be designed to withstand higher loads than would be the case outside mining subsidence regions, provided that this is not requested by and paid for by the mining company responsible. This also applies to reinforced concrete capping beams and their reinforcement, provided the beam remains above the water after the mining subsidence has occurred. Any damage can then be repaired or a beam completely renewed afterwards.

In order to minimise the susceptibility of quay walls to harmful mining subsidence effects, the portion of the sheet piling above the anchors should be as short as possible; anchor bars and walings should be positioned as close as possible to the top of the sheet piling. It is for this reason, and also because of wall deformations caused by tensile horizontal ground strains, that quay walls in mining subsidence areas should be designed to withstand the full active earth pressure and assume that the earth pressure is redistributed.

The steel grades for sheet piling in mine subsidence areas can be chosen according to R 67, section 8.1.6. For capping beams, walings and anchor bars, steel grades S 235 J2 and S 355 J2 as per DIN EN 10025 should be used. If walls are anchored with round steel tie rods, upset threaded ends are permissible provided the requirements of R 20, section 8.2.7.3, are fulfilled. Upset round steel tie rods offer the advantage of greater elongation and greater flexibility than tie rods without upset threaded ends; besides this, they are easier to install and cheaper.

Anchors for waterfront structures in mine subsidence areas should be attached to walings made from pairs of channels as the waling bolts of such walings can accommodate deformations of the wall more easily. The walings are to be designed so that all effects from mine subsidence can be accommodated without the need for subsequent strengthening. Concerning longitudinal movement of the wall, all walings and steel capping beams should be spliced via elongated holes or holes with sufficient play. If a wall has to be subsequently raised, this must be considered during the design of the capping beams, e.g. by ensuring that the beams can be easily removed. Anchor connections in a capping waling are to be avoided.

Horizontal or slightly inclined anchor rods are advantageous for waterfront structures in mining subsidence areas because only minimal additional stresses are caused in these in the event of differential settlement between anchorages and wall. For the same reason, anchor connections must be hinged. Wherever possible, anchor connections should be installed in the outboard troughs of the sheet piles so that they remain accessible and can be easily inspected.

When accepting delivery of sheet piles according to R 67, section 8.1.6, special attention should be paid to ensuring that the interlock tolerances have not been exceeded.

The deformation potential of sheet piling in the plane of the wall is not seriously impaired by welding the interlocks, in order to seal them, for example. However, welded interlocks do hinder vertical deformations. Therefore, for sheet piling in mining subsidence regions, do not weld all the interlocks connected on site.

When sheet piling in a mine subsidence area has to be watertight, the interlocks connected in the factory can therefore be sealed using an elastic sealing compound that does not hinder vertical movement, and interlocks connected on site should be welded.

Accessible interlocks can also be sealed on site by attaching a plate over an elastic sealing compound in front of the joint. Welding is generally to be avoided as much as possible if it impairs the flexibility of the sheet piling.

This requirement applies, in principle, to the interaction of reinforced concrete structural members and sheet piling. It is especially important that the flexibility of the sheet piling is not limited by the presence of heavyweight concrete members. Quay walls and craneways are to be kept separate, with separate foundations, so that they settle independently of each other and settlements can be directly compensated for. Where a craneway is not laid on sleepers, see R 120, section 6.17.2.1, but on reinforced concrete beams instead, the beams should be connected to each other by sturdy ties to maintain the gauge. Electric power should preferably be fed through trailing cables.

### 8.1.22.4 Monitoring of structures

Waterfront structures in regions at risk of mining subsidence require regular monitoring and reference measurements. Even if the mining company is liable for any damage, the owner of the facility still remains responsible for its safety.

## 8.1.23 Vibratory driving of U- and Z-section steel sheet piles (R 202)

### 8.1.23.1 General

Vibratory hammers (vibrators), which are rigidly clamped to the section being driven, can be used to drive U- and Z-section sheet piles into the ground. The vibrators generate vertically directed vibrations through eccentrics rotating synchronously in opposite directions which are then transferred to the pile section and cause the soil to resonate. This considerably reduces the soil's resistance to penetration (skin friction and toe resistance).

An expert must assess the interaction of vibrator, pile section and soil to determine whether vibratory driving can be successfully used in a given situation. Please refer to R 118, section 8.1.11.3, and R 154, section 1.5, for how soil and pile section influence vibration.

The effects of vibratory driving on the load-carrying capacity and settlement behaviour of foundation elements and the in situ density of the soil should be evaluated by a geotechnical expert when vibration is being considered. The loadbearing capacity should be assessed in advance by means of loading tests, especially in the case of foundation elements with alternating loads (tension and compression).

### 8.1.23.2 Terms and parameters for vibratory hammers

Important terms and parameters for vibratory hammers are as follows:

1. The type of drive; vibratory hammers can be driven electrically, hydraulically or electro-hydraulically.

2. The driving power $P$ [kW] determines the efficiency of the hammer. At least 2 kW should be available per 10 kN of centrifugal force.
3. Effective moment $M$ [kg m] is the total mass $m$ of the eccentrics multiplied by the spacing $r$ of the centre of gravity of an individual eccentric from its axis of rotation:

$$M = m \cdot r \, [\text{kg} \cdot \text{m}]$$

The effective moment also determines the stroke or amplitude of the vibratory hammer.
4. Revolutions per minute (rpm) $n$ [U min$^{-1}$] of the shafts on which the eccentrics are mounted. The rpm affects the centrifugal forces to the power of two. Electrical vibrators work with a constant rpm, whereas hydraulic vibrators have an infinitely variable rpm.
5. Centrifugal force (exciting force) $F$ [kN]. This is the product of the effective moment and the square of the angular velocity:

$$F = M \cdot 10^{-13} \cdot \omega^2 \, [kN] \quad \text{where} \quad \omega = \frac{2 \cdot \pi \cdot n}{60} \, [s^{-1}]$$

In practice, the centrifugal force is a benchmark for comparing different vibratory hammers. However, the rpm and effective moment at which the maximum centrifugal force is reached must be taken into account.

Modern vibrators include options for infinite adjustment of speed and eccentric moment during operation. The advantage of such vibrators is that they can be started with an amplitude of zero, free from all resonance. Only upon reaching the preselected speed are the weights extended and adjusted. This avoids undesirable amplitude peaks while starting and stopping.
6. Stroke $S$ [m] and amplitude $\bar{x}$ [m]. The stroke $S$ is the total vertical shift of the vibrating unit in the course of one revolution of the eccentrics. The amplitude $\bar{x}$ is half the stroke. Manufacturers' specifications sometimes list the stroke, sometimes the amplitude.

The amplitude is the quotient of the effective moment $M$ [kg m] and the dynamic mass $m$ [kg] of the hammer:

$$\bar{x} = \frac{M}{m_{\text{ham,dyn}}} \, [m]$$

On the other hand, the "working amplitude" $\bar{x}_A$ required in practice is an amplitude that is established during vibration. It is calculated as the quotient of the effective moment $M$ and the total resonating mass $m_{dyn}$:

$$\bar{x}_A = \frac{M}{m_{dyn}} \, [m]$$

The resonating mass $m_{dyn}$ is the sum of the dynamic mass $m_{ham,dyn}$ of the hammer, the mass $m_{pile}$ of the pile section and the mass $m_{soil}$ of the resonating soil volume. The latter is not usually known. Therefore, driving prognoses often use $m_{soil} \geq 0.7 \, (m_{ham,dyn} + m_{pile})$. To ensure optimum vibratory driving operations, the calculated working amplitude should be $\bar{x}_A \geq 0,003$ m.

7. Acceleration $a$ [m/s²]. The acceleration of the pile section acts on the granular structure of the surrounding soil. This influences the stresses between the individual grains of non-cohesive soils and, in the ideal case, even cancels them out completely so that the soil is turned into a "pseudo-liquid" state while the vibratory hammer is operating. This reduces the friction in the soil and soil's resistance to driving.

The product of the working amplitude and the square of the angular velocity yields the acceleration $a$ of the pile section:

$$a = \bar{x} \cdot \omega^2 \, [m/s^2] \quad \text{where} \quad \omega = \frac{2 \cdot \pi \cdot n}{60} \, [s^{-1}]$$

Experience shows that $a \geq 100$ m/s² is required for successful vibratory driving.

### 8.1.23.3 Connection between vibratory hammer and pile section

The vibrator must be connected to the pile section via hydraulic clamping jaws to create a connection that is as rigid as possible. This ensures that the energy from the vibratory hammer is ideally transferred to the pile section and from there to the soil and, consequently, that the vibratory driving is a success. As with impact driving, the vibrator should be positioned on the centroid axis of the driving resistance. Double clamping is recommended when driving double piles.

The number and positioning of the clamping jaws should be selected to suit the section.

### 8.1.23.4 Criteria for selecting a vibrator

For installing sheet pile sections in uniform, re-arrangeable (non-cohesive) and saturated soils, the vibratory hammer should generate at least 15 kN centrifugal force for each metre of driving depth and 30 kN centrifugal force for each 100 kg mass of pile section. The centrifugal force can therefore be calculated from

$$F = 15 \cdot \left(t + \frac{2 \cdot m_{pile}}{100}\right) [kN]$$

where

$F$ centrifugal force [kN]
$t$ embedment depth [m]
$m_{pile}$ mass of pile [kg]

Refer to the details of the equipment manufacturer or to computer-assisted prognosis models when selecting a vibrator. Suitability tests are recommended for larger construction projects.

### 8.1.23.5 Experience with the vibratory driving of U- and Z-section sheet piles

1. The effects of vibrations on nearby structures and other facilities arising from the vibratory driving of piles and sheet piles cannot be reliably predicted.
In principle, structures in the area affected by vibrations are loaded by oscillations, e.g. their foundations and also suspended floors (direct effects). DIN 4150 parts 2 and 3 contain reference values for permissible oscillation velocities.
Vibrations caused when operating vibratory hammers can also cause settlement of the soil, which then has an indirect effect on the structure.
Any direct and indirect effects expected to act on structures in the area affected by vibrations should be forecast in advance. Prognosis methods have been compiled by Achmus et al. [2], for example. Using these prognoses, machinery and operating data can be selected such that the reference values of DIN 4150 parts 2 and 3 are not exceeded and settlements due to indirect effects can be evaluated.
If penetration rates $\geq 1$ m/min are achieved, experience suggests that damaging effects on adjacent structures are unlikely. Penetration rates $\leq 0.5$ m/min over longer structures should be accompanied by measurements of the adjacent structures which are then examined in detail by a geotechnical expert together with a structural engineer to ascertain whether the oscillation velocities generated by the vibratory hammer are acceptable. Reduced penetration rates, e.g. when passing through compacted soil strata, are not usually problematic.

2. In soils that are not very amenable to re-arrangement (weakly cohesive soils, silt) or in dry soils, the effectiveness of vibration is significantly reduced. However, it can be improved by using jetting (R 203, section 8.1.23). Drilling to loosen the soil just in advance of the driving or soil replacement can also be considered as aids to improve driving.
3. Soil compaction through vibration as mentioned in R 154, section 1.5.2.3, is more likely to occur at high rotational speeds (section 8.1.23.2(7)). In order to avoid compaction, it can be expedient to carry out the work with a vibrator that generates the same centrifugal force but operates with a lower rpm and therefore with a lower acceleration.
4. More information on the watertightness of interlocks pre-sealed with synthetic materials can be found in R 117, section 8.1.21.3(1b).
5. R 118, section 8.1.11, applies for vibratory driving records. The driving logs should contain the time taken for each 0.5 m penetration in addition to the operational data of the vibrator. The vibration work should preferably be documented by continuous recording of operational data, vibration times and penetration.
6. Vibration is generally a low-noise driving method. Higher noise levels can occur with defective vibratory action as a result of the sheet pile wall vibrating as well and the clamps hitting each another. Vibration of the wall can be intensive in the case of tall piles and staggered or panel-by-panel driving. The use of driving aids in accordance with section 8.1.23.5(2), or padded clamping jaws can provide a remedy.
7. The risk of settlement in the vicinity of existing structures must be taken into account even when using modern high-frequency vibrators with variable eccentrics.
8. It should also be noted that low penetration rates and prolonged vibrating with powerful vibrators can lead to the interlocks being heated to such an extent that they fuse together. Where the penetration rate is reduced temporarily, water cooling of the pile, especially around the interlocks, can prevent this welding due to overheating.
9. Double clamps should always be used when driving double (or multiple) piles. To avoid cracks, the piles should have no lifting holes in the vicinity of the clamping jaw transferring the load.

## 8.1.24 Water-jetting to assist the driving of steel sheet piles (R 203)

### 8.1.24.1 General

R 104, section 8.1.12.5, R 118, section 8.1.11.4, R 149, section 8.1.14.4, R 154, section 1.5.2.5, and R 202, section 8.1.23, refer to water-jetting as an aid for driving piles. Jetting with water can be used with impact driving, vibrating and pressing in order to:

- generally facilitate installation,
- prevent overloading of the plant and overstressing of the pile sections,
- achieve the necessary embedment depth,
- reduce vibrations in the ground, and
- reduce costs through shortening installation times, reducing power requirements and/or enabling lighter equipment to be used.

Jetting pressure can be varied to suit the soil structure and strength. However, water-jetting is not permitted below the design depth.

### 8.1.24.2 Low-pressure jetting

Low-pressure jetting involves directing a jet of water at the base of the pile section. The subsoil is loosened by the water injected under pressure and the loosened material is carried away in the flow. Essentially, this reduces the toe resistance of the piles to be inserted. Depending on the soil structure, the skin friction and the interlock friction can also be reduced by the rising water.

The low-pressure method is limited by the strength of the subsoil, the number of jetting lances and the pressure and volume of water required. In order to establish the necessary operating data for low-pressure jetting, trial driving operations are recommended.

Low-pressure jetting can be used in non-cohesive, densely bedded soils, and especially in dry, uniform non-cohesive soils, sands and gravels.

Jetting lances have a diameter of 45–60 mm and the pressure at the pump is between 5 and 20 bar. Constricting the nozzle or using special nozzles can create a jet action with a correspondingly greater flushing effect. Water for jetting is usually supplied by centrifugal pumps, and water consumption can reach approx. 1000 l/min.

Depending on how difficult driving is, jetting lances are jetted in next to the pile section or fixed to the pile section and carried down with this into the ground. Reductions in the strength of the soil and settlement may occur as a result of introducing relatively large volumes of water.

Low-pressure jetting processes in which the jetting process is combined with vibration and restricting the jetting pressure to 15–40 bar have proven effective for driving sheet piling in very compact soils that would normally present extremely difficult driving conditions.

Owing to its environmental compatibility, the low-pressure jetting method can also be used in residential and inner-city areas.

The success of this method essentially depends on the optimum match between the jetting, the vibrator and the in situ subsoil. It is therefore important to assess local experience or, where this is unavailable, to carry out trials beforehand. Vibrators with variable effective moments and rpm control are ideal because their operating data can be adjusted to the specific penetration resistance.

Usually, between two and four lances are fixed to the pile section (double pile), with the tip of the lance flush with the base of the pile. The optimum arrangement is one pump per lance. Jetting begins simultaneously with vibration in order to prevent the nozzle from becoming clogged by the ingress of soil material.

At penetration rates $\geq 1$ m/min, jetting can continue until the embedment depth required by the calculations is reached. The soil properties previously determined for the sheet piling calculation then generally apply. However, the angles of inclination for active and passive earth pressures should be limited to $\delta_a = +\frac{1}{2}\varphi$ and $\delta_P = -\frac{1}{2}\varphi$ respectively. If high vertical loads are to be borne by a section inserted with the help of jetting, the vertical loadbearing capacity must be assessed in loading tests.

### 8.1.24.3 High-pressure jetting

The use of high-pressure jetting eases the driving of sheet piles in soils of varying compactness; indeed, in some cases, without jetting, driving would not be possible at all. Precision pipes 30 mm in diameter are used as jetting lances. The pressure at the (reciprocating) pump is between 250 and 500 bar. Jetting lances are fitted with screw-on nozzles, the cross-section of which can be adjusted to suit the surrounding soil. Water consumption lies between 30 and 150 l/min per nozzle.

High-pressure jetting is primarily suited to firm, overconsolidated cohesive soils such as silt and clay rocks and friable sandstone.

However, high-pressure jetting is only economical when the lances can be extracted and reused. To this end, the jetting lances cannot be permanently attached to the sections being driven. Instead, they are guided through clips welded to the section. The jetting nozzles must be about 5–10 mm above the bottom end of the pile and should be chosen to maximise their working life.

When using high-pressure jetting, the pressure, number of lances and type of nozzle must certainly be matched to the specific in situ soil. Where soil conditions vary, these adjustments must also be carried out during driving operations.

The angles of inclination for active and passive earth pressures should be limited to $\delta_a = +\frac{1}{2}\varphi$ and $\delta_P = -\frac{1}{2}\varphi$ respectively.

If high vertical loads are to be borne by a section inserted with the help of high-pressure jetting, the vertical loadbearing capacity must be assessed in loading tests.

### 8.1.25 Pressing of U- and Z-section steel sheet piles (R 212)

#### 8.1.25.1 General

Pressing meets the increasing demands for low noise levels and no vibrations during the driving of steel sheet piles. This method is often the only solution when sheet piling structures are to be erected economically in inner-city areas and/or adjacent to existing structures sensitive to settlement.

The cost of pressing is higher than that of methods employing hammers or vibrators. However, in many cases this is at least partly offset by the fact that, for example, soundproofing measures (see R 149, section 8.1.14) are then unnecessary.

#### 8.1.25.2 Pressing plant

We distinguish between presses suspended from cranes, presses with leader guidance and presses supported on sheet piles already driven.

Presses suspended from cranes require a frame that guides the piles at two levels. In the case of presses with leader guidance, the piles are guided at the top by the leader and at the bottom by a frame.

All presses for the above two methods are equipped with several adjacent rams on the axis of the wall which jack the piles into the ground in a given sequence. The reaction forces for the pressing forces are provided by the plant, the weight of the sheet piles and their skin friction in the soil.

Presses supported on sheet piles already driven are popular because of the small working space they require. They require no further guidance frames or means of support. The sheet pile to be pressed in is both aligned and inserted by the ram in the head of the jack. This type of press activates the reaction forces via the adjacent piles already driven.

Drilling to loosen the soil or water-jetting can be used to assist pressing. The fittings required for this are already integrated into some presses.

#### 8.1.25.3 Sheet piles for pressing

The majority of the machines available on the market today can only install single U- or Z-section sheet piles. Presses suspended from a crane and those with leader guidance can operate with loosely assembled double, triple or quadruple piles which are then driven as single piles. Sheet piling presses for installing double piles can be assisted during the driving procedure by drilling pilot holes in the trough of the pile.

In order to minimise interlock friction in non-cohesive, fine-grained soils, the open interlocks of the sheet piles should be filled with a material to displace the soil, e.g. hot bitumen or similar.

Sections to be pressed should be selected for both the structural requirements in the finished structure as well as the stresses during pressing. Experience has shown that sections being pressed in should not be too soft.

### 8.1.25.4 Pressing in sheet piles

Pressing in sheet piling is only successful when the penetration resistance of the piles is suitably estimated and the pressing forces and stiffness of the sections are matched to this.

Current sheet pile presses on the market are capable of generating a maximum pressing force of about 1500 kN. In the case of pile-supported presses, the reaction forces are provided solely by adjacent piles. Therefore, pressing forces of approx. 800 kN should not be exceeded with such presses as otherwise this could result in piles already driven being extracted again.

The use of driving aids such as low-pressure jetting (see R 203, section 8.1.24) or drilling to loosen the soil enables sheet piles to be pressed into difficult soils as well. When using low-pressure jetting, the angles of inclination for active and passive earth pressures should be limited to $\delta_a = +½\ \varphi$ and $\delta_P = -½\ \varphi$ respectively.

If necessary, any obstacles encountered in the soil must be removed.

## 8.2 Design of sheet piling

### 8.2.1 General

Eurocode EC 7-1 [85] offers three methods for analysing ultimate strength. In Germany, design approach 2 according to EC 7-1 is used with only one exception: when the partial safety factors are applied to loads (e.g. internal forces $M_k$, $V_k$, $N_k$) and resistances. To distinguish this from the variants also permissible under design approach 2, in which the partial safety factors are applied to the actions, the German design approach is denoted 2*.

The ultimate strength analyses for retaining structures include the following limit states:

- STR (structure failure, limit state for failure of structure or structural component),
- GEO 2 (geotechnical failure, limit state for failure of the subsoil), and
- GEO 3 (analysis of overall stability).

The analysis of the sheet piling component under bending moments and normal forces forms part of the STR limit state. Loadbearing capacity analyses for sheet piling made from

- reinforced concrete are carried out to DIN EN 1992,
- timber are carried out to DIN EN 1995, and
- steel are carried out to DIN EN 1993-5,

taking into account R 20, section 8.2.7.

DIN EN 1995-3 also permits plastic-plastic design and other methods for steel sheet piling structures. Such a design approach, which exploits both the plastic cross-section design and the plastic system loadbearing capacity (in the case of statically indeterminate systems) at the ultimate limit state, can also be appropriate for waterfront structures in certain cases. However, the elastic-elastic and elastic-plastic design approaches explained below are normally used. With regard to the plastic-plastic design of waterfront structures, there is still a lack of experience relating to, among other things, the active earth pressure distribution due to the redistribution of internal forces when utilising the system reserves to the full. The classic active earth pressure distribution and the earth pressure diagrams given in this book cannot be used as a basis for design using a plastic-plastic approach.

The following analyses are included in the GEO 2 limit state for retaining walls:

- Ground failure in the passive earth pressure zone due to horizontal action effects from soil support $B_{h,d}$ (analysis format $B_{h,d} \leq E_{ph,d}$, see DIN 1054, "note to 9.7.4")
- Axial sinking of the sheet pile wall in the subsoil due to vertical action effects $\Sigma V_{i,d}$ (analysis format $\Sigma V_{i,d} \leq \Sigma R_{i,d}$, see R 4, section 8.2.5.6
- Ground bearing capacity
- Stability of the lower failure plane for anchored retaining walls, see R 10, section 8.5

The analysis of safety against slope failure forms part of limit state GEO 3. DIN 1054, "note to 9.7.2", describes the minimum conditions for which the analysis of slope failure is to be carried out for retaining walls. The analyses of the horizontal base support for a retaining wall in the soil and the carrying of the vertical components of mobilised passive earth pressure are used to calculate the required embedment depth of the retaining wall in the subsoil. As part of both of these analyses, redistribution of the earth pressure on the loading side of the retaining wall can be considered as per section 8.2.3.1.

In the analysis of a sufficient horizontal base support for the retaining wall, the partial safety factor $\gamma_{R,e}$ may not be reduced for passive earth pressure. Reduced partial safety factors are discussed further in sections 8.2.1.2 and 8.2.1.3.

The serviceability limit state (SLS) embraces conditions that will cause the structure to become unusable but without it losing its loadbearing capacity. In the case of waterfront structures, this analysis must be carried out so that any wall deformation – possibly considering anchor elongation and the resulting settlement behind the wall – is not damaging for the structure or the surroundings. Further information on SLS analyses can be found in DIN 1054, "note to 9.8".

### 8.2.1.1 Partial safety factors for loads and resistances (R 214)

When designing sheet piling structures as well as anchor walls and anchor plates for round steel tie rods, the following partial safety factors govern for STR limit state analyses:

- $\gamma_G$ and $\gamma_Q$ for actions according to Table R 0-1
- $\gamma_{R,e}$ and $\gamma_{R,h}$ for resistances according to Table R 0-2
- $\gamma_{R,e,red}$ for passive earth pressure as per R 215, section 8.2.1.2
- $\gamma_{G,red}$ for excess water pressure actions according to R 216, section 8.2.1.3

For more information on applying the reduced partial safety factors $\gamma_{R,e,red}$ and $\gamma_{G,red}$, please see R 215, section 8.2.1.2, and R 216, section 8.2.1.3.

### 8.2.1.2 Determining the design values for the bending moments (R 215)

When calculating bending moments, a reduced partial safety factor $\gamma_{R,e,red}$ may be used for reducing earth pressure as per Table R 215-1 if there is non-cohesive soil with medium strength at least beneath the design bottom:

| Strength | In situ density $D$ | | Cone resistance |
|---|---|---|---|
| | $U \leq 3$ | $U > 3$ | $q_c$ [MN/m$^2$] |
| low | $0.15 \leq D < 0.30$ | $0.20 \leq D < 0.45$ | $5.0 \leq q_c < 7.5$ |
| medium | $0.30 \leq D < 0.50$ | $0.45 \leq D < 0.65$ | $7.5 \leq q_c < 15$ |
| high | $0.50 \leq D < 0.75$ | $0.65 \leq D < 0.90$ | $q_c \geq 15$ |

or cohesive soil with stiff consistency at least:

| State | Consistency index $I_C$ |
|---|---|
| soft | $0.50 \leq I_C < 0.75$ |
| stiff | $0.75 \leq I_C < 1.00$ |
| semi-firm to firm | $1.00 \leq I_C < 1.25$ |

Table R 215-1. Reduced partial safety factor $\gamma_{R,e,red}$ for passive earth pressure for determining bending moments

| GEO-2 | DS-P | DS-T | DS-A |
|---|---|---|---|
| $\gamma_{R,e,red}$ | 1.20 | 1.15 | 1.10 |

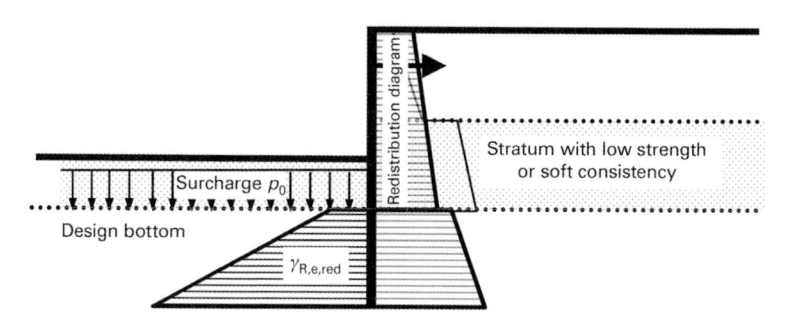

**Fig. R 215-1.** Loading diagram for determining bending moments with reduced partial safety factors in soils with low strength or soft consistency

Redistribution of the active earth pressure to R 77, section 8.2.3.1, is carried out down to the design bottom.

If there is initially soil with a soft consistency or low strength beneath the design depth (intended depth plus dredging tolerance, precautionary prior dredging, scour allowance, etc.), this cannot be used as horizontal support for the wall. Such strata can only be applied as a surcharge $p_0$ on the new design bottom, which is situated at the level of the beginning of the loadbearing, i.e. at least very stiff or firm, soil.

The reduced partial safety factor $\gamma_{R,e,red}$ for calculating bending moments may only be applied in the loadbearing strata (Fig. R 215-1). If the soil in the passive earth pressure zone is of low strength or soft consistency, then the action effects must be calculated using partial safety factors $\gamma_{R,e}$ that have not been reduced.

If the reduced partial safety factor $\gamma_{R,e,red}$ is not used, the active earth pressure redistribution diagram corresponding to Fig. R 215-2 need only be continued down to the bottom of the watercourse/excavation.

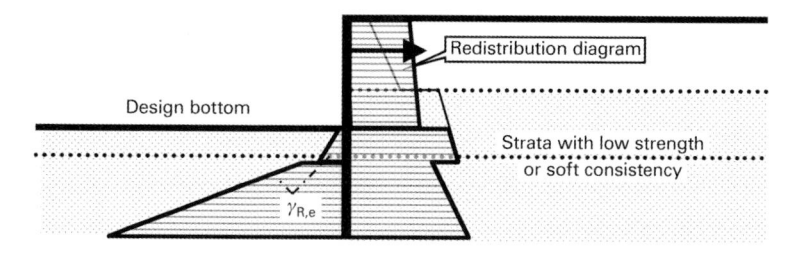

**Fig. R 215-2.** Loading diagram for determining bending moments without reduced partial safety factors in soils with low strength or soft consistency below the design bottom

Where the stipulated boundary conditions for using the reduced partial safety factor $\gamma_{R,e,red}$ are met, the degree of fixity for the full utilisation of the embedment depth/pile length calculated with non-reduced partial safety factors can be used for calculating bending moments. The internal forces (moments and shear forces) calculated with $\gamma_{R,e,red}$ play an important role in the sheet piling analysis.

For structures that fall within the remit of this book, anchor forces and the required embedment depth of the wall may only be determined using the full partial safety factor $\gamma_{R,e}$.

The redistribution of the active earth pressure according to R 77, section 8.2.3, is always carried out down as far as the design bottom defined beforehand.

### 8.2.1.3 Partial safety factor for hydrostatic pressure (R 216)

Actions due to hydrostatic pressure are to be calculated according to R 19, section 4.2, R 113, section 4.7, and R 114, section 2.9.

When the boundary conditions given below apply, it is possible to reduce the partial safety factor $\gamma_G$ used for calculating the design values of loads due to hydrostatic pressure. The partial safety factors $\gamma_{G,red}$ are given in Table R 216-1:

Reducing the partial safety factors for actions due to hydrostatic pressure is only permitted when at least one of the following three conditions is satisfied:

- Verified measurements are available regarding the positional and chronological relationships between groundwater and outer water levels to guarantee the hydrostatic pressure used in the calculations and also to serve as a basis for assigning to design situations DS-P, DS-T and DS-A.
- Numerical models of bandwidth and frequency of occurrence of the true water levels, and hence hydrostatic pressures, are analysed and lie on the safe side. These forecasts are to be checked through observations, beginning with the construction of the sheet pile wall. Where the measurements are larger than those predicted, the values on which the design was based must be guaranteed by appropriate measures such as drainage, pumping systems, etc. These must be monitored permanently.

**Table R 216-1.** Reduced partial safety factor $\gamma_{G,red}$ for calculating the design value of hydrostatic pressure actions

| STR | DS-P | DS-T | DS-A |
| --- | --- | --- | --- |
| $\gamma_{G,red}$ | 1.20 | 1.10 | 1.00 |

- There are geometric boundary conditions present which limit the water level to a maximum value – as is the case, for example, with the top edges of sheet pile walls designed as flood defence walls that limit the depth of the floodwaters. Drainage systems installed behind the sheet pile wall do not represent a clear geometrical limit to the water level in the meaning of this stipulation.

### 8.2.2  Free-standing sheet piling structures (R 161)

#### 8.2.2.1  General

Contingent on the bending resistance of the wall, sheet piling fully fixed in the ground and without an anchorage can be economical if there is comparatively little difference in ground levels. Such an arrangement can also be used for larger differences in level if the installation of an anchor or other pile head support would be very involved and if relatively large displacements of the pile head can be regarded as harmless in terms of serviceability.

#### 8.2.2.2  Design, calculations and construction

In order to attain the necessary stability of free-standing sheet pile walls (i.e. without anchors), their design, calculations and construction must satisfy the following requirements.

- Ascertain all actions as accurately as possible, e.g. including the compacted earth pressure in backfill according to DIN 4085. This applies in particular to those actions applied to the upper section of the sheet piling because these can substantially affect the design bending moment and the embedment depth required.
- It must be possible to establish an exact assignment to design situations DS-P, DS-T and DS-A taking into account, for example, unusually deep scouring and unusual excess water pressures.
- The design bottom level may not be exceeded in the passive earth pressure zone. Therefore, the design bottom level used in the calculations must include the additional depth required for any scouring and dredging work.
- The structural calculations may be carried out in accordance with Blum [22], with the active earth pressure applied in a classic distribution.
- The theoretical embedment depth required, taking into account R 56, section 8.2.9, and R 41, section 8.2.10, must be reached in the construction work.
- In the serviceability state – i.e. in an analysis with characteristic actions – the deformation of the wall is to be included in addition to the internal forces. The deformation values that occur must be

investigated to ensure their compatibility with the structure and the subsoil, e.g. with respect to the formation of gaps on the active earth pressure side in cohesive soils which could fill up with water. The compatibility of the deformations must also be checked with all other aspects of the project. This approach is particularly important for larger differences in ground levels.
- Free-standing, backfilled sheet piling is less critical with respect to deformations as the deformations already occur during backfilling and therefore generally have no detrimental effect on subsequent construction measures.
- Influences from wall deformations can be compensated for by suitable driving inclinations so as to avoid an unattractive overhang at the top of the wall.
- The top of a free-standing sheet piling wall, at least in a permanent structure, should be provided with a capping beam or waling of steel or reinforced concrete to distribute the actions and prevent non-uniform deformations as far as possible.

## 8.2.3 Design of sheet piling structures with fixity in the ground and a single anchor (R 77)

### 8.2.3.1 Active earth pressure

Calculations for anchored retaining structures in ports and harbours can be carried out using the active earth pressure. Under certain conditions the minimum earth pressure as per DIN 4085 is to be used in cohesive strata.

The resulting earth pressure force calculated with the classic distribution over the height $H_E$ – which, if necessary, is increased when considering the minimum earth pressure in cohesive strata – may be redistributed over the height $H_E$ for the ULS and SLS analyses. It must be redistributed for determining the anchoring force.

The ratio of the position of the anchor head $a$ to the redistribution depth $H_E$ serves as a criterion for deciding the case when selecting the redistribution diagrams (R 77, section 8.2.3.2). The distances $H_E$ and $a$ and the classic earth pressure distribution $e_{agh,k}$ due to soil dead load or minimum earth pressure from a variable surface load of up to $10\,\mathrm{kN/m^2}$ over a large area are defined in Figs. R 77-1 and R 77-2. Other actions (block loads or further surface loads over a large area) are to be redistributed while taking into account the actual loadbearing behaviour of the wall. In particular, it should be noted that stiffer building components attract loads.

Fig. R 77-2 shows an example of a superstructure with depth $H_Ü$ and a reinforced concrete slab to shield the earth pressure.

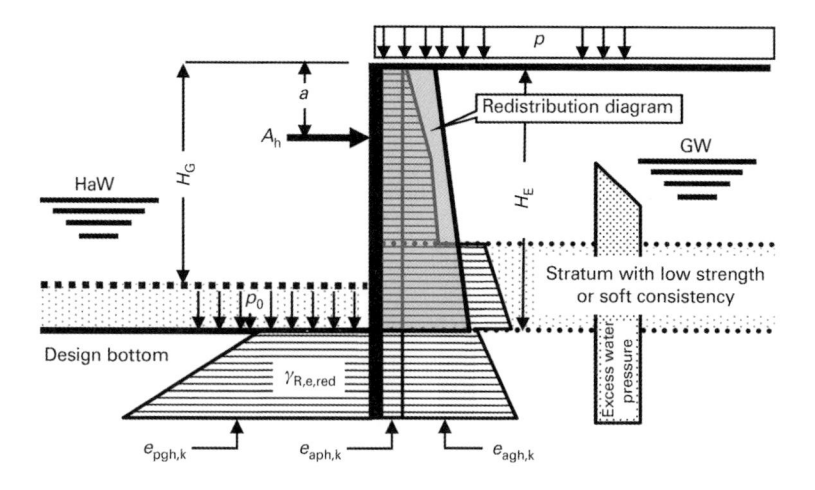

**Fig. R 77-1.** Example 1: Redistribution depth $H_E$ and anchor head position $a$ for calculating bending moments with $\gamma_{R,e,red}$

The following definitions apply:

$H_G$    total difference in ground levels

$H_Ü$    depth of superstructure from ground level to underside of shielding slab

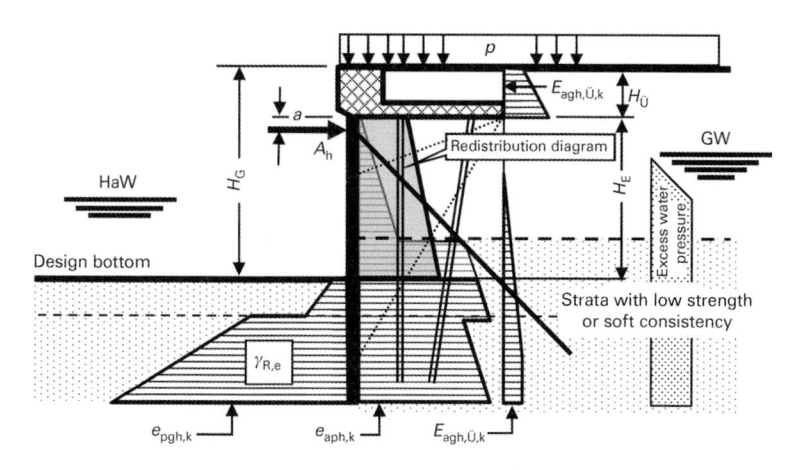

**Fig. R 77-2.** Example 2: Redistribution depth $H_E$ and anchor head position $a$ for calculating bending moments with $\gamma_{R,e}$

$H_E$     depth of earth pressure redistribution zone above design bottom (For a superstructure with a shielding slab, depth $H_E$ begins at the underside of the slab.)

$a$     distance of anchor head A from top edge of redistribution depth $H_E$

Beneath the design bottom, the non-redistributed active earth pressure is applied to the actions side.

### 8.2.3.2 Earth pressure redistribution

The earth pressure redistribution is selected depending on the method of construction:

- Trenching in front of wall    (cases 1 to 3, Fig. R 77-3)
- Backfilling behind wall    (cases 4 to 6, Fig. R 77-4)

We distinguish here between three ranges for anchor head distance $a$:

- $0 \leq a \leq 0.1 \cdot H_E$
- $0.1 \cdot H_E < a \leq 0.2 \cdot H_E$
- $0.2 \cdot H_E < a \leq 0.3 \cdot H_E$

Besides the designations shown in Figs. R 77-1 and R 77-2, in Figs. R 77-3 and R 77-4 the magnitude of the mean value $e_m$ of

**"Trenching in front of wall" method of construction**

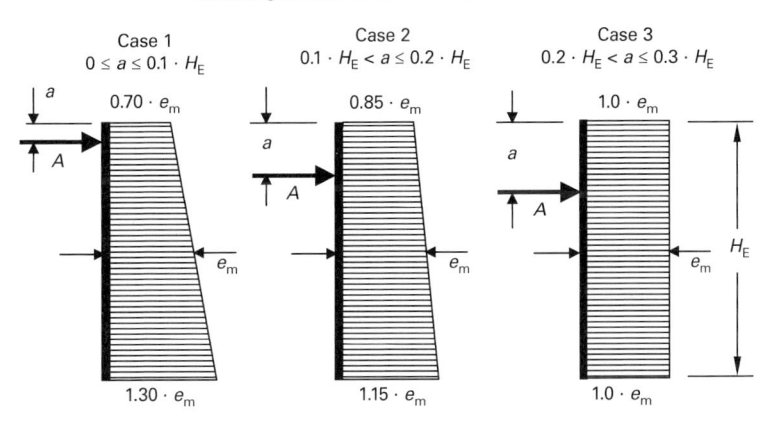

Case 1: $0 \leq a \leq 0.1 \cdot H_E$ — $0.70 \cdot e_m$ — $1.30 \cdot e_m$

Case 2: $0.1 \cdot H_E < a \leq 0.2 \cdot H_E$ — $0.85 \cdot e_m$ — $1.15 \cdot e_m$

Case 3: $0.2 \cdot H_E < a \leq 0.3 \cdot H_E$ — $1.0 \cdot e_m$ — $1.0 \cdot e_m$

**Fig. R 77-3.** Earth pressure redistribution for the "trenching in front of wall" method of construction

**"Backfilling behind wall" method of construction**

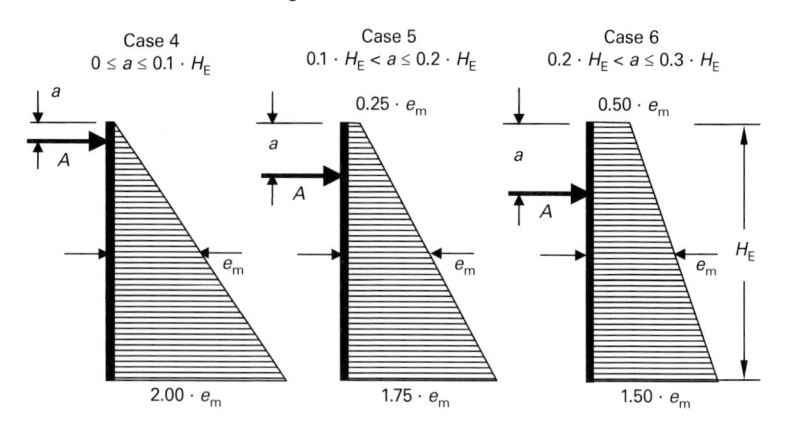

**Fig. R 77-4.** Earth pressure redistribution for the "backfilling behind wall" method of construction

the earth pressure distribution over the redistribution depth $H_E$ is as follows:

$$e_m = E_{ah,k}/H_E$$

The loading diagrams of Figs. R 77-3 and R 77-4 include all the anchor head positions $a$ in the range $a \leq 0.30 \cdot H_E$. These redistribution diagrams do not apply to anchors at lower levels, and in such cases the appropriate earth pressure diagrams must be determined separately.

If the ground surface is a short distance below the anchor, the earth pressure may be redistributed according to the value $a = 0$.

The loading diagrams for cases 1 to 3 are only valid on the condition that the earth pressure can redistribute to the stiffer support areas as a result of adequate wall deformation. A "vertical earth pressure vault" thus forms between anchor and soil support. Consequently, cases 1 to 3 may not be used when:

- the sheet pile wall is backfilled to a large extent between bottom of watercourse and anchorage and subsequent excavations in front of the wall are not deep enough for adequate additional deflection (guide value for adequate deflection is an excavation depth of approx. one-third the redistribution depth $H_{E,0}$ of the original system corresponding to Fig. R 77-5);

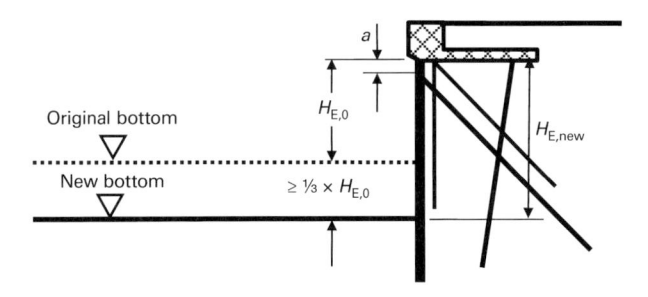

**Fig. R 77-5.** Additional excavation depth required for earth pressure redistribution according to the "trenching in front of wall" method of construction

- there is cohesive soil behind the sheet pile wall which is not yet sufficiently consolidated;
- a retaining wall with increasing flexural stiffness does not exhibit the wall deflections necessary for forming a vault, as is the case, for example, with reinforced concrete diaphragm walls. (In this situation the designer must check whether the displacement of the support at the base as a result of mobilisation of the passive earth pressure is sufficient for an earth pressure redistribution according to the "trenching" method cases 1 to 3.)

If loading diagrams cases 1 to 3 are not permitted for the above reasons, cases 4 to 6 belonging to the $a/H_E$ value for the "backfilling behind wall" method of construction may be used.

### 8.2.3.3 Passive earth pressure

When designing a sheet pile wall using the Blum method [22], the anticipated soil reaction is entered into the calculation with a linear increase opposite to the true progression. At the same time, and in the case of retaining walls with fixity, an equivalent force $C$ is required to maintain equilibrium.

Here, the characteristic soil support $B_{h,k}$ required for determining the embedment length is formed by the mobilised passive earth pressure $E_{ph,mob}$, which must exhibit a progression not unlike that of the characteristic passive earth pressure $E_{ph,k}$ and may not be redistributed.

Further information can be found in section 8.2.5.5.

### 8.2.3.4 Bedding

The soil support to a sheet piling wall can also be analysed using a horizontal bedding [66, 67, 130, 149, 198]. It should be noted that the soil reaction stress $\sigma_{h,k}$ at the design bottom as a result of characteristic actions may not be greater than the characteristic – i.e. maximum possible – passive earth pressure stress $e_{ph,k}$ (DIN 1054, Eqs. (A9.3) and (A9.4)). More information can be found in the 5th edition of the EAB.

### 8.2.4 Design of sheet pile walls with double anchors (R 134)

In contrast to R 133, section 8.4.7, which deals with issues regarding auxiliary anchors, R 134 covers sheet piling structures with a double row of anchors, i.e. anchors at two different levels.

The total actions acting on the sheet pile wall due to earth and hydrostatic pressures and variable loads are assigned to the two anchor positions $A_1$ and $A_2$ as well as soil support B. Owing to the distribution of the actions over the given structural system, the majority of the total anchorage force is carried by the lower anchor $A_2$.

Where round steel tie bars are used as anchors, it is advisable to connect both anchor positions to a common anchor wall at the same level and apply the direction of the resultant of anchor forces $A_1$ and $A_2$ as the anchor direction when checking stability at the lower failure plane according to R 10, section 8.5.

Anchors not connected to a common anchor wall (e.g. grouted anchors to DIN EN 1537) require the two rows of anchors to be analysed independently when checking stability.

#### 8.2.4.1 Active and passive earth pressures

Take into account the active and passive earth pressures in the same way as for a wall with a single row of anchors.

#### 8.2.4.2 Loading diagrams

The loading diagrams shown in R 77, section 8.2.3, for a wall with a single row of anchors are valid for determining the internal forces, support reactions and embedment length of a sheet pile wall with double anchors. Here, the level of the anchor head $a$ required for determining the $a/H_E$ value and for specifying the loading diagram is taken to be the average level between the two anchor positions $A_1$ and $A_2$. The earth pressure is then redistributed over the depth $H_E$ down to the design or model bottom in a similar fashion to the sheet pile wall with single anchor.

#### 8.2.4.3 Considering deformations due to previous excavations

As earlier deflections of sheet pile walls due to slippage of the soil on the earth pressure failure plane can be only partly reversed, the effects of temporary construction conditions on the stresses in the final condition must then be taken into account when they are critical for verifying serviceability. This might be the case, for example, when considering the wall deflection at the level of anchor $A_2$, which for a sheet pile wall temporarily anchored only at point $A_1$ should be considered as a yielding support for the structural system in the final condition.

### 8.2.4.4 Bedding

As for the sheet pile wall with a single row of anchors, a wall with two rows can be designed using horizontal bedding forces as a soil support (R 77, section 8.2.3.4).

### 8.2.4.5 Comparative calculations

As a comparison, a calculation according to R 133, section 8.4.7, must be carried out when designing the top of the wall and positioning the upper anchor $A_1$. If this calculation results in higher stresses, they govern the design.

## 8.2.5 Applying the angle of earth pressure and the analysis in the vertical direction (R 4)

The magnitude of the earth pressure angle selected, or permissible, depends on the maximum angle of internal friction between building material and subsoil (angle of wall friction) physically possible, the equilibrium conditions and the relative displacements of the sheet pile wall with respect to the soil.

The earth pressure angle has an influence on the analysis of the wall in the vertical direction. The following equilibrium and limit state conditions must be satisfied:

- Analysis of vertical component of mobilised passive earth pressure as per section 8.2.5.4
- Failure due to vertical movement as per section 8.2.5.5

In the case of untreated wall surfaces, the earth pressure angle to be applied depends on the properties of the wall. The following distinctions must be made:

- The back of a wall is classed as "joggled" when its form gives it such a large surface area that it is not the wall friction acting directly between soil and wall material which is critical, but rather the friction on a straight rupture surface in the soil which is only partly in contact with the wall. This is usually the case, for example, with walls constructed from bored cast-in-place piles. Diaphragm walls made from a hardened cement-bentonite suspension with suspended sheet piling or soldier piles can be classed as joggled. This also applies to driven, vibrated or pressed sheet pile walls.

- Untreated steel, concrete and timber surfaces can generally be regarded as "coarse", especially the surfaces of soldier piles and infill panels.
- The surface of a diaphragm wall can be classed as "less coarse" when there is only marginal filter cake formation, e.g. diaphragm walls in cohesive soil. Experience has shown that this also applies to diaphragm walls in non-cohesive soils if the length of time in which the trenches supported by the suspension is kept short (assuming excavation in accordance with the general rules for such work).
- The back of a wall is always classed as "smooth" when the clay content and consistency of the in situ soil are such that activation of a significant amount of friction cannot be expected.

The earth pressure angles can be applied depending on the wall properties as per sections 8.2.5.1 and 8.2.5.2.

With treated surfaces, soft subsoil that can form a lubricating layer on the surfaces of sections being driven and sections installed with the help of water-jetting, the friction can be reduced to such an extent that the angle of wall friction as per sections 8.2.5.1 and 8.2.5.2 cannot be established. In these cases the angle of wall friction must be restricted to a maximum of half the angle of internal friction, $|\delta_k| \leq \frac{1}{2} \cdot |\varphi'_k|$. Alternatively, a higher angle of wall friction will have to be verified by a geotechnical expert. The influence of reinforcement at the toe of the pile on the angle of wall friction must be assessed by a geotechnical expert.

### 8.2.5.1 Angle of inclination $\delta_{a,k}$ of active earth pressure

Active earth pressure is typically calculated assuming straight failure planes. The angle of inclination $\delta_{a,k}$ of the earth pressure can be used depending on the surface properties of the wall up to the following limits:

| Wall surface property | |
|---|---|
| Joggled wall | $|\delta_{a,k}| \leq (2/3) \cdot \varphi'_k$ |
| Coarse wall | $|\delta_{a,k}| \leq (2/3) \cdot \varphi'_k$ |
| Less coarse wall | $|\delta_{a,k}| \leq \frac{1}{2} \cdot \varphi'_k$ |
| Smooth wall | $\delta_{a,k} = 0$ |

### 8.2.5.2 Angle of inclination $\delta_{p,k}$ of passive earth pressure

Passive earth pressure is usually calculated for curved failure planes. In doing so, the angle of inclination $\delta_{p,k}$ of the passive earth pressure can be applied within the following limits:

$$-\varphi'_k \leq \delta_{p,k} \leq +\varphi'_k$$

To simplify the calculation, the approach using straight failure planes is permissible between the following limits:

$$-\tfrac{2}{3} \cdot \varphi'_k \leq \delta_{p,k} \leq +\tfrac{2}{3} \cdot \varphi'_k$$

provided the angle of internal friction $\varphi_k$, angle of wall inclination $\alpha_k$, angle of ground inclination $\beta_k$ and inclination angle $\delta_{p,k}$ lie within the following limits:

$\varphi'_k \leq 35°$
$\alpha_k \leq 0°$ (sign definitions as per Fig. R 4-1)
$\beta_k \geq 0$ for $\delta_{p,k} \geq 0°$ or $\beta_k \leq 0°$ for $\delta_{p,k} \leq 0°$

For these conditions there is no significant difference between the $K_{ph}$ values for the respective limit min $\delta_{p,k}$ from the methods with straight (min $\delta_{p,k} = (-2/3)\,\varphi'_k$) and curved (min $\delta_{p,k} = -\varphi'_k$) failure planes.

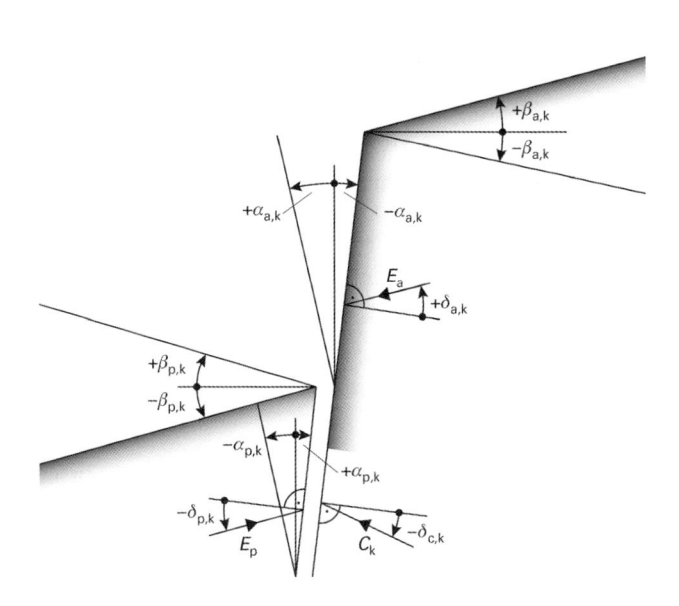

**Fig. R 4-1.** Sign definitions for angle of wall inclination $\alpha_k$, angle of ground inclination $\beta_k$, angles of inclination of active $\delta_{a,k}$ and passive $\delta_{p,k}$ earth pressures and angle of inclination $\delta_{C,k}$ of equivalent force $C$

Depending on the surface properties, the following additional limits on the passive earth pressure angle of inclination apply:

| Wall surface property | |
|---|---|
| Joggled wall | $|\delta_{p,k}| \leq \varphi'_k$ |
| Coarse wall | $|\delta_{p,k}| \leq \varphi'_k - 2.5° \leq 30°$ |
| Less coarse wall | $|\delta_{p,k}| \leq 1/2 \cdot \varphi'_k$ |
| Smooth wall | $\delta_{p,k} = 0°$ |

**8.2.5.3 Angle of inclination $\delta_{C,k}$ of equivalent force $C_k$**

When designing walls with full or partial fixity in the ground according to the Blum approach [22], the soil reaction below the theoretical base TF on the actions side is used to resist the equivalent force $C$. The extra depth required to resist this reaction force is calculated according to R 56, section 8.2.9, as a surcharge $\Delta t_1$ on the embedment depth $t_1$. As part of this approach, the direction of action of equivalent force $C$ is inclined at an angle $\delta_{C,k}$ to the horizontal.

The angle of inclination $\delta_{C,k}$ can be used within the following limits:

$$-\varphi'_k \leq \delta_{C,k} \leq +\tfrac{1}{3} \cdot \varphi'_k$$

However, depending on the surface properties of the wall, the angle cannot be larger than the limits for $\delta_{p,k}$ stipulated in section 8.2.5.2.

**8.2.5.4 Magnitude of equivalent force $C$**

The magnitude of the equivalent force $C_k$ for fixed walls, calculated using Blum's approach for passive earth pressure, is determined using the equilibrium condition $\sum H_k = 0$ for all characteristic actions and support reactions. For all analyses in the vertical direction, it must be remembered that this method results in an equivalent force $C$ that is too large, because the full passive earth pressure mobilised for soil support $B_k$ is assumed to act down to the theoretical base of the sheet pile wall TF. When the true progression of the soil reaction $B_k$ is taken into account, the equivalent force $C$ is only about half its theoretical magnitude. At the same time, the associated soil support $B_k$ is reduced by exactly this value (see Fig. R 4-2 and R 56, section 8.2.9, Fig. R 56-1a).

In order to compensate for this error, the horizontal components of the equivalent force and the soil support according to Blum ($C_{h,k}$ and $B_{h,k}$) are each reduced by $\tfrac{1}{2} \cdot C_{h,k}$ when calculating the associated vertical components.

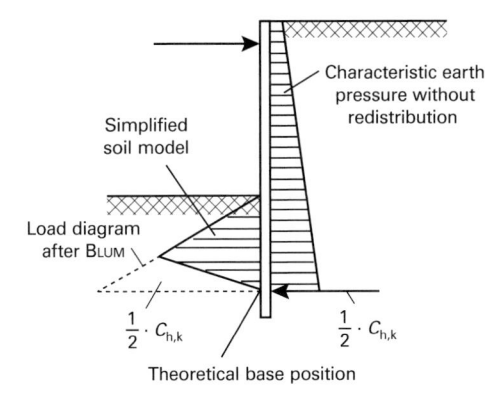

**Fig. R 4-2.** Effective portion of the soil reaction when fixed in the soil according to Blum

### 8.2.5.5 Analysis of vertical component of mobilised passive earth pressure

#### 8.2.5.5.1 Analysis format

This analysis, which must be carried out for each characteristic combination of actions, ensures that the angle of inclination $\delta_{p,k}$ selected for calculating passive earth pressure will actually be established in practice. The angle of inclination $\delta_{p,k}$ can only be applied to that negative value for which it has been proved that the downward characteristic actions $\Sigma V_{i,k}$ are greater than or equal to the upward characteristic actions $B_{v,k}$ (DIN 1054, section A 9.7.8). The analysis format required for this is as follows:

with simple support:

$$\Sigma V_{i,k} \geq |B_{v,k}|; \quad \text{where} \quad B_{v,k} = \Sigma B_{hi,k} \cdot \tan \delta_{pi,k}$$

with fixed wall:

$$\Sigma V_{i,k} \geq |B^*_{v,k}|; \quad \text{where} \quad B^*_{v,k} = \Sigma B_{hi,k} \cdot \tan \delta_{pi,k} - \tfrac{1}{2} C_{h,k} \cdot \tan \delta_{pr,k}$$

The analysis is to be carried out with the same angles of inclination used to calculate the active and passive earth pressures beforehand.
In order to carry out this analysis, the angle of inclination $\delta_{p,k}$ must be modified, if required, until the positive limit value for $\delta_{p,k}$ as per section 8.2.5.2 is reached. This results in a significant reduction in passive earth

pressure $E_{ph,k}$ and hence also in a greater required embedment depth $t_1$ for the wall.

The ground failure analysis in the passive earth pressure zone due to load $B_{h,d}$ from the soil support (see section 8.2.1) must be carried out for the modified angle of inclination $\delta_{p,k}$.

In the case of fixed walls, the soil support $B_k$ with $i = 1$ to $r$ strata down to the depth of the theoretical base TF is to be applied in the analysis according to section 8.2.9, Fig. R 56-1a.

### 8.2.5.5.2 Vertical component $V_{Q,k}$ from variable actions

The vertical component $V_{Q,k}$ due to variable actions $Q$ may not be used for this analysis if it does not make a significant contribution to the action effect of soil support $B_k$. This is the case, for instance, for actions that act directly at the top of the wall, e.g. the support reactions $F_{Qv,k}$ of the superstructure from crane and stack loads and the downward vertical components $\Delta A_{Qvi,k}$ of the anchor force due to horizontal, variable actions in the area around the top of the wall or above the anchor position, e.g.

- Crane lateral impact and storm locking device
- Line pull forces
- Earth pressure from variable actions on the wall area above the uppermost anchor position

### 8.2.5.5.3 Vertical force components $V_{i,k}$

The vertical force components $V_{i,k}$ are to be used with angle of inclination $\delta$ (positive upwards, negative downwards):

$V_{G,k} = \Sigma F_{G,k}$ due to *constant* axial effects $F$

$V_{Av,k} = P_{v,k,MIN}$ due to the anchor force for anchors inclined downwards

$P_{v,k,MIN} = P_{v,k} - \Delta P_{Qv,k}$ (according to last paragraph $= P_{v,k,MAX}$ for anchors inclined upwards or raking piles in compression)

$V_{Eav,k} = \Sigma(E_{ah,i,k} \cdot \tan \delta_{ai,k})$ due to earth pressure $E_{ah}$ with $i = 1$ to $r$ strata down to the depth of the theoretical base TF

$V_{Cv,k} = \frac{1}{2} C_{h,k} \cdot \tan \delta_{C,k}$ due to equivalent force $C_{h,k}$ (see also section 8.2.9)

### 8.2.5.6 Failure due to vertical movement

In addition to the analysis of the horizontal loadbearing capacity of the soil support and the vertical component of the mobilised passive earth pressure according to the model concept of active and passive sliding wedges of soil, an analysis must also be carried out with regard to the failure of soil-supported walls due to vertical movement as per section 8.2.1.

### 8.2.5.6.1 Model concept

The loading diagrams shown in Fig. R 4-3 are assumed. One of these two independent model concepts can be selected.

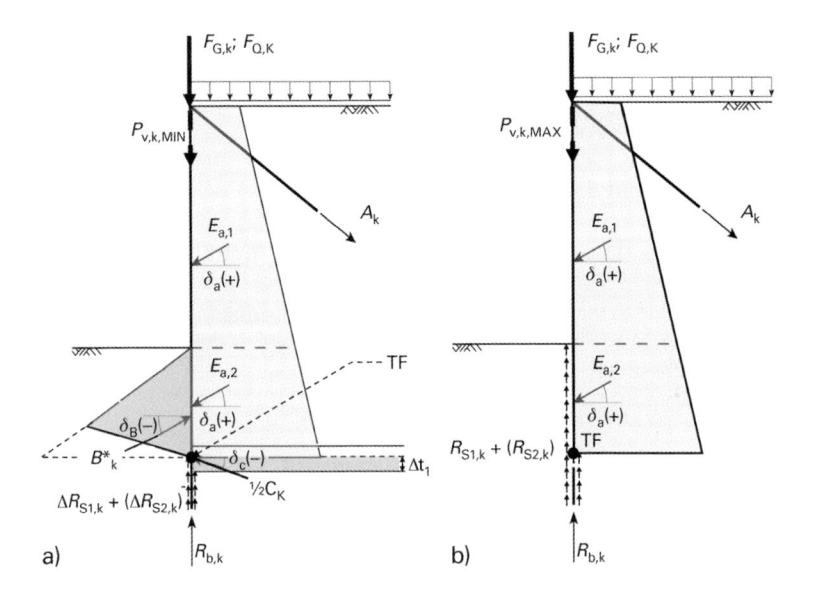

**Fig. R 4-3.** Approach for actions and resistances (example shows a fixed wall)

### 8.2.5.6.2 Analysis format

When analysing the safety of soil-supported walls against failure due to vertical movements in the subsoil (DIN EN 1997-1, section "note to 9.7.5"), all downward axial actions $\Sigma V_i$ and axial resistances $\Sigma R_i$ must be taken into account with their design values. The total load $V_d$ may not exceed the axial resistance $\Sigma Ri_{,d}$. The analysis for the limit state condition is

$$V_d = \Sigma V_{i,d} \leq \Sigma R_{i,d}$$

### 8.2.5.6.3 Vertical load $V_d$

In this case $V_d$ is the design value of all downward axial actions on a wall or soldier pile base according to DIN 1054:2010, section 9.7.5.

To calculate the load, all downward characteristic axial part-actions are multiplied by the partial safety factors for limit state GEO-2 as per Table R 0-1 for permanent ($G$) and variable ($Q$) actions for the respective design situation, separated within the combinations of actions according to cause.

$$V_{F,d} = \Sigma(V_{F,G,k} \cdot \gamma_G + V_{F,Q,k} \cdot \gamma_Q)$$

due to axial downward actions $F$

$$V_{Pv,d} = \Sigma(V_{Pv,G,k} \cdot \gamma_G + V_{Pv,Q,k} \cdot \gamma_Q)$$

due to anchor force component $P_v$

$$V_{Eav,d} = \Sigma(V_{Eav,G,k} \cdot \gamma_G + V_{Eav,Q,k} \cdot \gamma_Q)$$

due to the sum of stratum-by-stratum resultants $E_{av}$ arising from the earth pressure distribution of all strata down to the depth of the theoretical base TF.

### 8.2.5.6.4 Design values of axial resistances $R_{i,d}$

The design values $R_{i,d}$ of upward axial resistances are calculated by dividing the characteristic value $R_{i,k}$ of the individual resistance by the partial safety factors for limit state GEO-2 valid for the respective design situation.

In a similar way to the pile design, the partial safety factors for piles are used for skin friction and end bearing pressure.

The partial safety factor for passive earth pressure $\gamma_{R,e}$ is used for friction resistances $R_{Bv,k}$ or $R_B^*{}_{v,k}$ and $R_{Cv,K}$ due to the characteristic horizontal components of the soil support force $B_{h,k}$ or $B^*{}_{h,k}$ and half the equivalent force $\tfrac{1}{2}C_{h,k}$.

There are two options for the loadbearing capacity analysis:

a) Consideration of soil support (see Fig. R 4-3a)
   The following resistances are to be used:

$R_{Bv,d} = (B_{h,k} - \tfrac{1}{2}C_{h,k}) \cdot \tan \delta_B{}^{1)}/\gamma_{R,e}$    Wall friction resistance from mobilised soil support $B_{h,k}$

$R_{Cv,d} = \tfrac{1}{2}C_{h,k} \cdot \tan \delta_B/\gamma_{R,e}$    Wall friction resistance from half the equivalent force $C_{h,k}$

$R_{b,d} = R_{b,k}/\gamma_b{}^{3)}$    Base resistance due to end bearing pressure $R_{b,k}$
$R_{b,k} = A_{Pf} \cdot q_{b,soil}$

$\Delta R_{S1,d} = \Delta R_{s1,k}{}^{4)}/\gamma_S$ 

or
$R_{b,k} = A_W \cdot q_b$
Additional skin resistance due to skin friction (external)
$\Delta R_{s1,k} = U \cdot \Delta l \cdot q_S$
where $U = U_{PF}$ with plug formation
or $U = U_a$ without plug formation

$\Delta R_{S2,d} = \Delta R_{s2k}{}^{4)}/\gamma_S$

Additional skin resistance due to skin friction (internal)
$\Delta R_{s2,k} = U_i{}^{5)} \cdot 0.8(t + \Delta l) \, q_S$
only when plug formation at the base of the wall is impossible

b) Consideration of skin friction and end bearing pressure (see Fig. R 4-3b)
The following resistances are to be used:

$R_{s1,d} = R_{s1,k}{}^{4)}/\gamma_S$    Skin resistance due to skin friction (external)
$R_{s1,k} = U \cdot t \cdot q_{s,k}$
where $U = U_{PF}$ with plug formation
or $U = U_a$ without plug formation

$R_{s2,d} = R_{s2,k}{}^{4)}/\gamma_S$    Skin resistance due to skin friction (internal)
$R_{s2,k} = U_i \cdot 0.8 \cdot t \cdot q_s$
only when plug formation at the base of the wall is impossible

$R_{s,d} = Q_{s,k}/\gamma_S$    Skin resistance due to skin friction $Q_{s,k}$ when using $Q_{s,k}$ from loading tests

$R_{b,d} = R_{b,k}/\gamma_b{}^{3)}$    Base resistance
$R_{b,k} = A_{Pf} \cdot q_{b,soil}$ with plug formation
or
$R_{b,k} = A_W \cdot q_b$

$R_{b,d} = Q_{b,k}/\gamma_b$    Base resistance due to end bearing pressure $Q_{b,k}$ when using $Q_{b,k}$ from loading tests

[1] $\delta_B$ = magnitude of negative earth pressure angle of inclination during "failure due to vertical movement" analysis with $|\delta_B| = \varphi'_k$, irrespective of the analysis according to section 8.2.5.5.

[2] End bearing pressure $R_{b,d}$ is calculated by multiplying the end bearing pressure surface by the end bearing pressure at the base of the wall.

If the end bearing pressure is applied to the cross-sectional area $A_W$ of the wall section only, then the value from Table R 4-1 (which depends on $q_c$) can be used as an empirical value.

If the end bearing pressure is applied to a plug at base of the wall $A_{Pf}$, a geotechnical expert must be consulted to determine the area and the end bearing pressure within the scope of the design.

[3] The magnitude of the partial safety factor $\gamma_b$, which is independent of the design situation, depends on the end bearing pressure calculation. If the end bearing pressure $q_{b,k}$ is taken from
 - empirical values, then $\gamma_b = 1.40$,
 - loading tests, then $\gamma_b = 1.10$.

[4] To activate the additional skin friction resistances $\Delta R_{Si}$ required, the wall is to be extended beyond TF by $\Delta l$.

It is important to realise that, in the case of end bearing pressure that takes into account plug formation at the base of the wall, the skin friction applies solely to the peripheral area of the plug $U_{Pf}$. If there is no plug formation, $U_a$ (external) and $U_i$ (internal) may be taken as the developed surfaces of the wall section. The same values apply here for $\gamma_S$ as are listed for $\gamma_b$ under 3).

[5] In the case of I-sections in walls, the skin friction value as per Fig. R 4-4 can also be applied to the web and inner flange surfaces - only over 80% of the embedment depth, however.

### 8.2.5.6.5 Approaches for skin friction and end bearing pressure when analysing vertical loadbearing capacity

In non-cohesive soils, the characteristic empirical values of driven sheet pile walls and open bearing piles at the ultimate limit state for skin friction $q_{s,k}$ and end bearing pressure $q_{b,k}$ from Table R 4-1 can be used for preliminary calculations.

The values in the table are dependent on the resistance $q_c$, averaged over the depth, from cone penetration tests in non-cohesive soils. When determining the critical mean resistance $q_c$ in a cone penetration test, a distinction is made between the

  - zone critical for pile end bearing pressure ($1{*}D_{eq}$ above to $4{*}D_{eq}$ beneath toe of pile)
    and the

**Table R 4-1.** Empirical values to be used in preliminary calculations for the characteristic end bearing pressure $q_{b,k}$ and skin friction $q_{s,k}$ of open steel sections in non-cohesive soil

| Mean resistance $q_c$ in cone penetration test [MN/m²] | End bearing pressure $q_{b,k}$ at ultimate limit state [MN/m²] | Skin friction $q_{s,k}$ at ultimate limit state [kN/m²] |
|---|---|---|
| 7.5 | 7.5 | 20 |
| 15 | 15 | 40 |
| ≥25 | 20 | 50 |

- zone critical for pile skin friction (mean value of stratum concerned)

of the soil. If the stratification of the soil has a large influence on the resistance in the cone penetration test, then two or more middle zones are to be specified separately for pile skin friction.

Alternatively, the loadbearing capacity can be calculated using static and dynamic loading tests.

When mobilising axial resistances it is important to realise that the skin resistance is already effective after minor relative displacements, whereas base resistance requires large displacements – unless, on the basis of local experience, the pile sections have already been classed as sufficiently rigid during driving.

The empirical resistance and skin friction values given in Table R 4-1 have been derived from dynamic loading tests on pile and sheet pile sections driven over water. Where there is sufficient displacement of the wall due to backfilling, a higher horizontal stress state – compared with the state of the dynamic pile tests – ensues on the passive side of the wall. This increase is the result of the ratio of the mobilised horizontal passive earth pressure to the stress state after driving (see, for example, [137]). A higher skin friction value can be assumed on this surface provided a geotechnical expert is consulted. Based on geotechnical experience in Hamburg, the skin friction can be increased by up to a factor of 2.

A geotechnical expert is to be consulted during the design process when determining the resistances and surface areas.

The values given in Table R 4-1 apply to the sections commonly used in port and harbour construction:

- Sheet pile walls
- I-sections with $h \geq 0.50$ m
- II-sections with $h \geq 0.50$ m
- Pipe sections with $d \geq 0.80$ m

Resistance values for smaller section dimensions are given in *Recommendations on Piling* [55], section 5.4.4.

#### 8.2.5.6.6 Surface areas approach

Skin friction $q_{s,k}$ can be assumed on all the internal surfaces of open sections, see Fig. R 4-4, provided plug formation can be ruled out. Experience has shown that the inner skin height is 80% of the embedment depth in the loadbearing subsoil.

Skin friction $q_{s,k}$ cannot be used as a resistance on surfaces areas that are subjected to active earth pressure, see Figs. R 4-3b and R 4-4.

With section 8.2.5.6.4 in mind, either the vertical component of the soil support $B_{v,k}^*$ (Fig. R 4-3a) or skin friction $q_{s,k}$ (Fig. R 4-3b) can be applied on the passive earth pressure side.

In the case of combined sheet piling, the volumetric earth pressure is to be applied when calculating the soil support, e.g. in accordance with [225], when this is smaller than the continuous passive earth pressure for a selected embedment depth. As mentioned in section 8.1.4.2, for simplicity, the full passive earth pressure can assumed when using a clear bearing pile spacing of max. 1.80 m and a minimum embedment depth of 5.00 m in the passive earth pressure zone, even when the embedment depth of the intermediate piles is less than that of the bearing piles.

Where a plug forms, a reduced base resistance $R_{b,Pf,k}$ can be applied to the inner base surfaces of the open steel section in addition to the pile end bearing pressure, see section 8.2.5.6.7.

#### 8.2.5.6.7 Plug formation

In principle, plug formation is possible with open sections (tubes, U- and Z-section sheet piles, I-sections and box sections). Formation depends on the cross-section, the in situ density of the soil, the ratio of section diameter to embedment depth and the method of installation.

Plug formation is characterised by tension within the section which leads to the skin friction mobilised by the tension within the section being higher than the compressive force acting at the base on the column of soil inside the section. Consequently, a solid plug of soil forms inside the section. When using impact driving, dynamic effects can cause a continuous alternation between the formation and break-up of the solid plug, see [167], for example.

Findings in [141] assume plug formation at an embedment depth at least five times the pile diameter for driven tubes. An assessment of plug formation depending on section diameter and in situ density can be found in [111]. This applies to driven and pressed sections.

**Passive side**

·········· $R_{S,Bv}$ due to soil support force $B^*_{h,k}$ angle of inclination $\delta_B$

\*\*) In the case of a wall embedded below the theoretical base TF, an additional resistance $\Delta R_{S1}$ due to external skin friction can be applied to the passive side.

$R_{B,K}$ due to $q_{b,k}$ on cross-sectional area of wall section

$R_{S2}$ due to internal skin friction over 80 % of embedment from the bottom up

**Active side**

········ $R_{Cv}$ due to equivalent force $\frac{1}{2}C_{h,k}$ at angle of inclination $\delta_B$

\*) In the case of a wall embedded below the lowest point of the supplementary driving depth $\Delta t_1$, an additional resistance $\Delta R_{S1}$ due to external skin friction can be applied to the active side.

**Passive side**

— · — · — $R_{S,1}$ due to increased skin friction from the bottom up in $E_p$ zone

$R_{B,K}$ due to $q_{b,k}$ on cross-sectional area of wall section

$R_{S2}$ due to internal skin friction over 80 % of embedment from the bottom up

**Active side**

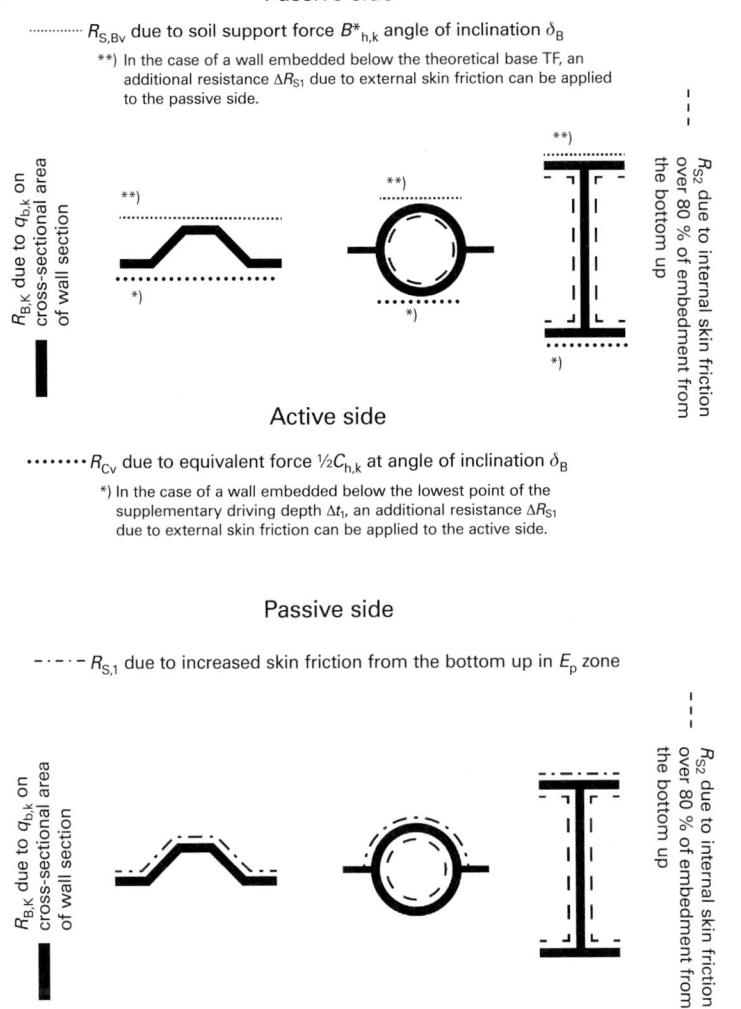

**Fig. R 4-4.** Resistant vertical components "$R$" from the design bottom up when analysing sinking; top: resistances due to earth pressure actions; bottom: resistances due to skin friction and end bearing pressure

385

Rausche et al. [169] state that plug formation cannot be expected in tubes with a diameter >1.5 m. Jardine et al. [111] conclude that plug formation is possible in tubular sections between 0.5 and 1.5 m in diameter.

Findings in [96] reveal that plug formation should not be assumed when using vibratory driving. The field measurements of Henke [94] confirm this for vibratory driving, whereas metrological observations of impact driving indicate plug formation. A high static axial load on the piles after driving considerably increases the probability of a plug forming.

Clausen et al. [40] present a method for calculating the end bearing pressure of driven tubular sections which is essentially dependent on the in situ density of the soil. In the case of loose bedding and assuming plug forming as per Clausen et al. [40], approx. 60% of the end bearing pressure of a solid cross-section can be assumed. In the case of dense bedding, the end bearing pressure should be reduced to 20%. A comparable method can be found in [131]. According to [40], the following applies for open sections:

$$\sigma_{b,\text{soil}} = 0,7 q_c / \left(1 + 3 I_D^2\right)$$

The provision of reinforcing plates at the base of the section to encourage plug formation should be agreed with a geotechnical expert in advance. The reinforcing plates can disrupt the internal skin friction and thus also reduce the probability of a plug forming, which leads to reduced loadbearing capacities, see [95], for example.

### 8.2.6 Taking account of unfavourable groundwater flows in the passive earth pressure zone (R 199)

Designs must allow for the influence of a flow around the sheet pile wall as a result of different water levels in front of and behind the wall (R 114, section 2.12.3.2).

Irrespective of this, it is necessary to verify the stability of the bottom against a hydraulic heave failure for limit state HYD.

Any risk to the stability caused by subsurface erosion of the bottom as a result of groundwater flows is to be investigated according to R 116, section 3.3. If necessary, the measures given in that section are to be initiated.

### 8.2.7 Verifying the loadbearing capacity of the elements of sheet piling structures (R 20)

#### 8.2.7.1 Quay wall

1. Predominantly constant loads
   Verifying the loadbearing capacity for all types of sheet pile wall is to be carried out according to DIN EN 1993-5. According to this

standard, the analysis format for safety against loss of the loadbearing capacity of the sheet pile wall section with the design value $E_d$ for internal forces and design value $R_d$ for the resistance of the section is

$$E_d \leq R_d$$

DIN EN 1993-5 refers to DIN EN 1997-1 for the method of calculation.

The torsional-flexural buckling analysis, which would only be carried out for I-section bearing piles in combined sheet piling, is unnecessary when the following boundary conditions apply:
- Fully backfilled combined sheet piling, or
- Combined sheet piling made from double I-section bearing piles (provided the bearing piles are embedded in loadbearing subsoil on at least three sides and the free-standing length is max. 7.5 m).

An analysis of skew bending is unnecessary for continuous U-section sheet pile walls that consist of double piles with shear-resistant connections, provided an elastic-elastic design method is used. Raking piles and all the components of the head details of piles and sheet piles for connecting walings, capping beams or reinforced concrete superstructures are designed according to DIN EN 1993-1-1. In any case, when designing the anchorage connection, the full characteristic loadbearing capacity $A_{pile} \cdot f_y$ of the actual installed anchor should be used as a design load for the connection.

2. Predominantly variable loads
Non-backfilled sheet pile walls free-standing in water are loaded primarily cyclically by the impact of waves. In this situation a large number of load cycles takes place over the lifetime of the wall so that verification of fatigue strength to DIN 19704-1 is required. In addition, please refer to DIN EN 1993-1-1. To prevent adverse effects caused by notch effects, such as from structural weld seams, tack welds, unavoidable irregularities on the surface due to the rolling process, pitting corrosion and the like, killed steels to DIN EN 10025 should be used in such cases.

### 8.2.7.2 Anchor wall, walings, capping beams and anchor head plates

1. Predominantly constant loads
Section 8.2.7.1, applies when verifying the loadbearing capacity of anchor plates and anchored sheet pile walls fixed in the ground. Walings, capping beams, bracing members and anchor head plates are designed according to DIN EN 1993-1-1. In doing so, it may be necessary to increase the partial safety factors for resistances for walings and capping beams according to R 30, section 8.4.2.3.

The resistance of sheet piles against the transfer of anchor and bracing forces must be verified according to DIN EN 1993-5, section 7.4.3.

2. Predominantly variable loads

Section 8.2.7.1(2), applies when verifying the loadbearing capacity. Close-tolerance bolts of grade 4.6 or higher are to be used for bolted connections in walings and capping beams. The fatigue strength analysis is to be carried out according to DIN EN 1993-1-1.

### 8.2.7.3 Round steel tie rods and waling bolts

The design of round steel tie rods and waling bolts is carried out in accordance with DIN EN 1993-5, section 7.2, using the notch factor $k_t$ and the core cross-sectional area $A_{core}$. (Therefore, the calculated design value of the section resistance lies on the safe side.)

#### 8.2.7.3.1 Predominantly static loads

The materials for round steel tie rods and waling bolts are given in section 8.1.22.3.

The analysis format for the ultimate limit state to DIN EN 1993-5 is

$$Z_d \leq R_d$$

The design values are calculated using the following variables:

| | |
|---|---|
| $Z_d$ | design value for anchor force, $Z_d = Z_{G,k} \cdot \gamma_G + Z_{Q,k} \cdot \gamma_Q$ |
| $R_d$ | design resistance of anchor, $R_d = \min [F_{tt,Rd}; F_{tg,Rd}]$ |
| $F_{tg,Rd}$ | $A_g \cdot f_y / \gamma_{M0}$ |
| $F_{tt,Rd}$ | $k_t \cdot A_{core} \cdot f_{ua,k} / \gamma_{M2}$ |
| $A_g$ | cross-sectional area of shaft |
| $A_{core}$ | core cross-sectional area of thread |
| $f_{y,k}$ | yield stress |
| $f_{ua,k}$ | tensile strength |
| $\gamma_{M0}$ | partial safety factor to DIN EN 1993-5 for anchor shaft |
| $\gamma_{M2}$ | partial safety factor to DIN EN 1993-5 for threaded segment |
| $k_t$ | notch factor ($k_t = 0.55$) |

DIN EN 1993-5/NA, section 3.14, takes into account the guidance of this book and has established the notch factor $k_t = 0.55$ for calculating the resistance of the threaded segment. This approach and using the core cross-section takes into account any additional stresses due to the installation of the anchor under the harsh conditions of a building site and any ensuing, unavoidable bending stresses in the threaded segment. Notwithstanding, it is still necessary to provide constructional measures to ensure that the anchor head can rotate sufficiently.

The additional serviceability analyses called for in DIN EN 1993-5 are already implied in the limit state condition $Z_d \leq R_d$ owing to the value selected for the notch factor $k_t$ and the customary upsetting relationships between shaft and thread diameters; the additional analyses are therefore unnecessary. Round steel tie rods can have cut, rolled or hot-rolled threads to R 184, section 8.4.8.

A prerequisite for a proper design is a suitable detail for the anchor connection, which means anchors must be connected via a form of hinge. Anchors must be installed at a higher level so that any settlement or subsidence does not cause any additional stresses.

Upsetting the ends of the anchor bars for the threaded segments and T-heads as well as tie rods with eyes are permissible when

- using steel grade groups J2 and K2, if necessary in the normalised/ normalising rolled condition (+N) – however, not thermomechanically rolled J2 and K2 group steels (see R 67, section 8.1.6.1);
- other steel grades, e.g. S 355 J0, are used and accompanying tests ensure that the strength values do not fall below those in DIN EN 10025 after the normalising procedure of the forging process;
- the upsetting and the provision of T-heads and eyes are carried out by specialist fabricators and it is ensured that the mechanical and technological values in all parts of the round steel tie rod are in accordance with the steel grade selected, the alignment of the fibres is not impaired during the machining process and detrimental microstructure disruptions are reliably avoided.

The "failure of an anchor" analysis does not need to be performed for round steel tie rods and anchor piles because the aforementioned verification of loadbearing capacity includes the notch factor $k_t$, the anchors are connected to ensure the full internal loadbearing capacity and the tie rods therefore have sufficient loadbearing reserves to avoid any possible failures.

As is mentioned in R 35, section 8.1.8.4(4), round steel tie rods should be installed without an anti-corrosion coating.

After installation in fill, round steel tie rods must be surrounded by a sufficiently thick layer of sand over their full length.

If it is necessary to coat round steel tie rods to prevent corrosion, measures will be necessary on site to ensure that the coating is not damaged. If, despite this, damage does occur, the coating must be repaired so that the original quality is restored.

The aforementioned measures reduce the risk of anodic areas on the round steel tie rods and any ensuing pitting.

The design and installation of sheet pile wall anchorages with grouted round steel tie rods is covered by DIN 1054 and DIN EN 1537.

#### 8.2.7.3.2 Predominantly fluctuating loads

Anchors are generally mainly subjected to static loads. Primarily fluctuating loads only occur in anchors in rare, special cases (section 8.2.7.1(2)), but more frequently in waling bolts.

Only fully killed steels to DIN EN 10025 may be used where fluctuating loads are likely.

Verification of fatigue strength is to be carried out according to DIN EN 1993-1-9.

If the basic static load is less than or equal to the reversed load amplitude, the recommendation is to apply a permanent, controlled prestress to the anchors or waling bolts which exceeds the stress amplitude. This ensures that the anchors or waling bolts remain under stress and do not fail abruptly when the stress increases again.

A prestress, which is not defined exactly, is applied to anchors and waling bolts, in many cases during the installation procedure. In cases without controlled prestressing, a stress of only $\sigma_{R,d} = 80 \text{ N/mm}^2$ may be assumed for the threads of anchors or waling bolts, regardless of design situation and steel grade, and neglecting the prestress.

Always ensure that the nuts of waling bolts cannot loosen due to repeated changes in stress.

### 8.2.8 Selection of embedment depth for sheet piling (R 55)

Structural, constructional, operational and economic matters may be relevant to the embedment depth of sheet piling in addition to the corresponding loadbearing analyses and the supplement required by R 56, section 8.2.9. Any foreseeable future deepening of the harbour bottom and any possible danger of scour below the design bottom should be considered to the same extent as the required margin of safety against slope, foundation, heave and erosion failures.

These latter requirements usually result in such a large minimum embedment depth that partial fixity is available at least – apart from the special case of foundations in rock. Even if a simple support would be adequate theoretically, it is often advisable to increase the embedment depth because this can have economic advantages as well. The resistance of the section is utilised more uniformly over the length of the sheet pile wall and therefore at least partial fixity of the sheet pile wall is advisable when using the Blum method [22].

If the sheet pile wall also has to transfer vertical loads into the subsoil, not all the piles have to continue as far as the loadbearing stratum. Instead, it can be sufficient to make the embedment length of only some of the piles long enough so that they are effective as vertical loadbearing piles provided it can be shown that this number of piles can carry the loads without sinking into the subsoil.

## 8.2.9 Determining the embedment depth for sheet pile walls with full or partial fixity in the soil (R 56)

If sheet piling is designed according to [22], with full fixity in the soil (degree of fixity $\tau_1 = 100$ %) the entire embedment length below the design bottom consists of the embedment depth $t_1$ down to the theoretical base plus the extra depth $\Delta t_1$ (additional driving depth). The extra length $\Delta t_1$ is necessary in order to accommodate the design value of the (mobilised) equivalent force $R_C$ (corresponding to the equivalent force $C$ of [22]) actually acting on the theoretical base TF as a soil reaction force distributed over depth $\Delta t_1$.

Notation for Figs. R 56-1a and R 56-1b:

| | |
|---|---|
| $t$ | required total embedment depth $t = t_1 + \Delta t_1$ for the sheet pile wall fixed in the soil [m] |
| TF | theoretical base of sheet pile wall (load application point for equivalent force $C$) |
| $t_1$ | distance between TF and design bottom [m] |
| $\Delta t_1$ | additional depth for accommodating equivalent force $\frac{1}{2} C_{h,d}$ via a soil reaction force below TF [m] |
| $\sigma_{z,C}$ | vertical soil stress at TF on equivalent force side [kN/m²] |
| $\delta_{p,k}$ | angle of inclination of passive earth pressure [°] |

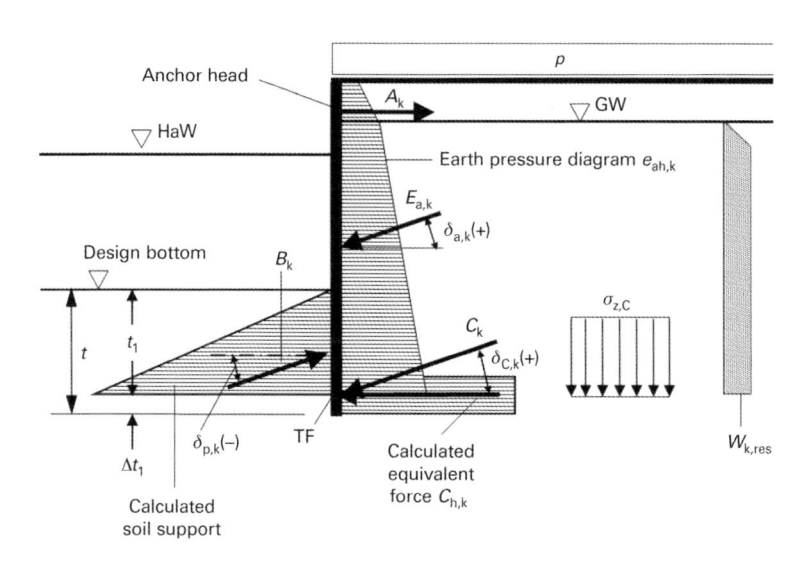

**Fig. R 56-1a.** Actions, calculated support and soil reactions of a sheet pile wall with full fixity in the soil

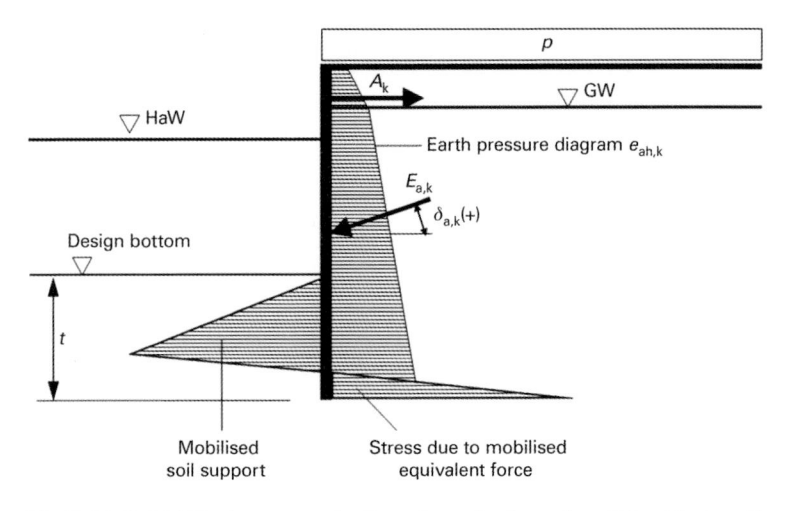

**Fig. R 56-1b.** Mobilised support and soil reactions of a sheet pile wall fixed in the soil

$K_{\text{pgh,C}}$     passive earth pressure factor at TF on equivalent force side for angle of inclination $\delta_{\text{C,k}}$

$\delta_{\text{C,k}}$     angle of inclination of equivalent force $C$ [°]

If the more exact analysis given below for calculating $\Delta t_1$ is not carried out, the additional depth for sheet pile walls fully fixed in the ground can be simplified to

$$\Delta t_1 = t_1/5$$

However, this is only possible when the actions do not include any significant hydrostatic pressure component.

The design value $C_{h,d}$ is

$$C_{h,d} = \Sigma(C_{\text{Gh,k}} \cdot \gamma_G + C_{\text{Qh,k}} \cdot \gamma_Q)$$

or

$$C_{h,d} = \Sigma(C_{\text{Gh,k}} \cdot \gamma_G + C_{\text{Gh,W,k}} \cdot \gamma_{G,\text{red}} + C_{\text{Qh,k}} \cdot \gamma_Q)$$

in the case of a reduced partial safety factor for the hydrostatic pressure component and separation according to equivalent force portions. The calculated equivalent force portions are:

$C_{Gh,k}$ due to permanent actions $G$
$C_{Gh,W,k}$ due to permanent hydrostatic pressure actions
$C_{Qh,k}$ due to variable actions $Q$

The associated partial safety factors are:

$\gamma_G$ for permanent actions
$\gamma_{G,red}$ for hydrostatic pressure with a permissible reduction
$\gamma_Q$ for variable actions

At failure, the characteristic value $E_{phC,k}$ of the soil support for accommodating the actual equivalent force $\frac{1}{2}C_{h,d}$ is the magnitude of the passive earth pressure on the equivalent force side beneath the theoretical base TF:

$$E_{phC,k} = \Delta t_1 \cdot e_{phC,k}$$

The characteristic value of the passive earth pressure stress $e_{phC,k}$ on the equivalent force side at the level of TF is

$e_{phC,k} = \sigma_{z,C} \cdot K_{pgh,C}$ in non-cohesive soils, and
$e_{phC,k} = \sigma_{z,C} \cdot K_{pgh,C} + c'_k \cdot K_{pch,C}$ in cohesive soils

(taking into account the respective consolidation state as a result of the shear parameters $c_{u,k}$ or $\varphi'_k$ and $c'_k$).
The vertical soil stress $\sigma_{z,C}$ is to be calculated at the level of the base TF on the equivalent force side.
The design value $E_{phC,d}$ of the soil support for accommodating the equivalent force $\frac{1}{2}C_{h,d}$ is calculated with the partial safety factor $\gamma_{R,e}$ for passive earth pressure:

$$E_{phC,d} = E_{phC,k}/\gamma_{R,e}$$

The analysis format for complying with the limit state condition for accommodating the equivalent force $C_{h,d}$ as a soil reaction is

$$\tfrac{1}{2}C_{h,d} \leq E_{phC,d}$$

From this limit state condition we get the magnitude of the required additional depth $\Delta t_1$ beneath the theoretical base TF for walls fully fixed in the soil:

$$\Delta t_1 \geq \tfrac{1}{2} C_{h,d} \cdot \gamma_{R,e} / e_{phC,k}$$

The above equation for the extra depth in the case of full fixity in the ground (degree of fixity $\tau_1 = 100\%$) is also used to determine the extra depth for sheet pile walls with only partial fixity in the ground, i.e. for any degree of fixity over a possible range $\tau_1 = 100\%$ to $\tau_0 = 0\%$ (simply supported in the soil).

The degree of fixity designated here with $\tau_{1-0}$ for a partially fixed sheet pile wall becomes $\tau_{1-0} = 100 \cdot (1 - \varepsilon/\max \varepsilon)$ [%] with the end tangent angle $\varepsilon$ of the deflection curve for the theoretical base TF selected and the end tangent angle max. $\varepsilon$ for a simple support in the soil. The embedment depth associated with the degree of fixity $\tau_{1-0}$ is designated $t_{1-0}$ and the extra depth $\Delta t_{1-0}$.

Partial fixity in the soil is associated with – compared with full fixity – smaller values for the equivalent force component $C_{h,d}$ and hence also additional depths $\Delta t_{1-0} < \Delta t_1$. In the case of a simple support for the sheet pile wall in the soil ($\tau_0 = 0\%$), $C_{h,d} = 0$ and $\Delta t_0 = 0$ apply.

A required minimum value $\Delta t_{MIN}$ must be maintained for the additional embedment depth, which is defined depending on the degree of fixity present ($100\% \geq \tau_{1-0} \geq 0\%$):

$$\Delta t_{MIN} = (\tau_{1-0}/100) \cdot t_{1-0}/10$$

### 8.2.10 Steel sheet piling with staggered embedment depths (R 41)

#### 8.2.10.1 Applications

Sheet piles (generally double piles) are frequently driven to different depths for technical and, in the case of fully fixed walls, for economic reasons, too. The permissible extent of these alternating embedment depths, known as a staggered arrangement, depends on the stresses in the bases of the longer piles and on construction issues. For driving reasons, staggering within a single pile unit is not recommended for sheet piles. At failure, a uniform, continuous passive earth pressure zone (which depends on the geometrical boundary conditions according to R 7, section 8.1.4.2) forms in the region of the soil support of staggered sheet piling (similar to the wedge of soil in front of closely spaced anchor plates). The full soil reaction down to the bottom of the deeper sheet piles therefore applies when determining the loads without taking into account the staggering. The bending moment at the bottom edge of the shorter piles must be resisted by the longer piles alone. Therefore, in order to limit the stresses in the deeper piles, only adjacent pile units (double piles at least) in sheet pile walls are staggered (Figs. R 41-1 and R 41-2). A length of 1.0 m is usual for staggered arrangements. In practice it has been found that a structural check of the longer piles is unnecessary. For a greater staggered length, it will be necessary to verify the loadbearing capacity of the deeper piles with respect to multiple stresses due to bending moments combined with axial and shear forces.

### 8.2.10.2 Sheet pile walls fixed in the soil

- According to the Blum method [22], sheet pile walls fully fixed in the soil ($\tau_1 = 100\%$) may exploit the entire stagger dimension $s$ to save steel: the longer sheet piles are driven to the depth determined according to R 56, section 8.2.9 (Fig. R 41-1), the shorter piles terminate at a level higher by the stagger dimension $s$.
- In the case of walls partially fixed in the soil ($100\% > \tau_{1\text{-}0} > 0\%$), the steel saving depends on the degree of fixity present. A corresponding saving in steel is achieved by driving the longer sheet piles below the level of the theoretical base of the wall by a certain fraction of the stagger dimension $s_U$, with the shorter piles again terminating at a level higher by the stagger dimension $s$.

The dimension $s_U$ depends on the degree of fixity $\tau_{1\text{-}0}$ [%]:

$$s_U = (100 - \tau_{1-0}) \cdot s / (2 \cdot 100)$$

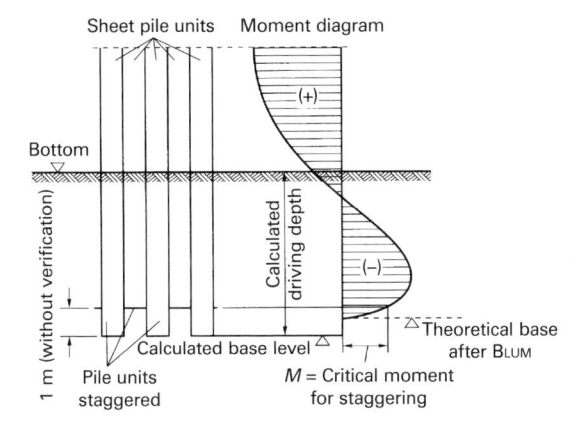

**Fig. R 41-1.** Staggered base of sheet piling for a sheet pile wall fully fixed in the soil

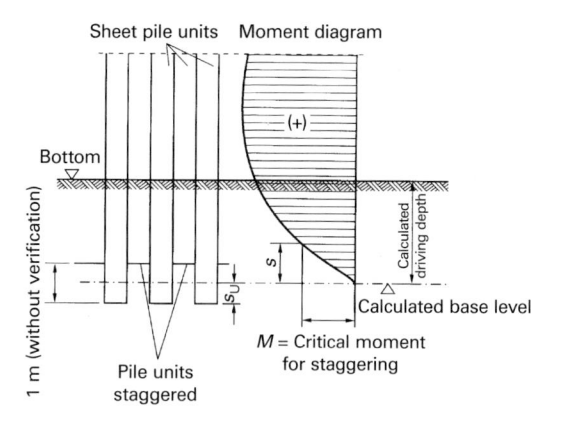

**Fig. R 41-2.** Staggered base of sheet piling for a sheet pile wall simply supported in the soil

### 8.2.10.3 Sheet pile walls simply supported in the soil

When the sheet piling is simply supported in the ground, the stagger dimension $s$ no longer leads to a saving in steel (owing to the equation for dimension $s_U$ from section 8.2.10.2, which also applies for degree of fixity $\tau_0 = 0\%$). Instead, it leads to an enlargement of the soil zone that can be mobilised for support B, but this cannot be used in the calculations.

In this case the longer sheet piles (see Fig. R 41-2) must be driven below the theoretical underside of the wall by a distance $s_U = s/2$.

With a stagger dimension $s > 1.0$ m, the loadbearing capacity of the longer piles must be verified according to Fig. R 41-2.

The same applies to reinforced concrete or timber sheet piling, provided the joints between the piles have sufficient strength to guarantee that the longer and shorter sheet piles work together.

### 8.2.10.4 Combined sheet piling

Sheet piling composed of bearing and intermediate piles (R 7, section 8.1.4) must take into account the excess water pressure due to flows around the wall such that the required margin of safety against hydraulic heave failure (R 155, section 3.2) is guaranteed in front of the shorter, intermediate piles. Longer piles should be used when there is a risk of scouring.

Where the subsoil at the bottom of the watercourse includes soft or very soft strata, the embedment depth of the shorter, intermediate piles should be determined by special investigations.

## 8.2.11 Horizontal actions on steel sheet pile walls in the longitudinal direction of the quay (R 132)

### 8.2.11.1 General

Combined piling and corrugated steel sheet pile walls react comparatively flexibly to horizontal actions in the longitudinal direction of the quay. If such actions occur, a check must be carried out to verify that the resulting horizontal action effects parallel to the quay can be accommodated by the sheet piling, or whether additional measures are required. In many cases, in-plane stresses in sheet piling structures due to earth and hydrostatic pressure can be avoided by choosing an appropriate design. Such designs can involve, for example, crossing anchors at quay wall corners according to R 31, section 8.4.11. Another example is radial arrangements of round steel tie rods, with an anchor plate at the centre of the curve, for curving sections of quay walls or pier heads; further tie rods then continue back from this central plate to an anchor wall of sheet pile sections. The anchor force resultants of the anchors connected here must act in the same direction as the anchor force resultant of the radial anchors so that the central anchor plate remains in equilibrium.

### 8.2.11.2 Transferring horizontal forces into the plane of the sheet piling

Available construction components such as capping beams and walings can be used to transfer horizontal forces provided they have been designed to do so. Otherwise, additional measures are required, e.g. the installation of diagonal bracing behind the wall. Welding the interlocks in the upper section of the wall will suffice in the case of smaller longitudinal forces.

The action components parallel to the quay due to line pull forces act at the mooring points, the maximum actions due to wind at crane wheel locking points and those due to ship friction at the fenders. The load application points of these friction forces can occur at any place along the wall. This also applies to horizontal loads as a result of crane braking, which must be transferred from the superstructure into the top of the wall. Longitudinal forces can be carried over a longer length of the wall provided the distributing construction components have been designed accordingly.

For this reason, the flanges of steel walings should be bolted or welded to the land-side sheet pile flanges (Fig. R 132-1).

Longitudinal forces can also be transferred via cleats that are welded to the waling and braced against the sheet pile webs (Fig. R 132-2).

When a waling consists of two channel sections, the waling bolts can only be used to transfer longitudinal forces when the two channels are joined by a vertical drilled plate welded in place on the land-side sheet piling flange. The force from the waling bolts is then accommodated by

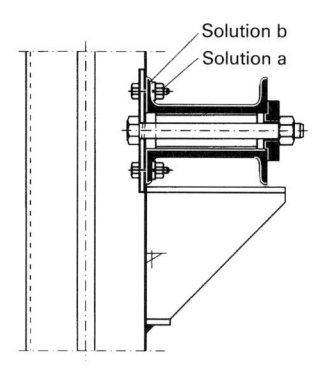

**Fig. R 132-1.** Transferring longitudinal forces by means of close-tolerance bolts in the waling flanges (solution a) or by welding (solution b)

the plate via bearing stresses, whereas the bolts are subjected to shearing stresses (Fig. R 132-3).

When loads act parallel to the plane of the sheet pile wall, the capping beams and walings, including their splices, should be designed for bending combined with axial and shear forces.

In order to transfer horizontal actions in the direction of the quay from a reinforced concrete capping beam to the top of a sheet pile wall, the latter must be adequately embedded in the beam. The design of the reinforced concrete cross-section must take account of all the global and local loads that occur in this area.

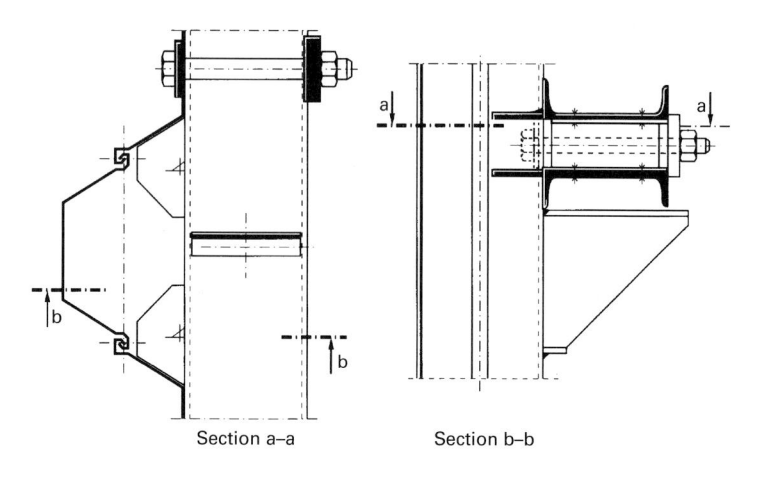

**Fig. R 132-2.** Transferring longitudinal forces by means of steel cleats welded to the waling

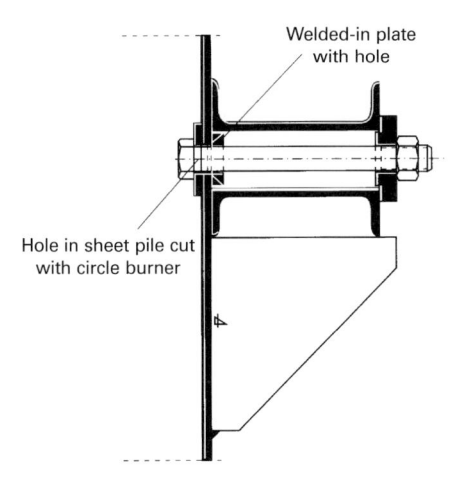

**Fig. R 132-3.** Transferring longitudinal forces by means of waling bolts and welded plate with hole drilled prior to welding

## 8.2.11.3 Transferring horizontal forces acting parallel to the line of the sheet piling from the plane of the wall into the subsoil

Horizontal longitudinal forces in the plane of the sheet pile wall are transferred into the ground by friction on the land-side sheet pile flanges and by resistance in front of the sheet pile webs. The latter, however, cannot be greater than the friction in the ground over the length of the sheet pile trough.

In non-cohesive soils the force can therefore be accommodated entirely by friction, for which a reasonable mean value of the friction coefficient between soil and steel as well as between soil and soil is used. The effectiveness of this force transfer in non-cohesive soils increases with the angle of internal friction and the strength of the backfill; in cohesive soils the force transfer improves with the shear strength and the consistency of the soil.

When transferring horizontal forces acting parallel to the plane of the sheet pile wall from a capping beam or waling into the subsoil, additional bending moments occur transverse to the principal loadbearing direction of the sheet pile wall. These bending moments can be calculated with the numerical model of a fixed or simply supported anchor wall. Instead of the resisting soil reaction as a result of the mobilised passive earth pressure, however, the aforementioned mean wall friction force, or a corresponding shear resistance, is assumed in this case.

As a rule, only double piles joined with shear-resistant welds should be considered as loadbearing elements carrying these additional loads.

Where welding is not used, the piles should be considered as single piles only.

When accommodating horizontal forces in the longitudinal direction of the quay, the sheet piles are stressed by bending in two planes. When superimposing the resulting stresses, DIN EN 1993-1-1 permits the design value for the equivalent tensile stress $\sigma_{v,d}$ to be used, which – for the boundary conditions for individual corner stresses given in the standard – may exceed the maximum permissible normal stress $\sigma_{R,d}$ by 10%.

By taking into account friction resistances in the horizontal direction, only a reduced angle of earth pressure inclination $\delta_{a,k,red}$ may be used in the sheet piling calculations. In doing so, the magnitude of the resultant of the vector addition of the two friction resistances orthogonal to each other may not exceed the maximum possible wall friction resistance value of the sheet pile wall with respect to the soil.

### 8.2.12 Design of anchor walls fixed in the ground (R 152)

Obstacles in the ground, such as ducts, pipes, cables, etc., sometimes mean that it is not possible to connect round steel tie rods to the centre of the anchor wall. In such cases the tie rods must then be positioned at a higher level and connected to the upper area of the anchor wall.

The calculations for the anchor wall in such cases are to be carried out as for a non-anchored sheet pile wall for limit state STR or GEO-2 (Fig. R 152-1). The anchor force $A_k$ of the sheet pile wall to be

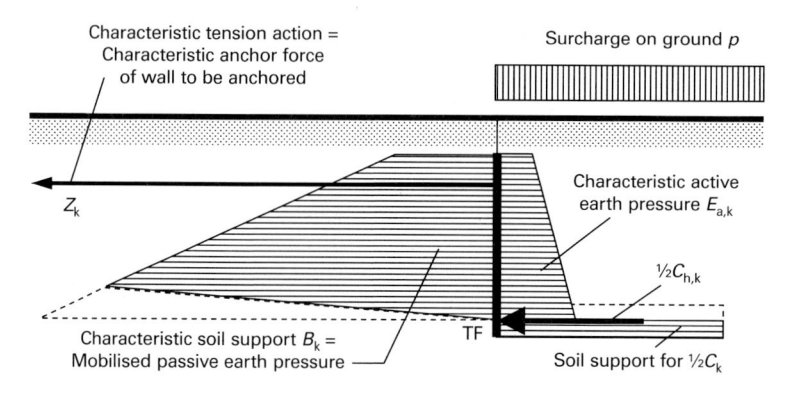

**Fig. R 152-1.** Actions, soil reaction and equivalent force for an anchor wall fully fixed in the ground as required for stability verification for limit state GEO-2

anchored is entered into the calculation as the characteristic value $F_k$ of the tensile force acting at the top of the anchor wall.

The partial safety factors for earth pressure and, if applicable, hydrostatic pressure actions are applied according to R 214, section 8.2.1.1. The factor for the tensile force $F_k$ is the quotient of the design value $A_d$ and the characteristic value $A_k$ of the anchor force, both of which are taken from the calculations for the sheet pile wall.

The extra embedment depth $\Delta t_1$ required is calculated according to R 56, section 8.2.9.

In this form of construction the sheet piles of the anchor wall are considerably longer than an anchor plate loaded centrally and also have a larger cross-section.

Staggering the anchor wall according to R 42, section 8.2.13, is only permissible at the base of the wall, but a stagger of up to 1.0 m for deep anchor walls is possible without any special verification.

In the case of predominantly horizontal groundwater flows, a number of weepholes must be provided in the anchor wall if the hydrostatic pressure acting on the wall needs to be reduced. The resulting hydrostatic pressure must be taken into account in the design of the sheet piles for the anchor wall.

## 8.2.13 Staggered arrangement of anchor walls (R 42)

In order to save materials, anchor walls may be staggered in the same manner as the waterfront sheet piling. Both ends may be staggered in the same wall. However, the stagger should not exceed 0.5 m top and bottom. When the wall is staggered top and bottom, all the double piles can be of the same length, i.e. 0.5 m shorter than the overall height of the anchor wall. The double piles are driven so that one double pile is driven to the bottom of the anchor wall, the next double pile driven so that its top matches the top of the anchor wall and so on, alternately. A stagger >0.5 m is permissible for deep anchor walls only, when the bearing capacity of the soil support and the loadbearing capacity of the section in terms of bending moment, shear force and axial force have been verified. Such verification is also required for a stagger of 0.5 m, however, if the overall height of the anchor wall is <2.5 m. In such a case it must be verified that the bending moments can be transferred from the deeper ends to the neighbouring piles.

The same applies to reinforced concrete or timber sheet piling, provided the joints between the piles have sufficient strength to guarantee that the longer and shorter sheet piles work together.

## 8.2.14 Steel sheet piling founded on bedrock (R 57)

When bedrock exhibits a fairly thick decomposed transition zone, with strength increasing with depth, or when the rock is soft, experience has shown that steel sheet piles can be driven deep enough into the rock to achieve at least a simple support.

In order to be able to drive sheet piles into bedrock, the piles must be modified and strengthened at the toe, and if need be, also at the top, depending on the pile section and the type of rock. Sheet piling of steel grade S 355 GP (R 67, section 8.1.6) is recommended, considering the high driving energy required. Heavy hammers and a correspondingly smaller drop height are very effective for this work. A similar effect can be achieved with hydraulic hammers whose impact energy can be controlled to suit the particular driving energy requirement (R 118, section 8.1.11.3).

Where there is strong, hard rock up to the surface, test piles and rock investigations are indispensable. If necessary, special measures must be taken to protect the toe of the pile and to ensure proper alignment. Holes 105–300 mm in diameter can be drilled at a spacing corresponding to the width of the sheet piles to perforate and relieve the subsoil and thus facilitate the driving of sheet piles.

The same effect can be achieved by high-pressure jetting in rock with changing strength and similar subsoils (see R 203, section 8.1.24.3).

If sheet piles must be driven deep into bedrock, accurate blasting can be used to loosen the rock along the line of the sheet piling and thus facilitate driving. When selecting the section and the grade of steel, possible irregularities in the subsoil and the ensuing driving stresses must be taken into account. Refer to R 183, section 8.1.10, for details of blasting.

The advantage of pre-drilling is that the properties of the rock in its undisturbed state remain intact. Compared with blasting, this has positive effects on the lower support reaction of the sheet piling. The pre-drilling depth is also less than the depth required for blasting.

## 8.2.15 Waterfront sheet piling in unconsolidated, soft cohesive soils, especially in connection with non-sway structures (R 43)

For various reasons, ports, harbours and industrial facilities with associated waterfront structures must sometimes be built in areas with poor subsoil. Alluvial cohesive soils, possibly with peat inclusions, are thus subjected to higher loads due to increases in the level of the terrain and hence placed in an unconsolidated state. The resulting settlement and horizontal displacements call for special construction features and a structural design treatment that is tailored to the particular site.

In unconsolidated, soft cohesive soils, sheet piling structures may only be designed as "floating" structures when neither the serviceability nor the stability of the entire structure and its parts will be endangered by the resulting (differential) settlement and horizontal displacements. To assess this and initiate the necessary measures, the expected settlement and displacements must be calculated.

If a quay wall is constructed in unconsolidated, soft cohesive soil in connection with a structure on a practically immovable foundation, e.g. on a pile trestle with vertical piles, the following solutions are possible: The sheet piling may be anchored or supported so that it is free to move in the vertical direction but such that the connection to the structure remains fully effective and loadbearing even in the case of the maximum theoretical displacements.

This solution is quite straightforward apart from the settlement and displacement calculations. For operational reasons, however, it can generally only be used for a rearward sheet pile wall for designs involving pile trestles. The vertical friction force at the support must be taken into account in the design of the trestle. Slotted holes are not sufficient at the anchor connections of sheet piling in front of the structure. In fact, a sliding anchorage is then required.

The sheet piling is supported against vertical displacements by driving a sufficient number of pile sections deep enough to reach the loadbearing subsoil deep in the ground. In this situation the loadbearing capacity of the sheet pile wall with respect to the following vertical loads must be guaranteed by the deeper piles alone:

- the self-weight of the wall,
- soil clinging to the sheet piling as a result of negative skin friction and adhesion, and,
- axial actions on the wall.

This solution is practicable in technical and operational terms when the sheet piling is located in front of the structure. Since the cohesive soil clings to the sheet piling during the settlement process, the active earth pressure decreases. If the supporting soil in front of the base of the sheet piling also settles, however, the characteristic passive earth pressure, and hence the potential soil support, also decreases as a result of negative skin friction. This must be taken into account in the sheet piling calculations. When calculating the vertical load on the sheet piling arising from soil settlement, then negative skin friction and adhesion for the initial and final states are taken into consideration.

Apart from the anchoring or support against horizontal forces, the sheet piling should be so suspended from the structure such that the aforementioned actions are transferred to the structure and from there to the loadbearing soil.

In this solution the sheet piling and its upper suspension are calculated using the aforementioned information.

If the loadbearing soil is located at a reasonable depth in construction terms, the entire wall can be driven down into the loadbearing stratum.

The passive earth pressure in this stratum is calculated with the usual earth pressure angles of inclination and the partial safety factors as per Table R 0-2. When calculating the soil reaction of the overlying soil with a lower strength or consistency, only a reduced characteristic passive earth pressure may be assumed. Serviceability must be verified.

## 8.2.16 Design of single-anchor sheet piling structures in earthquake zones (R 125)

### 8.2.16.1 General

First of all, the findings of soil investigations and soil mechanics tests must be carefully checked to establish what effects the vibrations caused by a critical earthquake may have on the shear strength of the subsoil. The results of these checks can be critical for the design of the structure. For example, when soil conditions are such that liquefaction is to be expected as per R 124, section 2.13, anchorages with high-level anchor walls or anchor plates are not permitted unless the mass of earth supporting the anchoring is adequately compacted in the course of the construction work and the danger of liquefaction thus eliminated. Please refer to R 124, section 2.13, for the magnitude of the seismic coefficient $k_h$ and other actions as well as the design values for loads and resistances plus the safety factors required.

### 8.2.16.2 Sheet piling calculations

Taking into account the sheet piling loads and support reactions determined according to R 124, sections 2.13.3, 2.13.4 and 2.13.5, the calculation can be carried out as per R 77, section 8.2.3, but without redistribution of the active earth pressure.

The characteristic active and passive earth pressures determined with the fictitious angles of inclination for the reference and ground surfaces are used as a basis for all calculations and analyses, although tests have revealed that the rise in the active earth pressure due to an earthquake does not increase linearly with the depth, instead, is higher near the surface. Anchorages must therefore be generously dimensioned.

### 8.2.16.3 Sheet piling anchorages

Verification of stability for anchorages at the lower failure plane is to be carried out according to R 10, section 8.4.9, taking into account the additional horizontal forces occurring at reduced variable loads due to acceleration of the mass of earth to be anchored and the pore water contained therein.

## 8.3 Calculation and design of cofferdams

### 8.3.1 Cellular cofferdams as excavation enclosures and waterfront structures (R 100)

#### 8.3.1.1 General
Cellular cofferdams are constructed from straight-web pile sections with a high interlock tensile strength of 2.0–5.5 MN/m. The advantage of such cofferdams is that, just by filling them with a suitable material, they can function as stable gravity walls without walings and anchors, even if embedment of the steel walls is impossible owing to rocky subsoil.

Cellular cofferdams can be economic where deeper water, i.e. large differences in ground levels, coincides with longer structures and when bracing or anchoring is impossible or uneconomic. In these situations the additional sheet piling surface area is compensated for in some circumstances by the weight-savings offered by the lighter and shorter sheet pile sections and the omission of walings and anchors.

#### 8.3.1.2 Types of cellular cofferdam
We distinguish between

- cellular cofferdams with circular cells (Fig. R 100-1a),
- cellular cofferdams with diaphragm cells (Fig. R 100-1b), and
- mono-cell cofferdams.

1. Circular cell cofferdams
   Circular cells, which are linked by small, connecting arcs, have the advantage that each main cell can be individually constructed and

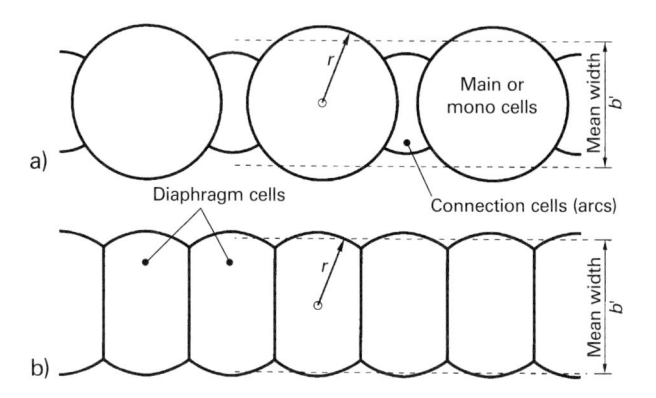

**Fig. R 100-1.** Schematic plans of cellular cofferdams: a) with circular cells, b) with diaphragm cells

filled, and is therefore stable in itself. The connecting arcs required to seal the structure can be installed later. Junction piles connect them to the stable circular cells. The junction piles generally consist of specially shaped rolled sections or welded or bolted sections in which the angle at the connection can be varied between 30° and 45°. To avoid lamellar tearing, only steel grades with the appropriate properties may be used for welded junction piles (see R 67, section 8.1.6). Construction details of welded junction piles can be found in DIN EN 12063.

In order to keep the unavoidable additional stresses in the junction piles low, the clear spacing of the circular cells and the radius of the connecting arcs should be kept to a minimum. If necessary, bent sheet piles may be used for the connecting arcs.

Information on design can be found in [39].

2. Diaphragm cell cofferdams

   Diaphragm cells with straight transverse walls and curved front walls are required when the design value of the circumferential tensile force $F_{t,Ed}$ in a circular cell is greater than the critical design value of the straight-web pile resistances $F_{ts,Rd}$.

   Since the individual diaphragm cells are not stable in themselves, a cofferdam of this type must be filled in stages unless other stabilisation measures are taken. For this reason, the ends of a diaphragm cell cofferdam must be designed as stable structures. The recommendation for a long structure is to incorporate intermediate fixed points, especially if there is a danger of ship collisions, because otherwise local damage could lead to a large part of the cofferdam collapsing. All other conditions being equal, diaphragm cell cofferdams require more steel per linear metre than circular cell cofferdams.

3. Mono-cell cofferdams

   Mono-cell cofferdams are individual cells that can be used as foundations in open water, e.g. for the ends of moles or guide and fender structures at harbour entrances, or foundations for navigation aids (beacons, etc.).

### 8.3.1.3 Verification of ultimate limit states STR, GEO-2 and GEO-3

#### 8.3.1.3.1 Verification of stability against failure of a cellular cofferdam in subsoil for ultimate limit states GEO-2 and GEO-3

1. Verification against failure due to overturning and sliding (GEO-2)
   Verification of stability against failure at the ultimate limit state GEO-2 for a cofferdam is to be carried out using the actions $W_ü$, $E_a$, any variable external actions, resistances $G$, $E_p$ and, if required, the

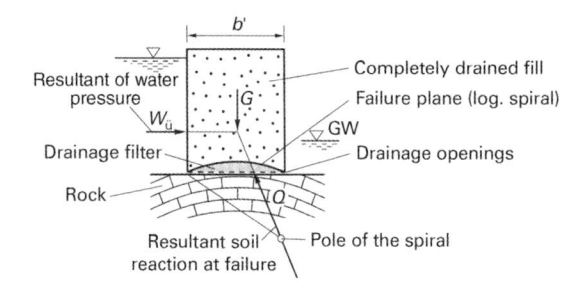

**Fig. R 100-2.** Cofferdam simply supported on rock, with drainage

cohesion at the failure plane, all of which are shown in Figs. R 100-2, R 100-3 and R 100-4. In the case of circular and diaphragm cell cofferdams, the mean width $b'$ as shown in Fig. R 100-1 is to be used as the theoretical width of the cellular cofferdam. This width results from converting the actual plan form into an equivalent rectangular area.

Where a cofferdam rests directly on rock (Fig. R 100-2), then upon failure, a convex failure plane forms between the bases of the cofferdam walls. As a first approximation, the curve for this failure plane can be generated by a logarithmic spiral for the characteristic value of the friction angle $\varphi_k$.

If the cofferdam is supported on rock that is overlain with other soil strata (Fig. R 100-3) or is embedded in loadbearing loose rock (Fig. R 100-4), the actions are increased by the additional active earth pressure of these soil strata and the resistance by the additional passive earth pressure. A reduced passive earth pressure is applied to take account of the minor shape changes – usually with $K_p = 1.0$ – and in the case of deeper embedment in loose rock, then with $K_p$ for $\delta_p = 0$.

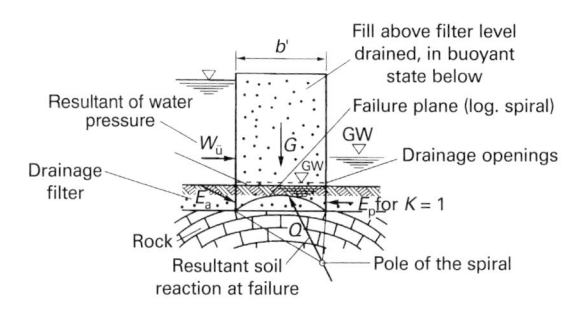

**Fig. R 100-3.** Cofferdam on rock overlain with other soil strata, with drainage

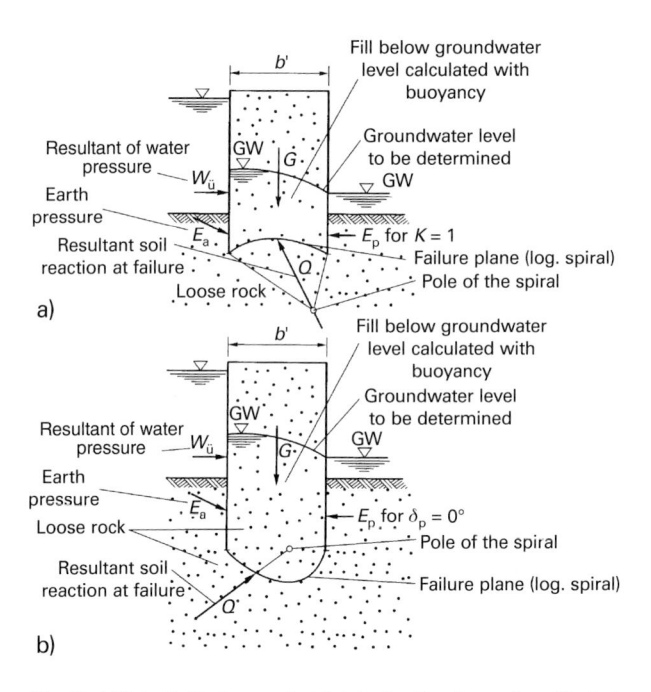

**Fig. R 100-4.** Cofferdam embedded in loadbearing soil, with drainage: a) shallow embedment, b) additional investigation for deep embedment

The resulting moments about the pole of the spiral are designated as a characteristic load due to actions and as characteristic resistances due to the resistant variables. To determine the design value of the moment $M_{Ed}$ as a result of horizontal actions $W_ü$, $E_a$ and variable external actions, e.g. line pull, the characteristic values of the individual moments are multiplied by the partial safety factors $\gamma_G$ and $\gamma_Q$ and added together.

The design value of the resisting moment $M_{Rd}$ as a result of the vertical action $G$ (dead load of cofferdam fill), passive earth pressure $E_p$ (for embedment in loose rock) and possible cohesion at the failure plane is determined by dividing the characteristic individual moments by the partial safety factors $\gamma_{R,h}$, $\gamma_{R,e}$ and $\gamma_c$ and adding them together.

$$M_{Ed} = M_{kG} \cdot \gamma_G + M_W \cdot \gamma_G + M_{kQ} \cdot \gamma_Q \leq \frac{M_{kG}^R}{\gamma_{R,h}} + \frac{M_{Ep}^R}{\gamma_{R,e}} + \frac{M_C^R}{\gamma_C} = M_{Rd}$$

where

$M_{kG}$ characteristic value of effective single moment due to earth pressure $E_a$

$M_W$ characteristic value of effective single moment due to resultant of water pressure $W_ü$

$M_{kQ}$ characteristic value of effective moment due to variable external force

$M_{kG}^R$ characteristic value of resisting single moment due to dead load of fill $G$

$M_{Ep}^R$ characteristic value of resisting single moment due to passive earth pressure $E_p$

$M_C^R$ characteristic value of resisting single moment due to cohesion in the logarithmic failure plane

Use the partial safety factors for the ultimate limit state GEO-2 according to section 0.2.1 for actions and resistances depending on the respective design situation.

Verification of stability against failure is satisfied when

$$M_{Ed} \leq M_{Rd}$$

The most unfavourable failure plane for the analysis is the logarithmic spiral that results in the smallest $M_{Rd}/M_{Ed}$ ratio.

The main action affecting cofferdams is generally the resultant of water pressure $W_ü$. This results from the difference in hydrostatic pressures acting on the outer and inner walls of the cofferdam, and is applied down to the bottom of the outer, i.e. loaded, wall. The water level within the cofferdam enclosure need not always correspond with the level of the bottom.

The resistance of the cofferdam against failure at the ultimate limit state GEO-2 "overturning and sliding" can be increased by
- widening the cofferdam,
- choosing a filling material with a higher unit weight and greater angle of internal friction,
- draining the cells, and
- in the case founding on loose rock, by embedding the cofferdam piles deeper in the subsoil.

Where a cofferdam with a deeper embedment is selected, the verification against failure is to be carried out with both a convex failure plane (Fig. R 100-4a) and a concave failure plane (Fig. R 100-4b). In the latter case the position of the spiral must be chosen such that its pole lies below the line of action of $E_p$ for $\delta_p = 0°$ (Fig. R 100-4), which is also a condition for deep embedment.

The above verification confirms both safety against failure due to overturning and also due to sliding.

2. Verification of safety against base failure (GEO-2)
   For cofferdams that are not founded on rock, verification of safety against heave failure is to be carried out to DIN 4017 on the basis of DIN 1054:2010, using the mean width $b'$ as the cofferdam width.
3. Verification of safety against failure due to loss of overall stability (GEO-3)
   According to DIN EN 1997-1, section 11, verification of failure due to loss of overall stability (ground failure) is to be carried out to DIN 4084 in the case of backfilled cofferdams that are part of a waterfront structure. To carry out the analysis, position the failure plane on the load-side, theoretical limit to the cofferdam width, which coincides with the above value for the mean width $b'$.
4. Additional analyses in the case of water flows:
   - Any flow force present is to be considered in the analyses called for in points 1 to 3.
   - Carry out an analysis to verify safety against failure of the subsoil as a result of a hydraulic heave failure.
   - Carry out an analysis to verify safety against failure of the subsoil as a result of piping.
   - In order to rule out the aforementioned failure modes, special sealing measures are required at the base of the sheet pile wall where cofferdams are founded on jointed rock or rock with varying strength.

### 8.3.1.3.2 Verification of safety against failure of straight-web piles due to circumferential tensile force at the ultimate limit state (STR)

When designing a cellular cofferdam it can be assumed that the stresses caused by external actions such as hydrostatic pressure and, if applicable, earth pressure can be accommodated by the monolithic block action of the cofferdam filling. To verify safety against failure of the straight-web section, it is sufficient in this case to determine the hoop tension at the level of the base of the excavation or watercourse because this is generally where the critical internal pressures occur.

However, in some circumstances it may be necessary to check the resistance to the hoop tension at several levels when the cofferdam is surrounded by or embedded in cohesive strata. Hoop tension values greater than those at the base can occur in these areas due to the abrupt rise in internal pressure or the smaller passive earth pressure as well as the possibility of pore water pressure.

Design values for hoop tension are calculated using the formula $F_{t,Ed} = \sum p_{i,d} \cdot r$. They are calculated by multiplying the characteristic

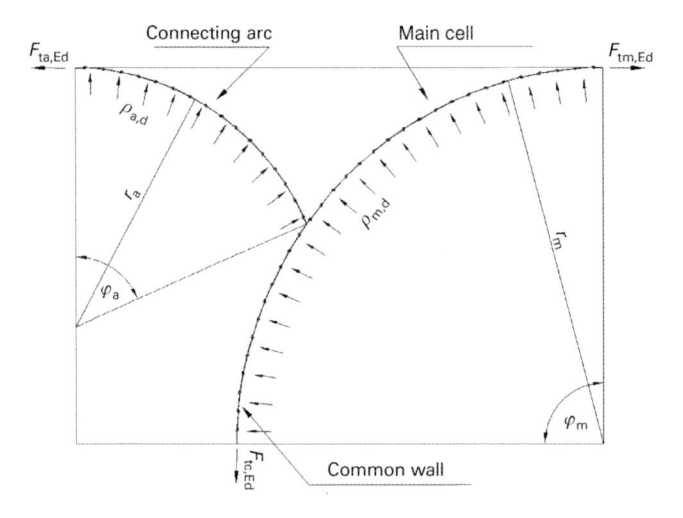

**Fig. R 100-5.** Circumferential tensile force $F_{t,Ed}$ in the individual wall elements of a cellular cofferdam

actions inside the cell due to excess water pressure, earth pressure at rest (where $K_0 = 1 - \sin\varphi_k$) and variable loads ($\sum p_{a,K}$, $\sum p_{m;k}$) by the appropriate partial safety factors (see section 0.2.1).

The design values for hoop tension ($F_{tc,Ed}$, $F_{tm,Ed}$, $F_{ta,Ed}$) in the individual wall elements (Fig. R100-5) may be determined using a simplified method according to DIN EN 1993-5, section 5.2.5(9), as follows:

In the common wall: $\quad F_{tc,Ed} = \sum p_{a,d} \cdot r_a \cdot \sin\varphi_a + \sum p_{m,d} \cdot r_m \cdot \sin\varphi_m$
In the main cell wall: $\quad F_{tm,Ed} = \sum p_{m,d} \cdot r_m$
In the connecting arc: $\quad F_{ta,Ed} = \sum p_{a,d} \cdot r_a$

The analysis of the sheet pile sections and the welded junction piles is carried out according to DIN EN 1993-5, section 5.2.5.

Verification of the wall sections is given when the resisting tensile strength $F_{ts,Rd}$ of the web and interlock is equal to or greater than the design values of the hoop tension ($F_{tc,Ed}$, $F_{tm,Ed}$, $F_{ta,Ed}$):

$$F_{ts,Rd} \geq F_{tc,Ed} \quad \text{or} \quad F_{tm,Ed} \quad \text{or} \quad F_{ta,Ed}$$

where

$$F_{ts,Rd} = \beta_R \cdot \frac{R_{k,S}}{\gamma_{MO}} \quad \text{(interlock)} \qquad F_{ts,Rd} = t_w \cdot \frac{f_{yk}}{\gamma_{MO}} \quad \text{(web)}$$

and

| | | |
|---|---|---|
| | $R_{k,S}$ | characteristic value of interlock tensile strength |
| | $f_{yk}$ | minimum yield stress of steel to R 67, section 8.1.6.2 |
| | $t_w$ | web thickness of the straight-web piles to R 67, section 8.1.6.2 |
| | $\gamma_{M0}$ | partial safety factor for sheet pile material |
| | $\beta_R$ | reduction factor for interlock tensile strength (DIN EN 1993-5: $\beta_R = 0.8$) |

Assuming that the junction pile is welded based on DIN EN 12063, safety is verified when the resisting tensile strength of the junction pile ($\beta_T \cdot F_{ts,Rd}$ for interlock and web) is greater than or equal to the circumferential tensile force design value $F_{tm,Ed}$:

$$\beta_T \cdot F_{ts,Rd} \geq F_{tm,Ed} = \sum p_{m,d} \cdot r_m$$

The reduction factor $\beta_T$ according to DIN EN 1993-5, section 5.2.5(14), can be taken as

$$\beta_T = 0,9 \cdot \left(1,3 - 0,8 \frac{r_a}{r_m}\right) \cdot (1 - 0,3 \tan \varphi_K)$$

### 8.3.1.4 Construction measures

Cellular cofferdams may be constructed on loadbearing subsoil only. Soft strata, especially if they occur near the bottom of the cofferdam, reduce the stability significantly due to the formation of fixed failure planes. Such soils should be replaced with sand inside the cofferdam or drained with vertical drains. If none of these measures are taken, the circumferential tensile force increases after filling as a result of the excess pore water pressure that occurs, which has a detrimental effect when analysing the failure of the sheet pile section (STR).

Fine-grained soil to DIN 18196 or DIN EN ISO 14688-2 may not be used for filling the cells.

The filling should be particularly water-permeable in the case of enclosures to excavations in order to guarantee the drawdown level for the dewatering.

Therefore, in order to minimise the dimensions of the cofferdam and to achieve adequate stability, a soil with a high unit weight $\gamma$ or $\gamma'$ and high internal angle of friction $\varphi_k$ should be used. Both these soil parameters can be improved by vibrating the filling.

1. Cellular cofferdams as excavation enclosures
   In excavation enclosures founded on rock it must be possible to lower the water in the cofferdam at any time with a drainage system monitored by observation wells to such an extent that it satisfies the stability analysis. Drainage openings at the bottom of the exposed wall, filters at the level of the base of the excavation and good permeability of the entire fill are essential.
   Experience has shown that the permeability of sheet piling interlocks subjected to tensile stresses is low, so no particular measures need to be taken here.
   The part of the excavation enclosure subjected to the external hydrostatic pressure must be adequately watertight. It can be practical in some cases to provide additional sealing measures on the outboard side of the cofferdam, e.g. underwater concrete.
2. Cellular cofferdams as waterfront structures
   In waterfront structures much of the cell fill is submerged. Therefore, deep drainage systems cannot be used. When there are large and rapid fluctuations in the level of the water table, however, providing drainage systems in the cell fill and the backfill to the structure can be advantageous because this prevents a larger resultant water pressure (Fig. R 100-6). In these situations the planned efficiency of the drainage measures is critical for the stability and useful life of the waterfront structure.

The superstructure should be designed and constructed so that

- the risk of local damage to the cofferdam cells due to ship impacts is prevented. This can be achieved, for example, through components designed to spread the load.
- the inclusion of appropriate measures, e.g. fenders with a high energy absorption capacity, reduces the magnitude of the global effect of a ship impact to such an extent that the stability of the cofferdam cells is not at risk.
- vertical loads from superstructures or other head details are fonded on the fill so that they generate only circumferential tensile force in the straight-web pile sections.

Fig. R 100-7 illustrates a corresponding wall head detail.
Components subjected to significant vertical actions, e.g. due to cranes, should be built on separate foundations, e.g. an additional pile foundation, which can be positioned either adjacent to or even within the cofferdam. This avoids an increase in the circumferential tensile force and larger eccentricities for the action resultants. Loadbearing piles

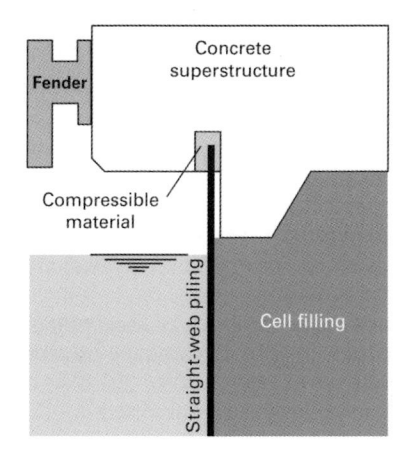

**Fig. R 100-6.** Detail of head of wall for circular cell cofferdam with superstructure

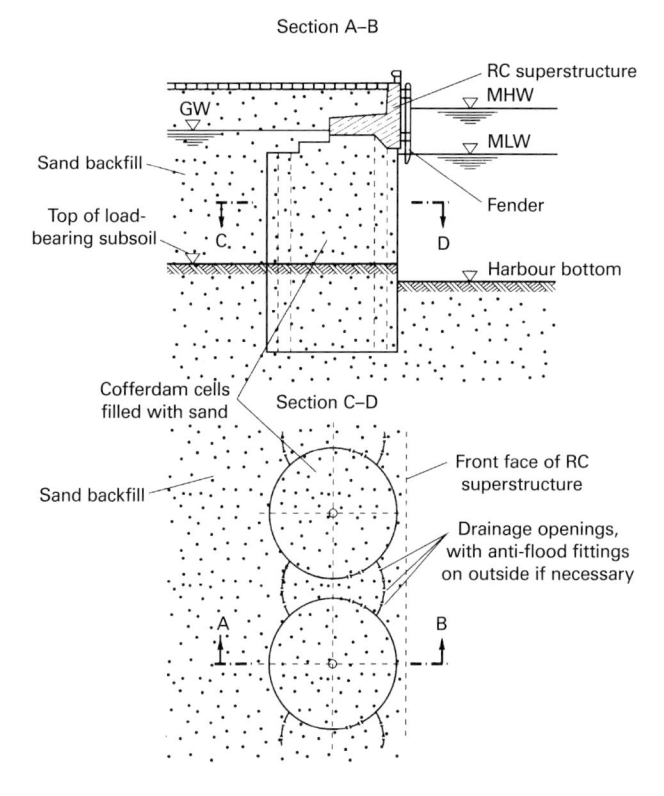

**Fig. R 100-7.** Schematic illustration of a waterfront structure constructed using circular cells, with drainage

should be driven into the loadbearing strata beneath the base of the cofferdam so that they do not need to be considered when analysing the stability of the cellular cofferdam.

### 8.3.1.5 Construction

1. Use of cofferdams
   Cofferdams are recommended for the conditions stated in section 8.3.1.1. They can be constructed both on land and in the water regardless of their particular purpose. Generally, a cofferdam built to provide protection against flooding or to enclose an excavation is built in the dry. By contrast, a cofferdam designed exclusively as a waterfront structure is usually built from a floating or jack-up platform over the water.
   - Construction on land
     The traditional case for constructing a cofferdam on land is for a weir or power station site on a river that is to be dammed. The cofferdam is built on dry ground in the floodplain during periods of low water in order to carry out excavations and building work within the protection of the cofferdam. The cofferdam elements (double-wall or cellular cofferdam, U-/Z-section or straight-web sheet piles) are set up on guide frames and, depending on the subsoil, driven as much as several metres into the ground. The inside of the cofferdam is filled with non-cohesive soil, which is subsequently compacted to a high degree and, if possible, drained. If the cofferdam is to be used to enclose an excavation, it must be driven into the ground to the appropriate depth. Filling of the cells is not, or not necessarily, required in this case as the surrounding soil represents the filling. However, it may be necessary to improve the soil in the cofferdam through compacting it by means of vibroflotation, or by installing vertical drains to relieve the pore water pressure and hence increase stability.
   - Construction in water
     In essence, the difference between construction in water and construction on land is that when working in water all the operations have to be carried out from a floating or jack-up platform. When building a double-wall cofferdam, pile-driving frames have to be set up first or otherwise jack-up platforms or pontoons will be required. Cellular cofferdams require working platforms to be built in the water. It may also be possible to pre-assemble a complete cell around a working platform erected on land. A floating crane is then used to lift the cell with the working

platform and place it in position. This avoids the need to assemble the individual pile sections on the open sea – often a laborious process.

Generally, superstructures in the form of cantilever retaining walls or wave chambers are built on cofferdams to incorporate all the necessary facilities for the berthing of ocean-going vessels.

The success or failure of the construction of waterfront structures comprising cellular cofferdams is very much influenced by the preparatory work. As individual cells without any filling are highly sensitive, non-loadbearing and easily damaged elements, a number of remarks concerning straight-web sections, the driving platform and driving methods are given below.

2. Diameter of circular cells

   Irrespective of the structural calculations, the actual diameter of a circular cell depends on the number of individual piles plus their rolling tolerances and the play at the individual interlocks. Therefore, the diameter can vary between two values, which are important for the size of the guide ring. A minimum diameter must be specified for fabricating guide rings.

3. Installation of straight-web sections

   The proper installation of the piles requires at least two, on high cofferdams three, guide rings. These are placed around the driving platform, which is generally constructed as a space frame and suspended from several driven piles or pile bents/trestles supported on a loadbearing seabed.

   All the piles are positioned around the driving platform and the rings tensioned accordingly. The piles are subsequently driven step by step, each about 50 cm per circuit of the vibrator or rapid-action hammer. It is also possible to lower the entire cell by using a multitude of vibrators, each one acting on several piles simultaneously.

   Prior to installation, the connections of all interlocks should be checked, with the help of divers if necessary.

   On very tall cells it may be necessary to divide the piles into several lengths for easier handling. Basically, as only tensile forces from the internal pressure or downward wall friction forces due to the earth pressure have to be accommodated, a corresponding stagger can be provided without suffering any drop in loadbearing capacity.

   After all the straight-web sections of the cell have been pitched, the individual sections should be driven step by step, e.g. using a driving procedure that makes numerous circuits of the cell, with the increment in the penetration being reduced each time. The angle of internal friction of the circular cell fill material is to be chosen for $\varphi_k$. One critical situation for the stability of a circular cell can occur after it has been set up and the driving platform has already been dismantled following installation of the piles. External actions, e.g. due to wave pressure, can cause a circular cell to collapse. Therefore, the interlocks at the top of the cell are welded and the inside filled to about two-thirds the height as quickly as possible. The recommendation is to fill the entire circular cell cofferdam at the same time as dismantling the driving platform.

## 8.3.2 Double-wall cofferdams as excavation enclosures and waterfront structures (R 101)

### 8.3.2.1 General

In double-wall cofferdams the parallel Z- or U-section steel sheet pile walls are driven into or otherwise placed on the bottom to suit the subsoil conditions and the hydraulic and structural requirements. They are then tied together. If the double-wall cofferdam is supported on rock or does not achieve a sufficient embedment, at least two rows of anchors are required.

The inclusion of cross-walls to form anchor cells as shown in Fig. R 101-1 can be expedient for the work on site. In long permanent structures they also limit any damage caused by vessel impact. The spacing of the cross-walls (or anchor cells) determines the length of the individual sections for constructing the cofferdam, including tie rods and fill.

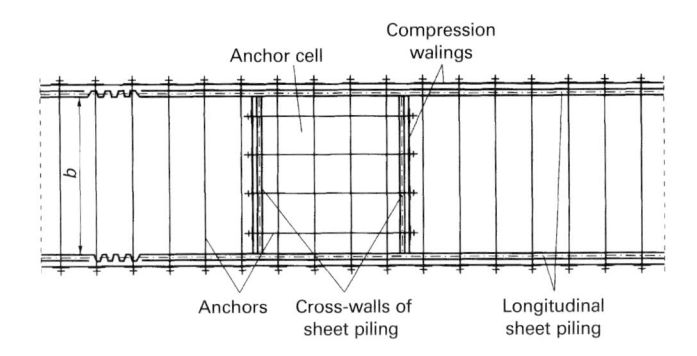

**Fig. R 101-1.** Plan of a double-wall cofferdam with anchor cells anchored in themselves

The remarks regarding the fill in R 100, section 8.3.1.4, apply here. When verifying the stability of a cofferdam with large external actions due to hydrostatic pressure, e.g. excavation enclosures in open water, the permanently effective drainage of its soil filling is crucial for minimising its dimensions. The fill is drained towards the excavation, for which weepholes according to R 51, section 4.4, are sufficient.

Drainage can also be useful when building a cofferdam to serve as a waterfront structure. Such cofferdams are drained towards the water side. However, if there is a danger of pollution, drainage systems with anti-flood fittings to R 32, section 4.5.2, must always be used.

In the following the external sheet pile wall to a cofferdam which is subjected to the actions due to earth and hydrostatic pressures plus variable actions is designated the load-side wall, whereas the opposite sheet pile wall not subjected to such loads is designated the air-, excavation- or water-side wall depending on the particular use of the cofferdam.

### 8.3.2.2 Calculations

#### 8.3.2.2.1 Verification of a sheet pile wall at the ultimate limit state (STR)

In a filled cofferdam, the load transfer into the subsoil relies on the cofferdam functioning as a compact block of soil. The moment as a result of horizontal actions (hydrostatic and earth pressures) with respect to the point of rotation is transferred into the loadbearing subsoil by means of vertical stresses in the soil through the cofferdam filling acting as a monolithic block. These vertical stresses in the soil change linearly over the width of the cofferdam and reach a maximum at the side opposite to the forces, i.e. the air- or water-side sheet pile wall, which is therefore subjected to a pressure higher than the active earth pressure. Experience has shown that this increase in active earth pressure can generally be taken into account with adequate accuracy by increasing the active earth pressure – calculated with $\delta_a = +2/3\, \varphi'_k$ – by 25%. Another action affecting the air side of the sheet pile wall is the excess water pressure, which results from a possible water level difference between the drawdown level within the fill and the air- or water-side water level.

If the cofferdam fill is installed by using hydraulic filling methods and compacted, the active earth pressure can rise to match the hydrostatic pressure as a result of the filling effect. When using in situ soil as the cofferdam filling, however, auxiliary failure planes can develop within the fill following excavation work. Redistribution of the active earth pressure according to R 77, section 8.2.3, is then permissible.

The air-side wall is designed as an anchored sheet pile wall taking into account all actions. If the sheet pile wall is embedded in loadbearing loose rock, the supporting passive earth pressure can be calculated with an angle of inclination as per R 4, section 8.2.5. The determination of the design values for action effects can be carried out according to R 77, section 8.2.3, for a sheet pile wall with one row of anchors and R 134, section 8.2.4, for two rows.

The load-side sheet piling can be built using a different section and can be shorter than the air- or water-side sheet piling provided this is checked for the individual construction phases or if the requirements regarding watertightness and limiting flow around the wall are satisfied.

Various actions and resistances must be considered when calculating the design values for the stresses in the load-side wall:

- Transferring the anchor force of the air-side sheet pile wall (or anchor forces with two rows of anchors)
- External hydrostatic pressure
- External active and, where applicable, also passive earth pressure
- Ship impacts, line pulls and other horizontal actions
- Support provided by the soil filling

The distribution and magnitude of the earth support over the height of the wall must be applied such that the equilibrium condition $\sum H = 0$ is satisfied. If the load-side wall is fixed in the ground, the equivalent force $C$ must be considered when checking the equilibrium.

### 8.3.2.2.2 Verification of safety against failure of a double-wall cofferdam at the ultimate limit state (GEO-2)

1. Verification of stability of the anchorage at the lower failure plane (internal stability)

    The stability of the anchorage at the lower failure plane is to be verified according to R 10, section 8.5. Here, the course of the lower failure plane can be approximated for a single row of anchors as follows:

    - Air- or water-side wall:

    from the theoretical base for a simply supported water-side wall, or from the point of zero shear in the zone of fixity for a fixed wall.

    - Load-side wall:

    At the base of an equivalent anchor wall assumed to be simply supported (Fig. R 101-2). The upper starting point for the lower failure plane can be placed deeper when the following aspects are verified:

    - The design value of the soil support induced by the forces above an imaginary dividing line is equal to or less than the design value of the partial passive earth resistance in the cofferdam above the dividing line.

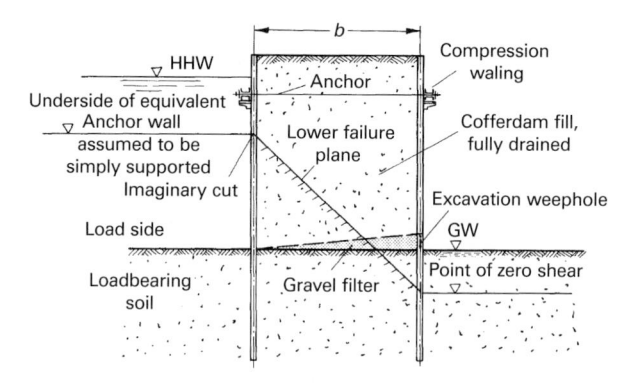

**Fig. R 101-2.** Verifying the stability of anchors for the lower failure plane to R 10, section 8.5

- The design value of the wall action effects in this area as a result of the aforementioned forces is less than or equal to the design value of the section resistance.

The upper starting point for the lower failure plane can then only be selected approximately as a point of zero shear for a fixed anchor wall when the following conditions apply:

- It must be possible to represent the design values of the soil support induced by the actions, and also equivalent force $C$, on the two opposite sides of the sheet pile wall as a passive earth pressure within the scope of the overall system. These values must be less than or equal to the design values for the passive earth pressures inside and outside the cofferdam on the two sides of the sheet pile wall.
- The design value of the wall action effects in this area as a result of functioning as a fixed anchor wall must be less than or equal to the design value of the section resistance.

If several anchors are used, an equivalent anchor wall can also be used in the calculations, but the imaginary dividing line below the bottommost anchor must be positioned in such a way that the lowest anchor cannot fail.

2. Verification against failure due to overturning and sliding (external stability)

The theoretical width of a double-wall cofferdam is taken as the centre-to-centre distance $b$ between the two sheet pile walls. For verification of the stability of double-wall cofferdams, essentially the same principles apply as for the stability of cellular cofferdams – see R 100, section 8.3.1.3.

In contrast to Figs. R 100-4a and R 100-4b, the passive earth pressure $E_p$ in front of the air-side sheet pile wall – owing to its greater deflection potential – may be assumed to correspond to that of a customary anchored sheet wall according to R 4, section 8.2.5, with an angle of inclination $\delta_p < 0°$. The following applies to the position of the failure planes to be investigated:
- Air-side sheet pile wall:
  - For a sheet pile wall simply supported in the ground, the logarithmic spiral of the failure plane intersects the base of this wall.
  - For a sheet pile fixed in the ground, the logarithmic spiral of the failure plane intersects the point of zero shear force.
- Load-side sheet pile wall:
  - The starting point of the logarithmic spiral is in this case generally at the same level as that of the air-side wall. If the load-side wall is shorter than the air-side wall, the failure plane on the load-side wall must continue to the existing base.

In a double-wall cofferdam, a concave failure plane usually governs because of the deeper embedment of the sheet pile walls in the subsoil. The resistance to failure of a double-wall cofferdam can be increased through one or more of the following measures:
- Widening the cofferdam.
- Selecting a fill material with better $\gamma_k$, $\gamma'_k$ and $\varphi'_k$.
- Compacting the cofferdam fill, possibly the subsoil as well.
- Driving the cofferdam sheet piles deeper if this generates a concave failure plane that satisfies the limit state conditions against failure of the soil.
- Providing further anchor levels (but the advisability of the installation of another anchor level with its associated difficulties, e.g. underwater with the help of divers, should be checked).

3. Base failure
   See R 100, section 8.3.1.3.1 (2).
4. Ground failure
   See R 100, section 8.3.1.3.1 (3).
5. Additional analyses in the case of water flow
   See R 100, section 8.3.1.3.1 (4).

**8.3.2.3 Constructional measures**

1. Excavation enclosures
   See R 100, section 8.3.1.4 (except the details regarding interlocks subjected to tension).

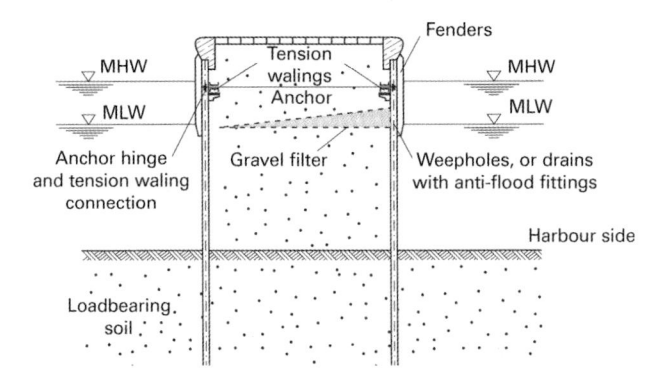

**Fig. R 101-3.** Schematic illustration of a mole structure built using a double-wall cofferdam

The weepholes near the base of the air- or water-side sheet pile wall should be located in the webs of the sheet pile sections.

The walings for transferring the anchor forces are mounted on the outer side of the sheet piling as compression walings, provided shipping operations do not preclude this. Waling bolts are not needed with this solution and there are also advantages for the installation of the anchors. The anchor penetration on the load side, i.e. water side, must be made watertight.

2. Waterfront structures, breakwaters and moles
   The designs in R 100, section 8.3.1.4, apply here accordingly (Fig. R 101-3).
3. Special aspects concerning the construction of double-wall cofferdams
   Special attention must be given to the installation of any lower anchors. As these are usually underwater and can only be installed with the help of divers, simple but effective connections to the sheet pile wall are vital.
   Prior to filling the cofferdam, the bottom surface inside the cofferdam should be cleaned in order to avoid increased earth pressure and hence higher loads on the low-level anchors.

## 8.3.3 Narrow moles in sheet piling (R 162)

### 8.3.3.1 General

Narrow moles in sheet piling are double-wall cofferdams in which the spacing of the sheet piling is only a few metres and thus considerably less than a customary double-wall cofferdam (R 101, section 8.3.2). The main loads on these moles are excess water pressure, vessel and ice impacts, line pull, etc.

The sheet pile walls are tensioned against each other at or near the top and also braced to resist compression and thus ensure joint transfer of the external actions.

The space between the sheet piles is filled with sand or gravelly sand with a medium density at least.

### 8.3.3.2 Calculation assumptions for a narrow mole as a non-anchored system at the ultimate limit state (GEO-2)

To accommodate the external forces acting perpendicular to the axis of the mole and transfer them into the subsoil, the mole should be considered as a free-standing structure fully fixed in the ground and consisting of two parallel, interconnected sheet pile walls. The influence of the soil fill between the two sheet piling walls as a result of silo action is neglected when determining the stiffness of the system. The two walls are interconnected by an essentially hinged connection at or near the top. A rigid connection is also possible but leads to large bending moments at the top of the sheet piling and thus the need for elaborate connections. In addition, it leads to axial forces in the sheet pile walls, which in some circumstances reduce the passive earth pressure that can be mobilised for a sheet pile wall in tension.

The full magnitude of the mobilisable passive earth pressure cannot be used because this is partly required to resist the active earth pressure and, possibly, excess water pressure due to the filling of the mole. This portion must be determined in advance and deducted from the total mobilisable passive earth pressure.

As both sheet pile walls exhibit an essentially parallel deflection curve due to the external actions, the bending moment in the total system can be distributed over both walls in the ratio of their flexural stiffnesses. The ultimate limit state conditions according to EC 3-5 are to be checked separately for each of the two sheet pile walls using the applicable design values of the loads ($M_{Ed}$, $V_{Ed}$, $N_{Ed}$) and the section resistances.

### 8.3.3.3 Calculation assumption for sheet pile walls tied to each other

The individual sheet pile walls are loaded by active earth pressure from the fill and the surcharge on the fill as well as by external actions. In addition, excess water pressures have to be considered if the water table can be higher inside the filling than in front of the sheet pile walls. Generally, a receding water level action should also be considered for the excess water pressure, which takes into account flooding of the mole with subsequent, brief lowering of the outer water level. In some circumstances an internal water level that drops at a much slower rate will cause a higher excess water pressure. The anchors tying together the sheet pile walls of the mole are to be designed for the tension loads resulting from the aforementioned actions.

#### 8.3.3.4 Design

Walings, anchors and bracing must be designed, detailed and installed in accordance with the relevant recommendations.

Special attention should be paid to the load transfer points for accommodating the external actions on the mole. An analysis of the forces acting parallel with the centre-line of the mole is to be carried out according to R 132, section 8.2.11.

Cross-walls or anchor cells are to be provided as per R 101, section 8.3.2.

## 8.4 Walings, capping beams and anchor connections

All components of anchor connections in sheet piling structures must be designed in such a way that the connection does not fail before the anchor.

### 8.4.1 Design of steel walings for sheet piling (R 29)

#### 8.4.1.1 Arrangement

Walings transfer the support reactions from the sheet piling to the anchors. Furthermore, they stiffen the sheet piling and the anchor wall and facilitate the alignment of the piling.

Usually, walings are installed as tension members on the inboard side of the sheet piling so that the outside remains smooth. At anchor walls they are generally used as compression members behind the wall.

#### 8.4.1.2 Design

It is expedient to provide walings in the form of two channels, the webs of which – perpendicular to the sheet piling – are spaced apart so that the anchors can be positioned freely between them (R 132, section 8.2). Where possible, the channels are positioned symmetrically around the anchor connection. Channels or plates maintain the spacing between the two channel. Additional stiffening of the waling is necessary in the area of the anchor in heavy-duty anchor systems or if there is a direct connection between anchor and waling.

Splices between waling sections should be positioned at points of minimum stress. A full cross-section splice is not required, but it must be capable to carry the calculated internal forces.

#### 8.4.1.3 Fixings

Walings are either supported on welded brackets (R 132, section 8.2.11, Figs. R 132-1, R 132-2 and R 132-3) or – especially with limited working space beneath the walings – suspended from the sheet piling. The fixings should be designed so that vertical loads on the walings can be transferred to the sheet piling. Brackets simplify the mounting of the walings. Suspension details must not weaken the walings and should therefore be welded to the walings or attached to the plate washers of the waling bolts.

The anchor force is transmitted through the waling bolts into the walings. Bolts are placed in the centre between the two waling channels and transfer their load through plate washers welded to the walings (R 132, section 8.2.11, Figs. R 132-1, R 132-2 and R 132-3). Overlong waling bolts are used to help align the sheet piling against the walings.

#### 8.4.1.4 Raking anchors

Connections for raking anchors must also be secured vertically.

#### 8.4.1.5 Extra waling

Sheet piling that has become severely misaligned during driving can be locally realigned with an extra waling, which remains as part of the structure.

### 8.4.2 Verification of steel walings (R 30)

Designing walings, waling bolts and plate washers are designed to DIN EN 1993-1 or DIN EN 1993-5. The load on the waling should be at least the loadbearing capacity of the tension anchor selected. Heavier walings made from steel grade S 235 JR are preferable to lighter ones made from S 235 J2 as they are more robust and can therefore be used to align the wall. Splices, stiffeners, bolts and connections must be designed in accordance with structural steelwork standards and the components should permit easy welding. To allow for possible corrosion, loadbearing weld seams must be at least 2 mm thicker than required by the structural calculations. Additionally, the welds must be designed to carry all anticipated horizontal and vertical actions to the anchors or the anchor wall. The following actions are to be considered for the design check:

#### 8.4.2.1 Horizontal actions

1. The design value of the horizontal component of the anchor tensile force taken from the sheet piling calculations or – as a minimum – the

horizontal component of the loadbearing capacity of the tension anchor selected.
2. The design values of line pull forces applied directly to the waling.
3. The design value of the berthing load, depending on the size of the vessel, the berthing manoeuvres and current and wind conditions. Ice impact may be neglected.

### 8.4.2.2 Vertical actions

1. The dead load of the waling members including stiffeners, bolts and plate washers.
2. The portion of the soil surcharge calculated from the rear surface of the sheet piling to a vertical plane through the rear edge of the waling.
3. The portion of the imposed load on the quay wall between the rear edge of the sheet piling capping beam and a vertical line through the rear edge of the waling.
4. The vertical component of the active earth pressure from underside of waling to ground surface and acting on the vertical plane passing through the rear edge of the waling.
5. For tension and compression walings, the vertical component of a raking anchor tensile force according to section 8.4.2.1(1).

The design values of the loads given in points 1 to 5 above are to be considered for limit state GEO-2.

### 8.4.2.3 Applying the actions

In the structural calculations for the walings, the only horizontal loads generally considered are the component of the anchor tensile force (section 8.4.2.1(1)) combined with the line pull forces (section 8.4.2.1 (2)). On the other hand, the vertical loads (section 8.4.2.2) are all included. The effects from berthing loads and the alignment of the wall are considered indirectly by ensuring that the waling is robust, i.e. by choosing the characteristic tensile resistance as the design value of the anchor tension load and increasing the partial safety factors for waling resistance by 15%. With several walings one above the other, the vertical loads are divided among the walings. In order to ensure the secure mounting of the waling brackets, the actions are applied at the rear edge of the waling.

### 8.4.2.4 Method of calculation

The actions are split into component forces vertical and parallel to the surface of the sheet pile wall (principal waling axes). In the calculations it should be assumed that the walings are supported by the anchors for carrying forces perpendicular to the plane of the wall and by the brackets or suspension details for actions parallel to the wall. If the anchors are connected to the sheet piling, the pressure of the piling on the waling in the areas around anchor connections has a supporting effect so that it is sufficient here to suspend the waling on the rear side – as is usually done for compression walings. The support and span moments resulting from the design value of the sheet piling support reaction are – considering the end bays – generally calculated using the equation $q \cdot l^2/10$.

### 8.4.2.5 Waling bolts

The waling bolts should be designed using the same principles as for the anchoring of the sheet piling (R 20, section 8.2.7.1). Due to the risk of corrosion, the loads triggered by the wall alignment and berthing loads, waling bolts must be at least 38 mm in diameter. Plate washers for waling bolts should be designed to provide the same loadbearing capacity as the waling bolts.

## 8.4.3 Sheet piling walings of reinforced concrete with driven steel anchor piles (R 59)

### 8.4.3.1 General

Anchors in the form of steel anchor piles driven with a rake of 1 : 1 have frequently proved practicable and quite economical for quay walls.

This is particularly the case when the upper strata behind a wall consist of soil types that make anchoring with round steel tie rods difficult or even impossible, but also when the wall must be anchored before backfilling.

If the anchor piles are driven before the sheet pile wall and the sheet piles suffer forward or backward lean during driving, the anchor piles already installed are not always in the right position relative to the sheet piling. However, inaccuracies of this kind can be compensated for with reinforced concrete walings in which the reinforcement considers the positions of the anchors (Fig. R 59-1).

It is advisable to mount a temporary steel waling on the sheet piling if the reinforced concrete waling is a considerable distance above the existing terrain. It is removed after the piles have been connected and the reinforced concrete waling has reached its full strength.

### 8.4.3.2 Construction of reinforced concrete walings

Reinforced concrete walings are connected to the sheet piling by round or square steel bars that are welded to the sheet pile webs (Fig. R 59-1, Nos. 4 and 5). Extra reinforcement is generally only used at expansion joints. The anchor forces are also transferred via round or square steel bars (Fig. R 59-1, Nos. 1 to 3).

The steel connecting bars welded to the sheet piles and the anchor piles are generally grade S 235 JR. Round steel bars of grade BSt 500 are also

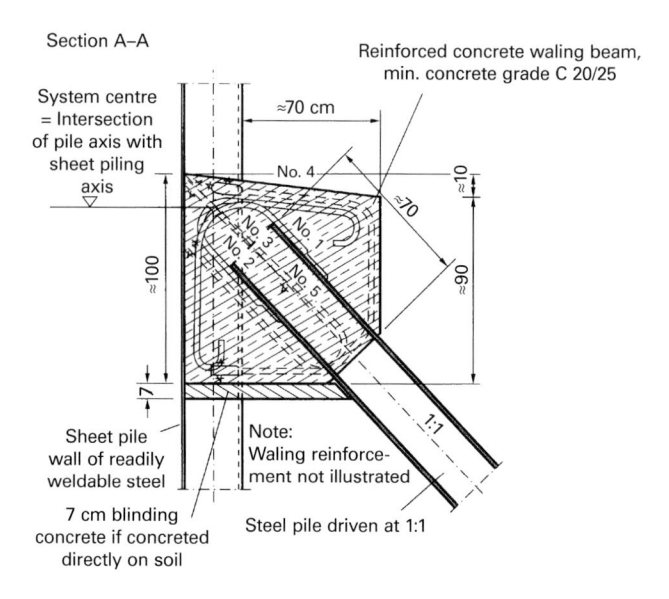

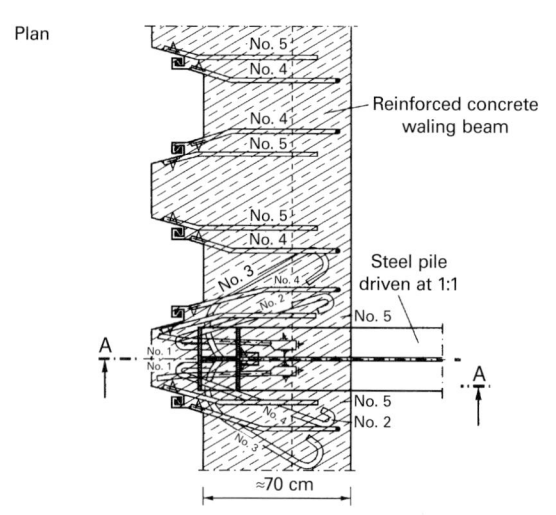

**Fig. R 59-1.** Reinforced concrete waling for steel sheet piling

used. Square steel bars can be welded directly to the wall and the anchor. Round steel bars must be forged flat at the ends to simplify welding.

Welding may only be carried out by qualified welders under the supervision of a welding engineer. Only use materials whose suitability for welding is known which are of uniformly good quality and are compatible with each other (see R 99, section 8.1.19).

### 8.4.3.3 Connection between pile and waling

If no large-scale settlement or subsidence are to be expected in the backfill behind the wall, the anchor piles can be directly fixed at the reinforced concrete waling. This economical solution can also be used when the restraint stresses resulting from minor settlement (e.g. thin soil strata liable to settle, or well-compacted backfill with non-cohesive soil) are included in the design of the waling. In these cases the fixity moment must be calculated from the restraint for the yield stress $f_{y,k}$ of the anchor steel used and the characteristic normal force $N_k$ in the anchor.

If only a short length of the piles is in a soil layer susceptible to settlement or if the depth of the backfill is shallow, a correspondingly smaller additional connection moment may be assumed.

The transfer of the internal forces in the steel pile to the connection at the reinforced concrete waling must be checked. In doing so, the combined stresses in the top of the pile due to normal force, shear force and bending moment should be taken into account. If necessary, strengthening plates may be welded to the sides of the steel pile to improve the transfer of the tensile forces. The anchoring bars, in the form of loops, are then connected to these plates. The voids that tend to form alongside the webs of the anchor piles as a consequence of this detail must be carefully filled with concrete to avoid corrosion.

Only specially killed steels resistant to brittle fracture, e.g. grade S 235 JG or S 355 JR, may be used for piles and their connections in quay walls anchored with piles subjected to larger bending stresses.

If anchors are placed in areas of soil with thicker strata of soil highly sensitive to settlement or if the backfill cannot be compacted, it is better to include a hinge in the pile connection detail.

### 8.4.3.4 Design of the anchor connection

The design value of the horizontal actions is the loadbearing capacity of the anchor selected; the anchor force from the sheet piling calculation must therefore be increased to this figure.

The anchor force is applied at the intersection between the axis of the sheet piling and the axis of the pile. The waling, including its connections to the sheet piling, is taken to be uniformly supported. The dead loads, vertical surcharges, pile forces, bending moments and shear forces from the anchor piles are actions and are included in the calculations as design values.

The forces acting at the pile connection, which result from the soil surcharges on the anchor pile due to backfill or settlement, are calculated for an equivalent beam assumed to be fixed at the waling and in the loadbearing soil. The fixity moment acting on the pile connection, and the shear force acting at the same position, must be considered when verifying the connection between waling and sheet pile wall. With regard to the analyses of the wall itself, these loads need only be taken into account when the anchor piles are considered to spread the load on the sheet piling.

A weakening of the pile cross-section at the point of fixity at the waling in order to reduce the connection moment and the associated shear force is not permissible because such a reduction can easily lead to pile failure, especially in the case of poor workmanship.

If anchor piles are connected via a hinge connections, then the hinge must also be verified for additional loads due to settlement and/or subsidence. Verification is carried out for the design values of the internal forces $Ed$, which may be reduced depending on the design situation according to section 8.2.1.2.

When a connection moment and the associated shear force are supposed to utilise the yield strength $fy,k$ in the anchor pile, then the connection components themselves may also be designed on the basis of the yield strength $fy,k$.

To ease construction, the dimensions of a reinforced concrete waling should not be smaller than those shown in Fig. R 59-1. In order to allow for variations in the anchor forces and the stresses in the waling, the reinforcement cross-section should be increased by at least 20% beyond the calculated requirements.

### 8.4.3.5 Expansion joints

Reinforced concrete walings can be cast with or without expansion joints. Their design is based on R 72, sections 10.2.4 and 10.2.5. See R 72, section 10.2.3, for details of construction joints.

Where expansion joints are specified, they should be arranged in such a way as to not hinder the changes in length of the sections.

To provide mutual support for the separate sections of the structure in the horizontal direction, expansion joints should have a joggled form, with dowels if necessary, as per Fig. R 59-2. The horizontal joggle is placed into the pile cap in the case of quay walls on pile trestles. Joints are to be designed to prevent the backfilling from leaking out.

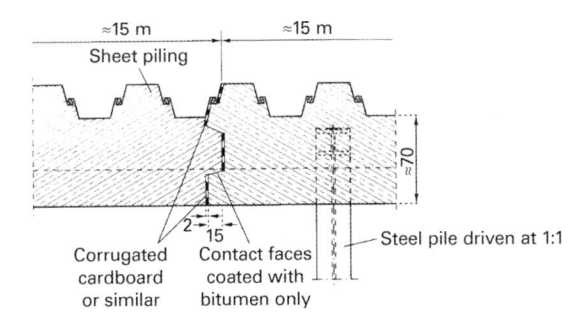

**Fig. R 59-2.** Joggle joint in a reinforced concrete waling

### 8.4.3.6 Typical connections of anchor piles to a reinforced concrete superstructure

The detail at the top of a steel anchor pile must be designed in such a way that the anchor force can be transmitted by the connection without exceeding the permissible stresses. Additional stresses from the anchor pile due to deflection and shear should be kept to a minimum in the connection zone. To this end, the embedment of the pile in the reinforced concrete should be about twice its depth (Fig. R 59-3). The connecting bars and their welds are designed in such a way that the full cross-sectional area of the anchor pile is connected.

With yielding subsoil under the anchor piles, stresses in the reinforced concrete superstructure are to be verified within the scope of the permissible stresses according to design situation DS-A. This applies not only to the full anchor pile force, but also to the loads from the shear force and bending moment at the anchor pile connection when the pile is stressed up to its yield strength.

Fig. R 59-3 shows a favourable connection solution with flat-head bars – as already used for many years for bollard anchor details. Here, one end of the round steel bar is upset to form a circular disc at the head which is up to three times the diameter of the bar. The end of the round steel bar to be welded to the tension pile is flattened to ensure a good weld.

End anchoring in the concrete can also be achieved, however, by welding cross-bars or plates of a suitable size to the round and square anchor bars.

### 8.4.4 Steel capping beams for sheet piling waterfront structures (R 95)

### 8.4.4.1 General

Steel capping beams for sheet piling walls are designed according to structural, operational and constructional requirements. R 94, section 8.4.6.1, applies for waterfront structures made from reinforced concrete.

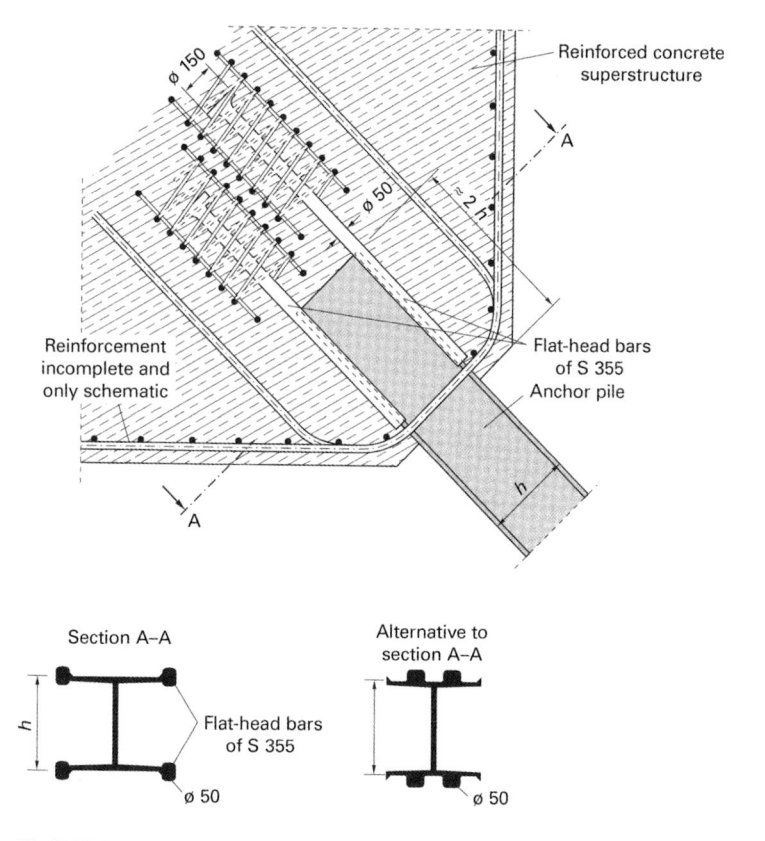

**Fig. R 59-3.** Example of an anchor pile connected to a reinforced concrete superstructure by means of flat-head bars

### 8.4.4.2 Structural and constructional requirements

A capping beam serves as an upper closure to sheet piling (Fig. R 95-1). With appropriate flexural stiffness (Fig. R 95-2), capping beams can also be used to accommodate forces arising during the alignment of the top of the sheet piling and loads from port operations. However, the top of the sheet piling can only be aligned with a capping beam if the length of the unembedded sheet piling is such that it can be deformed sufficiently.

With only a short distance between capping beam and waling, the alignment of the sheet piling will be successfully accomplished mostly with the stiffer waling.

During port operations, a capping beam distributes non-uniform loads to the top of the sheet piling and prevents uneven deflections of the head of the wall.

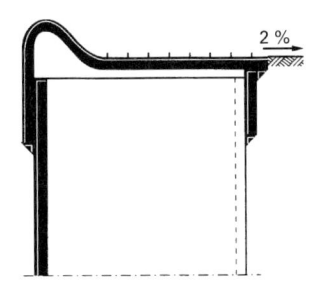

**Fig. R 95-1.** Rolled or pressed steel capping beam with bulb plate, welded to the steel sheet piling

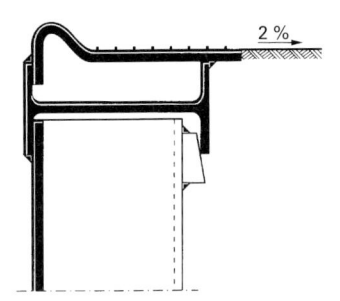

**Fig. R 95-2.** Welded capping beam waling with high section modulus, otherwise as Fig. R 95-1

Fig. R 95-1 shows a standard type of a steel capping beam. The greater the distance between capping beam and waling, the more important is the inherent stiffness to align the wall. Berthing loads must also be considered. Fig. R 95-2 shows a reinforced capping beam waling.

To prevent deflection or buckling, the capping beam shown in Fig. R 95-1 is strengthened with stiffeners in wide sheet pile troughs, which are welded to beam and pile.

If a capping beam also functions as a waling, it must be designed in accordance with R 29, section 8.4.1, and R 30, section 8.4.2.

### 8.4.4.3 Operational requirements

The top edge of a capping beam must be designed so that the hawsers trailing over it will not be damaged or damage the top of the quay. Furthermore, it must be ensured that hawsers and lines (e.g. also thin heaving lines) cannot become caught in gaps, joints, etc. For the safety of personnel working on the quay, the front section of the capping beam should be designed as a bulb plate. Horizontal surfaces of steel capping beams should have non-slip finishes (studs, chequered pattern) if possible (Figs. R 95-1 and R 95-2).

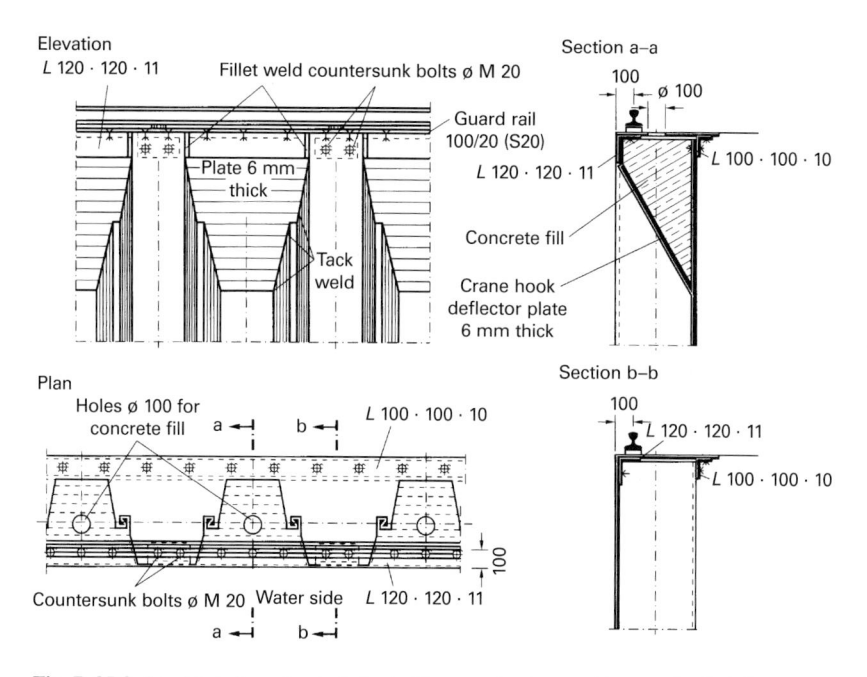

**Fig. R 95-3.** Special design of a steel sheet piling capping beam with crane hook deflector plates

Where heavy vehicular traffic operates, welding on a guard rail is recommended to protect the edge (Fig. R 95-3). If there is an outboard crane rail (R 74, section 6.3.4, Fig. R 74-3), this can serve as the guard rail.

The outboard side of a capping beam must be smooth. Unavoidable edges are to be chamfered if possible. The design must be such that ships cannot get under the capping beam and there is no danger of beam sections being ripped out by crane hooks (Fig. R 95-3).

### 8.4.4.4 Supply, mounting and corrosion protection

All steel capping beam parts are to be supplied without distortion and true to size. During fabrication in the workshop, the tolerances for the width and depth of the sheet pile sections and deviations during driving are to be taken into account. Where necessary, capping beams are to be modified and aligned on site. Capping beam splices are to be designed with the full cross-section.

After mounting the capping beam, it is to be backfilled from behind with compacted sand to avoid corrosion. The backfill is to be replenished when it has settled to such an extent that the beam is exposed.

If the capping beam is mounted so low that it can be flooded or water can flow over it, lies in the wave action zone, or, indeed, is intended to lie below the water level, there is a risk that the backfill designed to prevent corrosion will be washed out. To prevent this, an erosion-proof and watertight connection is required between the steel capping beam and the sheet piling, e.g. a backfilling of concrete behind the beam. This concrete backfill must be secured by means of fishtail anchors or bolts welded to the capping beam. A granular or geotextile filter must be laid beneath the paving of the port operations area behind the capping beam so that bedding of the paving cannot be washed away.

## 8.4.5 Reinforced concrete capping beams for waterfront structures with steel sheet piling (R 129)

### 8.4.5.1 General

Structural, constructional and operational considerations govern the design of reinforced concrete capping beams on steel sheet piling.

### 8.4.5.2 Structural requirements

In many cases the capping beam not only simply covers the top of the sheet piling, but also stiffens the wall and is therefore subjected to horizontal and vertical actions. If, in addition, it functions as a capping beam waling for transferring anchor forces, a sufficiently robust design is required, especially when it also has to support a crane rail directly.

Please refer to R 30, section 8.4.2, for more information concerning horizontal and vertical actions. Additional loads must be taken into account in areas with bollards or other mooring equipment (R 153, section 5.11, R 12, section 5.12, and R 102, section 5.13), in so far as such loads are not carried by special constructional measures. In addition, where a crane rail is supported directly on a reinforced concrete capping beam (section 8.4.5.3), the beam must be designed for horizontal and vertical crane wheel loads (R 84, section 5.14).

In the structural calculations it is expedient to treat the reinforced concrete capping beam – both horizontally and vertically – as a flexible beam on an elastic foundation (the sheet piling). In the case of heavy-duty capping beams on quay walls for sea-going vessels, the horizontal support can be calculated approximately using a modulus of subgrade reaction $k_{s,bh} = 25$ MN/m$^3$. The vertical support primarily depends on the pile sections, their length and the width of the capping beam. The modulus of subgrade reaction $k_s$,bv for the vertical support must therefore be calculated separately for each structure.

As an approximation during the conceptual design phase, the vertical modulus of subgrade reaction can be taken as $k_{s,bv} = 250\,\text{MN/m}^3$. Limit considerations are required for support conditions used in the detailed design of the capping beam. The design should be based on the least favourable case.

Sheet piling structure or bollard foundation anchors connected to the capping beam must be considered in the design.

Special attention should be paid to loads on the capping beam arising from changes in length due to shrinkage and temperature fluctuations. Changes in the length of the capping beam can be severely hampered by the connected sheet piling and the backfilling, resulting in the possibility of corresponding stresses due to shrinkage and temperature changes.

To take these actions, the different stiffnesses of the wall support for the capping beam and various anchor forces into account as a whole, the reinforcement cross-section as per R 59, section 8.4.3, should be at least 20% greater than that required by the calculations.

Please refer to R 72, section 10.2, for information on concrete types and reinforcement details for reinforced concrete capping beams.

The vertical loads in the plane of the sheet piling are generally transferred centrically into the top of the sheet pile. For this reason, the reinforced concrete capping beam should include sufficient tensile splitting reinforcement directly above the sheet piling. On steel sheet piling, the reinforced concrete capping beam can be supported on knife-edge bearings covered by national technical approvals. When transferring large concentrated loads, e.g. from a crane rail, via the beam into the sheet piling, it should be ensured that the sheet piles can carry the loads via plate action, e.g. by appropriate welding of the interlocks.

The geometric stipulations of port operations may render an eccentric support for the outboard craneway on the capping beam (see R 74, section 6.3).

Safe transfer of internal forces from the capping beam to the supporting wall must be verified.

### 8.4.5.3 Buckling and operational requirements

The top of the sheet pile wall must be aligned before the concrete beam is cast. A permanent or temporary steel waling can be used to do this. The reinforced concrete capping beam then forms the visually important alignment of the wall.

If it is necessary to ensure concrete cover of adequate depth over the top of the sheet piling, then this must be taken into account when choosing the width of the capping beam. Contingent on the design, the pile top should have a concrete cover of at least 15 cm on both the water and land side, and the depth of the capping beam should be at least 50 cm

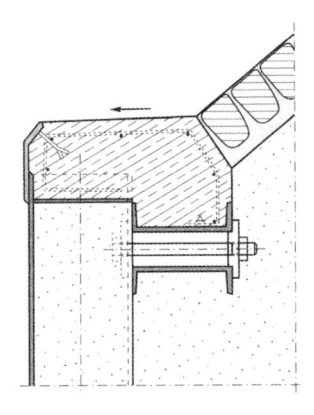

**Fig. R 129-1.** Reinforced concrete capping beam for sheet piling without water-side concrete cover on a partially sloped bank

(Figs. R 129-1 and R 129-2). The sheet piling should be embedded approx. 10–15 cm in the concrete capping beam.

The distance of a reinforced concrete capping beam above the water level should be sufficient to allow the sheet piles immediately below the beam to be inspected regularly for corrosion – and renewal of any corrosion protection measures if necessary.

For waterfront structures with an increased risk of corrosion (e.g. in seawater or brackish water), it is advisable to continue the sheet piles on the water side up to the top edge of the quay wall and position the

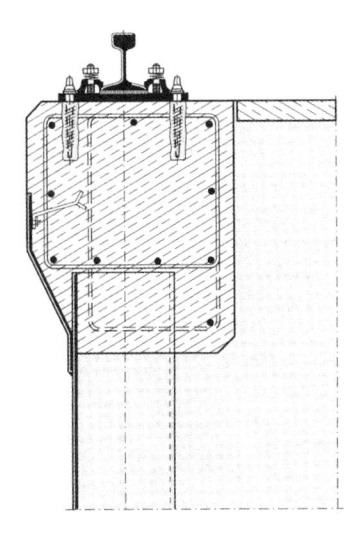

**Fig. R 129-2.** Reinforced concrete capping beam for sheet piling with concrete cover on both sides and craneway supported directly on the top surface

reinforced concrete capping beam behind them. This is an effective way of preventing corrosion at the transition from steel to concrete on the water side.

A capping beam may be designed as shown in Fig. R 129-2 in order to prevent a ship's hull from catching beneath it. On the water side, the beam is provided with a steel plate bent at 2 : 1 or steeper. The bottom edge of this plate is welded to the sheet piling or, as shown in Fig. R 95-3, welded into the outboard troughs of the sheet piles.

The outboard side of a reinforced concrete capping beam can be protected with a steel plate welded to the sheet piling (Fig. R 129-1). This solution is generally more economical than bolting the steel plate to the sheet piling. Fishtail anchors are then fitted above the sheet pile troughs to anchor the plate in the concrete. The edges of the plate should be chamfered to protect lines trailing over the capping beam. (Fig. R 129-1). Irregularities in the alignment of the top of the sheet piling of up to about 3 cm can be corrected by inserting small plates filling the gap. The capping beam is provided with edge protection and rubbing strips as per R 94, section 8.4.6, or DIN 19703.

Stirrup reinforcement (shear links) in reinforced concrete capping beams must be designed so that it provides a shear-resistant connection between the parts of the cross-section separated by the sheet piling. To this end, the stirrups should either be welded to the webs of the sheet piles or inserted through holes flame-cut in the piles or fitted into slots in the sheet piling. If the capping beam is reinforced for tensile splitting above the sheet pile wall to carry vertical loads, additional stirrups, e.g. in the troughs on both sides of the sheet piles, must be included so there is shear reinforcement on the underside of the capping beam as well.

Supporting the reinforced concrete capping beam on knife-edge bearings covered by a national technical approval enables a capping beam with a closed cross-section so that in this case special positioning of the stirrups is unnecessary.

Reinforced concrete capping beams on box sheet piling can also be protected with steel sections (Fig. R 129-3). In this case the reinforcement is moved into the sheet piling cells. To this end, the webs and flanges are cut away as necessary and holes flame-cut in them where required. Reinforced concrete capping beams on combined sheet piling can be arranged in the same way.

Reinforced concrete capping beams can be strengthened locally so that bollards can be mounted directly on them (Fig. R 129-4). In such cases, large line pull forces are best carried by heavy-duty round steel tie rods in order to minimise the elongation of the anchor and hence the bending moments in the capping beam.

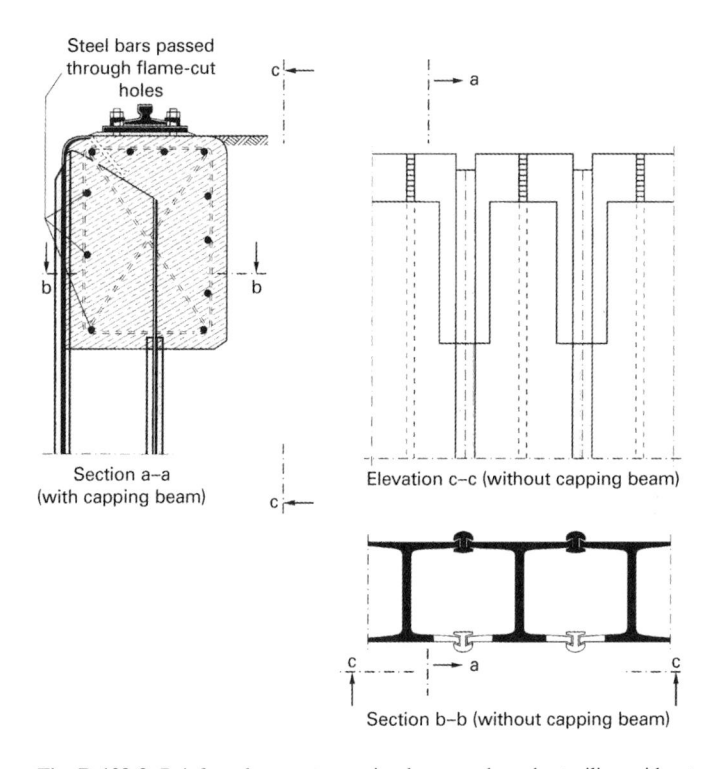

**Fig. R 129-3.** Reinforced concrete capping beam on box sheet piling without outboard concrete cover and craneway supported directly on the top surface

### 8.4.5.4 Expansion joints

Reinforced concrete capping beams can be constructed without joints when all the actions due to loads and restraint (shrinkage, creep, settlement, temperature) are taken into account (see R 70, section 10.2.4), provided the inevitable cracks are acceptable. The theoretical crack widths must be limited taking into account the environmental conditions (see R 70, section 10.2.5).

If expansion joints are planned, the lengths of the sections between them should be specified such that no significant restraint forces occur in the longitudinal direction of each section. Otherwise, the restraint forces should be taken into consideration in relation to the substructure or subsoil.

The joints themselves must also be designed so that the changes in length of the reinforced concrete capping beam at these points are not hampered by the sheet piling locally. To this end it is appropriate, e.g. for sheet

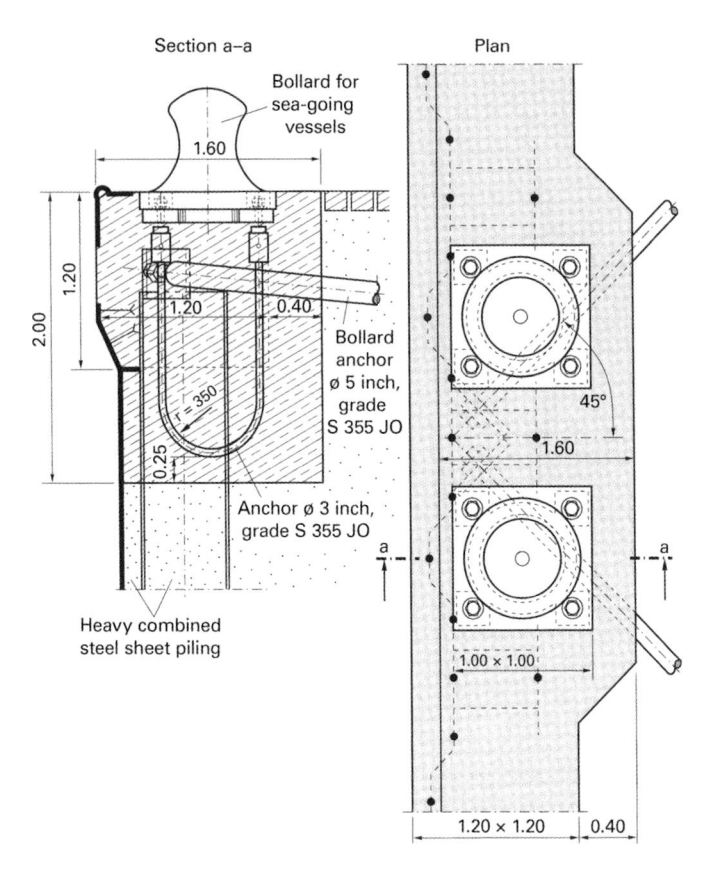

**Fig. R 129-4.** Heavy-duty reinforced concrete capping beam on a quay wall for sea-going vessels – details of an anchored bollard foundation

piling walls, to position an expansion joint directly above a sheet pile web. The web is then coated with elastic material that accommodates the changes in length of the capping beam without restraint stresses.

Where a capping beam expansion joint is located above a sheet pile trough, the embedment of this pile in the beam should be minimal. To ensure movement is possible, this pile must be covered with a substantial elastic coating that prevents direct force transfer between pile and beam and the same time guarantees a watertight joint between capping beam and sheet piling. Fig. R 129-5 shows an example of a capping beam joint above a sheet pile trough.

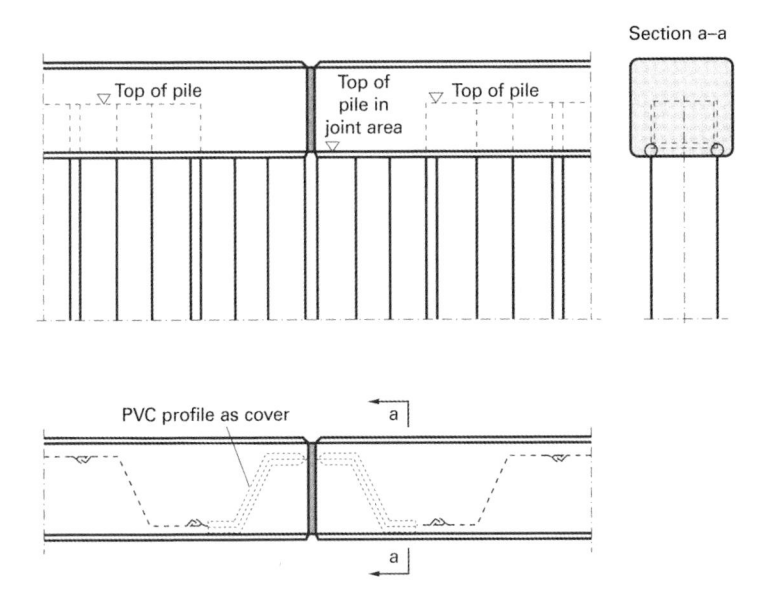

**Fig. R 129-5.** Expansion joint in a reinforced concrete capping beam

Reinforced concrete capping beams with a sufficiently large rectangular cross-section should include joggle joints to transfer horizontal forces across expansion joints. Steel dowels can be used in capping beams with smaller cross-section dimensions.

## 8.4.6 Steel nosings to protect reinforced concrete walls and capping beams on waterfront structures (R 94)

### 8.4.6.1 General

For practical purposes, the edges of reinforced concrete waterfront structures should be provided with carefully designed steel nosings on the water side. This is to protect both the edge and the hawsers running over it from damage caused by port operations. It also serves as a safety measure to prevent line handlers and other personnel working in this area from slipping over the edge. The nosing must be designed so that ships, or crane hooks, are caught on the underside (R 17, section 10.1.2).

If waterfront structures at inland ports can be flooded and there is a danger of ships grounding on the structures, the nosing must not have a raised edge.

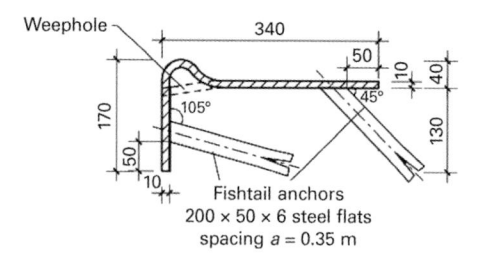

**Fig. R 94-1.** Nosing with weephole

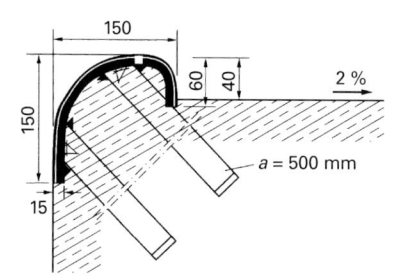

**Fig. R 94-2.** Special nosing section frequently used in The Netherlands

### 8.4.6.2 Examples

Fig. R 94-1 shows a nosing design frequently used for waterfront structures in ports and inland locks. The weephole can be omitted where precipitation is drained towards the land – which is required anyway for waterfront structures handling environmentally hazardous cargoes. The steel nosing shown in Fig. R 94-1 can also be supplied with angles other than 90° for fitting to waterfront structures with angled front or top surfaces. The separate parts of the nosing are welded together before mounting.

The design in Fig. R 94-2 depicts a special nosing developed in The Netherlands. It consists of relatively thick plates and strengthened fishtail anchors so the raised part does not need to be completely filled with concrete to guarantee the loadbearing capacity. However, the upper ventilation openings, the intention of which is to ensure that the section lies flat on the concrete during concreting, must be closed after concreting to prevent corrosion attacking the inner surface.

The designs shown in Figs. R 94-3 and R 94-4 have proved themselves on numerous German waterfront structures.

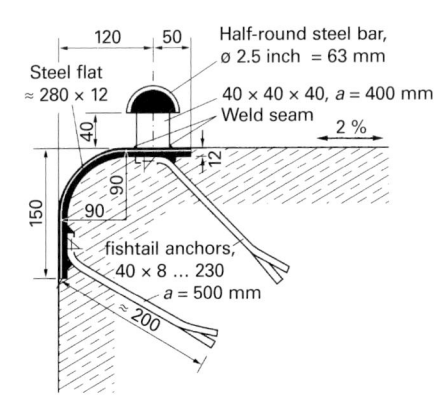

**Fig. R 94-3.** Nosing made from rounded plate, with foot railing in seaports, without in inland ports

The nosings in Figs. R 94-1 to R 94-4 must be carefully fixed in the formwork. The nosings in Figs. R 94-3 and R 94-4 must be well cast in without any voids. Any rust adhering to the inner surfaces of nosings must be removed with a wire brush before concreting.

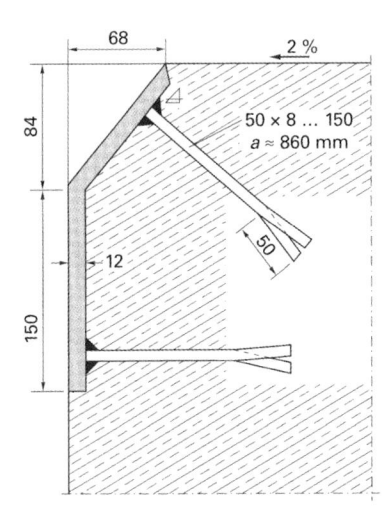

**Fig. R 94-4.** Nosing made from bent plate without foot railing for quaysides subject to flooding in inland ports

### 8.4.7 Auxiliary anchors at the top of steel sheet piling structures (R 133)

#### 8.4.7.1 General

For structural and economic reasons, the anchors to a waterfront sheet piling wall are, in general, not connected at the top of the wall, but rather at some distance below the top. This reduces the span of the wall between anchor and fixed support and thereby the bending moment. In addition, this contributes to redistribute the passive earth pressure.

In such cases the section above the anchors is frequently provided with auxiliary anchors at the top to secure the position of the top of the sheet piling and reduce its deflection. Auxiliary anchoring is not considered in the design of the sheet piling (R 77, section 8.2.3).

#### 8.4.7.2 Considerations for positioning auxiliary anchors

The height of the section above the main anchors determines whether auxiliary anchoring is necessary and depends on various factors such as the flexural stiffness of the sheet piling, the magnitude of the horizontal and vertical imposed loads and the demands that port operations place on the alignment of the top of the sheet pile wall.

When a waterfront structure is directly subjected to crane loads, auxiliary anchors should be added as near as possible to the top (unless it would be better to position the main anchors near the top of the wall so that they can carry the loads directly). As a rule, loads on the section above the anchor caused by mooring hooks also call for auxiliary anchoring. Bollards are usually anchored separately.

#### 8.4.7.3 Design of auxiliary anchors

Auxiliary anchors are calculated using an equivalent structural system in which the section above the anchor is considered to be fixed at the level of the main anchor. The loads as per R 5, section 5.5.5, are applied to this system. Auxiliary anchors are connected via a waling.

With regard to the stresses due to aligning the top of the sheet piling and the need to withstand lighter vessel impacts, the auxiliary anchor waling should be stronger than theoretically required, and is normally the same as the main anchor waling.

The auxiliary anchors are connected to anchor plates or continuous anchor walls, the stability of which must be checked for the both heave of the anchorage soil and the lower failure plane. The lower failure plane begins at the base of the anchor plate/wall and ends at the main anchor connection (Fig. R 133-1). The analysis is carried out as per R 10, section 8.5.

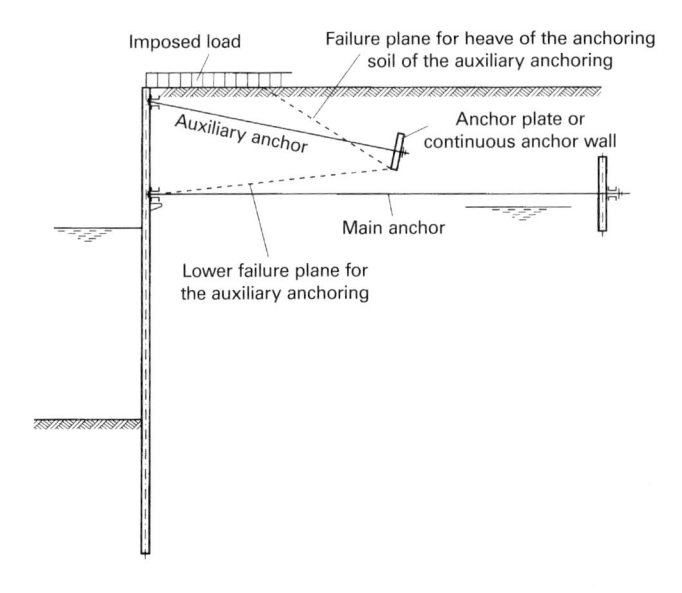

**Fig. R 133-1.** Sheet piling structure with single row of anchors plus auxiliary anchoring

### 8.4.7.4 Work on site

It is advisable to dredge the harbour bottom in front of the quay wall only after the auxiliary anchoring has been installed. If dredging is carried out beforehand, the top of the sheet pile wall may move uncontrollably, meaning that later adjustment with the auxiliary anchoring alone may not always be successful.

### 8.4.8 Screw threads for sheet piling anchors (R 184)

### 8.4.8.1 Types of thread

The following thread types are used for sheet piling anchors:

1. Cut thread (machined thread) to Fig. R 184-1
   The outside thread diameter is equal to the diameter of the round steel bar or the upsetting.
2. Rolled thread (non-cut thread produced in cold state) to Fig. R 184-2. After the thread has been rolled, the outer diameter of the thread is marginally greater than the diameter of the anchor rod. The diameter after upsetting, of tie rods with rolled threads can therefore be smaller than anchors with cut threads for the same loadbearing capacity.

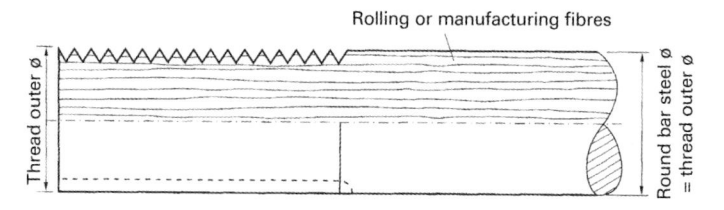

**Fig. R 184-1.** Cut thread

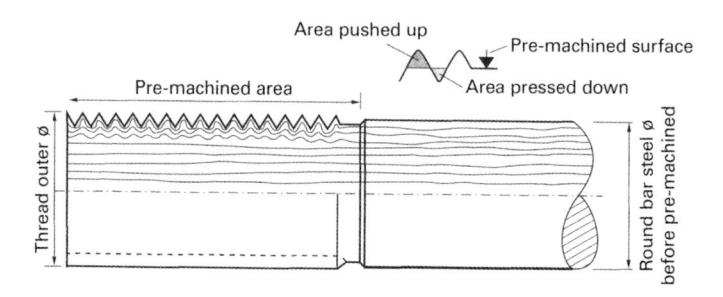

**Fig. R 184-2.** Rolled thread

When using steel grades S 235 JR and S 355 J2, before rolling the thread any upsetting must be turned down to the nominal thread diameter in order to achieve a thread conforming to the standard. Drawn steels (up to Ø36 mm) do not need to be pre-machined.

3. Hot-rolled thread (non-cut thread produced in hot state) to Fig. R 184-3

Hot-rolling produces two rows of thread flanks on the thread shaft which are opposite to each other but complement each other to form a continuous thread. The nominal diameter governs the loadbearing capacity, although the actual diameter can easily deviate from this. Components with the same thread type must be used for end anchorages and butt joints.

## 8.4.8.2 Required safety margins

For more information on verifying the loadbearing capacity of round steel tie rods with threads of various types plus the structural design of anchorages, see R 20, section 8.2.7.3.

# ANKER SCHROEDER HAVE SUPPLIED UPSET FORGED TIE BARS TO MAJOR PORTS OF THE WORLD SINCE 1920

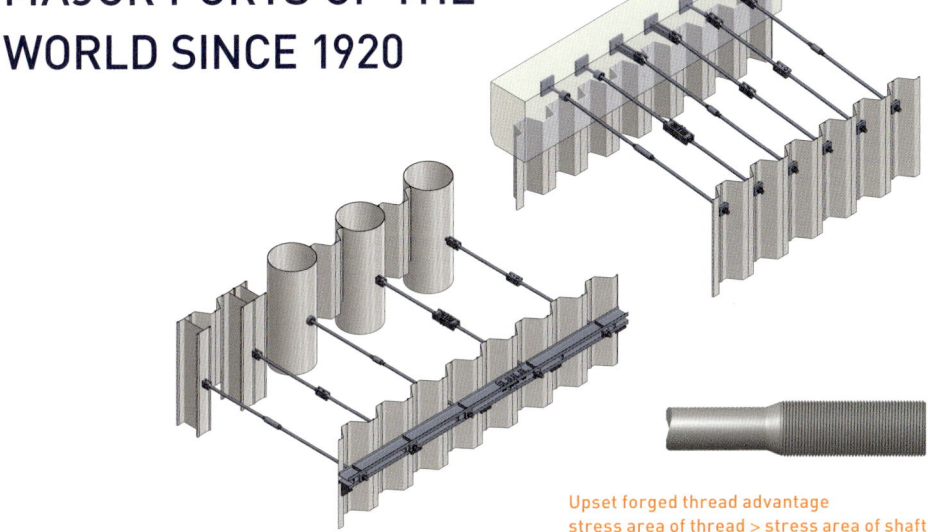

**ASDO**

**Upset forged thread advantage**
stress area of thread > stress area of shaft

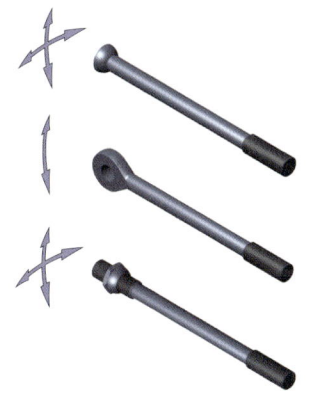

- Standard diameter range 48-180mm
- Steel grades ASDO350, 460, 500 and higher
- Load capacity up to 14,000kN
- Articulated and upset forged end connections to reduce bending stresses at wall connections caused by soil settlement & misalignment
- Individual bar lengths up to 22m
- Proven connection design & full scale testing facilities up to 20,000kN
- Design to EN1993-5 and manufacture to EN1090 Execution Class 2-4

Anker Schroeder ASDO GmbH
Hannöversche Straße 48 · 44143 Dortmund (Germany)
Phone   +49 231 51701-0
sales@anker.de · www.asdo.com

**ANKER SCHROEDER**
ASDO steel tension members

# Ernst & Sohn
A Wiley Brand

# Structural Concrete

Structural Concrete, the official journal of the *fib*, provides conceptual and procedural guidance in the field of concrete construction, and features peer-reviewed papers, keynote research and industry news covering all aspects of the design, construction, performance in service and demolition of concrete structures.

*fib* — fédération internationale du béton

Publisher: *fib* – International Federation for Structural Concrete
**Structural Concrete**
Journal of the fib
Volume 15, 2014
4 issues / year
Impact Factor 2013: 0,857
ISSN 1464-4177 print
ISSN 1751-7648 online
Also available as **ejournal**.

Other journals

- Beton und Stahlbetonbau
- Steel Construction

Order a free sample copy:
www.ernst-und-sohn.de/Structural-Concrete

**Ernst & Sohn**
Verlag für Architektur und technische
Wissenschaften GmbH & Co. KG

Customer Service: Wiley-VCH
Boschstraße 12
D-69469 Weinheim

Tel. +49 (0)800 1800-536
Fax +49 (0)6201 606-184
cs-germany@wiley.com

1003116_d

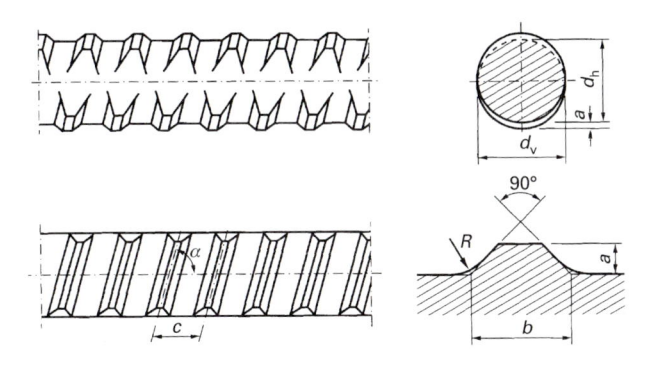

**Fig. R 184-3.** Hot-rolled thread

### 8.4.8.3 Further information on rolled threads

- Rolled threads have a high profile accuracy.
- Rolling a thread is a type of cold-forming. Cold-forming increases the strength and yield stress of the thread root and flanks, which has a favourable effect on transferring anchor forces concentrically via the thread.
- The thread root and flanks are particularly smooth on rolled threads and therefore have a higher fatigue strength under dynamic loads.
- In contrast to cut threads, the fibre orientation in the steel is not disrupted in rolled or hot-rolled threads.
- Rolled threads with larger diameters are primarily suited to concentrically loaded anchors with dynamic loads.
- In the case of round steel tie rods with rolled threads, it is important that the nuts, couplers and turnbuckles do not have rolled threads as well because internal threads are always stressed less than external ones. When the internal thread is loaded, hoop tension is generated, which supports the thread. Therefore, anchors with rolled external threads can be combined with nuts and turnbuckles having cut internal threads without hesitation.

## 8.4.9 Sheet piling anchors in unconsolidated, soft cohesive soils (R 50)

### 8.4.9.1 General

The calculation for sheet pile quay walls in unconsolidated, soft cohesive soil is covered in R 43, section 8.2.15. In principle, the settlement and wall deformations of sheet piling are greater in unconsolidated, soft cohesive soils than in soils with a higher strength.

The specific characteristics of soft cohesive soils must also be considered when anchoring such walls so that settlement and differential settlement do not result in unexpected loads.

As observations of completed structures have revealed, the shaft of a round steel tie rod is dragged downwards by the settlement of the backfill. The shaft itself however hardly cuts into soft cohesive soils.

Anchor connections at a quay wall whose base is situated in loadbearing soil (end bearing foundation) therefore undergo significant curvature when the soil behind the wall settles. Inclinations of anchor rods of up to 1 : 3 – where the anchor rods were originally installed horizontally – have already been noticed at quay walls of moderate height. The same applies to anchor connections on structures on deep foundations, anchor walls or pile bents/trestles.

Even when, in special cases, sheet pile quay walls are not embedded in loadbearing soil, it must nevertheless be assumed that the base area of such "floating" walls is located in soils that are stiffer than the soils above. It is to be expected that the soil around the anchorages of such walls settles and that the angles of anchor connections change.

In the case of anchor walls on a "floating" foundation, the differential settlement between "floating" main wall and anchor wall is generally small.

The settlement of unconsolidated cohesive soils can also be significantly different along the anchor itself. The anchor must therefore be able to bend without the normal stress due to the planned anchor load and the additional stresses from the bending exceeding the strength of the anchor.

Upset round steel tie rods with rolled threads have proved a viable solution for anchoring sheet pile walls in unconsolidated, soft cohesive soils because such tie rods exhibit a greater elongation and are more flexible than non-upset round steel tie rods for the same loadbearing capacity.

### 8.4.9.2 Connecting round steel tie rods to quay walls

Hinged connections are used to attach round steel tie rods to both quay and anchor walls. In order to provide the end of the anchor with sufficient flexibility for the increased rotational movement about the hinge, which is to be expected in soft cohesive soils, there must be a large enough gap between the webs of the channel sections forming the waling. However, the spacing required to accommodate the settlement frequently exceeds the structurally acceptable dimension. In such cases the anchor must be attached below the waling so that it can rotate freely regardless of the spacing of the channels. The unhindered flow of forces from the wall to

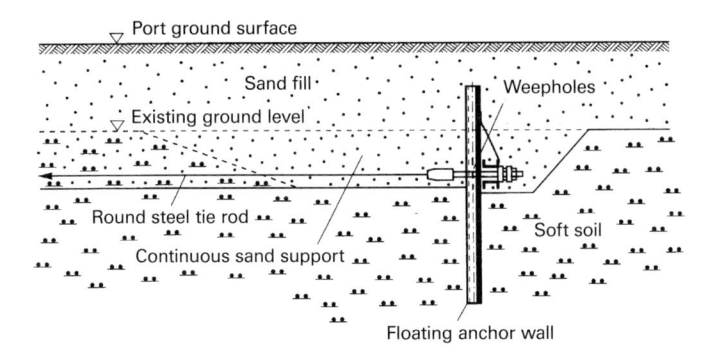

**Fig. R 50-1.** "Floating" anchor wall with eccentric anchor connection

the anchor must then be ensured by strengthening the sheet piling or additional measures on the waling.

**8.4.9.3 Connecting round steel tie rods to "floating" anchor walls**

In general, the usual gap between the two channel sections of a waling is sufficient to allow the free rotation of the anchor connection at a "floating" anchor wall when the anchor passes between the channels and is connected to a compression waling behind the wall with a hinged joint (Fig. R 50-1).

Hinged anchor connections are also required for structures and components on end bearing foundations.

**8.4.9.4 Waling design**

The properties of soft, unconsolidated cohesive soils can change dramatically over very short distances. Consequently, areas with a particularly low loadbearing capacity should not be excluded even if they have not been investigated by means of boreholes and penetrometer tests at the usual spacing (see R 1, section 1.2).

In order to accommodate the additional loads from differential settlement, walings on sheet pile walls in these soil types must be designed to be stronger than in other cases.

In general, U 400 channels of steel grade S 235 JR should be used for the walings, and S 355 J2 for larger structures, even if smaller sections will suffice structurally. Reinforced concrete walings must have at least the same load-carrying capacity as steel walings made from U 400 channels. RC walings are divided into sections 6.00–8.00 m long. They must be provided with joggle joints to prevent horizontal movement (R 59, section 8.4.3).

#### 8.4.9.5 Anchors at piles trestles or "floating" anchor walls

If displacement of the quay wall head is not permissible, the anchors must be fixed to pile trestles or non-sway structures or structural members.

If displacement of the quay wall can be accepted, the anchors can also be tied to a "floating" anchor wall. To this end, the horizontal pressures in front of the anchor wall must be limited in accordance with the permissible displacement of the wall. Where local experience relating to the displacement of the wall and the mobilised passive earth pressure is lacking, appropriate soil mechanics tests and, if necessary, loading tests should be carried out and the results assessed by a geotechnical expert.

#### 8.4.9.6 Design of anchors for "floating" anchor walls

If the terrain is filled with sand up to the level of the port operations area, a "floating" anchorage design as shown in Fig. R 50-1 is recommended. The soil in front of the anchor wall with a low loadbearing capacity is excavated to just below the anchor connection and replaced by a compacted bed of sand of sufficient width. The anchors can be laid in trenches that are either filled with carefully compacted sand or suitable excavated soil.

The anchor wall is then connected off-centre so that the permissible horizontal loads are not exceeded in either the sand fill or the area of the soft soil. A uniformly distributed bedding can be assumed both above and below the anchor connection, the magnitude of which is the result of the equilibrium conditions related to the anchor connection. The anchorage according to Fig. R 50-1 also compensates for irregularities in the bedding.

To prevent excess water pressure acting on the anchor wall, weepholes must be included in the anchor wall to equalise the hydrostatic pressure.

#### 8.4.9.7 Verifying the stability of "floating" anchor walls

For "floating" anchor walls, the stability at the lower failure plane must be verified for both the initial state of the unconsolidated soil as well as for all intermediate states and the final state of the consolidated soil as per R 10, section 8.2.6. In the case of soft cohesive soil whose shear strain in a triaxial test to DIN 18137-2 is >10%, the degree of utilisation is to be reduced according to DIN 4084.

### 8.4.10 Design of protruding quay wall corners with round steel tie rods (R 31)

#### 8.4.10.1 Guidance on anchorage design

A protruding corner in sheet piling held by anchors running diagonally from quay wall to quay wall leads to high tension loads being transferred to the walings. Waling splices are particularly problematic (R 132, section 8.2.11). The anchors located furthest from the quay wall corner are subjected to the highest tension loads.

The diagonal tension loads require a corresponding design of the walings and their connection to the wall. Steel sheet piling can handle the components of the anchor load in the plane of the wall but only with a large embedment depth.

Therefore, anchorage systems for protruding quay wall corners with diagonal anchors across the corner are technically awkward and also costly. It is also not easy to predict the effects of unintended loads, e.g. due to varying subsoil properties.

Anchorages for protruding quay wall corners are therefore not recommended.

### 8.4.10.2 Recommended crossing anchors

Tensile forces acting diagonally at protruding corners in sheet piling walls can be avoided through a crosswise arrangement of anchors tied to rear anchor walls (Fig. R 31-1). The quay wall and the anchor walls thus form a robust corner section that is stable by itself. However, the crossing anchors require the levels of the walings and anchor positions to be offset so that there is sufficient clearance between the anchors where they cross.

Edge bollards around protruding quay wall corners should have their own anchoring.

### 8.4.10.3 Walings

Walings for sheet piling are in the form of steel tension members and are shaped to match the alignment of the quay wall. The transition from the walings of a corner section to the walings of the quay wall should be arranged such as to allow the walling to move independently from each other. They are therefore connected by means of splices with slotted holes.

Freedom of movement is also necessary for anchor wall walings at the connections with the quay wall and the intersection with the other anchor wall.

Anchor wall walings are in the form of compression members made from steel or reinforced concrete.

### 8.4.10.4 Anchor walls

Anchor walls at the corner continue through to the quay wall (Fig. R 31-1). Staggered embedment lengths are used, gradually increasing to the depth of the harbour bottom alongside the quay wall. This arrangement reduces the risk that, in the case of a vessel impact at a corner (corners are always particularly vulnerable), the backfilling in the corner block may only be washed out as far as the anchor walls.

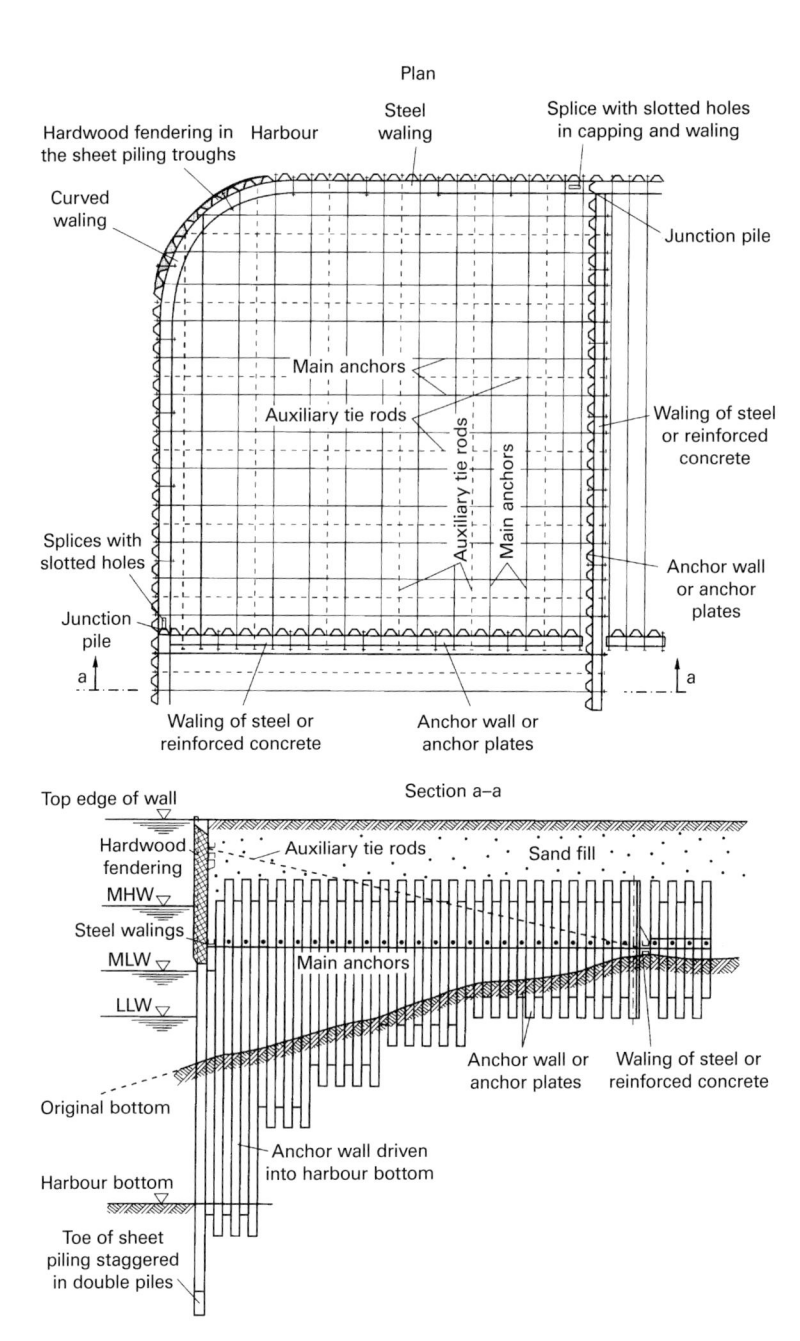

**Fig. R 31-1.** Anchoring a protruding corner in a sheet pile quay wall for a seaport

Embedment to the depth of the harbour bottom is also recommended when the round steel tie rods are anchored to anchor plates, e.g. made from reinforced concrete, instead of continuous anchor walls.

### 8.4.10.5 Fendering in sheet pile troughs
Ships and quay wall corners can be protected by fitting, for example, marine timbers or plastic sections into the sheet pile troughs at protruding corners. This fendering should project about 5 cm beyond the outer face of the sheet piling (Fig. R 31-1).

### 8.4.10.6 Rounded and reinforced concrete wall corners
Since protruding quay wall corners are particularly vulnerable to damage by waterborne traffic, they should be rounded off if possible and also strengthened by a strong reinforced concrete cap if necessary.

### 8.4.10.7 Protective dolphins at corners
If the waterborne traffic allows it, protruding quay wall corners can be protected against vessel impact by elastic dolphins or a guidance structure positioned in front of the exposed corner.

### 8.4.10.8 Stability verification
Verification of stability is carried out individually for each quay wall segment, in accordance with R 10, section 8.5. A special check for the corner section is not necessary if the anchors of the quay walls are carried through to the other wall as shown in Fig. R 31-1.

## 8.4.11 Design and calculation of protruding quay wall corners with raking anchor piles (R 146)

### 8.4.11.1 General
Protruding quay wall corners are particularly vulnerable to damage from vessel impacts. In many cases, quay wall corners are also equipped with heavy-duty bollards at the ends of the quay wall to secure large ships as per R 12, section 5.12.2. In addition, they are also equipped with the necessary fenders with a higher energy absorption capacity than the adjacent quay wall sections. Overall, protruding quay wall corners must be robust and as rigid as possible.

Supplementary to R 31, section 8.4.10, the anchorage of protruding quay wall corners with raking anchor piles is covered below, as this type of anchor essentially complies with the requirement for a particularly rigid and robust construction.

### 8.4.11.2 Design of the corner structure

The design of the corner structure is primarily determined by the structural design of the adjoining quay walls, the difference in ground levels and the angle enclosed by the wall sections forming the corner. In addition, the design to be selected is very much influenced by the existing water depth and the in situ soil conditions.

The raking piles to be used to anchor the corner structure should always be connected perpendicular to the quay wall. They must therefore overlap in the anchoring zone. In order to ensure that the anchor piles can be installed properly and without colliding, a minimum clearance must be maintained at the points where they cross. Whereas the clear spacing of crossing piles above the level of the harbour bottom can be kept comparatively small (usually about 25–50 cm), a similar spacing is not sufficient for long piles and difficult driving conditions. The clearance at crossing points below the level of the harbour bottom should therefore be at least 1.0 m, although 1.5 m is preferable. In soils in which a more significant deviation of longer piles is to be expected during driving, the clearance should be at least 2.5 m. Steel wings (if fitted) must always be taken into account when calculating the clearance between piles at crossing points.

In order to be able to comply with the minimum clearance requirements at crossing points, the spacing and inclination of the anchor piles must also be varied accordingly. However, the positions of the piles within a group should be kept fairly uniform because of the different loadbearing behaviour of piles at different angles.

Should deep foundations prove necessary at a quay wall corner for heavily loaded bollards or other items such as bracing structures for conveyor belts etc., the construction of a special reinforced concrete corner section with a pile cap on a deep foundation is recommended in most cases. The pile cap should be connected to the sheet pile wall via a hinge detail.

This solution is also suitable for quay wall corners where the minimum spacing of piles required for driving in the crossover area cannot be achieved by adjusting the positions of the piles (Fig. R 146-1). In such corner designs it is expedient to position the tension piles required in the corner area towards the rear of the pile cap. They then lie in a plane that is different from that of the tension piles of the neighbouring quay wall sections, thus making it easier to maintain adequate clearances at crossing points. As this solution requires additional compression piles at the rear edge of the pile cap to accommodate the vertical components of the anchor force, such designs are more costly. However, this is a reliable method of construction without any particular risks.

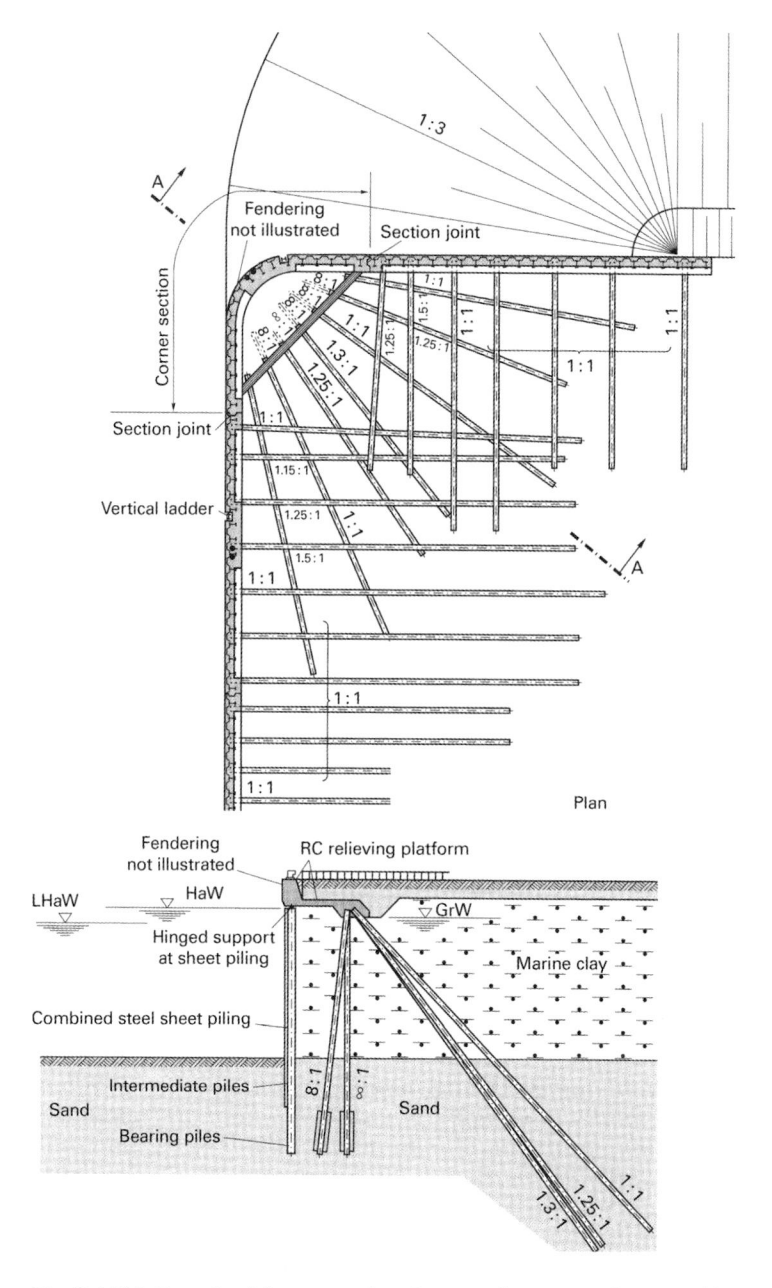

**Fig. R 146-1.** Example of the construction of a protruding quay wall corner with steel anchor piles

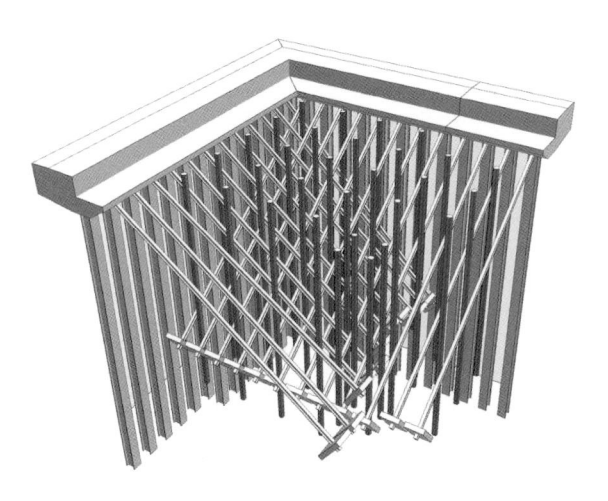

**Fig. R 146-2.** 3D image of protruding quay wall corner with raking anchors rotated down into position

The designs described in R 31, sections 8.4.10.5 and 8.4.10.7, for round steel tie rods also apply to protruding quay wall corners anchored with raking piles.

### 8.4.11.3 Use of 3D images for visualising pile positions

In order to anticipate driving difficulties, 3D images of difficult corner designs should be produced during the planning phase of a project (Fig. R 146-2). The colour coding of individual anchor positions means it is very easy to highlight the critical anchor crossing points and then find optimum solutions.

During construction on site, the 3D model should be updated with all the actual positions of anchors already installed to determine the necessary corrections due to driving deviations.

### 8.4.11.4 Verifying the stability of corner sections

The stability of corner sections must be verified for all structural parts of the corner anchoring. To this end, each wall at the corner must be considered separately. At corners with additional loads, e.g. due to corner structures, bollards, fenders and other equipment, it must be verified that the piles are also able to safely accommodate these additional forces.

If major changes to positions of piles are required during construction, their effects are to be verified by means of supplementary calculations.

## 8.4.12 High prestressing of high-strength steel anchors for waterfront structures (R 151)

### 8.4.12.1 General

Anchorages for sheet piling waterfront structures, also for subsequent securing of other structures such as walls on pile trestles, normally make use of non-prestressed anchors in steel grades S 235 JR, S 235 J2 or S 355 J2. However, in certain situations it can be beneficial to prestress anchors with a high proportion of their calculated anchor force. Such anchors must be made from high-strength steel.

Highly prestressed anchors can be practical for limiting displacements, particularly in the case of structures with long anchors, when building near structures sensitive to settlement or when subsequently connecting sheet pile walls driven beforehand.

Furthermore, the high prestressing can achieve a redistribution of the passive earth pressure when the sheet piling is backfilled with either medium density non-cohesive soil or stiff cohesive soil. Redistributing the passive earth pressure decreases the span moment and increases the anchor force.

Permanent anchors manufactured from high-strength steel must be protected against corrosion. Any existing special recommendations and standards e.g. for grouted anchors to DIN EN 1537, must be complied with.

### 8.4.12.2 Effects of high anchor prestressing on active earth pressure

Prestressing the anchors of sheet piling structures always reduces wall deformations towards the water side, especially in the upper part of the quay wall. A high prestress redistributes the active earth pressure upwards. Consequently, the resultant of the active earth pressure can shift from the bottom third point of the wall height $h$ above the harbour bottom to about $0.55h$ – with a corresponding increase in the anchor force. This active earth pressure redistribution is especially pronounced at quay walls that extend above the level of the anchors.

When the intention of prestressing is to achieve an active earth pressure distribution deviating from the classic distribution according to Coulomb, the anchors must be prestressed to about 80% of the characteristic anchor load determined for design situation DS-P.

### 8.4.12.3 Time for prestressing

The prestressing of the anchors may not begin until the respective prestressing forces can be accommodated without appreciable, undesirable movements of the structure or its members. This presumes appropriate backfilling behind the wall. The prestressing forces must then be transferred from the structure into the backfilling as intended.

Owing to the fact that when tensioning the anchors, the prestress in adjacent – already prestressed – anchors is at least partly lost due to the redistribution of loads in the soil, the anchors must be prestressed beyond the design value so that, after the adjacent anchors have been tensioned, they still possess their planned prestress. Random checks are to be made on prestressing forces during construction so that any required corrections can be made.

Prestressing in several steps can avoid the redistribution of loads and therefore also the short-term overloading of the anchors. However, this does complicate operations.

DIN EN 1537 is the main standard that applies to pressure-grouted anchors.

### 8.4.12.4 Further guidance

Where high anchor prestressing is only carried out for certain parts of a waterfront structure, it should be remembered that the freedom of movement of the wall as a whole varies locally. The prestressed zones act as fixed points, which are subjected to correspondingly higher spatial active earth pressures. The increased active earth pressure must be taken into account in the structural analyses for the respective wall areas and their anchors.

Prestressed anchors must remain permanently accessible so that, if required, the prestressing can be checked and corrected. Furthermore, anchor ends should have hinged connections.

Since bollards are loaded from time to time only, their anchors should not be made from prestressed high-strength steel, but rather from firmly tightened (but not prestressed) round steel tie rods made from grade S 235 JR, S 235 J2 or S 355 J2. The latter grades exhibit only a minor elongation under load.

### 8.4.13  Hinged connections between driven steel anchor piles and steel sheet piling structures (R 145)

### 8.4.13.1 General

Sheet pile walls are subjected to bending loads due to active earth pressure, which results in wall rotations around the anchor connections. These rotations are transferred to the anchors when the anchors are connected in accordance with R 59, section 8.4.3. The outcome is corresponding bending stresses in the anchor in addition to the intended stresses due to the anchor force. Settlement and/or subsidence may create additional bending loads on the anchor.

By contrast, a hinged connection for the driven steel anchor allows essentially non-restrained mutual rotation of the sheet piling and the anchor so that the anchor and the anchor connection are only subjected to the intended anchor loads and can therefore be designed economically. A hinged anchor connection detail must be designed in accordance with the principles of structural steelwork.

### 8.4.13.2 Guidance for designing a hinged connection

The anchor connection's ability to rotate can be achieved through the use of single or double hinge pins or through the plastic deformation of a structural member designed for this purpose (plastic hinge). A combination of hinge pin and plastic hinge is also possible. When designing for plastic hinges, the following points should be considered:

1. Plastic hinges should be located at a sufficient distance from butt and fillet welds so that the steel is not stressed to yield in the vicinity of weld seams. Lateral fillet welds should be located in the plane of the force or plane of the tension element so that they do not peel off. Otherwise, peeling-off is to be prevented by other measures.
2. Weld seams transverse to the tensile force of the anchor pile can act as metallurgical notches and should therefore be avoided.
3. Non-structural erection seams executed in difficult welding positions without observing proper welding techniques increase the risk of failure.
4. At difficult connections, also hinged connections, the recommendation is to calculate the likely plastic hinge cross-section for the intended normal forces in conjunction with possible additional stresses etc. (see R 59, section 8.4.3). When designing plastic hinges, DIN 18800 is to be taken into consideration.
5. Abrupt changes in stiffness, e.g. flame-cut notches in a pile and/or metallurgical notches due to cross-welds, as well as sudden increases in the steel cross-section, e.g. due to very thick, welded splice plates, are to be avoided, especially in the potential plastic hinge regions of anchor piles in tension, because they can lead to sudden failure without prior deformation.

Examples of hinged connections for steel anchor piles are shown in Figs. R 145-1 to R 145-8.

### 8.4.13.3 Work on site

Depending on the local conditions and the design, steel anchor piles may be installed either before or after the sheet piling. If the location of the connection depends on the geometry, as would be the case if the connection must be made in the trough of a sheet pile or on a bearing pile in combined sheet piling, it is important that the upper end of each anchor pile is as close as possible to its intended position. This is best

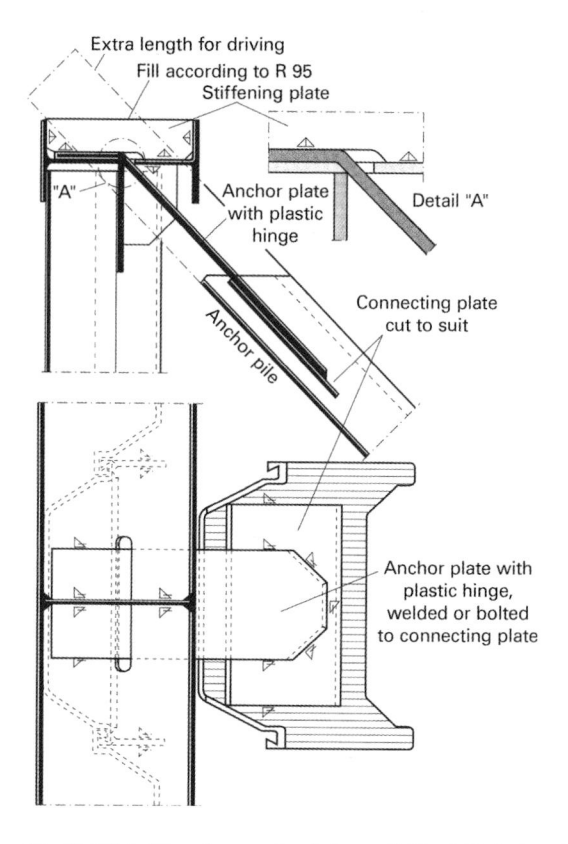

**Fig. R 145-1.** Hinged connection between lightweight steel anchor pile and lightweight steel sheet piling by means of splice plate plus plastic hinge

accomplished if the anchor piles are installed after the sheet piling. However, the design of the connection must always be such that certain deviations and rotations can be compensated for and accommodated.

If the steel anchor pile is driven directly above the top of the sheet piling, or through an opening in the sheet piling, the sheet piling can provide effective guidance for driving the anchor pile. A driving opening can also be provided by burning off the upper end of a double pile, hoisting it clear and later returning it to its original position and welding it in place. A certain degree of adjustment is possible at the head of steel piles whose upper end is not embedded in the soil at the time of making the connection to the sheet pile wall.

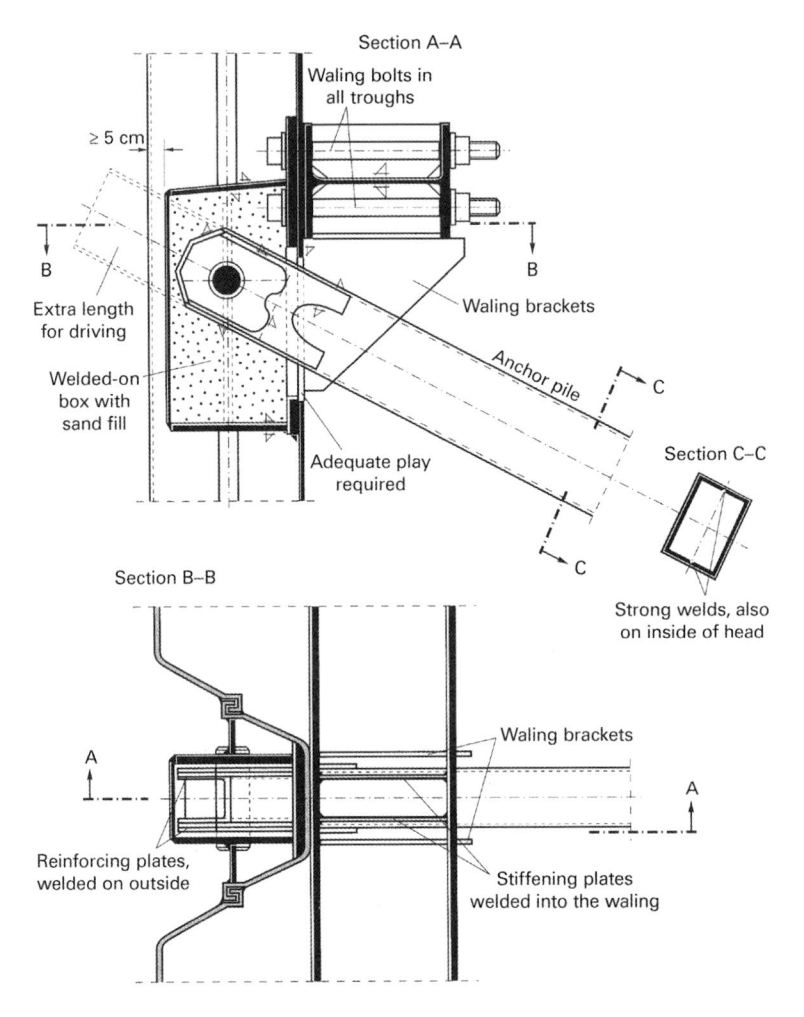

**Fig. R 145-2.** Hinged connection between steel anchor pile and heavy steel sheet piling by means of a hinge pin

Anchor piles should be provided longer than required to allow for cutting off the top end in the case of damage to the microstructure caused during driving.

Where possible, slots for connection plates should not be cut in sheet piles and/or anchor piles until after they have been driven to the final depth.

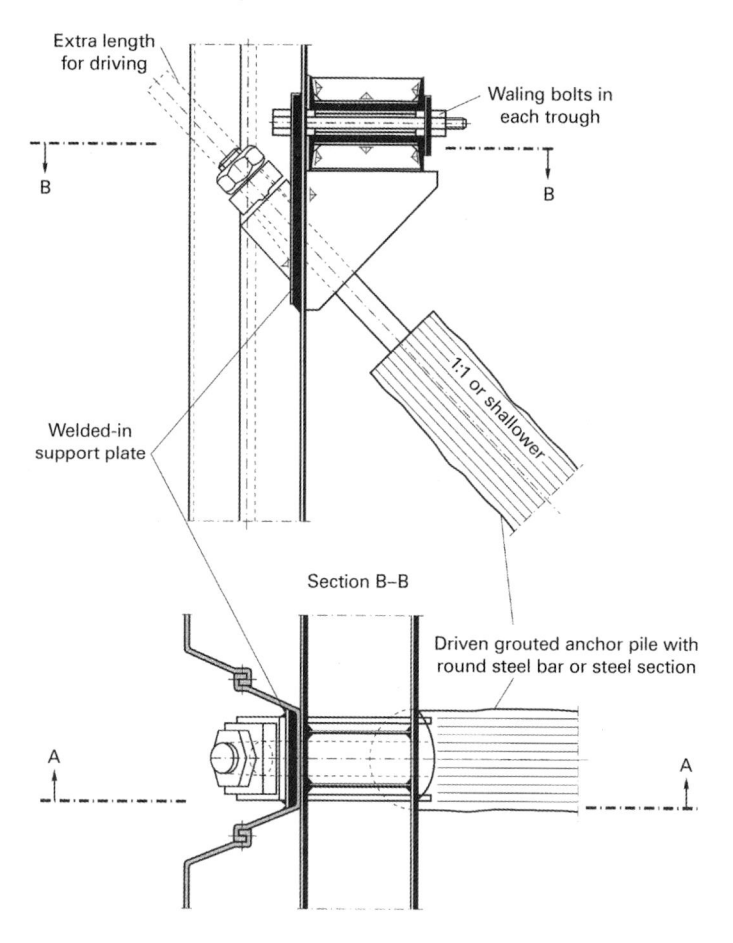

**Fig. R 145-3.** Hinged connection between driven grouted anchor pile and heavy steel sheet piling

### 8.4.13.4 Connection details

A hinged connection to sheet piling is generally placed in the trough, especially with interlocks on the centroid axis, or, for combined sheet piling, on the web of the bearing piles.

In the case of smaller anchor forces, the steel pile may also be connected to the capping beam on top of the sheet piling (Fig. R 145-1) or to a

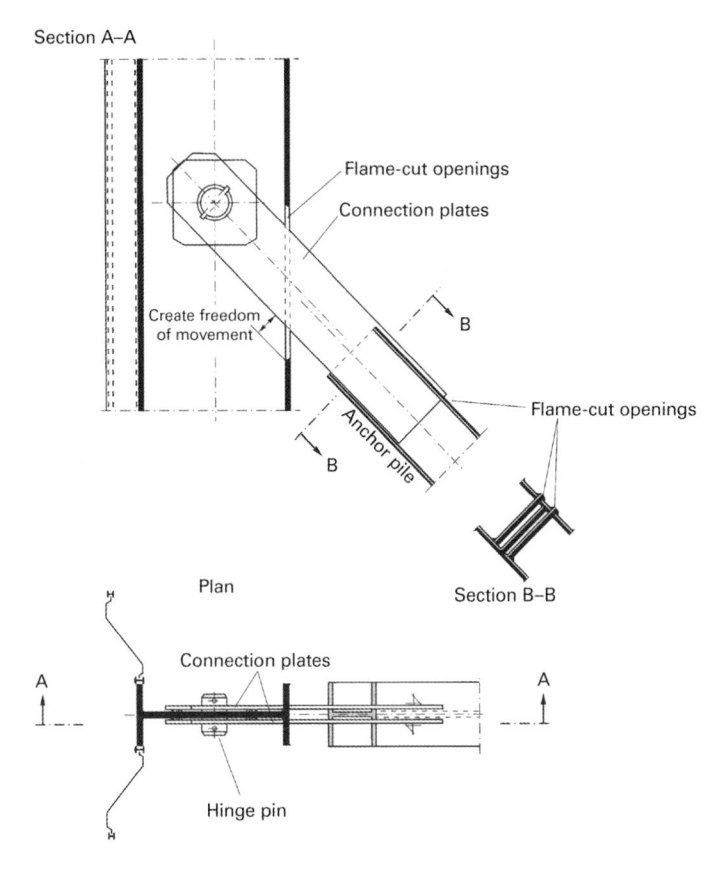

**Fig. R 145-4.** Hinged connection (hinge pin) between steel anchor pile and combined sheet piling with single bearing piles

waling behind the sheet piling by means of splice plate plus plastic hinge. Special attention must be paid to the danger of corrosion in such cases. Please refer to R 95, section 8.4.4, for berths and waterfront structures for cargo handling.

Tension elements made from round steel rods (Fig. R 145-3), steel plates or wide steel flats (tension splice plates) are frequently fitted between the connection in the sheet pile trough or bearing pile web and the upper end of the pile (Figs. R 145-4 and R 145-5). Connections consisting of a threaded steel rod, nut, plate washer and hinged plate have the advantage that they can be tensioned.

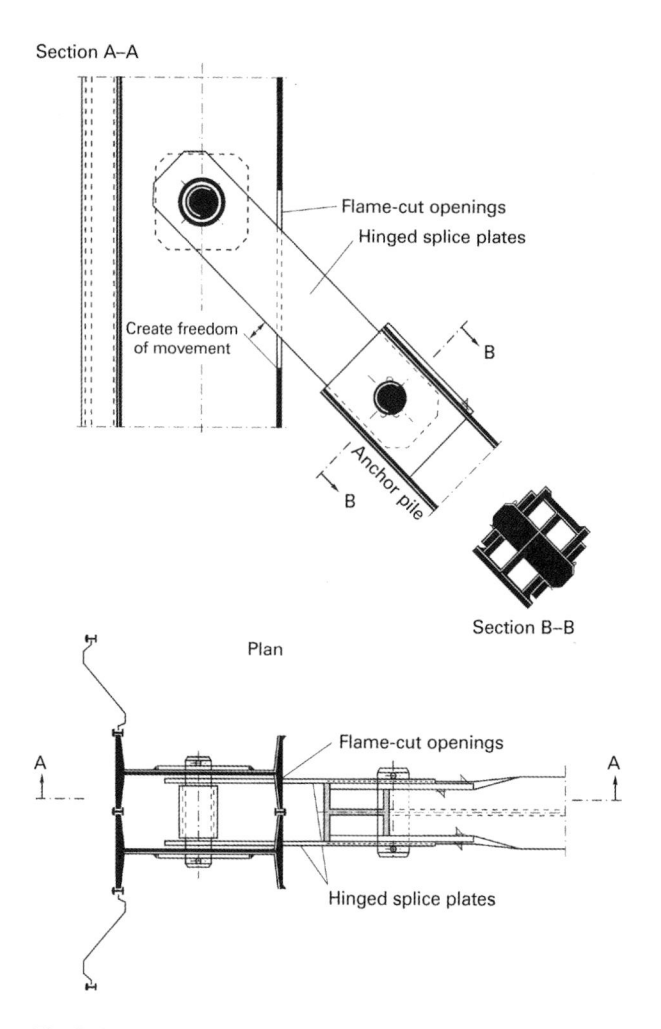

**Fig. R 145-5.** Hinged connection (hinged splice plates) between steel anchor pile and combined sheet piling with double bearing piles

In addition to the hinged connection in the sheet pile trough, at the capping beam or on the web of the bearing pile, an additional hinge may be included near the top end of the anchor pile in special cases.
This solution is depicted in Fig. R 145-5 for the case of double bearing piles. This type of connection can also be used for single bearing piles. The (flame-cut) slots in the flanges of the bearing piles are to be extended

far enough below the connection plates in order to allow sufficient freedom of movement for pile rotations and to rule out any restraint forces that can arise as a result of unwanted fixity should there be contact between plates and pile. Care is also required to ensure that the intended hinge effect is not impaired by incrustations, sintering or corrosion around the connection. This is to be checked for each individual case and considered in the structural design.

The anchor pile can also be driven through an opening in a sheet pile trough and connected via a hinge detail and supporting plates welded to the sheet piling (Fig. R 145-2).

If the connection is made on the water side trough of a sheet pile wall, all parts must terminate at least 5 cm behind the line of the sheet piling. Furthermore, the point at which the pile penetrates the sheet piling should be carefully protected against soil escaping and/or being washed out (e.g. by means of a sand-filled box as in Fig. R 145-2).

Depending on the design selected, preference should be given to those connections that can be largely fabricated in the workshop and which exhibit adequate tolerances (Fig. R 145-6). Extensive modifications on site are expensive and are therefore to be avoided if at all possible.

With a connection such as the one shown in Fig. R 145-6, all loadbearing weld seams on the bearing pile can be welded on in the horizontal flat

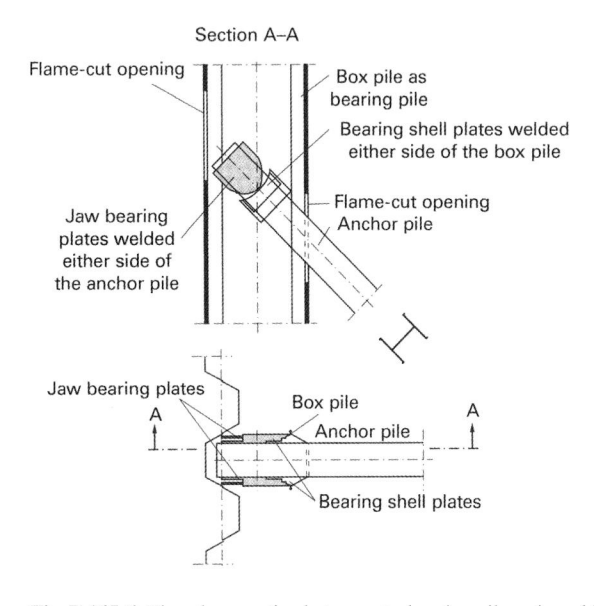

**Fig. R 145-6.** Hinged connection between steel anchor pile and combined steel sheet piling using jaw bearings/bearing shells

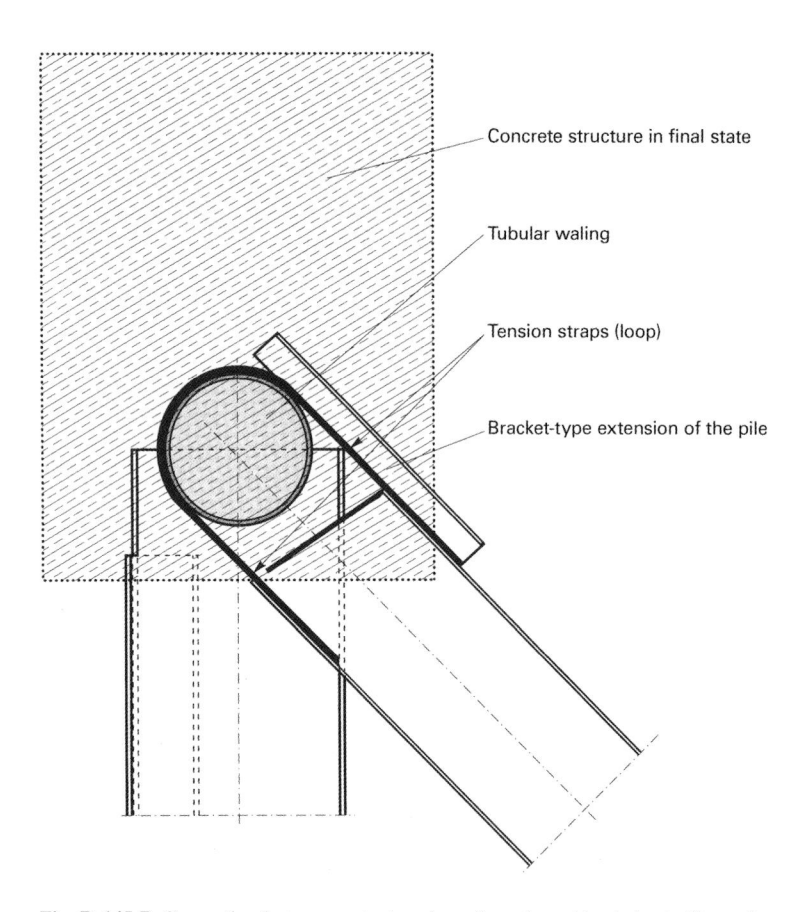

**Fig. R 145-7.** Connection between a steel anchor pile and combined sheet piling using loops

position in the workshop. However, this solution is only advisable when the jaw bearing plates are attached after driving the anchor piles and when the lengths of the bearing piles can be specified exactly. This sort of connection prepared in the workshop is therefore impossible where bearing piles have to be extended or shortened.

In the solution shown in Fig. R 145-7, the connection between anchor pile and sheet piling is created by loops that enclose a tubular waling welded into the sheet piling. Care is required here because the free rotation of the connection is prevented by friction between loops and tube. The loops are therefore to be dimensioned to allow for the adequate loadbearing reserves required for the resulting unequal load distribution.

As a rule, deflections of the anchor piles and displacement of the pile heads are to be expected, and so the connection has to be designed to transfer shear forces as well as anchor tensile forces. This can be achieved with a bracket-type extension of the piles through to the tubular waling (Fig. R 145-7).

Fig. R 145-8 depicts a hinged connection between a small injection pile and a steel sheet pile wall. As such connections are not generally covered by standards, a national technical approval may be required.

Corrosion protection of the steel tendon in the ground takes the form of a ribbed plastic tube into which cement mortar is injected. The outboard anchor connection is protected against corrosion by enclosing it in a fitting filled with a plastic compound. The connection to a reinforced concrete waling is regulated by the national technical approval for the injection pile.

### 8.4.13.5 Verifying the loadbearing capacity of the connection

All anchor connection parts are to be designed for the internal forces that can be transferred by the anchor system used. Loads from the water side, e.g. berthing pressure, ice pressure, mine subsidence, etc., can at times reduce the tensile force in the steel pile or even change it to a compressive force. If necessary, the stresses in the anchor connection and the anchor itself should be checked to ensure they can be accommodated. A buckling analysis is to be carried out on free-standing anchors. In some cases ice impact must also be taken into account.

Wherever possible, the connection should lie at the intersection of the sheet piling and pile axes (Figs. R 145-1 to R 145-8). If the anchor connection deviates substantially from this, additional moments in the sheet piling must be assumed.

The vertical and horizontal components of the anchor pile force are also to be taken into account in the connection to the sheet piling and – where not every loadbearing wall element is anchored – in the waling and its connections. If a vertical load due to soil surcharge is to be expected, this must also be allowed for in the reaction forces and in the verification of the loadbearing capacity of connections. This is always the case when deflection of the anchor piles is to be expected.

When the angle between anchor pile and sheet piling changes as a result of anchor deflection, then the ensuing changes to the tensile and shear forces at the anchor connection must also be considered in the analyses. Where anchor connections are located in the troughs of sheet pile walls, the horizontal force component is to be transferred into the sheet piling web via plate washers of sufficient width (Fig. R 145-3). The weakening of the sheet piling cross-section by the anchor penetration should be considered. It may be necessary to reinforce the sheet piling in the area around a connection.

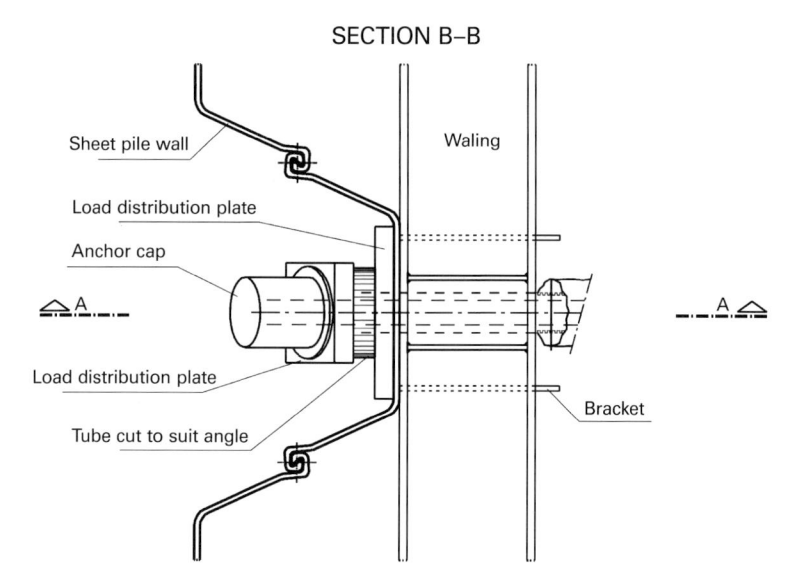

**Fig. R 145-8.** Hinged connection between small injection pile with double corrosion protection and steel sheet piling

Connections for hinged anchorages must be positioned such as to facilitate the transfer of tensile and shear forces in a constant flow. If the flow of forces is not obvious in the case of difficult, heavily loaded connection details, the calculated dimensions of the anchor connection should be tested by means of at least two loading tests to failure on full-size mock-ups.

## 8.5 Verification of stability for anchoring at the lower failure plane (R 10)

### 8.5.1 Stability at the lower failure plane for anchorages with anchor walls

Stability at the lower failure plane is verified using the method proposed by [127]. This method works on the basis of taking a section behind the retaining wall, along the lower failure plane and behind the anchor wall. The lower failure plane has a convex form and runs between the base of the anchor wall and the fulcrum of the sheet piling in the soil; it is approximated by a straight line in the analysis. In the case of quay walls simply supported in the soil, the fulcrum of the wall is the base of the wall.

The analysis of stability at the lower failure plane determines the minimum anchor length required for carrying the anchor force.

Fig. R 10-1 shows the forces in the body of soil between anchor wall, quay wall, lower failure plane and ground surface (sliding body FDBA). Notation for Fig. R 10-1:

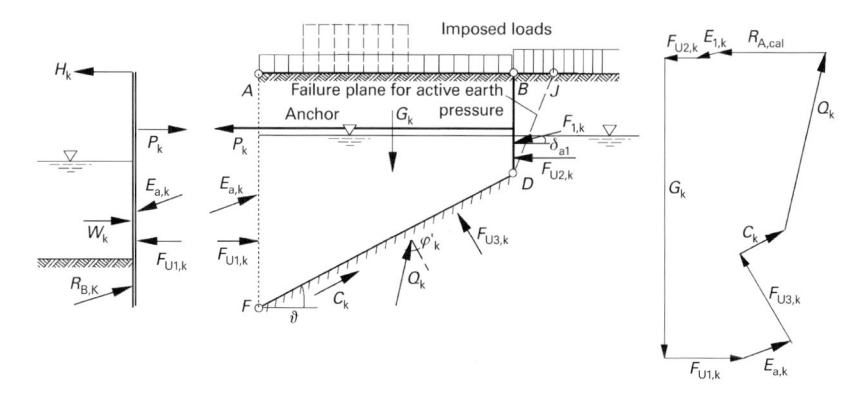

**Fig. R 10-1.** Analysis of stability at lower failure plane

$\vartheta$        inclination of lower failure plane
$G_k$      total characteristic weight of sliding body FDBA plus imposed load if applicable
$E_{a,k}$    characteristic active earth pressure (increased active earth pressure if applicable)

$F_{U1,k}$ characteristic hydrostatic pressure at section AF between soil and retaining wall
$F_{U2,k}$ characteristic hydrostatic pressure at anchor wall DB
$F_{U3,k}$ characteristic hydrostatic pressure at lower failure plane FD
$Q_k$ characteristic resultant force at lower failure plane due to normal force and maximum possible friction force (therefore inclined at angle $\varphi_k$ to normal to failure plane)
$C_k$ characteristic cohesion force at lower failure plane (with its magnitude dependent on the characteristic value of the cohesion and the length of the lower failure plane)
$E_{1,k}$ characteristic active earth pressure with imposed load on anchor wall DB
$P_k$ characteristic anchor force

In the case of the characteristic anchor force, we must distinguish between the component $P_{G,k}$ due to permanent actions and the component $P_{Q,k}$ due to variable actions.

The analysis must be carried out for exclusively permanent loads and also for permanent plus variable loads. In the latter case, the components due to variable loads are to be applied in an adverse position. The force $P_{Q,k}$ resulting from these components is to be identified separately. Stability at the lower failure plane is given when

$$P_{G,k} \cdot \gamma_G \leq R_{A,\mathrm{cal}}/\gamma_{Ep}$$

where $R_{A,\mathrm{cal}}$ is determined from the polygon of forces as per Fig. R 10-1 for exclusively permanent loads, and

$$P_{G,k} \cdot \gamma_G + P_{Q,k} \cdot \gamma_Q = R_{A,\mathrm{cal}}/\gamma_{Ep}$$

where $R_{A,\mathrm{cal}}$ is determined from the polygon of forces as per Fig. R 10-1 for permanent plus variable loads.

The following DIN 1054 partial safety factors are applied when analysing stability at the lower failure plane:

$\gamma_G$ partial safety factor for permanent actions
$\gamma_Q$ partial safety factor for variable actions
$\gamma_{Ep}$ partial safety factor for passive earth pressure

The analysis of stability at the lower failure plane is based on the concept that transferring the anchor force into the soil causes a failure body to become established behind the retaining wall and that this body is bounded by the quay wall, the anchor wall and the lower failure plane. In this situation the maximum possible shear resistance at the lower failure plane is exploited, whereas the limiting value for the reaction force at the base is not achieved. Force $R_{A,cal}$ is the characteristic anchor force that can be accommodated by the sliding body FDBA assuming the full shear strength of the soil. The definition of the exploitation of the anchor force is equivalent to the exploitation of the shear strength of the soil.

Equilibrium of the applied moments is not considered in the verification of stability at the lower failure plane either. This is because only the resultants of the actions transferred via the boundaries of the sliding body are included in the analysis. It is sufficiently accurate to replace the lower failure plane by the straight line DF as the critical failure plane.

If the flow force of any groundwater flowing in the sliding body (water table descending downwards towards sheet piling) is to be taken into account, the hydrostatic pressures on quay wall, anchor wall and lower failure plane must be determined using a flow net according to R 113, section 4.7. The pressures are added together to form resultants at the respective boundary surface of the sliding body.

## 8.5.2 Stability at the lower failure plane in unconsolidated, saturated cohesive soils

Verification of stability at the lower failure plane for quay walls and their anchors in unconsolidated cohesive soils is carried out as per section 8.5.1. The active earth pressure is determined for the unconsolidated, saturated case as per R 130, section 2.5. The characteristic cohesion force $C_{u,k}$ is effective in the lower failure plane. The angle of internal friction is to be taken as $\varphi_u = 0$ for unconsolidated, saturated, virgin cohesive soils.

## 8.5.3 Stability at the lower failure plane with varying soil strata

Verification of stability at the lower failure plane in varying soil strata is carried out as per section 8.5.1. The sliding body as shown in Fig. R 10-2 is divided by imaginary vertical planes passing through the intersections of the lower failure plane with the boundaries of the strata (Fig. R 10-2). This splits the sliding body FDBA into as many parts as there are strata intersected by the lower failure plane. Equilibrium of forces is then carried out for all individual parts in turn (Fig. R 10-2). If parts of the lower failure plane pass through cohesive soil, a cohesion force is applied to these sections (cohesion is not taken into account in the polygon of forces in Fig. R 10-2).

The active earth pressures in the vertical sections between the individual parts are assumed to act parallel to the surface.

The following characteristic forces act on the sliding bodies in Fig. R 10-2:

$G_{1,k}$     total weight of sliding body $F_1DBB_1$, plus imposed load if applicable

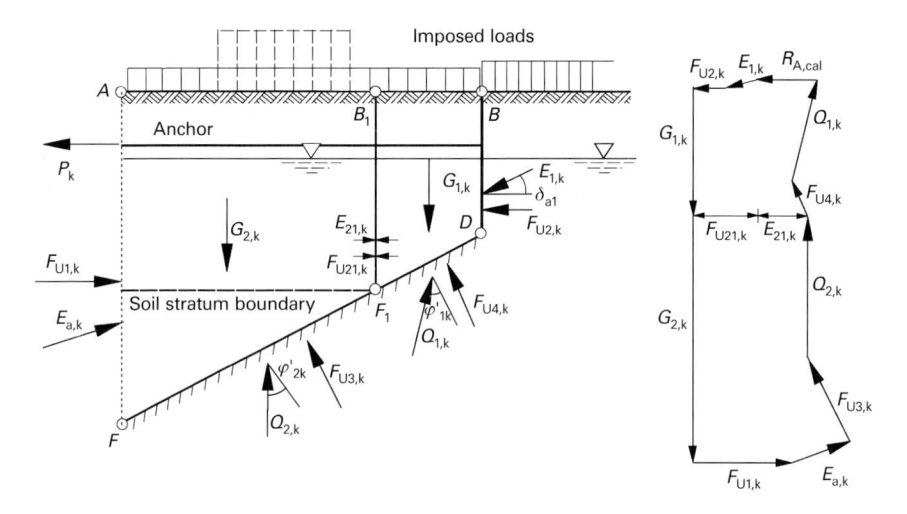

**Fig. R 10-2.** Verification of stability for the lower failure plane in stratified soil

| | |
|---|---|
| $G_{2,k}$ | total weight of sliding body $FF_1B_1A$, plus imposed load if applicable |
| $E_{a,k}$ | active earth pressure (across all soil strata) |
| $P_k$ | anchor force |
| $F_{U1,k}$ | hydrostatic pressure between soil and retaining wall AF |
| $F_{U2,k}$ | hydrostatic pressure on anchor wall DB |
| $F_{U3,k}$ | hydrostatic pressure on lower failure plane in section $FF_1$ |
| $F_{U4,k}$ | hydrostatic pressure on lower failure plane in section $F_1D$ |
| $F_{U21,k}$ | hydrostatic pressure on vertical separation plane $F_1B_1$ |
| $E_{1,k}$ | active earth pressure with imposed load on anchor wall DB |
| $E_{21,k}$ | earth pressure in vertical separation plane $F_1B_1$ |

The stability at the lower failure plane is determined using the inequalities in section 8.5.1.

## 8.5.4 Verification of stability at the lower failure for a quay wall fixed in the soil

The aforementioned analysis of stability at the lower failure plane is also sufficiently accurate for sheet pile walls fixed in soil. With this type of wall, the lower failure plane runs between the base of the anchor wall and the point of zero shear in the wall's fixity zone. This point coincides with the position of the greatest fixed-end moment. Its position can therefore be taken from the sheet piling calculation.

The active earth pressure in this case is determined only down to the theoretical base of the sheet piling and the anchor force is taken from the sheet piling analysis for the fixed wall.

**8.5.5 Stability at the lower failure plane for an anchor wall fixed in the soil**
If an anchor wall is fixed in the ground, then section 8.5.4 suggests that the lower failure plane should continue to the theoretical base at the level of the point of zero shear in the fixed-end zone of the anchor wall.

**8.5.6 Stability at the lower failure plane for anchors with anchor plates**
If anchors are connected to separate anchor plates with a clear spacing $a$, an imaginary equivalent anchor wall placed at a distance $½\,a$ in front of the anchor plates is assumed when analysing stability at the lower failure plane.

**8.5.7 Verification of safety against failure of anchoring soil**
Analysing the safety against failure of the anchoring soil is intended to verify that the design values of the resisting horizontal forces in front of anchor plates/wall from the underside of the anchor plates/wall to the surface of the ground are at least equal to or greater than the sum of the horizontal components of the design value of the anchor force, the design value of the active earth pressure acting on the anchor wall and any excess water pressure present.

The active and passive earth pressures acting on the anchor wall or individual anchor plates is determined according to DIN 4085. A non-permanent action (imposed load on ground) may only be applied when it has an unfavourable effect. This is usually the case with imposed loads applied behind the anchor wall or anchor plates. The groundwater level is to be applied at its most unfavourable level.

When calculating the passive earth pressure on the anchor wall, the only angle of inclination to be considered should be the one for which the condition $\Sigma V = 0$ is satisfied at the anchor wall (sum of all applied vertical forces including dead load, with earth surcharge corresponding to vertical component of passive earth pressure).

In the case of simply supported anchor plates and walls, the anchor connection is generally located in the middle of the height of the plate/wall. For further details, see R 152, section 8.2.14, and R 50, section 8.4.9.

### 8.5.8 Stability at the lower failure plane for quay walls anchored with anchor piles or grouted anchors at one level

Quay walls are also anchored with anchor piles or grouted anchors that transfer the anchor force into the soil via skin friction. Basically, we distinguish between three groups of anchor elements:

- Anchor piles with and without grouted skin according to section 9.2
- Grouted micropiles to DIN EN 14199 and DIN EN 1536
- Grouted anchors to DIN EN 1537

The required lengths of these anchor elements are determined as part of the stability verification at the lower failure plane as per Fig. R 10-3. Notation for Fig. R 10-3 (see Fig. R 10-1 for forces on failure body):

$l_a$    length of anchor element

$l_r$    required minimum anchor length or nominal grouting length for a grouted anchor calculated from the design value of anchor force $P_d$ and the design value of skin friction $T_d$ of the pile ($l_r = P_d/T_d$)

$T_d$    design value for skin friction, calculated from the design value of the pull-out resistance $R_{t,d}$ and the force transfer length $l_0$ in a tensile test ($T_d = R_{t,d}/l_0$)

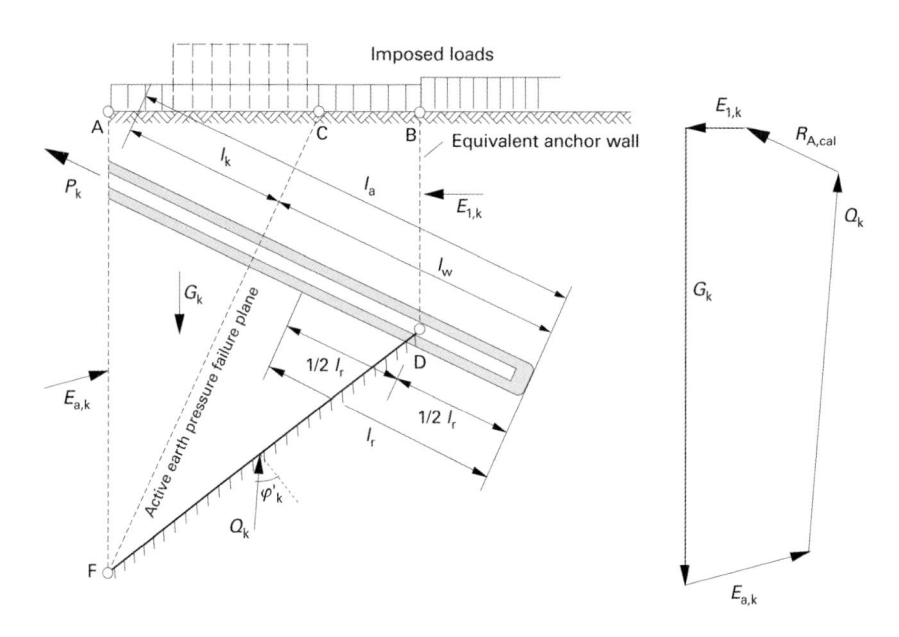

**Fig. R 10-3.** Verification of stability for the lower failure plane when using piles and grouted anchors

$l_k$ upper anchor pile length not effective structurally (which begins at the head of the anchor pile and ends at the active earth pressure failure plane or at the upper edge of the loadbearing soil if this is deeper)

$l_w$ structurally effective anchor length (which extends from the active earth pressure failure plane or the upper edge of the loadbearing soil to the end of the anchor – excluding toe of anchor; $l_w \geq l_r$ must always apply, and $l_w \geq 5.00$ m for grouted skin anchors)

In this case the sliding body is limited by an equivalent anchor wall that is assumed to be vertical in the centre of the force transfer length $l_r$ of the anchor element. In the case of grouted anchors, the force transfer length $l_r$ is the nominal length of the grout.

The active earth pressure $E_{1,k}$ at the equivalent anchor wall is always assumed to act parallel with the surface.

If the anchor spacing $a_a$ is greater than half the calculated force transfer length ($a_a < \frac{1}{2} \cdot l_r$), the possible anchor force $R_{A,cal}$ must be reduced by the ratio of half the calculated force transfer length and the anchor spacing:

$$R^*_{A,cal} = R_{A,cal} \cdot (1/2 \cdot l_r)/a_A$$

The possible anchor force reduced in this way is usually smaller than the anchor force that must be actually accommodated when the pull-out resistance of the anchor elements mobilised behind the base of the equivalent wall is not applied in the analyses.

This analysis represents a considerable simplification. A more accurate analysis is required when the anchors are so short that the calculated force transfer length $l_r$ is the same as the anchor length $l_w$ outside the active sliding wedge ($l_w = l_r$) [90].

A more accurate – but also more complex – analysis is possible by varying the inclination of the failure plane to find the most unfavourable failure plane while taking into account the remaining potential pull-out resistance behind the base of the equivalent anchor wall (in line with DIN 4084, [89]). In doing so, the characteristic value of the pull-out force that is transferred from the force transfer length behind the failure plane to the unaffected soil should be used.

**8.5.9 Stability at the lower failure plane for quay walls with anchors at more than one level**

If several rows of anchor piles or grouted piles are to be used, the sliding bodies to be analysed for stability at the lower failure plane are bounded

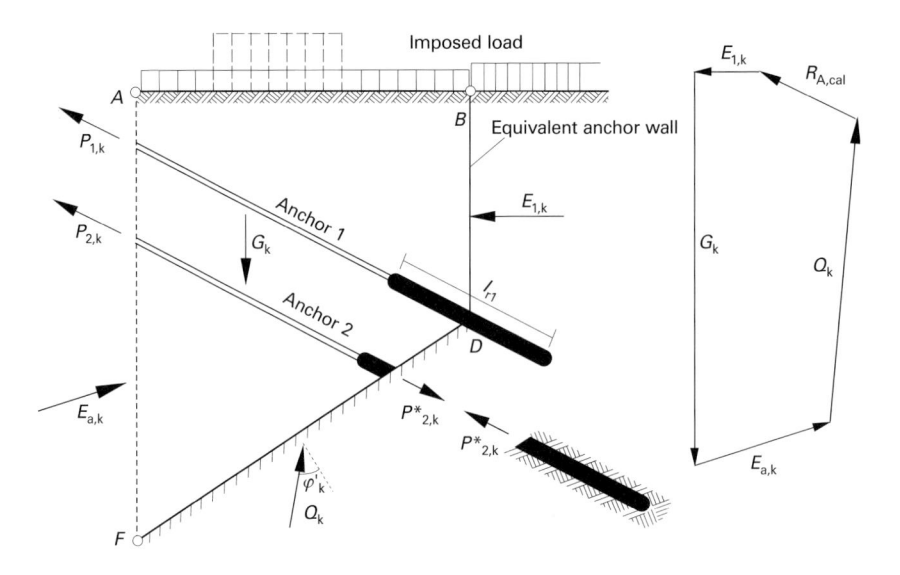

**Fig. R 10-4.** Verification of stability at the lower failure plane when using several anchors

by a failure plane through each of the centroids of the force transfer paths. In a design with several anchor levels, if the lower failure plane intersects an anchor before or in the force transfer path, the force that can be transferred behind the failure plane in undisturbed soil may be considered (force $P^*_{2,k}$ in Fig. R 10-4). The separated anchor force $P^*_{2,k}$ may be determined from a uniform distribution of the anchor force $P_{2,k}$ over transfer path $l_r$.

In the analysis we must distinguish between the permanent components $P_{G,k}$ and the variable components $P_{Q,k}$ of the anchor force.

Stability at the lower failure plane is assured when

$$\Sigma(P_{G,k} - P^*_{G,k}) \cdot \gamma_G \leq R_{A,cal}/\gamma_{Ep}$$

where $R_{A,cal}$ is determined from the polygon of forces using exclusively permanent loads, and

$$\Sigma(P_{G,k} - P^*_{G,k}) \cdot \gamma_G + \Sigma(P_{Q,k} - P^*_{Q,k}) \cdot \gamma_Q \leq R_{A,cal}/\gamma_{Ep}$$

where $R_{A,cal}$ is determined from the polygon of forces using permanent and variable loads, where

$\Sigma(P_{G,k} - P^*_{G,k})$ sum of all permanent components of characteristic anchor forces minus the forces transferred to the undisturbed soil behind the failure plane

$\Sigma(P_{Q,k} - P^*_{Q,k})$ sum of all variable components of characteristic anchor forces minus the forces transferred to the undisturbed soil behind the failure plane

(Safety factors $\gamma_i$ as given in section 8.5.1.)

## 8.5.10 Safety against slope failure

The verification of stability at the lower failure plane in conjunction with the analyses required for anchor walls and plates to verify adequate safety against the failure of the anchoring soil replace the slope failure investigation to DIN 4084 normally required.

Irrespective of the stability at the lower failure plane, the overall stability according to DIN 4084 must still be verified if there are unfavourable soil strata (soft strata beneath the anchoring zone) or high loads behind the anchor wall or equivalent anchor wall or particularly long anchors are being used.

# 9 Tension piles and anchors (R 217)

## 9.1 General

As a rule, waterfront structures need appropriate anchoring elements to prevent overturning and sliding, to carry horizontal loads from earth and water pressure as well as loads from the superstructure such as line pull and vessel impact. For smaller changes in ground level, these loads can be carried by appropriately designed structures supported on pile bents or trestles if necessary. Larger changes in ground level, such as those in modern seaports and inland ports, require special anchoring solutions. The main components used for anchoring waterfront structures are as follows:

- Displacement piles
- Micropiles
- Special piles
- Anchors

## 9.2 Displacement piles

### 9.2.1 Installation

The installation of displacement piles is regulated by DIN EN 12699 "Displacement piles" in conjunction with DIN SPEC 18539. No soil is removed during the installation of displacement piles; it remains in the ground, directly around the pile and, as far as possible, soil is compacted by the displacement process. Displacement piles are impact-driven, vibratory driven or screwed into the subsoil. Installing raking piles at a shallow angle requires secure guidance. In principle, owing to the longer time per impact, but also environmental considerations (noise, vibrations), slow-action pile-drivers are preferred over fast-action types. However, the "vibratory effect" of fast-action plant can compact non-cohesive soils and hence increase the load-carrying capacity. The loss of energy due to the inclined position of raking piles should be taken into consideration when calculating the weight of the hammer. The exposed length of pile below the driving guide should be limited so the permissible bending stresses in the pile are not exceeded during installation. The consequences of any water-jetting used to assist driving should be taken into consideration.

---

*Recommendations of the Committee for Waterfront Structures Harbours and Waterways – EAU 2012,* 9th Edition. Issued by the Committee for Waterfront Structures of the German Port Technology Association and the German Geotechnical Society.
© 2015 Ernst & Sohn GmbH & Co. KG. Published 2015 by Ernst & Sohn GmbH & Co. KG.

# Stump

## Ground Engineering in Europe

### Consulting • Engineering • Execution

Micro-Piles • Jet Grouting (Stump Jetting) • Permanent anchoring • Temporary anchoring • Grouting • Renovation of masonry/Conservation technology • Shotcrete/Nailing Excavation pits • Embankment protection • Slope stabilization • Stump HLV®-Composite Pile System • New foundations • Foundation restoration • Shoring against uplift Bored piling works • Pile driving works • Diaphragm walls Foundation underpinnings • Soil freezing • Exploratory drillings • Soil stabilization • Berlin-type support system

*Construction of the 5th lock chamber in Brunsbüttel*
*Test piling for the anchoring of the chamber walls with jet grouted piles, max. length approx 60 m*

**Berlin – Headquarters**
Tel. +49 (0) 30 754904-0
Fax +49 (0) 30 754904-420

**Chemnitz**
Tel. +49 (0) 371 262519-0
Fax +49 (0) 371 262519-30

**Hanover**
Tel. +49 (0) 511 94999-0
Fax +49 (0) 511 499498

**Langenfeld**
Tel. +49 (0) 2173 27197-0
Fax +49 (0) 2173 27197-990

**Munich**
Tel. +49 (0) 89 960701-0
Fax +49 (0) 89 960701-4297

**Colbitz**
Tel. +49 (0) 39207 856-0
Fax +49 (0) 39207 856-50

**www.stump.de**

... also in the Czech Republic and Poland

---

# Ernst & Sohn
A Wiley Brand

## The most comprehensive code on concrete structures

fib – International Federation
**fib Model Code for Concrete Structures 2010**
2013. 500 pages.
€ 199,–*
ISBN 978-3-433-03061-5
Also available as ebook.

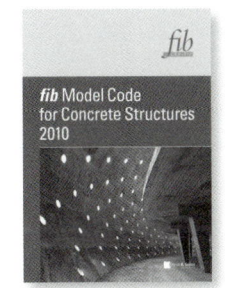

The fib Model Code 2010 is now the most comprehensive code on concrete structures including their complete life cycle. It represents an important document for both national and international code committees, practitioners and researchers.

Journals recommendation
■ Structural Concrete

Order online:
www.ernst-und-sohn.de

**Ernst & Sohn**
Verlag für Architektur und technische
Wissenschaften GmbH & Co. KG

Customer Service: Wiley-VCH
Boschstraße 12
D-69469 Weinheim

Tel. +49 (0)6201 606-400
Fax +49 (0)6201 606-184
service@wiley-vch.de

*€ Prices are valid in Germany, exclusively, and subject to alterations. Prices incl. VAT. excl. shipping. 1007116_dp

# Ernst & Sohn
A Wiley Brand

# Geomechanics and Tunnelling

The contributions published in Geomechanics and Tunnelling deal with tunnelling, rock engineering and applications of rock and soil mechanics as well as engineering geology in practice. Each issue focuses on a current topic or specific project. Brief news, reports from construction sites and news on conferences round off the content. An internationally renowned Editorial Board assures a highly interesting selection of topics and guarantees the high standard of the contributions.

ÖGG
ÖSTERREICHISCHE GESELLSCHAFT FÜR GEOMECHANIK

Editor: ÖGG – Österreichische Gesellschaft für Geomechanik (Austrian Society for Geomechanics)
**Geomechanics and Tunnelling**
Geomechanik und Tunnelbau
Volume 8, 2015.
6 issues / year.
Language of Publication: English/German
ISSN 1865-7362 print
ISSN 1865-7389 online
Also available as ejournal.

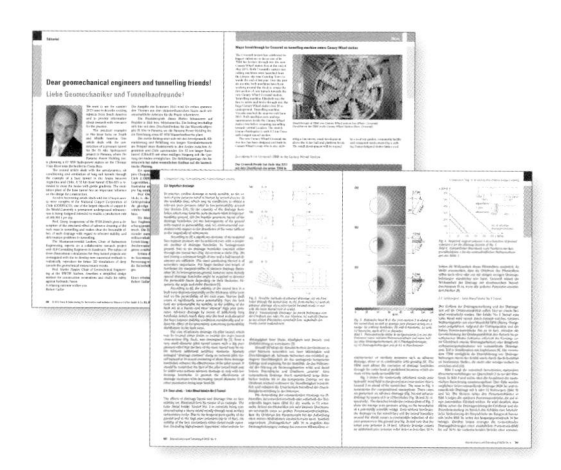

Other journals
- Steel Construction
- Structural Concrete
- geotechnik

Order a free sample copy:
www.ernst-und-sohn.de/Geomechanics-and-Tunnelling

**Ernst & Sohn**
Verlag für Architektur und technische Wissenschaften GmbH & Co. KG

Customer Service: Wiley-VCH
Boschstraße 12
D-69469 Weinheim

Tel. +49 (0)800 1800-536
Fax +49 (0)6201 606-184
cs-germany@wiley.com

## 9.2.2 Types

### 9.2.2.1 Steel piles (without grouting)

Steel piles are rolled products that can be supplied as I-sections in long lengths. They are characterised by their ability to adapt well to the particular structural, geotechnical and installation conditions. They can be driven or vibrated as "plain" piles and are relatively unaffected by obstructions in the ground and difficult driving conditions. Where required by the local soil conditions, they can be lengthened at the pile head. Steel piles are easy to weld to other structures made from steel or reinforced concrete.

The steel grades used should comply with DIN EN 1993-1-1 and DIN EN 10025.

Under special circumstances, steel wings can be welded to steel piles to improve their loadbearing capacity. However, such piles should only be used in soils – preferably non-cohesive – free from all obstacles, and must be embedded sufficiently deep in loadbearing subsoil. Where there are cohesive strata, the wings should be located below these and any open driving channels should be closed off, e.g. by grouting. Wings should be designed and positioned such that they do not render driving too difficult and can themselves resist the driving process intact. The shape of the wings and level at which they are attached should therefore be carefully adjusted to suit the particular soil conditions. It should be noted that saturated cohesive soils are indeed displaced during driving, but not compacted. In non-cohesive soil, a highly compacted solid plug can form, primarily around the wings, as a result of the vibrations of driving, which then makes further driving more difficult.

Wings should be positioned symmetrical about the pile axis just above the toe of the pile so that there is enough space to fit a min. 8 mm thick weld seam between wing and end of pile. The upper end of the wing must also have a correspondingly strong transverse weld. These weld seams link up with approx. 500 mm long joints on both sides of the wing in the longitudinal direction of the pile. Between these, intermittent weld seams suffice.

Considering the restraint forces, the wing attachment area should be sufficiently wide (generally at least 100 mm). Depending on the soil stratification, wings can also be positioned higher up the pile shaft.

### 9.2.2.2 Grouted piles

This term also includes driven piles with a grouted skin and grouted displacement piles. Grouted piles consist of steel sections as described in section 9.2.2.1. However, they have a special toe form for introducing the grout at the base of the pile. A box-shaped cutting shoe made from welded sheet steel is attached to the toe of the pile.

This cutting shoe, acting as a "full displacement tool", creates a prismatic cavity in the soil around the entire steel section of the driven pile which is filled with cement mortar/fine-grained concrete while the pile is being driven. The filling material is pumped through a pipe attached to the pile down to the pile toe where it is forced out and fills the cavity left behind by the cutting shoe.

The hardened cement mortar/fine-grained concrete in the cavity creates the bond between the steel pile and the subsoil. Depending on the subsoil, the skin friction that can be activated can be three to five times greater than that of a non-grouted steel pile.

A further increase in the loadbearing capacity of the grouted pile results from the displacement of the soil during driving, which creates three-dimensional tension in the subsoil. Grouted piles are generally installed at rakes between 2:1 and 1:1.

Grouted piles are particularly suited to non-cohesive soils with a relatively high pore volume.

### 9.2.2.3 Vibratory driven grouted piles

Vibratory driven grouted piles have certain similarities with grouted piles. They consist of steel sections as described in section 9.2.1.1. At the base, to widen the I-shaped pile cross-section, approx. 20 mm thick steel plates are welded to the web and flanges on all sides. These create a cavity along the pile shaft equal to the plate thickness, which is filled with cement mortar during vibratory driving in a similar way to the grouted pile to increase the shaft resistance.

Vibratory driven grouted piles can be installed vertically and at rakes of up to 1:1. However, it is difficult to achieve a satisfactory effective degree of penetration (pile penetration per unit of time related to the vibration energy applied) for vibratory driven raking piles. It has proved beneficial to apply a "prestress" of approx. 100 kN axial compression to raking vibratory driven grouted piles. This enables the pile to maintain continuous frictional contact with the subsoil and the energy input can then be transferred more effectively to improve the penetration rate.

## 9.2.3 Loadbearing capacity of displacement piles

### 9.2.3.1 Internal loadbearing capacity

The internal loadbearing capacity of a pile should be verified depending on the materials used. *Recommendations on Piling* [155], section 5.10, contains a detailed description of the analyses required.

### 9.2.3.2 External loadbearing capacity

The resistance of a displacement pile in tension is calculated according to DIN EN 1997-1. Two different types of failure must be investigated for tension piles: pull-out of the piles from the ground and uplift of the block of soil containing the piles. According to DIN EN 1997-1, the external loadbearing capacity of a tension pile should always be verified by loading tests. Estimating the tension pile resistance using empirical data is only permitted in exceptional circumstances.

The procedure for verifying the external loadbearing capacity in accordance with DIN EN 1997-1 is set out in detail in [155], section 6.3. That publication also contains full information on pile loading tests.

For preliminary design, the external loadbearing capacity of tension piles can also be derived approximately from the resistance measured in cone penetration tests. To ensure that the in situ soils are correctly identified, the local surface friction should always be measured as well during penetrometer tests. The unequivocal allocation of the values obtained from penetrometer tests to the in situ soil types requires the help of at least one borehole to calibrate the results of the penetrometer tests. Reference values for characteristic pile resistances depending on the types of soil and their strengths can be found in [155].

### 9.2.3.3 Bond stresses in grouted displacement piles

According to DIN SPEC 18538, it is necessary to verify the bond stresses at the interface between the pile shaft and the grout of a grouted displacement pile. A distinction is made between prestressed and non-prestressed areas. The result is noticeably higher bond stresses for grouted piles (section 9.2.2.2) than for vibratory driven grouted piles (section 9.2.2.3).

### 9.2.3.4 Additional stresses

A wide variety of additional stresses may be encountered when anchoring waterfront structures with raking piles, e.g. due to consolidation settlement, compaction settlement, block loadbearing behaviour, ground deformation due to the deflection of a sheet pile wall, etc., especially transverse to the axis of the pile. With ductile piles, no verification of the internal stability in the span is required for the additional stresses referred to above and when the wall connection detail is sufficiently flexible or ductile. Instructions on how to take account of these additional stresses can be found in [155], section 4.6.2.

## 9.3 Micropiles

### 9.3.1 Installation

The installation of micropiles is regulated by DIN EN 14199 in conjunction with DIN SPEC 18539. Micropiles can be installed in the form of bored piles (maximum shaft diameter 300 mm) or as displacement piles (maximum cross-sectional measurement 150 mm). The tension member (tendon) is either inserted into a pre-drilled hole or, in the case of self-drilling micropiles, itself functions as a sacrificial drilling tool. Micropiles are generally not prestressed. Unlike grouted anchors, micropiles transfer the load into the subsoil over their entire length, provided there is no unbonded segment.

### 9.3.2 Types

#### 9.3.2.1 Composite piles

Micropiles in the form of composite piles have a continuous, prefabricated steel loadbearing tendon filled and surrounded by pressure-injected cement mortar. Grouting takes place under increased hydrostatic pressure (min. 5 bar) on the grouting material. With appropriate equipment it is possible to post-grout the micropiles using sleeve pipes or tubes à manchette. Post-grouting is not possible with self-drilling micropile systems. Composite piles require an approval. The tendons can be made from ribbed steel reinforcing bars, solid steel bars or steel pipes with cut or rolled threads or welded-on wires. This makes it easy to create connections to reinforced concrete or steel wall structures. Approvals are also required for these connection details.

#### 9.3.2.2 Cast-in-place concrete piles

Micropiles in the form of cast-in-place concrete piles have a shaft diameter of 150–300 mm. They are inserted into pre-drilled holes, include longitudinal reinforcing bars over their full length and are filled with concrete or cement mortar. Grouting uses compressed air on the exposed cast-in-place concrete surface. Post-grouting is possible by inserting sleeve pipes with each pile.

### 9.3.3 Loadbearing capacity of micropiles

#### 9.3.3.1 Internal loadbearing capacity

The internal loadbearing capacity of micropiles in the form of composite piles in accordance with section 9.3.2.1 is determined by the material and cross-section parameters of the steel used and regulated by the approval that is required.

The internal loadbearing capacity of micropiles in the form of cast-in-place piles in accordance with section 9.3.2.2 should be verified in line with the design rules for reinforced concrete.

#### 9.3.3.2 External loadbearing capacity

The resistance of micropiles in tension is calculated according to DIN EN 1997-1. In order to verify the external loadbearing capacity, loading tests should always be carried out afterwards.

The external loadbearing capacity of micropiles for preliminary designs may be calculated using the empirical values given in [155]. However, these values, too, should be verified in loading tests. According to DIN EN 1997-1, loading tests for micropiles in tension should be carried out on a minimum of 3% of the piles (but no fewer than two).

### 9.3.3.3 Bond stresses in grouted micropiles

Verifying the bond stresses at the pile shaft/grout interface of a grouted micropile is always carried out for each particular type during investigations for the approval.

### 9.3.3.4 Additional stresses

Basically, the information given in section 9.2.3.4 also applies to micropiles. According to [155], section 4.6.2(4), the corrosion protection to the loadbearing tendon in a micropile is not impaired by additional stresses applied transverse to the pile axis when the tendon is housed in a continuous plastic sleeve filled with cement mortar.

## 9.4 Special piles

### 9.4.1 General

Special solutions for anchoring quay walls are always appearing on the market. These primarily involve a combination of different, well-known building processes and methods that have been developed for a specific method of construction, e.g. jet grouting combined with driven or bored steel sections. These piles have proved worthwhile in individual cases; however, there is insufficient experience that can be generalised to discuss them separately in the context of this recommendation. Where such special solutions are to be used, their feasibility should always be verified on a case-by-case basis.

An exception is the prefabricated raking pile, which has proved itself in many building processes due to its robust construction and is described below.

### 9.4.2 Prefabricated raking piles

This form of raking pile is a fully prefabricated steel tension element consisting of a steel pile section plus an anchor plate. The angle of installation is between 0° and 45° from the horizontal.

During prefabrication, a hinged steel fitting is mounted on the pile head so that the anchor pile can later be attached to the quay wall. In order to accommodate this pile head detail, each bearing pile is fitted with a compatible steel mounting in which the raking anchor pile is free to rotate. The anchor plate is welded to the pile toe at 90° to the pile axis.

Lifting gear is required to install these anchor piles. While suspended from a crane, the pile head is attached to the sheet pile wall so that it can rotate but still remains a fully structural connection. The pile toe, still supported by the crane, is then lowered so that the pile shaft pivots about the hinge at the wall and the pile toe with its fixed anchor plate can be placed in position on the bottom of the watercourse. Once this "pivoting procedure" is complete, the pile is then in its final position. Backfilling of the area can then commence. It is imperative that the first loads of fill material be placed directly in front of and on the anchor plate so that pile resistance can be mobilised right from an early stage. The loadbearing capacity of such prefabricated raking anchor piles can be improved by vibrating each anchor plate into the bed of the watercourse for a distance of half its height with the help of an underwater vibrator.

## 9.5 Anchors

### 9.5.1 Construction

Unlike displacement piles, anchors always have a well-defined anchorage length or anchoring point via which the load is transferred into the ground as well as an equally clearly defined unbonded length over which no load is intended to be transferred into the ground. Anchors are installed by inserting them into cased or uncased holes (grouted anchors). Alternatively, they can be placed close to the surface, fixed to anchor plates and then covered (e.g. anchors made from round steel rods or rolled sections). The head details of grouted anchors are regulated by approvals. See section 8.4 for anchor connection details.

### 9.5.2 Types

#### 9.5.2.1 Grouted anchors

The construction of grouted anchors is regulated by DIN EN 1537 in conjunction with DIN SPEC 18537. Grouted anchors consist of an anchor head and a tension member (tendon) with a defined unbonded anchor length and an anchorage length. Structural steel sections, reinforcing bars and prestressing steels (bars or strands) can be used for the tendons. Grouted anchors in accordance with DIN EN 1537 can be prestressed. All grouted anchors for long-term use (>2 years) require an approval. The approval covers all aspects from the factory production of the tendon, the corrosion protection, the insertion of the tendon through to testing and prestressing the anchors.

#### 9.5.2.2 Anchoring with anchor plate ("dead-man" anchors)

As with the grouted anchors referred to in section 9.5.2.1, the "dead-man" anchor consists of an anchor head and a defined unbonded anchor length. The load is not transmitted into the subsoil via the bond between anchor and soil, but via a large anchor plate (anchor wall). Upset T-heads and threaded ends, anchor chairs, coupling sleeves and turnbuckles as well as anchor plates made from sheet pile sections or precast concrete elements allow for various structural solutions to address a wide variety of problems. The angle of installation for these "dead-man" anchors is normally limited to a maximum of 8–10° because otherwise the earthworks required to excavate the anchor trench and then backfill it once again are uneconomic.

Horizontal anchors in the form of "dead-man" anchors are suitable for special structures such as pier heads, quay walls and headlands provided the anchor can be tied to the opposite quay wall. In such cases it is not usually possible to install anchors in the form of raking piles.

## 9.5.3 Loadbearing capacity of anchors

### 9.5.3.1 Internal loadbearing capacity

The internal loadbearing capacity of anchors is determined by the material and cross-sectional parameters of the steel used.

### 9.5.3.2 External loadbearing capacity

Suitability tests should always be carried out to determine the external loadbearing capacity of grouted anchors. The characteristic pull-out resistance is determined from the results of the tests according to DIN EN 1997-1.

In addition, all grouted anchors must pass an acceptance test in accordance with the German Eurocode 7 manual [85], [86] prior to being used. Random tests, as used for piles and micropiles, are not enough.

### 9.5.3.3 Loadbearing capacity of anchor plates (anchor walls)

Verification of the stability of anchorages with anchor plates is dealt with in section 8.2.

# 10 Quay walls and superstructures in concrete

## 10.1 Design principles for quay walls and superstructures in concrete (R 17)

### 10.1.1 General principles

Durability and robustness are particularly important factors to consider when designing waterfront structures made from plain concrete, reinforced concrete or prestressed concrete. The proposed lifetime must be taken into account in this respect, which for port and harbour structures can be shorter than for general or waterway engineering structures. Waterfront structures are exposed to the effects of changing water levels, waters and soils harmful to concrete, ice, ship impacts (berthing pressures or accidents), chemicals from goods being shipped or stored, etc. It is therefore not sufficient to design the reinforced concrete parts of waterfront structures simply according to the structural requirements.

In addition, the requirements regarding simple construction methods without difficult formwork, for the straightforward incorporation of sheet pile walls, piles, etc. are also important. This gives rise to structural and constructional measures that can go beyond the minimum requirements laid down in DIN EN 1992. Those measures must be agreed between the client, the design engineer, the engineer in charge of checking the structural design and the building authority responsible.

The following sections take into account the loads and stresses in hydraulic engineering and the ensuing increased requirements regarding the construction and maintenance of concrete components in contact with water.

### 10.1.2 Edge protection

Concrete walls should have a $5 \times 5$ cm chamfer along their upper edges or be correspondingly rounded and/or protected on the water side by steel angles in the case of transhipment operations; if necessary, R 94, section 8.4.6, should be observed. Any special edge protection affixed to protect the wall and as a safeguard to prevent port personnel slipping over the edge must be designed in such a way that water can drain away easily. In the case of quay walls with steel sheet piling and a reinforced concrete superstructure, the reinforced concrete cross-section must project about 15 cm beyond the front line of the sheet pile wall.

---

*Recommendations of the Committee for Waterfront Structures Harbours and Waterways – EAU 2012,*
9th Edition. Issued by the Committee for Waterfront Structures of the German Port Technology Association and the German Geotechnical Society.
© 2015 Ernst & Sohn GmbH & Co. KG. Published 2015 by Ernst & Sohn GmbH & Co. KG.

The lower edge of the reinforced concrete superstructure on the water side should, however, be at least 1 m above the high tide or mean water level in order to prevent the corrosion that would otherwise occur to an increased extent at this point. The transition should be at a slope of roughly 2:1 so that ships and crane hooks cannot be caught underneath, and ideally should include a protective bent steel plate that fits flush with both the sheet pile wall below and the concrete wall above.

## 10.1.3 Facing

Where facing is appropriate as protection against particular mechanical or chemical effects, or for architectural reasons, the use of basalt, granite or engineering/facing bricks is recommended. Ashlar stones or slabs forming a coping to the front top edge of a wall must be protected against displacement and lifting.

## 10.2 Design and construction of reinforced concrete components in waterfront structures (R 72)

### 10.2.1 Preliminary remarks

The recommendations regarding structural design calculations given in this section are based on the safety concept according to DIN EN 1990-1 and DIN EN 1991-1.

To ensure sufficient reliability, numerical analyses are carried out for the ultimate and serviceability limit states, and the structures are designed in accordance with the rules of structural design and the concepts for ensuring durability.

In general, DIN EN 1992-1-1 in conjunction with the NAD shall apply, plus national standards not replaced by Eurocodes and the applicable regulations cited in those standards.

### 10.2.2 Concrete

The strength and deformation characteristics must be taken from DIN EN 1992-1-1, section 3.1. The concrete properties must be specified in compliance with DIN EN 1992-1-1, section 4, and according to DIN EN 206-1 in conjunction with DIN 1045-2.

When choosing concretes, the relevant exposure classes are critical, taking the ambient conditions into account. The exposure classes govern the minimum requirements for concrete quality, concrete cover and the criteria regarding maximum crack widths.

The exposure class for chemical attack by natural soils and groundwater is determined according to DIN EN 206-1, Table 2, in conjunction with DIN 4030.

Where significant expansion of concrete components is expected (see section 10.2.4) and large cross-sections are in use, attention must be paid to employing concrete mixes with low shrinkage and low heat of hydration development. It is also important to use a concrete that is as dense as possible, to employ intensive curing and to ensure sufficient concrete cover to the steel reinforcement.

The concrete cover should be larger than that given in DIN EN 1992-1-1 and at least $c_{min} = 50$ mm, with a nominal cover $c_{nom} = 60$ mm. Regarding limiting crack widths under serviceability loads, section 10.2.5 must be observed.

The strength categories to be chosen depending on exposure class, minimum cement content, maximum water/cement ratio and further requirements are derived from DIN EN 1992-1-1, the additional technical contractual conditions for hydraulic engineering ZTV-W (LB 215) [234], and, in the case of bulky structural components (smallest dimensions $\geq 80$ cm), in accordance with the guideline on such bulky components published by the German Committee for Structural Concrete [44]. For typical components in waterfront structures, allocations to the exposure classes are shown in Figs. R 72-1 and R 72-2 for both seawater and freshwater.

### 10.2.3 Construction joints

Construction joints must be avoided wherever possible. Unavoidable construction joints should be planned before beginning concreting

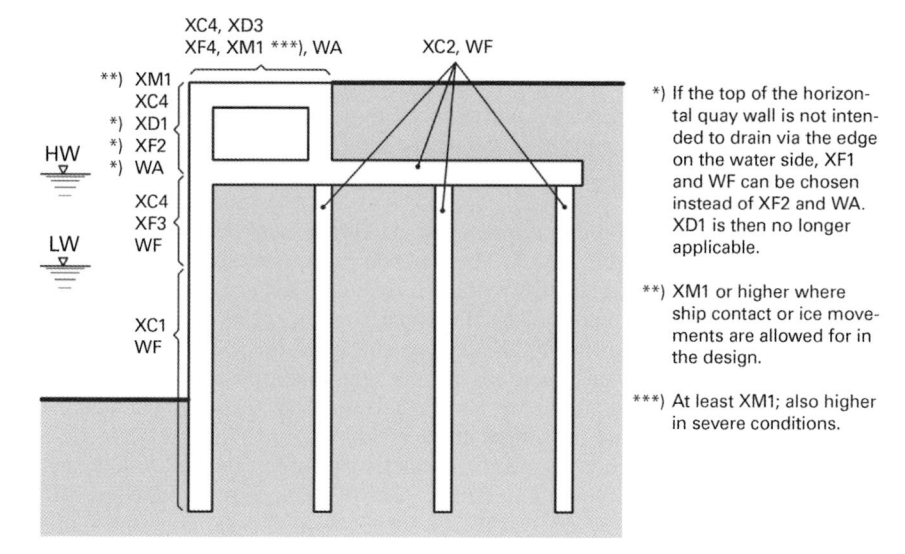

**Fig. R 72-1.** Example of exposure classes for a quay wall in tidal freshwater

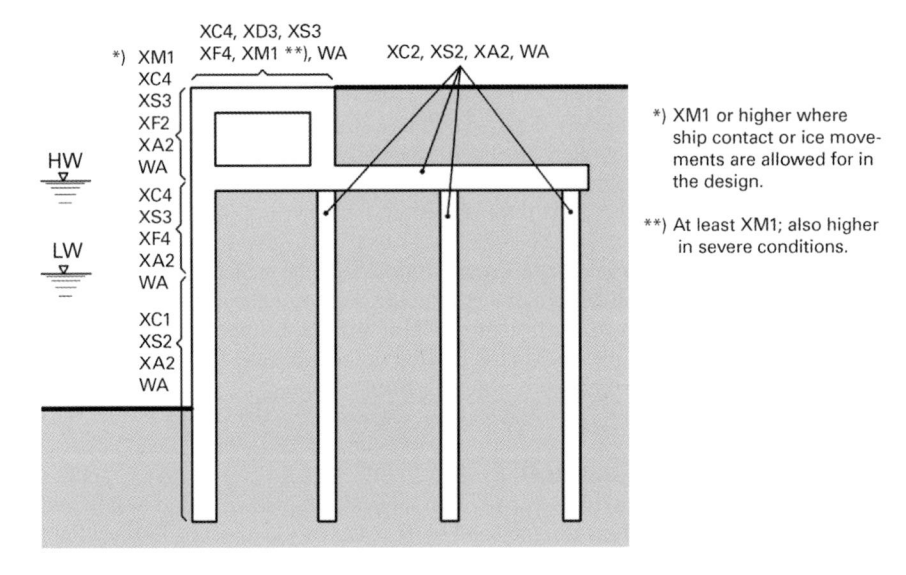

**Fig. R 72-2.** Example of exposure classes for a quay wall in seawater

works and designed in such a way that all the loads occurring can be accommodated (DIN EN 1992-1-1, section 6.2.5).

A construction joint must not impair the durability of a concrete component. To this end, careful planning and preparation of construction joints (see also DIN EN 13670/DIN 1045-3, section 8.2) and further works (e.g. grouting) might be necessary. Joint locations must be suitable for such work. Ribbed expanded metal or similar materials along with cement slurry must be completely removed prior to concreting the adjoining component.

It is good practice to expose the grains of aggregate in the concrete component cast first by means of high-pressure water jets and to apply a preliminary mix containing a high percentage of cement prior to concreting the second component to serve as a cushion and to improve adhesion.

## 10.2.4 Structures with large longitudinal dimensions

Long linear structures can be built with or without movement joints. The decision regarding the location and number of such joints must be made with a view to finding the best solution in terms of economy, durability and the robustness of the structure. In this context, consideration must be given to the influences of the subsoil, the substructure and the construction details.

Where waterfront structures are built without joints, in addition to actions caused by loads, the additional actions due to restraint (shrinkage, creep, settlement, temperature) must also be taken into account in the calculations. Verifying maximum crack widths under loading and restraint stresses then becomes especially important if damage due to excessive cracking is to be avoided (see also section 10.2.5).

See section 10.2.3 for information on construction joints.

Movement joints must be detailed in such a way that they do not impede the changes in length of separate blocks of the structure.

Joggle or dowelled movement joints ensure mutual support between the separate blocks of the structure. Where walls are supported on pile bents or trestles, the interlocking is located in the pile cap. Joint gaps must be secured to prevent the backfilling from escaping.

### 10.2.5 Crack width limitation

Given the increased risk of corrosion, reinforced concrete waterfront structures must be designed and built in such a way that cracks that could affect durability do not occur. If concrete technology measures are not sufficient in themselves, the maximum crack widths must be verified taking into account the ambient conditions and actions (DIN EN 1992-1-1, section 7.3).

The crack width must be chosen in such a way that self-healing can occur. Generally speaking, this can be assumed for theoretical crack widths $w_k \leq 0.25$ mm.

If there is an increased risk of corrosion, e.g. in the tropics and when using prestressing steel, tougher demands must be specified for maximum crack widths.

All cracks >0.25 mm must be permanently injected in accordance with the additional technical contractual conditions for engineered structures ZTV-ING [231].

Restraint stresses can be reduced by choosing a logical concreting sequence in which the inherent deformations (due to loss of heat of hydration and shrinkage) of subsequent concreting sections are not overly impeded by concrete components cast earlier and now largely cool. This can be achieved for larger elements, e.g. a pier structure, if beams and slabs are cast in a single concreting process.

Crack widths can be limited in the case of large cross-sections by using minimum reinforcement in accordance with the code of practice supplied by the Federal Waterways Engineering & Research Institute (MFZ [16]). Further principles are given in Rostasy [183].

In addition, or as an alternative, the reinforcement can be protected by a cathodic corrosion protection system. Advice on such systems can be found in HTG [103].

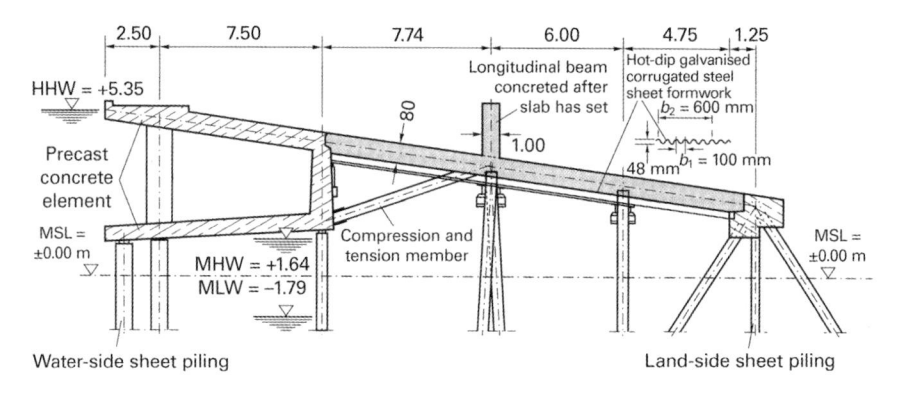

**Fig. R 169-1.** Example of the construction of a quay wall in seawater using a precast concrete element and corrugated metal sheeting as formwork

## 10.3 Formwork in areas affected by tides and waves (R 169)

Formwork should be avoided if possible in areas affected by the tide and/or waves, e.g. by using precast concrete elements, raising the underside of the concrete structure or similar means.

Concreting works in areas affected by the tide and/or waves should, as far as possible, be carried out in periods of calm weather.

Areas with difficult access, such as the underside of a pile cap, should be cast on permanent formwork whenever possible, e.g. concrete paving slabs, corrugated metal sheeting, etc.

Formwork in the sea should be able to absorb wave impacts essentially elastically. This is the case in the example shown in Fig. R 169-1, where the slab is cast on corrugated metal sheeting raised and positioned to suit the conditions.

Permanent formwork in the form of corrugated metal sheeting must be secured against uplift and fitted, for example, with galvanised wires or other anchor systems for subsequent connection to the concrete.

Where corrugated metal sheeting is used as formwork, the joints between the individual sheets preclude this from being used as part of the corrosion protection for the slab reinforcement.

## 10.4 Box caissons as waterfront structures in seaports (R 79)

### 10.4.1 General

Box caissons can represent economic solutions when enclosing heavily loaded, high vertical banks in areas with loadbearing soils, especially in the case of structures projecting into the open water.

Box caissons consist of rows of reinforced concrete elements open at the top which are generally stabilised through the addition of ballast. After they have been floated into position and sunk onto loadbearing soil, they are filled and backfilled with sand, stones or other suitable materials. The chambers on the water side are often left unfilled in order to reduce compression along the edges. Once installed, a box caisson projects only a little above the lowest working water level. Above this, a reinforced concrete superstructure is built to provide additional stiffening for the structure and to form the top of the front wall. Choosing a suitable form for the reinforced concrete superstructure compensates for the differential settlement and horizontal displacements caused during sinking and backfilling the caisson.

### 10.4.2 Design

Apart from verifying the stability for the final state, it is also necessary to investigate the states during construction such as the stability of the box caisson during launching, floating, sinking and backfilling. For the final state, proof of safety against scour must also be furnished.

Deviating from DIN 1054, the joint between the underside of the box caisson and the soil may not open up under any combination of actions due to characteristic loads.

### 10.4.3 Safety against sliding

It is very important to check whether mud can build up on the foundation bearing surface in the period between completing the bearing area and sinking the box caisson. If this is possible, it must be verified that there is still sufficient safety against sliding of the caisson on the contaminated bearing area. The same applies to the interface between existing subsoil and the backfilling to an excavation.

An economic way of improving safety against sliding is to provide a base slab with a rough underside. In this regard, the degree of roughness must match the average grain size of the material directly beneath the caisson. With an appropriate roughness, the angle of friction between the concrete and the foundation bearing area is assumed to be equal to the angle of internal friction $\varphi'$ of the foundation material, but is only $\frac{2}{3} \varphi'$ of the foundation material if the underside of the caisson is smooth. Safety against sliding can also be improved by increasing the depth at which the box caisson is founded. The analysis of safety against sliding should be based on the least favourable combination of water pressures on the base and sides of the caisson. These pressures can be caused by backfilling operations, changing tides, precipitation, etc. In addition, line pull must also be taken into account.

**10.4.4　Construction details**

The joint between two adjacent box caissons must be formed in such a way that the differential settlement to be expected between the caissons as a result of positioning, filling and backfilling can be accommodated without any risk of damage. On the other hand, in the final state, the joint must constitute a reliable seal to prevent any backfill material from being washed out.

Even if sealing is perfect, the use of a continuous tongue and groove joint over the whole height is only permitted if the movement between adjacent box caissons remains small.

The solution shown in Fig. R 79-1 has proven to be practical. Here, four vertical reinforced concrete nibs are included on the side walls of each caisson in such a way that they are situated opposite each other on either side of the joint and form three chambers after the box caissons have been installed. As soon as the neighbouring caisson has been sunk, the outer chambers are filled with graded gravel of a suitable granulometric composition to seal the joint. After backfilling the caissons and once settlement has largely abated, the middle chamber is flushed empty and carefully filled with underwater concrete or concrete in geotextile tubes.

Where there is a great variation in water levels between the front and back of a box caisson, there is a risk that soil beneath the foundation may be washed out. In such cases the granulometric composition of the strata of the bearing area must constitute a stable filter in relation to each other and in relation to the underlying subsoil. High excess water pressures can be eliminated successfully with the help of drainage systems with anti-flood fittings as described in R 32, section 4.5.

The risk of scour caused by current and wave forces can be countered by including adequate safeguards as described in R 83, section 12.4.

**10.4.5　Construction work**

Box caissons must be positioned on a well-levelled bed of stones, gravel or sand. If the bearing area contains soils with a low bearing capacity, they must be dredged out beforehand and replaced by sand or gravel (R 109, section 7.9).

**10.5　Compressed-air caissons as waterfront structures (R 87)**

**10.5.1　General**

Compressed-air caissons can provide advantageous solutions for enclosing high banks if their installation can be carried out from the land. In this case the compressed-air caissons are firstly installed from the existing terrain and the dredging works in the port or harbour basin carried out afterwards.

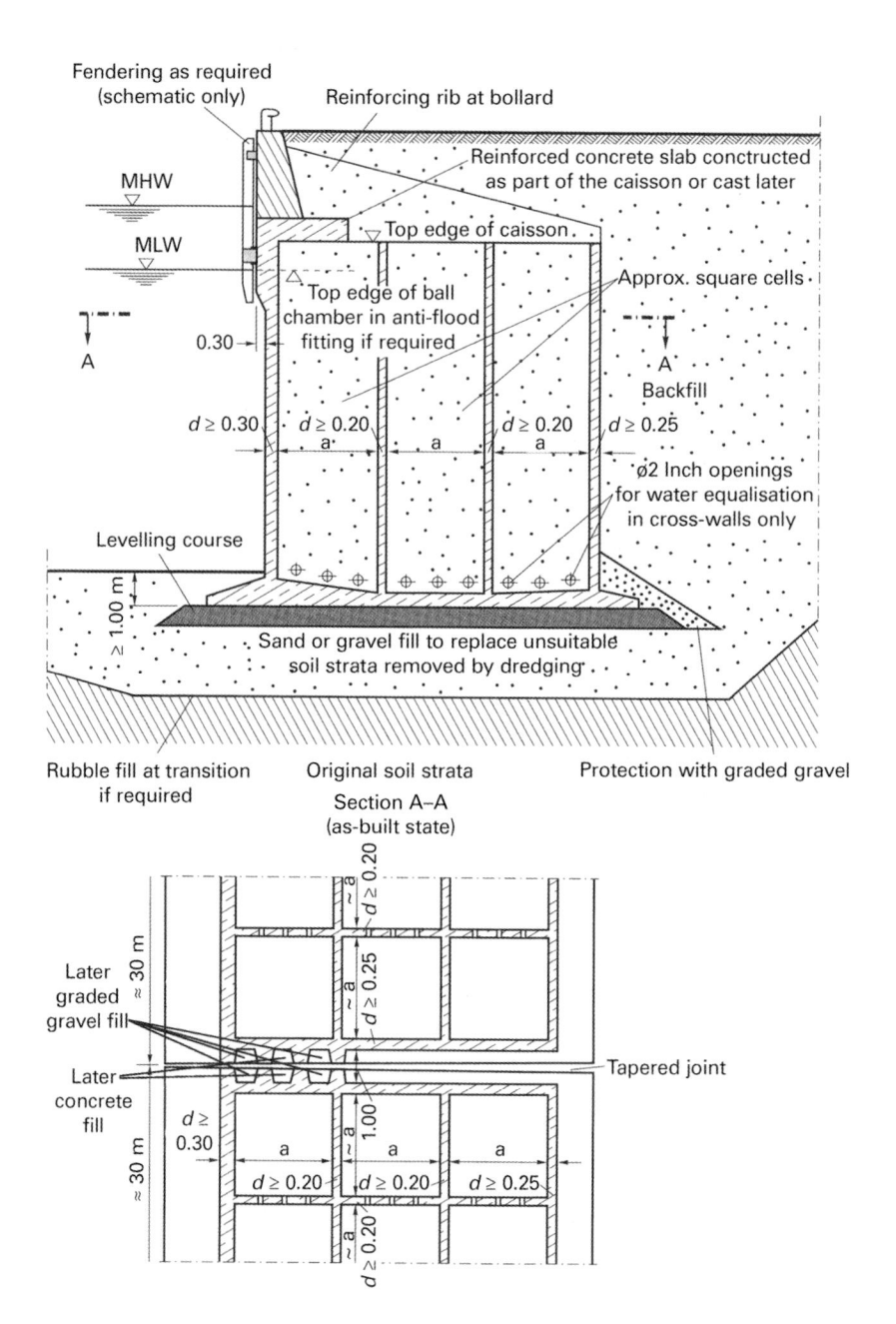

**Fig. R 79-1.** Construction of a waterfront structure with box caissons

Compressed-air caissons can also be designed as box caissons if a substrate with adequate bearing capacity is not available or cannot be established in the area of the caissons or if the levelling of the bearing surface is particularly difficult, as is the case with rocky ground, for example. The principles given in R 79, section 10.4.1, then also apply similarly to compressed-air caissons.

### 10.5.2 Verification

R 79, section 10.4.2, applies here. In addition, for the lowering states in the soil, the usual analyses for bending and shear forces in the vertical direction as a result of uneven support for the caisson edges and bending and shear forces in the horizontal direction due to non-uniform earth pressures are also required.

Since compressed-air caissons are regarded as normal shallow foundations with regard to the location and form of the bearing area and because of the good dovetailing of the caisson edges and the working chamber concrete with the subsoil, here – in contrast to R 79, section 10.4.2 – the joint between the underside of the caisson and the soil may open up. However, for actions due to characteristic loads, the minimum distance between the resultants and the front edge of the caisson may not be less than ¼ the width of the foundation.

In the case of high excess water pressure, it is necessary to investigate the danger of soil being washed out from in front of and beneath the base of the caisson. Where necessary, special safeguards should be taken to prevent such scour, e.g. soil consolidation carried out from the working chamber or similar. Founding at a lower depth or widening the base of the caisson may be more economical, however.

In the final state, compressed-air caissons require no special proof of stresses as a result of uneven support in the longitudinal direction. If the dimensions are particularly large, however, the recommendation is to check the stresses on the structure in relation to a bearing pressure distribution after Boussinesq.

### 10.5.3 Safety against sliding

R 79, section 10.4.3, applies.

### 10.5.4 Construction details

R 79, section 10.4.4 applies. Experience has shown that the joint shown in Fig. R 87-1 is a good solution for compressed-air caissons. After the caissons have been sunk, elastic make-up piles (threaded into cast-in sheet pile interlocks) are driven into the 40–50 cm wide joints to close the gaps. The space between the front and rear sheet piles is then cleared and, if the subsoil is firm, backfilled with underwater concrete or, if the

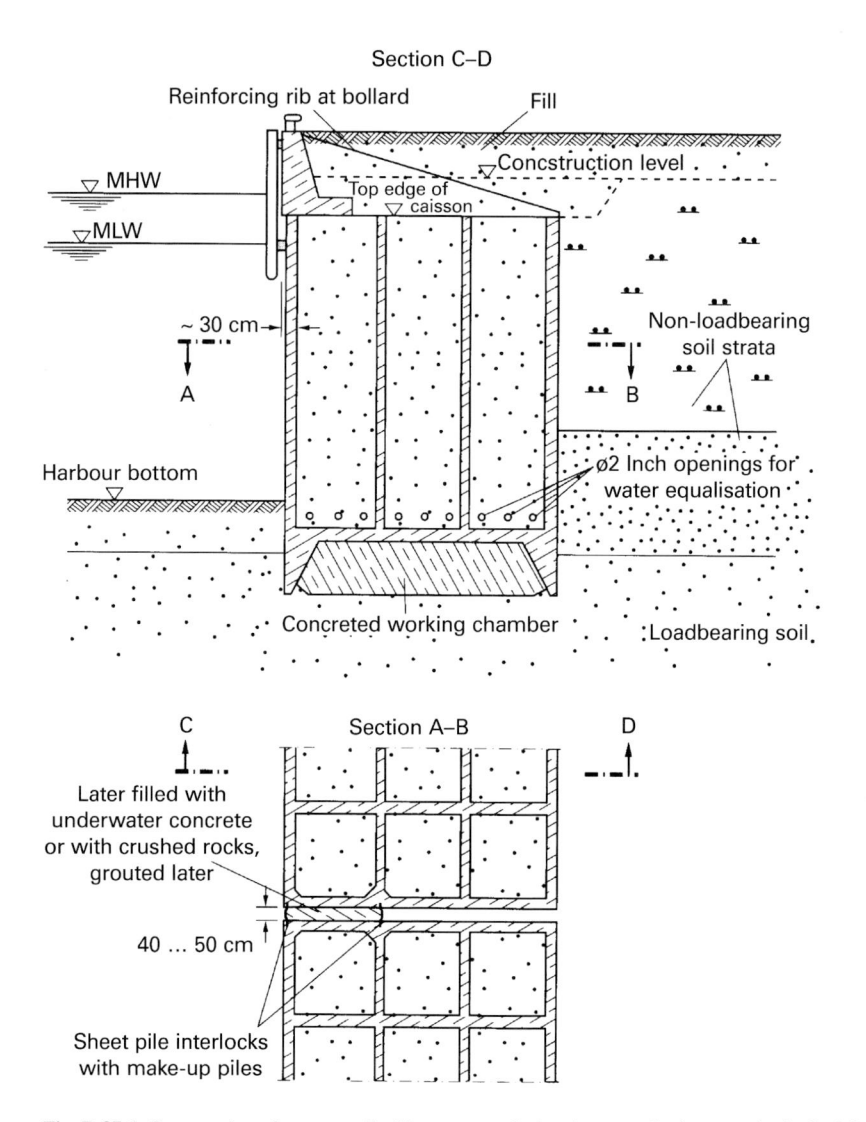

**Fig. R 87-1.** Construction of a quay wall with compressed-air caissons and subsequent basin dredging

subsoil is soft, with stones that can be pressure-grouted later. The back of the front pile designed to close a gap can be fitted flush with the front edge of the caissons. However, it can also be set back slightly to form a shallow recess for a ladder or similar.

If there are major differences in water levels, the foundation level should be such that there is sufficient protection against scour. Alternatively, suitable backfill material and drainage systems should be provided to balance out the differences in water pressure.

### 10.5.5 Work on site

Compressed-air caissons installed from the land are sunk from the ground level on which they had been constructed previously. The soil in the working chamber is generally excavated working almost exclusively in compressed air, or is loosened by jetting and pumped out. If the soil at the proposed foundation depth does not have a sufficient bearing capacity, the caisson must be sunk to a correspondingly greater depth.

Once the necessary founding depth has been reached, the bearing surface is levelled and the working chamber filled with concrete, again working in compressed air.

Compressed-air caissons that are floated into position must first be sunk to the existing or deepened bottom. Generally, the bearing area only needs to be roughly levelled since the narrow caisson edges easily penetrate the soil, which renders minor unevenness in the surface unimportant. The caissons are then lowered and filled with concrete as described above.

### 10.5.6 Frictional resistance during sinking

The frictional resistance depends on the properties of the subsoil, the position of the caisson in relation to the groundwater and the design of the caisson.

Friction is affected by:

- Soil type, density and strength of the in situ strata (non-cohesive and cohesive soils)
- Groundwater level
- Founding depth of the caisson
- Plan form and size of the caisson
- Geometry of the bottom edges and the outer wall surfaces

Determining the necessary "surplus sinking weight" for the respective sinking state is more a matter of experience than precise calculation. In general, it is sufficient when the "surplus weight" (total of all vertical forces, ignoring friction) is enough to overcome a skin friction of $20\,kN/m^2$ on the outside of the caisson to be embedded. With a lower surplus weight (modern reinforced concrete caissons), the recommendation is to employ additional measures to reduce friction, such as the use of lubricants, e.g. bentonite.

## 10.6 Design and construction of block-type quay walls (R 123)

### 10.6.1 Basic principles

Waterfront structures employing a block-type construction can only be built successfully where there is loadbearing subsoil beneath foundation level. If necessary, the bearing capacity of the in situ soil must be improved (e.g. by compaction), or inadequate soil replaced.

The dimensions and weights of the individual blocks are determined according to the building materials available, the production and transport options, the plant available for placing the blocks on site and the likely conditions in terms of site location, wind, weather and wave impacts during construction and after completion. From the point of view of economy, the lowest possible number of large blocks should be chosen, since the duration of individual activities during the building and striking of formwork, the transport and placing is not dependent on the size of the blocks. During transport to the site, buoyancy can be used to relieve the load on the means of transport if the blocks can be transported in the immersed state.

Buoyancy is frequently exploited used during installation of the blocks to reduce the effective weight and improve the utilisation of the crane. But in any event, the blocks must be sufficiently large and/or heavy enough to be able to withstand wave impacts. The plant to be used must be chosen on this basis. In particular, when a floating crane is used for installation, blocks with an effective installation load of 600–800 kN are often chosen.

The blocks must be shaped and laid in such a way that they are not damaged during installation. If the blocks are simply stacked vertically, one on top of the other, which is recommended if the subsoil is susceptible to settlement, it is very difficult to avoid larger joint widths. However, wider joints are acceptable if suitable backfilling material is used. Generally speaking, the aim should be to achieve an economic optimum with regard to the permissible joint width on the one hand and the choice of backfilling on the other. Blocks can be joined using tongue and groove joints or dovetailed with one another in an "I" shape. If vertical joints are to be avoided, this can be achieved, for example, by laying the blocks an angle of 10–20° from the vertical. The support can be in the form of, for example, blocks laid horizontally, a sunken caisson or similar, with wedge-shaped blocks forming the transition. The latter can also be used if it is necessary to correct the angle. Blocks installed at an angle results in the smallest possible joint width between the individual blocks, even though there is a greater number of block types. Using this form of construction, all the blocks will need tongue and groove joints along their sides. The projecting tongue should be on the outside of the

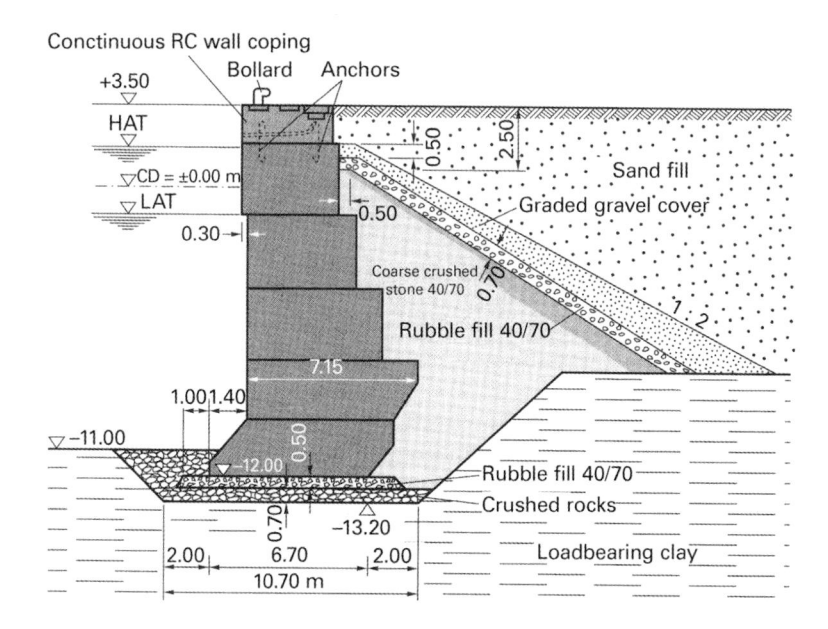

**Fig. R 123-1.** Section through a quay wall built from large blocks

block already laid so that, during installation, the next block can slide down, guided on this tongue.

A bedding of rubble and hard ballast at least 1.0 m thick should be installed between the loadbearing subsoil and the underside of the block wall (Fig. R 123-1). The surface must be levelled carefully, generally using special plant and the help of divers. In water containing sedimentation, the surface must also be cleaned particularly carefully prior to the positioning of the blocks in order to ensure that the soil-structure interface does not become a failure plane. This is particularly important in the case of blocks that are stacked vertically on top of one another.

In order to prevent the bedding from sinking into the subsoil (in fine-grained, non-cohesive subsoil in particular), the pores must be filled with graded gravel with a suitable particle distribution. In addition, a graded gravel filter can be installed between the bedding and the subsoil. If the subsoil is very fine-grained but not cohesive, a layer of non-woven material should be laid under the graded gravel filter to secure the positions of both subsoil and gravel filter.

Depending on the plant available, construction with blocks is particularly suitable for areas with more powerful waves and in countries that lack a skilled workforce. In addition to the use of heavy equipment, however, the method also requires, above all, costly diving operations in order to ensure and monitor the careful completion of the bedding, the positioning of the blocks and backfilling work. Further advice on execution can be found in Zdansky [231].

### 10.6.2 Forces acting on a block wall

#### 10.6.2.1 Active and passive earth pressures

Active earth pressure may be assumed since the movements of the wall can be presumed for its mobilisation. As the foundations for block walls are generally not very deep, passive earth pressure should not be included in the calculations.

#### 10.6.2.2 Excess water pressure

If the joints between the individual blocks are readily permeable and if the choice of backfilling material (Fig. R 123-1) ensures rapid equalisation of the water levels, the excess water pressure on the quay wall only needs to be assessed for half the height of the largest waves to be expected and at the least favourable level according to R 19, section 4.2. Otherwise, the excess water pressure as per R 19 must be added to the half wave height. In cases of doubt, drainage systems with anti-flood fittings should be installed, but they must also work reliably even under the action of waves. Conversely, experience shows that it is not possible to achieve a perfect seal in the joints between blocks.

A permanently effective filter must be positioned between the quay wall, or a backfilling of coarse material, and a subsequent fill consisting of sand, etc. in order to prevent the backfill material being washed out (Fig. R 123-1).

#### 10.6.2.3 Stresses resulting from waves

If waterfront structures are to be built from blocks in areas where high waves are likely, then special studies must be carried out to assess stability. In particular, it is important – using model trials in cases of doubt – to establish whether breaking waves could occur. If this is the case, the risks in terms of the stability and service life of a block wall are so great that this method of construction can no longer be recommended. The ratio between the water depth $d$ in front of the wall and the wave height $H$ can be used to establish whether breaking or reflected waves occur. With a water depth $d \geq 1.5 \cdot H$, it can generally be assumed that only reflected waves will occur (see also R 135, section 5.7.2, and R 136, section 5.6).

The wave pressure not only affects the front side of a block wall, but also travels into the joints between individual blocks. The joint water pressure can temporarily reduce the effective block weight more than the buoyancy does, reducing the friction between individual blocks to such an extent that the stability of the wall is at risk. As the wave recedes, the pressure drop in the narrow joints, which is also affected by the groundwater, takes place more slowly than that along the outer surface of the quay wall, resulting in a higher water pressure in the joints than that corresponding to the water level in front of the wall. At the same time, however, the active earth pressure and excess water pressure from behind are still fully effective. This state, too, can be relevant in terms of stability.

### 10.6.2.4 Line pull, vessel impact and crane loads

The relevant recommendations apply here, including R 12, section 5.12, R 38, section 5.2, and R 84, section 5.14.

## 10.6.3 Design

### 10.6.3.1 Base of wall, bearing pressures, stability

The block wall cross-section must be designed in such a way that the bearing pressures in the soil beneath the wall due to permanent loads are distributed as uniformly as possible. This can generally be achieved relatively easily by designing the base of the wall with a toe projecting beyond the line of the wall on the water side and by allowing the wall to overhang the base on the land side ("knapsack") (Fig. R 123-1).

If voids beneath overhanging blocks at the rear of the wall are to be avoided, the blocks must be undercut at the back. The angle of batter must be steeper than the angle of friction of the backfill (Fig. R 123-1). Bearing pressures must be verified for all the important phases during construction. If necessary, the quay wall must be backfilled more or less at the same time as laying the blocks in order to counter any overturning in the direction of the land or excess bearing pressures at the land end of the foundation (Fig. R 123-2). In addition to the permissible bearing pressures, it is also necessary to check safety against sliding, ground failure and slope failure. When it comes to sliding, the most relevant information is to be found in R 79, section 10.4.3.

Possible changes in the harbour bottom as a result of scouring, but primarily as a result of foreseeable deepening, must be taken into consideration. During subsequent port or harbour operations, inspections of the level of the bottom in front of the wall must be carried out at regular intervals and suitable protective measures initiated immediately if necessary.

In order to allow for any overturning of the quay wall in the direction of the water side during port operations, the wall must be designed with a slight batter towards the land side. The crane track gauge may change as a result of unavoidable wall movements and must therefore be adjustable.

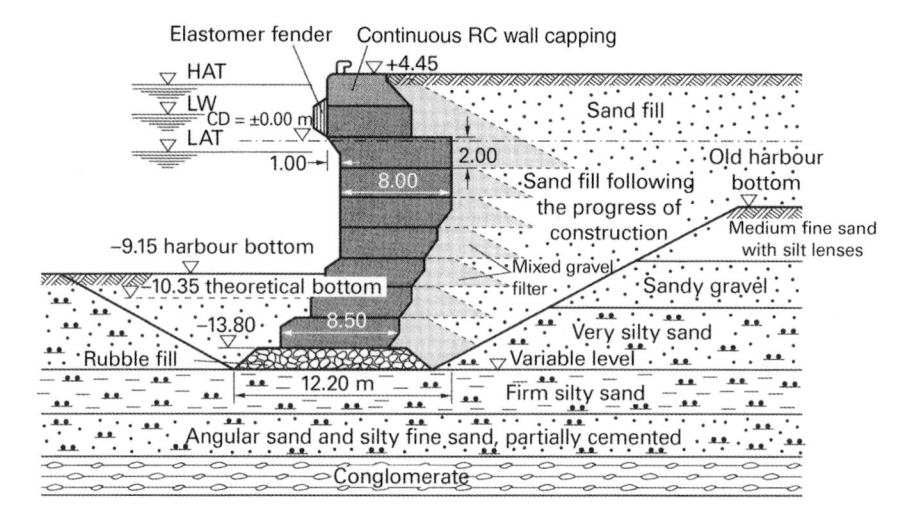

Fig. R 123-2. Design for a quay wall built from large blocks in an earthquake zone

**10.6.3.2 Horizontal joints in the block wall**

Safety against sliding and the location of the resultant of the applied forces must also be verified for the horizontal joints in the block wall for all important construction states and the final state. In contrast to the foundation joint, a theoretical opening up of the joints as far as the centroid axis may be permitted with the simultaneous application of all the forces acting unfavourably.

**10.6.3.3 Reinforced concrete capping beam**

The in situ reinforced concrete beam cast on the top of every block wall is needed to even out inaccuracies associated with laying the blocks and the construction in general, to spread concentrated horizontal and vertical loads and to even out local variations in earth pressures and support conditions at the base. On account of the differential settlement in the block wall, the RC capping beam may only be cast after all settlement has abated. To speed up the settlement process, a temporarily higher loading on the wall is useful, e.g. by applying additional loads in the form of concrete blocks. The settlement behaviour must be monitored constantly in this case.

Loads relieved by excavation work in cohesive soil do not generally need to be considered in the structural design calculations, since they are compensated for by the increasing wall load.

When calculating the internal forces at the top of the wall due to vessel impact, line pull and lateral crane impacts, it can generally be assumed that the capping beam is rigid in comparison with the block wall supporting it. This assumption is generally on the safe side.

When designing the capping beam for vertical forces, primarily crane wheel loads, the modulus of subgrade reaction method can generally be used. If greater differential settlement or subsidence of the block wall is to be expected, the internal forces at the top of the wall must be limited through comparative studies involving various support conditions – "riding" in the middle or at the ends. The dead load of the capping beam must also be taken into account in these studies. Verification of maximum crack widths to R 72, section 10.2.5, is required. If necessary, block joints must be included.

At joints between blocks, capping beams are only interlocked for the transmission of horizontal forces. Interlocking for vertical forces should be avoided because of the unpredictable settlement behaviour of block walls.

At block joints, rail supports should be protected against differential settlement by construction details that make use of short intermediate bridges; the crane rails must be able to pass through without joints.

To transmit horizontal forces between capping beam and block wall, both should be effectively interlocked with one another. Anchoring can also be used instead of interlocking.

## 10.7 Design of quay walls using open caissons (R 147)

### 10.7.1 General

Open caissons are used in seaports for waterfront structures and mooring points, but also as foundations for other structures, albeit very rarely. As with compressed-air caissons according to R 87, section 10.5, they can be built on the ground above the water level in the sinking zone or in a driven or floating scaffold or floated into position as finished units on jack-up platforms or floats and then sunk. The sinking of open caissons involves lower labour and site setup costs than sinking using compressed air and can be carried out at significantly greater depths. However, it cannot achieve the same degree of positional accuracy. In addition, it does not lead to similarly reliable support conditions at the foundation. Obstacles encountered during sinking can only be passed or removed with difficulty. Positioning on sloping rock surfaces always requires additional measures.

The construction principles specified in relation to box caissons for quay walls in R 79, section 10.4.1, also apply similarly to open caissons.

Otherwise, particular reference is made to section 3.3 of the *Geotechnical Engineering Handbook* [77].

### 10.7.2 Verification

R 79, sections 10.4.2 and section 10.4.3, and R 87, section 10.5.2, must be observed.

### 10.7.3 Construction details

Open caissons may be square, rectangular or circular on plan. When choosing the plan form, both operating and construction-related considerations are crucial.

Owing to the funnel-shaped excavation, open caissons with a rectangular plan form do not stand as evenly on their edges as circular open caissons. There is therefore an increased risk of deviations from the target position. Where a rectangular shape is necessary, it should therefore be realised in a compact form. Since the excavation, and hence the sinking process, is difficult to monitor and open caissons can only be ballasted to a limited extent, thicker walls should be chosen so that, allowing for buoyancy, the dead load of the caisson definitely exceeds the expected wall friction force.

The bases of outer walls are provided with a stiff steel cutting edge. An edge made from high-strength concrete (at least C80/95) or steel fibre-reinforced concrete is also conceivable. Jetting lances set in the cutting edge, with outlets facing inwards, can facilitate the removal of non-cohesive soil (Fig. R 147-1, section C-D, shown on the water side).

The bottom edges of cross-walls must terminate at least 0.5 m above the bottom edge of the caisson so that no loads can be diverted into the subsoil.

Outer walls and cross-walls are provided with reliable bearing recesses, which can easily be cleaned after sinking, for transferring the loads to the underwater concrete base.

The loosening of the soil beneath the caisson, unavoidable when sinking open caissons, leads to settlement and tilting of the finished structure. This must be taken into account in the design and construction, also during construction operations.

The information given in R 79, section 10.4.4, applies without exception. For the joints, a solution according to R 87, section 10.5, Fig. R 87-1, is recommended. However, filling the space between the sheet piles with graded gravel to create a filter should be preferred over a rigid filling because gravel can adjust to any settlement that occurs without causing damage.

Considering the method of excavation, the distance of 40–50 cm between caissons, which is given in R 87, section 10.5.4, for compressed-air caissons, is only sufficient for open caissons if the actual sinking depth is small and obstacles – including those caused by embedded, firm, cohesive soil strata – are not anticipated. If the sinking process is difficult, a distance of 60–80 cm should be selected. Suitable

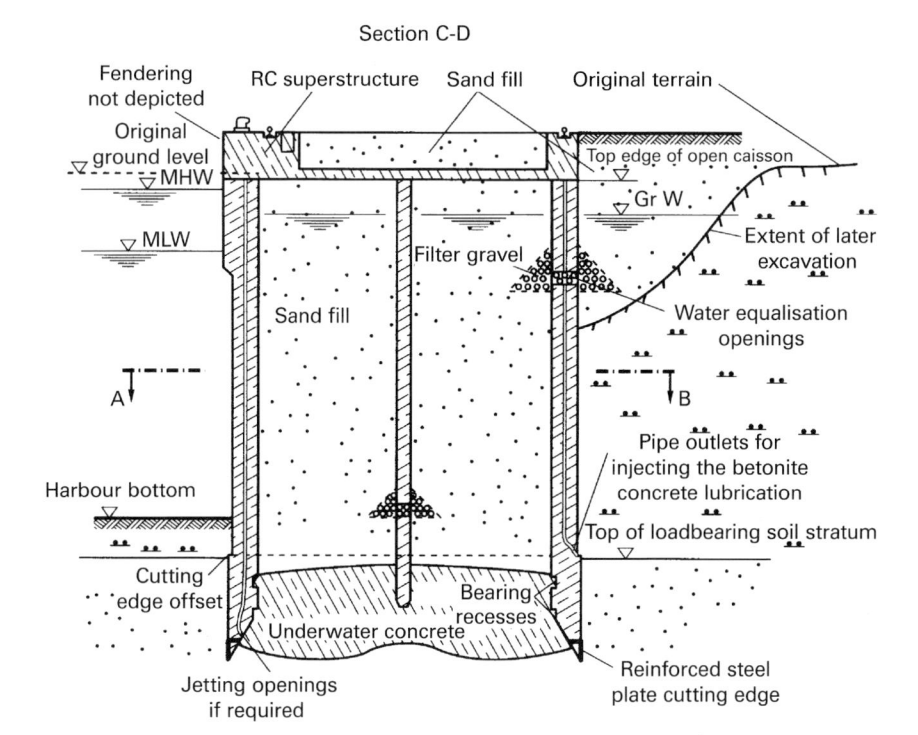

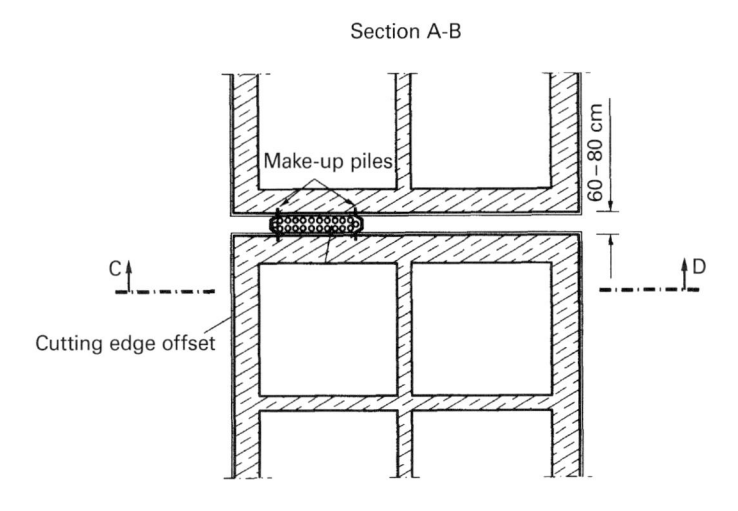

**Fig. R 147-1.** Design for a quay wall built from open caissons with subsequent harbour dredging

closures would then be correspondingly wide make-up piles or groups of sheet piles arranged in a looped shape to accommodate deformations.

**10.7.4** **Work on site**

If built on land, the bearing capacity of the subsoil beneath the caisson must be particularly carefully checked and observed so that the soil under the cutting edge does not give way too much or too unevenly. The latter can also cause fracturing of the cutting edge. The soil within the caisson is excavated with grabs or pumps, with the water level inside the caisson always being kept at least at the level of the external water in order to avoid hydraulic heave failure.

When sinking a row of caissons, the sequence 1, 3, 5 . . . 2, 4, 6 can be appropriate because this sequence ensures that the active earth pressures on both end faces of each caisson are balanced.

Sinking a caisson can be made considerably easier by lubricating the outside surfaces above the base with a thixotropic liquid, e.g. a bentonite suspension. To ensure that the entire outer surface is actually lubricated, the lubricant should not be poured in from the top but instead forced in through pipes that are cast into the caisson and end immediately above the step at the base, protected by a sheet steel spreader if necessary (Fig. R 147-1, shown on land side). However, injecting the lubricant must be carried out carefully so that the thixotropic liquid cannot break through into the excavation area below and flow away. The base part of the open caisson up to the set-back outer face must therefore be correspondingly high. Particular care is required if, as a result of vibration of the loosened sand under the base of the excavation, the top surface of the sand sinks to any significant extent.

After the planned foundation depth has been reached, the bearing area should be carefully cleaned. Only then should the base slab made from underwater or grouted aggregate concrete be cast.

**10.7.5** **Frictional resistance during sinking**

The information given in R 87, section 10.5.6, concerning compressed-air caissons also applies to open caissons. However, as an open caisson cannot be ballasted to the same extent as a compressed-air caisson, the thixotropic lubrication of the outer surface acquires particular significance with greater sinking depths. Experience has shown that it reduces the mean skin friction to $<10 \text{ kN/m}^2$.

**10.7.6** **Preparation of the subsoil**

Non-cohesive soil susceptible to liquefaction must be compacted or replaced beyond the area of the actual foundation. On account of the loosening associated with soil excavation in open caissons, subsequent compaction of the soil under the base of the excavation is necessary when using open caissons.

## 10.8 Design and construction of solid waterfront structures (e.g. blocks, box caissons, compressed-air caissons) in earthquake zones (R 126)

### 10.8.1 General
R 124, section 2.16, must be taken into account.

When determining the horizontal inertial forces of the waterfront structure, it is important to realise that they must be derived from the mass of the respective structural components plus the surcharges resulting from the associated backfill. In doing so, the mass of the pore water in the soil must also be considered.

### 10.8.2 Active and passive earth pressures, excess water pressure, variable loads
The information given in sections 2.16.3, 2.16.4 and 2.16.5 under R 124 applies accordingly.

### 10.8.3 Safety
The main advice to observe is given in R 124, section 2.16.6.

With block-type construction, even if the effects of earthquakes are taken into account, the eccentricity of the resultant forces in the horizontal joints between individual blocks may only be so large that there is no theoretical widening of the joints beyond the centroid axis under characteristic loads.

### 10.8.4 Base of the wall
At the soil-structure interface, where opening of the joint is not permitted even in non-earthquake situations, there should be no widening of the joints beyond the centroid axis under characteristic loads.

## 10.9 Use and design of bored cast-in-place piles (R 86)

### 10.9.1 General
Bored cast-in-place piles can also be used for waterfront structures if suitably designed and constructed. Besides economic and technical reasons, they are ideal when a reliable, largely vibration-free and/or low-noise method of construction is required.

### 10.9.2 Design
By arranging bored cast-in-place piles in a row, straight or curved walls can be constructed to suit the desired plan layout.

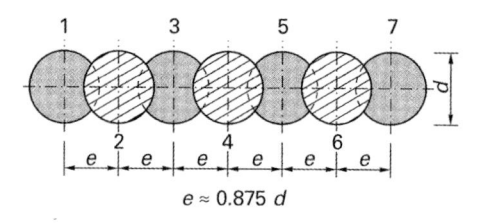

**Fig. R 86-1.** Interlocking bored cast-in-place pile wall

Two types of bored cast-in-place pile wall, distinguished by the distance between the piles, are suitable for waterfront structures.

### 10.9.2.1 Interlocking bored cast-in-place pile wall

The centre-to-centre spacing of the bored cast-in-place piles in this case is smaller than the pile diameter. Firstly, the primary piles (1, 3, 5, ... ) in plain concrete are installed. These are cut during the installation of the intermediate, secondary piles (2, 4, 6 ... ) in reinforced concrete (in special cases, three plain concrete piles can also be installed next to each other). The overlap is generally 10–15% of the diameter, but at least 10 cm. The amount of overlap must be matched to the construction tolerances of the bored cast-in-place piles in such a way that a reliable overlap is also achieved at the required depth (Fig. R 86-1).

The finished wall is thus generally practically watertight. A structural interaction of the piles can be at least partially assumed for loads perpendicular to and in the plane of the wall, but this is generally not taken into account when designing the loadbearing, secondary piles. Vertical loads can also be distributed to a certain extent – apart from the use of a capping beam to spread the load – by shear forces between the adjoining piles provided the cut surfaces are sufficiently rough and clean. In the case of walls short on plan, any outward displacement of the pile toes in the plane of the wall must be adequately prevented by embedding the bases of the piles in a subsoil with a particularly good bearing capacity.

### 10.9.2.2 Secant bored cast-in-place pile wall

The centre-to-centre spacing of the bored cast-in-place piles is, for construction reasons, about 5 cm larger than the pile diameter. Normally, every pile is reinforced. This type of wall can only be made watertight if additional measures are taken, e.g. by creating overlapping columns using the jet-grouting method or other injection processes. No plate-type interaction of the piles can be expected in the plane of the wall.

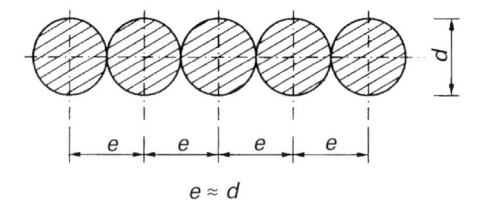

**Fig. R 86-2.** Secant bored cast-in-place pile wall

### 10.9.3 Construction of bored cast-in-place pile walls

The construction of bored cast-in-place pile walls according to Figs. R 86-1 and R 86-2 presumes a high degree of boring precision, which calls for good guidance of the casing. The accuracy of the drilling position is guaranteed by using a template.

Where possible, bored cast-in-place pile walls are built from the natural terrain or else from a platform made from dumped material.

The borehole is formed using rotary drilling or grab excavation methods. Obstacles are removed by breaking them up or core-drilling through them. Care must be taken to ensure that there is sufficient overpressure from the water filling in the casing, which generally has to be kept at least 1.5 m above the groundwater level.

Interlocking bored pile walls are mostly constructed using plant in which the casing is driven into the soil by rotating and/or pressing as the hole is drilled with the rotary cutting head or with the help of a casing oscillator. The bottom edge of the casing functions as a cutting edge.

With appropriate guidance, cased holes can also be executed at an angle, although raking bored cast-in-place pile walls should generally be limited to exceptional, unavoidable cases.

Obviously, the concrete used in the plain primary piles should include a retarder to slow setting. The sequence of operations should be selected such that the strength of the concrete during cutting by the secondary piles – depending on the capacity of the rotary drilling plant – does not exceed 3–10 MPa under normal circumstances. The difference in strength between adjacent primary piles to be cut should be kept as small as possible in order to avoid any changes in the direction of the secondary piles. Concrete with a strength class lower than C 30/37 is permitted for the plain primary piles in interlocking bored cast-in-place pile walls.

DIN EN 1536 and DIN SPEC 18140 apply regarding cleaning the base, concreting, concrete cover and reinforcement details.

### 10.9.4 Construction guidance

A radially symmetric arrangement is required for the reinforcement in each pile. Only in exceptional cases – where operations and supervision are carried out very meticulously – may a different arrangement be used. This is because accidental twisting of the reinforcement cage cannot be ruled out when extracting the casing.

Accidental extraction of the reinforcement cage can be avoided by applying fresh concrete to a plate built into the base of the cage and/or corresponding coordination between the largest grain size of the concrete and the space between reinforcement cage and casing.

The pile reinforcement must be sufficiently stiff in order to maintain the necessary concrete cover and prevent any deformation of the reinforcement cage.

Welded stiffening rings in accordance with ZTV-ING [231] have proved effective. The minimum measures specified in that document are only sufficient for piles up to about 1 m in diameter. For piles with a large diameter ($d=$ approx. 1.30 m), at a spacing of 1.60 m, for example, two $\varnothing 28$ mm stiffening rings of steel grade Bst 420 S with eight $\varnothing 22$ mm spacers at a spacing $l = 400$ mm, are recommended. These are welded to each other and to the longitudinal reinforcement of the pile.

The piles are designed on the basis of DIN EN 1997-1-1 and DIN 1054. If the piles are not incorporated into a sufficiently rigid superstructure with a short distance to the anchoring plane, walings are generally required to distribute the loads and accommodate the anchoring forces. Walings are unnecessary for anchored bored cast-in-place pile walls, interlocking and secant types – at least in moderately densely bedded non-cohesive or semi-firm cohesive soils, and provided at least every second pile in interlocking walls or every second gusset between piles in secant walls is held by an anchor. At the same time, however, both ends of the wall must be provided with tension walings over a sufficient length.

Connections to adjacent building elements should preferably only be made through the reinforcement at the pile head; in the rest of the wall in special cases only and even then only via openings or connection fittings cast in specifically for this.

Since a standardised crack width limitation (R 72, section 10.2.5) is not possible for bored cast-in-place pile walls, additional measures must be taken in consultation with the client (e.g. facings, cathodic corrosion protection, increased concrete cover).

## 10.10 Use and design of diaphragm walls (R 144)

### 10.10.1 General

Regarding the use of diaphragm walls, R 86, section 10.9.1, applies accordingly.

The term "diaphragm wall" is applied to in situ concrete walls that are constructed in sections (panels) in trenches. The trenches are excavated between guide walls with a special grab bucket or a continuous cutting and recirculating machine and then filled with a supporting slurry on a continuous basis. After the supporting slurry has been cleaned and homogenised, the reinforcement is suspended in the trench and the concrete introduced via a tremie pipe, which pushes the supporting slurry upwards and allows it to be pumped out.

The design of diaphragm walls is regulated in DIN EN 1997-1-1 and DIN 1054. Construction is described in detail in DIN EN 1538, which primarily provides information regarding:

- the excavation of the trench,
- the manufacture, mixing, swelling, storage, introduction, homogenisation and reprocessing of the supporting slurry, and
- reinforcement and concreting.

The stability of the liquid-supported trench is regulated in DIN 4126, and DIN 4127 lays down the test procedures for supporting slurries and their raw materials.

The following literature can also be consulted: [223], [145], [63], [133], [213], [214], and [215].

Diaphragm walls are practically watertight and are constructed continuously in thicknesses of 60, 80 and 100 cm; thicknesses of 120 and 150 cm are also used for quay walls with large steps in the terrain. Where loads are high, instead of a single wall, a wall consisting of T-shaped elements arranged in a row can be constructed. With very loosely bedded or soft soils, constructing such T-shaped elements is only recommended if additional measures are taken, e.g. improvement of the soil beforehand.

A wall that is curved on plan is replaced by a polygonal sequence of straight panels. The possible length of a panel excavated in one operation is limited by the stability of the liquid-supported trench. The minimum length of a trench section is determined by the opening size of the grab or the cutting and recirculating machine. In the case of a high groundwater level, lack of cohesion in the soil, adjacent heavily loaded foundations, sensitive buried services, etc., the maximum dimension of approx. 10 m must be reduced to the aforementioned minimum opening where necessary. Diaphragm walls can transfer high horizontal and vertical loads into the subsoil. Connections to other horizontal or vertical parts of the structure are possible using connecting elements cast in or connected with dowels and housed in recesses if necessary. Good fair-face concrete surfaces can be achieved by installing suspended precast concrete elements, the use of which is limited, however, to depths of 12–15 m on account of their high self-weight.

### 10.10.2 Verifying the stability of the open trench

The analysis of the stability of the open trench is dealt with in DIN 4126. Three verifications are normally necessary:

a) safety against the entry of groundwater into the trench and against displacement of the supporting slurry,
b) safety against sliding of individual grains or groups of grains, and
c) safety against failure planes in the soil which endanger the trench.

For verification a), the hydrostatic pressure of the supporting slurry is compared with that of the groundwater. From this it is possible to derive the necessary density for the supporting slurry.

Verification b) requires the minimum liquid limit of the supporting slurry to be calculated.

To verify safety against failure planes in the soil which endanger the trench, verification c) requires the equilibrium of a sliding wedge of soil to be investigated. The loads are the dead load of the soil plus any additional loads from neighbouring structures, construction plant or other variable loads together with the external water pressure. On the resistance side there is the pressure of the supporting slurry, the full friction at the failure planes, which leads to the active earth pressure, and friction on the lateral surfaces of the sliding wedge of soil plus any cohesion. In addition, the braced guide wall can be taken into account. This is particularly important for high-level failure planes because the shear stress is still not very effective here in the case of non-cohesive soils. With low-level failure planes, the influence of the guide wall is negligible.

In tidal areas the critical external water level must be determined and stipulated based on the proposed supporting slurry level. If it is anticipated that the permissible external water level will be exceeded, e.g. as a result of storm floods, an open trench must be backfilled in good time.

### 10.10.3 Composition of the supporting slurry

A clay or bentonite suspension is used as the supporting slurry. Particular attention must be paid to the fact that, in the case of structures in seawater or very salty groundwater, the ion balance of the clay suspension is altered unfavourably as a result of the ingress of salts. Flocculation is the result, which can lead to a reduction in the supporting capacity of the suspension. For this reason, saltwater-resistant bentonite suspensions must be used when diaphragm walls are built in such areas. Many different mixes can be used. Suitability tests must certainly be carried out prior to commencing building work. Such tests must cover the salt content of the water, the soil conditions and any other relevant aspects (e.g. passing through corals). The contamination of a suspension under saltwater conditions manifests itself most obviously by the increase in filtrate water discharge.

Special caution is required with soil impurities (polluted areas), soil constituents composed of peat or brown coal, etc. Appropriate additives can be used to partly compensate for unfavourable influences. Suitability tests of the supporting slurries are urgently recommended in such cases.

### 10.10.4 Diaphragm wall construction

Generally speaking, a section of diaphragm wall is excavated in the ground between guide walls. The walls are usually 1.0–1.5 m high and made from reinforced concrete. Depending on the soil conditions and the loads exerted by the excavation plant and the extraction equipment for the stop end pipes as well as the suspended reinforcement, they are designed as continuous wall strips propped against each other beyond the excavation areas or as cantilever walls.

The supporting slurry becomes enriched with ultrafine particles during the excavation work. It should therefore be checked at regular intervals and replaced if it no longer complies with the specification. As a rule, the supporting slurry can be used several times. The density, filtrate water discharge, sand content and liquid limit of the supporting slurry must be checked on site.

The trench and, if applicable, the adjacent joints must be cleaned prior to incorporating the reinforcement. Before concreting, the properties of the supporting suspension must be checked. Where the limits for the liquid limit and sand content are exceeded, the suspension must be recycled, or replaced if necessary. The interval between the end of the excavation work and the start of the concreting must be kept as short as possible.

Details regarding diaphragm wall construction are shown in Fig. R 144-1. In some cases the step-by-step construction of the panels in the order 1, 2, 3, etc. is preferable. The formwork elements should be kept as narrow as possible in order to keep the unreinforced zone to a minimum.

### 10.10.5 Concrete and reinforcement

First and foremost, the reader should consult the detailed information given in DIN EN 1538.

When designing the reinforcement, it is important to avoid concentrations of and gaps in the reinforcement that hinder the flow. Ribbed reinforcing bars should be preferred on account of their superior bonding properties. To ensure the necessary concrete cover, a sufficient number of generously sized spacers must be provided.

Adequate bracing of the cages should be ensured, especially in the case of minimum reinforcement.

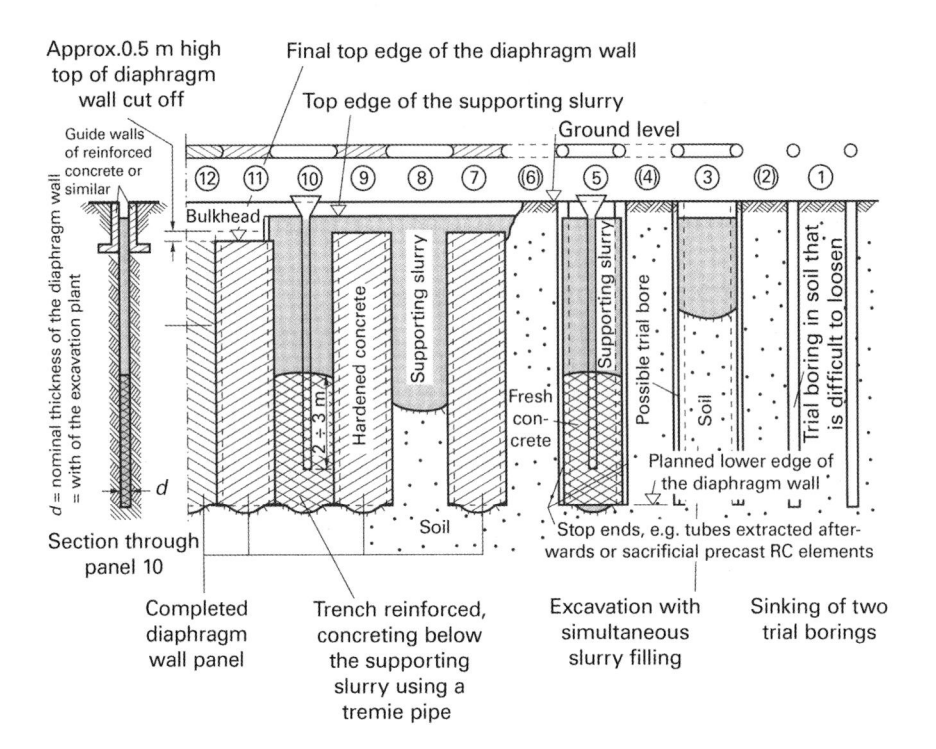

**Fig. R 144-1.** Example of diaphragm wall construction

### 10.10.6 Guidance for the design of diaphragm walls

On account of their high flexural stiffness and low deformations, diaphragm walls generally have to be designed for an increased active earth pressure. The application of the active earth pressure can only be justified where – as a result of sufficient flexibility at the base of the wall and the supports (with sufficiently flexible anchoring) and the horizontal deflection as well – the necessary displacement is possible for full mobilisation of the shear stresses at the failure planes.

With large differences in terrain levels and head displacements in the centimetre range, e.g. a change in level $\geq 20$ m in the case of quay walls for sea-going ships, a distribution of the active earth pressure according to R 77, section 8.2.3, is possible. The possible deformation behaviour must be taken into account in every individual case.

Generally speaking, full fixity at the base of the wall in the soil cannot be achieved with anchors or support at the top of the wall because of the wall's high flexural stiffness. It is therefore a good idea, when designing the wall according to Blum, to allow for partial fixity or else to carry out the design using the modulus of subgrade reaction or modulus of coefficient of compressibility method with elastic fixity at the base. If the finite element method is used, the most important thing is to use appropriate material models for the soil. The angles of inclination in the active and passive zones essentially depend on the type of soil, the progress of the work and the time the trench is left open. Coarse-grained soils result in a greater roughness to the wall of the excavation, whereas fine-grained soils will lead to relatively smooth excavation faces. Slow progress and trenches left open for longer periods will encourage deposits from the supporting slurry (formation of filter cakes). On account of their dependency on the factors listed above, the angles of inclination can generally be assumed to lie within the following limits:

$$0 \leq \delta_{a,k} \leq \tfrac{1}{2} \cdot \phi'_k \quad \text{or} \quad -\tfrac{1}{2} \cdot \phi'_k \leq \delta_{p,k} \leq 0$$

Walings for supports or anchors can be formed by additional transverse reinforcement within the panels. If the panels are narrow, a central support or anchor will suffice; wider panels will need two or more positioned symmetrically in relation to the panel. Where necessary, a punching shear analysis must be carried out.

Design is carried out in accordance with DIN EN 1992-1-1 and DIN 19702. Checks of maximum crack width are based on R 72, section 10.2.5. Supplementary measures including, for example, increased concrete cover or cathodic corrosion protection (see [103]) may be necessary in this respect. The measures relating to corrosion protection of the reinforcement must be agreed with the client in each individual case.

In the light of the unfavourable effect of supporting slurry residue on the steel or fine sand deposits, the bond stresses for horizontal reinforcing bars should correspond to the moderate bond conditions of DIN 1045, section 12.4. Good bond conditions can generally be assumed for vertical steel bars, but it is recommended that the anchorage lengths be increased at the top and bottom of the diaphragm wall.

## 10.11 Survey prior to repairing concrete components in hydraulic engineering structures (R 194)

### 10.11.1 General

Constructional measures concerning the repair of concrete components will only be successful if they correctly take into account the causes of the defects or damage. Since several causes are generally involved, a systematic inspection of the actual situation must be carried out beforehand by a qualified engineer.

As the correct assessment of the causes of the defects and damage is an essential prerequisite for a permanent repair, a number of recommendations for assessing the actual condition and determining the causes are given below.

The following provisions also essentially apply to the testing of concrete parts as part of the survey according to R 193, section 14.1. However, the individual inspections listed will in some cases go beyond the extent of a normal survey and are therefore listed here separately.

The following data must be collected beforehand:

1. Description of the structure
    - year of construction
    - loads arising from use, operation and the environment
    - existing stability analyses
    - information on the subsoil
    - construction drawings
    - specific factors during the construction of the structure
2. Assessment of the components affected by the damage:
    - type, position and dimensions of the components
    - building materials used (type and quality class)
    - description of the damage (nature and extent of the damage with dimensions of damaged areas)
    - documentation (photos and sketches)
3. Tests required:
   The type and number of tests needed to determine the causes are defined on the basis of the findings according to section 10.11.1(1) and (2).

### 10.11.2 Tests performed on the structure

More detailed information regarding the actual condition of the structure can be obtained by carrying out the following tests on the structure:

1. Concrete
    - discolouration, dampness, organic growths, efflorescence/sintering, concrete spalling, voids
    - surface roughness
    - adhesion
    - watertightness, pay coming
    - depth of carbonisation
    - chloride content (quantitative)
    - cracking: positions, widths, depths, lengths
    - crack movements
    - conditions of joints

2. Reinforcement
   - concrete cover
   - corrosion, rate of corrosion
   - reduction in cross-section
3. Prestressing tendons
   - concrete cover
   - condition of grouting (using ultrasound, radiography tests, endoscopy if necessary)
   - condition of prestressing steel
   - actual degree of prestressing
   - grouting mortar ($SO_3$ content)
4. Components
   - deformations
   - forces
   - vibration behaviour
5. Sampling carried out on the structure
   - products of efflorescence/sintering
   - concrete parts
   - drilled cores
   - drilling debris
   - reinforcing elements

## 10.11.3 Tests performed in the laboratory

1. Concrete
   - density
   - porosity/capillarity
   - water absorption
   - water penetration depth
   - abrasion resistance (to DIN 52108)
   - micro air pore content
   - chloride content (quantitatively in various depth zones)
   - sulphate content
   - compressive strength (to DIN EN 12504)
   - modulus of elasticity (to DIN 1048)
   - mix composition (to DIN 52170)
   - granulometric composition
   - splitting tensile strength (to DIN EN 12390)
   - depth of carbonisation
   - near-surface tensile strength (to DIN EN 1542) at various depths
2. Steel
   - tensile test
   - fatigue test

### 10.11.4 Theoretical investigations

- Structural calculations for the structural safety and deformation behaviour of the structure or individual parts thereof before and after repair
- Estimate of the progress of carbonisation and/or of the temporary chloride accumulation with and without repair

## 10.12 Repairing concrete components in hydraulic engineering structures (R 195)

### 10.12.1 General

Hydraulic engineering structures are subjected to certain environmental stresses that can arise from physical, chemical and biological factors. In addition, in port and harbour installations, for example, and other areas that must be kept accessible for operational reasons, the effect of de-icing salts and other harmful impurities must also be expected, occasionally involving transhipped goods that attack the concrete.

Apart from imposed loads and impact and frictional forces from ships, the physical actions are mainly the result of the repeated wetting and drying of the concrete, constant temperature changes with the harsh effects of frost on water-saturated concrete and the effects of ice. Chemical loads, in addition to the effects of de-icing salts and transhipped goods in individual cases as a result of port operations, are primarily caused by the salts in seawater. Chlorides that infiltrate the concrete can destroy the passive coating on the reinforcement. The result can be corrosion of the reinforcement in areas where there is an adequate supply of oxygen and moisture simultaneously in the concrete, e.g. above the low water zone. The primary causes of biological loads are plants and the products of their metabolic processes.

The factors listed can lead to concrete with cracks and surface damage plus corroded reinforcement. Components in the splash and tidal zones are particularly at risk, especially those structural components exposed to seawater or those in the immediate vicinity of the coast where the air is very salty. Fig. R 195-1 shows in diagrammatic form how seawater can attack reinforced concrete.

If concrete components have to be repaired as a result of damage, an experienced engineer must always be brought in to ascertain the actual condition, assess the damage and plan the repair measures, in accordance with the guidelines laid down by the German Committee for Structural Concrete concerning the protection and repair of concrete components [43]. The lasting success of the repairs essentially depends on the qualifications of the personnel carrying out the work, the quality and suitability of the building materials used and the care taken in executing and supervising the work.

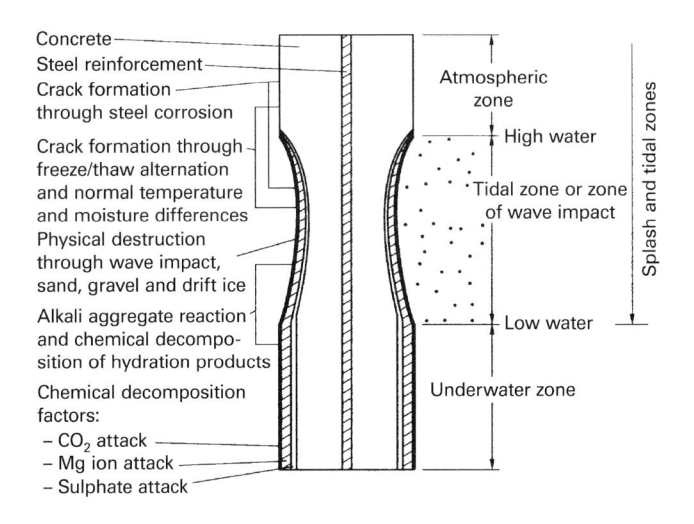

**Fig. R 195-1.** Diagram of the effect of seawater on reinforced concrete [145].

Protection and repair work on components that can be protected against the ingress of water – at least during the execution of the work (working above water) – should be carried out on the basis of [233]. When filling cracks and cavities in such components [231] should be used as a basis. In the case of protective and repair work that has to be carried out underwater, trial repairs should be carried out to prove that the envisaged measures are expedient under the given boundary conditions for each individual case. Verification of the quality of the building materials should be provided by way of suitability and quality tests tailored to the relevant component conditions.

Measures in connection with cathodic corrosion protection should be planned on the basis of [99].

Repair work should only be carried out by contractors with sufficient expertise and experience in this area and who satisfy the requirements in terms of staff and equipment as laid down by the German Committee for Structural Concrete [43], ZTV-W LB 219 [233] and ZTV-ING [231].

### 10.12.2 Assessing the actual condition

The effect of the defects and damage on the stability, serviceability and durability of the structure must be assessed on the basis of a thorough survey of the structure according to R 194, section 10.11.

The actions and stresses to which the component to be repaired is subjected should be ascertained as precisely as possible because these are then used as a basis for the requirements to be met by the building materials and building methods being used.

### 10.12.3 Planning the repair works

#### 10.12.3.1 General

Comparing the actual condition in accordance with section 10.12.2 and the envisaged target condition following completion of the repair measures results in the repair requirements. The objectives of the repair measures should be defined as precisely as possible. In this respect, according to the guidelines of the German Committee for Structural Concrete [43] and ZTV-W LB 219 [233], a distinction should be drawn between measures aimed at protecting and repairing the concrete itself, and measures designed to restore or maintain the corrosion protection of the reinforcement. When filling cracks, it should be made clear whether a closing or sealing of the crack is to be achieved or whether the crack edges are to be connected together in order to transfer forces or allow expansion.

The possible effects of protective and repair measures on the durability and structural behaviour of a component or the entire structure must be examined. In particular, any changes that are unfavourable in terms of building physics as well as changes to the structural behaviour (increase in dead load, load redistribution, etc.), should be considered.

When drawing up the repair plan, the fundamentally different boundary conditions for reinforcement corrosion in components above or below water must be taken into account, primarily the different ways in supplying the oxygen necessary for corrosion.

The causes of cracking must be investigated along with the associated stresses or deformations likely in the future.

#### 10.12.3.2 Repair plan

An experienced engineer should prepare a repair plan for every repair measure. The plan should describe all the details relevant to the execution of the repair measures – from the preparation of the substrate, the types and grades of the materials to be used, the building methods to be selected and curing measures right up to quality assurance.

As far as possible, the repair plan should be drawn up on the basis of the German Committee for Structural Concrete guidelines [43] and ZTV W LB 219 [233].

The repair plan should contain details of the following at least:

1. Repair principles/basic solutions according to the German Committee for Structural Concrete guidelines [43] and ZTV-W LB 219 [233].

2. Requirements in terms of contractors/personnel, e.g.
   - proof of suitability for nozzle operators performing sprayed concrete work,
   - proof of suitability in terms of handling polymers used in concrete construction.
3. Substrate pre-treatment
   - the aim of substrate pre-treatment and the nature of the pre-treatment process,
   - the extent of concrete removal and/or exposure of reinforcement,
   - the degree of rust removal from the reinforcement.
4. Replacement of concrete
   - the nature and quality of the materials and processes to be used, e.g.
   concrete,
   sprayed concrete,
   injected mortar/sprayed concrete with polymer additive (SPCC),
   cement mortar/concrete with polymer additive (PCC),
   - formwork,
   - layer thicknesses,
   - additional reinforcement,
   - construction joints.
5. Cracks
   - filling materials,
   - types of filling.
6. Joints
   - preparatory work,
   - type of joint sealing material,
   - execution.
7. Surface protection systems
   - type of system,
   - layer thicknesses.
8. Curing
   - type,
   - duration.
9. Quality
   - basic inspections,
   - suitability tests,
   - quality control.

## 10.12.4 Execution of the repair works

### 10.12.4.1 General

The execution of repair works above water is described in detail in [231]. Prior to the application of cement-bonded concrete substitutes (concrete, sprayed concrete, SPCC, PCC), the concrete substrate should be sufficiently pre-wetted (initially 24 hours beforehand). However, prior to applying the concrete substitute the surfaces affected must have dried to the point that they are only slightly moist.

Adequate curing is crucial to the success of repair works. Cement-bonded concrete substitutes should be cured in the first few days after application by means of measures that aim to supply water. This applies in particular to thinner layers where the concrete is replaced with PCC. Given the different raw materials, differences in colour between the old concrete and the concrete repair must always be expected for local repairs with cement-bonded concrete substitutes.

The following advice does not generally apply to measures involving cathodic corrosion protection.

### 10.12.4.2 Substrate pre-treatment

1. General
   The aim of the substrate pre-treatment, rather than its nature, should be specified.
   Following completion of the substrate pre-treatment, tests should be carried out to establish whether the concrete substrate has the adhesive strength necessary for the proposed repair works.
2. Working above water
   To achieve a good bond, the concrete substrate must be uniformly firm and free from intrinsic or other separating substances. Loose, friable concrete and all foreign matter such as plants, shells, oil or paint residue must be removed. The concrete removal required in addition and the exposure of reinforcement in order to achieve the repair objective depends on the method selected in accordance with the German Committee for Structural Concrete guidelines [43] and ZTV-W LB 219 [233], and should be taken from the repair plan.
   Prior to the application of a cement-bonded concrete substitute, and after completion of the substrate pre-treatment, grains of aggregate with a diameter $\geq 4$ mm firmly embedded near the surface should generally be exposed as small peaks.
   Following completion of the substrate pre-treatment, loose rust on exposed reinforcement and, if applicable, exposed cast-in elements, must be removed. The degree of rust removal in the case of corrosion protection through restoration of the alkaline environment according to the German Committee for Structural Concrete guidelines [43] and ZTV-W LB 219 [233] must at least correspond to standard Sa 2, and for corrosion protection through coating of the reinforcement at least standard Sa 2½. Only high-pressure water-jetting ($\geq 600$ bar) is permitted when removing rust from reinforcement in the case of chloride-induced reinforcement corrosion.

The following methods of substrate pre-treatment are suitable depending on the intended purpose:
- chiselling,
- milling,
- grinding,
- blasting with solid blasting media or a water/sand mix or high-pressure water.

Any material removed during substrate pre-treatment and any mixtures created by the process must be disposed of in accordance with the applicable waste disposal legislation.

3. Working underwater

The advice given in section 10.12.4.2(2) applies accordingly. The following methods of substrate pre-treatment are suitable depending on the intended purpose:
- hydraulically powered cleaning equipment,
- underwater blasting with solid blasting media or high-pressure water.

### 10.12.4.3 Repairs using concrete

1. General

    Repairs with concrete are preferable – for both technical and economic reasons – where repairs cover large areas and involve thicker layers.

2. Working above water

    Repairs with concrete should be carried out on the basis of [233], which specifies particular requirements regarding the composition and characteristics of the concrete depending on the loads acting on the structural component.

3. Working underwater

    The work must be carried out in accordance with section 10.12.4.3(2). Proper placement and compaction of the concrete without segregation must be ensured by adding a suitable DIBt-approved stabiliser or by working to the provisions governing underwater concrete given in DIN EN 206 and DIN 1045-2.

### 10.12.4.4 Repairs using sprayed concrete

1. General

    Sprayed concrete has proved successful for the repair of concrete components used in hydraulic engineering when work is carried out above water. It is probably the most commonly used repair method.

2. Working above water

    Refer to section 10.12.4.3(2). ZTV-W LB 219 [233] specifies certain requirements regarding the composition of the mix and the properties of the finished sprayed concrete depending on the loads acting on the component.

In essence, ZTV-W LB 219 [233] makes a distinction between sprayed concrete with a layer thickness of up to about 5 cm, which can be applied without reinforcement, and sprayed concrete with a layer thickness >5 cm, which must be additionally reinforced and connected to the component using anchors.

The surface of the sprayed concrete must be left rough, as sprayed. If a smooth or specially textured surface is required, a mortar or sprayed mortar must be applied, and worked accordingly, in a separate operation once the sprayed concrete has hardened.

3. Working underwater
   Sprayed concrete cannot be applied underwater.

### 10.12.4.5 Repairs using sprayed polymer-modified concrete (SPCC)

1. General
   The use of SPCC can be advantageous, especially with thinner layers, since the polymer additives improve certain properties of the concrete, including the water retention capacity, adhesive strength or watertightness. In addition, the use of polymer additives means that a deformation behaviour similar to that of the old concrete can be achieved. Only polymer additives unaffected by moisture may be used.

2. Working above water
   Refer to section 10.12.4.3(2). The layer thickness of the SPCC should be between 2 and 5 cm when applied over an area. Additional reinforcement is unnecessary in layers of this thickness.
   The surface of the SPCC should be left rough, as sprayed. If a smooth or specially textured surface is required, then:
   - in the case of single-layer application, after the SPCC has hardened, apply a mortar compatible with the SPCC in a separate operation and then process accordingly.
   - in the case of multiple-layer application, the final layer must be processed accordingly.

3. Working underwater
   SPCC cannot be applied underwater.

### 10.12.4.6 Repairs using cement mortar/concrete with a polymer additive (PCC)

1. General
   PCC is particularly suitable for repairing small areas of damage. It is applied by hand or machine. However, in contrast to sprayed concrete or SPCC, compaction is by hand in both cases. Only polymer additives unaffected by moisture may be used.

2. Working above water
   Refer to section 10.12.4.3(2). The thickness of the layer of PCC can be up to about 10 cm for local repairs.
3. Working underwater
   Special products are available for applications underwater.

**10.12.4.7 Repairs using polymer concrete (PC)**

1. General
   Polymer concrete is almost impermeable to water vapour. Not least for this reason, the use of polymer concrete is limited to local repairs and to working underwater.
2. Working above water
   Polymer concrete should only be used in exceptional cases and only for local repairs [231].
3. Working underwater
   Special products are available for applications underwater.

**10.12.4.8 Encasing the concrete component**

1. General
   The damaged concrete component is encased in a watertight casing that is sufficiently resistant to the anticipated mechanical, chemical and biological attacks. The protective casing can be attached with or without being bonded to the component to be protected. The aim of the process is to prevent the ingress of water, oxygen or other substances between the casing and the component. This method can be used both above and below water.
2. Cleaning and pre-treating the substrate
   The work should be carried out in accordance with section 10.12.4.2(2) or 10.12.4.2(3).
3. Encasing the concrete with a precast concrete shell
   Requirements regarding the precast concrete element:
   - dense concrete free from capillary pores,
   - a high level of corrosion protection for the reinforcement, e.g. by means of a coating.
   
   The space between the shell and the pre-treated concrete is filled by injecting a low-shrinkage cement mortar with a high frost resistance.
4. Encasing the concrete with a prefabricated fibre-reinforced concrete shell
   Suitable fibres:
   - steel fibres,
   - alkali-resistant glass fibres.
   
   Requirements and execution in accordance with section 10.12.4.8(3).

5. Encasing the concrete on site with fibre-reinforced concrete – see section 10.12.4.8(4)
6. Encasing the concrete with a polymer shell – for columns
   Requirements regarding the polymer shell:
   - resistant to UV radiation (only above water),
   - resistant to the in situ water,
   - watertight and sufficiently diffusion-tight,
   - if necessary, with adequate mechanical resistance to the anticipated impacts, including ice loads, debris and vessel impacts.

   Filling between shell and pre-treated concrete as described in section 10.12.4.8(3).
7. Wrapping the concrete component in flexible sheeting – for columns
   Cleaning and pre-treatment of the substrate in accordance with section 10.12.4.2(2) or 10.12.4.2(3).
   Provision of corrosion protection for the reinforcement and filling the damaged areas as per sections 10.12.4.3 to 10.12.4.7, wrapping the columns in flexible sheeting.
   Requirements regarding the system:
   - resistant to UV radiation,
   - resistant to the in situ water,
   - watertight and gastight,
   - adequate mechanical resistance to the anticipated external factors, e.g. ice loads,
   - leakproof closures between the edges of the polymer sheeting and leakproof upper and lower connections to the column so neither liquid nor gaseous substances can infiltrate between sheeting and substrate.

### 10.12.4.9 Coating the concrete component

1. General
   As an additional measure, a coating can be applied to the cleaned, pre-treated component (which might also have been repaired beforehand with a concrete substitute) to prevent the infiltration of harmful substances into the concrete, especially chlorides and carbon dioxide in the case of reinforced and prestressed concrete structural components (if no casing as per section 10.12.4.8 is envisaged).
2. Working above water
   Refer to section 10.12.4.3(2).
   Coatings may only be used if the risk of moisture penetrating from the rear can be excluded.
3. Working underwater
   Special products are available for applications underwater.

**10.12.4.10 Filling cracks**

As far as possible, the work should be carried out on the basis of ZTV-ING [231]. Regarding the filling of cracks in hydraulic engineering concrete components with their frequently high degrees of saturation, cement pastes/suspensions have proved to be successful where loads have to be transferred across cracks, while polyurethane has proved itself where expansion is necessary.

**10.12.4.11 Preparing joints and joint seals**

- Clean the joints and, if necessary, widen the existing joint opening.
- Repair damaged edges with epoxy resin mortar, for example,
- Incorporate joint sealing in accordance with the relevant regulations and guidelines.

Joint closing and filling can be carried out in accordance with DIN 18540. As regards joints in trafficked areas, information sheet FGSV-820 [66] must be observed.

# 11 Pile bents and trestles

## 11.1 General

The horizontal head deformations of the pile bents or trestles (= interconnected pile bents) covered in the following sections should always be determined as well and checked for compatibility with serviceability requirements. For large changes in ground level, raking anchors can be used in addition to pile trestles. In this respect it is important to ensure that the anchor forces correspond to the deformation options of the structure.

Piles can be exposed to reversed loads (tension/compression) caused by line pull, lateral crane impacts, tidal effects, etc. The suitability of the chosen pile system should therefore be verified for such loads.

## 11.2 Calculating subsequently strengthened pile bents/trestles (R 45)

### 11.2.1 General

Pile bents/trestles with existing water- and/or land-side sheet pile walls often have to be strengthened by building an additional sheet pile wall driven in front of the old one and driven deeper in order to cope with greater water depths (see Fig. R 45-1). The new sheet pile wall is then loaded with the earth support reaction from the existing water-side sheet pile wall and often also loaded by soil stresses from the existing pile loads just below the deepened harbour bottom.

Even with new pile trestles, a water-side sheet pile wall can also be within the area of influence of pile loads (see R 78, section 11.3).

The loads acting on the sliding wedge of soil and the new sheet pile wall can only be determined approximately. The following information applies primarily to determining the internal forces in non-cohesive soil. To do this, it is presumed that the overall stability is verified in accordance with DIN 4084 and that the embedment depth of the new water-side sheet pile wall is determined as part of this analysis. DIN 4084 states that in such cases combined failure mechanisms with straight failure planes are suitable for verifying the overall stability.

---

*Recommendations of the Committee for Waterfront Structures Harbours and Waterways – EAU 2012,*
9th Edition. Issued by the Committee for Waterfront Structures of the German Port Technology Association and the German Geotechnical Society.
© 2015 Ernst & Sohn GmbH & Co. KG. Published 2015 by Ernst & Sohn GmbH & Co. KG.

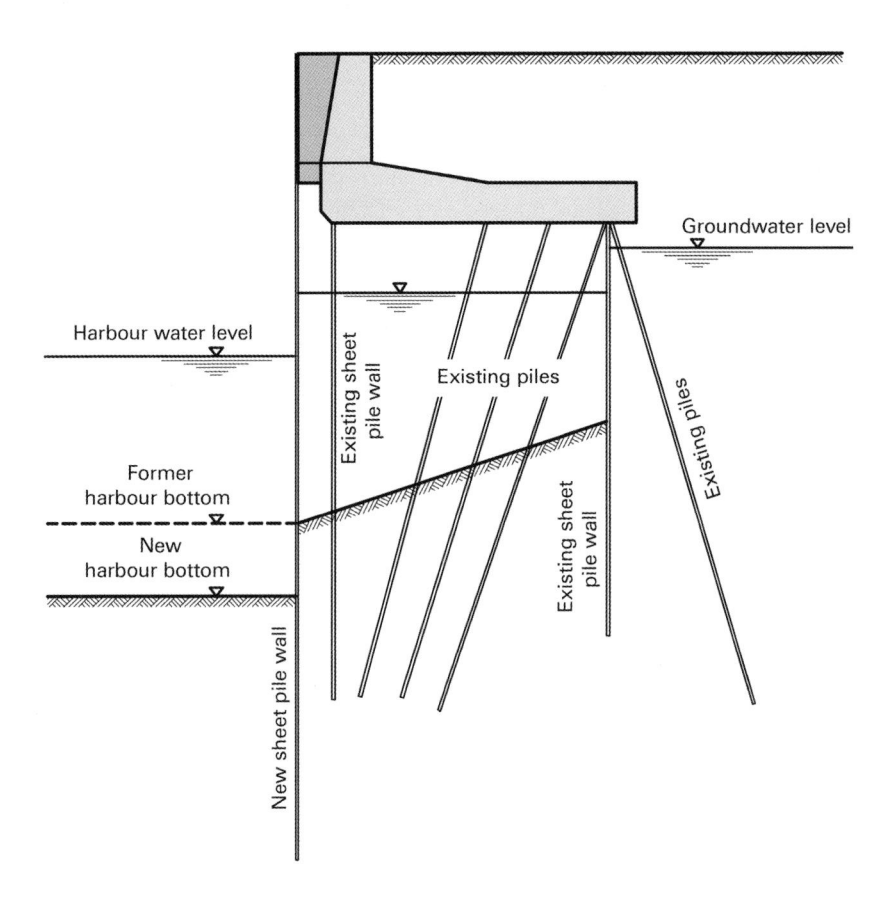

**Fig. R 45-1.** Example of a subsequently strengthened pile trestle

### 11.2.2 Loads

The active earth pressure on the new water-side sheet pile wall is affected by:

1. The active earth pressure from the soil behind the structure. It normally relates to the plane of any existing land-side sheet pile wall or to the vertical plane through the rear edge of the superstructure (land-side reference plane). It is calculated with straight failure planes for the existing height of the terrain and the surcharge. The active earth pressure friction angle should be estimated in accordance with DIN 4085.

2. The pile toe reaction force of any existing land-side sheet pile wall.
3. The flow force in the subsoil behind the existing water-side sheet pile wall, caused by the difference between the water table and the water level in the harbour.
4. The dead load of the masses of soil between the existing water-side sheet pile wall and the land-side reference plane in conjunction with the active earth pressure referred to in (1) above. Where there is a land-side sheet pile wall, this is the pile toe reaction force required for the equilibrium of the land-side sheet pile wall which is transferred into the soil between the two sheet pile walls.
5. The pile forces resulting from the vertical and horizontal loads from the superstructure. In order to calculate the pile forces, the upper support reaction forces of the existing water-side sheet pile wall must be included unless additional anchoring independent of the pile trestle is used.
6. The resistance of the ground between the water-side sheet pile wall and the land-side reference level as per (1) above where the structure moves in the direction of the harbour water.

### 11.2.3 Calculation for cohesive substrata

The procedure here is similar to the one for non-cohesive soil. For soils with cohesion $c'$, the cohesive force $C' = c' \cdot l$ along the particular failure plane being investigated is also taken into account. For virgin saturated soil, the value $c_u$ replaces $c'$, where $\varphi' = 0$.

### 11.2.4 Load from excess water pressure

The excess water pressure acting on the water-side sheet pile wall depends on the soil conditions, the depth of soil behind the wall and the presence of a drainage system as well as other factors. In the case of new structures with a water-side sheet pile wall only and soil extending up to the underside of the pile cap, the excess water pressure is assumed to act directly on the sheet pile wall according to R 19, section 4.2. Where the ground surface is below the outer water level for land-side sheet pile walls and the new wall has permanently effective openings for equalising the water level, the excess water pressure is again according to R 19, section 4.2. The flow force to be estimated in accordance with section 11.2.2(3) is the difference between the water level behind the wall and the outer water level. As a precaution, an excess water pressure from the outer water level plus half the height of the waves expected in the harbour acting directly on the water-side sheet pile wall should be assumed. As a rule, a water level difference of 0.5 m is sufficient as a characteristic value.

## 11.3 Design of plane pile bents (R 78)

There are two main types of plane pile bent for dealing with changes in ground level:

1. Exposed pile bents above an underwater slope with land- and water-side steel pile walls to deal with a change in ground level (Fig. R 78-1). To determine the largest bending moments and anchor forces in each case, s gradients below the slab of 1:4 as well as 1:10 should be analysed.

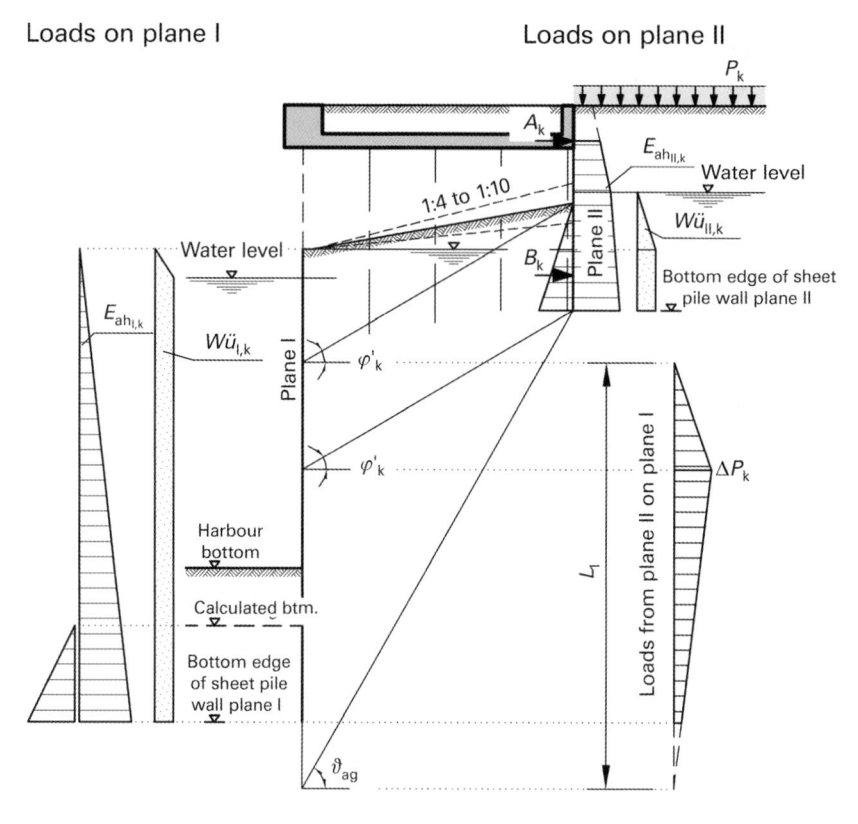

**Fig. R 78-1.** Pile bent with land- and water-side sheet pile walls

The load of the land-side sheet pile wall (plane II) on the water-side sheet pile wall (plane I) can be derived from the following equation:

$$B_k = E_{ah_{II,k}} + W\ddot{u}_{II,k} - A_k$$
$$\Delta P_k = \frac{2 \cdot B_k}{L_1}$$

The corresponding partial safety factors should be applied to dead loads, variable loads and water pressure.

2. A pile bent with water-side sheet pile wall . . .
   - either as a strengthening structure in front of or above the existing pile bent to increase the design depth, with an existing embankment generally remaining intact. The sheet pile wall should continue to a certain depth, but at least down to the base of the embankment, and
     – to prevent excess water pressure from building up behind it –
     should not be impermeable.
   - or as a structure for a new pile bent with a fully backfilled sheet pile wall. The superstructure slab supported on the piles shields the earth pressure on the sheet pile wall against surcharges (Fig. R 78-2).

For economic reasons, the relatively high compression pile loads below a superstructure slab are often carried by cast-in-place driven piles. Where they carry the loads primarily by way of end bearing, the actions from the pile loads on the quay wall can be neglected if the piles are installed at a rake of 1:2 or less as per Fig. R 78-2. This applies to both single pile bents as well as several rows of compression piles.

Should the pile loads be carried via surface friction and end bearing, they generate actions on the quay wall which must then be multiplied by partial safety factors. These, too, can be neglected if the centroid of the point of application of the pile forces is below a straight line sloping at 1:2 as per Fig. R 78-2.

The magnitude of the actions from the pile forces acting on the quay wall is calculated in accordance with DIN 4085 and depends, in particular, on the location of the piles in relation to the wall, the rake of the piles and the properties of the in situ subsoil.

When designing the compression piles, it should be remembered that they are subjected to lateral pressure due to deflection of the quay wall and thus also subjected to bending. The load results from the deformation of the wall as a result of backfilling or excavating, and the deformations decrease as the distance from the wall increases.

In this case the backfill soil – if sufficiently compactable – can be additionally compacted by the construction of the piles. Such compaction increases the strength of the backfill soil and the active earth pressure is correspondingly lower. However, during construction of

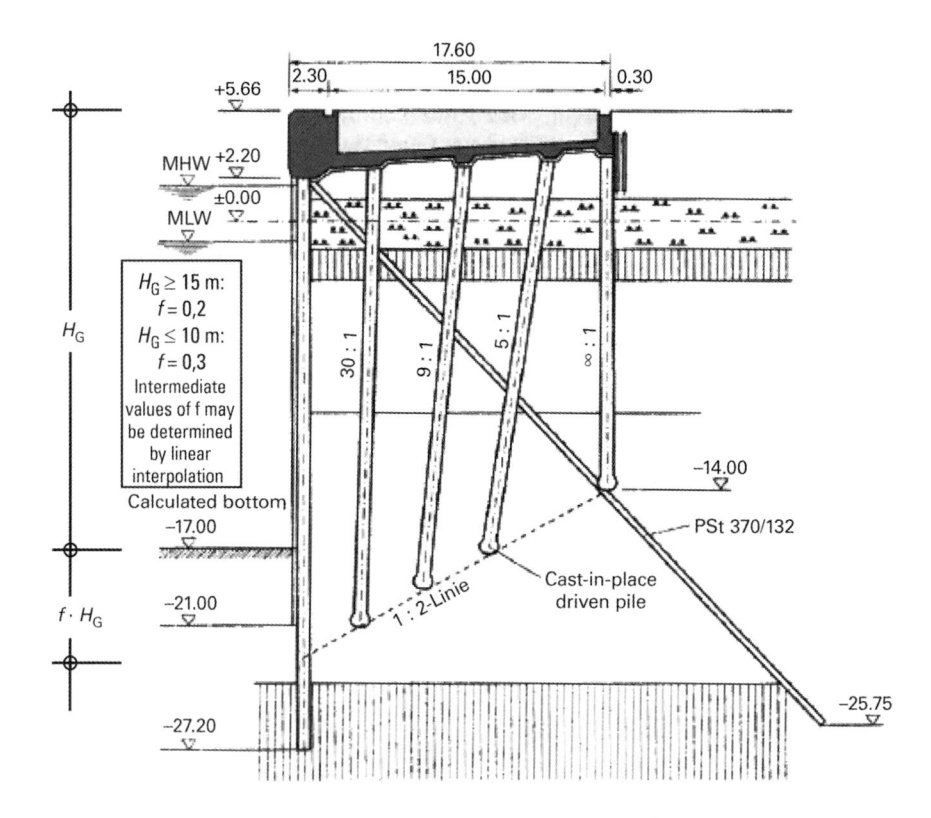

**Fig. R 78-2.** Example of a pile bent with a water-side sheet pile wall

the piles in the direct vicinity of the wall, an additional local compaction earth pressure acts, which can cause further deflection of the wall.

Pile bents are one-dimensional linear structures in the longitudinal direction and are thus planar structures for the purpose of structural design. Consequently, actions and resistances can be defined per linear meter or per system module length.

Concentrated actions from bollards and fenders can be distributed proportionately over the design cross-section within a section of the structure due to the plate effect of the pile cap. Where high loads acting at a point are carried by a group of piles for structural reasons, the loadbearing capacity of the group of piles should also be verified.

Determining the internal forces for piles and superstructure for the planar case is relatively straightforward:

The structural system of the slab strip with the piles placed in its plane can best be modelled by an elastically supported continuous beam. The piles are thus represented here by springs in the direction of the pile axis, and the associated spring stiffness is derived from the pile data. It must be guaranteed that the distance between the piles is large enough to avoid them influencing each other when carrying loads and that the interaction between pile cap and piles is negligible.

Where piles have a very high axial stiffness, e.g. cast-in-place driven piles, then a rigid support for the pile cap can be assumed as well.

As an equivalent to the elastically supported continuous beam, the system can be modelled as a plane frame consisting of several pile legs and a cross-member representing the superstructure.

Fixed or hinged supports for the piles in the soil and at the superstructure slab, lateral bedding (and hence bending of the pile) and axial bedding can be handled with standard 2D software.

Calculations for waterfront structures with a "rigid" superstructure are therefore only appropriate in special cases, e.g. where old quays with solid pile head blocks must be re-analysed.

In general, anchor forces should be transferred into the anchor elements via the shortest route. Provided the subsoil conditions are suitable, this design principle is best achieved with horizontal anchors fixed to anchor plates (or fixed to pile bents in the case of deeper loadbearing strata) or by anchoring with raking piles.

Special construction methods may be required in difficult driving conditions or where obstacles in the subsoil suggest that the planned driving of the quay wall and pile bents may not appear to be feasible. When verifying the anchorage of a sheet pile wall driven afterwards in front of an existing structure, it should be taken into account that the anchoring of the existing structure is also loaded by additional earth, water and line pull loads. How the loads are distributed between the old and new anchors should then be analysed. In order to keep the displacements small and to avoid overloading the existing anchors, the anchors for a sheet pile wall driven afterwards in front of an existing structure can be prestressed. The displacement of the anchor connection should be specified based on the local circumstances.

## 11.4 Design of spatial pile trestles (R 157)

A spatial pile trestle can be modelled as a three-dimensional frame on which a pile cap is supported as a superstructure. The loads acting on the pile cap are distributed to all the piles in the pile trestle through the loadbearing capacity of the superstructure acting as both plate (= in-plane forces) and slab (= out-of-plane forces) to achieve an effective load transfer.

### 11.4.1 Special structures designed as spatial pile trestles

Sharp bend and corner structures as well as moles and jetty heads in the form of pile trestles are special structures for which the foundation piles have to be positioned in accordance with the in situ site conditions. This is also the case where, for example, the layout of ro/ro ramps requires projections or different heights for the front edge of the quay.

In such cases it is normally no longer possible to arrange the piles in the plane of the pile bent. Instead, the piles must be installed with a large number of intersections and rakes intersecting in three dimensions. In addition, the connections to the superstructure can be at different heights.

### 11.4.2 Free-standing pile trestles

Tall, free-standing pile trestles with a superstructure slab are mostly used where particular boundary conditions dictate this. This can be where, for example, loadbearing subsoil is only found at a great depth. Another reason for choosing a free-standing pile trestle may be that the passage of waves should not be interrupted or where it should be ensured that waves are not reflected at vertical quay walls. And, indeed, free-standing pile trestles can also be very economic compared with other forms of construction.

Pile caps supported on free-standing pile trestles are generally made from reinforced concrete, whereas driven steel sections are normally chosen for the foundation piles. Foundation piles made from steel can also meet the requirements for icy conditions (Fig. R 157-1).

Some of the foundation piles should be arranged as pile trestles so that the load on the slab can be accommodated horizontally. This simultaneously reduces the bending moments in all piles where otherwise no further external horizontal actions with a large influence occur, e.g. lateral pressure on piles from flowing soil masses, strong currents, ice pressure, ice impact, etc.

Should the aforementioned assumptions apply, assuming that the piles are hinged at the slab and at their base is sufficiently accurate for the calculations, even if the piles are not actually constructed as such. In order to deal with any unwanted fixity, special construction details will be needed at the head of the pile. This may be necessary where large changes in the length of the superstructure slab occur due to temperature changes.

Displacements of the head of a pile due to shrinkage during construction can therefore be kept under control by constructing sections of the superstructure slab with wide contraction joints in between. These joints are closed off with structural concrete after the shrinkage has abated. Restraint stresses due to shrinkage are generally kept within safe limits by employing such joints.

The depth of the pile cap is determined by calculating the effective bending moments and shear forces at the supports on the foundation

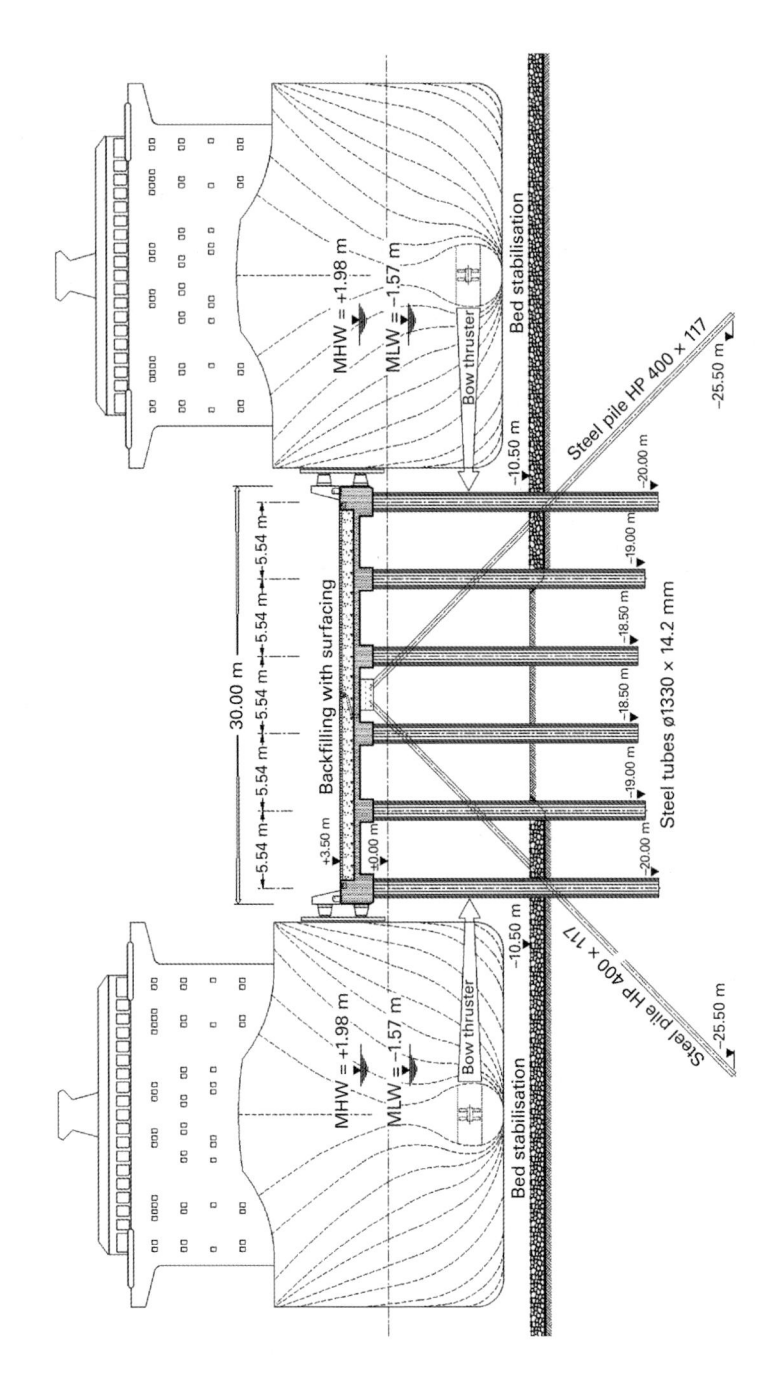

Fig. R 157-1. Example of a pier

piles. Such pile caps are normally only 50–75 cm thick, and so should be considered flexible with respect to the spring stiffness of the piles.

Transferring larger bending moments caused by loads on the pile cap can be avoided by positioning the piles to suit the load transfer. The actions applied at the respective loading position should largely be carried by the piles directly adjacent.

Pile caps should not be directly traversed by vehicles. Covering the slab with a sand fill and paving the surface to suit the traffic loads has operational and constructional benefits because pipes, cables, ducts, etc. can be accommodated within the sand layer and the loading on the slab and the piles due to localised actions are lower due to the load being spread within the sand compared with slabs traversed directly. At the same time, the substructure need not be designed for any dynamic actions caused by vehicles.

A sand fill with a thickness of about 1.0 m or more will in most cases provide adequate depth for all the services within the filling.

If the actions due to road and rail traffic are assumed to be uniformly distributed (basic situation 2 or 3 according to R 5, section 5.5), the depth of the sand fill must be at least 1.0 or 1.5 m respectively.

### 11.4.3 Structural system and calculations

A pile cap supported on a spatial pile trestle can be properly modelled as an elastic slab on elastic supports. From a structural point of view and with regard to the spread of the loads, the pile cap is an elastic plane frame that is supported on – likewise – elastic piles with or without a column head. According to R 78, section 11.3, each pile can be represented by an elastic spring acting in the direction of the pile axis and the pile lengths set as elastic lengths between the hinge points. Such structural systems are designed with the help of standard commercial software.

The wave pressure on the piles can be calculated in accordance with R 159, section 5.10. Loading on the pile cap from below due to "wave slamming" can also be assumed as set out in section 5.10.9.

The most beneficial arrangement of the piles below the pile cap is when all the support moments and the pile loads on one pile grid-line are roughly equal. However, this cannot be achieved in all situations due to the actions from cranes, line pull and vessel impact.

### 11.4.4 Construction guidance

The following are just some points that should be complied with to ensure that the pile trestles are as economical as possible:

- As far as possible, the berthing forces of large vessels should be absorbed completely by fenders plus heavy-duty fender panels

- positioned in front of the pile trestle. If necessary, the fenders can also be supported off the pile cap.
- In the vicinity of the fenders, a heavy-duty mooring bollard at the top of the wall can be combined with the fender structure.
- Local horizontal loads such as a line pull or vessel impact are distributed through the superstructure slab, which is very stiff in its plane, to all piles in a block.
- Fender piles should be specified for small waterborne traffic in order to protect the structural piles and the hulls of the vessels.
- Craneway beams can be included as structural components in the reinforced concrete pile cap.
- Vertical loads from crane operations can be carried by additional piles along the line of the crane if necessary.
- In order to minimise the effect on the bending moments, cranks in the pile cap slab should only be positioned above rows of piles.
- In tidal areas it is expedient to place the pile cap at a sufficient height above mean high tide in order to be independent from the normal tide levels during the construction of the pile trestle.
- When designing the formwork for the piled slab, the effects of waves should also be taken into account.
- Rows of vertical and raking piles should be arranged offset with respect to each other.
- If the pile cap is divided into blocks, these are generally connected to one another through horizontal joggle joints.
- Horizontal actions in the longitudinal direction are carried by pile bents in the middle of the block with raking piles as shallow as possible.
- Horizontal actions in the transverse direction are carried by pile bents on the longitudinal axis of the structure with raking piles as shallow as possible.
- Resisting all the horizontal actions in the manner described above minimises the stresses in the pile cap and the piles.
- Pile caps for large transhipment bridges can be concreted on sliding or moving formwork mainly supported on vertical piles.
- The fixity zones of steel foundation piles at the pile caps must be protected against corrosion, especially in the areas with a high risk of corrosion (saltwater, brackish water).

Further measures are required for parts that deviate from the standard cross-section and thus disrupt progress on site, e.g. connections for the aforementioned raking pile bents.

Where possible, raking piles should only be driven from the pile cap through dedicated driving openings after concreting and then connected to the pile bent with local reinforced concrete plugs in a second concreting operation.

The contraction joints referred to in section 11.4.2 can also be used as driving openings, which for this purpose might need to be widened locally. The stability of the blocks divided by joints must be verified. To ensure stability, temporary bracing across the openings and contraction joints may be required under certain circumstances.

## 11.5 Design of piled structures in earthquake zones (R 127)

### 11.5.1 General

When designing pile trestles for earthquake zones, it should be remembered that the superstructure – including fill materials, imposed loads and supported structures – is accelerated by the effects of the earthquake in such a way that additional, horizontal inertial forces arise which load the structure and its foundations. Therefore, in principle, the weight of pile trestles should be kept as small as possible in seismic regions. In the case of pile trestles with a slab to shield the earth pressure, it is necessary to check whether the benefit of such shielding could possibly be negated during an earthquake by the ensuing horizontal inertial forces due to the large mass of concrete at a relatively high level which is associated with this form of construction.

Please refer to R 124, section 2.16, for more information on how earthquakes can affect pile trestles, the permissible stresses and the safety measures required. Structures that are particularly tall and narrow must also be checked for the resonance that amplifies seismic amplitudes.

### 11.5.2 Active and passive earth pressures, excess water pressure, variable loads

The information in R 124 (sections 2.16.3, 2.16.4 and 2.16.5) applies accordingly. However, in the event of an earthquake it should be remembered that variable loads and soil dead loads assumed to act behind the slab should be applied at a shallower angle due to the horizontal earthquake acceleration and that the shielding is therefore less effective.

### 11.5.3 Resisting the horizontal inertial forces of the superstructure

The horizontal inertial forces due to an earthquake can act in any direction. At right-angles to the waterfront structure they can generally be readily accommodated by raking piles. Arranging additional pile bents in the longitudinal direction of the structure can, however, present problems in certain circumstances.

If the backfill soil to a water-side sheet pile wall extends up to the underside of the pile cap, it is advantageous to resist the horizontal loads acting in the longitudinal direction through the bending of the piles in the subsoil. However, this has to be mobilised by displacing the piles against the soil. To ensure their serviceability and for constructional reasons, the displacements should not be greater than approx. 3 cm and the bedding should be limited correspondingly.

Where superstructures are built over embankments, the earthquake load due to active earth pressure is significantly lower than for backfilled structures. Pile caps above such slopes should be designed to be as light as possible in order to minimise the horizontal inertia forces from earthquakes.

# 12 Protection and stabilisation structures

## 12.1 Embankment stabilisation on inland waterways (R 211)

### 12.1.1 General

In the presence of unsteady hydraulic loads, soil banks are permanently stable only at very shallow angles (1:8 to 1:15). Steeper slopes require armour that can ensure stability against hydraulic loads and adequate overall bank stability.

When choosing the slope angle, the technical benefits of a shallower bank should be compared with the disadvantage of the larger armoured area and the greater amount of land consumed. The construction and maintenance costs should therefore not be disproportionate to the economic and ecological benefits.

With greater hydraulic loads (lock entrances/exits, berths), the bottoms of waterways can also be protected by revetments. Where the bottom of a waterway comprises an impervious lining, a suitable revetment will be required to protect it.

The following conditions apply for the design:

- the angle of the embankment stabilisation should be as steep as possible without compromising stability, and
- it should, if possible, be able to install the armour with mechanical plant.

The following information refers primarily to inland waterways but in principle also applies in other situations. Further guidance on the design and construction of revetments can be found in the document published by the Federal Waterways Engineering & Research Institute [139].

### 12.1.2 Loads on inland waterways

The banks of man-made inland waterways (canals) are essentially subjected to hydraulic actions caused by waterborne traffic, and those of impounded or free-flowing rivers from natural currents as well.

The hydraulic loads caused by waterborne traffic can be divided into propeller wash, return current and the resulting drawdown alongside a vessel, a transverse stern wave with slope supply flow at the level of the ship's stern and the secondary wave system. Every action results in a separate load on the waterfront. The return current primarily affects the bank below still water level, while the transverse stern wave and the secondary wave system essentially affect the zone around still water level. The critical hydraulic actions for designing revetments on inland waterways are generally the transverse stern wave with the slope supply flow and the drawdown. Information about wave loads due to the inflow and outflow of water as well as waterborne traffic can be found in R 185, section 5.8, and R 186, section 5.9.

Waterfront revetments must be designed so that they can withstand the hydraulic shear and flow forces. Which loading component is critical for the design depends on the types of vessel expected (propulsion power and cross-section) and the cross-section of the waterway.

Water level differences arising from waterborne traffic and tides or those occurring naturally also lead to loads on the waterfront. A distinction must be made here between the upward water pressure below the revetment, which increases as the permeability of the revetment decreases, and the hydraulic gradient in the filter layers and the subsoil. With permeable revetments, the open water interacts with the groundwater. With a restricted navigable cross-section, the drop in the water table as a vessel passes takes place faster than the corresponding pressure drop in the pore water, but is dependent on the permeability of the subsoil. The result is an excess pore water pressure in the subsoil, the decrease in which in the direction of the surface can be represented with good approximation as an exponential function [120]. Excess pore water pressure in the soil reduces its shear strength and thus leads to a loss of stability that must be taken into account when designing the bank and its revetment. Soils with low permeability and no or only very little cohesion, e.g. silty fine sands, are particularly at risk.

### 12.1.3 Construction of bank protection

Revetments are the most common form of bank protection. The various components used, which depend on the specific requirements, are – from top (external) to bottom (internal) – as follows (see Fig. R 211-1):

- armour layer,
- filter/separating layer,
- impervious lining.

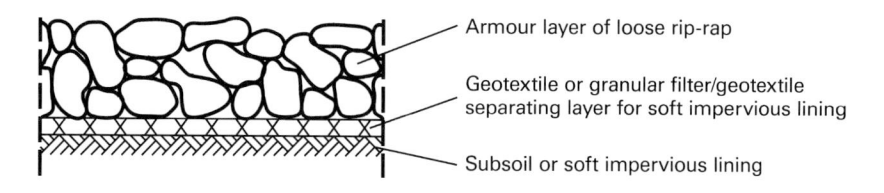

**Fig. R 211-1.** Construction of a revetment

In order to create a smooth subgrade, a levelling layer may also be needed below the revetment. Likewise, a cushioning layer can be used to protect individual components of the revetment, e.g. a impervious lining or geotextile filter, against loads.

Other parts of a revetment are the toe protection and, where applicable, connections to other elements.

### 12.1.3.1 Armour layer

The armour layer is the uppermost, erosion-resistant layer of bank protection measures and can be either permeable or impermeable. It is designed to resist hydraulic and geotechnical aspects: currents, wave attack and the drop in the water table must not cause displacement of parts of the construction or the subsoil, and pressure impacts must be absorbed without damage. At the same time, wave energy should be dissipated. The guidelines for scour protection given in R 83, section 12.4, also apply in principle to armour layers of bank protection.

The armour layer is often in the form of loose armourstones (rip-rap). The armourstones must be stable as well as light-resistant, and frost- and weather-resistant (DIN EN 13383-1). They should also have the highest possible density to ensure their stability over the long term. This type of construction is very flexible and easily adapts to deformations in the subsoil. Rip-rap is relatively simple to install and also easy to repair in the event of damage. The stones are mostly natural stone but can also be made industrially during metal production. For loose armour layers, hydraulic actions would normally cause the stones to move, but this can be kept to a minimum through suitable design. Fundamental research backed up by experiments has led to the development of design approaches for such revetments in recent years [158], [159], [71].

In order to increase the resistance to hydraulic loading, rip-rap can be either partially or fully grouted with a hydraulically bonded mortar. This form of construction is recommended for heavily loaded areas, e.g. manoeuvring areas such as berths or outer basins in particular. Partial grouting must ensure that the armour layer retains sufficient flexibility and permeability. Ideally, partial grouting produces conglomerates that are securely interlocked with each other and possess the necessary adaptability and erosion resistance of large individual stones. Owing to the high hydraulic resistance of partially grouted armour layers, it is in some cases possible to use much smaller armourstones than would be the case with loose rip-rap.

A similar effect can be achieved using precast concrete blocks laid like paving provided the armour layer is sufficiently permeable and the blocks are linked together. The weight per unit area that can be achieved with concrete blocks is, however, limited.

Impermeable (fully grouted) armour layers act both as lining and protective layer, and exhibit better resistance to erosion and other mechanical damage. In general, they can be thinner than permeable revetments; however, they are inflexible and ecologically controversial.

### 12.1.3.2 Filter

In principle, any bank protection scheme must be built to prevent materials being washed out from the subsoil (winnowing). To this end, filter layers should be placed between the subsoil and the armour layer, as required, and must have mutually compatible properties. All layers must constitute a stable filter with respect to the adjacent layer. Where a special clay lining is integrated into the revetment, a geotextile or mineral filter should be laid between the armour layer and the lining as a separating layer without filter function (Fig. R 211-1).

The filter should be designed according to geohydraulic aspects (i.e. pore water flows and their interaction with the granular structure). Besides granular filters, geotextile filters are also suitable (R 189, section 12.3). Water permeability and filter stability (filter and separating functions) are typical design requirements for both types of construction.

Owing to the turbulent inflows and alternating throughflows, both granular and geotextile filters in bank and bottom stabilisation are subjected to high unsteady hydraulic loads in contrast to their use in drainage, where they are exposed to flows in one direction only. The properties of the filters should therefore match with the subsoil, the armour layer and the loading conditions.

### 12.1.3.3 Impervious lining

Impervious lining is required for reasons of water management or – in the case of embankment dams – to improve stability. In principle, a distinction should be made between watertight revetments (surfacing or full grouting with asphalt or hydraulically bonded mortar) that act as both lining and protection and those revetments with a separate impervious lining such as natural clay, earth composites or geosynthetic clay liners (GCL, bentonite mat). R 204, section 12.6, contains information on mineral linings.

Unlike a permeable revetment, an excess water pressure can act on the underside of a watertight revetment, which should be taken into account when designing the revetment. The magnitude of the excess water pressure depends on the magnitude of the changes to the water level on the bank and the simultaneous groundwater levels behind the revetment. The excess water pressure reduces the potential friction force between the revetment and the soil below it.

Asphalt is a viscous material. Roots and rhizomes can thus penetrate revetments with asphalt grouting and the revetment can creep. Revetments made from asphalt or those with full asphalt grouting should be regarded as rigid solutions if the not insignificant ground deformations below the revetment caused by erosion, suffosion or subsidence take place more quickly than the creep process in the asphalt, which proceeds very slowly. Detailed guidelines on asphalt linings can be found in [56].

It is a requirement for all load cases that the self-weight component of the revetment perpendicular to the bank should always be greater than the maximum water pressure arising directly underneath it so that the lining is never subjected to uplift.

### 12.1.3.4 Levelling layer

Levelling layers are used to create a smooth subgrade where this cannot be achieved through excavation.

A levelling layer can be designed as a filter layer where there is very inhomogeneous soil underneath. In this case the particle distribution of the levelling layer is adjusted to the coarser in situ soil and acts as a filter with respect to the finer material. This solution avoids having to construct filters differently in different areas. A levelling layer is also an alternative to soil replacement, e.g. where an embankment cannot be shaped properly because the in situ soil is unstable.

### 12.1.3.5 Cushioning layer

In certain cases, cushioning layers are used to protect against very heavy loads on the underlying layers (e.g. particularly large rip-rap stones on a geotextile filter). A cushioning layer must comply with the filter criteria of the adjacent layers.

## 12.1.4 Toe protection

On steep banks the downslope forces from the bank protection cannot be fully resisted by the friction between revetment and subsoil. The portion of the downslope forces exceeding the friction must be resisted by toe protection. The requirement to resist the downslope forces entirely through friction would lead to revetments that are too thick or the inclination of the bank is too shallow, both of which are uneconomical solutions.

In the case of slopes that continue down to the bottom of the watercourse, the revetment normally continues at the same angle into the subsoil (embedded toe, Fig. R 211-2a). For subsoil at risk of erosion (sands and non-cohesive sand/silt mixtures), the embedment depth should not be less than 1.5 m [139], [71]. Alternatively, the revetment can also continue horizontally across the bottom (toe apron, Fig. R 211-2b). However, such a toe apron should not be used if there are soils in the

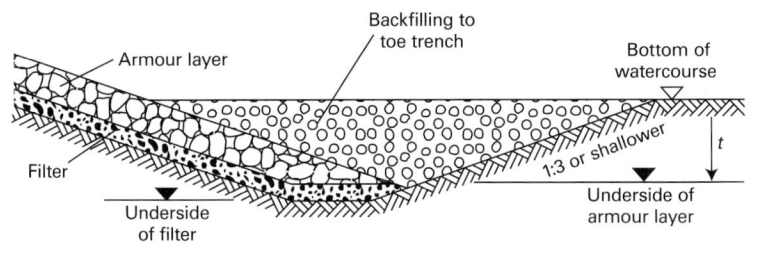

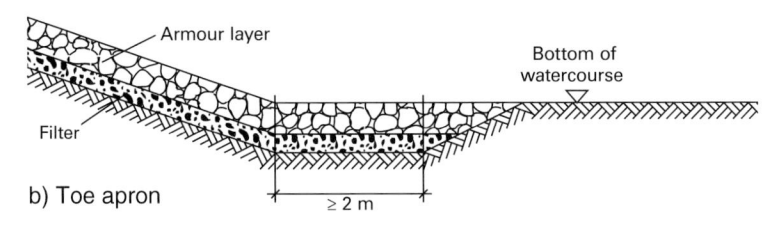

**Fig. R 211-2.** Toe protection by means of embedded toe or toe apron (to BAW 2008 [139])

bottom at risk of erosion (sands and finer materials). Where there is a revetment on the bottom as well, the revetment is supported by this.

The toe of a revetment can also be supported by a sheet pile wall at the toe of the bank, provided the soil is suitable for driving. In terms of its design and construction, this solution corresponds to supporting revetments on partial slopes behind sheet piling (R 106, section 6.4, and R 119, section 6.5).

### 12.1.5 Junctions

Junctions with structures and covering materials or the subsoil require particular attention. In practice, many cases of damage can be traced back to design and/or construction errors at these junctions. When connecting a revetment to a sheet pile wall or other component, care should be taken to ensure that there is a good transfer of forces and stable filter action, and that the joint is protected against erosion.

It can often make sense to specify full grouting for a strip of the loose riprap 0.5–1.0 m wide at the junction with a rigid structure and partial grouting adjacent to this with a decreasing quantity of grout.

The connections between filter layers or impervious linings and structures should also be designed and built with particular care and attention.

## 12.1.6 Design of revetments

The following aspects must be considered when designing revetments:

- Stability of individual stones with respect to hydraulic attack
  The size of stone required is essentially determined by the height of the transverse stern wave and the velocities of the return current and the slope supply flow as well as – disproportionately – by the density of the stone.
- Sufficient revetment weight to avoid bank failure where there is a rapid drop in water level
  The weight of revetment required is provided by the thickness of the armour layer. The key influencing factors are the density of the stone, the magnitude and speed of the drop in water level, the type of filter and the permeability of the subsoil.

For inland waterways the design can be carried out in accordance with [71]. Under certain boundary conditions, standard forms of construction for revetments according to [139] can be applied without the need for numerical verification. Stones of class $LMB_{5/40}$ with a minimum density of $2650 \, kg/m^3$ are recommended for class V waterways with modern waterborne traffic. Typical armour layer thicknesses lie between 60 and 80 cm and depend on the subsoil and the type of filter.

## 12.2 Slopes in seaports and tidal inland ports (R 107)

### 12.2.1 General

The waterfronts in ports where bulk cargoes are handled, at berths and at port/harbour entrances and turning basins can be constructed as permanently stable sloped banks even where the tidal range is large and the water level is subjected to wide fluctuations. Certain design principles should be followed to avoid the need for more extensive maintenance. Large sea-going vessels generally enter ports under their own power with tug assistance. Embankments can suffer considerable damage from propeller wash during berthing and deberthing. In addition, the propeller wash as well as the bow and stern waves of large tugboats, inland vessels, small sea-going ships and coastal vessels can also attack the slope over a depth of about 6–7 m below the respective water level (R 83, section 12.4). Bow and stern thrusters or Azipod propulsion units (azimuth thrusters) cause certain actions that require specially tailored solutions in each case (e.g. ferry ports).

#### 12.2.1.1 Examples of permeable revetments

Fig. R 107-1 shows a solution employed in Bremen.
The transition from the protected to the unprotected area of the bank is in the form of a 3.00 m wide horizontal berm covered with rip-rap. Above

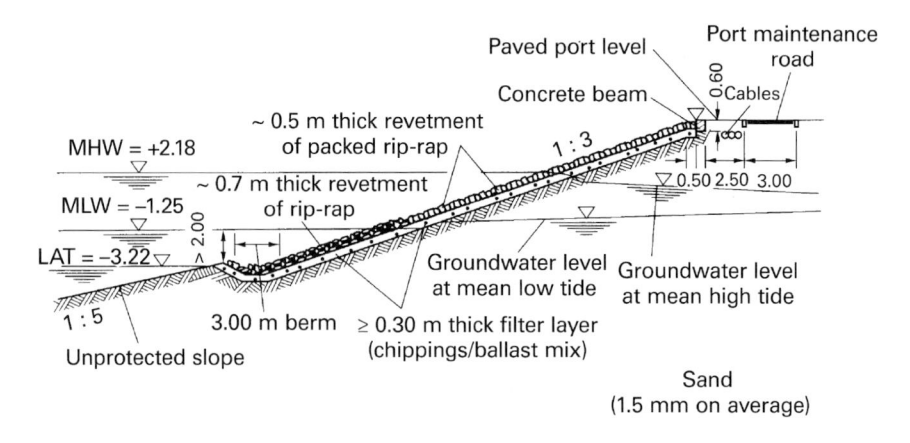

**Fig. R 107-1.** Port embankment in Bremen with permeable revetment (example)

this berm the stone revetment has a gradient of 1:3. A concrete beam, 0.5 m wide x 0.6 m deep, located on the level of the port operations area, serves as the upper boundary to the revetment. The revetment consists of heavy rip-rap dumped in a layer approx. 0.7 m thick just above mean low tide level. On the slope above this, the armour layer consists of an approx. 0.5 m thick layer of rip-rap tightly packed together. During laying, it should be ensured that the stones are properly fixed and that there is sufficient mutual support between stones to prevent individual stones being washed away by waves.

For the maintenance of the revetment and to provide access to berths, a 3.00 m wide maintenance road suitable for heavy vehicles should be included 2.50 m behind the upper concrete beam (see Fig. R 107-1). Electricity cables for the port facilities and the port's navigation lights as well as telephone cables, etc. are laid in a strip between the concrete beam and the maintenance road.

Fig. R 107-2 shows a section through an embankment in the Port of Hamburg. In this solution and abutment made from brick rubble, approx. 3.5 $m^3$/m, is built on top of geotextile at the toe of the revetment. Above this a uniform double-layer revetment covers the majority of the slope. Taking into account the in situ soil conditions, bank protection generally only continues down to 0.7 m below mean tide low water level. Where washout occurs below this level, this is easily rectified by dumping further brick rubble. In this form of construction the rip-rap can be torn away when ice forms, although the cost for supplementing the rip-rap is seen as relatively low in Hamburg.

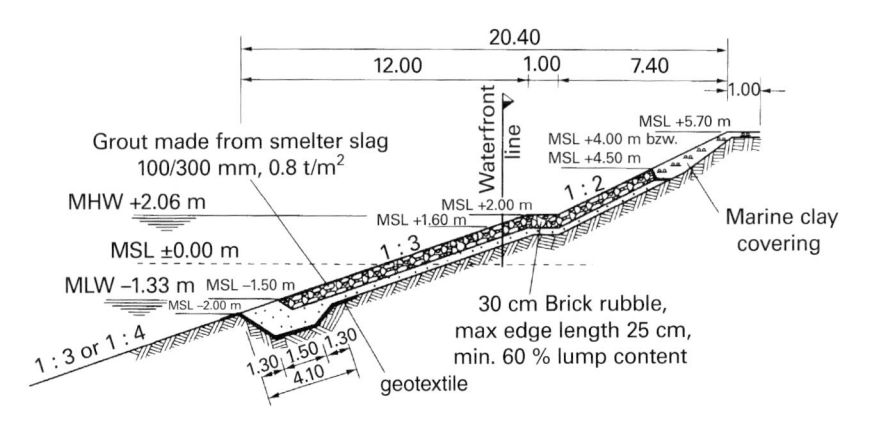

**Fig. R 107-2.** Port embankment in Hamburg with permeable revetment (example)

In order that banks blend into the natural landscape as much as possible, rip-rap covered banks in Hamburg include pockets for planting (Fig. R 107-3). In the standard cross-section as per Fig. R 107-2, a horizontal strip approx. 8.00–12.00 m wide, depending on the space available, is included in the region of AMSL +0.4 m. This strip is filled with marine clay 0.4–0.5 m thick to create a zone where vegetation can grow. The brick rubble substructure is thickened to 0.5 m and between the marine clay and the brick rubble there is a 0.15 m thick layer of Elbe sand as a ventilation zone.

Bulrushes (*Schoenoplectus tabernaemontani*), sedges (*Carex gracilis*) and reeds (*Phragmitis australis*) are planted in this pocket, staggered

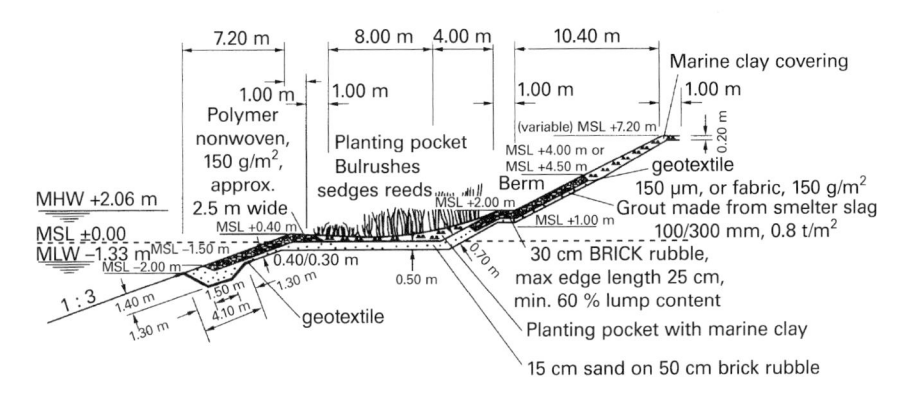

**Fig. R 107-3.** Port embankment in Hamburg with permeable revetment and planting pocket (example)

549

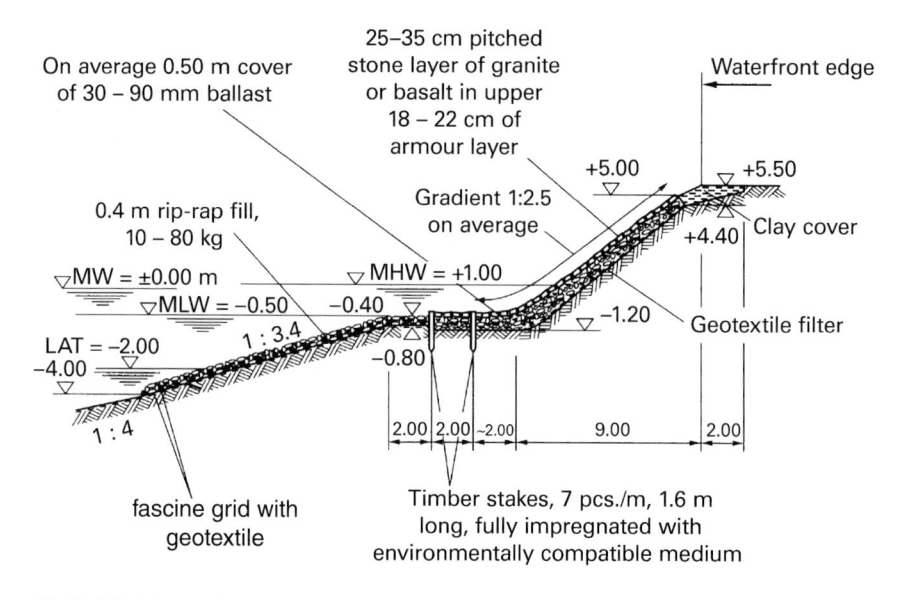

**Fig. R 107-4.** Port embankment with permeable revetment in Rotterdam (example)

according to the steepness of the location. Willow cuttings are used above the berm at AMSL +2.0 m, i.e. also within the standard rip-rap revetment. Planting (in NW Europe) should take place in April/May due to the better growing conditions.

However, the special shape of the planting pocket can only be used on banks with sufficient space and little surge from waterborne traffic and wave impacts.

Fig. R 107-4 shows a solution with a permeable revetment in the Port of Rotterdam. Apart from the revetment itself, its construction is largely similar to the Rotterdam solution with impermeable revetment. Further design details can therefore be found in section 12.2.3. Fig. R 107-5 shows another solution for a permeable port embankment.

### 12.2.2 Examples of impermeable revetments

Fig. R 107-6 shows a revetment developed and tested in Rotterdam. It has an "open toe" to reduce the excess water pressure. This toe consists of coarse gravel fill $d_{50} \geq 30$ mm secured by two rows of closely spaced timber stakes (2.00 m long x 0.2 m thick) that are fully impregnated with an environmentally compatible medium. At the bottom end of the asphalt-grouted quarry stone covering, the coarse gravel is then covered by a permeable layer of 25–30 cm of large granite or basalt stones.

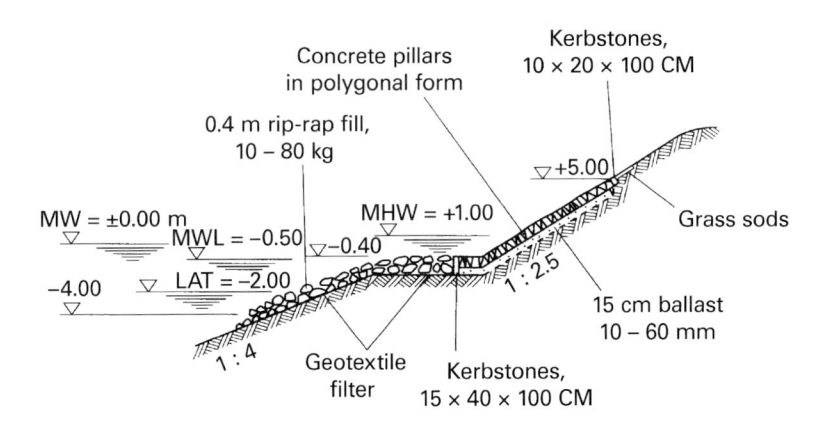

**Fig. R 107-5.** Port embankment with permeable revetment in Rotterdam (example)

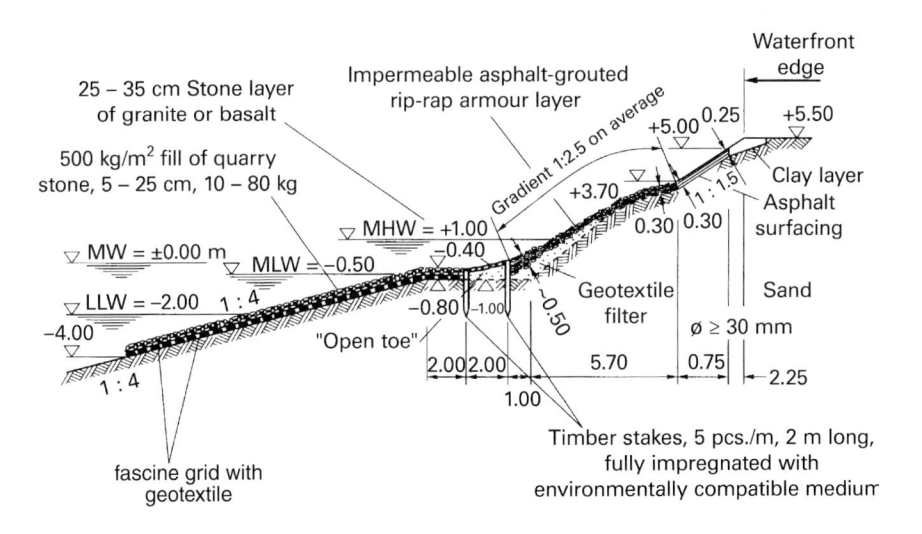

**Fig. R 107-6.** Port embankment with impermeable revetment in Rotterdam (example)

Under this layer of coarse gravel, which also extends below a significant part of the impermeable armour layer, there is a geotextile filter. There is a 2.0 m wide berm next to the "open toe", below that underwater stabilisation in the form of a timber mattress with a – in sand – 1:4 gradient down to about 3.5 m below mean low tide level. The bundles in the fascine grid are laid horizontally and in the direction of the embankment. On top of

551

that there is a 0.30–0.50 m deep fill of quarry stones because the load from the armour layer is intended to be about 3–5 kN/m$^2$.

The asphalt-grouted quarry stone revetment extends from the land-side row of timber stakes up to about 3.7 m above mean low tide level and has an average gradient of 1:2.5. Its weight per unit area (thickness) must be designed for each situation depending on the magnitude of the critical water pressure acting on its underside. In the standard case the thickness reduces from bottom to top from approx. 0.5 m to approx. 0.3 m.

The individual quarry stones weigh between 10 and 80 kg.

Adjoining the revetment at a level of 1.3 m is a 0.25–0.30 m thick asphalt surfacing layer with a gradient of 1:1.5 and above this at the same gradient is a clay cover at a level of 0.5 m. This is designed to allow pipes and cables to be subsequently laid without having to disturb the bank protection.

## 12.3 Use of geotextile filters in bank and bottom protection (R 189)

### 12.3.1 General

Geotextiles in the form of wovens, nonwovens and composites are used for bank and bottom protection.

To date the following polymers have proved to be effective rot-resistant materials for geotextile filters: acrylic, polyamide, polyester, polyvinyl alcohol, polyethylene and polypropylene. Details of the properties of these materials can be found in [158].

Where geosynthetics are to be used in bank and bottom protection, their properties must comply with DIN EN 13253. Threshold values for these properties depend on the specific application. Examples of threshold values can be found in [159] and [209].

The benefit of geotextile filters over mineral filters is the factory prefabrication, which results in very consistent properties. Geotextile filters can also be used underwater as long as certain installation rules and product requirements are met. The geotextile filters themselves weigh very little. A thicker armour layer can therefore sometimes be required in comparison to mineral filters. For non-cohesive fine-grained soils, the actions of waves in the zone between high and low water levels and below can give rise to a risk of liquefaction and a shifting of the soil below the revetment. To prevent this, the filter must meet geometrical filter criteria and the revetment must be sufficiently heavy [136].

### 12.3.2 Design principles

Geotextile filters in bank and bottom protection can be designed in terms of the mechanical and hydraulic effectiveness of the filter, the in situ loads such as punching and tensile forces and their durability with respect to abrasion in non-bonded armour layers in accordance with the rules set out in [158], [136], [53]. PIANC [58] and DVWK [53] contain design rules for unsteady hydraulic loads which are based on past experience with steady loads. MAG [136] contains design rules based on throughflow tests ("soil-type method") which are designed for unsteady hydraulic loads. Both methods are essentially based on German domestic experience. International experience and design principles can be found, for example, in [216], [117], and [37].

Besides the mechanical and hydraulic filter effectiveness, geotextile filters must be mainly designed to resist in situ loads. In this respect, relatively thick ($d \geq 4.5$ mm) or heavy ($g \geq 500$ g/m$^2$) geotextiles have proved effective for installation underwater while waterborne traffic operations continue.

### 12.3.3 Requirements

The tensile strength of the geotextile filters at breaking point should be at least 1200 N/10 cm in the longitudinal and transverse directions when laid wet.

For armour layers made from dumped stone material, the perforation resistance should be verified [185].

If abrasive wear movements of the armour layer stones can occur as a result of wave or current loads, the abrasive resistance of the geotextile should be verified [185].

### 12.3.4 Additional measures

Where required, the properties of the geotextile can be improved through the use of additional measures. Examples of such are set out below.

By including coarser additional layers on the underside of the geotextile adjusted to the subgrade grain sizes, it is possible to achieve an interlock with the subsoil, thus stabilising the boundary layer between subsoil and filter. However, in most cases it is better to increase the load on the geotextile filter, especially as such additional layers can also give rise to negative effects (e.g. lifting of the geotextile, unintended drainage effect). A fascine grid on the top of a geotextile can prevent creases forming in the geotextile during installation and enhance the stability of the fill material on the geotextile. Such grids have long been successfully used in the construction of fascine mattresses. A combination of woven fabric and needle-punched nonwovens can be used to increase the friction between the woven geotextile and the subsoil and to improve the filter action.

A factory-produced mineral fill consisting of sand or other granulates ("sand mat") increases the weight per unit area of the geotextile, which improves stability during installation. Furthermore, the filling reduces the risk of creases during laying.

### 12.3.5 General installation guidelines

Prior to laying geotextile filters, it is important to verify that the geotextile supplied complies with the contract and the relevant terms of supply, e.g. [185] and [209]. Once on site, the geotextile should be carefully stored and protected against UV radiation, the weather and other detrimental effects.

In order to prevent functional defects, care should be taken to ensure that multi-layer geotextile filters (composites) with filter layers graded according to pores are laid the right way up (top and underside are different).

Geotextiles must be laid without creases or folds in order to avoid creating any water channels and thus the possibility of soil particle migration.

Nailing to the subsoil at the top of the embankment is only permitted when this does not cause any restraint stresses in the geotextile as construction progresses. A better alternative to rigid fixing by nailing is to embed the geotextile filter in a trench at the top of the embankment. This allows the geotextile to give way in a controlled manner if there are any high loads during subsequent stages of construction. As geotextiles float during wet installation, they must be held in position by installing either the armour layer or a cushioning layer on top directly after being laid. Geotextiles should not be laid at temperatures below +5 °C.

Careful stitching or overlapping when joining together individual sheets of geotextile filter material is particularly important for their soil retention capacity. With stitching, the strength of the stitches should meet the required minimum strength for that particular geotextile. When installing in dry conditions on an embankment at a gradient of 1:3 or shallower, the planned overlaps must be at least 0.5 m. In wet conditions and for all steeper embankments, the overlap should be at least 1 m wide. Where the subsoil is soft, a check should be carried out to determine whether larger overlaps are required so that any movement of the geotextile sheets does not result in uncovered areas when installing rip-rap.

In principle, site stitching and overlapping should always follow the slope of the bank. If in exceptional cases overlapping in the longitudinal direction cannot be avoided, the sheet further down the slope must overlap the sheet further up in order to avoid downslope erosion of the bank through the overlapping.

When laying above water, care should be taken to avoid the relatively light geotextile from becoming displaced by wind.

The following points should be taken into consideration when installing geotextile filters underwater in areas with waterborne traffic in order to ensure that the geotextile filter is laid on the subgrade with sufficient overlap and without creases, folds, gaps or distortion:

- Warning signs must be set up to indicate the construction site in such a way that all vessels are warned to pass by at slow speed only.
- The subgrade must be carefully prepared and cleared of all stones.
- The installation plant must be positioned so that laying is not impaired by currents and the drawdown of passing vessels and so that the geotextile is not subjected to any inadmissible forces (mounted on stilts is preferable for installation).
- The risk of the geotextile sheets floating should be countered by using appropriate installation methods. It is advantageous to press the geotextile onto the subsoil when laying. There should only be a short interval between laying the geotextile and dumping rip-rap, and the dumping height should be kept small.
- Mechanisms for fixing the geotextile sheets to the installation plant must be released upon dumping rip-rap.
- Installing rip-rap on slopes with geotextiles must proceed from bottom to top.
- Laying underwater is only permitted when the contractor can prove that all requirements can be met.
- Diver inspections are essential.

## 12.4 Scour and protection against scour in front of waterfront structures (R 83)

### 12.4.1 General

Waterfront structures, especially vertical ones, divert and concentrate currents, which can result in bed material being eroded, a phenomenon known as "scouring". Its causes are essentially two-fold:

1. Natural currents carrying material away from the base of the waterfront structure. This occurs, for example, around pierheads at entrances to seaports and inland ports, which are subjected to strong cross-currents, or on the outer banks of river bends, where port facilities are often located due to the greater depth of water.
2. Wash from ship propellers and other manoeuvring aids such as bow thrusters or tugboats. This carries material away from the bed of the watercourse in front of the (in most cases) vertical waterfront structure.

Scouring is a process, which means that scour does not appear immediately, but develops over time, as the current carries away a certain amount of subsoil from the bed only gradually. This is particularly important for scour caused by vessel manoeuvres because the flow forces they cause are high but act for a short time only, whereas natural currents, while often relatively small forces, are long-term, ongoing effects.

The two causes of scour can be superimposed. However, it is often also the case that natural sedimentation processes can cause the mooring basins in front of quay facilities to silt up, e.g. in basins with no through-current. The localised scouring resulting from vessels berthing and deberthing as well as vessels moored at a berth once again carries off the accretions and thus reduces maintenance requirements.

When designing waterfront structures, it is necessary to resolve the question of whether scouring is likely in that particular case. Experience of other waterfront structures in a similar location can be very helpful when making an assessment. In addition, the following boundary conditions should also be given consideration:

- Features of the particular body of water, e.g. the strength of the natural current along the waterfront structure and the concentration of sediments in the water.
- The type and properties of the in situ soil on the bed. Non-cohesive, fine-grained soils are particularly prone to scour, whereas cohesive soils with a semi-firm to firm consistency are mostly resistant to erosion and hence less prone to scouring.
- The manner in which vessels berth and deberth. How are manoeuvring aids such as tugs or bow thrusters used, but also the vessel's main engines? Do strong currents or winds making berthing or deberthing more difficult?
- Type of vessel: ro-ro ships and ferries generally always dock at the same place and so cause scouring at the same place. Container quays are subjected to different conditions; vessels varying significantly in size dock at different locations each time so the effects from different berthing and deberthing procedures are superimposed. This can cause the removal of soil from new scouring to fill older scouring, partially or completely.

The risk of scouring can be counteracted either by a deepening the basin or watercourse or by protecting the bottom.

**12.4.2 Choosing a greater design depth (allowance for scouring)**

Opting for a greater design depth (allowance for scouring) ensures that scour does not endanger the stability of the waterfront structure down to the depth of the scouring allowance. In conjunction with regular soundings, the allowance for scouring allows the effects on the bed associated with berthing and deberthing procedures to be monitored without endangering the stability of the facility. Decisions can then be made on the basis of this experience as to whether the effects of port operations make it necessary to protect the bed in the long term.

The size of the allowance for scouring depends on the local conditions. Experience of facilities with similar environmental conditions is the best guide in this respect.

Further indications as to the size of the scouring allowance can also be gained by estimating the maximum likely scouring depth expected with particular types of vessel or particular currents. However, the empirical approaches available should be applied with care, and the calculated scouring depths can vary significantly. Drewes et al. [51] describe model experiments that allow an estimate of the scouring depth arising solely from vessel manoeuvres.

Where the allowance for scouring exceeds 10–20 % of the height of the waterfront structure, the recommendation is to verify whether it is more economical to cover the bed in front of the waterfront structure to protect against scour.

**12.4.3 Covering the bottom (scour protection)**

To protect the bottom of a watercourse against scour, the following measures should be considered:

1. Covering the bottom with loose stone fill
2. Covering the bottom with a grouted (stable) stone fill
3. Covering the bottom with a flexible composite system
4. Underwater concrete bottom (e.g. in ferry berths)
5. Designing the quay wall to deflect wash, including covering the bottom where applicable

1. Covering the bottom with loose stone filling
   Loose stone fill (natural stones and waste material such as slag) is one of the mostly commonly used protection systems. The requirements for this solution are as follows:
   – Adequate stability against damage caused by propeller wash properly covering the subsoil, i.e. two or three layers of stones
   – A filter-type installation, i.e. on a granular or geotextile filter that is tailored to the particular subsoil, see [136] and [138]
   – Undercurrent- and hence erosion-proof connection to the waterfront structure, especially in the case of sheet pile quay walls
   R 211, section 12.1, contains further guidance on design.
   The thickness of the layer of stones, or the weight of an individual stone, depends on the current loads and/or velocities at the bottom.

For currents induced by propeller wash, section 12.4.4 provides a method of calculating the current velocity near the bed. The formula for this is shown below.

The mean stone diameter required for loose stone fill according to Römisch [180] is as follows:

$$d_{reqd} \geq \frac{v_{btm}^2}{B^2 \cdot g \cdot \Delta'}$$

$d_{reqd}$ = mean stone diameter required for protection [m] (armour layer)
$v_{bed}$ = bed velocity according to section 12.4.4 [m/s]
$B$ = stability coefficient [1] after Römisch [180]
 = 0.90 for stern thrusters without central rudder
 = 1.25 for stern thrusters with central rudder
 = 1.20 for bow thrusters
$g$ = 9.81 (acceleration due to gravity) [m/s$^2$]
$\Delta'$ = relative density of bed material underwater [1]
 = $(\rho s - \rho 0)/\rho 0$
$\rho_s, \rho_0$ = density of fill material and water respectively [t/m$^3$]

The particle diameter $d_{50}$ is normally used for $d_{reqd}$, sometimes also the particle diameter $d_{75}$ of a stone size class. The stability coefficient $B$ determined experimentally takes into account the different turbulence intensity (erosive effect of current) that can arise due to the various configurations of propellers and rudders.

For bottom velocities > 3 m/s, loose stone fill becomes increasingly uneconomic as the associated mean diameters become greater than about 0.5 m, making the armour layer disproportionately deep. For higher bottom velocities, grouted stone fills, special forms of construction with flexible covers or underwater concrete are necessary.

2. Covering the bottom with a grouted (stable) stone fill

    A distinction is made between partial and full grouting for grouted stone fills or revetments. With full grouting, all the voids in the stone fill are filled with grout, which results in an armour layer similar to an plain concrete bottom. Normally, the grout is applied in such a way that the tips of the stones still protrude and contribute to dissipating the energy of the current. With partial grouting, only as much grout is added to the stone structure as is needed to fix the individual stones in position; the stone fill still has sufficient permeability to prevent excess water pressure below the armour layer. R 211, section 12.1, contains further guidance.

Owing to the interlocking effect, partially grouted stone fills remain stable up to bed velocities of 6–8 m/s (see [182]). Propeller wash does not usually cause faster current velocities at the bottom.

3. Covering the bottom with a flexible composite system.

   Composite systems are designed to create a planar protective system by coupling together individual elements. An important principle is that the coupling should be flexible enough to adapt well to edge scour and thus stabilise it. The following technical coupling forms are known:
   - Concrete elements coupled together with ropes or chains
   - Interlocking precast concrete blocks
   - Wire mesh containers filled with quarry stone (stone or scrap mattresses or gabions)
   - Mortar-filled geotextile mats
   - Geotextile mats with permanently connected concrete blocks
   - Mats made from fabric-reinforced heavy rubber
   - Sandbags or sand-filled geotextile nonwoven bags

   These systems have excellent stabilising properties provided they are adequately dimensioned. Owing to the wide variety of systems on offer, a universal flow-mechanics design approach is only available for special cases, see [179]. Dimensions are often therefore based on the empirical values of suppliers.

   Where the coupling is sufficiently flexible, these systems can themselves stabilise edge scour and thus prevent any regressive erosion. The wire mesh of stone mattresses is, however, prone to corrosion, sand abrasion and mechanical damage despite its good stabilising and protective properties regarding edge scour. If the wire mesh is damaged, the gabions lose their mechanical stability. Mattresses or gabions must be joined together with tension-resistant connections.

4. Underwater concrete bottom (e.g. at ferry berths)

   An underwater concrete bottom, which can be constructed to a far higher degree of precision in terms of its thickness when compared with stone fill, constitutes very effective erosion protection in certain situations (e.g. ferry berths). Owing to the homogeneous structure of the concrete, the thrust transmitted locally onto the bed by ship propellers is distributed over a wide area, so a bed protected by a carefully cast concrete slab remains stable even with very severe wash actions.

   One disadvantage is that differential settlement can fracture the rigid concrete slab. Further, on its own it is not capable of stabilising edge scour, which calls for special solutions. Cut-off walls have proved to be effective around the edges of underwater concrete slabs. Such slabs are cast in depths between 0.3 and 1.0 m depending on the flow forces and the installation technology.

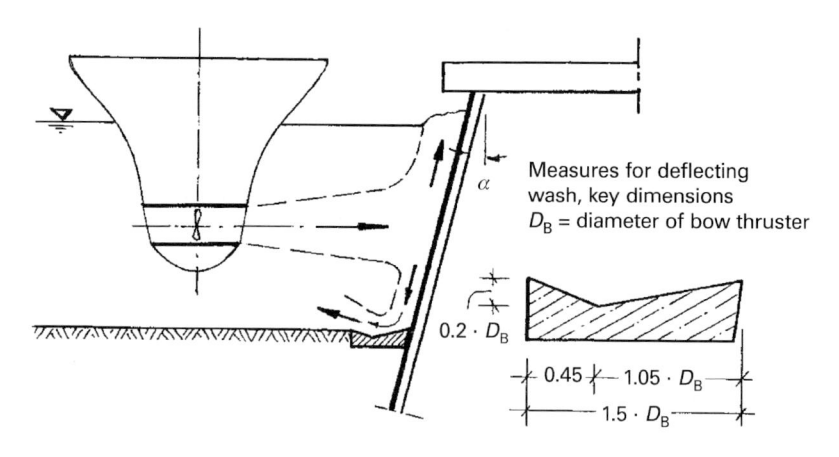

**Fig. R 83-1.** Measures for deflecting wash at a quay wall to reduce scouring [181], minimum dimensions

Key dimensions in figure: $D_B$ = diameter of bow thruster; $0.2 \cdot D_B$; $0.45$; $1.05 \cdot D_B$; $1.5 \cdot D_B$; angle $\alpha$.

Underwater concrete should only be installed by a specialist contractor able to demonstrate the necessary expertise in this field of work. As the layers are generally relatively thin, it is not possible to place the concrete with a tremie pipe. Instead, it is necessary to use so-called erosion-resistant underwater concrete, which does not segregate as it falls through the water.

5. Designing the quay wall to deflect wash, including covering the bottom where applicable

   Shifting the sheet pile wall as per Fig. R 83-1 to create a cushion of water between the front edge of the quay and the side of a vessel, possibly in combination with measures to deflect the wash, can be an effective way of avoiding or minimising loads on the bottom. Such approaches are particularly suitable for reducing scouring due to wash erosion caused by bow or stern thrusters.

   Where the measures for deflecting the wash, e.g. sloping the wall at $\alpha = 10°$ and placing a concrete apron on the bed as per Fig. R 83-1, are adequate, the scouring effect of the propeller wash is reduced to such an extent that no further measures are needed to protect the bottom (see [181]).

## 12.4.4 Current velocity at revetment due to propeller wash

Designing a loose stone fill to act as a revetment protecting against scour from propeller wash requires the wash-induced current velocity near the bottom to be estimated. Certain types of structure also require this approach. A procedure for estimating this velocity is set out below.

### 12.4.4.1 Wash caused by stern thrusters

According to [180], the wash velocity caused by a rotating propeller, the so-called induced wash velocity (occurring directly behind the propeller) can be calculated as follows:

$$v_0 = 1.6 \cdot n \cdot D \cdot \sqrt{k_T}$$

where

$n$   rotational speed of propeller [l/s]
$D$   propeller diameter [m]
$k_T$   thrust coefficient of propeller [1], $k_T = 0.25 \ldots 0.50$

The thrust coefficient takes into account the different types of propeller, the number of propeller blades and their pitch, which all depend on the type of ship. A simplified way of arriving at a mean value for the thrust coefficient is to use the following equation:

$$v_0 = 0.95 \cdot n \cdot D$$

The wash velocity is therefore essentially the product of the rotational speed of the propeller $n$ and the propeller diameter $D$. The rotational speed of the propeller when berthing and deberthing is crucial when designing measures to protect against scour at a waterfront structure. Practical experience indicates that the rotational speed of the propeller when manoeuvring is between 30 and 50 % of the rated speed (ship speeds "dead slow ahead" and "slow ahead" according to [28]). As the required diameter of the stones for a loose stone fill is calculated from the square of the rotational speed (see section 12.4.3), the estimate of this value is very important. The data in the literature varies considerably. The rated speed of the propeller and its diameter are key design features of a vessel's propulsion. The larger the propeller, the lower its rotational speed must be to avoid cavitation on the tips of the propellers. Table R 83-1 lists customary dimensions. It shows that the product of the rated speed and the propeller diameter are relatively constant for a wide range of sizes and types of vessel.

The wash continues to spread out in a conical shape due to turbulent exchange and mixing processes (Fig. R 83-2) and its velocity decreases as the distance increases.

According to [180], the maximum wash velocity found in the region of the bottom, and which is primarily responsible for scouring, can be

**Table R 83-1.** Customary values for propeller diameter and rated propeller speed

| type of vessel | propeller diameter $D$ [m] | rated rotational speed $n$ [min$^{-1}$] | peripheral propeller speed $n*D$ [m/s] |
|---|---|---|---|
| **container ship** | | | |
| 800 TEU | 5.2 | 135 | 11.5 |
| 2500 TEU | 7.2 | 105 | 12.5 |
| 5000 TEU | 8.4 | 100 | 14 |
| 8000 TEU | 9.2 | 100 | 15 |
| **multi-purpose cargo vessel** | | | |
| 5000 dwt | 3.4 | 200 | 11 |
| 12 000 dwt | 5.2 | 150 | 13 |
| 25 000 dwt | 6.1 | 120 | 12 |
| **bulk carrier** | | | |
| 20 000 dwt | 4.8 | 140 | 11.5 |
| 50 000 dwt | 6.3 | 115 | 12 |
| 75 000 dwt | 6.8 | 105 | 12 |
| 180 000 dwt | 8.1 | 82 | 11 |
| **tanker** | | | |
| 10 000 dwt | 4.4 | 180 | 13 |
| 20 000 dwt | 5.2 | 140 | 12 |
| 44 000 dwt | 6.4 | 115 | 12 |
| 120 000 dwt | 7.8 | 90 | 11.5 |
| 300 000 dwt | 9.6 | 75 | 12 |

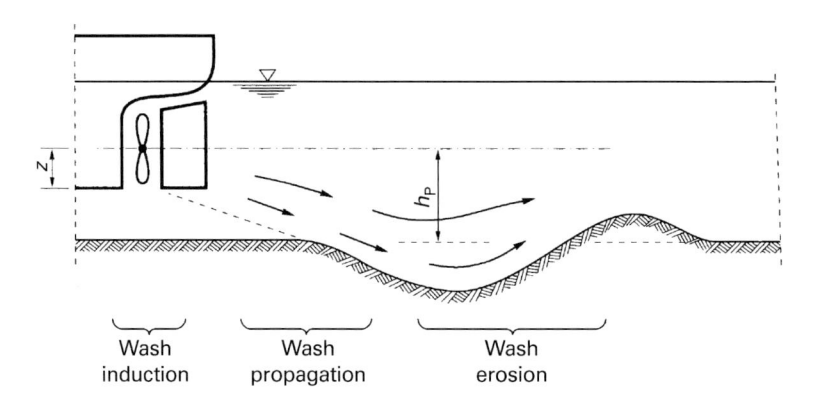

**Fig. R 83-2.** Wash caused by stern thrusters

calculated as follows:

$$\frac{\max v_{btm}}{v_0} = E \cdot \left(\frac{h_P}{D}\right)^a$$

$E$ = 0.71 for single-propeller ship with central rudder
  = 0.42 for single-propeller ship without central rudder
  = 0.42 for twin-propeller ship with central rudder, valid for $0.9 < h_P/D < 3.0$
  = 0.52 for twin-propeller ship with two rudders behind the propellers, valid for $0.9 < h_P/D < 3.0$
$a$ = $-1.00$ for single-propeller ships
  = $-0.28$ for twin-propeller ships
$h_P$ height of propeller axis above bed [m] (Fig. R 83-2)
$D$ propeller diameter [m]

### 12.4.4.2 Wash caused by bow thrusters

A bow thruster is a propeller that operates in a pipe placed transverse to the ship's longitudinal axis. It is designed to perform manoeuvres from a standstill and is therefore installed near the bow, less commonly near the stern. When using the bow thruster in the vicinity of a quay, the wash it produces strikes the quay wall directly and is deflected in all directions. Crucial for the quay wall is the proportion of the wash directed towards the bed, which can cause scouring directly adjacent to the wall when it strikes the bed, see Fig. R 83-3.

According to [180], the wash velocity $v_{0,B}$ at the bow thruster exit can be calculated as follows:

$$v_{0.B} = 1.04 \cdot \left[\frac{P_B}{\rho_0 \cdot D_B^2}\right]^{1/3}$$

where

$P_B$ power of bow thruster [kW]
$D_B$ inside diameter of bow thruster opening [m]
$\rho_0$ density of water [t/m³]

The bow thrusters of large container ships ($P_B = 2500$ kW, $D_B = 3.00$ m) are likely to produce wash velocities of 6.5–7.0 m/s.

The part of the wash velocity responsible for bed erosion max $v_{bed}$ is calculated as follows:

$$\frac{\max v_{btm}}{v_{0,B}} = 2.0 \cdot \left(\frac{L}{D_B}\right)^{-1.0}$$

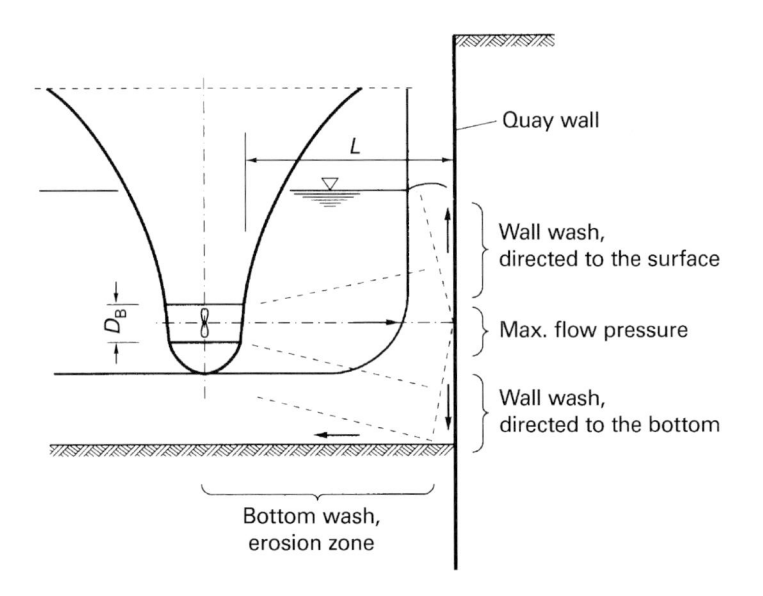

**Fig. R 83-3.** Wash loads on harbour caused by bow thrusters

where $L$ is the distance [m] between the bow thruster opening and the quay wall (Fig. R 83-3).

A bow or stern thruster is normally operated at full power.

## 12.4.5 Designing bottom protection

Determining the dimensions of any bottom protection should take account of flow mechanics factors, ensuring that the wash velocities near the edges of the protection are reduced to such an extent that there is no risk of the protection being undermined by edge scour. However, this requirement can lead to the dimensions of the bottom protection being very large, which can increase costs significantly.

For economic reasons and following the principle that the bottom and not the structure (quay wall or similar) should be protected, the dimensions of the bottom protection should ensure that at least the intensive current loading can be accommodated. Furthermore, the minimum dimensions of the protection should ensure that the area of the structurally effective passive earth pressure wedge at the base of the quay wall is protected against edge scour.

The values shown in Fig. R 83-4 are recommended as an initial approximation for the minimum dimensions. It should be remembered that about 70–80 % of the maximum bottom velocity is still present at the edges of the bottom protection when using these dimensions for scour

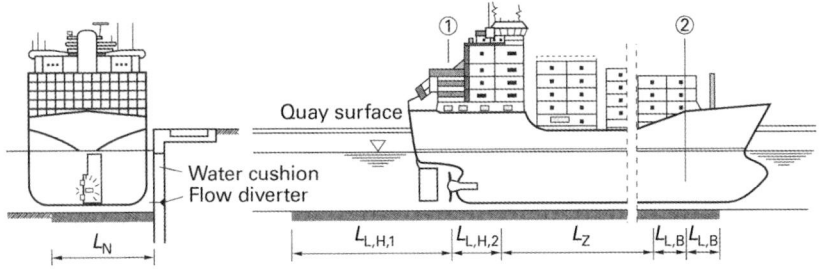

**Fig. R 83-4.** Dimensions of protected areas in front of a quay wall

① Extreme position of stern of ship
② Extreme position of bow of ship

protection measures. For soils on the bottom which are prone to erosion, the protection should be suitably designed along the edges so that it can adapt flexibly to edge scour and thus stabilise it.

The recommended minimum dimensions of bottom protection for single-propeller ships are as follows: perpendicular to the quay:

$$L_N = (3\ldots 4 \cdot D) + \Delta EP$$

parallel to the quay:

$$L_{L,H,1} = (6\ldots 8 \cdot D) + \Delta RS$$
$$L_{L,H,2} = 3 \cdot D + \Delta EP$$
$$L_{L,B} = (3\ldots 4 \cdot D_B) + \Delta EP$$

where

$D$ propeller diameter
$\Delta EP$ allowance for edge protection, approx. 3–5 m

For twin-propeller ships, the dimensions given above should be doubled. The total extent of the protection along the quay depends on the anticipated variation in berthing positions. For berths with precisely defined ship positions, the intermediate length $L_Z$ can be left unprotected.

For berths in frequent use, e.g. ferry docks, and for quay structures particularly susceptible to settlement, the extent of the protection should be investigated in greater detail – going beyond the scope of the above recommendation (minimum dimensions) – by analysing the reach of the propeller wash.

## 12.5 Scour protection at piers and dolphins

In principle, scour protection at piers and dolphins should be built like a revetment (section 12.1). The armour layer should withstand the maximum loading and a filter layer should guarantee the long-term stability of the scour protection (preventing erosion of the soil through the voids in the revetment). The installation of scour protection is made more difficult by currents and waves. Solutions are therefore needed which combine the necessary filter function with sufficient weight to withstand the hydraulic loading. Geosynthetic containers offer such properties.

Geosynthetic containers can also be laid to form a filter layer in those situations where mineral or geotextile filters can no longer be reliably installed because of fast currents. With a suitable choice of size and filling, geosynthetic containers can even be laid at high current velocities. Experiments involving a barrier of stacked geocontainers (three layers up to a height of 1.8 m) in a hydraulic test channel resulted in stability with a maximum flow velocity of approx. 4 m/s and average of 1.5–2 m/s perpendicular to the barrier [165]. Adequate stability is also to be expected at significantly higher velocities where containers are placed over a larger area.

To ensure that the filter layer functions properly, there should be no gaps between the elements. Therefore, two layers of containers are normally installed. Furthermore, the filling should not exceed 80 % of the theoretical volume as fully filled geocontainers cannot adapt to the subsoil, structures or adjacent geocontainers; and with lower fill rates, the geotextile can become displaced by oscillating motions (flapping) caused by the current, which can lead to fatigue failure of the geotextile material.

Owing to their good extensibility, there is a very low risk of the geotextile sustaining mechanical damage during installation. The nonwoven containment is very flexible and so the container can absorb the high impact loads that can occur when the container hits the bottom or when rip-rap is dumped on top. A minimum weight per unit area of 500 g/m$^2$ for the nonwoven geotextile and a minimum tensile strength of 25 kN/m are recommended to ensure sufficient robustness during installation and usage. As the friction angle of a nonwoven geotextile is greater than that of a woven fabric, geotextile containers with nonwoven containment are also suitable for protecting relatively steep embankments.

An armour layer of rip-rap (loose or partially grouted) is normally laid on top of a filter layer of geotextile containers. However, the containers themselves can also be used as a permanent armour layer. Ultraviolet radiation limits the design life of geotextile containers above water unless additional measures are taken. The UV radiation is only of minor importance underwater, so containers can be used, for example, as permanent protection against scouring around bridge piers and dolphins. Sufficient resistance to abrasion is required, which can be verified, for example, by corresponding RPG tests [185].

## 12.6 Installation of mineral impervious linings underwater and their connection to waterfront structures (R 204)

### 12.6.1 Concept

A mineral impervious lining consists of a natural, fine-grained soil whose composition or pretreatment gives it a very low permeability without the need for additional materials to achieve the sealing effect, or which achieves the necessary properties through suitable additives. (Impervious linings in the form of fully grouted rip-rap revetments are covered in section 12.1.3).

### 12.6.2 Installation in dry conditions

Mineral linings installed in dry conditions are covered in detail in [52]. Geosynthetic clay liners are dealt with in [58] and [61].

### 12.6.3 Installation in wet conditions

#### 12.6.3.1 General

When deepening or extending lined basins or waterways, impervious linings often have to be installed underwater, sometimes with waterborne traffic still operating. In this situation it is inevitable that part of the bed is temporarily without a lining. The resulting effects in terms of the water levels to be assumed and the quality of the groundwater should be considered during the detailed design work. Depending on the installation procedure, the sealing material will have to satisfy certain requirements.

#### 12.6.3.2 Requirements

Mineral lining materials installed underwater cannot be compacted mechanically, or at best to a limited degree only. They should therefore be homogenised beforehand and installed in a consistency that provides a uniform sealing effect from the outset, ensures that the material used adapts to any unevenness in the subgrade without splitting, can withstand the erosion forces of waterborne traffic during installation and can guarantee that junctions with waterfront structures are sealed, even if these structures deform.

Impervious linings installed on slopes must be strong enough to ensure that the sealing material remains stable on the slope.

When proposing a mineral impervious lining, it should be verified that it has adequate resistance to:

- the risk of the newly installed lining material disintegrating underwater,
- erosion due to the return currents of waterborne traffic adjusting to the conditions imposed by the building works,
- the lining developing cracks or holes in the form of narrow channels on coarse-grained subsoil (piping),
- sliding on embankments with a gradient of up to 1:3, and
- the loads due to dumping filter materials and rip-rap on the lining.

Impervious linings made from natural soils without additives generally meet these requirements provided the sealing material fulfils the following criteria (geotextile clay liners must be considered as a special case):

- Proportion of sand ($d \geq 0.063$ mm) < 20 %
- Proportion of clay ($d \leq 0.002$ mm) > 30 %
- Permeability $k \leq 10^{-9}$ m/s
- Undrained shear strength 15 kN/m² $\leq c_u \leq$ 25 kN/m²
- Thickness (for 4 m water depth) $d \geq 0.20$ m

When specifying mixes that include certain additives plus a proportion of cement and which solidify after installation, it is important that the flexibility of the lining in its final state is not compromised. This must be verified through testing, e.g. according to [97].

Special investigations are required when installing mineral impervious linings at greater depths, in gravelly soil, where the subsoil has large pores, on embankments steeper than 1:3 and when designing the lining material in terms of its self-healing properties at any cracks and the sealing effect at butt joints [192], [193].

See ZTV-W 210 [235] for suitability and monitoring tests.

There are now several methods available (some patented) for single-layer soft mineral impervious linings [61]. Mounting the laying equipment on stilts is recommended for all methods.

### 12.6.4 Connections

Mineral impervious linings are normally connected to structures via butt joints. The sealing material is generally pressed on with equipment that ensures that the lining adapts to the shape of the joint (e.g. a sheet pile section). An adequate quantity of sealant is applied to the line of the joint beforehand using suitable equipment. As the sealing effect comes from the perpendicular contact stress between lining material and joint [192], [193], the pressing procedure should be carried out with great care.

For lining materials with an undrained shear strength $c_u <$ 25 kN/m², the contact length between a mineral lining and a sheet pile wall or component should be at least 0.5 m, and at least 0.8 m for a higher strength. The shear strength of a mineral lining in the area of the wall connection should not exceed $c_u = 50\,\text{kN/m}^2$. A geosynthetic clay liner is connected via a sealing wedge made from suitable sealing material with a contact length to the lining material of at least 0.8 m.

## 12.7 Flood defence walls in seaports (R 165)

### 12.7.1 General

Flood defence walls protect land against flooding. They require considerably less space than dykes and are therefore often used in and around ports and harbours. The particular requirements to be met by such walls are described in the following sections.

### 12.7.2 Critical water levels

#### 12.7.2.1 Critical water levels for flooding

##### 12.7.2.1.1 Outer water level and nominal level

The nominal level of a flood defence wall is derived from the critical still water level (design water level corresponding to highest expected storm flood level, highest astronomical tide) plus freeboard allowances for the localised sea state effects (waves, section 5.7) and surge where applicable.

Owing to the larger run-up of waves on walls, the top of a flood defence wall is set higher than for dykes, unless short-term overflow over the walls is acceptable. In this case it should therefore be ensured that the floodwater causes no scouring behind the wall and can drain away without causing any damage (section 12.7.7.1).

The following values for the permitted wave overtopping rate are recommended in Table A 4.2.3 of EAK [60]:

$q_T < 0.5\,\text{l/(s·m)}$ for flat terrain with stationary traffic
$q_T < 5$ to $10\,\text{l/(s·m)}$ for paved but empty areas

Where applicable, a different value should be assumed for areas around ports and harbour facilities depending on the potential for damage.

The wave overflow can be effectively reduced by building an overflow deflector along the top edge of the wall [84].

##### 12.7.2.1.2 Inner water level

The inner water level (groundwater) should normally be taken as the level of the top of the terrain directly behind the wall (Fig. R 165-1).

Design situation DS-T (transient, R 18) can be used to verify stability for flooding, taking the mean wave height into account. If the maximum wave pressure or a special load according to section 12.7.5 is considered, verification can be carried out for DS-A (accidental). For extremely rare load combinations, the extreme case can be applied.

#### 12.7.2.2 Critical water levels for low water

##### 12.7.2.2.1 Outer water level

Mean low tide should be considered as the standard low water level for DS-P (permanent).

Exceptionally low outer water levels occurring only once a year should be allocated to DS-T (transient).

The lowest astronomical tide ever measured or lowest outer water level expected in the future should be categorised in DS-A (accidental).

##### 12.7.2.2.2 Inner water level

Generally, the inner water level should be assumed to be the top of the terrain, unless a lower water level can be demonstrated through more precise studies of the flows or can be permanently ensured through construction measures such as drainage. There should nevertheless still be a safety margin $\geq 1.0$ for the event of failure of the drainage (extreme case). The critical inner water level for an individual case can also be determined by observing the groundwater levels – presuming detailed knowledge of the local conditions.

##### 12.7.2.2.3 Flood run-off

Flood run-off can give rise to water level differences that correspond to the low water situation (excess pressure on land side) but also lead to a higher loading on the wall, e.g. with a water level above the terrain on the inner side.

### 12.7.3 Excess water pressure and unit weight of soil

The progression of the excess water pressure coordinates can be calculated with the help of a flow net in accordance with R 133, section 4.7, or based on R 11, section 2.12. The change in the effective unit weight due to the flowing groundwater can be taken into account according to R 114, section 2.12.

If a gap forms between the wall and a less permeable stratum due to deflection of the wall, this stratum is to be considered as fully permeable. Further information on calculations for walls in flowing groundwater can be found in R 113, section 4.7.

### 12.7.4 Minimum embedment depths for flood defence walls

The minimum depth of embedment for a flood defence wall can be derived from the structural calculations and the required verification of safety against slope failure. The reduction in the unit weight as a result of the upward vertical flow through the soil in the passive earth pressure zone must be taken into account (see also R 114, section 2.12). The risk to the subsoil and the operations on site due to potential leaks (interlock failure) should also be considered, and it should be noted that:

- just one flaw in a flood defence wall can lead to failure of the entire structure,
- it is not possible to carry out a suitability test for the flooding design load case, and
- the driving depth allowance, taking into account that an embankment could potentially act in an unfavourable manner, should be determined in accordance with R 56, section 8.2.9.

Values for the flow path in the soil should therefore not drop below the following for the flooding load case:

- For homogeneous soils with a relatively permeable soil structure and where a gap has formed due to deflection of the wall: four times the difference between the design water level and the top of the terrain on the land side (regardless of the actual inner water level).
- Stratified soils with permeability differences exceeding two powers of 10: three times the difference between the design water level and the top of the terrain on the land side (regardless of the actual inner water level); horizontal seepage paths due to, for example, settlement beneath a superstructure slab cannot be included.

### 12.7.5 Special loads on flood defence walls

Apart from the normal imposed loads, other loads (min. 30 kN) caused by impacts from floating objects (including vessels) during flooding and from the collision of land-based vehicles should be allowed for (Fig. R 165-1). A reasonable line of action of the load must be specified. However, a much higher impact load should be assumed where the location is vulnerable due to unfavourable current and wind conditions or is easily accessible. Using suitable constructional measures to distribute the loads is permitted provided the functionality of the flood defence wall is not impaired.

Special loads also include ice pressure. For special loads, verification of stability may be conducted according to DS-A (accidental).

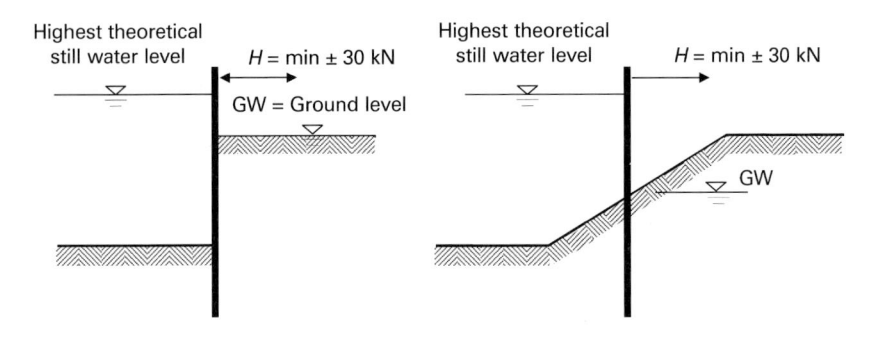

**Fig. R 165-1.** Critical water levels for flooding

### 12.7.6 Guidance on designing flood defence walls in slopes

The low water levels of the outer water are generally critical when designing flood defence walls in or in the vicinity of slopes.

Increasing the load due to the inner excess water pressure and increased unit weights, also due to flow pressure, is accompanied by lower passive earth pressure on the outside. The different water levels often also lead to less stability against slope failure. It is therefore recommended that when determining the seepage flow, only half of the horizontal seepage path should be assumed.

In stratified subsoils (cohesive intermediate strata) that are not secured with sufficiently long sheet piling, in addition to the normal analysis of slope failure, the stability of the wedge of soil in front of the flood defence wall should also be checked (slip circle failure, safety against sliding).

The outer slope should be protected against scouring using rocky material or similar measures. Stability against slope failure should be verified for DS-T (transient) at least in accordance with DIN 4084. These slopes should be inspected regularly.

### 12.7.7 Constructional measures

#### 12.7.7.1 Surface protection on land side of flood defence wall

In order to avoid land-side scouring caused by overflows during flooding, the surface should be protected. The width of the protection should be at least equal to the exposed land-side height of the wall.

#### 12.7.7.2 Flood defence road

It is recommended that an asphalted road be built near the flood defence wall. It should be min. 2.50 m wide and at the same time protect the surface in accordance with section 12.7.7.1.

### 12.7.7.3 Pressure relief filter

A 0.3–0.5 m wide pressure relief filter should be placed directly adjacent to the flood defence wall on the land side to prevent greater water pressure building up beneath the flood defence road.

For sheet piling structures, it is sufficient to fill the land-side troughs with an appropriate filter material (e.g. 35/55 smelter slag).

### 12.7.7.4 Imperviousness of the sheet pile wall

Sections of sheet pile wall projecting above ground level are generally provided with synthetic interlock seals as set out in section R 117, section 8.1.21.

## 12.7.8 Buried services in the region of flood defence walls

### 12.7.8.1 General

For various reasons, buried services in the region of flood defence walls can represent weak spots. The main reasons are:

- Leaks from pipes carrying liquids reduce, through washout, the seepage paths in the soil that would otherwise exist.
- Digging trenches to replace damaged/faulty pipes or cables reduces the supporting effect of the passive earth pressure and again shortens seepage paths.
- Decommissioned pipes can leave behind uncontrolled cavities; such pipes should therefore be removed, but at the very least filled.

Wherever possible, work on buried pipes or cables should be avoided at times when storm floods are likely. Where this is unavoidable, the construction work should take account of potential flood situations.

### 12.7.8.2 Buried services parallel to a flood defence wall

Pipes or cables parallel to a flood defence wall must not be laid within an adequately wide (> 15 m) safety strip on both sides of the flood defence wall. Existing pipes/cables should be repositioned or decommissioned. Any ensuing voids must be properly backfilled.

Any pipes/cables remaining in the safety strip should be paid particular attention. It must be possible to close off pipes that transport liquids with suitable shut-off valves at the points where they enter and leave the safety strip.

### 12.7.8.3 Buried services crossing a flood defence wall

Pipes or cables passing through a flood defence wall are also potential weak spots and should therefore be avoided wherever possible. As such,

- services, especially high-pressure pipes or high-voltage cables, should therefore be routed over the flood defence wall wherever possible,
- individual pipes/cables located in the subsoil outside the safety strip should be combined and routed through the safety strip and the flood defence wall as a single pipe/cable or bundle of pipes/cables, and
- services should cross the wall at 90° as far as possible.

The different settlement behaviour of buried services and flood defence walls should be taken into account by way of constructional measures (flexible penetrations, articulated pipe joints). Rigid penetrations are not permitted.

The design of pipe/cable intersections depends on the type of buried service and should take account of any relevant regulations.

## 12.8   Dumped moles and breakwaters (R 137)

### 12.8.1   General

Moles differ from breakwaters primarily in the way they are used; it is possible to drive or at least walk along moles. They are therefore generally higher than breakwaters, which in some cases do not even project above still water level. Further, breakwaters are not always connected to the land.

Besides careful determination of the wind and wave conditions, currents and any potential sand drift, it is essential to have accurate information about the subsoil when constructing moles and breakwaters. For the sake of simplicity, only dumped breakwaters are covered here.

The positions and cross-sections of large dumped breakwaters are determined not just by their intended purpose, but also by their constructional feasibility.

### 12.8.2   Stability analyses, settlement and subsidence, guidance on construction

Loosely bedded non-cohesive soils below the footprint of an intended mole or breakwater must first be compacted; soils with a low bearing capacity must be replaced.

It is also possible to displace soft cohesive layers by deliberately exceeding their bearing capacity so that the dumped material is embedded in the subsoil. Blasting below the dumped material is also possible so that the subsoil is displaced. However, both procedures result in the finished structure experiencing greater differential settlement because the displacement achieved in this way is never uniform.

Silt strata are displaced by pushing them ahead of the dumped material. The ensuing build-up of silt should be removed because otherwise it can infiltrate the dumped material and have a long-term negative effect on its properties.

In the case of dumped breakwaters, safety against ground and slope failure should be checked. Here, the effect of waves is taken into account with the characteristic value of the design wave. In earthquake zones the risk of soil liquefaction should be assessed.

The total settlement due to the load of the dumped material, subsidence under the effect of waves and the dumped material becoming embedded in the subsoil, or vice versa, can amount to several metres and must be compensated for by specifying a greater height.

Dumped breakwaters are permeable and are thus subjected to through-flows. Where their construction is inhomogeneous, the stability of the filter action of adjacent strata must be guaranteed.

### 12.8.3 Specifying the geometry of the structure

The key input parameters for determining the cross-section of a breakwater are:

- design water levels
- significant wave heights, wave periods (frequencies), approach direction of waves
- subsoil conditions
- construction materials available

The crest height is set so that once settlement has ceased, any wave overtopping is kept to a minimum.

The recommendation is to calculate the overall height of a breakwater as follows:

$$R_c = 1.2 H_S + s$$

where

$R_c$    freeboard height (overall height above still water level) [m]
$H_S$    significant wave height $H_{1/3}$ of design sea state [m]
$t$    total expected final settlement, subsidence and embedment [m]

When $R_c < H_S$, significant wave overtopping can be expected.
When $R_c = 1.5 \cdot H_S$, wave overtopping is negligible.
Wave overtopping $q$ can be calculated as set out by [210] or [152].
The stone size for the armour layer is determined based on tried-and-tested empirical equations, e.g. after Hudson. The approach used by Hudson is described in the following section.
The sizes of blocks that can be obtained economically from quarries are frequently inadequate for the armour layer. Precast concrete blocks can

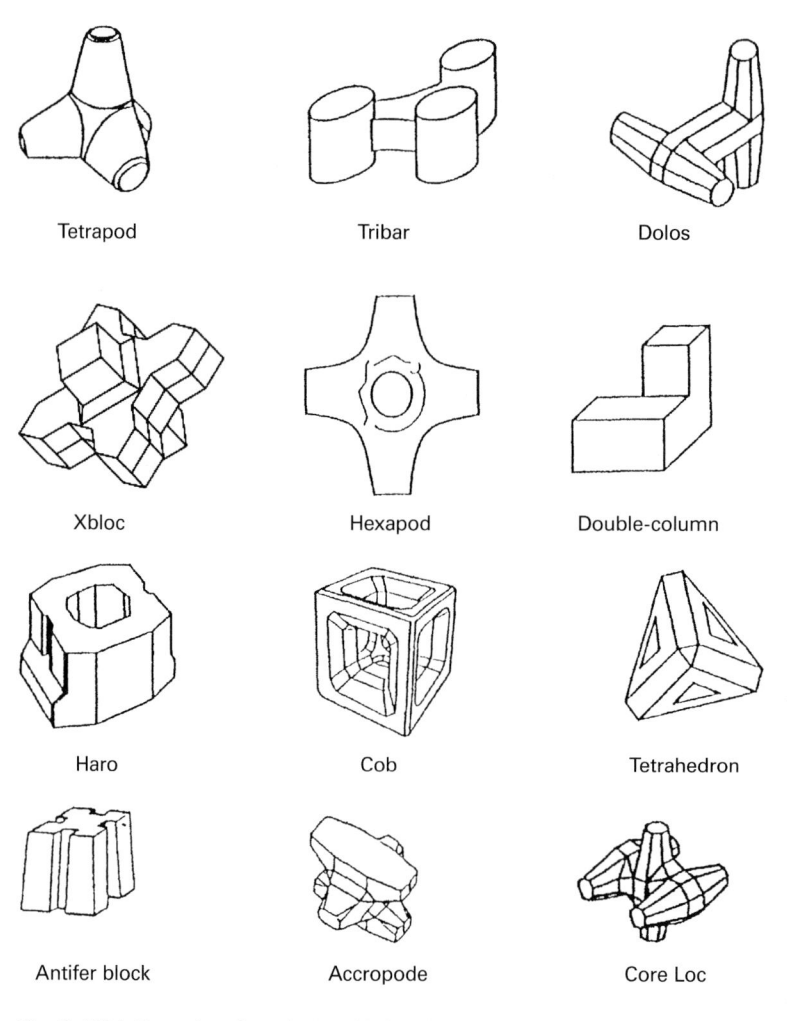

**Fig. R 137-1.** Examples of standard prefabricated elements

be used instead, e.g. the standard precast blocks listed in Table R 137-1 and illustrated in Fig. R 137-1.

Reference values for the seaward embankment are given in Table R 137-1.

**Table R 137-1.** Recommended $K_D$ values for designing the armour layer for a permissible destruction of up to 5% and only negligible wave overtopping (extract from [2021])

| Armour layer elements (examples) | No. of layers | Type of arrange-ment | Breakwater side $K_D$ [1] | | Breakwater crest $K_D$ | | |
|---|---|---|---|---|---|---|---|
| | | | breaking waves[5] | non-breaking waves[5] | breaking waves | non-breaking waves | gradient |
| Smooth, rounded natural stones | 2<br>3 | random<br>random | 1.2<br>1.6 | 2.4<br>3.2 | 1.1<br>1.4 | 1.9<br>2.3 | 1:1.5 to 1:3<br>1:1.5 to 1:3 |
| Angular quarry stones | 2 | random | 2.0 | 4.0 | 1.9<br>1.6<br>1.3 | 3.2<br>2.8<br>2.3 | 1:1.5<br>1:2<br>1:3 |
| | 3<br>2 | random special arrange-ment[2] | 2.2<br>5.8 | 4.5<br>7.0 | 2.1<br>5.3 | 4.2<br>6.4 | 1:1.5 to 1:3<br>1:1.5 to 1:3 |
| Tetrapod | 2 | random | 7.0 | 8.0 | 5.0<br>4.5<br>3.5 | 6.0<br>5.5<br>4.0 | 1:1.5<br>1:2<br>1:3 |
| Antifer block | 2 | random | 8.0 | – | – | – | 1:2 |
| Accropode | 1 | | 12.0 | 15.0 | 9.5 | 11.5 | up to 1:1.33 |
| Core Loc | 1 | | 16.0 | 16.0 | 13.0 | 13.0 | up to 1:1.33 |
| Tribar | 2 | random | 9.0 | 10.0 | 8.3<br>7.8<br>6.0 | 9.0<br>8.5<br>6.5 | 1:1.5<br>1:2<br>1:3 |

**Table R 137-1.** (*Continued*)

| Armour layer elements (examples) | No. of layers | Type of arrange-ment | Breakwater side $K_D$ [1] | | | Breakwater crest $K_D$ | | |
|---|---|---|---|---|---|---|---|---|
| | | | breaking waves[5] | non-breaking waves[5] | gradient | breaking waves | non-breaking waves | gradient |
| Tribar | 1 | arranged uniform-ly | 12.0 | 15.0 | | 7.5 | 9.5 | 1:1.5 to 1:3 |
| Dolos | 2 | random | 15.8[3] | 31.8[3] | | 8.0 7.0 | 16.0 14.0 | 1:2[4] 1:3 |
| Xbloc | 1 | random | 16 | 16 | | 13 | 13 | up to 1:1.33 |

[1] For gradients of 1:1.5 to 1:5.
[2] Longitudinal axis of stones perpendicular to surface.
[3] $K_D$ values only confirmed experimentally for a 1:2 gradient; for higher requirements (destruction < 2 %), $K_D$ values should be halved.
[4] Gradients steeper than 1:2 are not recommended.
[5] Breaking waves occur increasingly where the still water depth is less than the wave height in front of the breakwater.

The minimum width of the crest is calculated from

$$W_{min} = (3 \text{ to } 4) D_m$$

or

$$D_m = \sqrt[3]{\frac{W}{\rho_s}} \qquad D_m = \sqrt[3]{\frac{W_{50}}{\rho_s}}$$

where

| | |
|---|---|
| $W_{min}$ | minimum width of breakwater crest [m] |
| $D_m$ | mean diameter of individual stone or block in armour layer [m] |
| $W$, $W_{50}$, $\rho_s$ | see section 12.8.4 |

As fine-grained material is in most cases considerably less expensive than the coarse armour layer material, most breakwaters have the traditional structure as shown in Fig. R 137-2:

- core,
- filter layer, and
- armour layer.

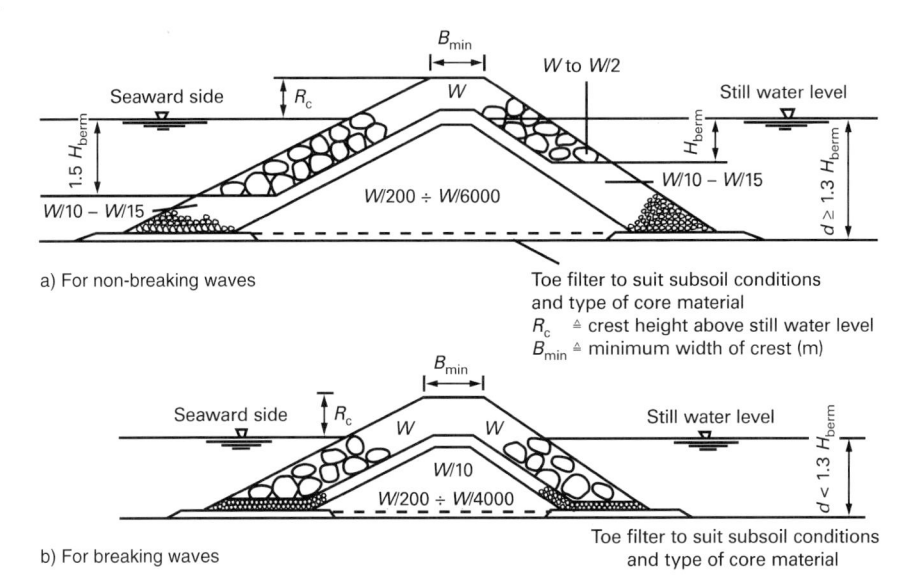

**Fig. R 137-2.** Filter structure of breakwater in three gradations

$R_c$ ≙ crest height above still water level
$B_{min}$ ≙ minimum width of crest (m)

a) For non-breaking waves
b) For breaking waves

However, it is possible that the difference in cost of the core and armour layer materials can sometimes vanish in the case of large transport distances, for example. The breakwater can then be constructed from uniform block sizes, especially when being built with sea-based.

Installing a toe filter should be given particular consideration where coarse-grained cores are used.

The stones forming mole crests are often covered with concrete to make them suitable for vehicular traffic.

Walls on top of dumped breakwaters are very often used to repel wave overtopping and spray as well as for access on moles. However, a wall constitutes a "foreign body" that reveals the considerable settlement and differential settlement. Cracks in such walls and tilting are therefore not uncommon.

### 12.8.4 Designing the armour layer

With given wave conditions, the stability of the armour layer depends on the size, weight and form of the constructional elements as well as the gradient of the armour layer.

Following a series of tests over many years, Hudson developed the following equation for the required block weight ([202], [159], [29]), which has proved effective in practice:

$$W = \frac{\rho_s \cdot H_{des}^3}{K_D \cdot \left(\frac{\rho_s}{\rho_w} - 1\right) \cdot \cot\alpha}$$

where

| | |
|---|---|
| $W$ | block weight [t] |
| $\rho_s$ | density of block material [t/m$^3$] |
| $\rho_w$ | density of water [t/m$^3$] |
| $H_{des}$ | characteristic height of "design wave" multiplied by partial safety factor [m] |
| $\alpha$ | gradient of armour layer [°] |
| $K_D$ | shape and stability coefficient [l] |

This equation applies to an armour layer built from stones with a roughly uniform weight. The most common form and stability coefficients $K_D$ for quarry stones and moulded blocks for inclined breakwater armour layers according to [202] are summarised in Table R 137-1. Fig. R 137-1 shows examples of standard prefabricated elements.

When selecting elements for the armour layer, it should be taken into account that, depending on the form of the element, additional tensile, bending, shear and torsion loads can occur with the possible settlement or subsidence movements in accordance with section 12.8.2. Owing to the high sudden loading, the $K_D$ values should be halved for larger Dolos elements.

According to [202], the following amended equation is recommended for designing an armour layer of graded natural stone sizes with design wave heights of up to about 1.5 m:

$$W_{50} = \frac{\rho_s \cdot H_{des}^3}{K_{RR} \cdot \left(\dfrac{\rho_s}{\rho_w} - 1\right) \cdot \cot\alpha}$$

where

$W_{50}$ weight of average-size stone [t]
$K_{RR}$ shape and stability coefficient [1]
= 2.2 for breaking waves
= 2.5 for non-breaking waves

Here, the weight of the largest stones should be $3.5 \cdot W_{50}$ and the smallest $0.22 \cdot W_{50}$. According to [202], owing to the complex processes involved, the block weight should in general not be reduced where waves approach the structure at an angle.

Incidentally, [159] recommends assuming the characteristic value of the "design wave" to be at least $H_{des} = H_s$ for all wave heights when using the Hudson equation. This value is generally extrapolated with the help of statistics of extreme values to cover a longer period (e.g. 100-year return). For extrapolation to be reliable, there must be sufficient data on wave measurements available (see also R 136, section 5.6).

The significance of the design wave for the structure is that the required weight of an individual block $W$ increases in proportion to the wave height to the power of three.

When planning a dumped breakwater, economic considerations can lead to different criteria for the lowest possible destruction rate where extreme sea state loads occur only very rarely or at the land end of the breakwater where silting-up occurs on the seaward side to such extent that the armour layer is no longer needed. The more economical options should be chosen if the capitalised repair costs and the likely costs of rectifying any other damage in the port area are lower than the increased capital expenditure when designing the block weight for a particularly high design wave that occurs only rarely. The feasibility of carrying out general maintenance in situ as well as the likely duration of the work should be considered separately in each case.

Further calculation methods are given in [153], [154], and [159]. Abromeit [1] contains fundamental information on how the size, installed thickness and dry bulk density of the rip-rap used influence the stability of a grouted armour layer with respect to current and wave loads as well as suggestions for calculating technically equivalent armour layers.

In addition to the Hudson formula for designing the armour layers of dumped breakwaters, the report by PIANC Working Group 12 of the Permanent Technical Committee II for Coastal and Ocean Waterways as well as [159] includes, in particular, the Van der Meer formula.

These equations take into account the breaking form of the waves (plunging and surging breakers), which is calculated based on the height and period (frequency) of the wave according to the Iribarren number. However, they also take into account the duration of the storm, the degree of damage and the porosity of the breakwater. The formulas were derived from model experiments with waves that corresponded to the natural wave spectrum in terms of wave height and wavelength distribution. Hudson [106], [107], on the other hand, used only regular waves in his experiments. Van der Meer's method of calculation presumes advance knowledge of many detailed relationships, as can be found in [159]. According to Hudson and Van der Meer, the results of calculations for armour layer sizes vary significantly for extreme cases. For large mole or breakwater structures, the recommendation is therefore to appoint an approved hydraulic engineering institute to investigate the chosen cross-section as a whole with the help of hydraulic models. Such models can also reveal how a crest wall might influence the overall stability of the breakwater.

### 12.8.5 Construction of breakwaters

According to the recommendations of [202], breakwaters in three-layer gradations according to Fig. R 137-2 have proved effective in practice. Notation for Fig. R 137-2:

$W$ weight of individual block [t]
$H_{des}$ height of "design wave" [m]

A single-layered structure made from quarry stones should not be used. The general recommendation for the slope inclination on the seaward side is no steeper than 1:1.5.

Particular attention should be paid to the toe support of the armour layer, especially when it does not reach as far down as the base of the seaward slope. Stability requirements might indicate that the slope requires a suitable berm (Fig. R 137-3).

Filter principles should be observed with respect to the subsoil, too. This can be achieved primarily by including a special filter layer (granular filter, geotextile, sand mat, geotextile containers with filter function) especially underneath block-type outer layers at the base because the installation of these layers is more reliable.

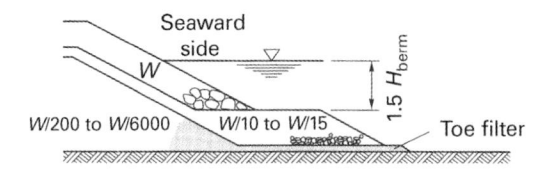

**Fig. R 137-3.** Seaward toe stabilisation of a breakwater

## 12.8.6 Construction and use of plant

### 12.8.6.1 General

The construction of dumped moles and breakwaters often calls for large quantities of materials to be installed in a relatively short time and under difficult local conditions caused by weather, tides, sea states and currents. The mutual dependency of individual operations in such construction conditions requires very careful planning of the construction schedule and the use of the plant.

The design engineer and the contractor should find out about the wave heights expected during the construction period. To do this they require information about the prevailing sea state during operations as well as very rare wave events. The duration of wave height $H_s$ and $H_{max}$ occurring in one year, for example, can be estimated in accordance with Fig. R 137-4. Observations must be made over significant periods of time in order to obtain a reliable description of the wave climate by way of a wave height duration curve.

Breakwaters must be designed so that serious damage, even in the case of a sudden storm, can be avoided, e.g. by using a layered construction with few gradations.

When specifying the productivity of the construction site, or when choosing the size of plant, realistic approaches must be applied that take account of work being interrupted due to bad weather.

Depending on the particular construction, dumping works are carried out

1. with floating plant,
2. with land-based plant building ahead of itself,
3. with fixed plant, jack-up platforms, etc.,
4. with cableways, or
5. any combination of these methods.

For particularly exposed sites seriously affected by winds, tides, sea states and currents, building methods with fixed plant, jack-up platforms, etc. are preferred. This is especially the case if there is no sheltered harbour at or near the construction site.

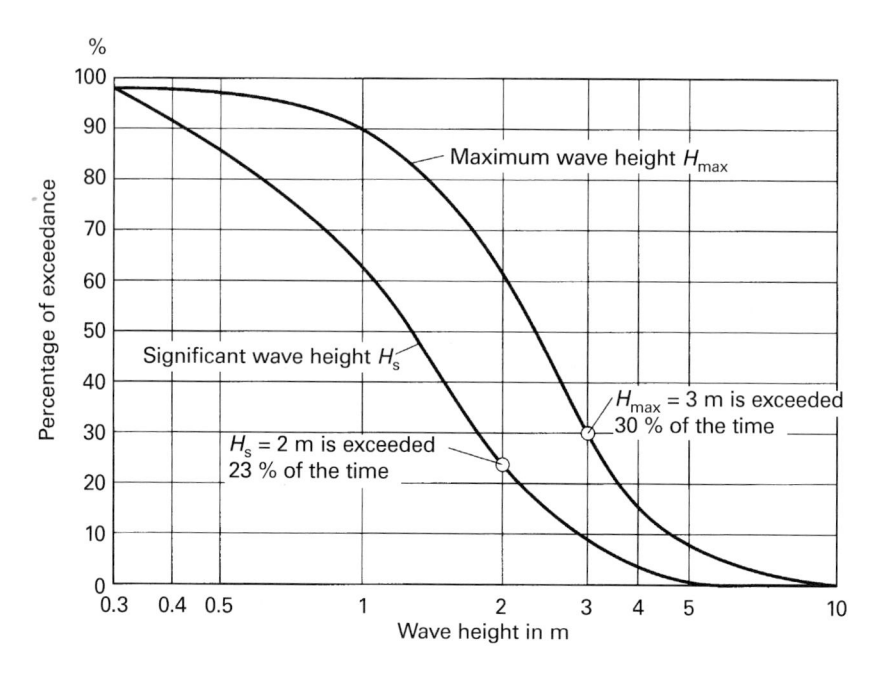

**Fig. R 137-4.** Wave height duration curve: period of time over which a particular wave height is exceeded in a year, e.g. $H_s = 2\,\text{m}$; $H_{max} = 3\,\text{m}$

### 12.8.6.2 Provision of fill and other construction materials

The provision of fill and construction materials requires careful planning depending on the options available for their acquisition and transportation. Procuring coarse materials is often the main problem.

### 12.8.6.3 Installing materials with floating plant

When building with floating plant, the cross-section of the breakwater must be adapted to the plant. Split hopper barges always require a sufficient depth of water. Side stone dumping vessels can be moved sideways even with low water depths. By exploiting today's computer-controlled positioning procedures, floating plant can achieve the accuracy that in the past was only possible with land-based plant.

### 12.8.6.4 Installing fill material with land-based plant

The working level of land-based plant should normally be above the effects of the normal sea state and surf. The minimum width of this working level should be adjusted to suit the requirements of the plant being used.

Land-based plant works by gradually extending the construction ahead of itself, the materials being delivered to site dump trucks. This construction method therefore generally requires a core protruding out of the water with an extra wide crest. The core serves as a road and therefore must be removed again to certain depth before the armour layers can be placed on top, so that sufficient interlocking and the hydraulic homogeneity can be restored.

With a narrow working area it is often advantageous to use a gantry crane for the installation work as the materials for the ongoing work can be transported underneath it.

Stones to be installed by the crane are mostly delivered in skips on flatbed trailers, trucks with a special loading area or low-loaders.

Where the roadway is narrow, trailers that can be driven backwards without needing to turn around are used. Large stones and precast concrete elements are placed using orange peel or other, special grabs. Electronic monitors in the crane cabin make it easier to install materials in accordance with the design profile, even underwater.

Covering the core should follow quickly after dumping just a short length, especially when building the structure progressively ahead of the plant. This avoids unprotected core material from being washed away. Further information can be found in [38].

### 12.8.6.5 Installing material from fixed scaffolds, jack-up platforms, etc

Installing from fixed scaffolds, jack-up platforms, etc., also with a cableway, is primarily considered for bridging a zone with continuous, powerful surf.

When using a jack-up platform, construction progress generally depends on the capacity of the installation crane. The crane chosen should therefore have a large safe working load for the required reach.

The design should indicate clearly which parts of the breakwater cross-section must be installed when the sea is calm and which may still be constructed in certain wave conditions. This applies to both the core material and the precast concrete elements of the armour layer. Even in low swell, precast concrete elements can still suffer impacts underwater due to their large weight, which can lead to cracks and fractures.

### 12.8.7 Settlement and subsidence

Consistent and minor settlement of dumped breakwaters can be allowed for by additional height. Once the settlement has abated, which can always be monitored by taking settlement levels, the concrete on top of the crest should be cast in sections that are not overlong.

Where large differential settlement is expected, walls on the crest should be avoided because their settlement can subsequently lead to a visually unappealing overall appearance of the breakwater even though neither function nor stability are at risk.

### 12.8.8 Invoicing for installed quantities

As the settlement and subsidence behaviour of such structures is very difficult to predict, the recommendation is to specify realistic tolerances (±) from the outset for the purpose of invoicing based on the drawings. These should take into account the form of the mole and the installation layers in order to compensate for the settlement and, where applicable, embedded volumes and depending on the technical procedure selected. The tender should certainly specify whether invoicing is based on measured or actually installed quantities of materials. Where actually installed quantities are to be invoiced, settlement levels should be included in the tender.

If soil investigations reveal that it is likely to prove particularly difficult to invoice actually installed quantities, it is recommended that the invoice be based on weight [38] if no other solution specific to the given subsoil conditions is possible. The measurement procedure (highest points of a stone layer or the use of a sphere/hemisphere at the bottom of a measuring stick) should be specified.

# 13 Dolphins (R 218)

## 13.1 General principles

### 13.1.1 Dolphins – purposes and types

Dolphins are designed to allow ships to berth and moor safely. They are also used to protect waterfront structures and guide waterborne traffic. Breasting dolphins must absorb the impacts of the berthing process. Mooring dolphins should be designed to take account of the effects of line pull forces, wind loads and currents. Breasting dolphins generally also serve as mooring dolphins. Dolphins as part of a crash barrier are sacrificial structures and are designed to meet specific requirements. They are not included in this recommendation.

Dolphins can be designed as single piles, row of piles (also named fender rack) or group of piles (cluster dolphins), see Fig. R 218-1. They can be fitted with fenders to reduce the contact forces between a vessel and the dolphin.

### 13.1.2 Stiffness of the system

For the purpose of design, a distinction should be made between rigid and flexible dolphins. Rigid dolphins accommodate actions without any significant deformation, whereas flexible dolphins exhibit greater deformations under the effect of the loads.

The stiffness of a dolphin is therefore an important aspect in its design. It is derived from the interaction between the pile (or group of piles), the fender (where fitted) and the subsoil. The resulting overall stiffness can therefore be significantly non-linear. Fig. R 218-2 shows the design model for a dolphin as well as typical stress–strain curves for individual components and the overall system.

### 13.1.3 Loads on dolphins and design principles

Mooring dolphins are loaded by line pull and contact forces, which are in turn significantly affected by wind and wave loads. These loads can be assumed to be static for the purpose of design.

Breasting dolphins, on the other hand, are loaded by forces arising from ship berthing manoeuvres. They are not designed by prescribing a load, but by specifying the berthing energy of a vessel, which corresponds to the integral of the stress–strain diagram of the dolphin deformation (Fig. R 218-2). The magnitude of the resulting horizontal reaction force

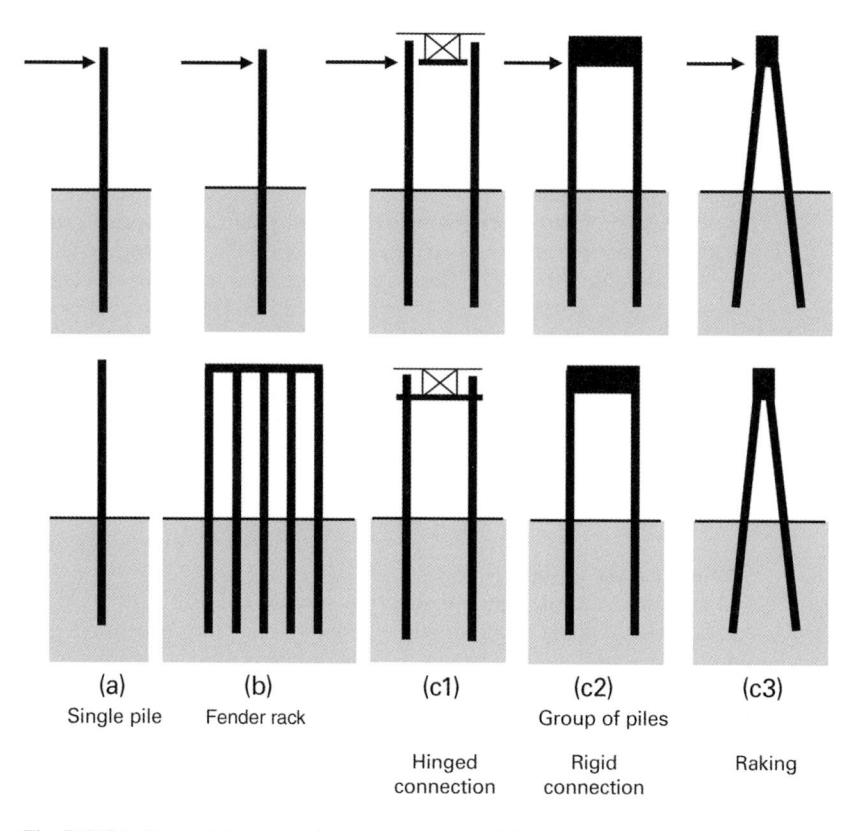

**Fig. R 218-1.** Types of dolphin, side elevations (top) and front elevations (bottom)

(a) Single pile  (b) Fender rack  (c1) Hinged connection  (c2) Rigid connection  (c3) Raking — Group of piles

$F_R$ is essentially determined by the system stiffness. A greater stiffness results in smaller deformations and larger forces; less stiffness has the opposite effect. There is no explicit solution to this design problem. It is the responsibility of the engineer to reach an optimum compromise for each particular situation, taking into account that the berthing force is limited by the permissible pressure on the hull of the vessel (section 13.2.3.1) and the dolphin deformation should not exceed approx. 1.5 m (section 13.2.3.2).

Breasting dolphins in particular are often designed as flexible elements because of the need to limit the reaction force from berthing to avoid the vessel being damaged. Breasting dolphins must nevertheless absorb the kinetic energy of the berthing vessel either fully or partially. The energy

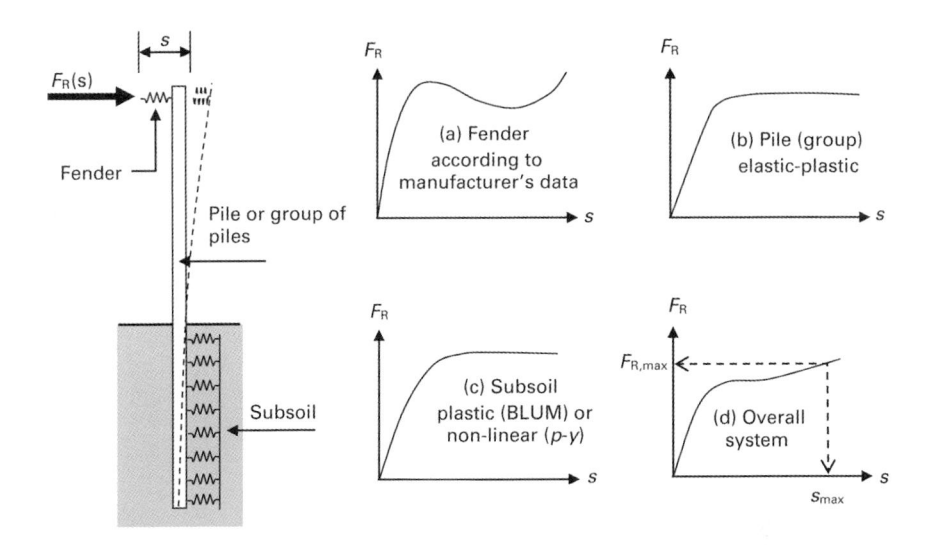

**Fig. R 218-2.** Overview of the structural system of a dolphin with fender and typical stress–strain relationships of individual components (a to c) as well as the overall system (d)

absorbed by the deformed dolphin is called the energy absorption capacity and according to Fig. R 218-2d is as follows:

$$A = \int_0^{s_{max}} F_R(s) \cdot ds$$

where

- $R$      energy absorption capacity (internal work) of dolphin [kNm]
- $F_R(s)$      horizontal reaction force (berthing force) between vessel and dolphin as a function of the deflection $s$ at the level of the force application point [kN]
- $s_{max}$      maximum deflection of dolphin at the level of the force application point [m]

For linear load–deformation behaviour, this equation can be simplified to

$$A = \frac{1}{2} \cdot F_{R,max} \cdot s_{max}$$

where $F_{R,max}$ [kN] is the horizontal reaction force (characteristic berthing force) related to the deflection $s_{max}$.

The horizontal reaction force is transferred to the dolphin pile and in turn to the subsoil via the embedment of the pile. For raking piles or groups of piles with a rigid connection (Fig. R 218-1, c2 and c3), normal forces can also arise, which can be very large if the spacing of the piles is small compared with the vertical distance from the bottom to the force application point.

The magnitude of the horizontal soil bearing pressure due to the loads to be transferred to the soil depends on the deformations and can increase up to the point where the soil fails. The horizontal soil bearing pressure can be calculated using different models (section 13.2.1).

### 13.1.4 Actions

#### 13.1.4.1 Loads due to berthing manoeuvres

The energy of berthing manoeuvres which must be resisted by the dolphin can be determined as shown in R 60, section 6.15. The berth configuration factor of the waterfront structure should be taken as $C_c = 1.0$ (open structure).

The level of the vessel impact force transferred to the dolphin depends on its form of construction, the vessel dimensions and the water levels, and may differ depending on the design parameters (force, deflection, stresses).

The berthing force arising from berthing manoeuvres is determined by the stiffness of the dolphin (section 13.2.1). For precise calculations, the flexibility of the vessel's hull must be taken into account. This can be allowed for by reducing the kinetic energy of the vessel vessel $C_s$ factor (s is the softness factor), section 6.15.4.2.

As the stiffness of the system can only be determined appropriately for characteristic loads and resistances, breasting dolphins are designed using characteristic values. Table R 218-1 lists the partial safety factors to be used.

#### 13.1.4.2 Mooring and contact forces

Vessels moored to dolphins are subjected to wind, current and wave loads, which have to be resisted by the dolphin. This gives rise to tension loads (mooring forces) or compression loads (lean-on forces) depending on the location of the dolphin in relation to the vessel. The tensile forces can be directed upwards at an angle of up to 45°.

Wind loads acting on vessels can be calculated as shown in R 153, section 5.11. Recommendations for estimating current loads can be found in DNV (2010).

**Table R 218-1.** Partial safety factors for verifying the ultimate limit state of a dolphin

|  | Actions | Resistances | |
|---|---|---|---|
|  |  | soil | steel |
|  | $\gamma_Q$ | $\gamma_{R,e}$ | $\gamma_M$ |
| Loads from berthing manoeuvres | 1.00 | 1.00 | 1.00 |
| Mooring forces (line pull) and contact forces | 1.20 | 1.15 | 1.10 |
| Wave, wind and current loads | 1.20 | 1.15 | 1.10 |
| Ice loads (see also section 5.16.1) | 1.00 | 1.10 | 1.10 |

Mooring and contact forces in sheltered harbour/port areas, i.e. locations with no significant wave effects (sea state and swell) can be determined using the information set out in R 12, section 5.12, and R 102, section 5.13.

The wave loads on vessels moored at unsheltered berths can become critical. In these cases the design loads for the dolphin can be determined using a related to time series simulation of the vessel movements induced by the sea state. The effects on the fatigue strength must be taken into account (see section 13.2.3.3).

Unsheltered jetties alongside navigation channels can be subjected to additional loads caused by passing vessels. Seelig and Flory have set out design approaches for determining these loads (Naval Facilities Engineering Service Center, 2005).

The resulting mooring and lean-on forces on individual dolphins are derived from the applicable equilibria of forces and moments. Mooring forces are limited by the loadbearing capacity of the vessel's on-board mooring equipment (ropes and winches). The winches normally fail at 60 % of the rope loadbearing capacity. This load must be accommodated by mooring dolphins:

$$F_T = 0.6 \cdot n \cdot F_{rope}$$

where

$F_T$     critical tensile force on bollard [kN]
$n$     number of ropes pulling on dolphin simultaneously in same direction [-]
$F_{rope}$     tensile strength of ropes of critical vessel, also called minimum breaking load (MBL) [kN]

#### 13.1.4.3 Other actions

If there is no vessel berthed at the dolphin, current and wave forces act directly on the dolphin pile or dolphin topside/deck. These forces can be determined as shown in R 159, section 5.10. Even if these actions are generally smaller than the actions due to berthing manoeuvres or the mooring and contact forces of moored vessels, in exceptional cases they can affect the fatigue strength of the dolphin due to their cyclic nature.

Ice loading on dolphins can be estimated as described in R 205, section 5.16, with any vertical loads due to any adhering ice being included in the calculations as well. Ice loading should be given particular attention if pontoons are permanently attached to a dolphin. In this case the ice loading acts on the dolphin through the pontoon and can therefore be considerably greater than an ice loading acting on the dolphin only.

Mooring dolphins also accommodate vertical loads that are transmitted through friction between vessel and dolphin due to the effects of waves, tides, loading/unloading, etc. These loads can become critical for the design of the dolphins and the stresses in the vessel's hull.

### 13.1.5 Safety concept

In principle, elastic design is used for dolphins; see section 13.2.3 for more details. When analysing the ultimate limit state of the dolphin, the partial safety factors given in Table R 218-1 are recommended. For the serviceability limit state, the characteristic actions and resistances should be used.

## 13.2 Design of dolphins

### 13.2.1 Soil–structure interaction and the resulting design variables

#### 13.2.1.1 Overview

The design variables for dolphins are derived from the deformation-dependent interaction between the soil and the dolphin. Earth pressure at rest acting on a non-loaded dolphin is rotationally symmetric and is thus cancelled out for the dolphin itself. The dolphin is pushed against the soil by the loads of berthing or moored vessels. This increases the lateral soil stresses in the loading direction beyond the steady-state earth pressure until a critical tensile stress is reached. At the same time, the lateral soil stresses on the opposite side of the dolphin are reduced. The dolphin is thus subjected to bending.

Two fundamentally different approaches have proved effective for modelling the deformation-dependent interaction between dolphin and soil.

The traditional method according to [22] estimates the spatial passive earth pressure $E_{ph}$ in front of the dolphin as a lateral stress (Fig. R 218-3a).

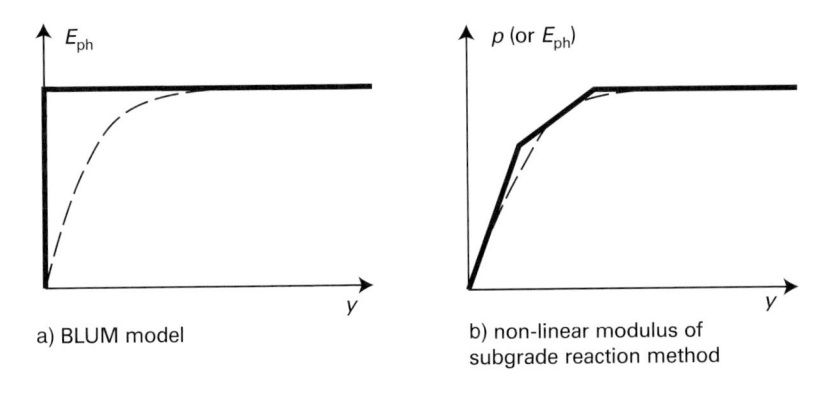

**Fig. R 218-3.** Idealised load–displacement diagrams for the dolphin foundation (soil–structure interaction)

a) BLUM model

b) non-linear modulus of subgrade reaction method

For a homogeneous non-cohesive subsoil in the embedment area, this stress can be calculated using traditional earth pressure theory. Cohesive and/or stratified non-cohesive subsoils require further factors to be taken into account when determining the spatial passive earth pressure. The Blum method therefore assumes the lateral soil stress to be the critical stress. This method thus normally provides an upper bound for the bending and shear loads on a dolphin. However, the method is of only limited use when verifying whether a dolphin complies with serviceability requirements (deformations). Section 13.2.1.2 contains further details of Blum's method.

The $p$-$y$ method is a modulus of subgrade reaction based on non-linear load–deflection curves (Fig. R 218-3b). Here, the lateral soil stresses are assumed to correspond to the deformations of the dolphin over the embedment depth, which vary with the load. The $p$-$y$ method thus supplies more realistic values for the load on the soil and the dolphin, and also allows the deflections of the dolphin to be calculated. Section 13.2.1.3 contains further guidance on the $p$-$y$ method.

Comparative calculations [186] have shown that both methods yield fundamentally comparable results for breasting and mooring dolphins in cohesive and non-cohesive subsoils with a strength $c_u < 96$ kN/m². In general, when modelling the soil with a lower stiffness, the $p$-$y$ method supplies larger deformations than the Blum method. Consequently, for breasting and lead-in dolphins, lower forces – and thus generally more economic structures – are the outcome.

However, for stiff cohesive subsoils with $c_u > 96$ kN/m², the two methods can yield significantly different results. In this case smaller structural member dimensions are calculated with the Blum method compared

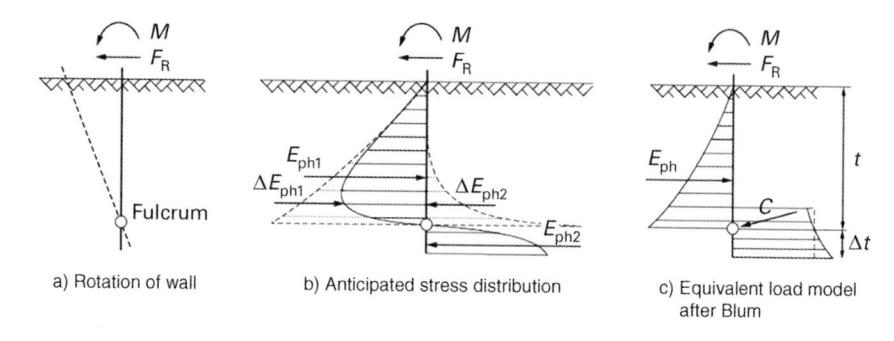

**Fig. R 218-4.** System idealisation for spatial passage of pressure according to Blum

with the p-y method. Therefore, for dolphins in stiff cohesive subsoils, the recommendation is to use the p-y for design.

The embedment depth determined for a dolphin consisting of a group of piles is greater than that for a comparable number of individual piles. Section 13.2.1.4 contains further guidance on this.

### 13.2.1.2 The Blum method

According to Blum, the embedment of the dolphin in the subsoil should be modelled as a fixed support with a parabolic distribution of the passive earth pressure and an equivalent force $C$ applied at the theoretical base (Fig. R 218-4). Contrary to the original approach according to Blum, the recommendation is to apply the characteristic passive earth pressure as a spatial passive earth pressure $E_{ph}^r$ in accordance with DIN 4085:

$$E_{ph,k}^r = E_{pgh,k}^r + E_{pch,k}^r + E_{pph,k}^r$$

Here, $E_{ph,k}^r$ is the total characteristic passive earth pressure due to dead load, cohesion and surcharges acting from the bottom of the watercourse down to the calculated embedment depth $t$. The higher spatial passive earth pressure, compared with the approach set out above, is taken into account through equivalent dolphin widths or shape factors. The equivalent dolphin width is calculated depending on the angle of internal friction of the soil and the embedment depth of the dolphin as shown in Eqs. 74–77 of DIN 4085.

The curved line of the spatial passive earth pressure is calculated from the differentiation of the spatial active earth pressure in accordance with Eq. 78 (DIN 4085) for depth $z$, with Eqs. 74–77 (DIN 4085) being applied and derived. In contrast to the calculation for the resultant active earth pressure, a distinction must be made between a position near the surface ($d/z < 0.3$) and a deeper position ($d/z \geq 0.3$), where

- $d$    diameter of dolphin or width of dolphin perpendicular to direction in which the loads act (for cluster dolphins the distance between outside edges of outer piles)
- $z$    depth in the soil

The approach as well as the result of this differentiation can be found, for example, in [186].

In cohesive subsoils the passive earth pressure should be calculated using the undrained shear parameters $\varphi_u$ and $c_u$ if the load is applied quickly in comparison to the consolidation of the soil. The submerged unit weight of the relevant soil strata is assumed to be $\gamma'_{k,i}$.

The process is more complex in stratified subsoils as the different shear parameters and unit weights of the upper strata must be taken into account. Further guidance on determining the spatial earth pressure coordinates for homogeneous and stratified subsoils can be found in, for example, [186].

The first step in the design is to calculate the required embedment depth $t$ from the moment equilibrium at the base ($\Sigma M_{base} = 0$). Equivalent force $C$ is derived from the equilibrium of the horizontal forces as the difference between the mobilised spatial passive earth pressure and the forces acting while neglecting the active earth pressure:

$$C_{h,k,BLUM} = E^r_{ph,mob} - \Sigma F_{h,k,i}$$

where

- $C_{h,k,BLUM}$    horizontal component of Blum equivalent force
- $\Sigma F_{h,k,i}$    total of characteristic horizontal actions
- $E^r_{ph,mob}$    $= E^r_{ph,k}/(\gamma_Q \cdot \gamma_{Ep})$; horizontal component of mobilised spatial passive earth pressure
- $\gamma_Q$    partial safety factor for actions
- $\gamma_{R,e}$    partial safety factor for passive earth pressure

The equivalent force can be inclined at an angle of up to $\delta_{c,k} = +\frac{1}{3}\varphi$ with respect to a normal to the dolphin as long as equilibrium of vertical forces is assured [186]. Verifying that the vertical forces can be accommodated is carried out with the reduced $C$ force value by including an allowance on the embedment depth in accordance with R 4, section 8.2.4.3:

$$\Delta t = \frac{1}{2} \cdot C_{h,k,BLUM} \cdot \gamma_Q \cdot \frac{\gamma_{R,e}}{e^r_{ph,k}}$$

transverse to the pile axis. No local analysis of the tension in the soil ($\sigma_{h,k} \leq e_{ph,k}$) is required here because it is automatically satisfied by the *p-y* curves. In a global analysis of the tension in the soil, the design value of the mobilised lateral soil force $F_{h,d}$ up to the point of zero displacement may not be greater than the design value of the maximum passive earth pressure $E_{ph,d}$ to be mobilised:

$$F_{h,d} = \int p \cdot dz \cdot \gamma_Q \leq E_{ph,d} = \int p_u^* \cdot dz / \gamma_{R,e}$$

Here, $p_u^*$ is the maximum value of the *p-y* curve. Where the shape of the *p-y* curve follows Fig. R 218-6b, $p_u^* = p_{max}$ must be applied if the displacement *y* is smaller than that required to achieve $p_{max}$, otherwise $p_u^* = p_{rest}$.

Unlike the Blum method, the *p-y* method does not supply the embedment depth of the dolphin directly. This must normally be determined iteratively. To do this, a secondary condition must be introduced. This could be, for example, the embedment depth that satisfies the ultimate limit state analysis according to EC 7. Section 7.7 of the Eurocode 7 Manual sets out the procedure for this.

Alternatively, the embedment depth can be determined by limiting the horizontal deformation of the dolphin. A common and proven approach in practice is to determine the deformations of the dolphin for different embedment depths. The deflection reduces degressively as the depth of embedment increases. The embedment depth is then set at the depth at which the deflection no longer decreases significantly. Occasionally, the embedment depth is set at the depth at which the deflection curve becomes vertical. This leads to very large embedment depths that are unnecessary for stability.

The *p-y* curves can generally be used for designing dolphins for static loads, but a check should be carried out to establish whether the lateral support must be reduced for repetitive cyclic loads (see section 13.2.3.3). Further guidance on designing dolphins using the *p-y* method plus application examples can be found in, for example, [186].

#### 13.2.1.4 Soil–structure interaction for cluster dolphins

Cluster dolphins are groups of piles with a pile spacing that is less than six times the pile diameter. When designing such dolphins, the overlapping of the areas from which the spatial passive earth pressure is generated means that the passive earth pressure of an individual pile cannot be used.

Reducing the active earth pressure compared with that of the individual pile depends on the pile spacing, the pile diameter and the arrangement of the piles. It can be estimated in accordance with, for example,

*Recommendations on Piling* [55] or DIN 4085. Further methods are described in the literature, e.g. [143].

### 13.2.2 Required energy absorption capacity of breasting dolphins

#### 13.2.2.1 General remarks on the energy absorption capacity of dolphins

The energy absorption capacity of a dolphin, or rather the deformation energy of the dolphin (plus fender if fitted), must be as large as the kinetic energy of the berthing vessel (R 60, section 6.15). At the same time, neither the permissible stresses in the dolphin components nor the deformation limits may be exceeded. Please refer to section 13.2.3 for more information.

Vessels cannot always be manoeuvred to the middle of a berth. When designing breasting dolphins, the energy absorption capacity should therefore always be calculated for a distance $e = 0.1 \cdot 1 \leq 15$ m (parallel to the row of fenders) between the vessel's centre of gravity and the centre of the berth. For tankers, an eccentricity of the transfer connections (manifold) with respect to a vessel's centre of gravity may need to be allowed for as well.

#### 13.2.2.2 Special guidance for seaports

Where a ship is manoeuvred to a dolphin berth with the help of tugboats, it can be assumed that it will hardly move in the direction of its longitudinal axis. Therefore, the longitudinal component of the berthing velocity can generally be ignored when calculating the berthing energy. As an approximation, the velocity vector υ for large vessels can be assumed to be perpendicular to distance $r$ ($\alpha = 90°$) (Fig. R 60-2, section 6.15.4.1).

In practice, all dolphins in a berth are generally of the same type. In principle, the inner dolphins could also be designed for lower loads than the outer ones. In this case it should then be taken into account that the berthing angle for large vessels docking with tug assistance is generally smaller than that for smaller vessels that dock without tugs.

#### 13.2.2.3 Special guidance for inland ports

For dolphins in inland ports, using the approach of R 60, section 6.15, often leads to the kinetic energy of the vessel being overestimated. This can result in dolphin dimensions that are uneconomic.

The required energy absorption capacity of dolphins in inland ports should therefore be governed by economic considerations. A low energy absorption capacity leads to lightweight dolphins with lower investment costs but also to potentially higher maintenance costs for any damage to the dolphins. In the end this can be more economical than specifying a heavy structure with a large energy absorption capacity. Choosing

**Table R 218-2.** Recommended maximum berthing pressures at berths with fender panels

| Capacity (dwt) | ≤20 000 | 40 000 | 60 000 | 80 000 | 100 000 | ≥120 000 |
|---|---|---|---|---|---|---|
| Pressure (kN/m²) | 400 | 350 | 300 | 250 | 200 | 150 |

between the two options must take account of local conditions (likelihood of damage to the dolphins).

For the standard vessel sizes in use on German inland waterways, according to TAB (1996) [64], an energy absorption capacity of 70–100 kNm, occasionally also 120 kNm, has proved expedient.

### 13.2.3 Other calculations

#### 13.2.3.1 Berthing/Hull pressure

The maximum permissible berthing force $F_{R,max}$ between a vessel and a breasting dolphin is determined by the permissible berthing pressure of the vessel. Where no more detailed information is available, the maximum berthing pressures (under unfavourable berthing conditions) for dolphins with fender panels can be taken from Table R 218-2 depending on the loading capacity of the ship.

Higher berthing pressures may be permitted if it can be proved that the load can be accommodated by the shells and frames of the vessels using the berth.

For gas tankers the permissible berthing pressure may be lower than the values given in Table R 218-2. Vessels with rubbing strips and steel belts all round, e.g. ferries, require special constructional measures for fenders.

#### 13.2.3.2 Deformations

Unless particular circumstances dictate otherwise, e.g. a greater energy absorption capacity than actually required at berths for large vessels, the maximum dolphin deflection $s_{max}$ should not exceed approx. 1.5 m because otherwise the impact between vessel and dolphin becomes so soft that the skipper can no longer recognise clearly enough when the vessel touches the dolphin.

Furthermore, the advice given in section 13.3.2 should be followed.

#### 13.2.3.3 Structural design of dolphins

Breasting dolphins are generally designed elastically, exploiting the yield strength. The unused plastic reserve acts as a safety margin for unforeseen berthing procedures.

Owing to the wind and wave loads frequently acting on a vessel, mooring and breasting dolphins may need to be designed for fatigue as well (see section 13.1.4.2). At least twice the number of load cycles for the anticipated design life should be assumed here (EN ISO 19902). Local scour (e.g. due to currents flowing around the dolphin or propeller wash) should be taken into account in the design. Please refer to [100] for more information on depth of scouring. The scour depth should be taken as the design depth over the whole area.

## 13.3 Construction and arrangement of dolphins (R128)

### 13.3.1 Type of dolphin structure

A dolphin can be designed as a single pile, a row of piles or a group of piles (Fig. R 218-1). The type chosen depends on the loads to be accommodated, the sections and materials available, the subsoil, the function of the dolphin, etc.

### 13.3.2 Layout of dolphins

At least two breasting dolphins are needed to moor a vessel. The distances between dolphins should be chosen so that the straight part of the hull (called the parallel body) of the smallest vessel using the berth is in contact with at least two dolphins. At the same time, the distance between dolphins should ensure that the ensuing lever arms for the mooring lines are not too short. In practice, a dolphin spacing of 25–40 % of the vessel's length has proved suitable.

To cover all likely vessel lengths, more than two dolphins are normally required at a berth. The outer dolphins are normally more heavily loaded than the inner ones as the largest vessels always berth at the outer dolphins. The inner dolphins can therefore be designed for lower loads. However, their design should still protect them against overloading. This can be done, for example, by setting the line of the inner dolphins back from that of the outer dolphins.

The spacing of dolphins designed to protect waterfront structures should be close enough to prevent any contact between vessel and structure. A dolphin spacing $\leq 15\%$ of the length of the shortest vessel is generally sufficient. The distance between a fully deflected dolphin and the quay wall should be at least 0.5 m. The same distance should be maintained for each part of the vessel's hull except for the point of contact, with an unfavourable listing angle of min. 3° being assumed. Providing a dolphin pile with a small outward rake can help to meet both these requirements. If the dolphin is fitted with a fender, this should be placed at a height that allows all vessels to safely berth for all combinations of water level and loading conditions. With a view to the simple construction and maintenance of the key parts of the structure, such parts should be above the water table for as much of the time as is possible.

### 13.3.3 Equipment for dolphins

Breasting dolphins are fitted with bollards, mooring hooks and/or quick release hooks. Bollards are proven and reliable mooring devices that require almost no maintenance. Slip hooks have the advantage that mooring lines – even those under tension – can be released in emergencies if required, which is impossible with bollards. This can be especially necessary at berths where hazardous materials are being loaded/unloaded, which is why slip hooks are often specified in such cases. However, as the mechanism can also be released accidentally, which can result in damage, e.g. the rope becoming entangled in a vessel's propeller, slip hooks should really only be used where the safety requirements of particular transhipment operations dictate this.

Where bollards with break-off bolts are mounted on dolphins, high dynamic loads occur once the bollard is torn off. These loads must be taken into account when designing the dolphin.

The edges of dolphin top side deck should have nosing (rope protection) to prevent chaffing hawsers. Such edge protection can, for example, consist of steel pipes placed wherever the ropes could make contact with the dolphin.

Dolphins are often connected by walkways (called catwalks) to ensure quick and safe operations. Walkways must be designed to absorb safely any potential deformation of the dolphins during berthing manoeuvres. Ladders that reach down to the lowest water level should also be fitted so that anyone falling into the water can climb out again. The ladders can at the same time provide access to a pilot boat, for instance. It should be possible to access dolphins safely from the land via a walkway that can also serve as an escape route.

Breasting dolphins are often fitted with fenders. These are attached with bolts on a steel plate mounted on the dolphin cap. The connection should ensure that precipitation and spray water can easily run off and that mooring lines cannot get caught.

Where the lighting from the adjacent quayside is inadequate, each dolphin should have its own lighting on the top to ensure safe working at night as well. However, the lighting should not have an adverse effect on navigation lights.

Systems to control and monitor the berthing and mooring process on the dolphins are available. These include systems for measuring the berthing speed, which activate an alarm when a critical speed is exceeded, or slip hooks that continually measure the tensile forces and regulate the tensioning of the ropes, also dolphin position loggers that provide conclusive information in the event of a dolphin being damaged. Such systems have proved particularly beneficial for berths that handle hazardous goods as well as in unprotected areas. The relevant figures are displayed on large illuminated panels. The current trend is towards GPS-controlled portable systems that do not require expensive installations on the berths and can be read regardless of location.

**13.3.4 Advice for selecting materials**

The energy absorption capacity of tubular pile dolphins can be increased while saving materials by manufacturing the dolphin from tubular sections with different wall thicknesses and grades of material. The use of fine-grained structural steel also increases the energy absorption capacity. The high permissible contact loads as well as the large deformations associated with such high loads result in an increased absorption of energy. Data on materials is given in R 67, section 8.1.6.

It is worthwhile building the topmost part of a dolphin from weldable low-strength fine-grained structural steel or structural steel with a strength $\leq 355\,\text{N/mm}^2$ so that it is easy to weld on bracing members and other structural components.

R 99, section 1.18, applies to all welding work accordingly. The information given in R 35, section 8.1.8, applies to the corrosion protection of steel piles accordingly.

# 14 Inspection and monitoring of waterfront structures (R 193)

## 14.1 General

As with other hydraulic engineering structures, regular inspections of waterfront structures are necessary under consideration of the various potential risks and the robustness of the structure, so that any damage that could influence the loadbearing capacity and serviceability of the structure is identified in good time. On the basis of such inspections, the persons responsible for maintenance can thus assume responsibility for guaranteeing the safety and soundness of the structure and its fitness for purpose.

Careful, regular and consistent inspections of the structure also help in the planning and management of maintenance work. Higher refurbishment costs or the premature replacement of a structure can therefore also be avoided.

Structures can be categorised according to their structural and constructional design, or rather their structural properties (robustness), and the consequences of any damage. This is set out in more detail in the guidance contained in VDI Guideline 6200 [210] and Argebau [9]. Using this classification, it is then possible to plan and specify the type and regularity of inspections and examinations of waterfront structures or port/harbour structures that do not need to comply with DIN 1076 or other provisions (see also Table R 193-1).

Irrespective of the potential hazards or a structure's robustness or structural/constructional design, taking into account the individual case based on DIN 1076 for civil engineering structures, a *structural inspection of* a *waterway structure* consists of:

- *structural check* – a close inspection of all the parts of the structure, including those difficult to access, by experienced engineers who assess the structural, constructional and hydromechanical conditions in relation to the requirements imposed by the structure's use and can instruct divers where necessary,
- *structural monitoring* – intensive, comprehensive visual inspections of the structures by experienced engineers, and
- *structural survey* – examining the structure for any obvious damage by experienced personnel familiar with the structural behaviour and functions of the structure.

---

*Recommendations of the Committee for Waterfront Structures Harbours and Waterways – EAU 2012,*
9th Edition. Issued by the Committee for Waterfront Structures of the German Port Technology Association and the German Geotechnical Society.
© 2015 Ernst & Sohn GmbH & Co. KG. Published 2015 by Ernst & Sohn GmbH & Co. KG.

**Table R 193-1.** Inspection intervals for waterfront structures

| Robustness to [210] | Risk potential to [210] | Structural survey[1] | Structural monitoring[2] | Structural check[3] |
|---|---|---|---|---|
| RC 1 to RC 4 | low (CC1) | annually | – | – |
| RC 3 and RC 4 | moderate (CC2) | annually | – | – |
| RC1 and RC 2 | | annually | 3 years after structural check | every 6 years |

[1] Corresponds to structural monitoring according to DIN 1076 and "inspection by the owner" according to VDI 6200 [210].
[2] Corresponds to the simple check according to DIN 1076 and the "inspection by an experienced person" according to VDI 6200 [210].
[3] Corresponds to the main check according to DIN 1076 and the "detailed inspection by a very experienced person" according to VDI 6200 [210].

The following provisions and regulations applicable to certain areas should be referred to when planning and carrying out structural inspections (listed in alphabetical order):

- Argebau, Conference of German Building Ministers: "Hinweise für die Überprüfung der Standsicherheit von baulichen Anlagen durch den Eigentümer/Verfügungsberechtigten" [9]
- BAW Code of Practice: Asset Inspection (MBI), Federal Waterway Engineering & Research Institute [13]
- BAW Code of Practice: Classifying Waterway Construction Damages (MSV), Federal Waterway Engineering & Research Institute [14]
- DIN 1076:1999-11 – Engineering structures in connection with roads – Inspection and test
- DIN 19702:2013-02 – Solid structures in hydraulic engineering – Bearing capacity, serviceability and durability (in German)
- ETAB: Technische Empfehlungen und Berichte, Association of German Public Inland Ports [61]
- PIANC Report 2004: "Inspection, maintenance and repair of maritime structures exposed to material degradation caused by a salt water environment" MarCom Report of WG 17 [160]
- PIANC Report 1998: "Life cycle management of port structures – General principles", Report of WG 31 [157]
- PIANC-Report 2006: "Maintenance and renovation of navigation infrastructure", InCom Report of WG 25 [161]
- RÜV-Richtlinie für die Überwachung der Verkehrssicherheit von baulichen Anlagen des Bundes [186]
- VDI Guideline 6200: "Structural safety of buildings – Regular inspections" [210]

- VV-WSV 2101 Bauwerksinspektion [216]
- VV-WSV 2301 Damminspektion [217]

## 14.2 Documentation

Structural inspection files and logbooks form the basis of the structural inspections. They contain all the information needed for an inspection, including the structure's key design data together with sketches, as-built drawings, as-built calculations, measurement software, measurement results (e.g. soundings, measurements taken during construction, reference measurements after completion), information on corrosion protection, reports, information on repairs and the results of previous inspections (examination, monitoring and survey reports).

Checklists or task sheets are suitable for systematically recording the condition of a structure, especially if the as-built structure is very heterogeneous in terms of its construction, design and the materials used. Task sheets should contain guidance on preparing and carrying out structural inspections and should also describe typical signs of damage for each component and material used. The structural, constructional and hydromechanical conditions should be taken into account. As preparation for inspections, the task sheets can also include important information about the inspection intervals and activities specific to that structure such as the use of divers, lifting gear, vehicles, etc.

The recommendation is to set standard criteria and assessment principles to ensure maximum consistency when assessing damage. To assess the condition of waterway structures, these principles are summarised in, for example, the BAW Code of Practice "Classifying Waterway Construction Damages" [14]. VDI Guideline 6200 "Structural safety of buildings – Regular inspections" [210] also provides guidance on changes to and ascertaining the properties of the building materials used.

Using specially developed software to record the structural inspections also supports the standard presentation of reports and gives those responsible for maintenance the opportunity to investigate further any damage found. Examples include the WSV-Pruf software developed by the Wasser- und Schifffahrtsverwaltung des Bundes (Federal Waterways & Shipping Authority) and the SIB-Bauwerke software developed by the Federal Highway Research Institute, although this is very much geared towards highway structures (bridges etc.).

## 14.3 Carrying out structural inspections

### 14.3.1 Structural check/Principle check

During a structural check, all parts of the structure, including those difficult to access, should be examined closely with the help of any equipment required to do so. Where necessary, individual parts of the structure should be carefully cleaned before the structural check so that any hidden damage can be identified. The nature and scope of the structural check should be specified for each structure in accordance with the local conditions, e.g. with the help of task sheets.

Depending on their relevance and the boundary conditions plus the robustness of the structure and its design (loadbearing systems, damage processes, advance warning of failure), inspections of waterfront structures particularly at risk should include checking and measuring the following:

- Location and extent of damage to a quay wall, using divers where necessary
- Condition and functionality of drainage systems
- Condition of previous repairs
- Condition of corrosion protection coatings
- Condition of cathodic protection system
- Settlement and subsidence behind quay wall
- Soundings of watercourse bed in front of quay wall
- Seals at joints and connections
- Movements at joints and supports
- Damage to concrete (including steel reinforcing bars)
- Measurements of horizontal movements (including head deformations) and vertical movements (settlement, uplift)
- Measurements of residual wall thickness (mean and maximum corrosion rates)

Further measurements may be necessary for particular structures, e.g. measurement of anchor forces, inclinometer readings, potential field measurements, etc.

The inspection report should describe any damage found and assess the suspected or confirmed causes as well as the loadbearing capacity and serviceability of the structure. The structural check may also involve structural analyses depending on the residual wall thicknesses of sheet piling. The inspection report should also set out any further action required.

## 14.3.2 Structural monitoring/Intermediate inspection

Structural monitoring involves carrying out an intensive, comprehensive visual inspection of all accessible parts of the structure where this is feasible without draining or using inspection equipment. However, use can be made of inspection and access facilities already available on the structure. The nature and scope of the structural monitoring should be specified for each particular structure depending on the local boundary conditions, e.g. with the help of task sheets. As a minimum, the water-side of the structure should be subjected to an intensive visual inspection from a boat during low water. For waterfront structures particularly at risk, the following abnormalities are generally relevant for structural monitoring, depending on the structural and constructional design of the structure (Table R 193-1):

- Damage or changes to surfaces
- Subsidence, settlement, displacements
- Changes to joints and connections
- Damage to or missing equipment
- Improper usage
- Functionality of drainage system
- Damage to sheet pile walls
- Scouring or accretions in front of a sheet pile wall

The monitoring report should describe any damage found and assess the suspected or confirmed causes as well as the loadbearing capacity and serviceability of the structure. The monitoring report should also set out any further action required.

### 14.3.3 Structural survey/Routine inspection

During a structural survey, the structure should be closely inspected without any significant aids as such survey vehicles, scaffolding, etc. However, any available survey equipment can be used and any accessible voids in the structure subjected to a visual inspection. The fitness for purpose, the overall condition of the structure and any abnormalities should be investigated.

As a minimum, the water-side of the structure should be subjected to an intensive visual inspection from a boat during low water.

The following items in particular should be verified during a structural survey:

- Unusual changes to the structure, significant changes to damage
- Significant damage to or missing equipment
- Significant concrete spalling, noticeable cracks
- Deformations to or displacements of the structure visible to the naked eye
- Unusual water discharges
- Damage to embankments or slopes, scouring, accretions

Any significant damage or unusual changes identified and any further actions that should follow are to be described in the structural survey report.

## 14.4 Inspection intervals

Carrying out inspections no less frequently than every six years has proved effective for waterway structures (see [221]) and engineered highway structures. For waterfront structures particularly at risk, the recommendation is therefore to carry out inspections at least every six years depending on the structural and constructional design of the structure (see Table R 193-1). The structure should be checked for the first time during the acceptance procedure for the structure. A further check should then be carried out before the end of the warranty period for any claims relating to defects in accordance with VOB/B, the German construction contract procedures [217]. Levelling and alignment should be included and reference levels and positions taken. Further follow-up measurements should be specified as part of the structural checks.

At the very latest three years after a structural check, all waterfront structures that require such checks (see Table R 193-1) should undergo structural monitoring.

The structural survey set out in section 14.3.3 should normally be carried out annually for all waterfront structures. A structural survey is normally sufficient for most waterfront structures due to the robustness of such structures and the consequences of any damage caused (see [212]). This recommendation is based on experience of ports, harbours and quay facilities acquired over many years. Reference values relating to the inspection intervals can be found in Table R 193-1.

## 14.5 Maintenance management systems

Regular inspections of the structure documented in a uniform manner provide a sound overview of the constructional condition of the structures for which maintenance is necessary.

These inspections therefore form the main foundation for maintenance management systems, which systematise and optimise decision-making for maintenance strategies and measures. Financial resources can then be allocated effectively in good time and used efficiently.

Using this type of maintenance management system, appropriate intervention times can be specified, for example, on the basis of scores assigned to conditions and specific damage development models. Funding for maintenance activities should then be allocated which – along with other factors taken into account when prioritising maintenance activities – enables objective sequencing of maintenance activities.

For a number of years, different national and international management systems have been in development across different areas of the transport infrastructure. The first applications intended for waterway infrastructures of the Wasser- und Schifffahrtsverwaltung des Bundes (Federal Waterways & Shipping Authority) have already appeared.

# Annex I   Bibliography

## I.1   Annual technical reports

The basis of this compendium are the Annual Technical Reports of the Committee for Waterfront Structures published in the journals *Die Bautechnik* (*Bautechnik* from 1984 onwards) and *HANSA*; in particular the following:

| | | |
|---|---|---|
| *HANSA* | 87 (1950), No. 46/47, p. 1524 | |
| *Die Bautechnik* | 28 (1951), No. 11, p. 279 | 29 (1952), No. 12, p. 345 |
| | 30 (1953), No. 12, p. 369 | 31 (1954), No. 12, p. 406 |
| | 32 (1955), No. 12, p. 416 | 33 (1956), No. 12, p. 429 |
| | 34 (1957), No. 12, p. 471 | 35 (1958), No. 12, p. 482 |
| | 36 (1959), No. 12, p. 468 | 37 (1960), No. 12, p. 472 |
| | 38 (1961), No. 12, p. 416 | 39 (1962), No. 12, p. 426 |
| | 40 (1963), No. 12, p. 431 | 41 (1964), No. 12, p. 426 |
| | 42 (1965), No. 12, p. 431 | 43 (1966), No. 12, p. 425 |
| | 44 (1967), No. 12, p. 429 | 45 (1968), No. 12, p. 416 |
| | 46 (1969), No. 12, p. 418 | 47 (1970), No. 12, p. 403 |
| | 48 (1971), No. 12, p. 409 | 49 (1972), No. 12, p. 405 |
| | 50 (1973), No. 12, p. 397 | 51 (1974), No. 12, p. 420 |
| | 52 (1975), No. 12, p. 410 | 53 (1976), No. 12, p. 397 |
| | 54 (1977), No. 12, p. 397 | 55 (1978), No. 12, p. 406 |
| | 56 (1979), No. 12, p. 397 | 57 (1980), No. 12, p. 397 |
| | 58 (1981), No. 12, p. 397 | 59 (1982), No. 12, p. 397 |
| | 60 (1983), No. 12, p. 405 | |
| *Bautechnik* | 61 (1984), No. 12, p. 402 | 62 (1985), No. 12, p. 397 |
| | 63 (1986), No. 12, p. 397 | 64 (1987), No. 12, p. 397 |
| | 65 (1988), No. 12, p. 397 | 66 (1989), No. 12, p. 401 |
| | 67 (1990), No. 12, p. 397 | 68 (1991), No. 12, p. 398 |
| | 69 (1992), No. 12, p. 710 | 70 (1993), No. 12, p. 755 |
| | 71 (1994), No. 12, p. 763 | 72 (1995), No. 12, p. 817 |
| | 73 (1996), No. 12, p. 844 | 75 (1998), No. 12, p. 992 |
| | 76 (1999), No. 12, p. 1062 | 77 (2000), No. 12, p. 909 |
| | 78 (2001), No. 12, p. 872 | 79 (2002), No. 12, p. 850 |
| | 80 (2003), No. 12, p. 903 | 81 (2004), No. 12, p. 980 |
| | 82 (2005), No. 12, p. 857 | 83 (2006), No. 12, p. 842 |
| | 84 (2007), No. 7, p. 496 | 84 (2007), No. 12, p. 849 |
| | 85 (2008), No. 8, p. 512 | 85 (2008), No. 12, p. 812 |
| | 86 (2009), No. 8, p. 465 | 86 (2009), No. 12, p. 780 |
| | 87 (2010), No. 2, p. 124 | 87 (2010), No. 12, p. 761 |
| | 88 (2011), No. 12, p. 848 | |

*Recommendations of the Committee for Waterfront Structures Harbours and Waterways – EAU 2012*, 9th Edition. Issued by the Committee for Waterfront Structures of the German Port Technology Association and the German Geotechnical Society.
© 2015 Ernst & Sohn GmbH & Co. KG. Published 2015 by Ernst & Sohn GmbH & Co. KG.

## I.2 Books and papers

[1] Abromeit, H.-U. (1997): Ermittlung technisch gleichwertiger Deckwerke an Wasserstraßen und im Küstenbereich in Abhängigkeit von der Trockenrohdichte der verwendeten Wasserbausteine. Mitteilungsblatt der Bundesanstalt für Wasserbau, Heft 75, Karlsruhe.
[2] Achmus, M.; Kaiser, J. und Wörden, F. T. (2005): Bauwerkserschütterungen durch Tiefbauarbeiten, Grundlagen, Messergebnisse, Prognosen. Mitteilungen Institut für Grundbau, Bodenmechanik und Energiewasserbau (IGBE), Universität Hannover, Heft 61.
[3] AHU der HTG: Empfehlungen und Berichte des Ausschusses für Hafenumschlagtechnik (AHU) der Hafenbautechnischen Gesellschaft e. V, Hamburg.
[4] AK Numerik (1991): Empfehlungen des AK Numerik in der Geotechnik. Deutsche Gesellschaft für Geotechnik. Geotechnik 14, S. 1–10.
[5] Alberts, D. (2001): Korrosionsschäden und Nutzungsdauerabschätzung an Stahlspundwänden und –pfählen im Wasserbau. 1. Tagung „Korrosionsschutz in der maritimen Technik", Germanischer Lloyd, Hamburg.
[6] Alberts, D. und Heeling, A. (1996): Wanddickenmessungen an korrodierten Stahlspundwänden; statistische Datenauswertung. Mitteilungsblatt der BAW Nr. 75, Karlsruhe.
[7] Alberts, D. und Schuppener, B. (1991): Comparison of ultrasonic probes for the measurement of the thickness of sheet-pile walls. Field Measurements in Geotechnics, Sørum (ed.), Balkema, Rotterdam.
[8] Andrews, J. D. und Moss, T. R. (1993): Reliability and Risk Assessment, Verlag Longman Scientific & Technical, Burnt Mill (UK).
[9] API (2000): Recommended Practice for Planning, Designing and Constructing Fixed Offshore Platforms - Working Stress Design, API RP2A-WSO, 21, edition 2000.
[10] Argebau: Hinweise für die Überprüfung der Standsicherheit von baulichen Anlagen durch den Eigentümer/Verfügungsberechtigten. Bauministerkonferenz (2006).
[11] Barron, R. A. (1948): Consolidation of fine-grained soils by drain wells. Trans. ASCE, Vol. 113, Paper No 2346.
[12] Battjes, J. A. (1975): Surf Similarity. Proc. of the 14th International Conference on Coastal Engineering. Copenhagen 1974, Vol. I.
[13] Baumaschinen-LärmVO: 15. Verordnung zur Durchführung des BImSchG vom 10.11.1986 (Baumaschinen-LärmVO).
[14] BAW-Merkblatt: Bauwerksinspektion (MBI). Bundesanstalt für Wasserbau (2010).
[15] BAW-Merkblatt: Schadensklassifizierung an Verkehrswasserbauwerken (MSV). Bundesanstalt für Wasserbau (2011).
[16] BAW-Merkblatt: Standsicherheit von Dämmen der Bundeswasserstraßen (MSD). Bundesanstalt für Wasserbau (2011).
[17] BAW-Merkblatt: Rissbreitenbegrenzung für frühen Zwang in massiven Wasserbauwerken (MFZ). Bundesanstalt für Wasserbau (2011).
[18] BAW (2004): Grundlagen zur Bemessung von Böschungs- und Sohlensicherungen an Binnenwasserstraßen. Bundesanstalt für Wasserbau, Mitteilungsheft 87, Karlsruhe.
[19] Binder, G. (2001): Probleme der Bauwerkserhaltung – eine Wirtschaftlichkeitsberechnung. BAW-Brief Nr. 1, Karlsruhe.
[20] Binder, G. und Graff, M. (1995): Mikrobiell verursachte Korrosion an Stahlbauten. Materials and Corrosion **46**, S. 639–648.

[21] Bjerrum, L. (1973): Problems of soil mechanics and constructions on soft clays and structurally unstable soils (collapsible, expansive and others). Proc. of 8th ICSMFE, Moscow, Vol. 3, pp. 111–155.
[22] Blum (1932): Wirtschaftliche Dalbenformen und deren Bemessung. Bautechnik, **10** (5), 1932, Seiten 50–55.
[23] Blum, H. (1931): Einspannverhältnisse bei Bohlwerken. Verlag Ernst & Sohn, Berlin 1931.
[24] Brennecke, L. und Lohmeyer, E. (1930): Der Grundbau. 4. Auflage, **II.** Bd.,Verlag Ernst & Sohn, Berlin 1930.
[25] Brinch Hansen, J. (1953): Earth pressure calculations. The Danish Technical Press, Kopenhagen.
[26] Brinch Hansen, J. und Lundgren, H. (1960): Hauptprobleme der Bodenmechanik. Springer Verlag, Berlin 1960.
[27] BRL 1120 (1997): Nationale Beoordelingsrichtlijn, Geokunststoffe: Geprefabriceerde verticale drains. Rijswijk, Kiwa.
[28] Broughton, P. und Horn, E. (1987): Ekofisk platform 2/4C: Re-analysis due to subsidence. Proc. Inst. Civ. Engrs.
[29] Bruderreck, L.; Rökisch, K.; Schmidt, E. (2011): Kritische Propellerdrehzahl bei Hafenmanövern als Basis zur Bemessung von Sohlsicherungen. Hansa, **148** Jhg. Nr. 5.
[30] Brunn, P. (1980): Port Engineering. London.
[31] BSH (2001): Eiskarten der deutschen Nord- und Ostseeküste (seit 1879). Bundesamt für Seeschifffahrt und Hydrographie.
[32] Bundesverband öffentlicher Binnenhäfen: Empfehlungen des Technischen Ausschusses Binnenhäfen. Neuss.
[33] Burkhardt, O. (1967): Über den Wellendruck auf senkrechte Kreiszylinder. Mitteilungen des Franzius-Instituts Hannover, Heft **29**.
[34] Busse, M. (2009): Einpressen von Spundwänden – Stand der Verfahrens- und Maschinentechnik, bodenmechanische Voraussetzungen. Tagungsband zum Workshop Spundwände – Profile, Tragverhalten, Bemessung, Einbringung und Wiederverwendung, Veröffentlichungen des Instituts für Geotechnik und Baubetrieb TU Hamburg Harburg, Heft **19**, S. 27–45.
[35] Bydin, F. I. (1959): Development of certain questions in area of river's winter regime. III. Hydrologic Congress, Leningrad.
[36] Camfield, F. E. (1991): Wave Forces on a Wall. J. Waterway, Port, Coastal and Ocean Engineering **117** (1), 76–79, ASCE, New York.
[37] CEM (2001): Costal Engineering Manual Part VI. Design of Coastal Projects Elements. US Army Corps of Engineers, Washington D. C.
[38] CFEM (2006): Canadian Foundation Engineering Manual. 4th Edition, Canadian Geotechnical Society.
[39] CIRIA/CUR (1991): Manual on the use of rock in coastal and shoreline engineering. Ciria Special Publication 83, Cur Report 154, Rotterdam, A. A. Balkema 1991.
[40] Clasmeier, H.-D. (1996): Ein Beitrag zur erdstatischen Berechnung von Kreiszellenfangedämmen. Mitteilung des Instituts für Grundbau und Bodenmechanik, Universität Hannover, Heft **44**.
[41] Clausen, C. J. F.; Aas, P. M. and Karlsrud, K. (2005): Bearing capacity of driven piles in sand, the NGI approach. International Symposium on Frontiers in Offshore Geotechnics (ISFOG2005), Perth, pp. 677–681.

[42] Construction and Survey Accuracies for the execution of dredging and stone dumping works. Rotterdam Public Works Engineering Department, Port of Rotterdam, The Netherlands Association of Dredging, Shore and Bank Protection Contractors (VBKO). International Association of Dredging Companies (IADC).
[43] CUR (2005): Handbook of Quay Walls. Centre for Civil Engineering Research and Codes, Taylor & Travess, Leiden.
[44] DAfStb (1990, 1991, 1992): Richtlinie für Schutz und Instandsetzung von Betonbauteilen. Teile 1 bis 4, Deutscher Ausschuss für Stahlbeton (DAfStb).
[45] DAfStb (2012): Richtlinie ,,Massige Bauteile aus Beton". Deutscher Ausschuss für Stahlbeton (DAfStb).
[46] Davidenkoff, R. (1964): Deiche und Erddämme, Sickerwasser-Standsicherheit. Werner Verlag Düsseldorf.
[47] Davidenkoff, R. (1970): Unterläufigkeit von Bauwerken. Werner Verlag Düsseldorf.
[48] Davidenkoff, R. und Franke, O. L. (1965): Untersuchungen der räumlichen Sickerströmung in einer umspundeten Baugrube im Grundwasser. Bautechnik **42**, Heft 9, 1965.
[49] Det Norske Veritas (1991): Environmental conditions and environmental loads. Classification Notes No. 30.5.
[50] Dietze, W. (1964): Seegangskräfte nichtbrechender Wellen auf senkrechte Pfähle. Bauingenieur **39** (9): 354.
[51] DNV (2010): Design of Offshore Wind Turbine Structures. Offshore Standard DNV-OS-1101. Det Norske Veritas.
[52] Drewes, U.; Römisch, K.; Schmidt E. (1995): Propellerstrahlbedingte Erosionen im Hafenbau und Möglichkeiten zum Schutz für den Ausbau des Burchardkais im Hafen Hamburg. Mitteilungen des Leichtweiß-Instituts für Wasserbau der Technischen Universität Braunschweig, H. **134**.
[53] DVWK (Deutscher Verband für Wasserwirtschaft und Kulturbau e. V.) (1990): Dichtungselemente im Wasserbau. DK 626/627 Wasserbau; DK 69.034.93 Abdichtung, Verlag Paul Parey, Hamburg, Berlin.
[54] DVWK (Deutscher Verband für Wasserwirtschaft und Kulturbau e. V.) (1992): Anwendung von Geotextilien im Wasserbau. Merkblatt 221.
[55] Dynamit Nobel AG (1993): Sprengtechnisches Handbuch. Dynamit Nobel Aktiengesellschaft (Hrsg.), Troisdorf.
[56] EA Pfähle (2012): Empfehlungen des Arbeitskreises ,,Pfähle". Deutsche Gesellschaft für Geotechnik e. V (Hrsg.), Verlag Ernst & Sohn, Berlin 2012.
[57] EAAW (2008): Empfehlungen für die Ausführung von Asphaltarbeiten im Wasserbau. Deutsche Gesellschaft für Geotechnik, 5. Ausgabe, nur digital verfügbar auf http://www.dggt.de/images/PDF-Dokumente/eaaw2008.pdf.
[58] EAB (2006): Empfehlungen des Arbeitskreises ,,Baugruben". Deutsche Gesellschaft für Geotechnik e. V (Hrsg.), Verlag Ernst & Sohn, Berlin 2006.
[59] EAG-GTD (2002): Empfehlungen zur Anwendung geosynthetischer Tondichtungsbahnen. Deutsche Gesellschaft für Geotechnik (Hrsg.), Verlag Ernst & Sohn, Berlin 2002.
[60] EAK (2002): Empfehlungen für Küstenschutzbauwerke. Ausschuss für Küstenschutzwerke der DGGT und der HTG, ,,Die Küste" Heft 65-2002, Westholsteinische Verlagsanstalt Boyens & Co., Heide i. Holst.

[61] EAK (2002): Empfehlungen für Küstenschutzbauwerke. Ausschuss für Küstenschutzwerke der DGGT und der HTG, „Die Küste" Heft 65-2002, Westholsteinische Verlagsanstalt Boyens & Co., Heide i. Holst.
[62] EAO (2004): Empfehlungen zur Anwendung von Oberflächendichtungen an Sohle und Böschung von Wasserstraßen. Mitteilung Nr. 85, Bundesanstalt für Wasserbau, Karlsruhe 2004, digital verfügbar auf www.baw.de.
[63] EBGEO (2010): Empfehlungen für den Entwurf und die Berechnung von Erdkörpern mit Bewehrungen aus Geokunststoffen (EBGEO). Deutsche Gesellschaft für Geotechnik e. V. (Hrsg.), 2. Auflage, April 2010, 327 Seiten.
[64] Edil, T. B.; Roblee, C. J. und Wortley, C. A. (1988): Design approach for piles subject to ice jacking. Journal of Cold Regions Engineering Vol. **2**, Nr. 2, Paper 22508, American Society of Civil Engineers.
[65] ETAB: Technische Empfehlungen und Berichte. Bundesverband öffentlicher Binnenhäfen, Neuss, siehe: http://www.binnenhafen.de/die-themen/infothek/technische-empfehlungen-und-berichte.
[66] F. E. M. (1987): Federation Européenne de la Manutention, Section I, Rules for the design of hoisting appliances, Booklet 2: Classification and loading on structures and mechanisms F. E. M. 1.001. 3rd Edition, Deutsches National-Komitee Frankfurt/Main.
[67] Fages, R. und Gallet, M. (1973): Calculations for Sheet Piled or Cast in Situ Diaphragm Walls. Civil Engineering and Public Works Review, Dec.
[68] Fages, R. und Bouyat, C. (1971): Calcul de rideaux de paroismouileeset de palplanches. Travaux Nr. **439**, S. 49–51 und Nr. **441**, S. 38–46.
[69] Feile, W. (1975): Konstruktion und Bau der Schleuse Regensburg mit Hilfe von Schlitzwänden. Bauingenieur **50**, Heft 5, S. 168.
[70] FGSV-820 (1982): Merkblatt für die Fugenfüllung in Verkehrsflächen aus Beton. Forschungsgesellschaft für Straßen- und Verkehrswesen e. V.
[71] Galvin, C. H. Ir. (1972): Wave Breaking in Shallow Water, in Waves on Beaches. New York, Ed. R. E. Meyer,Academic Press.
[72] GBB (2010): Grundlagen zur Bemessung von Böschungs- und Sohlensicherungen an Binnenwasserstraßen. Merkblatt der Bundesanstalt für Wasserbau, Karlsruhe, digital verfügbar auf www.baw.de.
[73] Gebreselassie, B. (2003): Experimental, analytical and numerical investigations in normally consolidated soft soils. Schriftenreihe Geotechnik der Universität Kassel, Heft **14**.
[74] Germanischer Lloyd (1976): Vorschriften für Konstruktion und Prüfung von Meerestechnischen Einrichtungen, Band I – Meerestechnische Einheiten – (Seebauwerke). Hamburg, Eigenverlag des Germanischen Lloyd.
[75] Germanischer Lloyd (2005): Guideline for the construction of fixed offshore installations in ice infested waters, Rules and Guidelines IV-Industrial Services (Part 6, Chapter 7).
[76] Goda, Y. (2000): Random Seas and Design of Maritime Structures. University of Tokyo Press. 1985; auch 2. geänderte Auflage, Advanced Series of Ocean Engineering – Volume 15, World Scientific Singapore.
[77] Graff, M.; Klages, D. und Binder, G. (2000): Mikrobiell induzierte Korrosion (MIC) in marinem Milieu. Materials and Corrosion **51**, S. 247–254.
[78] Grundbau–Taschenbuch (2001). 6. Auflage, Teil 1, 2 und 3, Verlag Ernst & Sohn, Berlin 2001.

[79] Gudehus, G. (1981): Bodenmechanik. Ferdinand Enke Verlag, Stuttgart 1981.
[80] Hafner, E. (1977): Kraftwirkung der Wellen auf Pfähle. Wasserwirtschaft **67**, H. 12, S. 385.
[81] Hafner, E. (1978): Bemessungsdiagramme zur Bestimmung von Wellenkräften auf vertikale Kreiszylinder. Wasserwirtschaft **68** (7/8): 227.
[82] Hager, M. (1975): Untersuchungen über Mach-Reflexion an senkrechter Wand. Mitteilungen des Franzius-Instituts für Wasserbau und Küsteningenieurwesen der Technischen Universität Hannover, H. **42**.
[83] Hager, M. (1996): Eisdruck. Kap. 1.14 im Grundbau-Taschenbuch, 5. Auflage, Teil 1, Verlag Ernst & Sohn, Berlin 1996.
[84] Hager, M. (2002): Ice loading actions. Geotechnical Engineering Handbook, Vol **1**: Fundamentals. Chap. 1.14, Verlag Ernst & Sohn, Berlin 2002.
[85] Hamburg (2007): Freie und Hansestadt Hamburg – Berechnungsgrundsätze für Hochwasserschutzwände, Flutschutzanlagen und Uferbauwerke im Bereich der Freien und Hansestadt Hamburg.
[86] Handbuch Eurocode 7-1 (2011): Handbuch Eurocode 7 – Geotechnische Bemessung, Band **1**: Allgemeine Regeln.1. Auflage, Herausgeber: DIN Deutsches Institut für Normung e. V., Beuth Verlag, Berlin.
[87] Handbuch Eurocode 7-2 (2011): Handbuch Eurocode 7 – Geotechnische Bemessung, Band **2**: Erkundung und Untersuchung. 1. Auflage, Herausgeber: DIN Deutsches Institut für Normung e. V., Beuth Verlag, Berlin.
[88] Hansbo, S. (1976): Consolidation of clay by band-shaped prefabricated drains, Ground Engineering, Foundation Publications Ltd., July 1976.
[89] Hansbo, S. (1981): Consolidation of fine-grained soils by prefabricated Drains, 10th International Conference on Soil Mechanics and Foundation Engineering, Stockholm.
[90] Heibaum, M. (1987): Zur Frage der Standsicherheit verankerter Stützwände auf der tiefen Gleitfuge. Technische Hochschule Darmstadt, Fachbereich Konstruktiver Ingenieurbau, Diss., 1987. Erschienen in: Franke, E. (Hrsg.): Mitteilungen des Instituts für Grundbau, Boden- und Felsmechanik der Technischen Hochschule Darmstadt, Heft **27**.
[91] Heibaum, M. (1991): Kleinbohrpfähle als Zugverankerung – Überlegungen zur Systemstandsicherheit und zur Ermittlung der erforderlichen Länge. In: Institut für Bodenmechanik, Felsmechanik und Grundbau der Technischen Universität Graz (Veranst.): Bohrpfähle und Kleinpfähle – Neue Entwicklungen (6. Christian Veder Kolloquium). Graz, Institut für Bodenmechanik der Technischen Universität.
[92] Heil, H.; Kruppe, J. und Möller, B. (1997): Berechnungsansätze für HWS-Wände und Uferbauwerke. Hansa, **134** (5): 77 ff.
[93] Hein, W. (1990): Zur Korrosion von Stahlspundwänden in Wasser. Mitteilungsblatt der BAW Nr. **67**, Karlsruhe.
[94] Heiß, P.; Möhlmann, F. und Röder, H. (1992): Korrosionsprobleme im Hafenbau am Übergang Spundwandkopf zum Betonüberbau. HTG-Jahrbuch, 47.Bd.
[95] Henke, S. (2011): Numerical and experimental investigations of soil plugging in open-ended piles. Tagungsband zum Workshop Ports for Container Ships of Future Generations,Veröffentlichungen des Instituts für Geotechnik und Baubetrieb der TU Hamburg-Harburg, Heft **22**, S. 97–122.
[96] Henke, S. (2012): Large deformation numerical simulations regarding the soil plugging behaviour inside open-ended piles. 31st International Conference on

Ocean, Offshore and Arctic Engineering, Rio de Janeiro, Brasilien, digitally published under OMAE2012-83039.
[97] Henke, S. und Grabe J. (2008): Numerische Untersuchungen zur Pfropfenbildung in offenen Profilen in Abhängigkeit des Einbringverfahrens. Bautechnik **85** (8): 521–529.
[98] Henne, J. (1989): Versuchsgerät zur Ermittlung der Biegezugfestigkeit von bindigen Böden. Geotechnik, H. **2**, S. 96ff.
[99] Herdt, W.; Arndts, E. (1973): Theorie und Praxis der Grundwasserabsenkung. Verlag Ernst & Sohn, Berlin 1973.
[100] Hirayama, K. I.; Schwarz, J. und Wu, H. C. (1974): An investigation of ice forces on vertical structures. Iowa Institute of Hydraulic Research, IIHR Report No. **158**.
[101] Hoffmans, G. J. C. M. and Verheij, H. J. (1997): Scour Manual 1997. Published by Balkema, Rotterdam/Brookfield.
[102] Horn, A. (1984): Vorbelastung als Mittel zur schnellen Konsolidierung weicher Böden. Geotechnik, Heft **3**, S. 189.
[103] HTG (1985): Beziehung zwischen Kranbahn und Kransystem. Ausschuss für Hafenumschlagtechnik der Hafenbautechnischen Gesellschaft e. V., Hansa **122**, H. 21, S. 2215 und H. 22, S. 2319.
[104] HTG (1994): Kathodischer Korrosionsschutz für Stahlbeton. Hafenbautechnische Gesellschaft e. V. (HTG), Hamburg.
[105] HTG (1996): Hochwasserschutz in Häfen – Neue Bemessungsansätze. Tagungsband zum HTG-Sprechtag Oktober 1996, Hafenbautechnische Gesellschaft (HTG) e. V., Hamburg.
[106] HTG (2010): Empfehlungen des Arbeitsausschusses Sportboothäfen und wassertouristische Anlagen. Handlungsempfehlungen für Planung, Bau und Betrieb von Sportboothäfen und wassertouristischen Anlagen. Entwurf Mai 2010, in Bearbeitung.
[107] Hudson, R. Y. (1958): Design of quarry stone cover layers for rubble mound breakwaters. Waterway Experiment Station, Research Report No. 2-2, Vicksburg, USA.
[108] Hudson, R. Y. (1959): Laboratory investigations of rubble mound breakwaters. Waterway Experiment Station Report, Vicksburg, USA.
[109] Idriss, I. M. und Boulanger, R. W. (2004): Semi-Empirical Procedures For Evaluating Liquefaction Potential During Earthquakes. Joint 11th Int. Conference on Soil Dynamics & Earthquake Engineering (ICSDEE) and The 3rd International Conference on Earthquake Geotechnical Engineering (ICEGE) January 7–9, 2004, Berkeley CA, USA.
[110] ISO/FDIS 19906 (2010) (E): Petroleum and natural gas industries. Arctic offshore structures.
[111] Jamiolkowski, M.; Ladd, C. C.; Germaine, J. T. und Lancellotta, R. (1985): New developments in field and laboratory testing of soils. Proc. of 11th Int. Conf. on Soil Mechanics and Foundation Engineering in San Francisco (USA), Vol. 1, pp. 57–153.
[112] Jardine, R.; Chow, F.; Overy, R. and Standing, J. (2005): ICP design methods for driven piles in sand and clays. London, Thomas Telford.
[113] JSCE (1996) –Japan Society of Civil Engineering: The 1995 Hyogoken-Nanbu Earthquake – Investigation into Damage to Civil Engineering Structures – Committee of Earthquake Engineering, Tokyo 1996.
[114] Kaplan, P. (1992): Wave impact forces on offshore structures: re-examination and new Interpretations. Offshore Technology Conf., OTC 6814, Houston, USA.
[115] Kempfert, H.-G. (1996): Embankment foundation on geotextile-coated sand columns in soft ground. Proceedings of the 1st European geosynthetics conference EurGeo 1, Maastricht, Netherlands.

[116] Kempfert, H.-G. und Stadel, M. (1997 c): Berechnungsgrundlagen für Baugruben in normalkonsolidierten weichen bindigen Böden. Bauingenieur **72**, S. 207–213.
[117] Kirsch, K.; Sondermann, W. (2001): Baugrundverbesserung. In: Grundbautaschenbuch, Teil 2, 6. Auflage 2001.Verlag Ernst & Sohn, Berlin 2001.
[118] Kjellmann, W. (1948): Accelerating consolidation of fine grained soils by means of card-board wicks. 2nd International Conference on Soil Mechanics and Foundation Engineering.
[119] Koerner, R. M. (2005): Designing with Geosynthetics. Prentice-Hall, Englewood Cliffs, N. Y.
[120] Köhler, H.-J. (1997): Porenwasserdruckausbreitung im Boden, Messverfahren und Berechnungsansätze. Mitteilungen des Instituts für Grundbau und Bodenmechanik der Universität Braunschweig, Braunschweig, Heft **50**, S. 247–258.
[121] Köhler, H.-J. und Haarer, R. (1995): Development of excess pore water pressure in over-consolidated clay, induced by hydraulic head changes and its effect on sheet pile wall stability of a navigable lock. Of the 4th Int. Symp. on Field Measurements in Geomechanics (FMGM 95), Bergamo, SG Editorial Padua, pp. 519–526.
[122] Köhler, H.-J. und Schulz, H. (1986): Bemessung von Deckwerken unter Berücksichtigung von Geotextilien. 3. Internationale Konferenz über Geotextilien in Wien 1986, Rotterdam, A. A. Balkema.
[123] Kohlhase, S.; Dede, Ch.; Weichbrodt, F. und Radomski, J. (2006): Empfehlungen zur Bemessung der Einbindelänge von Holzpfählen im Buhnenbau, Ergebnisse des BMBF-Forschungsvorhabens Buhnenbau. Universität Rostock.
[124] Kokkinowrachos, K. (1980) in: „Handbuch der Werften", Bd. **15**, Hamburg.
[125] Koppejan, A. W. (1948): A Formular combining the Therzaghi Load-compression relationship and the Buisman secular time effect. Proceedings 2nd Int. Conf. On Soil Mech. And Found. Eng.
[126] Kortenhaus, A. und Oumeraci, H. (1997): Lastansätze für Wellendruck. Hansa, **134** (5), S. 77 ff.
[127] Korzhavin, K. N. (1962): Action of ice on engineering structures. English translation, U.S. Cold Region Research and Engineering Laboratory, Trans. T. L. 260.
[128] Kovacs, A. (1996): Sea-Ice Part II. Estimating the Full-Scale Tensile, Flexural, and Compressive Strength of First-Year Ice, US Army Corps of Engineers,CRREL Report **96**– 11.
[129] Kranz, E. (1953): Über die Verankerung von Spundwänden. 2. Auflage,Verlag Ernst & Sohn, Berlin 1953.
[130] Kriebel, D. (2005): Mooring loads due to parallel passing ships. Technical Report TR-6056-OCN. Naval Facilities Engineering Center.
[131] Ladd, C. C. und DeGroot, D. J. (2003): Recommended practise for soft ground site characterisation. Proc. of 12th Panam. CSMGE, Arthur Casagrande Lecture, Cambridge (USA).
[132] Ladd, C. C. und Foott, R. (1974): New design procedure for stability of soft clays. Journal of the Geotech. Eng. Div., ASCE, GT7, **100** (1): 763–786.
[133] Laumans, Q. (1977): Verhalten einer ebenen, in Sand eingespannten Wand bei nichtlinearen Stoffeigenschaften des Bodens. Baugrundinstitut Stuttgart, Mitteilung **7**.
[134] Lehane, B. M.; Schneider, J. A. and Xu, X. (2005): The UWA-05 method for prediction of axial capacity of driven piles in sand. International Symposium on Frontiers in Offshore Geotechnics (ISFOG2005), Perth, 683–689.

[135] Leinenkugel, H. J. (1976): Deformations- und Festigkeitsverhalten bindiger Erdstoffe. Veröffentlichungen des Instituts für Bodenmechanik und Felsmechanik der Universität Karlsruhe, Heft **66**.
[136] Loers, G. und Pause, H. (1976): Die Schlitzwandbauweise – große und tiefe Baugruben in Städten. Bauingenieur **51**, Heft 2, S. 41.
[137] Longuet-Higgins, M. S. (1952): On the Statistical Distribution of the Heights of Sea Waves. Journal of Marine Research, Vol. **XI**, No. 3.
[138] MacCamy, R. C. und Fuchs, R. A. (1954): Wave Forces on Piles: A Diffraction Theory. Techn. Memorandum 69, US Army, Corps of Engineers, Beach Erosion Board, Washington D.C., Dec. 1954.
[139] MAG (1993): Merkblatt „Anwendung von geotextilen Filtern an Wasserstraßen". Ausgabe 1993, Bundesanstalt für Wasserbau, Karlsruhe.
[140] Mahutka, K.-P.; König, F. und Grabe, J. (2006): Numerical modelling of pile jacking, driving and vibro driving. Proceedings of International Conference on Numerical Simulation of Construction Processes in Geotechnical Engineering for Urban Environment (NSC06), Bochum, ed. by T. Triantafyllidis, Balkema, Rotterdam, pp. 235–246.
[141] MAK (1989): Merkblatt „Anwendung von Kornfiltern an Wasserstraßen". Ausgabe 1989, Bundesanstalt für Wasserbau, Karlsruhe.
[142] MAR (2008): Merkblatt „Anwendung von Regelbauweisen für Böschungs- und Sohlensicherungen an Binnenwasserstraßen". Bundesanstalt für Wasserbau, Karlsruhe, digital verfügbar auf www.baw.de.
[143] Matlock, H. (1970): Correlations for Design of Laterally Loaded Piles in Soft Clay. OTC 1204.
[144] Meek, J. W. (1995): Der Spitzenwiderstand von Stahlrohrpfählen. Bautechnik **72** (5): 305–309.
[145] Metha, P. and Gerwick, B. (1982): Cracking-corrosion interaction in concrete exposed to marine environment. Concrete International Design & Construction, **4** (10): 45–51.
[146] Möbius, W.; Wallis, P.; Raithel, M.; Kempfert, H.-G. und Geduhn, M. (2002): Deichgründung auf geokunststoffummantelten Sandsäulen. Hansa, **139**. Jhg., Heft 12, S. 49–53.
[147] Mokwa, Robert L. (1999): Investigation of the resistance of pile caps to lateral loading; PhD Thesis. Faculty of the Virginia Polytechnical Institute and State University.
[148] Morison, J. R.; O'Brien, M. P.; Johnson, J. W. und Schaaf, S. A. (1950): The force exerted by surface waves on piles. Petroleum Transactions, AIME, Vol. **189**.
[149] Müller-Kirchenbauer, H., Walz, B. und Kilchert, M. (1979): Vergleichende Untersuchung der Berechnungsverfahren zum Nachweis der Sicherheit gegen Gleitflächenbildung bei suspensionsgestützten Erdwänden. Veröffentlichungen des Grundbauinstituts der TU Berlin, Heft **5**.
[150] Muttray, M. (2000): Wellenbewegung an und in einem geschütteten Wellenbrecher. Dissertation, TU Braunschweig.
[151] O'Neill and Murchinson (1983): Fan Evaluation of p-y Relationships in Sands. By M. W.: A report to the American Petroleum Institute.
[152] Odenwald, B. und Herten, M. (2008): Hydraulischer Grundbruch: neue Erkenntnisse. Bautechnik, **85** (9): 585–595.
[153] Os, P. J. van (1976): Damwandberekening: Computermodel of Blum. Polytechnisch Tijdschrift, Editie B, Nr. **6**, S. 367–378.

[154] Oumeraci, H. (2001): Küsteningenieurwesen. In: Lecher, K. et al.: Taschenbuch der Wasserwirtschaft. 8. völlig neu bearbeitete Auflage, Berlin, Paul Parey Verlag, Kap. 12, S. 657–743.
[155] Oumeraci, H. und Kortenhaus, A. (1997): Anforderungen an ein Bemessungskonzept. Hansa, **134**. Jhg., Seite 71 ff.
[156] Owen, M. W. (1980): Design of seawalls allowing for wave overtopping. Hydraulic Research, Wallingford, Report No. 783, Delft.
[157] PIANC (1973): Report of the International Waves Commission, PIANC-Bulletin No 15, Brüssel.
[158] PIANC (1976): Report of the International Waves Commission, PIANC-Bulletin No 25 (1976), Brüssel.
[159] PIANC (1980): Report of the 3rd International Wave Commission. Supplement of the Bulletin No 36, Brüssel.
[160] PIANC (1986): IAHR/PIANC: Intern. Ass. For Hydr. Research/Permanent Intern. Ass. Of Navigation Congresses. List of Sea State Parameters. Supplement to Bulletin No. 52, Brüssel.
[161] PIANC (1987): Report of the Pianc Working Group II-9 ,,Development of modern Marine Terminals". Supplement to Pianc-Bulletin No 56, Brüssel.
[162] PIANC (1987): Report of Working Group I-4 ,,Guidelines for the design and construction of flexible revements incorporating geotextiles for inland waterways". Supplement to Pianc-Bulletin No 57, Brüssel 1987.
[163] PIANC (1992): Report of Working Group II-21 ,,Guidelines for the design and construction of flexible revetments incorporating geotextiles in marine environment". Supplement to Pianc-Bulletin No 78/79, Brüssel 1992.
[164] PIANC Report (1998): Life cycle management of port structures – General principles. Pianc-Bulletin No 99, Brüssel (1998).
[165] PIANC (2001): PIANC-Report ,,Effect of Earthquakes on Port Structures". Report of MarCom, WG 34.
[166] PIANC (2002): PIANC-Report ,,Guidelines for the Design of Fender Systems: 2002". Report of MarCom, WG 33.
[167] PIANC Report (2004): Inspection, maintenance and repair of maritime structures exposed to material degradation caused by a salt water environment. MarCom Report of WG 17 (2004).
[168] PIANC Report (2006): Maintenance and renovation of navigation infrastructure. InCom Report of WG 25 (2006).
[169] Pilarczyk, K. und Zeidler, R. (1996): Offshore Breakwaters and Shore Evolution Control. Balkema, Rotterdam, The Netherlands.
[170] Raithel, M. (1999): Zum Trag- und Verformungsverhalten von geokunststoffummantelten Sandsäulen. Schriftenreihe Geotechnik, Universität Kassel, Heft **6**.
[171] Randolph, M. F. (2003): Science and empiricism pile foundation design. Géotechnique **53** (10), 847–875.
[172] Randolph, M. F. (2004): Characterisation of soft sediments for offshore applications. 2nd International Site Characterisation Conference, Port, Portugal, Vol. 1, pp. 209–232.
[173] Rausche, F.; Likins, G. und Klingmüller, O. (2011): Zur Auswertung dynamischer Messungen an großen offenen Stahlrohrpfählen. Pfahl-Symposium 2011, Braunschweig, Mitteilungen des Instituts für Grundbau und Bodenmechanik, TU Braunschweig, Heft **94**, S. 491–507.

[174] Reese, L. C. and Cox, W. R. (1975): Field Testing and Analysis of Laterally Loaded Piles in Stiff Clay, OTC 2312.
[175] Richtlinie 79/113/EWG vom 19.12.1978 zur Angleichung der Rechtsvorschriften der Mitgliedsstaaten betreffend die Ermittlung des Geräuschemissionspegels von Baumaschinen und Baugeräten (Amtsbl. EG 1979 Nr. L 33 S. 15).
[176] Richtlinie „Berechnungsgrundsätze für private Hochwasserschutzwände und Uferbauwerke im Bereich der Freien und Hansestadt Hamburg" (Mai 1997). Amtlicher Anzeiger, Teil II des Hamburgischen Gesetz- und Verordnungsblattes, Nr. 33, 1998.
[177] Richwien, W. und Lesny, K. (2003): Risikobewertung als Schlüssel des Sicherheitskonzepts – Ein probabilistisches Nachweiskonzept für die Gründung von Offshore-Windenergieanlagen. In: Erneuerbare Energien **13**, Heft 2, S. 30–35.
[178] Ridderbos, N. L. (1999): Risicoanalyse met behulp van foutenboom en golfbelasting ten gevolge van 'slamming' op horizontale constructie. Master thesis, TU Delft, Faculty of Civil Engineering and Geoscience, Delft.
[179] Rienecker, M. M. und Fenton, J. D. (1981): A Fourier approximation method for steady water waves. Journal of Fluid Mechanics, Vol. **104**.
[180] Rollberg, D. (1976): Bestimmung des Verhaltens von Pfählen aus Sondier- und Rammergebnissen. Forschungsberichte aus Bodenmechanik und Grundbau FBG **4**, Techn. Hochschule Aachen.
[181] Rollberg, D. (1977): Bestimmung der Tragfähigkeit und des Rammwiderstands von Pfählen und Sondierungen. Veröffentlichungen des Instituts für Grundbau, Bodenmechanik, Felsmechanik und Verkehrswasserbau der Techn. Hochschule Aachen, Heft **3**, S. 43–224.
[182] ROM (1990): Recomendaciones para Obras Maritimas. (Englische Fassung), Maritime Works Recommendations (MWR): Actions in the design of maritime and Harbor Works (ROM 0.2-90), Ministerio de Obras Publicas y Transportes, Madrid.
[183] Römisch, K. (1993): Propellerstrahlinduzierte Erosionserscheinungen in Häfen. Hansa, **130**. Jhg., Nr. 8.
[184] Römisch, K. (1994): Propellerstrahlinduzierte Erosionserscheinungen – Spezielle Probleme. Hansa, **131**. Jhg., Nr. 9.
[185] Römisch, K.: Scouring in Front of Quay Walls Caused by Bow Thruster and New Measures for its Reduction. V. International Seminar on Renovation and Improvements to Existing Quay Structures, TU Gdansk (Poland), May 28–30, 2001.
[186] Römisch, K.: Strömungsstabilität vergossener Steinschüttungen. Wasserwirtschaft **90**, 2000, Heft 7–8, S. 356–361.
[187] Rostasy, F. S., Onken, P. (1995): Wirksame Betonzugfestigkeit bei früh einsetzendem Temperaturzwang. Deutscher Ausschuss für Stahlbeton – Heft 449, Beuth Verlag, Berlin.
[188] RPB (2001): Richtlinie für die Prüfung von Beschichtungssystemen für den Korrosionsschutz im Stahlwasserbau (RPB). Ausgabe 2001, Bundesanstalt für Wasserbau, Karlsruhe.
[189] RPG (1994): Richtlinien für die Prüfung von geotextilen Filtern im Verkehrswasserbau. Bundesanstalt für Wasserbau, Karlsruhe.
[190] Rudolph, C.; Mardfeldt, B. und Dührkop, J. (2011): Vergleichsberechnungen zur Dalbenbemessung nach Blum und mit der p-y-Methode. Geotechnik, Heft **4**.
[191] RÜV: Richtlinie für die Überwachung der Verkehrssicherheit von baulichen Anlagen des Bundes. Bundesministerium für Verkehr, Bau und Stadtentwicklung, Abteilung Bauwesen, Bauwirtschaft und Bundesbauten (2006).

[192]  Sainflou, M. (1928): Essai sur les digues maritimes verticales. Annales des Ponts et Chaussées, tome 98 II (1928), übersetzt: Treatise on vertical breakwaters. US Army, Corps of Engineers.
[193]  Schenk, W. (1968): Verfahren beim Rammen besonders langer, flachgeneigter Schrägpfähle. Bauingenieur **43**, Heft 5.
[194]  Scherzinger, T. (1991): Materialverhalten von Seetonen – Ergebnisse von Laboruntersuchungen und ihre Bedeutung für das Bauen in weichem Baugrund. Veröffentlichungen des Institutes für Bodenmechanik und Felsmechanik der Universität Karlsruhe, Heft **122**.
[195]  Schüller, G. I. (1981): Einführung in die Sicherheit und Zuverlässigkeit von Tragwerken, Verlag Ernst und Sohn, Berlin 1981.
[196]  Schulz, H. (1987a): Mineralische Dichtungen für Wasserstraßen. Fachseminar "Dichtungswände und Dichtsohlen", Juni 1987 in Braunschweig, Mitteilungen des Instituts für Grundbau und Bodenmechanik, Techn. Universität Braunschweig, H. **23**.
[197]  Schulz, H. (1987b): Conditions for day sealings at joints. Proc. of the IX. Europ. Conf. on Soil MecH. and Found. Eng., Dublin.
[198]  Schüttrumpf, R. (1973): Über die Bestimmung von Bemessungswellen für den Seebau am Beispiel der südlichen Nordsee. Mitteilungen des Franzius-Instituts für Wasserbau und Küsteningenieurwesen der Technischen Universität Hannover, Heft **39**.
[199]  Schwarz, J. (1970): Treibeisdruck auf Pfähle. Mitteilung des Franzius-Instituts für Grund- und Wasserbau der Technischen Universität Hannover, Heft **34**.
[200]  Schwarz, J.; Hirayama, K.; und Wu, H. C. (1974): Effect of Ice Thickness on Ice Forces. Proceedings Sixth Annual Offshore Technology Conference, Houston, Texas, USA.
[201]  Seah, T. H. und Lai, K. C. (2003): Strength and deformation behavior of soft Bangkok clay. Geotechnical Testing Journal, 26 (4).
[202]  Sherif, G. (1974): Elastisch eingespannte Bauwerke. Tafeln zur Berechnung nach dem Bettungsmodulverfahren mit variablen Bettungsmoduli. Verlag Ernst & Sohn, Berlin/München/Düsseldorf 1974.
[203]  Siefert, W. (1974): Über den Seegang in Flachwassergebieten. Mitteilungen des Leichtweiß-Instituts für Wasserbau der Technischen Universität Braunschweig, Heft **40**.
[204]  SNiP (1995): SNiP 2.06.04-82: Ministry of Russia. Bautechnische Normen und Regeln – Belastung und Einflüsse aus Wellen, Eis und von Schiffen auf hydrotechnische Anlagen. Moskau.
[205]  Sparboom, U. (1986): Über die Seegangsbelastung lotrechter zylindrischer Pfähle im Flachwasserbereich. Mitteilungen des Leichtweiß-Instituts der TU Braunschweig, Heft **93**, Braunschweig.
[206]  SPM (1984): Shore Protection Manual. US Army Corps of Engineers, Coastal Engineering Research Center, Vicksburg, USA.
[207]  Streeter, V. L. (1961): Handbook of Fluid Dynamics. New York.
[208]  Takahashi, S. (1996): Design of Breakwaters. Port and Harbour Research Institute, Yokosuka, Japan.
[209]  Tanimoto, K. und Takahashi, S. (1978): Wave forces on horizontal platforms. Proc. of 5th Int. Ocean Development Conf.
[210]  Terzaghi, K. und Peck, R. B. (1961): Die Bodenmechanik in der Baupraxis. Springer Verlag, Berlin 1961.

[211] TGL 35983/02 (1983): Sicherungen von Baugruben und Leitungsgräben. Böschung im Lockergestein. Fachbereichsstandard der Deutschen Demokratischen Republik.
[212] TLG (2003): Technische Lieferbedingungen für Geotextilien und geotextilverwandte Produkte an Wasserstraßen – Ausgabe 2003 – des Bundesministeriums für Verkehr, Bau- und Wohnungswesen. Verkehrsblatt 2003, Heft **18**.
[213] Van der Meer, J. W.; Janssen, J. P. F. M. (1994): Wave run-up and waver overtopping at dikes and revetments. Delft Hydraulics Publication No. **485**, Delft.
[214] VDI-Richtlinie 3576: Schienen für Krananlagen, Schienenverbindungen, Schienenbefestigungen, Toleranzen. Verein Deutscher Ingenieure, 2011.
[215] VDI-Richtlinie 6200: Standsicherheit von Bauwerken – Regelmäßige Überprüfung. VDI-Richtlinienausschuss 6200, 2010.
[216] Veder, Ch. (1975): Die Schlitzwandbauweise – Entwicklung, Gegenwart und Zukunft. Österreichischer Ing. Z. **18**, Heft 8, S. 247.
[217] Veder, Ch. (1976): Beispiele neuzeitlicher Tiefgründungen. Bauingenieur **51**, Heft 3, S. 89.
[218] Veder, Ch. (1981): Einige Ursachen von Misserfolgen bei der Herstellung von Schlitzwänden und Vorschläge zu ihrer Vermeidung. Bauingenieur 56, Heft 8, S. 299.
[219] Veldhuyzen van Zanten, R. (1994): Geotextiles and Geomembranes in Civil Engineering. Balkema, Rotterdam/Boston.
[220] VOB/B: Allgemeine Vertragsbedingungen für die Ausführung von Bauleistungen. Hrsg.:DIN Deutsches Institut für Normung e. V., Beuth Verlag, Berlin2010.
[221] VV-WSV 2101: Bauwerksinspektion. Herausgegeben vomBundesminister für Verkehr, Bonn, erhältlich bei der Drucksachenstelle der Wasser- und Schifffahrtsdirektion Mitte, Hannover 2009.
[222] VV-WSV 2301: Damminspektion. Herausgegeben vom Bundesminister für Verkehr, Bonn, erhältlich bei der Drucksachenstelle der Wasser- und Schifffahrtsdirektion Mitte, Hannover 1981.
[223] Walden, H. und Schäfer, P. J. (1969): Die winderzeugten Meereswellen, Teil II, Flachwasserwellen. Heft 1 und 2, Einzelveröffentlichungen des Deutschen Wetterdienstes, Seewetteramt Hamburg.
[224] Wehnert, M. (2006): Ein Beitrag zur drainierten und undrainierten Analyse in der Geotechnik. Mitteilung **53**. Institut für Geotechnik, Universität Stuttgart.
[225] Weichbrodt, F. (2008): Entwicklung eines Bemessungsverfahrens für Holzpfahlbuhnen im Küstenwasserbau. Veröffentl. Dissertation, Rostock.
[226] Weiss, F. (1967): Die Standfestigkeit flüssigkeitsgestützter Erdwände. Bauingenieur-Praxis, Heft **70**,Verlag Ernst & Sohn, Berlin/München/Düsseldorf 1967.
[227] Weißenbach, A. (1961): Der Erdwiderstand vor schmalen Druckflächen. Mitteilung des Franzius-Instituts TH Hannover **1961**, Heft 19, S. 220.
[228] Weißenbach, A. (1985): Baugruben, Teil II, Berechnungsgrundlagen. 1. Nachdruck, Verlag Ernst & Sohn, Berlin 1985.
[229] Wiegel, R. L. (1964): Oceanographical Engineering. Prentice Hall Series in Fluid Mechanics.
[230] Wirsbitzki, B. (1981): Kathodischer Korrosionsschutz im Wasserbau. Hafenbautechnische Gesellschaft e. V., Hamburg.
[231] Wroth, C. P. (1984): The interpretation of in situ soil tests. Géotechnique **34** (4): 449–489.

[232] Zaeske, D. (2001): Zur Wirkungsweise von unbewehrten und bewehrten mineralischen Tragschichten über pfahlartigen Gründungselementen. Schriftenreihe Geotechnik, Universität Kassel, Heft 10.
[233] Zdansky, V. (2002): Kaimauern in Blockbauweise, Bautechnik 79, Heft **12**, S. 857–864, Ernst & Sohn 2002.
[234] Ziegler, M.; Aulbach, B.; Heller, H. und Kuhlmann, D. (2009): Der Hydraulische Grundbruch – Bemessungsdiagramme zur Ermittlung der erforderlichen Einbindetiefe. Bautechnik **86**, Heft 9, S. 529–541.
[235] ZTV-ING (2003): Zusätzliche Technische Vertragsbedingungen und Richtlinien für Ingenieurbauten (ZTV-ING). Bundesanstalt für Straßenwesen (Hrsg.), Verkehrsblatt-Sammlung Nr. S 1056, Verkehrsblatt-Verlag, Dortmund.
[236] ZTV-W LB 220 (1999): Zusätzliche Technische Vertragsbedingungen – Wasserbau (ZTV-W) für kathodischen Korrosionsschutz im Stahlwasserbau (Leistungsbereich 220). Bundesministerium für Verkehr, Bau und Stadtentwicklung.
[237] ZTV-W LB 219 (2012): Zusätzliche Technische Vertragsbedingungen – Wasserbau (ZTV-W) für Schutz und Instandsetzung der Betonbauteile von Wasserbauwerken (Leistungsbereich 219). Bundesministerium für Verkehr, Bau und Stadtentwicklung.
[238] ZTV-W LB 215 (2012): Zusätzliche Technische Vertragsbedingungen – Wasserbau (ZTV-W) für Wasserbauwerke aus Beton und Stahlbeton (Leistungsbereich 215). Bundesministerium für Verkehr, Bau und Stadtentwicklung.
[239] ZTV-W LB 210 (2006): Zusätzliche Technische Vertragsbedingungen – Wasserbau (ZTV-W) für Böschungs- und Sohlensicherungen (Leistungsbereich 210). Bundesministerium für Verkehr, Bau und Stadtentwicklung.

## I.3 Technical standards

BS 6349-1:2000: Maritime Structures – Part 1: Code of practice for general criteria, Section 5.
DIN EN 206 Concrete – Specification, performance, production and conformity.
DIN 536-1 Crane rails; dimensions, sectional properties, steel grades for crane rails with foot flange, form A.
DIN EN 1090-2 Execution of steel structures and aluminium structures – Part 2: Technical requirements for steel structures.
DIN 1045 Concrete, reinforced and prestressed concrete structures.
DIN 1048 Testing concrete.
DIN 1052 Design of timber structures.
DIN 1054 Subsoil – Verification of the safety of earthworks and foundations – Supplementary rules to DIN EN 1997-1.
DIN 1076 Engineering structures in connection with roads – inspection and test.
DIN EN 1536 Execution of special geotechnical work – Bored piles.
DIN EN 1537 Execution of special geotechnical works – Ground anchors.
DIN EN 1538 Execution of special geotechnical work – Diaphragm walls.
DIN EN 1542 Products and systems for the protection and repair of concrete structures – Test methods – Measurement of bond strength by pull-off.
DIN EN ISO 1872 Plastics – Polyethylene (PE) moulding and extrusion materials.
DIN EN 1990 Eurocode: Basis of structural design.

DIN EN 1991 Eurocode 1: Actions on structures.
DIN EN 1992 Eurocode 2: Design of concrete structures.
DIN EN 1992-1-1: 2011-01(E) Eurocode 2: Design of concrete structures- Part 1-1: General rules and rules for buildings (includes Corrigendum AC:2010).
DIN EN 1993 Eurocode 3: Design of steel structures.
DIN EN 1994 Eurocode 4: Design of composite steel and concrete structures.
DIN EN 1995 Eurocode 5: Design of timber structures.
DIN EN 1996 Eurocode 6: Design of masonry structures.
DIN EN 1997 Eurocode 7: Geotechnical design.
DIN EN 1997-1:2014-03 Eurocode 7: Geotechnical design - Part 1: General rules.
DIN EN 1997-2:2010-10 Eurocode 7: Geotechnical design - Part 2: Ground investigation and testing (includes Corrigendum AC:2010).
DIN EN 1998 Eurocode 8: Design of structures for earthquake resistance.
DIN EN 1998-1:2010-12.
DIN EN 1998-1/NA:2011-01.
DIN EN 1998-5:2010-12.
DIN EN 1998-5/NA:2011-07.
DIN EN 1998-5 Eurocode 8: Design of structures for earthquake resistance – Part 5: Foundations, retaining structures and geotechnical aspects.
DIN EN 1999 Eurocode 9: Design of aluminium structures.
DIN EN ISO 2560 Welding consumables – Covered electrodes for manual metal arc welding of non-alloy and fine grain steels – Classification.
DIN 4017 Soil – Calculation of design bearing capacity of soil beneath shallow foundations.
DIN 4019 Soil – Analysis of settlement.
DIN 4020 Geotechnical investigations for civil engineering purposes – Supplementary rules to DIN EN 1997-2.
DIN 4030 Assessment of water, soil and gases for their aggressiveness to concrete.
DIN 4084 Soil – Calculation of embankment failure and overall stability of retaining structures.
DIN 4085 Subsoil – Calculation of earth-pressure.
DIN 4094-2 Subsoil – Field testing – Part 2: Borehole dynamic probing.
DIN 4094-4 Geotechnical field investigations – Part 4: Field vane test.
DIN 4126 Stability analysis of diaphragm walls.
DIN 4127 Earthworks and foundation engineering – Test methods for supporting fluids used in the construction of diaphragm walls and their constituent products.
DIN 4150 Vibrations in buildings.
DIN ISO 9613 Acoustics – Attenuation of sound during propagation outdoors.
DIN EN 10025 Hot-rolled products of structural steels.
DIN EN 10219 Cold-formed welded structural hollow sections of non-alloy and fine grain steels.
DIN EN 10248 Hot-rolled steel sheet piling.
DIN EN 10249 Cold-formed steel sheet piling.
DIN EN 12063 Execution of special geotechnical work – Sheet pile walls.
DIN EN 12390 Testing hardened concrete.
DIN EN 12504 Testing concrete in structures.
DIN EN 12699 Execution of special geotechnical work – Displacement piles.
DIN EN ISO 12944 Paints and varnishes – Corrosion protection of steel structures by protective paint systems.

DIN EN 13253 Geotextiles and geotextile-related products – Characteristics required for use in erosion control works (coastal protection, bank revetments).
DIN EN 13383-1 Armour stone – Part 1: Specification.
DIN EN 13670 Execution of concrete structures.
DIN EN ISO 14171 Welding consumables – Solid wire electrodes, tubular cored electrodes and electrode/flux combinations for submerged arc welding of non-alloy and fine-grain steels.
DIN EN 14199 Execution of special geotechnical work – Micropiles.
DIN EN ISO 14341 Welding consumables – Wire electrodes and weld deposits for gas shielded metal arc welding of non alloy and fine-grain steels – Classification.
DIN 14504 Inland navigation vessels – Floating landing stages – Requirements, tests.
DIN EN ISO 14688 Geotechnical investigation and testing – Identification and classification of soil.
DIN EN ISO 15527 Plastics – Compression-moulded sheets of polyethylene (PE-UHMW, PE-HD) – Requirements and test methods.
DIN EN ISO 15614 Specification and qualification of welding procedures for metallic materials – Welding procedure test.
DIN EN 16228-1 Drilling and foundation equipment – Safety – Part 1: Common requirements.
DIN 18134 Soil – Testing procedures and testing equipment.
DIN 18137-1:2010-07 Soil, investigation and testing - Determination of shear strenght - Part 1: Concepts and general testing conditions.
DIN 18137 Soil, investigation and testing – Determination of shear strength.
DIN SPEC 18140 Supplementary provisions to DIN EN 1536:2010-12, Execution of special geotechnical works – Bored piles.
DIN 18196 Earthworks and foundations – Soil classification for civil engineering purposes.
DIN SPEC 18537 Supplementary provisions to DIN EN 1537:2001-01, Execution of special geotechnical works – Ground anchors.
DIN SPEC 18538 Supplementary provisions to DIN EN 12699:2001-05, Execution of special geotechnical work – Displacement piles.
DIN SPEC 18539 Supplementary provisions to DIN EN 14199:2012-01, Execution of special geotechnical works – Micropiles.
DIN 18540 Sealing of exterior wall joints in building using joint sealants.
DIN 19666 Drain pipes and percolation pipes – General requirements.
DIN 19702 Solid structures in hydraulic engineering – Bearing capacity, serviceability and durability.
DIN 19703 Locks for waterways for inland navigation – Principles for dimensioning and equipment.
DIN 19704-1 Hydraulic steel structures – Part 1: Criteria for design and calculation.
DIN EN ISO 19902:2007: Petroleum and natural gas industries – Fixed structures.
DIN EN ISO 22475-1 Geotechnical investigation and testing – Sampling methods and groundwater measurements – Part 1: Technical principles for execution (ISO 22475-1:2006).
DIN EN ISO 22476-2 Geotechnical investigation and testing – Field testing – Part 2: Dynamic probing (ISO 22476-2:2005 + Amd 1:2011).
DIN 45669 Measurement of vibration immission.
DIN 52108 Testing of inorganic non-metallic materials – Wear test using the grinding wheel according to Böhme – Grinding wheel method.
DIN 52170 Determination of composition of hardened concrete.

# Annex II  Notation

II.1    Symbols for variables, sorted according to . . .
II.1a   Latin lower-case letters
II.1b   Latin upper-case letters
II.1c   Greek letters
II.2    Subscripts and indexes
II.3    Abbreviations
II.4    Designations for water levels and wave heights

The most important symbols and abbreviations used in the text, figures and equations are listed below. As far as possible they are as used in the Eurocode. All symbols are also explained in the respective passages of text.

## II.1a    Latin lower-case letters

| Symbol | Definition | Unit |
|---|---|---|
| $a$ | geometric information, length, tidal range, etc. | m |
|  | acceleration | m/s$^2$ |
| $b$ | geometric information, width | m |
|  | pore water pressure parameter | 1/m |
| $c$ | wave propagation velocity | m/s |
|  | spring constant | kN/m, MN/m |
|  | concrete cover | mm |
|  | cohesion, e.g.: | kN/m$^2$, MN/m$^2$ |
| $c_c$ | apparent cohesion, capillary cohesion | kN/m$^2$, MN/m$^2$ |
| $c_u$ | undrained cohesion | kN/m$^2$, MN/m$^2$ |
|  | geometric factor, e.g.: | 1 |
| $c_B$ | shape factor | 1 |
| $d$ | geometric information, thickness | m |
|  | diameter | m |
|  | particle size | mm |
|  | embedment depth | m |
| $e$ | void ratio | 1 |
|  | active earth pressure | kN/m$^2$, MN/m$^2$ |
|  | eccentricity, e.g.: | m |
| $e_r$ | permissible eccentricity | m |
| $f$ | deflection | m |
|  | frequency | 1/s |
|  | material strength, e.g.: | kN/m$^2$, MN/m$^2$ |

*Recommendations of the Committee for Waterfront Structures Harbours and Waterways – EAU 2012*, 9th Edition. Issued by the Committee for Waterfront Structures of the German Port Technology Association and the German Geotechnical Society.
© 2015 Ernst & Sohn GmbH & Co. KG. Published 2015 by Ernst & Sohn GmbH & Co. KG.

| Symbol | Definition | Unit |
|---|---|---|
| $f_u$ | tensile strength | $kN/m^2$, $MN/m^2$ |
| $f_y$ | yield strength/point/stress | $kN/m^2$, $MN/m^2$ |
| $g$ | gravitational acceleration | $m/s^2$ |
| $h$ | geometric information, height (beam: depth) | m |
| $i$ | hydraulic gradient | 1 |
| $k$ | permeability coefficient | m/s |
| | wave number | 1/m |
| | radius of gyration of a ship | m |
| $k_s$ | modulus of subgrade reaction | $kN/m^2$, $MN/m^2$ |
| | creep rate | mm |
| $l$ | geometric information, length | m |
| $m$ | mass | t |
| $n$ | porosity | 1 |
| | number | 1 |
| $p$ | permanent load (per unit area/length) | $kN/m^2$, $kN/m$ |
| $q$ | flow rate | $m^3/(s \cdot m)$ |
| | variable load (per unit area/length) | $kN/m^2$, $kN/m$ |
| | compressive strength, e.g.: | $kN/m^2$, $MN/m^2$ |
| $q_b$ | end bearing pressure | $kN/m^2$, $MN/m^2$ |
| $q_s$ | skin friction | $kN/m^2$, $MN/m^2$ |
| $q_u$ | uniaxial compressive strength | $kN/m^2$, $MN/m^2$ |
| $r$ | radius | m |
| $s$ | geometric information, e.g. displacement, settlement | cm |
| $t$ | depth | m |
| | time | s, h, d, a |
| $u$ | horizontal component of velocity of water particles | m/s |
| | pore water pressure | $kN/m^2$, $MN/m^2$ |
| $v$ | velocity | m/s |
| $w$ | water/hydrostatic pressure | $kN/m^2$, $MN/m^2$ |
| $x$ | ordinate/geometric information | m |
| $y$ | ordinate/geometric information | m |
| $z$ | ordinate/geometric information | m |

## II.1b   Latin upper-case letters

| Symbol | Definition | Unit |
|---|---|---|
| $A$ | energy capacity (dolphin) | kNm, MNm |
| | area | $m^2$ |
| | accidental action | kN |
| $B$ | reaction force in soil | kN, MN, kN/m, MN/m |
| $C$ | equivalent force for soil reaction (after Blum) | kN/m |
| | coefficient, factor, e.g.: | 1 |
| $C_D$ | resistance factor for flow pressure | 1 |

| Symbol | Definition | Unit |
|---|---|---|
| $C_e$ | eccentricity factor | 1 |
| $C_m$ | mass factor | 1 |
| $C_M$ | resistance factor for flow acceleration | 1 |
| $C_S$ | stiffness factor | 1 |
| $D$ | in situ density (soil), e.g.: | 1 |
| $D_{pr}$ | Proctor density | 1 |
| $E$ | modulus of elasticity/Young's modulus | $kN/m^2$, $MN/m^2$ |
| | energy | kJ |
| | earth pressure force | kN, MN, kN/m, MN/m |
| | action effect, load | kN, MN, kN/m, MN/m |
| $E_s$ | stiffness modulus | $kN/m^2$, $MN/m^2$ |
| $F$ | force | kN, MN, kN/m, MN/m |
| | action, e.g.: | kN, MN, kN/m, MN/m |
| $F_c$ | cohesive force | kN, MN, kN/m, MN/m |
| $F_s$ | flow force | kN, MN, kN/m, MN/m |
| $F_u$ | pore water pressure force | kN, MN, kN/m, MN/m |
| $F_w$ | wind load | kN, MN, kN/m, MN/m |
| $F_z$ | tensile force | kN, MN, kN/m, MN/m |
| $G$ | dead load, self-weight, permanent vertical action | kN, MN, kN/m, MN/m |
| $H$ | horizontal load | kN, MN, kN/m, MN/m |
| | wave height | m |
| $I$ | second moment of area | $m^4$ |
| | condition index, e.g.: | 1 |
| $I_c$ | consistency index | 1 |
| $I_D$ | relative in situ density | 1 |
| $I_P$ | plasticity index | 1 |
| $K$ | earth pressure coefficient | 1 |
| $M$ | moment | kNm, MNm, kNm/m, MNm/m |
| $N$ | normal force | kN, MN, kN/m, MN/m |
| $N_{10}$ | number of blows per 10 cm penetration (penetrometer test) | 1 |
| $P$ | anchor force, load | kN, MN, kN/m, MN/m |
| | probability | 1 |
| $Q$ | variable action | kN, MN, kN/m, MN/m |
| | flow rate | $m^3/s$ |
| $R$ | reaction force, resistance force | kN, MN, kN/m, MN/m |
| | resistance | kN, MN, kN/m, MN/m |
| $R_B$ | reaction force in soil | kN, MN, kN/m, MN/m |
| $R_C$ | equivalent force for soil reaction (after Blum) | kN/m |
| $R_e$ | Reynolds number | 1 |
| $S$ | parameter, e.g.: standard surface finish | 1 |
| | first moment of area | $m^3$ |
| $T$ | wave period | s |
| | shear force | $kN/m^2$, $MN/m^2$ |
| | temperature | °C, K |

| Symbol | Definition | Unit |
|---|---|---|
| $T_c$ | cohesive force | $kN/m^2$, $MN/m^2$ |
| $U$ | uniformity coefficient | 1 |
| $V$ | vertical load | kN, MN, kN/m, MN/m |
| | volume | $m^3$ |
| $W$ | hydrostatic force | kN, MN, kN/m, MN/m |
| | section modulus | $m^3$ |

## II.1c  Greek letters

| Symbol | Definition | Unit |
|---|---|---|
| $\alpha$ | angle of inclination of bottom | degree |
| | angle of wall inclination | degree |
| | moment reduction factor | 1 |
| $\alpha_T$ | coefficient of thermal expansion | 1/°C |
| $\beta$ | angle of slope | degree |
| | angle of inclination of terrain | degree |
| $\gamma$ | partial safety factor | 1 |
| | unit weight, e.g.: | $kN/m^3$ |
| $\gamma'$ | buoyant/submerged unit weight | $kN/m^3$ |
| $\Delta$ | increase, decrease, change | 1 |
| $\delta$ | angle of interface (e.g. wall) friction, e.g.: | degree |
| $\delta_a$ | angle of inclination of active earth pressure | degree |
| $\delta_p$ | angle of inclination of passive earth pressure | degree |
| $\varepsilon$ | strain | 1 |
| $\eta$ | modification factor | 1 |
| $\vartheta$ | angle of inclination of failure plane | degree |
| $\kappa$ | reflection coefficient | 1 |
| $\lambda$ | hoop tension factor, coefficient | 1 |
| $\mu$ | coefficient of friction | 1 |
| | correction factor | 1 |
| $\nu$ | stiffness factor | 1 |
| $\xi$ | breaker index | 1 |
| | reduction factor | 1 |
| $\rho$ | density | $t/m^3$ |
| $\Sigma$ | total, sum | |
| $\sigma$ | stress, normal stress, e.g.: | $kN/m^2$, $MN/m^2$ |
| $\sigma'$ | effective (normal) stress | $kN/m^2$, $MN/m^2$ |
| $\sigma_v$ | equivalent stress | $kN/m^2$, $MN/m^2$ |
| $\tau$ | (transverse) stress, shear stress | $kN/m^2$, $MN/m^2$ |
| $\Phi$ | impact factor, dynamic factor | 1 |
| $\varphi$ | angle of internal friction | degree |
| $\psi$ | combination factor | 1 |
| $\omega$ | angular frequency | 1/s |

## II.2 Subscripts and indexes

| Symbol | Definition |
|---|---|
| a | active |
| abs | absolute |
| at | atmospheric |
| b | base |
| btm | bottom |
| c | cohesion |
|  | compression |
| cal | calculation |
| crit | critical |
| d | design value |
|  | dry |
| dyn | dynamic |
| dst | destabilising |
| e | eccentric |
| eff | effective |
| E | fixity |
| f | fracture |
| g | due to permanent actions |
| h | horizontal |
| k | characteristic value |
| kin | kinetic |
| m | mean, average |
| max | maximum |
| min | minimum |
| mob | mobilised |
| p | passive |
|  | persistent |
| pl | plastic |
| poss | possible |
| pr | Proctor |
| q | due to variable actions |
| r | resultant |
|  | resistance |
| red | reduced |
| rep | representative |
| reqd | required |
| s | layer, stratum |
|  | shaft |
|  | flow, current |
| stat | static |
| stb | stabilising |

| Symbol | Definition |
|---|---|
| t | tension |
|   | transient |
| tot | total |
| tr | transverse (load) |
| u | undrained |
|   | bottom |
|   | uniaxial |
| upl | uplift |
| v | vertical |
|   | comparative |
| w | water, water level |
| y | yield point |

## II.3  Abbreviations

| Symbol | Definition |
|---|---|
| AMSL | above mean sea level |
| dwt | deadweight tonnage |
| GRT | gross register tonnage |
| DS | design situation, e.g.: |
| DS-P | persistent design situation |
| DS-T | transient design situation |
| DS-A | accidental design situation |
| DS-E | earthquake design situation |
| EQU | limit state for a loss of equilibrium of the structure or ground regarded as a rigid body, where the strength of the material and the subsoil are not critical for the resistance |
| GEO | limit state of the ground: |
| GEO-2 | limit state of the ground for which design approach 2 is used |
| GEO-3 | limit state of the ground for which design approach 3 is used |
| HYD | limit state of failure caused by hydraulic gradients in the ground, e.g. hydraulic heave, internal erosion or piping |
| MSL | mean sea level |
| SLS | serviceability limit state |
| STR | limit state of failure or very large deformations of the structure or its individual parts, including the foundations, piles, basement walls, etc., where the strength of the material is critical for the resistance |
| TEU | twenty feet equivalent unit |
| ULS | ultimate limit state |
| UPL | limit state for a loss of equilibrium of the structure or the ground as a result of uplift by water pressure (buoyancy) or other vertical actions |

## II.4 Designations for water levels and wave heights

Non-tidal water levels

| | |
|---|---|
| GW, GrW | groundwater level, water table |
| HaW | normal harbour water level |
| LHaW | lowest harbour water level |
| HHW | highest high water |
| HW | high water |
| MHW | mean high water |
| MW | mean water level |
| MLW | mean low water |
| LW | low water |
| LLW | lowest low water |
| HNW | highest navigable water level |

Tidal water levels

| | |
|---|---|
| HAT | highest astronomical tide |
| MHWS | mean high water spring |
| MHW | mean high tide |
| MW | mean tide level |
| T½W | half tide |
| MT½W | mean half tide |
| MLW | mean low tide |
| MLWS | mean low water spring |
| LAT | lowest astronomical tide |
| CD | chart datum (roughly corresponds to MLWS) |

Wave heights

| | |
|---|---|
| $H_b$ | height of breaking wave |
| $H_d$ | design wave height |
| $H_m$ | mean wave height |
| $H_{max}$ | maximum wave height |
| $H_{rms}$ | root-mean-square wave height |
| $H_{1/3}$ | mean of 33% highest wave heights |
| $H_{1/10}$ | mean of 10% highest wave heights |
| $H_{1/100}$ | mean of 1% highest wave heights |

# Annex III  List of keywords

**A** — Section

Acceptance conditions for steel sheet piles and steel piles on site ............ 8.1.7
Access ladders ................................................................................................ 6.12
    design ........................................................................................................ 6.12.2
    layout ......................................................................................................... 6.12.1
Active earth pressure ..................................................................................... 2.12.3.1, 2.16.3
    in stratified soil ......................................................................................... 2.5
Active noise control measures ...................................................................... 8.1.14.4
Against failure of a cellular cofferdam ........................................................ 8.3.1.3.1
Allowance for scouring ................................................................................. 12.4.2
Analysing ultimate strength .......................................................................... 8.2.1
Anchor cells ................................................................................................... 8.3.2.1
Anchoring
    elements .................................................................................................... 9.1
        Displacement piles ............................................................................... 9.1
        Micropiles .............................................................................................. 9.1
        Special piles ........................................................................................... 9.1
        Anchors .................................................................................................. 9.1
    point .......................................................................................................... 9.5.1
Anchoring with anchor plate ("dead-man" anchors) ................................. 9.5.2.2
Anchor plates ................................................................................................ 8.5.6, 9.5.3.3
Anchors .......................................................................................................... 9, 9.5
    internal ...................................................................................................... 9.5.3.1
    load bearing capacity ............................................................................... 9.5.3
Anchor wall ................................................................................................... 8.2.7.2
    design ........................................................................................................ 8.2.12
    staggered arrangement ............................................................................. 8.2.13
Angle of earth pressure ................................................................................. 8.2.5
Armour layer ................................................................................................. 12.1.3.1
    design ........................................................................................................ 12.8.4
Armourstones ................................................................................................ 12.1.3.1
Arrangement of tops of quay walls at container terminals ....................... 6.1.5
Artesian water pressure ................................................................................. 2.10
Attenuation factor ......................................................................................... 13.1.4.1
Auxiliary anchors .......................................................................................... 8.4.7
Auxiliary driving measures .......................................................................... 1.5.2.5

**B**

Base failure .................................................................................................... 8.3.1.3.1
Backfilling ..................................................................................................... 7.4
Bearing capacity ............................................................................................ 7.12.3, 7.13

*Recommendations of the Committee for Waterfront Structures Harbours and Waterways – EAU 2012,*
9th Edition. Issued by the Committee for Waterfront Structures of the German Port Technology
Association and the German Geotechnical Society.
© 2015 Ernst & Sohn GmbH & Co. KG. Published 2015 by Ernst & Sohn GmbH & Co. KG.

Bearing pressure/base resistance ............................................................. 8.1.4.5
Bedding .................................................................................................... 8.2.3.4
Berthing force of ships at quays ............................................................. 5.2
Berthing manoeuvres ............................................................................... 13.1.4.1
Berthing pressure
    berthing force ................................................................................... 5.2, 13.2.3.1
Berthing process, control and monitor ...................................................... 13.3.3
Berthing velocities .................................................................................... 5.3
Berths for large vessels ............................................................................. 6.11
Blasting guidance ..................................................................................... 8.1.10.3
Blasting to assist the driving of steel sheet piles .................................... 8.1.10
Blum method ............................................................................................ 13.2.1.1
Bollards .................................................................................................... 5.13, 13.3.3
    layout ................................................................................................. 5.12
Bored cast-in-place piles .......................................................................... 10.9
Boring
    depth .................................................................................................. 1.2
    intermediate borehole ....................................................................... 1.2.3
    layout ................................................................................................. 1.2
    pressing in sheet piles ...................................................................... 8.1.25.4
    principal borehole ............................................................................. 1.2.2
Bow thrusters ............................................................................................ 12.4.4.2
Box caissons ............................................................................................. 10.4
Box piles, H-piles or tubular piles ............................................................ 8.1.4.4
Breaking waves ........................................................................................ 5.6.5
Break-off bolts .......................................................................................... 13.3.3
Breakwaters .............................................................................................. 12.8
    buckle ................................................................................................ 8.1.4.4

## C

Calculating bending moments .................................................................. 8.2.1.2
Calculating subsequently strengthened pile bents/trestles ....................... 11.2
Calculations .............................................................................................. 0.3
(Capillary cohesion) in sand ..................................................................... 2.3
Capping beams ......................................................................................... 8.2.7.2
Capping beams on waterfront structures .................................................. 8.4.6
Cast-in-place concrete piles ...................................................................... 9.3.2.2
Cathodic corrosion protection system ...................................................... 10.2.5
Cofferdams ............................................................................................... 8.3.1.5
    diaphragm ......................................................................................... 8.3.1.2
    double-wall ....................................................................................... 8.3.2, 8.3.2.2.2
        overturning and sliding ............................................................... 8.3.2.2.2
    excavation enclosures ....................................................................... 8.3.1.3.1
    mono-cell .......................................................................................... 8.3.1.2
    waterfront structure .......................................................................... 8.3.1.3.11
Cellular cofferdams .................................................................................. 8.3.1, 8.3.1.2
    construction measures ...................................................................... 8.3.1.4
    ground improvement ......................................................................... 8.2.3

| | |
|---|---|
| Choice of steel grade | 8.1.5.4 |
| Circular cells | 8.3.1.2 |
| Classification of the welded joints | 8.1.19.3 |
| Cluster dolphins | 13.2.1.4 |
| Cohesion $c_u$ | 1.1.1, 1.4.1, 1.4.2, 1.4.3 |
| Cohesion in cohesive soils | 1.1.1, 2.2 |
| Combination factors | 0.2.2 |
| Combined sheet piling | 8.1.4.2 |
|     steel | 8.1.4 |
| Composite piles | 9.3.2.1 |
| Compressed-air caissons | 10.5 |
| Compression members | 8.4.1.1 |
| Compressive strength of the ice | 5.16.3 |
| Concrete | 10.2.2 |
|     cover | 10.2.2 |
|     exposure class | 10.2.2 |
| Connecting steel sheet piling to a concrete structure | 6.20 |
| Connecting areas | 8.3.1.2 |
| Connection between pile and waling | 8.4.3.3 |
| Connection of expansion joint seal in reinforced concrete bottom to loadbearing steel sheet pile wall | 6.19 |
| Consolidation | 1.1.1, 7.11, 7.12 |
|     settlement | 7.11.1 |
| Construction guidance | 11.4.4 |
| Construction in water | 8.3.1.5 |
| Construction joints | 10.2.3 |
| Construction on land | 8.3.1.5 |
| Contact forces | 13.1.4.2 |
| Container cranes | 5.14.2 |
| Control and monitor the berthing process | 13.3.3 |
| Conversion of the sea state | 5.6.5 |
| Convex failure plane | 8.3.1.3 |
| Corner sections | 8.4.11.4 |
| Corrosion of steel sheet piling | 8.1.8 |
|     influence | 8.1.8.2 |
| Corrosion protection | 8.1.1.1, 8.1.8.4 |
| Corrosion zones | 8.1.8.1 |
| Cost of reinforcement | 8.1.17.10 |
| Crack width limitation | 10.2.5 |
| Creep settlement | 7.11.1 |
| Culmann method | 2.4 |
| Cushioning layer | 12.1.3.5 |

**D**

| | |
|---|---|
| Damaged during driving (declutching) | 7.4.4 |
| Damage to sheet pile interlocks | 7.1 |
| Darcy's law | 4.7.1 |

Declutching ............................................................................... 7.4.4, 8.1.12.3, 8.1.13.2, 8.1.16.1
   repairing interlock declutching ......................................... 8.1.6
Decrease in wall thickness .................................................... 8.1.8.1
Depths ................................................................................... 1.2
   of preload fill ...................................................................... 7.12.5
Description of the sea state .................................................. 5.6.2
Designing .............................................................................. 12.8.4
   bottom protection ............................................................... 12.4.5
   and construction of block-type quay walls ........................ 10.6
   depth in front of quay wall ................................................ 6.7.3
   depth of harbour bottom ..................................................... 6.7
   of plane pile bents .............................................................. 11.3
   of quay walls using open caissons ..................................... 10.7
   sea state ............................................................................... 5.6
   of spatial pile trestles ......................................................... 11.4
Density
   in situ ................................................................................... 7.5
   of waterfront areas in inland ports according to operational aspects ..... 6.6, 13.2.2.2
Determination of design situations (DS) ............................... 0.2.1
Determining active earth pressure ......................................... 2.4, 2.7
   in saturated, non-or partially consolidated, soft cohesive soils ........... 1.1.1, 2.9
   for a steep, paved embankment in a partially sloping
      waterfront structure ....................................................... 2.6
Determining excess water pressure ....................................... 2.11.3, 2.12.2
Determining the compressive strength of the ice ................. 5.15.2
Determining the effects of earthquakes ................................ 2.16.3
Diaphragm cells ..................................................................... 8.3.1.2
Determining the sea state parameters ................................... 5.6.3
Diaphragm walls .................................................................... 10.10
Diffraction ............................................................................... 5.6.5
Displacement .......................................................................... 5.1.4
Displacement piles ................................................................. 9.2, 9.2.1
   additional stresses ............................................................... 9.2.3.4
   bond stresses ....................................................................... 9.2.3.3
   load bearing capacity
      external ............................................................................ 9.2.3.2
      internal ............................................................................ 9.2.3.1
Disposal systems .................................................................... 6.14
Disturbed base of excavation ................................................. 2.11
Dolphin caps .......................................................................... 13.3.3
   lighting ................................................................................ 13.3.3
   nosings ................................................................................ 13.3.3
Dolphins ................................................................................. 13
   absorption capacity ............................................................. 13.2.3.2
   capacity ............................................................................... 13.1.3, 13.2.2.1, 13.2.2.3

| | |
|---|---|
| cluster dolphins | 13.2.1.4 |
| cyclic loads | 13.2.1.3 |
| embedment depth | 13.2.1.2, 13.2.1.3 |
| energy absorption capacity | 13.2.2.1 |
| equipment | 13.3.3 |
| fromtubular sections | 13.3.4 |
| ice loading | 13.1.4.3 |
| impact and pressure | 5.15 |
| inland ports | 13.2.2.3 |
| maximum permissible berthing force | 13.2.3.1 |
| passive earth pressure can be assumed to have the maximum possible angle of inclination | 13.2.1.2 |
| position | 13.3.2 |
| protection dolphin | 13.1.1 |
| protect waterfront structures | 13.3.2 |
| scour protection | 12.5 |
| serviceability limit state | 13.1.5 |
| spatial passive earth pressure | 13.2.1.2 |
| stiffness of the system | 13.1.2 |
| stress–strain diagram | 13.1.3 |
| from tubular sections | 13.3.4 |
| ultimate limit state | 13.1.5 |
| walkways | 13.3.3 |
| Double-wall cofferdam | 8.3.2.2 |
| as excavation enclosures | 8.3.2 |
| Drainage systems | 4.5, 4.5.1, 4.5.2, 4.5.3 |
| Dredging in front of quay walls in seaports | 7.1 |
| Dregders | 7.1 |
| bucket-ladder dredgers | 7.1 |
| cutter-suction dredgers | 7.1 |
| cutter-wheel suction dredgers | 7.1 |
| grab dredgers | 7.1 |
| hopper suction dredgers | 7.1 |
| plain suction dredgers | |
| Drains | |
| sand | 7.11.5 |
| plastic | 7.11.5 |
| Dredging tolerances | 7.2.2 |
| Dredging underwater slopes | 2.6, 7.7 |
| Driving bearing piles | 8.1.13.4 |
| Driving | |
| plant | 8.1.11.3 |
| at low temperatures | 8.1.15 |
| combined steel sheet piling | 8.1.12 |
| deviations and tolerances | 8.1.13.3 |
| difficult conditions | 8.1.5.3 |

| | |
|---|---|
| easy | 8.1.5.3 |
| low-noise | 8.1.14 |
| planing a drive site | 8.1.14.5 |
| pile sections and installation methods | 1.5.2.6 |
| procedure | 8.1.12.5 |
| sheet piles | 8.1.11.4 |
|     steel | 8.1.11 |
| tolerances | 8.1.13.2 |
| vibrations and noise expected | 8.1.11.3 |
| Drop hammer | 8.1.11.3 |
| Dumped moles | 12.8 |
| Dynamic compaction | 2.14.3, 7.10 |

## E

| | |
|---|---|
| Earth pressures | 8.3.2.2.1 |
| distribution under limited loads | 2.8 |
| redistribution | 8.2.3.2 |
| Earthworks and dredging | 7 |
| Edge bollards | 6.1.4 |
| Edge protection | 6.15.8, 10.1.2 |
| Effective compaction | 7.10 |
| Effects of waves due to ship movements | 5.9 |
| Elastomer fenders | 6.15.5.1 |
| Embankment | |
| construction | 12.1.3 |
| depth | 8.2.8, 8.2.9 |
| determining active earth pressure for a steep, paved embankment in a partially sloping waterfront structure | 2.6 |
| dredging underwater sloes | 7.7 |
| protection and stabilisation on inland waterways | 12.1 |
|     construction | 12.1.3 |
| stabilization | 12.1, 12.1.1 |
| unsteady hydraulic loads | 12.1.1 |
| Energy absorption | |
| actions | 13.1.4 |
| actions | 13.2.2 |
| maximum permissible berthing force | 13.2.3.1 |
| Equipment for waterfront structures | 6.14 |
| Equipotential line | 2.12.2 |
| Excavation | 7.9.4 |
| depths | 8.1.3.2 |
| Excess water pressure | 4.2, 4.8.1 |
| on sheet piling | 4.3 |
| on waterfront structure | 2.12.2 |
| Expansion joints | 8.4.5.4 |
| Exposed locations | 8.1.12.5 |

External loadbearing capacity .................................................. 9.5.3.2
Extra waling ............................................................................ 8.4.1.5

**F**
Facing .................................................................................... 10.1.3
Failure due to vertical movement ........................................... 8.2.5
Failure plane ........................................................................... 8.5.5
   in unconsolidated, saturated cohesive soils ....................... 8.5.2
Fenders ................................................................................... 6.15.1, 13.1.1,
                                                                                                                                                                                13.1.2
   absorption capacity ........................................................... 6.15.4.1
   chains ................................................................................ 6.15.7
   design principle ................................................................ 6.15.3
   elastomer fender ............................................................... 6.15.5.1
   inland ports ...................................................................... 6.16
   large vessels ..................................................................... 6.15
   natural materials .............................................................. 6.15.5.2
   in sheet pile troughs ......................................................... 8.4.10.5
   types ................................................................................. 6.15.5
Filled cofferdam ..................................................................... 8.3.2.2.1
Filling should be particularly water-permeable ..................... 8.3.1.4
Filter ....................................................................................... 12.1.3.2
   geotextile .......................................................................... 12.3
Fine-grained sand ................................................................... 7.3.3.2
Fixings ................................................................................... 8.4.1.3
   crane rails to concrete ...................................................... 6.18
Floating anchor walls ............................................................. 8.4.9.3
Floating berths in seaports ..................................................... 6.21
Flood defence walls in seaports ............................................. 12.7
   buried services ................................................................. 12.7.8
   flood defence road ........................................................... 12.7.7.2
   inner water level .............................................................. 12.7.2.1.1
   minimum embedment depth ............................................ 12.7.4
Flow channel .......................................................................... 4.7.7.1
Flow force .............................................................................. 2.12.3.1
Flow net ................................................................................. 2.12.3.1, 4.7.3
Fluidization ............................................................................ 3.1
Forces from breaking waves .................................................. 5.10.5
Form of bored piles ............................................................... 9.3.1
Foundations to craneways on waterfront structures .............. 6.17
Free-standing pile trestles ...................................................... 11.4.2
Free-standing sheet piling structures ..................................... 8.2.2
Full portal cranes ................................................................... 5.14.1.1

**G**
General requirements placed on sheet pile wall sections ....... 8.1.12.4
Geotechnical categories ......................................................... 0.2.5

639

Geotechnical report ............. 1.3
Geotextile ............. 12.3
   geotextile-encased columns ............. 7.13.4
Ground failure ............. 3.2
Groundwater flow ............. 4.7
   on excess water pressure ............. 2.12
Groundwater level ............. 4.1
Groundwater lowering ............. 4.8
Groundwater models ............. 4.7.5
Grouted anchors ............. 9.5.2.1
Grouted piles ............. 9.2.2.2
Guide piles ............. 6.5.2
Guiding devices ............. 6.15.8

# H
Hardened cement mortar/fine-grained concrete ............. 9.2.2.2
Head details of steel anchor piles ............. 8.4.3.6
Head of a bollard ............. 6.1.4
Heavy weights ............. 7.10
Height of the wall ............. 8.3.2.2
Hinged connections between driven steel anchor piles ............. 8.4.13
Hoop tension ............. 8.3.1.3, 8.3.1.4
Horizontal actions ............. 8.2.11
Horizontal ice load on group of piles ............. 5.15.5
Hydraulic filling ............. 7.3
   tolerances ............. 7.2
   conservation ............. 7.3.1
   dyke ............. 7.3.2
   fine materials ............. 7.3.2
   settlement ............. 7.3.2
   suspended particles ............. 7.3.2
Hydraulic heave failure, ground failure ............. 8.3.1.3.1
Hydrostatic pressure ............. 8.2.1.3, 8.3.2.2.1

# I
Ice loads
   on narrow structures ............. 5.16.5
   on vertical piles ............. 5.15.4
   on waterfront structures and other structures of greater extent ............. 5.16.4
Impact and pressure of ice on waterfront structures ............. 5.15
   piers and dolphins at inland facilities ............. 5.16
Ice surcharges ............. 5.15.6
Ice thickness ............. 5.16.2
Impact driving ............. 1.5.2.2, 8.1.4.4
   of steel tubes ............. 8.1.4.4
Impervious lining ............. 12.1.3.3
Improvement through vacuum consolidation ............. 7.12.8

| | |
|---|---|
| Improving the bearing capacity | 1.1.1, 7.13 |
| Initially loaded soil | 1.1.1 |
| Inland ports | 6.16, 13.2.2.3 |
|     bollards | 5.13 |
|     design of waterfront areas in inland ports according to operational aspect | 6.6 |
|     dolphins | 13.2.2.3 |
|     fenders | 6.16 |
|     upgrading partially sloped waterfronts | 6.5 |
|         redesign | 6.10 |
|         standard cross-sections | 6.3 |
| Inland waterways | 6.4 |
|     classification | 5.1.3, 12.1.2 |
|     embankment stabilization | 12.1 |
|         construction | 12.1.3 |
|     loads | 12.1.2 |
|     sheet piling | 6.4 |
|     unfavorable canal and groundwater levels | 6.4.3 |
|     vessels | 5.1.3 |
| In situ density | |
|     of dumped non-cohesive soils | 1.1.1, 7.6 |
|     of hydraulically filled non-cohesive soils | 1.1.1, 7.5 |
| Inspection | 14 |
| Installation | 8.1.1, 8.1.4.1 |
|     of reinforced concrete sheet pile walls | 8.1.2 |
|     of sheet piles | 8.1.13 |
|     of steel sheet pile walls | 8.1.3 |
| Installation methods | 1.5.2 |
| Interlock | 8.1.4.1 |
|     connections | 8.1.4.4 |
|     declutching | 8.1.16.2 |
|     distortion | 8.1.4.4 |
|     effect using welding | 8.1.5.2 |
|     forms | 8.1.6.5 |
|     welding | 8.1.5.3 |
| Intermediate boreholes | 1.2.3 |
| Intermediate piles | 8.1.4.1 |
|     with interlock types | 8.1.12.3 |

## J

| | |
|---|---|
| Jack-up platforms | 8.1.12.5, 8.3.1.5 |
| Jagged wall | 8.1.5.1 |
| jet grouting | 9.4.1 |
| Jetting | |
|     high-pressure | 8.1.24.3 |
|     low-pressure | 8.1.24.2 |
|     water-jetting | 8.1.24 |

## L

| | |
|---|---|
| Landings | 6.13.3 |
| Layout | 1.2 |
| Levelling layer | 12.1.3.4 |
| Level of port operations area | 6.2.2 |
| Lighting | 13.3.3 |
| Linear load–deformation behaviour | 13.1.3 |
| Loadbearing capacity of the vessel's own onboard mooring equipment (ropes and winches) | 9.2.3, 13.1.4.2 |
| Loading assumptions for quay surfaces | 5.5.5 |
| Loads | |
|    arising from surging and receding waves due to inflow or outflow of water | 5.8 |
|    on mooring and fender equipment | 5.11.4 |
|    vertically imposed | 5.5 |
|    vertical wave load | 5.10.9 |
|    due to non-breaking waves | 5.7.2 |
|    due to waves breaking | 5.7.3 |
|    on a single vertical pile | 5.10.3 |
|    on dolphins | 13.1.3 |
| Load specifications for port cranes | 5.14.3 |
| Load transfer | 9.5.1 |
| Local scour | 13.2.3.3 |
| Logarithmic spiral | 8.3.1.3 |
| Lowly stressed | 8.1.19.5 |
| Low-pressure jetting | 8.1.24.2 |

## M

| | |
|---|---|
| Making weld joints | 8.1.19.4 |
| Mean characteristic values of soil parameters | 1.1 |
| Measures for increasing the passive earth pressure | 2.14 |
| Measure the degree of corrosion | 8.1.8.1 |
| Mechanical ice pressure | 5.15.3.1 |
| Microbiologically induced corrosion | 8.1.8.3 |
| Micropiles | 9.3 |
|    additional stresses | 9.3.3.4 |
|    bond stresses | 9.3.3.3 |
|    external | 9.3.3.2 |
|    internal | 9.3.3.1 |
|    loadbearing capacity | 9.3.3 |
| Mineral impervious linings | 12.6 |
| Monitoring | |
|    during the installation | 8.1.13 |
|    of structures | 8.1.22.4 |
| Mono-cell | 8.3.1.2 |
| Mooring | |
|    equipment | 6.3.5 |

forces .................................................................................... 13.1.4.2
hooks ................................................................................... 13.3.3
Movement joints ................................................................. 10.2.4

## N
Narrow moles ..................................................................... 8.3.3
Noise control ...................................................................... 8.1.11.3, 8.1.14.4
   low-noise driving ........................................................ 8.1.14
Nominal depth .................................................................... 6.7.1
   and design depth of harbour bottom ........................... 6.7
Nosings ............................................................................... 13.3.3
No unbonded ...................................................................... 9.3.1

## O
Outer water levels .............................................................. 4.3.1
Overturning ........................................................................ 8.3.1.3.1

## P
Partial grouting .................................................................. 12.1.3.1
Partially sloped upgrades .................................................. 6.5.1
Partial safety factors ......................................................... 0.2.1, 8.2.1.1
   for the material resistance ......................................... 0.2.1
      of steel tension members ...................................... 0.2.1
   for shear strength ....................................................... 0.2.1
Passive earth pressures ..................................................... 2.16.3, 8.2.3.3
   in front of abrupt changes in ground level in soft cohesive soils with
      rapid load application on land side ................... 2.15
Passive noise control measures ....................................... 8.1.14.3
Penetrometer tests ............................................................. 1.2.4
Percolation ......................................................................... 5.6.5
Permissible dimensional deviations for interlocks ....... 8.1.6.6
Armoured steel piling ....................................................... 8.1.17
Bearing piles ...................................................................... 8.1.3.2
   installing ........................................................................ 8.1.4.4
tubular bearing piles ......................................................... 8.1.4.3
spirally welded .................................................................. 8.1.4.3
longitudinally welded ....................................................... 8.1.4.3
helical line ......................................................................... 8.1.4.3
Piles
   bents and trestles ......................................................... 11
   cast-in-place concrete piles ........................................ 9.3.2.2
   cement mortar piles .................................................... 9.2.2.2
   composite piles ............................................................ 9.3.2.1
   driving frames ............................................................. 8.3.1.5
   form and reinforcement fabrication ........................... 8.1.17.5
   grouted piles ................................................................ 9.2.2.2
   guided piles ................................................................. 6.5.2

643

ice loads on vertical piles ... 5.15.4
impact driving ... 1.5.2.2
junction piles ... 8.3.1.2
   assessing the sub soil ... 15.2
   auxiliary driving measures ... 12.5.2.5
   installation methods ... 15.2.6
   piles section ... 15.2.6
intermediate piles ... 8.1.3.2
prefabricated raking piles ... 9.4.2
sheet pile wall ... 8.3.2.2
   logarithmic spiral ... 8.3.2.2
single pile, wave load ... 5.10.3
special piles ... 9.4
steel wings ... 9.2.2.1
structures in earthquake zones ... 11.5
tension piles ... 9
timber fenders and fender piles ... 6.15.8.2
vibratory driven grouted piles ... 9.2.2.3
Piling frames ... 8.1.18
Piping ... 8.3.1.3.1
Platforms
   driving ... 8.3.1.5
   jack-up platform ... 8.3.1.5
   working ... 8.3.1.5
Planning and executing dredging work ... 7.1
Plastic drains ... 7.11.4
Plug formation ... 8.2.5.6.7
Position of outboard crane rail ... 6.3.4
Potential differential equation ... 4.7.2
Precast concrete blocks ... 12.1.3.1, 12.8.3
Prefabricated raking piles ... 9.4.2
Prefabricated steel tension element ... 9.4.2
Preload fill ... 7.12.5.2
Preloading ... 7.12
Preparation of joint ends ... 8.1.19.4
Pressing ... 1.5.2.4
   in sheet piles ... 8.1.25.4
   of U- and Z-section steel sheet piles ... 8.1.25
   systems ... 8.1.11.3
Pressure ... 2.12.1
Prestressing steels ... 9.5.2.1
Primary settlement ... 7.11.1
Principal boreholes ... 1.2.2
Probabilistic analysis ... 0.2.6
Propeller wash ... 12.4.4
Protected against corrosion ... 8.1.3.1
Protection and stabilisation structures ... 12
$p$-$y$ method ... 13.2.1.1, 13.2.1.3

## Q

Quality requirements for steels ... 8.1.6
Quay loads ... 5.14
Quay walls ... 8.4.9.2
  fixed in the soil ... 8.5.4
Quick-release hooks ... 6.11

## R

Railings ... 6.13.4
Raking piles ... 5.10.7
Raking anchor piles ... 8.4.11
Recommended crossing anchors ... 8.4.10.2
Records of the driving ... 8.1.13.5
Redesign ... 6.10
Reflections from the structure ... 5.6.5
Refraction ... 5.6.5
Regions with mining subsidence ... 8.1.22
Regulations and directives for noise control ... 8.1.14.2
Reinforced concrete capping beams ... 8.1.22.3, 8.4.5
Reinforced steel sheet piling ... 8.1.17
Reinforcing bars ... 9.5.2.1
Relief wells ... 4.6.3
Relieving artesian pressure ... 4.6
Repairing concrete components in hydraulic engineering structures ... 10.12
Repairing interlock declutching ... 8.1.16.2
  on driven steel sheet piling ... 8.1.16
Required energy absorption capacity ... 13.2.2
Resinous pine wood ... 8.1.1.2
Revetments ... 12.1.3
Rotting ... 8.1.1.5
Round steel tie rods ... 8.2.7.3, 8.4.10
Rubbing strips of polyethylene ... 6.15.8.4

## S

Safety against hydraulic heave failure ... 3.1
Safety concept ... 0.2
Sand
  abrasion ... 8.1.9
  columns ... 7.13.1
  fill ... 7.9.5
Scour and protection ... 12.4
Scour protection ... 12.4.3, 12.5
  dolphins ... 13
Screw threads for sheet piling anchors ... 8.4.8
  cut threat ... 8.4.8.1
  hot-rolled threat ... 8.4.8.1
  rolled threat ... 8.4.8.1

645

seal
   artificial .................................................................................................. 8.1.21.3
Sea state parameters ................................................................................... 5.6.3
Secondary settlement ................................................................................. 7.11.1, 7.12.10
Seepage line with wellpoint dewatering ..................................................... 4.7.5.1
Seismic regions .......................................................................................... 2.16
   design situation ...................................................................................... 2.16.6
   effects of earthquakes ............................................................................ 2.16.2
   piled structures in earthquake zones ..................................................... 11.5
Selecting type of section and grade of steel ............................................... 8.1.3.2
Self-sealing ................................................................................................. 8.1.21.2
Semi-portal cranes ...................................................................................... 5.14.1.3
Serviceability .............................................................................................. 8.1.3.2
Settlement ................................................................................................... 7.3.3.4
Shear-resistant interlock connections ......................................................... 8.1.5
Shear strength ............................................................................................. 1.1.1, 1.4
Sheet piling ................................................................................................. 6.4, 8.1.16.2
   concrete sheet pile ................................................................................. 8.1.2
   drainage systems .................................................................................... 4.5.1
   interlocks have been verified ................................................................. 8.1.4.2
   sections for walls ................................................................................... 8.1.12.3
   structures with fixity in the ground and a single anchor ....................... 8.2.3
Ship dimensions ......................................................................................... 5
Shoaling ...................................................................................................... 5.6.5
Shock blasting to assist the driving ............................................................ 8.1.10
Shorter, lighter intermediate piles .............................................................. 8.1.4.1
Shorter sheet pile ........................................................................................ 8.3.1.1
Skin friction ................................................................................................ 8.2.5.6.5
Sliding ......................................................................................................... 8.3.1.3.1
Slip hooks ................................................................................................... 13.3.3
Slope failure ............................................................................................... 8.5.10
   dredging ................................................................................................. 7.7.4.2
Slow-action pile-drivers ............................................................................. 9.2.1
Soft cohesive soils ...................................................................................... 7.11, 7.12
Soft sediment deposits ................................................................................ 7.3.3.3
   cohesive .................................................................................................. 7.13
Soil
   assessing the subsoil for the installation of piles and sheet piles .......... 1.5, 2.16.2
   compaction ............................................................................................. 2.14.3
      dynamic compaction ......................................................................... 7.10
   effects of earthquakes ............................................................................ 2.16.2
   improvement through vacuum consolidation ........................................ 7.12.7
Soil replacement ......................................................................................... 2.11, 2.14.2
   along a line of piles ............................................................................... 2.11, 7.9
   along a line of piles ............................................................................... 7.9
   stabilization ............................................................................................ 2.14.5
   surcharge ................................................................................................ 2.14.4
Sound level and sound propagation ........................................................... 8.1.14.1

| | |
|---|---|
| Splices | 8.4.1.2 |
| Spring piles | 8.1.13.4 |
| overall stability | 8.3.1.3.1 |
| Stability at the lower failure plane for anchorages with anchor walls | 8.5 |
| Stairs in seaports | 6.13 |
| Standard cross-sections | 6.1.1, 6.3, 13.2.2.3 |
|    dimensions | 6.1 |

Steel
| | |
|---|---|
|    capping beams | 8.4.4 |
|    cutting of steel sections | 8.1.20 |
|    grades | 8.1.6.1 |
|    quality requirements | 8.1.6 |
|    in concrete | 8.1.8.4 |
|    piles | 9.2.2.1 |
|    welded joints | 8.1.19 |
|    sheet piling with staggered embedment depths | 8.2.10 |
|    walings | 8.4.1 |
|    verification of steel walings | 8.4.2 |
|    wings | 9.2.2.1 |

| | |
|---|---|
| Stiffening | 8.1.4.4 |
| Stiff piling frames | 8.1.12.5 |
| Straight-web pile sections | 8.3.1.1 |
| Strengthening waterfront structures for deepening harbor bottoms in seaports | 6.8 |
| Strengthening plates | 8.1.5.7 |
| Stress–strain curves | 13.1.2 |
| Structural design calculations | 10.2.1 |
| Structural design of dolphins | 13.2.3.3 |
| Structural requirements | 8.4.5.2 |
| Structural steel sections | 9.5.2.1 |
| Structures in mining subsidence regions | 8.1.22.2 |
| Structural system for combined sheet piling | 8.1.4.2 |
| Subsidence of non-cohesive soils | 1.1.1, 7.8 |
| Subsoil | 1 |
|    active earth pressure in stratified soil | 2.5 |
|    cohesion in cohesive soil | 2.2 |
|    determining active earth pressure in saturated, non-or partially consolidated, soft cohesive soils | 2.9 |
|       cohesion | 1.4.1, 1.4.2, 1.4.3 |
|       sear strength | 1.4 |
|    in situ density | |
|       of dumped non-cohesive soils | 7.6 |
|       of hydraulically filled non-cohesive soils | 7.5 |
|    soft cohesive | |
|       consolidation | 7.11, 7.12 |
|       improving bearing capacity | 7.13 |

subsidence of non-cohesive soils ............................................. 7.8
undrained initially loaded soil ................................................ 1.1.1
Suitable for welding ............................................................... 8.1.19.2
Superstructure ....................................................................... 8.3.1.4
Supply, mounting and corrosion protection ............................ 8.4.4.4
Surging and receding waves .................................................. 5.8, 5.8.1
Survey prior to repairing concrete components ..................... 10.11

## T
Taper piles ............................................................................. 8.1.11.4
Terzaghi's consolidation theory ............................................. 7.113
Tendons can be made from ribbed steel ................................. 9.3.2.1
Tension members ................................................................... 8.4.1.1
Tension piles ......................................................................... 9
Thermal ice pressure .............................................................. 5.15.3.2
Three cases to verify ultimate limit state of bearing capacity ... 0.2.1
Timber fenders and fender piles ............................................ 6.15.8.2
Timber sheet pile walls ......................................................... 8.1.1.1
    corrosion protection ........................................................ 8.1.1.1
    installation ....................................................................... 8.1.1
    seal .................................................................................. 8.1.1.4
Toe protection ....................................................................... 12.1.4
Tolerances for steel sheet piles .............................................. 8.1.6
Tolerances to be maintained when driving combined sheet piling ........... 8.1.13.4
Top edges of waterfront structures ........................................ 6.2
Tropical hardwoods ............................................................... 8.1.1.2, 8.1.1.5
Tubes made from fine-grained structural steels ..................... 8.1.4.3
Tubular bearing piles ............................................................. 8.1.4.3

## U
Ultimate limit state (ULS) capacity ....................................... 0.2.1, 8.3.2.2.1
    analysis ............................................................................ 0.2.4
        for dolphins ................................................................ 13.1.5
    bearing capacity .............................................................. 0.2.1
Ultrasound measurements ..................................................... 8.1.8.2
Uncased holes ........................................................................ 9.5.1
    underwater slopes ............................................................ 7.7
Underseepage ........................................................................ 4.7.7.1
Undrained .............................................................................. 1.1.1
Unsteady hydraulic loads, soil banks ..................................... 12.1.1
Upgrading partially sloped waterfronts ................................. 6.5, 13.2.2.3

## V
Vacuum consolidation ........................................................... 7.12.7
Varying soil strata ................................................................. 8.5.3
Verification of safety ............................................................. 8.5.7
Verifying the loadbearing capacity of the elements .............. 8.2.7

| | |
|---|---|
| Vertical drains | 7.11, 7.12.7, 7.12.8 |
| Vertical elements | 7.13 |
| Vibrator | |
| selection | 8.1.23.4 |
| Vertical imposed loads | 5.5 |
| Vertical wave load | 5.10.9 |
| Vibratory driven grouted piles | 9.2.2.3 |
| Vibratory hammer | 8.1.11.3, 8.1.23.2 |
| Vibratory | |
| driving | 1.5.2.3, 8.1.4.4, 8.1.23.5 |
| experience with | 8.1.23.5 |
| of U- and Z-section steel sheet piles | 8.1.23 |
| Vibro-displacement stone columns | 7.13.3 |

## W

| | |
|---|---|
| Waling bolts | 8.2.7.3, 8.4.2.5 |
| Walings | 8.2.7.2 |
| Wall, loss of wall thickness | 8.1.8.3 |
| Walkway (towpath) | 6.1.2 |
| Wall head | 8.3.1.4 |
| backfilling | 7.4 |
| Waterfront structures | 13.1.4.1 |
| calculations | 0.3 |
| configuration | 6 |
| ice loads | 5.16.4 |
| impact of ice | 5.15, 5.16 |
| in inland ports | |
| redesign | 6.10 |
| standards cross-sections | 6.3 |
| inspection | 14 |
| made from plain concrete, reinforced concrete or prestressed concrete | 10.1.1 |
| press stressed concrete | 10.1.1 |
| pressure of ice | 5.15, 5.16 |
| redesign, inland ports | 6.10 |
| in sea ports | |
| standards cross-sections | 6.1.1 |
| strengthening | 6.8 |
| supply and disposal systems | 6.14 |
| top edges | 6.2 |
| seismic regions | 2.16 |
| Water-jetting to assist the driving of steel sheet piles | 8.1.24 |
| Water pressure | |
| artesian | 2.10 |
| hydrostatic | 8.2.1.3 |
| Watertightness | 8.1.21 |

649

| | |
|---|---|
| Wave forces act directly on the dolphin pile | 13.1.4.3 |
| Wave heights | 5.9.2 |
| Wave loads | 5.10.3 |
|    on a group of piles | 5.10.6 |
| Wave pressure | |
|    on piled structures | 5.10 |
|    on vertical quay walls in coastal areas | 5.7 |
| Wave slamming | 5.10.9 |
| Weepholes | 4.4 |
| Weldability | 8.1.6.4 |
| Weld seams | 8.1.5.3 |
| Well-defined anchorage length | 9.5.1 |
| Well-graded sand | 7.3 |
| Well outlets | 4.6.1 |
| Wind loads | |
|    on moored ships | 5.11 |
|    on moored vessels | 5.11.3 |
| Wind speed, critical | 5.11.2 |